Geology of the Malay Peninsula

a volume in the
REGIONAL GEOLOGY SERIES

edited by
L. U. DE SITTER

Cover photograph: Hill of Kinta Limestone with cave temple, Ipoh, Perak
Photograph: C. S. Hutchison

Published

GEOLOGY OF THE HIMALAYAS
Augusto Gansser

THE TECTONICS OF THE APPALACHIANS
John Rodgers

THE EAST GREENLAND CALEDONIDES
J. Haller

GEOLOGY OF THE MALAY PENINSULA
Edited by D. J. Gobbett and C. S. Hutchison

THE SCANDINAVIAN CALEDONIDES
T. Strand and O. Kulling

GEOLOGY OF DONEGAL
W. S. Pitcher and A. R. Berger

GEOLOGY OF THE MALAY PENINSULA

(WEST MALAYSIA AND SINGAPORE)

EDITED FOR

THE GEOLOGICAL SOCIETY of MALAYSIA

BY

D. J. GOBBETT and C. S. HUTCHISON

FROM THE CONTRIBUTIONS OF

C. K. Burton
D. J. Gobbett
K. F. G. Hosking
C. S. Hutchison
C. R. Jones
P. H. Stauffer
H. D. Tjia

WILEY-INTERSCIENCE, a Division of John Wiley & Sons, Inc.
New York • London • Sydney • Toronto

Copyright © 1973, by John Wiley & Sons, Inc.

All rights reserved. Published simultaneously in Canada.

No part of this book may be reproduced by any means, nor transmitted, nor translated into a machine language without the written permission of the publisher.

Library of Congress Cataloging in Publication Data

Gobbett, Derek John
    Geology of the Malay Peninsula.

   (Regional geology series)
   Includes bibliographies.
    1. Geology—Malay Peninsula.   I. Hutchison, Charles Strachan, joint author.   II. Burton, Cedric Keith.   III. Geological Society of Malaysia. IV. Title.

QE299.G62   555.95'1   72-8490
ISBN 0-471-30850-1

Printed in the United States of America

10-9 8 7 6 5 4 3 2 1

Editor's Note to Regional Geology Series

The aim of this series on Regional Geology is to add to the available geological literature concise descriptions of large structural units, independent of national boundaries.

It is important that the personal opinion of an author, formed by his work and experience in the structural unit he describes, comes clearly to the attention of the reader. Theorizing about the geological history of a particular kind of structure too often does not take into account the great diversity of the observed phenomena, and then generalizes in an unwarranted way. We aim to give a better basis to these general concepts and thereby stimulate a deeper understanding of the relations between different kinds of structures.

Some of the books describe classical territory where new work has brought new conceptions, others are concerned with hitherto relatively unknown regions, but always the surveys are presented from a fresh aspect.

L. U. de Sitter

JOHN BROOKE SCRIVENOR, pioneer of Malayan geology
Photograph by courtesy of Sir Thomas Scrivenor.

Preface

This book presents a detailed summary of our present knowledge of Malayan geology. It is a successor to Scrivenor's book *The Geology of Malaya,* published in 1931, and is based on the work instigated by Scrivenor and subsequently largely carried out by him and other members of the Geological Survey of Malaya. Our concept of the geology of Malaya (now West Malaysia) has broadened immeasurably since 1930, and especially during the last two decades, yet no connected general account has been published, and this realization prompted some members of the Geological Society of Malaysia, which was formed in 1967, to collaborate in preparing this book. Inevitably, the process has been lengthy. Several of the contributors had left Malaya, and manuscripts, figures, and plates had to be collated and welded into a connected unit largely by correspondence.

ACKNOWLEDGMENTS

This volume has been sponsored by the Geological Society of Malaysia, c/o Department of Geology, University of Malaya, Kuala Lumpur, Malaysia.

The editors wish to thank all the contributors and also Dr. B. N. Koopmans and Mr. S. K. Chung who originally agreed to contribute but unfortunately were unable to continue to do so. We are grateful to the Council of the Geological Society of Malaysia for providing funds to cover a voluminous air mail correspondence and for expenses incurred in the preparation of the book.

Individuals who greatly aided the preparation of the manuscript include those who drafted the figures: Mr. S. Sriniwass, Mr. Mohammed bin Haji Majid, Mr. Ching Yu Hay, and Mrs. Haidar bt. Ludin in Kuala Lumpur, and Mr. John Lewis in Cambridge, and those who did photographic work: Mr. Jaafar bin Haji Abdullah in Kuala Lumpur and Mr. David Bursill in Cambridge. Special thanks are due to Miss Anne Chong in Kuala Lumpur and Miss Maureen Templeton in Cambridge for typing the edited manuscripts.

We are indebted to Sir Thomas Scrivenor for the loan of a photograph of his father and permission to publish it, to the Director of the Malayan Geological Survey for permission to reproduce text Figures 8.8, 9.19, 10.7, 10.8, 10.9, 10.17, 10.18, and 10.21, and to *Overseas Geology and Mineral Resources* for permission to reproduce text Figure 10.16.

Dr. P. H. Stauffer wishes to acknowledge the generous cooperation of the Geological Survey of Malaysia and its director, S. K. Chung, who made available much material in unpublished manuscripts and gave permission to use it in compiling Chapter 6; C. K. Burton, P. V. O. Drummond, N. S. Haile, C. R. Jones, D. A. Hooijer, R. A. Newell, H. C. Olander, B. A. V. Peacock, and S. P. Sivam freely shared other unpublished information.

Professor K. F. G. Hosking wishes to acknowledge his gratitude to his erstwhile colleague Mr. J. H. Leow and to his present one Mr. E. B. Yeap for long discussions on the Malaysian ore deposits. He is also most grateful to Dr. C. S. Hutchison and Mr. J. D. Bignell for data and discussions concerning the Malaysian igneous rocks. He wishes to thank the Chief Inspector of Mines, the Director of the Geological Survey, and the Assistant Director, Mr. Santokh Singh, for always providing assistance. He is particularly grateful to Mr. Aw Peck Chin for information and specimens from the Sungei Aring skarn deposit and to Mr. Lee Chong Yan for information and specimens from the Sungei Lah section of Chenderiang Tin Ltd., Perak.

We are grateful to Miss Koh Yen Foong for considerable help in the compilation of the index.

THE EDITORS

Dr. Derek John Gobbett Ph.D., M.A., F.G.S., is presently Senior Assistant in Research at the Sedgwick Museum, Downing Street, Cambridge, England. From 1961 to 1968 he was Lecturer in Geology (Palaeontology and Stratigraphy) at the University of Malaya, Kuala Lumpur.

Dr. Charles Strachan Hutchison Ph.D., B.Sc., F.G.S., is presently Senior Lecturer in Geology (Mineralogy and Petrology) at the University of Malaya, Kuala Lumpur. He has been on the staff of the University of Malaya since 1957, for the first three years in Singapore, and since 1960 in Kuala Lumpur. His previous experience (1955 to 1957) was as oilfield geologist in Trinidad. He has been a visiting professor in petrology at the University of Kansas, Lawrence, on two occasions—1963 and 1970.

THE CONTRIBUTORS

Dr. Cedric Keith Burton, D.Sc., B.Sc., F.G.S., is presently employed by Kennecott Exploration Inc. From 1966 to 1969 he was Senior Lecturer in Geology (Historical and Stratigraphic Palaeontology) at Chulalongkorn University, Bangkok, Thailand. From 1956 to 1966 he was geologist in the Geological Survey of the Federation of Malaya. He did detailed field mapping in Kedah and Perak and south Johore, as well as extensive field work through West Malaysia and Singapore.

Dr. D. J. Gobbett (see under editors)

Professor Kenneth F. G. Hosking, Ph.D., D.Sc., M.I.M.M., F.G.S., A.S.C.M., is presently professor of Applied Geology at the University of Malaya, Kuala Lumpur. He was formerly head of the Department of Geology and Applied Geochemistry of the Camborne School of Mines, Cornwall, England.

Dr. C. S. Hutchison (see under editors)

Dr. Clive Roderick Jones, B.Sc., Ph.D., is presently Principal Senior Scientific Officer, Overseas Division, Institute of Geological Sciences, London S.W.7: on secondment to the Geological Survey of Botswana. Previously he was a field geologist with the Geological Survey of Malaya from 1954 to 1962 and Principal Geologist in charge of the Field Mapping Division from 1962 to 1966.

Dr. Peter Herman Stauffer, Ph.D., B.S., M.S., has been lecturer in Geology (Stratigraphy and Sedimentology) in the Department of Geology, University of Malaya, Kuala Lumpur, since 1965. Previously he was Assistant Instructor, Stanford University Summer Geology Field Course, 1964.

Dr. Hong Djin Tjia, Cand., Drs., D.Sc., is presently head of the Geology Department of the University Kebangsaan, Kuala Lumpur. From 1968 to 1970 he was Lecturer in the Geology Department of the University of Malaya, Kuala Lumpur. From 1962 to 1968 he was Research Geologist, National Institute for Geology and Mining (Indonesia) and Assistant to Senior Lecturer in the Department of Geology, Institute Teknologi Bandung, Bandung, Indonesia.

D. J. GOBBETT
C. S. HUTCHISON

Cambridge, England
Kuala Lumpur, West Malaysia
May 1972

Contents

Plates	xv
Figures	xvii
Tables	xix
Glossary of Commonly Used Malay Terms	xxi

Chapter 1 INTRODUCTION by D. J. Gobbett 1

HISTORY OF GEOLOGICAL INVESTIGATIONS IN MALAYA, 1

    Scattered Observations before the Expansion of the Tin Industry—pre-1870, 1
    1870 to 1903—Mining Geologists, Travelers, and Prospectors before Scrivenor, 3
    1903 to 1930—The Scrivenor Era, 4
    1930 to 1955—Systematic Mapping by the Geological Survey, 7
    1955 to the Present—New Discoveries and Revisions, 8

SUMMARY OF THE GEOLOGY AS PRESENTLY UNDERSTOOD, 9

    Late Cambrian to Triassic, 9
    Late Triassic Orogeny, 11
    Jurassic to Paleogene, 11
    Neogene to Recent, 11

STRATIGRAPHIC NOMENCLATURE, 12

Chapter 2 GEOMORPHOLOGY by H. D. Tjia 13

HIGHLAND AREAS, 15
LOWLAND AREAS, 16
LIMESTONE AREAS, 16
COASTAL FEATURES, 19

    Coastal Plain Bordering the Malacca Strait, 19
    Southern Coastal Plain, 20
    Eastern Coastal Plain, 22
    Quaternary Sea Levels, 23

Chapter 3 LOWER PALEOZOIC by C. R. Jones 25

DISTRIBUTION AND CLASSIFICATION, 25
ROCKS OF THE MIOGEOSYNCLINE, 27

    Shelf Deposits, 27
        Deltaic Facies, 30
            Machinchang formation, 30
            Jerai formation, 33
            Correlatives in Thailand, 33
        Shelf Limestone, 33
            Setul formation, 33
        Shelf Limestone Elsewhere in Malaya, 36
        Correlatives in Thailand and Burma, 39
    Basin Deposits, 40
        Mahang Formation, 41
        Detrital Members of the Setul Formation, 43
        Hawthornden Schist, 45
        Correlatives in Thailand and Burma, 45

ROCKS OF THE GEANTICLINE, 46

    Baling Group in East Kedah, 47
    Baling Group in North Perak, 47
    Other Formations, 49
    Correlatives in Thailand and Burma, 50

ROCKS OF THE EUGEOSYNCLINE, 50

    Bentong Group in West Pahang, 51
    Other Formations, 53
    Volcanic Facies, 54
        Tuff, 54
        Ophiolites, 54

PALEOGEOGRAPHY AND GEOLOGICAL HISTORY, 55

    Cambrian, 55
    Ordovician, 58
    Silurian and Early Devonian, 59
    Mid-Paleozoic Igneous Activity, 60

Chapter 4 UPPER PALEOZOIC
by D. J. Gobbett   61

NORTHWEST MALAYA, 63

    Singa Formation, 63
    Kubang Pasu Formation, 68
    Chuping Formation, 68
    Summary, 70

KINTA VALLEY, 71

    Devonian, 75
    Carboniferous, 76
    Permian, 76
    Summary, 78

CENTRAL MALAYA, 78

    North Pahang and South Kelantan, 80
        Calcareous Facies, 80
        Shale Facies, 82
        Volcanic Facies, 82
        Sub Facies, 82
    North Kelantan, 82
    Central Pahang, 83
    Summary, 83

EAST MALAYA, 84

    Kuantan Area, 84
    Trengganu, 86
    Southeast Pahang and Johore, 87
    Summary, 87

KENNY HILL FORMATION by P. H. Stauffer, 87

    Occurrence, 87
    Lithology, 87
    Structure, 90
    Fossils and Age, 90
    Origin of Sediments, 91

CORRELATION AND SYNTHESIS, 91

    Correlation with Neighboring Countries, 91
    Geological History, 94

Chapter 5 MESOZOIC by C. K. Burton   97

LOWER TRIASSIC, 97

    Occurrence and Nomenclature, 97
    Gua Musang Formation, 101
    Gunong Pulai member, 103
    Synthesis, 104

MIDDLE AND UPPER TRIASSIC, 105

    Occurrence and Nomenclature, 105
    Axial Malayan Province, 106
        Jelai Formation, 106
        Jurong Formation, 108
        Pasir Panjang member, 109
        Kerdau Formation, 111
        Gunong Rabong Formation, 113
    West Malayan Province, 114
        Semanggol Formation, 114
    Chert Member, 115
    Rhythmite Member, 117
    Conglomerate Member, 118
    Paleontology, 118
        Kodiang Limestone, 120
    Middle to Upper Triassic Granite, 121
    Synthesis, 121

JURASSIC AND CRETACEOUS, 122

    Tembeling Formation, 123
        Occurrence and Nomenclature, 123
        Lithology, 123
        Paleontology and Age, 127
        Summary, 127
    Gagau Group, 128
        Occurrence and Nomenclature, 128
        Lithology, 129
    Badong Conglomerate, 129
    Lotong Sandstone, 130
        Paleontology and Age, 131
        Summary, 133
    Tebak Formation, 134
        Occurrence and Nomenclature, 134
        Lithology, 134
        Paleontology and Age, 136
        Summary, 136
    Synthesis, 137

Chapter 6 CENOZOIC by P. H. Stauffer   143

TERTIARY SEDIMENTARY ROCKS, 143

    Bukit Arang-Betong, 144
        Extent and Thickness, 144
        Lithology, 144
        Fossils and Age, 144
    Enggor, 146
        Extent and Thickness, 146
        Lithology, 146
    Batu Arang, 146
        Extent and Thickness, 146
        Lithology, 148
        Fossils and Age, 149
    Kepong, 150
        Extent and Thickness, 150
        Lithology, 150
        Fossils and Age, 150
    Kluang-Niyor, 151

Extent and Thickness, 151
Lithology, 151
Fossils and Age, 151
Other Areas, 151
Tanjong Rambutan, 151
Lawin, 151
Layang Layang, 152
Summary and Discussion, 152

QUATERNARY DEPOSITS, 153

Volcanic Deposits, 153
Basaltic Flows, 153
Rhyolitic Ash, 153
Soils and Residual Deposits, 155
Soils, 155
Residual Deposits, 156
Fluvial Terrace Deposits, 157
Inland Valley Fill, 157
Boulder Beds, 158
Old Alluvium, 159
Kinta Valley, 160
Kuala Lumpur Area, 165
Johore and Singapore, 166
Age and Origin, 167
Young Alluvium, 168
Age and Origin, 170
Cave Deposits, 171
Coastal Plains, 171
Thickness of Sediments, 172
Subsurface Deposits, 172
Surficial Deposits, 173
Beach Ridges, 173
Tektites, 174
Archaeology, 175
Concluding Remarks, 175

Chapter 7 VOLCANIC ACTIVITY
by C. S. Hutchison 177

LOWER PALEOZOIC VOLCANISM, 177

West Kedah, 180
Upper Perak-Grik Area, 181
Central Perak—Northeast and South of Sungei Siput North, 182
Selangor, 182
Selangor-Pahang Boundary, 182
East and South of the Main Range, 185
Chemistry, 185

UPPER PALEOZOIC TO LOWER MESOZOIC VOLCANISM, 185

Central and North Pahang, 187
Petrology of the Volcanic Rocks, 187
Petrology of the Pyroclastic Rocks, 188

North Kelantan and North Trengganu, 190
Petrology of the Volcanic Rocks, 191
Petrology of the Pyroclastic Rocks, 192
The Temangan Ignimbrite, 194
East Pahang, 195
Tuff, 195
Rhyolite, 195
The Jengka Triangle, 195
West Pahang, 197
Bentong Area, 197
Raub Area, 198
South Johore, 198
Lava, 199
Tuff, 199
Central Johore, 200
The Sedili Volcanic Rocks, 200
Chemendong Volcanic Rocks, 200
East Johore, 201
Chemistry, 201

MIDDLE TO UPPER TRIASSIC VOLCANISM, 201
JURASSIC TO CRETACEOUS VOLCANISM, 201

CENOZOIC VOLCANISM, 203
Kuantan, 204
Petrology of the Basalt, 204
Petrology of the Dolerite, 204
Segamat, 206
Petrology of the Basalt, 206
Petrology of the Intrusive Rocks, 206
Petrogenesis, 206
Chemistry, 207
Pleistocene Rhyolite Ash, 207

SUMMARY OF VOLCANIC ACTIVITY IN MALAYA, 207
REGIONAL CORRELATION, 209

Lower Paleozoic, 209
Upper Paleozoic to Lower Mesozoic, 211
Cenozoic, 213

Chapter 8 PLUTONIC ACTIVITY
by C. S. Hutchison 215

GRANITE, 215

Epizone, 215
Petrography, 218
Mesozone, 218
Petrography, 219
Shearing of the Granite, 222
Contamination of the Granite, 222
Modification of the Granite, 224
Microgranite, 224
Porphyry, 225
Aplite, 225

Pegmatite, 225
Vein Quartz, 227
Tourmaline Greisenization, 229
Catazone, 230
Chemistry, 230
Age, 232
   Stratigraphic Evidence, 232
   Radiometric Evidence, 235
Correlation with Thailand, 240
Correlation with Indonesia, 242

BASIC AND ULTRABASIC ROCKS, 242

Diorite, 243
Grabbro and Eucrite, 243
Serpentinite, 245
Dyke Rocks, 245
   Dolerite, 245
   Lamprophyre, 247
Chemistry, 249

PETROGENESIS, 249

Chapter 9 METAMORPHISM
by C. S. Hutchison 253

ZONAL CLASSIFICATION OF THE MALAYAN OROGEN, 253

Epizone, 254
Mesozone, 254
Catazone, 254

ROCKS OF THE CATAZONE, 254

Taku Schist, 254
   Distribution, 256
   Petrology, 256
   Age, 260
   Geochemistry, 260
   Conditions of Metamorphism, 260
   Structure, 261
   Relation to Overlying Rocks, 262
The Stong Magmatite Complex, 262
   Petrology, 262
   Geochemistry, 266
   Conditions of Metamorphism, 266
   Structure, 267
   Relation to Overlying Rocks, 267
The Benta Migmatite Complex, 267
   Distribution, 269
   Petrography, 269
   Chemistry, 269
   Petrogenesis, 273
   Southward Extension of the Complex, 275
   General Considerations, 277

The Bukit Ranjut Complex, 277
   Petrology, 278
   Petrogenesis, 279
The Bukit Berentin Complex, 279
   Petrogenesis, 279
   Metamorphism of Associated Sedimentary Rocks, 281
   Petrogenesis, 281
   Conditions of Metamorphism, 281
The Gunong Jerai Massif, 283
   Age, 283
   Petrology, 283
   Structure, 284
   Conditions of Metamorphism, 285
   Metamorphic Facies Series, 285
The Kupang Gneiss, 287

ROCKS OF THE MESOZONE, 287

Northwest Malaya, 288
   Langkawi, Perlis, West Kedah, 288
   East Kedah, 289
   North Perak, 289
Kinta Valley, 290
   Calcareous Rocks, 290
   Pelitic Rocks, 290
   Tourmaline-Corundum Rocks, 291
Selangor, 293
   Amphibole Schists, 293
   Amphibolite, 294
Kuala Lumpur Area, 294
Pahang, East of Kuala Lumpur, 295
   Schist Series, 295
Raub Area of Pahang, 295
   Schist Series, 295
   Amphibole Schist, 296
   Upper Paleozoic and Triassic, 296
North Pahang, 296
Northeast Malaya, 297
   Volcanic Rocks, 297
East Pahang, 298
East Johore, 298
South Johore, 298
   Metabasites, 298
Physical Effect of Metamorphism on Calcareous Rocks, 299
   Dolomitization of Calcareous Formations, 300

ROCKS OF THE EPIZONE, 301

Dislocation Metamorphism, 301

SUMMARY OF METAMORPHISM IN THE MALAYAN OROGEN, 301

REGIONAL CORRELATION, 302

Thailand, 302
Indonesia, 303

Chapter 10 TECTONIC HISTORY
by D. J. Gobbett and H. D. Tjia   305

THE POSITION OF THE MALAY PENINSULA IN SOUTH-EAST ASIA, 305
STRUCTURAL OUTLINE OF THE MALAY PENINSULA, 306
MID-PALEOZOIC TECTONISM, 308

Langkawi Islands, 308
Kuala Lumpur, 309
West Pahang, 310
Plutonism, 310

UPPER PALEOZOIC TECTONISM, 310

Western Zone, 312
Langkawi, 312
Main Range, 312
Axial Zone, 312
Eastern Zone, 312

LATE TRIASSIC TECTONISM, 313

Western Zone, 314
Axial Zone, 315
Eastern Zone, 317

POST-TRIASSIC FOLDING, 317
POST-TRIASSIC FAULTING, 318

Bok Bak Fault, Kedah, 318
Lebir Fault Zone, 318
Kuala Lumpur-Endau Fault Zone, 322
Bukit Tinggi Fault, 326
Kledang Fault, 326
Chegar Perah-Benta Fault, 327
Faulting at Sungei Lembing, Pahang, 327
Conclusions, 328

STRUCTURE OF THE MAIN RANGE GRANITE, 328

Southern Part, 328
Batang Padang, Perak, 330
North Perak, 330

SYNTHESIS OF CRUSTAL EVOLUTION IN THE MALAY PENINSULA, by C. S. Hutchison, 330

Chapter 11 PRIMARY MINERAL DEPOSITS
by K. F. G. Hosking   335

REGIONAL DISTRIBUTION OF PRIMARY MINERALIZATION, 336

MAGMATIC DISSEMINATIONS AND SEGREGATIONS, 341

PEGMATITES AND APLITES, 343

PYROMETASOMATIC (SKARN) DEPOSITS, 346

Ferriferous-Stanniferous Types, 347
Stanniferous Types, 351
Mode of Occurrence of Tin in Stanniferous Skarns, 352
Other Types, 356

HYDROTHERMAL DEPOSITS, 357

Tin and Tungsten Deposits, 363
Tin and Tungsten Minerals, 365
Mineralogically Simple Veins and Vein Systems, 371
Simple Replacement Deposits, 373
Moderately Complex Veins, 373
Mineralogically Complex Pipes and Veins, 374
Ban Hock Hin Lode, 374
Manson Lode, 378
Lodes of the Cornish Type, 379
Iron Deposits, 381
Gold Deposits, 381
Base Metal Sulfide Deposits, 382
The Maran Lead Ore, 383
The Segamat Deposits, 383
Barite Deposits, 385

IN SITU EFFECTS OF WEATHERING OF PRIMARY DEPOSITS, 385

TIME RELATIONSHIPS BETWEEN THE PRIMARY MINERAL DEPOSITS AND IGNEOUS ROCKS, 387
PROSPECTING, 388

REFERENCES, 391

INDEX   403

Plates

DEDICATION **John Brooke Scrivenor,** vi

1. Limestone hills in Perlis, **18**
2. Coastal plain of south Perlis, **20**
3. Machinchang formation of Pulau Terutau, **32**
4. Setul limestone, Pulau Langkawi, **36**
5. Lower Paleozoic fossils, **38**
6. Gunong Kendrong, north Perak, **48**
7. Devonian and Carboniferous fossils, **69**
8. Bukit Ketri, Perlis, **72**
9. Pulau Jong, Langkawi, **73**
10. Permian fossils, **77**
11. Triassic fossils, **100**
12. Bukit Kodiang, Kedah, **120**
13. Tembeling sandstone, Sungei Jempol, Pahang, **126**
14. Gunong Gagau plateau, **128**
15. Badong conglomerate unconformable on Upper Paleozoic shale, Ulu Pahang, **130**
16. Gagau flora, **132**
17. Hills capped by the Tebak formation, south Pahang, **135**
18. Klang Gates ridge, Kuala Lumpur, **228**
19. Dolerite dykes cutting granite, Kuantan, **246**
20. Recumbent fold in Triassic limestone, Bukit Kalong, Kedah. **315**

Figures

1.1	Structural elements of southeast Asia, 2	5.2	Distribution of Mesozoic rocks in the Malay Peninsula, 99
2.1	Drainage and major divides of Malaya, 14	5.3	Bedded chert of the Semanggol formation, 116
2.2	Weathering of granitic rocks, 15	5.4	Rhythmite member of the Semanggol formation, 117
2.3	Topographic map of northeast Pahang, 17	5.5	Stratigraphic sections in the Tembeling formation, 124
2.4	Lapiés, Bukit Takun, Selangor, 18	5.6	Murau Conglomerate, 125
2.5	Cave in limestone at Bukit Kepala, Kedah, 19	5.7	Tembeling Formation sandstone, Sungei Jempol, Pahang, 125
2.6	Displacement of river mouths along the Malayan coast, 21	5.8	Summit profile of the Gagau Group and Tebak Formation, 140
2.7	Migration of the lower Perak River, 22	6.1	Distribution of Cenozoic rocks in the Malay Peninsula, 145
3.1	Distribution of Lower Paleozoic rocks in the Malay Peninsula, 26	6.2	Geological map of the Batu Arang basin, Selangor, 147
3.2	Lithofacies map of the Lower Paleozoic rocks, 28	6.3	Columnar section through the Tertiary sediments at Batu Arang, 148
3.3	Idealized section through the Malayan Lower Paleozoic geosyncline, 30	6.4	Soil and weathering profile developed on granite, 155
3.4	Lower Paleozoic formations in northwest Malaya, 31	6.5	Coarse granite colluvium at Kuala Dipang, Perak, 156
3.5	Marmorized and stylolitic Setul limestone, 35	6.6	Boulder beds (early Quaternary?), 158
3.6	Kuala Lumpur Limestone at Segambut, Kuala Lumpur, 39	6.7	Open-cast tin mine near Bidor, Perak, 159
3.7	Bukit Anak Takun, Selangor, 40	6.8	Old Alluvium from near Gunong Rapat, Kinta Valley, 161
3.8	Geological sketch map of the northwest coast of Pulau Langgun, Langkawi, 44	6.9	Old Alluvium near Tanjong Rambutan, 161
3.9	Sheared conglomerate in the Bentong Group, 52	6.10	Diagrammatic sections in Old Alluvium in the Kinta Valley, 162
3.10	Paleographic maps of the Malay Peninsula during the Lower Paleozoic, 56	6.11	Section in Old Alluvium, near Tanjong Rambutan, 163
3.11	Schematic sections through the Lower Paleozoic geosyncline, 57	6.12	Bedding structures in Old Alluvium near Kampar, 164
4.1	Distribution of Upper Paleozoic rocks in Malaya, 62	6.13	Limestone bedrock near Batu Gajah, 164
4.2	Geological map of northwest Malaya, 64	6.14	Old Alluvium at Batu Gajah, 165
4.3	Relationships of the Upper Paleozoic formations in northwest Malaya, 65	6.15	Diagrammatic section of Old Alluvium filling "pothole," 165
4.4	Geological sketch map of southwest Langkawi, 66	6.16	Rudaceous Old Alluvium near Bedak, Singapore, 166
4.5	Singa formation, Langkawi Islands, 67	6.17	Unconformity between Young and Old Alluvium, 169
4.6	Section through north Perlis, 67	6.18	Section in Young Alluvium near Tanjong Rambutan, 169
4.7	Section at Bukit Temiang, Perlis, 70	7.1	Distribution of volcanic rocks in the Malay Peninsula, 178
4.8	Geological sketch map of the Kinta Valley, 74	7.2	Photomicrographs of rhyolitic rocks, 181
4.9	Hypothetical section through Kinta in the Lower Permian, 78	7.3	Photomicrographs of ignimbrite and acid tuff, 193
4.10	Geological sketch map of central Malaya, 79	7.4	Geological map of the Jengka triangle, Pahang, 196
4.11	Structure of the Upper Paleozoic and Triassic in central Malaya, 80	7.5	Flow-banded rhyolite, Ulu Endau, Johore, 203
4.12	Section at the Jenka Pass, Pahang, 83	7.6	Map and section of the Kuantan basalt, 205
4.13	Geological sketch map of east Malaya, 85	7.7	Variation diagram of Malayan volcanic rocks, 210
4.14	Slump folds in Carboniferous (?) sediments, 86	7.8	Distribution of volcanic rocks in Southeast Asia, 212
4.15	Geological sketch map of the Kenny Hill formation, 88		
4.16	Kenny Hill formation, Petaling, Selangor, 89		
4.17	Upper Paleozoic outcrops in Southeast Asia, 93		
5.1	Geosynclinal organization in Malaya, 98		

FIGURES

8.1 Distribution of granite in the Malay Peninsula, 216
8.2 Geology of the area around Mount Ophir, 217
8.3 Weathered porphyritic granite, 219
8.4 Flow alignment of feldspar phenocrysts, 219
8.5 Triangular diagram for modal quartz, plagioclase, and K-feldspar of Malayan granites, 221
8.6 Photomicrographs of biotite granites, 223
8.7 Granodiorite dykes cutting gabbro, Singapore, 224
8.8 Details of granite–gabbro contact, Singapore, 225
8.9 Distribution of tonalite in the Boundary Range Granite, 226
8.10 Tourmaline clots in granite, Langkawi, 229
8.11 Normative quartz, albite, orthoclase, and plagioclase in Malayan granites, 233
8.12 Frequency distribution of radiometric ages, 239
8.13 Distribution of granite plutons in Southeast Asia, 241
8.14 Geological map of the southwest margin of the Singapore granodiorite, 244
8.15 Variation diagrams of plutonic rocks from Malay Peninsula, 250
9.1 Distribution of metamorphic grades in the Malay Peninsula, 255
9.2 Geological map of Ulu Kelantan, 257
9.3 Metamorphic facies diagrams for the Taku Schist, 261
9.4 Rocks of the Stong Migmatite Complex, 264
9.5 Metamorphic facies diagrams for the Stong Migmatite Complex, 266
9.6 Geological map of the Stong Migmatite Complex, 268
9.7 Geological map of the Benta migmatite, 270
9.8 Photomicrograph drawings of the Benta migmatite, 271
9.9 Foliated gneiss near Benta, Pahang, 271
9.10 Contacts in the Benta Migmatite Complex, 273
9.11 Triangular variation diagram, 274
9.12 Foliated venite psammite, 275
9.13 Geological map of part of the Benta Migmatite Complex, 276
9.14 Geological map of the Gunong Jerai area, Kedah, 282
9.15 Photomicrograph of quartz-mica schist, 283
9.16 Analysis of foliation planes of the Jerai Formation, 285
9.17 Distribution of index metamorphic minerals in the Gunong Jerai area, 286
9.18 Calc-silicate hornfels, 290
9.19 Section through tourmaline-corundum rock deposit in Kinta, 292
9.20 Distribution of thermoluminescence ages of marble, 300
10.1 Structure of the Sunda area, 306
10.2 Structural outline of the Malay Peninsula, 307
10.3 Geological sketch map of the Langkawi Islands, 309
10.4 Geological sketch map of the Kuala Lumpur area, 311
10.5 Diagrams representing the Kisap Thrust, Langkawi, 313
10.6 Geological sketch map of southeast Kedah and north Perak, 314
10.7 Structural map of north Pahang, 316
10.8 Tectonic style in Pahang, 317
10.9 Disharmonic folds at Raub Gold Mine, 317
10.10 Structural map of the Tembeling Formation, 319
10.11 Major faults in Malaya, 320
10.12 Lebir fault zone near Kuala Krai, Kelantan, 321
10.13 Wrench faults and dolorite dykes in Trengganu, 323
10.14 Structural map of part of south Pahang, 324
10.15 Structure of the southern part of the Main Range Granite, 325
10.16 Major joints at Klang Gates Gorge dam site, 326
10.17 The Kabang fault at Sungei Lembing, 327
10.18 Strike frequencies of tin lodes at Sungei Lembing, 328
10.19 Lineaments in the Fraser's Hill area, 329
10.20 Lineament pattern in the Rasa area, 330
10.21 Strike frequencies of joints in the Batang Padang area, 331
10.22 Major lineaments in north Perak and west Kelantan, 332
10.23 Schematic tectonic sections through Malaya, 333
11.1 Distribution of tin and tantalum/niobium deposits, 337
11.2 Distribution of tungsten deposits, 338
11.3 Distribution of iron and manganese deposits, 339
11.4 Distribution of gold deposits, 340
11.5 Photomicrographs of tin ore, 349
11.6 Photomicrograph of galena-rich vein, 350
11.7 Photomicrograph of a garnet-bearing tin ore, 352
11.8 Photomicrograph of malayaite, 353
11.9 Zoned malayaite, 354
11.10 Photomicrographs of tin ore from Sungei Gow, 354
11.11 Photomicrograph of tin ore from Telok Kruen Pipe, 355
11.12 Section of the Beatrice Mine, 356
11.13 Photomicrograph of streaky-bacon ore, 357
11.14 Metal zones in the Kinta Valley, 358
11.15 Plan of the Kramat Pulai Mine, 359
11.16 Plan of part of the Gakak Mines, 361
11.17 Photomicrographs of cassiterite, 364
11.18 Photomicrographs of sulfide ores, 368
11.19 Photomicrographs of ore from the Manson Lode, 369
11.20 Photomicrographs of tin ore from Ampang, Perak, 370
11.21 Photomicrographs of tin oblique tungsten ores, 371
11.22 Tin ore from Setapak, Selangor, 375
11.23 Evolution of the Setapak "telescoped" ore, 377
11.24 Photomicrograph of sulfide ore, 378
11.25 Section through Pahang Consolidated Mines, 380
11.26 Photomicrographs of lead ore from Maran Lode, 384
11.27 Photomicrograph of sulfide ore in basalt, 385
11.28 Lineaments and mineralization near Gambang, Pahang, 389

Tables

1.1 Time correlation of the major rock formations of the Malay Peninsula, 10
2.1 Evidence for Quaternary sea levels in Malaya and Singapore, 23
2.2 Correlation chart of known and interpreted Quaternary shorelines in Malaya and Singapore, 24
3.1 Schematic classification and correlation chart of Malayan Lower Paleozoic formations, 29
3.2 Chemical analyses of Lower Paleozoic limestones, 34
4.1 Chemical analyses of Upper Paleozoic limestones, 71
4.2 Upper Paleozoic sequence west of Kampar, 75
4.3 Upper Paleozoic and Triassic of north Pahang, 81
4.4 Comparison of the four zones of the Malayan Upper Paleozoic, 92
5.1 Chemical analyses of Triassic limestone, 102
5.2 Comparison of granodiorite and rhyodacite from Ulu Endau, 103
5.3 Correlation of the Mesozoic formations, 107
7.1 Correlation chart of volcanic and pyroclastic activity in the Malay Peninsula, 179
7.2 Chemical analyses and Niggli norms of Lower Paleozoic rocks, 186
7.3 Chemical analyses and Niggli norms of Upper Paleozoic and Lower Mesozoic rocks, 202
7.4 Chemical analyses and Niggli norms of post-orogenic basalt and dolerite, 208
8.1 Modal analyses of selected Malayan granites, 220
8.2 Chemical analyses and Niggli norms of Malayan granites, 231
8.3 Normative corundum and acmite in salic rocks whose normative $Q + Ab + Or > 70\%$, 232
8.4 Chemical analyses and Niggli norms of granitic rocks, 234
8.5 Radiometric dates of Malayan granites, 236
8.6 Chemical analyses and Niggli norms of Malayan mafic rocks, 248
9.1 Chemical analyses of Malayan metamorphic rocks, 258
9.2 Chemical analyses and Niggli norms of rocks from the Benta area, 272
11.1 Analyses for tin (ppm Sn) of "fresh" vein-free granites, 343
11.2 Niobium content (ppm) of West Malaysia granites, 344
11.3 Paragenetic sequence of the minerals of Pelepah Kanan, 351
11.4 General paragenesis of the mineral deposits of West Malaysia, 362
11.5 Chemical analyses of some West Malaysian cassiterites, 366
11.6 Recovery of cassiterite by the Frantz Isodynamic Separator, 367
11.7 Paragenesis of the Bukit Lentor deposits, 373
11.8 Paragenesis of the Manson Lode, 379
11.9 Spectrographic analyses of ore from the Manson Lode, 379
11.10 Production of gold from Raub, 382
11.11 Paragenesis of the Maran lead ore, 383

Glossary of Commonly Used Malay Terms

Malay (Bahasa Malaysia)	English Equivalent
Batu	Rock or milestone
Besar	Large
Batang	Main river
Bukit	Hill
Changkul	Hoelike implement used for cutting earth
Dulang	Large shallow wooden bowl; a prospector's pan for concentrating tin-ore or gold
Gunong	Mountain but may also mean a high prominent hill in generally low-lying country
Jabatan Kerja Raya (J.K.R.)	Public Works Department
Kampong	Village or small settlement
Kechil	Small
Kuala	The mouth of a main river or a tributary stream
Lampan	A method of extracting ore from weathered rock by bringing a stream of water to it and concentrating the ore without lifting the containing rock
Pekan	Town
Pulau	Island
Sungei (or Sungai)	River or stream
Tanjong	Headland or promontory
Tasek	Lake
Telok	Bay
Ulu	Upper waters of a river; up-country; the interior of a country or state

CHAPTER 1

Introduction

D. J. GOBBETT

The Malay Peninsula forms the southeastern extremity of the Eurasian landmass, to which it is connected by the isthmus of southernmost Thailand and Burma. It straddles latitudes 1 to 7°N and is thus wholly within the equatorial region; under natural conditions, its hot, wet, equable climate allows a luxuriant rain forest to cover the entire land surface.

Structurally the peninsula forms part of a larger area, now partly submerged beneath the Straits of Malacca and the South China Sea, curving southeastward through the Indonesian Rhio Archipelago and the islands of Bangka and Billiton and east-northeastward through west and north central Borneo (Kalimantan). This larger region, which for convenience we may call "Chersonesia"[1] has been a positive structural element since the Jurassic period. It is flanked to the west, south, and east by Cenozoic geosynclines (Figure 1.1). The geology of the southeastern part of Chersonesia is relatively poorly known, but we may expect it to be essentially similar to that of the northwestern part, the Malay Peninsula, as described in this book.

The Malay Peninsula, formerly the Federation of Malaya, now includes two nations, West Malaysia and Singapore. West Malaysia has a central government but has evolved historically by the compounding of several states, each still headed by a Sultan or Chief Minister. The state boundaries largely coincide with the major watersheds. In general, economic development and the destruction of the natural environment have proceeded furthest in the states bordering the Malacca Strait and least in Upper Perak, Ulu Kelantan, and Ulu Pahang. Our knowledge of the geology reflects this pattern of development, as the following historical account shows.

HISTORY OF GEOLOGICAL INVESTIGATIONS IN MALAYA

This book results from a long series of investigations on Malayan geology, beginning in the first quarter of the nineteenth century. Both the kind of geological work and its emphasis varied, so it is convenient to divide the history into five periods.

Scattered Observations before the Expansion of the Tin Industry—pre-1870

Prior to 1870, half a dozen voyagers recorded their observations on the topography and rock strata. Amateur geologists working by themselves, these men were disinclined or unable to follow up their own or previous work. Their observations were confined to the coastal areas, particularly of Singapore and the west coast. Tin deposits were mentioned, but no descriptions of them were published.

The earliest publication (Jack 1822) was a letter giving a brief account of the rocks forming Penang

---

[1] The term "Sundaland," which has been used in this context, correctly refers to the "land" embracing the whole of Sumatra, Java, and Borneo at times of low sea level during the Pleistocene Epoch.

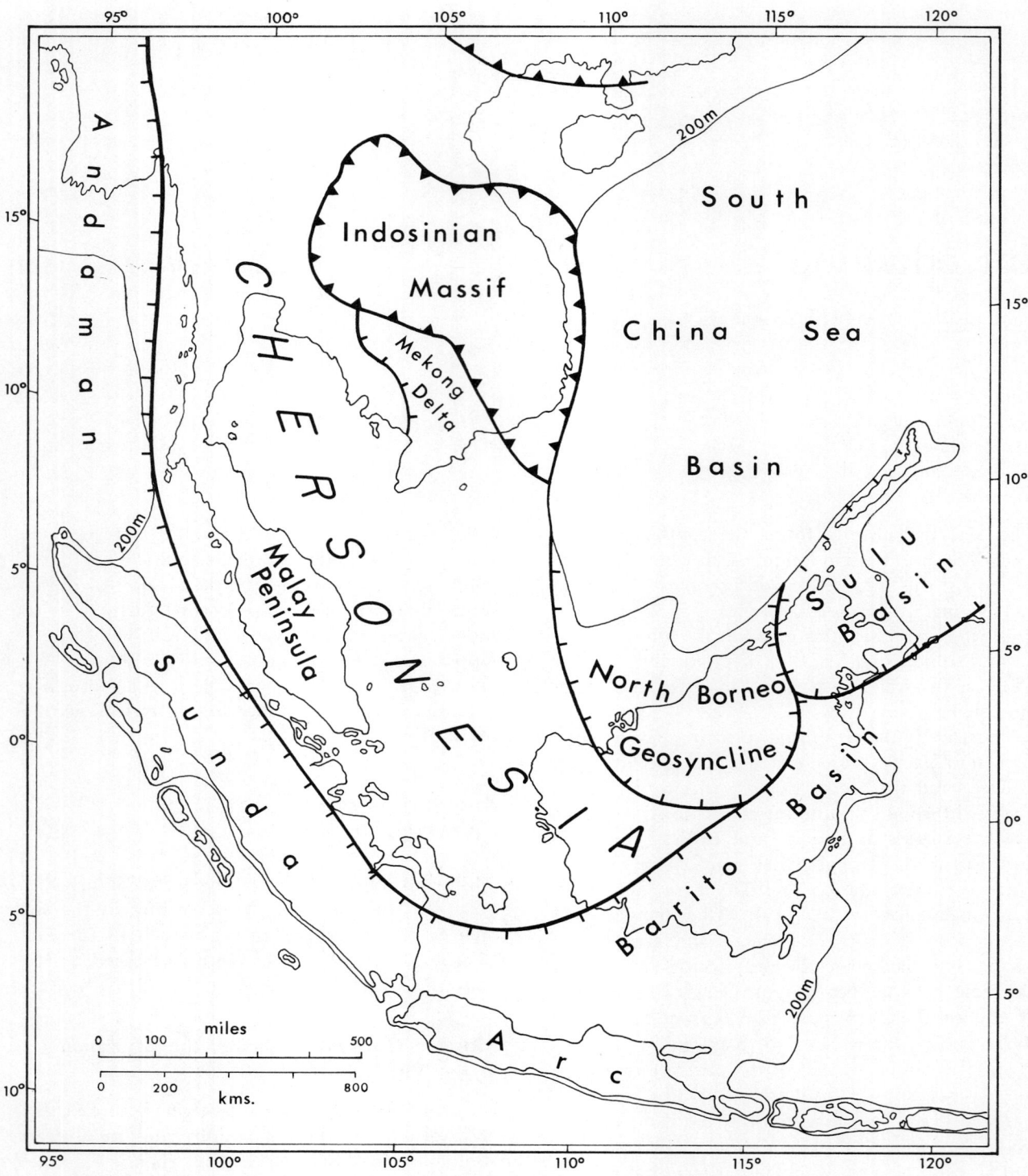

FIGURE 1.1 *Structural elements of Southeast Asia showing the position of the Malay Peninsula. Partly after Van Bemmelen (1949).*

and Singapore. Crawford (1824) mentioned the rocks of Singapore, and a slightly longer account of the geology of this island was given by Low (1847). The geomorphology of the Perak, Penang, and Kedah coast was described earlier by Low (1833), and in the same year Ward (1833) published a geological map and sections showing the geological composition of Penang and neighboring islands. In 1851 Thompson gave a clear account of the metasediments, granite, and volcanic rock of the coast of east Johore and southeast Pahang, illustrated by a colored map. Lateritized quartz-mica schist and phyllite from this area had been described a few years earlier by Logan (1848b).

Logan probably resided longer in Malaya than the other authors mentioned. He recorded detailed descriptions of the rocks of Singapore and Pulau Ubin (Logan 1851, 1847b), journeyed inland in Malacca, and was able to outline the general structure of the Peninsula (Logan 1848a). Besides merely describing geological features, he presented a hypothesis of laterization and cited evidence for Recent sea-level changes (Logan 1848a).

One of the most impressive features of Malayan geology, and one that certainly struck the early workers, was the extreme depth of weathering and laterization of all rock types. This is particularly evident in Malacca and Johore, where the lateritized rock was often considered to be a distinct formation. Logan (1847a) thought that this laterization must be due to the addition of iron from a volcanic or plutonic source, metasomatizing the country rock.

After 1851 no geological publications appeared for twenty years. In the last three decades of the century, almost all geological investigations were directly associated with tin mining.

1870 to 1903—Mining Geologists, Travelers, and Prospectors before Scrivenor

Between 1870 and 1903 Europeans began to participate in tin mining, and geological activities, restricted as they were, were largely carried out by miners and prospectors. In the 1870s Larut, near Taiping, was the main tin-mining centre, and it continued to be prominent at least until the end of the century. In the 1880s, however, the Kinta Valley became the site of a rapidly expanding tin industry, and by 1889 its production had surpassed that of Larut.

Descriptions of the occurrence of alluvial ore and of the nature of the alluvium in Larut were given by Doyle (1878, 1879a,b) and Tenison-Woods (1884a). However, the workings were shallow and did not expose the bedrock. The introduction of the steam engine and the centrifugal pump (from 1877 onward) allowed the alluvium to be worked to a greater depth and resulted in more frequent exposure of the bedrock.

The French were among the first Europeans to open mines in the Kinta Valley, and general descriptions of the geology of the area were given by de la Croix (1882) and de Morgan (1886). Most of these early investigators were content merely to describe the rocks, but de Morgan also interpreted their age relationships. He considered the granite, which he termed "gneiss," to be of Precambrian age and to be overlain by Silurian schist and quartzite and Upper Silurian or Devonian marble, the latter with the brachiopod *Platystrophia*. Although this was probably a misidentification, it seems to be the first record of a fossil from Malaya. The tin-bearing alluvium he thought to be post-Pliocene.

Collet (1903) gave a general account of tin in Malaya. He, however, based his geology on Kinta and compared a supposed succession of Archeaen gneiss, Silurian and Devonian schist and quartzite, and Carboniferous limestone with that of Burma.

Other descriptions of the tin deposits were given by Hampton (1887, 1899) and Penrose (1903). The latter provided a clear and accurate account of the manner of occurrence of alluvial and lode tin ore in the Kinta Valley.

The authors just mentioned were all connected with the established mining industry. However, a small number of prospectors were opening up new areas during this period. In Ulu Rompin, Swinney (1891) was engaged in mineral prospecting and made notes on the local geology of an area little frequented by Europeans. Gold mining received some attention from prospectors in central Pahang (Swann 1900, Becher 1893), and in the 1890s lode tin mining at Sungei Lembing encouraged a small amount of geological work in that area (Stephens 1901).

Travelers and explorers in the Malay Peninsula occasionally published their geological observations which, like those made earlier in the century, remained isolated records. Tenison-Woods wrote several papers mainly on Perak geology. He described (1884b) the limestone of Kinta as overlain by Paleozoic clay slate but mistakenly identified what were probably pebbles of tourmaline-corundum rock as Recent basalt (Tenison-Woods, 1885). Explorations in the southern part of Malaya were made by Lake (1894) and by Snow (1902), who optimistically reported gold occurrences at Mount Ophir to the Sultan of Johore. Daly, employed as a topographic surveyor, made extensive journeys into the interior of the peninsula. He recorded the distribution of gold and tin on either side of the Main Range (Daly 1878) and in 1882 described the tin mines at Ampang, Selangor, and the Chindrass gold mines near Mount Ophir.

In 1883 Leonard Wray, Jr., was commissioned by the British Resident in Perak to collect specimens for a state museum. Wray became curator of the Perak Museum, which was completed in 1886 and contained a section on local geology and mineralogy. Wray published a number of geological papers on Perak between 1885 and 1898. The geology of Perak, and particularly that of the Kinta Valley, provided a basis for the interpretation of other areas and so for many years epitomized Malayan geology.

Fossils were collected and sent to the British Museum by at least two members of the public. H. F. Bellamy collected some pieces of fossiliferous sandstone from near Kuala Lipis in 1897, and these were described by Newton (1900) as bivalve molluscs of Triassic age. Limestone with crinoid remains was collected by H. M. Becher from Gua Sai, near Kuala Lipis; this occurrence was also published by Newton.

Up to 1903 all geological work had been of a temporary and haphazard nature. In 1903 the government appointed a geologist, mainly to assess the country's potentiality for the further exploitation of tin and gold. Thus J. B. Scrivenor came to Malaya and eventually exerted a profound influence on the history being traced here.

1903 to 1930—The Scrivenor Era

During the period from 1903 to 1930, European tin mining expanded greatly. Indeed most of the tin production was in European hands by 1930, whereas before tin mining (as opposed to tin smelting) had been dominated by the Chinese. In 1912 the first tin dredge was introduced, but the main development of dredging did not begin until after World War I. European companies also ventured into gold mining but without success.

It was against this background that the Government Geologist and later the Geological Survey Department worked. The geologist, J. B. Scrivenor, arrived in Malaya in 1903 on a three-year contract. He returned to Malaya, after leave, in 1907 and subsequently devoted most of his working life to Malayan geology. A geological and chemical laboratory and an office, set up at Batu Gajah in the Kinta Valley, became his headquarters.

Scrivenor was an energetic man who pursued geology in the field with great enthusiasm and determination. Skillfully interpreting his observations with little previous work to guide him, he held his theories firmly and defended them vigorously in fluent speech and writing. From the time of Scrivenor's arrival until recent years, almost all geological publications emanated from the Survey, and until about 1930 most were from Scrivenor himself.

During this period, brief accounts of mining geology in Trengganu (Fawns 1904, Brelich 1914), the Kinta tin deposits (Rumbold 1906, Roux-Brahis 1910), the Rahman tin mines in Upper Perak (Harris 1924), and the Batu Arang coal field (McCall 1922) comprised most of the published geological observations unconnected with Scrivenor's work. A mass of geological data gradually accumulated in his field notebooks, which were kept at Batu Gajah. Much of this was published without delay, in some cases rather precipitately, but at least the information was made available to the miners and the public at large.

During his first tour of duty Scrivenor was directed to study the occurrence of tin and gold, since the Government was concerned about the future of these resources. Accordingly, he familiarized himself with the tin deposits of Larut, Kinta, and Seremban, and visited Singapore and Ulu Pahang, where he made extensive geological observations. His report of progress (Scrivenor 1907) indicates quite astonishing activity during three and and a half years in a scarcely developed tropical country, without topographic maps, where the elephant and the ox cart provided the main forms of travel and communication, and the climate was not inducive to prolonged physical work in the field.

In Ulu Pahang four rock "Series" were designated. The Chert Series, of carbonaceous shale and radiolarian chert, was developed in west Pahang. The Raub Series was formed of limestone and shale with "Permo-Carboniferous" fossils. The Tembeling Series, of sandstone and shale (frequently red or yellow), included conglomerates with chert from the Chert Series. Sandstone of the Tembeling Series contained Triassic fossils found near Kuala Lipis. Finally, the Pahang Volcanic Series was mainly of pyroclastic rocks, andesitic to trachytic in composition, interbedded with the Raub

Series. Hornblende granite and augite syenite were recognized in the Benom area, and several metamorphic and skarn rocks were described. Scrivenor reviewed gold and tin mining and had to report pessimistically about the former.

Scrivenor's observations on Pahang geology were somewhat expanded and modified in a paper published in 1911, *The geology and mining industries of Ulu Pahang*, which included a colored geological sketch map. The Raub, Chert, and Pahang Volcanic Series were retained, but the rocks of the Tembeling Series were renamed "Gondwana rocks." They were thought to have been formed littoral to the Gondwanna continent, then recently interpreted from the Upper Paleozoic and early Mesozoic continental rocks of peninsular India.

The phenomena of rock weathering in the equatorial climate and the formation of laterite arrested Scrivenor's attention, as they had that of earlier observers, and he published several communications on laterite, arguing against overly strict definition of the term. Some Malayan laterites, notably those of Malacca, were recognized to have properties similar to those of the typical Indian variety.

Much of Scrivenor's work was centered on the most important tin-mining area, the Kinta Valley, and problems of Kinta geology overshadowed those of other areas for many years. In 1912, at a meeting of the Geological Society of London, he proposed that certain rocks, resembling tillite and containing boulders of a variety of rocks and also tin ore, were older than the granite. He called them the Gopeng Beds and assumed that they had been deposited by Gondwana ice that had eroded tin-bearing rocks. The Beds themselves, he asserted, were later further enriched in tin by the Mesozoic granite. He also suggested that the limestone hills of Kinta were mainly fault-bounded.

These theories were advanced in 1913 in *The geology and mining industry of the Kinta district, Perak*. The succession of events in the Kinta Valley was stated to be as follows. The deposition of the limestone was succeeded by that of the Gondwana clays and boulder clays with detrital cassiterite, along with younger Gondwana phyllites and quartzites. Granite was later intruded, introducing more tin ore to enrich the Recent alluvium. Scrivenor (1914a) thought that the Gondwana clays were older than the granite because they were enriched by tourmaline near the granite and carried thin veins of muscovite, fluorite, and corundum. Their soft condition was thought to be due to intense chemical weathering.

In 1913 W. R. Jones joined Scrivenor as a geological assistant. He worked mainly in North Selangor and partly on the rich cassiterite-bearing rocks, which had newly been discovered on Gunong Bakau. These had been described by Scrivenor (1914b) as primary topaz-quartz rocks. Jones (1915b), 1916) criticized this interpretation and stated his opinion that the topaz was formed by the alteration of feldspar within the granite by fluorine vapors. The topaz was similar in occurrence and mineral content to the tin-bearing rocks of the Erzgebirge, which were of secondary origin. This led to some controversy between the two geologists, and Scrivenor argued (1916, 1919) that not all the topaz could be accounted for by the secondary introduction of fluorine.

However, a much greater controversy began on the geological history of the Kinta Valley. Scrivenor's theories were criticized by Jones (1917), who refuted the Government Geologist's interpretations of a glacial origin and a Gondwana age for the tin-bearing "clays and boulder clays." Jones reasoned that these were partly alluvial and partly elluvial deposits deformed by collapse, subsequent on the solution of the underlying limestone. Tin ore was associated with the Mesozoic granite instrusion only. Jones also criticized Scrivenor's ideas on the formation of the limestone hills by faulting, attributing this to stream erosion. Scrivenor (1918) argued for the original unbedded nature of the clays and their contained boulders. He discussed the evidence against a glacial origin but concluded that the glacial theory best fitted all the facts.

In 1919 *The geology of south Perak, north Selangor and the Dindings* was published under the joint authorship of Scrivenor and Jones. However, this work was mainly written by Scrivenor and supported his theory of Kinta geology. The Gondwana deposits of Kinta were here divided into the Western Clays and Boulder Clays, characterized by the presence of tourmaline-corundum boulders, and the Eastern Clays and Boulder Clays (formerly Gopeng Beds), characterized by the presence of pure corundum pebbles and discoidal granite boulders. The conclusions were that both deposits were of Gondwana age and were probably glacial or fluvio-glacial in origin.

In 1924 Cameron published a paper on Kinta geology and supported Scrivenor's ideas of an older set of pregranite tin-bearing deposits (Gondwana boulder clays), which he called "deep leads." However, he postulated two unconformable limestones and two tin-bearing granites of different ages, and he named the following succession: schist, older limestone, granite, deep leads, younger limestone, granite.

Cameron (1925a) supported W. R. Jones on the origin of the limestone hills, considering them to have been formed by differential erosion. Jones (1925a) criticized Cameron's ideas of two granites and believed his older granite to be a marginal differentiate. Cameron countered this and asked (1925b) for more detailed surveys of Kinta to settle differences of opinion on the geological succession, particularly with regard to the stratigraphical relationships of the schist and limestone. Scrivenor (1925) replied that a bore, put down on the spot postulated by Cameron to be underlain by schist only, actually reached limestone underlying the schist.

Finally Jones (1925b) reaffirmed the sequence as limestone, schist, granite, and alluvium with detrial tin ore. Scrivenor agreed with Jones, although he still doubted the theory of the origin of the boulder clays and the tin ore they contained. In spite of this, a new contributor to the problem (Marriot 1927) argued that the schist really underlay the limestone and, being less resistant to erosion, caused undermining of the limestone hills, which were thus cliffed. The intrusion of granite domes caused gravity collapse structures in the overlying schist, which then locally overlay the limestone tectonically.

An independent view was sought, and to this end R. H. Rastall visited Kinta. He published his interpretation in 1927. The succession according to Rastall was limestone, schist (including his Tekka Clays formed by collapse of the schist over the dissolving limestone), granite, Western Boulder Clays, and Gopeng Beds. Rastall considered the distribution of the limestone hills to be due to three symmetrical anticlines striking north-south and with steep eastern limbs. Scrivenor (1928, 1931) finally acknowledged that the Gopeng Beds were younger than the granite, particularly because they were seen to contain lignite veins, and he compared them with the high-level alluvium of Johore. However, he refused to accept Rastall's interpretation of the age and origin of the Western Boulder Clays.

The geological survey of other parts of the country was continuing during this time. Aided by E. S. Willbourn, Scrivenor traveled extensively and laid the foundations for a general appraisal of Malayan geology. Specialist help was forthcoming from the Imperial Institute and the British Museum in London. Members of the Imperial Institute investigated minerals from the alluvium and *amang* and reported on corundum, monazite, struverite, and tungsten ores with a view to their possible economic importance. Upper Paleozoic and Triassic fossils were described by the British Museum staff, notably by Newton (1906, 1923, 1925, 1926), who took a general interest in the geology of Malaya and published reviews of some of Scrivenor's more important papers.

By the 1920s a large amount of data had been assembled and Scrivenor proceeded with a series of syntheses—on the physical geology (1921), on the structure (1923), and on the paleontology (1926) of Malaya—leading to the production of two summary works, *The geology of the Malayan ore deposits* (1928) and *The geology of Malaya* (1931). In connection with these, a mineral distribution map was published in 1926, and general geological maps of Malaya appeared in 1927 and 1930. The stratigraphy was poorly recognized at this time. The lack of fossil discoveries in many areas permitted false correlations to be made based solely on general similarities in lithology. Thus all limestones were grouped together as Carboniferous and Permian. They were underlain by assumed Lower Carboniferous clastics in the Langkawi Islands, but elsewhere they formed the oldest rocks, including the Raub shales in Pahang. Above these were clastic rocks, everywhere referred to the Triassic even though Triassic fossils had been discovered only in a few scattered localities in Perak, Pahang, and Singapore. Likewise all the granites were assumed to have been intruded in the later Mesozoic.

Regional geological studies were published by Scrivenor on Singapore (1924), Malacca (1927), with Jones on south Perak and north Selangor (1919), and with Willbourn on Langkawi (1923). Willbourn also contributed a review of the Pahang Volcanic Series (1917) and papers on the geology of south Selangor and Negri Sembilan (1922), Kedah and Perlis (1926), and Johore (1928). In 1925 a third member of the Geological Survey Department, H. E. F. Savage, summarized the geology of Kelantan as it was then known.

Thus at the end of Scrivenor's career in Malaya, considerable areas of the country had been covered by geological reconnaissance and it was time to proceed to the next stage in its exploration, that of more detailed surveys and mapping on a scale of one inch to one mile (1:63,360).

1930 to 1955—Systematic Mapping by the Geological Survey

On Scrivenor's retirement, Willbourn took over the directorship of the Geological Survey Department. Savage, his assistant, undertook a detailed survey of the Sungei Siput area north of the Kinta Valley during the years 1931 to 1933. The publication of this work was delayed until 1937 for economic reasons. It included a one inch to the mile geological map, the first to be compiled for Malaya. Willbourn himself began a detailed survey of the gold belt north of Raub, and F. T. Ingham mapped the Tapah and Telok Anson area to the south of the Kinta Valley during 1934 to 1937. Ingham's work was published in 1938 as a second memoir. In 1937 the Geological Survey Department was enlarged and two more geologists, J. A. Richardson and F. W. Roe, arrived in Malaya. They were followed by F. H. Fitch in 1938, and H. Service and J. B. Alexander in 1939.

With this increase in staff, the organization of Survey work was decentralized, and district offices were opened in various parts of the country where detailed surveys were being made. Richardson took over Willbourn's field work in the Raub area, and a third memoir on this area was published in 1939. Fitch's work was based on Kuantan, Alexander's on Bentong, Service's on Kuala Lipis, and Roe's on northeast Selangor.

The rocks were divided lithologically into limestone and associated rocks, and quartzites and associated rocks. Richardson used the terms "Calcareous Formation" and "Arenaceous Formation" and included in the former an "Amphibole Schist Series." A few more fossil faunas were discovered, which tended to confirm the Carboniferous age of the limestone and Triassic age of the quartzite. They were identified by paleontologists at the British Museum (Edwards 1933, Cox 1936). A relatively large Lower Carboniferous fauna, already known since about 1920, was collected from the vicinity of Sungei Lembing and described by Muir-Wood (1948).

By 1941, however, field work had almost ceased, since all the Survey officers were engaged in some form of local military service. During the Japanese occupation (1942–1945), all except Roe were prisoners of war or internees. The chemical laboratory at the Batu Gajah headquarters was used by the Japanese. There was a total loss of scientific equipment, although the records and library were largely saved, thanks mainly to the efforts of the clerical staff. However, the district offices suffered heavy losses of note books, manuscripts, and maps.

Because of this, after the war it was necessary to resurvey much of the area previously covered. Willbourn retired from the directorship in 1945 and Ingham took his place. Savage was reappointed, first working in south Selangor and later in the Kuala Lipis area, where he discovered Middle Triassic ammonoids (Savage 1950). Alexander, Fitch, and Service continued their mapping at Bentong, Kuantan, and Kuala Lipis, respectively. Richardson resigned in 1946 but his work in northwest Pahang (Chegar Perah and Merapoh) was published in 1950 as a memoir. In this, three facies of the Calcareous Formation (now termed Calcareous Series) were recognized, and the Amphibole Schist Series, Arenaceous Formation (now termed Arenaceous Series), and Pahang Volcanic Series were extended along the strike from the Raub area. Granitic and syenitic igneous suites were differentiated, but the structure and age relationships of these remained obscure.

Soon after the war had ended, field work again became curtailed because of terrorist activities, which continued as a serious menace until the late 1950s. A. C. Amies, who was appointed as a Survey geologist and began work in southern Trengganu in 1948, was killed by terrorists the following year. A Geological Survey Department of the British Territories in Borneo was initiated in 1949, and Fitch and Roe left Malaya to head the new establishment.

Roe's work in north Selangor was published as two memoirs in 1951 and 1953. The records of his original survey in the years 1937 to 1941 were lost during the Japanese occupation, and the area had to be rapidly resurveyed in 1947 and 1948. The limited time available for the resurvey and the activity of bandits restricted the scope of the publication, emphasis being given to the mineral resources. Richardson's rock units were extended to include the pregranite rocks of Selangor. They were also used by Fitch in describing the rocks of the Kuantan district (Fitch 1952).

Alexander's studies of the Bentong area, prepared for publication in 1952, remained unpublished until 1968. He recognized a Schist Series formed of the oldest rocks in the area, followed by an Older Arenaceous Series, and the Calcareous Series (here mainly of shale), all supposedly of Upper Paleozoic age: he also noted a Younger Arenaceous Series, supposedly Triassic. No diagnostic fossils had been found in the area before 1960. The final memoir to be prepared during this period, but not published until later, was on the Kinta Valley (Ingham and Bradford 1960). In this the Tekka Clay and Western Boulder Clays were regarded as mainly being of weathered schist. The limestone beneath the alluvial floor of the valley was seen to be continuous with that of the hills, and the authors found no evidence of block faulting.

Work outside the Geological Survey Department was little in evidence except for that done in Singapore by Mrs. F. E. S. Alexander, who produced a notable synthesis of Singapore geology (1950). This work has since been neglected but remains a sound basis for subsequent research.

During the period 1930 to 1955, ideas on Malayan geology in general had become rather stereotyped, based on Scrivenor's earlier interpretation. Factors that were to change this attitude began to appear in about 1956 and led to the expansion of ideas and eventually to the present outlook.

1955 to the Present—New Discoveries and Revisions

The easing and eventual cessation of terrorism allowed geological work to be extended greatly between 1955 and 1960. Efforts were also intensified because the Geological Survey Department was greatly enlarged; an average of about 20 geologists worked during this period under the directorship, for most of the time, of J. B. Alexander.

Certain anomalies were noted between the geological map of Malaya and that of Thailand, and a survey of the border areas by a joint Thai–Malayan geological party was arranged and carried out in 1955 and 1956. This led to the recognition of Lower Paleozoic fossils in Langkawi and Perlis and to the extension of the known geological history of the peninsula back to the Upper Cambrian (Alexander 1959). Regional mapping on a scale of one inch to one mile was started in Perlis, Langkawi, Kedah, Upper Perak, Kelantan, Trengganu, Malacca, south Johore, and Pulau Tioman, in addition to work continuing in the Kuala Lipis area and in southwest Pahang.

Rapid surveys by a number of Survey geologists in coordination were carried out in remote areas of south Pahang and around the junction of the States of Kelantan, Trengganu, and Pahang. In the latter area, late Jurassic or early Cretaceous plant remains were discovered in the postgranite continental sediments of Gunong Gagau (Paton 1959). Correlative beds were later found in Johore. Detailed mapping in south Kedah led to the discovery of abundant graptolites and tentaculitids (Burton 1967a), which were at first thought to be Lower Silurian only but were later recognized to include a Lower Devonian fauna (Jones 1967a). Silurian, Devonian, Carboniferous, and Permian fossils were found in the limestone of the Kinta Valley (Alexander and Muller 1963, Gobbett 1966, Suntharalingham 1968), and Silurian fossils were discoverd in the limestone of Kuala Lumpar (Thomas 1963, Boucot et al. 1966). Lower Devonian fossils were collected near the top of the "Older Arenaceous Series" of Alexander in southwest Pahang (Jones 1967b), and Triassic fossils were found to be widespread by Ja'afar bin Ahmad in the Raub Series of central Pahang. These occurrences and many others, summarized by Jones et al. (1966), resulted in a complete review of Malayan stratigraphy.

In 1959 the Geology Department of the University of Malaya moved to Kuala Lumpur from Singapore, and it has increased considerably in staff, accommodation, and equipment since then. This department had professional geologists who were able to pursue purely academic lines of research, and this resulted in an improvement of our understanding of the stratigraphy (Gobbett 1965a, Suntharalingham 1968), the structure (Koopmans 1965, 1966), and the plutonism (Hutchison 1964a) of parts of the country.

Many of the fossil faunas were investigated by Japanese paleontologists who also visited Malaya and made their own collections. Thus the Lower Paleozoic shelly faunas were described by Kobayashi (1957, 1958, 1959); the Upper Paleozoic faunas by Igo (1964, 1966), Ishii (1966), and Sakagami (1963); Triassic fossils by Tokuyama (1961), Hada (1966), Sato (1963), and Kobayashi (1963a,b); and the Gagau flora by Kon'no (1966, 1968).

Considerable progress was also made in geophysical studies. An airborne magnetometer and scintillation counter survey was financed under the Colombo Plan (Agocs 1966). This study provided information for a general structural interpretation of the area covered and indicated possible economic deposits of iron ore, as well. Work on the radioisotope dating of Malayan granites was begun by the Institute of Geological Sciences (Snelling 1965,

1967; Anon. 1966). A range of granite ages from Carboniferous to Tertiary indicated a complex plutonic history. Pioneer work in delimiting boundaries in the alluvium and bedrock by electrical resistivity methods was carried out by Chan (1967).

Although aerial photographs had been available previously, a new aerial photographic coverage of Malaya (flown in 1966 under the auspices of the Colombo Plan) encouraged their use in interpreting the geological structure of the country.

Pleistocene geology and geomorphology were studied mainly by workers outside the institutions just mentioned. An important study on the Kinta Valley was undertaken by Walker (1956), who divided the alluvial deposits into Boulder beds (including the Gopeng Beds, the Western Boulder Clays, and the Tekka Clays), Old Alluvium, and Young Alluvium. He interpreted earlier sea levels at about 70 m (230 ft), and 15 to 30 m (50 to 100 ft). Nossin (1961 to 1965) published a series of studies on the geomorphology of the east coast, attributing the beach ridges to a gradual fall in sea level of about 6 m during the Holocene.

In the last few years several oil companies have become interested in the possibility of oil reserves underlying the offshore areas in the South China Sea. Interest has also been shown in the offshore areas of the west coast by tin-mining concerns. The Geological Society of Malaysia was founded in 1967 and has benefited from the generosity of oil and mining companies in providing funds for publications.

The present time is thus one of very active interest in Malayan geology and it is appropriate to review our knowledge.

SUMMARY OF THE GEOLOGY AS PRESENTLY UNDERSTOOD

The Malay Peninsula records a Phanerozoic history without major gaps from the late Cambrian to the late Triassic. The record of the succeeding Jurassic and Cretaceous is less complete, and Tertiary rocks have a very limited distribution and are probably all late Tertiary. Quaternary deposits are extensive and have economic importance. The Peninsula is composed of a great variety of rock types reflecting a spectrum of environments in space and time, and in this respect it is typical of a continental margin that has experienced a complex set of geosynclinal, orogenic, and postorogenic events.

The geological history of Malaya is summarized in Table 1.1.

Late Cambrian to Triassic

A period of marine geosynclinal sedimentation, plutonism, and volcanism with intermittent and more or less localized earth movements occurred from the Late Cambrian to the Triassic. From the Lower Paleozoic rocks it is possible to interpret a major geosyncline occupying at least the western half of the peninsula, continuous northward with the Shan States of Burma and with Yunnan. For most of its history this geosyncline had a well-defined miogeosynclinal trough to the west and a eugeosynclinal trough to the east, separated by a geanticlinal ridge. Volcanism was confined to the eugeosyncline and the flanks of the geanticline, where it was widespread. A foreland to the west has been identified with Gondwanaland but may be represented by a basement, as yet unknown, in Sumatra.

Lower Paleozoic sedimentation apparently continued without a break into the Lower Devonian within the miogeosyncline, geanticline, and eugeosyncline. In the northern part of the miogeosyncline, however, there is clear evidence of a period of folding between the Siegenian and the end of the Devonian. This folding may not have had a very wide extent. In Perak no apparent unconformity is present within the Kinta Limestone, which ranges in age from Silurian to Permian. Elsewhere the Lower Paleozoic has a tectonic boundary or is overlain by poorly dated Upper Paleozoic or Triassic rocks. However, in Selangor the Lower Paleozoic rocks were metamorphosed prior to the deposition of the Upper Paleozoic, and in west Pahang the Lower Paleozoic metasediments contrast with the unmetamorphosed Triassic rocks overlying them. Granite clasts from rocks overlying the unconformity in Langkawi cannot be related to any known plutonism of mid-Paleozoic age within the geosyncline, and these clasts have been interpreted as derived from the foreland.

Upper Paleozoic sediments occupy two distinct tracts, one on the west of the peninsula, the other on the east. In the center, limited outcrops of mainly Permian rocks occupy anticlinal cores, but most of the Upper Paleozoic material is buried beneath the Triassic. The earlier geosynclinal pattern is more obscure, although we know that shallow-water sedi-

TABLE 1.1 *Time correlation of the major rock formations of the Malay Peninsula (terrestrial formations are stippled).*

Period	Northwest Malaya	Lower Perak	Selangor, Negri Sembilan, Malacca	Kelantan and Trengganu	Pahang	Johore and Singapore
TERTIARY	Bukit Arang beds	Enggor beds	Batu Arang beds		Kuantan basalt	Niyor and Kepong beds
CRETACEOUS			Mt. Ophir Granite	Gagau Group	Tebak Formation	
JURASSIC U				Tembeling Formation		
JURASSIC M	Kodiang lst.		Major Period of Folding	and Granite Intrusion		Jurong Formation
JURASSIC L	Semanggol Formation			Gunong Rabong Formation / Jelai Formation / Kerdau Formation		
TRIASSIC U						
TRIASSIC M	Thrusting and Granite intrusion			Gua Musang Formation	Granite intrusion	
TRIASSIC L						
PERMIAN U	Chuping Formation / Kubang Pasu Formation	Kinta	Kenny Hill Formation	Verbeekinid limestones and andesites		
PERMIAN M				Granite intrusion	Pecopteris beds	
PERMIAN L	Singa Formation		Folding	Metasediments, andesites, and rhyolites	Kuantan Group	
CARBONIFEROUS U		Limestone	Kuala Lumpur Limestone / Dinding Schist			Metasediments of east Johore
CARBONIFEROUS L	Folding		Metasediments of Negri Sembilan and Malacca			
DEVONIAN	Setul Formation / Mahang Formation / Baling Formation			Taku Schist and migmatite complexes	Bentong Group	
SILURIAN U						
SILURIAN L						
ORDOVICIAN	Machinchang Formation					
CAMBRIAN						

10

ments are important in the west and volcanic rocks are limited to the central and eastern parts of Malaya. In the northwest, considerable relief, which could have been produced by the Devonian movements, is suggested by the presence of slump bedding and associated poorly sorted pebbly mudstones and sandstones. Shallow-water carbonate deposition was important in the west and central parts of the peninsula, particularly in the Permian. The presence of land plants in the later Paleozoic sediments of Pahang and Kelantan is evidence for the presence of a land mass to the east, possibly connected with "Indosinia," which dates from the middle Carboniferous.

In central Malaya the Permian appears to be conformable with the Lower Triassic. The Lower Scythian bivalve *Claraia* is found in mudstones similar in lithology to Upper Permian mudstones with the brachiopod *Leptodus*. Elsewhere the Lower Triassic and probably much of the Upper Permian are missing, and Anisian or younger Triassic strata overlie the Paleozoic. In Langkawi, thrusting occurred in the Permian, and this was followed by late Permian granite intrusion. Radiometric dating implies that granite intrusion into the mesozone began in the Upper Carboniferous and continued intermittently into the Triassic. Phases of this plutonic activity were probably accompanied by uplift, and an important phase of uplift seems to have occurred in the late Permian. It is probable that catazonal plutonism and metamorphism also occurred at about this time, but some may have been earlier and some later. This deep-seated metamorphism produced migmatite complexes—now exposed mainly in central Malaya where, at a later stage, they have been pushed up farther into the upper levels of the crust to be exposed by Tertiary denudation.

The final period of geosynclinal marine sedimentation is recorded in Middle and Upper Triassic rocks, which are widespread in the northwest and central parts of Malaya. In these, three facies are widely developed. Coarse clastic sediments with a benthonic bivalve fauna, mainly Carnian in age, are found in central and south Malaya. Shales, in places associated with conglomerate and chert, and containing a nektonic fauna of ammonoids and the bivalves *Daonella* and *Halobia,* range in age from Anisian to Carnian and are widespread in the northwest, and in Pahang and Kelantan. A third facies is represented by mainly pyroclastic andesitic to rhyolitic rocks in Kelantan, Pahang, and Johore.

Late Triassic Orogeny

In late Triassic times a major orogenic phase terminated marine sedimentation in the peninsula, and the Paleozoic and Triassic rocks were tightly folded generally about north-south axes. The major fold pattern was a synclinorium flanked by two anticlinoria. The Upper Paleozoic rocks were metamorphosed generally in the greenschist facies, although along the eastern side of the peninsula they were more strongly metamorphosed and cleaved than elsewhere. The Triassic rocks were not so deeply buried and did not suffer regional metamorphism. At a late stage in the orogeny, granite was intruded on a large scale, conforming generally to the regional strike but discordant to individual sedimentary formations. This granite effected tin mineralization in the vicinity of its contact with the country rock.

Jurassic to Paleogene

Succeeding the Late Triassic orogeny was a long period of erosion, repeated uplift, and rejuvenation of granite intrusion. This was accompanied by a molasse type sedimentation in continental basins associated with normal faulting and local volcanism. Molasse sediments were widespread in the Jurassic. These were folded into open flexure folds trending north-northwest before the deposition of late Jurassic or early Cretaceous plant-bearing fluviatile and lacustrine sediments in the eastern part of the peninsula. These sediments were later subjected to uplift and erosion, and they now form very restricted outcrops. In Malacca minor granite bosses were intruded into the epizone at the end of the Cretaceous.

There is no record of Lower Tertiary sediments. Indeed, throughout this period Malaya must have been a positive region, contrasting with the developing Tertiary geosynclines of Indonesia.

Neogene to Recent

During the Neogene, major transcurrent faults developed in Malaya in response to large-scale earth movements in Indonesia and the formation of the Sunda Arc. A general east-southeasterly stretching caused transcurrent faulting of a dominantly sinistral type and imparted an arcuate form to the regional strike and the axis of the Main Range Granite in the western half of the peninsula. Small sedimentary basins with coal measures are probably of late Tertiary of Quaternary age, and some appear to be localized in depressions aligned with the major faults. The basaltic lavas around Kuantan indicate the presence of crustal tension in that area. Quaternary erosion, accentuated by climatic and sea-level changes, has resulted in widespread and thick alluvial deposits in the coastal areas and offshore. Of less bulk but of great economic importance are the tin-bearing alluvial deposits of the major valleys draining the Triassic granites.

STRATIGRAPHIC NOMENCLATURE

In this work the names employed for rock units are taken mainly from previous names, regardless of whether they have been published, although some have been modified to conform with modern practice and with the Geological Society of Malaysia's code of stratigraphical nomenclature. Only the names of well-documented rock units are capitalized; other names are treated as informal.

CHAPTER 2

Geomorphology

H. D. TJIA

Eight prominent north-south to northwest-southeast trending mountain ranges alternating with valleys dominate the topography of the Malay Peninsula. Most of the ranges are formed of granitic rocks and possess peaks around 2000 m (6600 ft) high. The highest summit of the peninsula is formed by Gunong Tahan (2178 m, 7186 ft), which consists of resistant sedimentary rocks. Figure 2.1 shows the drainage and main divides of the peninsula. The trunk streams and major tributaries run parallel as well as transverse to the structural grain of the peninsula. Two of the large rivers, the Sungei Pahang (420 km, 262 miles) and the Sungei Kelantan (280 km, 175 miles) debouch into the South China Sea; the third large river, the Sungei Perak (350 km, 218 miles) drains into the Malacca Strait.

As in other humid tropical regions, chemical weathering processes predominate in Malaya and Singapore because of the constant high temperatures, the availability of water, and the resulting dense vegetation (Figure 2.2). Various landscape elements demonstrate the influence of chemical disintegration. A talus deposit at the foot of a scarp is rare. This is because the debris that should have formed the talus is usually weathered or achieves this state rapidly, so that particles of the decomposed rock are sufficiently small to be transported into the rivers by sheetwash during heavy rains. Torrential rains and deep zones of weathered rock further result in the development of more drainage lines per unit area than in similar rock types under different climatic conditions. Denudation rates are probably very high and may be of the order of a few millimeters each year, as the figures from Java and Indonesia suggest (Rutten 1938); the average denudation rates in Europe and North America, on the other hand, are only a millimeter or less annually.

Under tropical climatic conditions, iron-containing rocks generally weather into a typical red-brown hardpan of nodular to spongy structure, or laterite. Laterization is envisaged as a process of leaching of silica and basic oxides while concentrating hydrated iron and aluminum oxides. The formation of laterite appears to require prolonged geologic periods of subaerial weathering, as its absence upon Quaternary deposits implies (Mohr and Van Baren 1954). These authors have also presented an excellent compilation of the various stages of laterization.

In Malaya, laterite has been used as building material in Malacca (Scrivenor 1931), whereas in Johore this weathering process has accounted for the formation of economic bauxite deposits. Grubb (1968) described the bauxite occurrences of Johore. Alexander (1959) presented her observations on tropical weathering of rocks on Singapore Island. She found that pure iron oxide concretions were associated with igneous rocks and shales and that, as expected, secondary quartz occurred in quartzites.

Richardson (1947c) suggested the following morphologic development. The drainage system was thought to have evolved since the Cretaceous. Initially it consisted of north-south subsequent streams and east-west consequent streams. By stream cap-

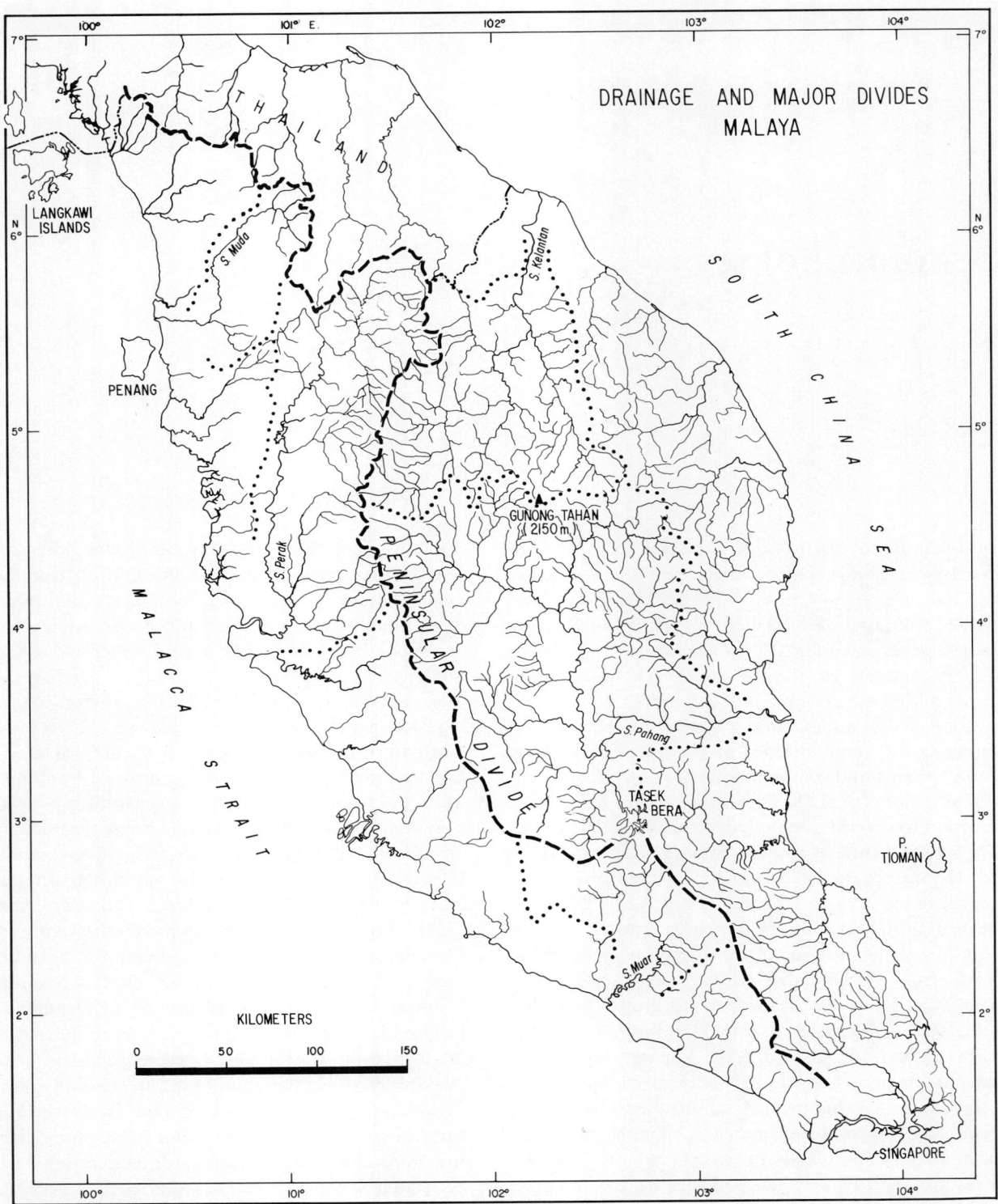

FIGURE 2.1 *Drainage and major divides of Malaya. Heavy dashes indicate the peninsula divide; dots denote the major watersheds. Formerly the Sungei Pahang may have drained into the Malacca Strait through the swampy area of Tasek Bera and the Sungei Muar.*

FIGURE 2.2 *Weathering of granitic rocks:* (a) *Earth pillars formed by heavy rain on granite colluvium, Petaling Jaya, Selangor; Photograph D. J. Gobbett.* (b) *Colluvium with granite core boulders at Tanjong Tokong, Penang. Note sharp junction of hard, relatively fresh granite and deeply weathered rock. Photograph P. H. Stauffer.*

ture, rectangular drainage patterns were formed. During the interglacial high sea levels, the lower courses of the rivers became aggraded and they were re-eroded in each following emergence: there are now a number of depositional stream terraces and knickpoints in the valley profiles. It is improbable, however, that the higher located evidence of rejuvenation—up to 1500 m (5000 ft) in Pahang, according to Richardson—is related to Quaternary sea levels. If we take into consideration the tectonically stable character of the area during this period, a 30-m (100 ft) upper limit of the Quaternary sea level is likely, as found elsewhere (Flint 1957).

Special mention should be made of the limestone topography of Malaya. The karst features consist of tower or mogote karst, which is a typical limestone topography for the tropics and is indicative of a deeply seated water table. Other karst landforms are rectilinear valleys, elongated *uvalas* or *poljes*, and solutional depressions or dolines.

Eyles (1966, 1968a,b) introduced geomorphometry into the study of Malayan landforms. Quantitative studies are claimed to be in good agreement with qualitative descriptions of various morphological features. Generally, high and low mountains have been found in association with granitic rocks; isolated steep, high hills characterize metamorphic limestone and quartzite, and low convex hill country is formed on shale or deeply weathered igneous rock.

HIGHLAND AREAS

The greater part of the Malay Peninsula is formed of a deeply dissected highland area which has had a long erosional history through successive uplifts during the Cenozoic. Its morphology is largely controlled by rock type and structure.

Granitic rocks dominate the topography by forming all but one of the major mountain ranges with summits exceeding 2000 m (6600 ft). Eight granite ranges are present in the latitude of Penang Island. The topography of these granites is characterized by steep valley walls, numerous waterfalls and rapids, and small remnant upland plateaux now providing hill resorts like the Cameron Highlands. These features indicate a late youthful stage of erosion. The drainage pattern in granitic regions is mainly rectangular or angulate, indicating the influence of joints and faults.

Lineament patterns on aerial photographs show agreement with fracture patterns as predicted by fracture analysis. Certain lineament directions may be predominant in these granite areas. For instance, the Penoh and upper Kinta valleys, east of Ipoh, Perak, run as an extraordinarily straight line that is at least 15 km (10 miles) long with a bearing N125°E. Other examples are the Tanglir–Benus valleys west of Bentong, Pahang; the strike is N125°E, and the lineament can be followed over a distance of almost 25 km (16 miles). Many of the linear stream valleys in granite areas are associated with sheared rock (Alexander 1968) and obviously mark important fault zones. Radial drainage may occur on more or less isolated granite peaks.

A different morphology is expressed by the granitic rocks of Johore and Singapore. In these regions granite is marked by undulating hill topography with elevations of a few hundreds of meters. Apparently lower initial elevation of the granite masses and deep weathering account for this. Burton (1969) has inferred from the summit distribution of the nonmarine Gagau Group and Tebak formation (late Jurassic or early Cretaceous) in central and south Malaya, a southward tilt of their surface amounting to a net gradient of 1 in 453 (Figure 5.8). This summit gradation plane is considered to be depositional as well as erosional in character. The time of warping is interpreted as late Cretaceous to early Tertiary. Furthermore, the vertical range of the presumably early Pleistocene "Older Alluvium" in Singapore and Johore, from its base at 46 m (150 ft) below sea level to its surface at 70 m (230 ft) above sea level, may represent a younger manifestation of the previous tilting movement.

Large quartz dykes with precipitous walls are conspicuous ridge formers. A well-known landmark is the large Klang Gates quartz ridge north of Kuala Lumpur (Plate 18). This ridge is traceable for a distance of 15 km (10 miles); it has summits as high as 200 m (650 ft) above the surrounding country, and locally its width may exceed 150 m (500 ft). Other quartz ridges are usually 0.5 to 2 km long. The quartz ridges of Malaya are believed to have been fracture fillings. Some ridges consist of strained and sheared quartz that indicate post-crystallization movements in these zones of weakness. One prominent dyke of ignimbrite character occurs in Kelantan and forms a row of elongated hills that show alignment for a distance of more than 20 km (12.5 miles) (Aw 1967).

The morphology of the Taku Schist in Kelantan is marked by the prevalence of slopes over other topographic features, by steep valley walls and sharp-crested ridges with mean summits at elevations of 375 m (1238 ft), and by one more or less centrally located peak of 500 m (1650 ft) high. The Taku Schist area is now in its mature stage of erosional development. The drainage basins are distinctly elongated, and streams radiate from the central peak area. Aerial photographs show that stream segments are linear with recurrent bearings; in other words, the river pattern seems to be influenced by foliation or fractures.

The topography of areas formed of sedimentary rocks is controlled by their strike. The sediments frequently dip steeply and give rise to a series of strike ridges and valleys in successive beds of varying competence. Such features are most clearly displayed in the Triassic formations, the Tembeling Formation, and the Bentong Group, which contain competent sandstone and conglomerate beds alternating with shales. Apart from the general north or northnorthwest trends, some of the strike ridges also show, by zigzag ridge and valley patterns, the location of fold plunges (Figure 2.3). Streams draining the sedimentary regions are well adjusted to the structure.

Where strike ridges are present, trellis drainage is the rule, and drainage patterns in the Tembeling Formation of Pahang provide excellent examples. The medium-sized tributaries in these regions are mostly synclinal or homoclinal streams. Anticlinal rivers are of smaller dimension and relatively unimportant morphologic elements. Semiannular drainage patterns occur in brachy-synclinal areas, that are underlain by the Tembeling and Gagau beds.

The well-adjusted drainage and the predominance of slopes over other topographic elements suggest a mature morphologic stage. On the other hand, stream profiles still possess knick points. Accordingly, the combined morphologic features indicate late youth.

LOWLAND AREAS

Most of the low-lying part of the Malay Peninsula forms the coastal plain and is discussed later, in connection with the coast. However, a broad low-lying inland marshy area is present around Tasek Bera in South Pahang.

A few investigators have hinted that in the not so remote past the Pahang River flowed through the Tasek Bera area and the present Muar River to debouch into the Malacca Straits (Scrivenor 1931). Later, an eastward-flowing stream appeared to have captured the upper reaches of the Pahang River in the vicinity of this marshy area, and thus the present configuration of the Pahang River became established.

LIMESTONE AREAS

Towerlike limestone hills occur in many localities in the northern half of Malaya (Gobbett 1965b). Their precipitous cliffs rise abruptly above the surrounding country and form local relief up to 600 m (2000 ft) (Plate 1). In their lowest 20 m (65 ft) these cliffs often carry notches and deep horizontal grooves. The hills are usually honeycombed by caves of various dimensions. The nature and origin of the caves have been fully discussed by Wilford (1964) and Gobbett (1965b). Other karst features are lapiés (Figure 2.4), dolines (locally known as *wangs*), dead-end valleys, and subterranean rivers. On aerial photographs these features may show distinct alignments and may

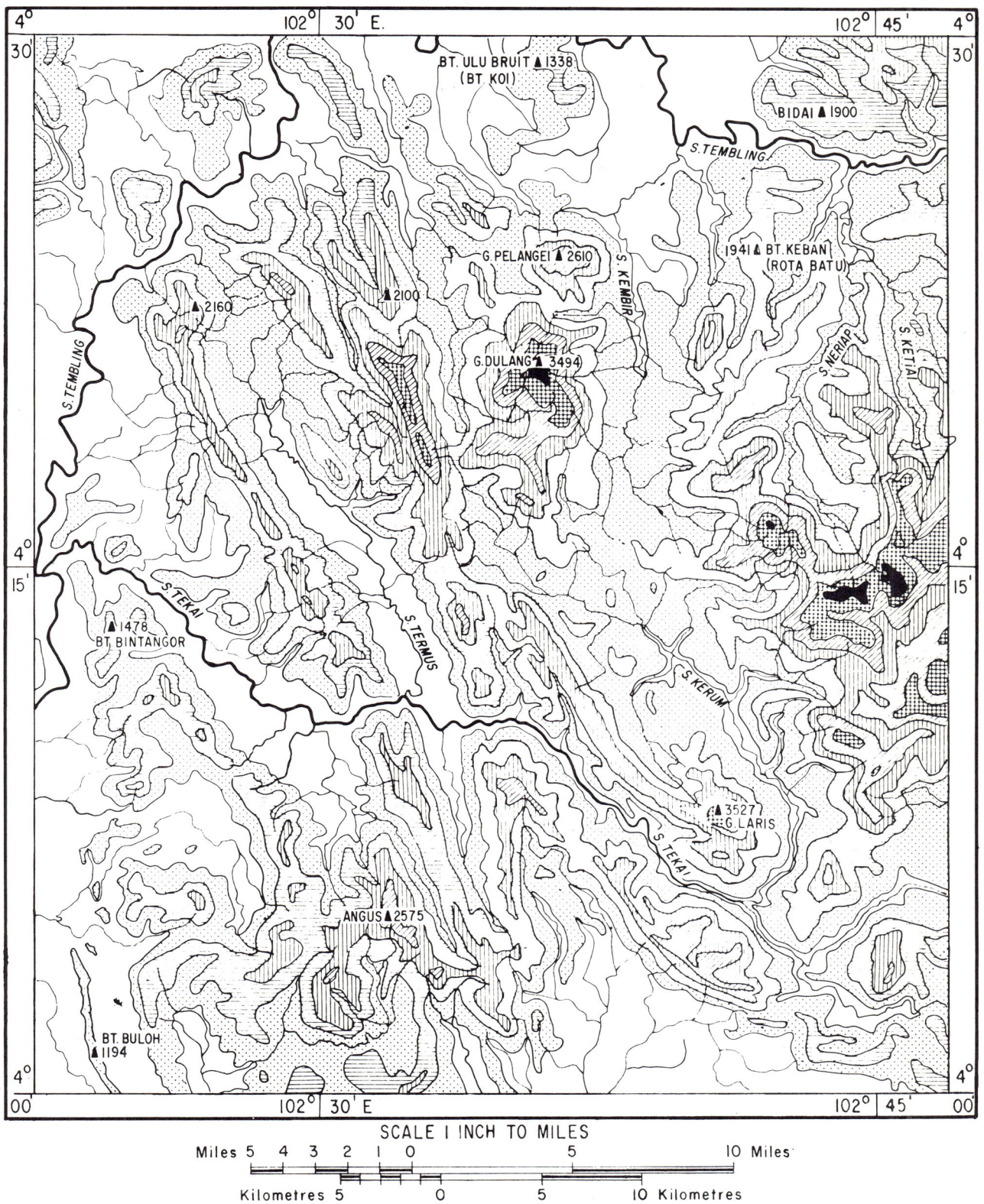

FIGURE 2.3 *Topographic map of northeast Pahang based on a 1:253,440 map published by the Directorate of National Mapping, Malaysia, sheet B-48S. Zigzag ridges are composed of Tembeling Formation and indicate plunging folds striking north-northwest. The contour interval is 152 m (500 ft). The Sungei Termus occupies a synclinal valley; the segment of the Sungei Tekai west of Gunong Laris flows in a homoclinal valley.*

PLATE 1 *Limestone hills in Perlis, of the Chuping Formation (Permian). View north, from left to right Bukit Chuping, Bukit Ketri, Bukit Chabang, and Bukit Jerneh. Photograph C. R. Jones.*

FIGURE 2.4 *Lapiés, Bukit Takun, Selangor. Photograph P. H. Stauffer.*

demonstrate their relation to regional fractures. In the Kinta Valley, around Kuala Lumpur, and in other areas, limestone underlies the alluvial plains which surround the hills. Its surface is generally deeply pitted, but the peaks of the pinnacles often define a flat surface (Figure 3.6).

Several theories have been advanced to explain the formation of limestone hills in Malaya. These invoke structural control, with or without the aid of various erosional processes including ablation, abrasion, and subaerial erosion. One of these theories ascribes the patchy distribution of the hills to their association with interpreted horsts, crests of anticlines, or even tectonic outliers. Paton (1964) reassessed the various hypotheses on the formation of the calcareous hills and concluded that lithology appears to be the dominating factor. He decided that the steep slopes were the result of subaerial erosion along joint planes and that the platformlike surface that is indicated by the pinnacles in calcareous bedrock was formed by marine abrasion. The deep horizontal basal grooves in the slopes seem to be due to solution by acid swamp water or river water. Some of the notches on limestone hills in Perlis and Kedah resemble present-day sea-level notches that are so common along calcareous sea cliffs and were probably formed in the same way (Figure 2.5).

FIGURE 2.5 *Cave in limestone at Bukit Kepala, Kedah. Note the prominent notch and remnant of smooth platform around the notch on the right. Evidently these notches in the Kedah limestone hills were produced by marine erosion when the sea level was higher. Photograph C. S. Hutchison.*

Other workers on tropical karst morphology are convinced that tower or mogote karst, like the karst landscapes that occur in Malaya, develop if the water table is deep. On the other hand, sinoid-karst, which is composed of numerous low hillocks, is formed if the water table is shallow. Flathe and Pfeiffer (1965) have described the development of the sinoid karst in the classical area of central Java in this way. The present author has proposed that the development of straight, convex, or concave slopes on sinoid hillocks depends on the ratio between rate of vertical erosion and rate of lateral erosion. In this way, the precipitous mogote slopes are indicative of extremely rapid vertical erosion, in addition to low water tables and availability of vertical joint planes (Tjia 1969a).

COASTAL FEATURES

The result of the last Quaternary sea-level rise is still apparent along the coasts of Malaya. Estuaries, offshore islands, and irregular shorelines are especially clear along the south and west coasts. On these coasts most deltas at the large river mouths appear to be estuarine, and they seem to have resulted from the aggradation of bays. In contrast, the eastern shoreline possesses many straight coastal stretches with well-developed beach ridges. A fan-shaped delta occurs at the mouth of the Kelantan River, but the largest river of Malaya, the Sungei Pahang, has a cuspate delta. The straight eastern coastal tracts are the result of their alignment with the geologic structures, aided by strong abrasion by the South China Sea. On the whole, however, aggradation on this shoreline seems to be predominant, particularly in the vicinity of the large rivers. The storm waves of the South China Sea also account for the development of beach ridges. Weaker wave action in the Malacca Strait is reflected by the occurrence of smaller beach ridges and narrow beach ridge zones along the west coast. The irregular outline of the west coast has been enhanced because in many localities it obliquely cuts the structural grain of the peninsula.

Coastal Plain Bordering the Malacca Strait

The Langkawi island group mainly exhibits irregular ria coasts consisting of numerous embayments, offshore islands, sea stacks, and sea arches, especially along the limestone coasts of the southern islands. The drowned character is further revealed by wide estuaries at the mouths of the larger rivers and by narrow sea straits, indicating submerged former river valleys. Most shorelines in the Langkawi Islands consist of steep cliffs. Narrow sandy beaches with low beach ridges may be exceptionally present, as along the southwest coast of Pulau Langkawi. Here a tombolo is being formed between Pulau Rebak and the main island.

The coastal plain of Kedah and Perlis (Plate 2) shows evidence of recent aggradation. Wave-cut notches and marine beach deposits around the isolated hill of Bukit Keriang, now 8 km (5 miles) from the coast in Kedah, show how recently the hill was an offshore rocky islet.

On Penang Island, only the west and east coasts show evidence of recent aggradation. In these places the coastal plain may be 2 km wide. However, steep cliffs and irregular shorelines of submergence predominate in the coastal regions.

The western coastal plain of the Malay Peninsula is widest in the vicinity of the Perak River mouth, where it is about 45 km (27 miles) wide. On the average, the coastal plain has widths varying between 20 and 30 km (12–18 miles). Locally, cliffs mark the shoreline. Small cuspate deltas have developed at the mouths of the Kedah and Muda rivers; but in other localities of the west coast, deltas are of the estuarine type. Indeed, rather than indicating aggradation, some of the delta-like regions may even

PLATE 2 *Coastal plain of south Perlis. View west toward the Langkawi Islands. Photograph C. R. Jones.*

represent drowned topography. In several places extensive drowning of the coastal region is indicated by swamps through which short meandering tidal rivers flow: examples are the Trong region, where the coastal swamps are 10 km (6 miles) wide, and the Sungei Panjang area in north Selangor, where coastal swamps measure 28 km (16 miles) across. Occasionally a beach ridge zone, up to a few kilometers wide, lines the coastal plain. Near the Perak River such ridges are 3 to 6 m (10 to 20 ft) above the present sea level (Koopmans 1964).

East and north of Butterworth, Penang, beach ridges have been recorded for at least 6 km (4 miles) inland (Courtier 1962). These sandy ridges rise a few meters above the surrounding plain. Raised beaches occur at 3 and 6 m above sea level south of the town. Farther inland this coastal plain is fringed by alluvial terraces at about 35, 25, 16, and probably also at 8 m above the present sea level. The degree of dissection of these terraces increases progressively with increasing height. Granitic clasts and sand grains are the main constituents of the terrace alluvium; some of the better sorted sands probably reflect a marine environment of deposition.

The prevailing west-northwest to northwest longshore currents between 0 and 6°N latitudes (U.S. Navy Hydrographic Office 1960) have resulted in the deflection of most river mouths to the northwest (Figure 2.6). For the Langat River such traceable displacement amounts to 10 km (6 miles). Some southeasterly displacements of lower courses on the west coast have also been observed, but in all instances, except one, these displacements have occurred in areas sheltered by former headlands. An exception is the Perak River, which was traced by Koopmans (1964); this river shifted its mouth to the south over a distance of about 30 km (18 miles). However, at one point some 20 km (12 miles) inland from the present shoreline, a progressive northwesterly displacement occurred during the river's more recent history (Figure 2.7). The evidence indicates that in previous times southeasterly directed currents were dominant along this stretch of the Malayan shoreline, whereas the northwesterly currents have only become important in more recent times. Beyond 6°N latitude, the resultant longshore current is toward the southeast; accordingly, it has deflected the Perlis and Kedah river mouths in the same direction.

A study of historical maps shows a rate of coastal accretion of 100 m (330 ft) in 80 years, or 16 m (50 ft) annually, in the Dindings area of Perak (Koopmans 1964).

Southern Coastal Plain

The southern coastal plain of Malaya has a typical submerged character. Two wide estuaries are the

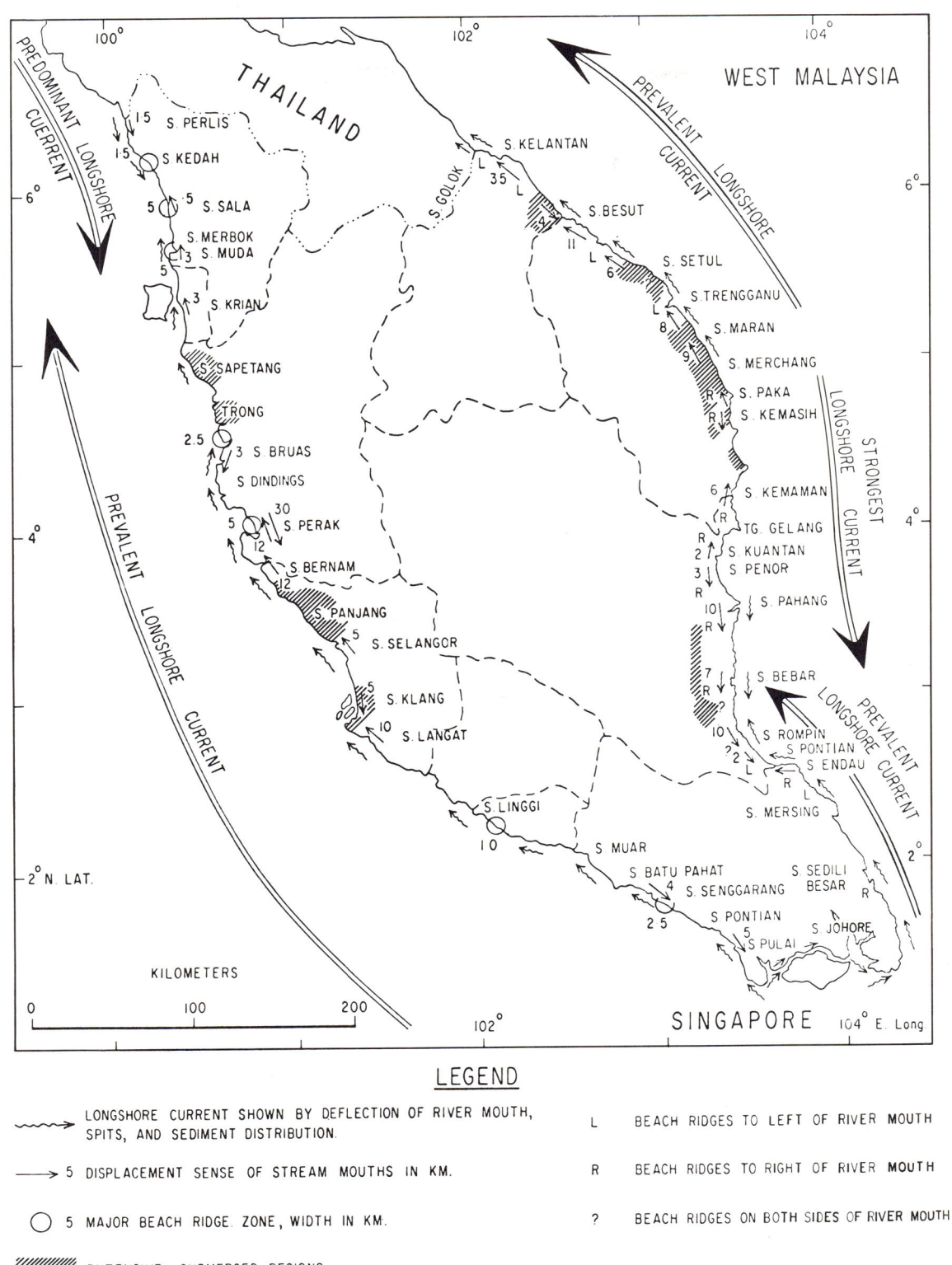

FIGURE 2.6 *Displacement of river mouths along the Malayan coast. Major beach ridge zones are common along the east coast but localized along the west coast at the circled localities. Longshore current movements have been derived from Van der Stok (1922) for the east coast and from the U.S. Navy Hydrographic Office (1960) for the west coast.*

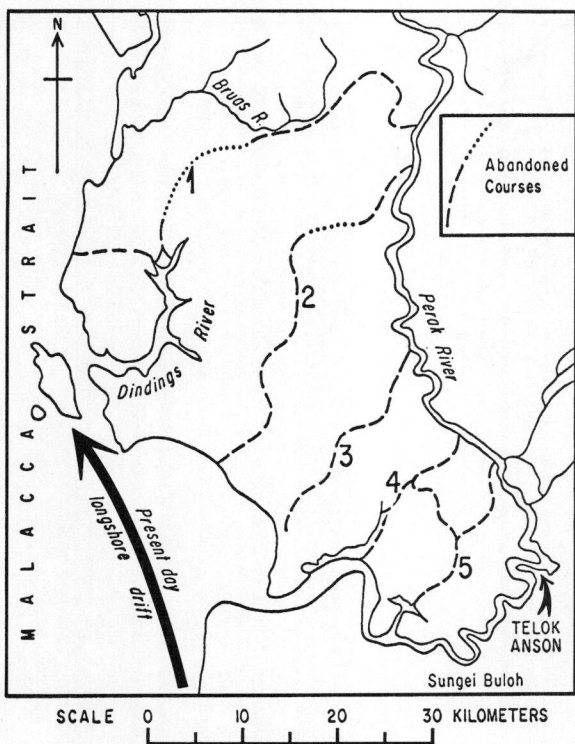

FIGURE 2.7 *Migration of the lower Perak River. The abandoned courses were traced from aerial photographs and show progressive southward migration of the mouth of the Perak River. When the prograding shoreline was at the locality of Sungei Buloh village, the river commenced to shift back to the north, probably influenced by the net longshore drift. After Koopmans (1964), with a few additions.*

Pilai and Johore river mouths. The deflection of stream mouths to the east corresponds to the prevailing direction of longshore currents (Figure 2.6). The presence of estuaries, irregular shorelines, and rare and narrow beaches, reveals that Singapore Island has a shoreline of submergence. Burton (1964) reported that the presumably lower Pleistocene "Older Alluvium" in Johore and Singapore extends from a depth of 45 m (150 ft) to an elevation of 69 m (226 ft) above the present sea level.

Eastern Coastal Plain

Off the east coast of Malaya the islands exhibit drowned coastlines that generally consist of steep cliffs and small embayments with narrow coastal flats.

Irregular shorelines also predominate along certain parts of the eastern coastal stretch of the mainland. Such shorelines are especially evident where structural ridges are cut obliquely by the shoreline, as between the Sungei Kemasih and the Sungei Kuantan, and between the Sungei Endau and the Sungei Mersing. On the whole, however, the eastern coastal plain bears marked evidence of aggradation, and straight sandy shorelines predominate, accompanied by wide coastal plains. The widest coastal plain along this coast is in the Kelantan River area, where it is 60 km (37 miles) across. In many other localities the coastal plain is 20 to 30 km (12 to 19 miles) wide and is usually extensively covered by fresh-water swamps (Nossin 1964b).

Two wide swampy areas are in the lower Merchang and lower Pahang regions, where these tracts are more than 30 km (48 miles) wide. On the seaward side of the swamps there is generally a zone of beach ridges that is 5 km (8 miles) or less across. Traces of abandoned beach ridges can be noted as far as 20 km (12 miles) inland in the region of the Kelantan and Bebar rivers.

Most investigators have recognized two beach ridge series, one with crests at 12 m (40 ft) and another at 5 to 6 m (16 to 20 ft) above sea level (Fitch 1952, Nossin 1961a, 1962, 1964a, 1965b). Nossin (1962) considered coastal accretion of the east coast as the result of infilling of bays that had been protected from the open sea by offshore bars. The general presence of extensive swampy areas behind a comparatively narrow zone of beach ridges is thought to support the assumption. However, Hill (1966) observed on a smaller scale that beach ridges may also extend seaward by direct accretion.

The wide extent of the beach ridge zone and the presence of high hills among the coastal swamps suggest that the swamps have developed by silting and inadequate drainage of low-lying areas rather than by downwarping, as is implied for the western coastal plain.

The thickness of the alluvium in the Kuantan area reaches 50 m (165 ft) in the Sungei Soi area, where the base lies 47 m (155 ft) below sea level. In Kuantan town fluviatile alluvium exceeds 30 m (100 ft) in thickness (Fitch 1952).

Nossin (1964b, 1965a) learned from ancient maps that the coastal area south of Kuantan became land only after A.D. 1800. It has been estimated that, within the last 1500 years, the same coastline has prograded at a rate of 16 m (53 ft) annually.

The movement pattern of various topographic elements along the east coast of Malaya corresponds to the direction of prevailing longshore currents of the South China Sea (Figure 2.6). Monsoon-controlled currents are predominantly southward throughout the year for the eastern coastal tract that lies between 3 to 5°N. latitudes, whereas the other coastal stretches are subjected to mainly north-northwesterly longshore current (Van der Stok 1922. The prevailing direction of longshore current influences the directional displacements of the lower stream courses and is also shown by trailing spit-ends. Beach ridges tend to develop on the down-current sides of river outlets with respect to the longshore current during the wet monsoon, when the streams are transporting the bulk of their annual load (Tjia 1970a).

Quaternary Sea Levels

The fluctuations of Pleistocene glaciations left records of sea level changes in the Sunda area. It is widely accepted that the lowest Quaternary sea level was about 100 m (330 ft) below and the highest sea level about 30 to 50 m (100-165 ft) above the present one (Umbgrove 1929, Flint 1957). Lower shorelines of Quaternary age have also been interpreted. However, these −130 m (−430 ft) sea levels have been recorded in such tectonically unstable areas as Japan (Fujii and Fuji 1967).

Tables 2.1 and 2.2 show the various levels that have been interpreted as indicating shorelines in the Malay Peninsula and Singapore Island. Because of the tectonically stable nature of the area during the Upper Tertiary and Quaternary age, the elevated shorelines may be correlated by height alone with shorelines of the Mediterranean area. Accordingly, tentative ages have been ascribed to these levels (Tjia 1970b).

Several Quaternary sea-level indicators in Malaya point to shorelines more than 30 m (100 ft) higher

TABLE 2.1 *Evidence for Quaternary sea levels in Malaya and Singapore*

Locality	Reference to Present Sea Level (m)	Type	Source
1. Perlis	+3	Terrace, 5200 ± 200 yr old shells	Alexander, 1961
2. Langkawi Islands	> +60	Elevated shorelines	Scrivenor and Willbourn, 1923
3. Dindings, Perak	+3	Beach ridge	Courtier, 1962 and Koopmans, 1964
	+6	Beach ridge	
4. Sitiawan, Perak	−108*	Alluvium upon phyllite	Willbourn, 1941
5. Lapan Utan, Selangor	−66*	Shells and unconsolidated sediments	Roe, 1951
6. Selangor	> −30*	Alluvium	Roe, 1949
7. Sungei Bernam, Selangor	−161*	Alluvium	Willbourn, 1923
8. West coast Malaya	−120*	Bedrock under alluvium, Kinta Valley;	Walker, 1956
	+75	Ubiquitous slope break	Walker, 1956
	+30 to +33	Coastal bench	Walker, 1956
	+3 to +5	Raised shoreline	Walker, 1956
9. Malacca Strait	−26.5	Peat, 10,000 yr B. P.	Keller, 1967
10. Chukai, Trengganu	+5	Beach ridge	Fitch, 1949
	+11	Beach ridge	Fitch, 1949
11. East coast Kuala Trengganu to Kuala Pahang	+10	Beach ridge	Nossin, 1965b
12. Sungei Soi, Kuantan, Pahang	−51*	Alluvium	Fitch, 1951
13. Kuantan, Pahang	+16	Alluvial flat	Fitch, 1949
	+5	Beach ridge	Fitch, 1951
	+11	Beach ridge	Fitch, 1951
14. Northeast Johore	+6	Beach ridge	Nossin, 1962
15. Johore	−46*	Marine clastics	Burton, 1964
	+70	Marine clastics	Burton, 1964
16. Singapore Island	+16 to +33	Alluvium	Scrivenor, 1924

* Reference plane is land surface that is usually a few meters above sea level.

TABLE 2.2 *Correlation chart of known and interpreted Quaternary shorelines in Malaya and Singapore (last column gives the tentative ages).*

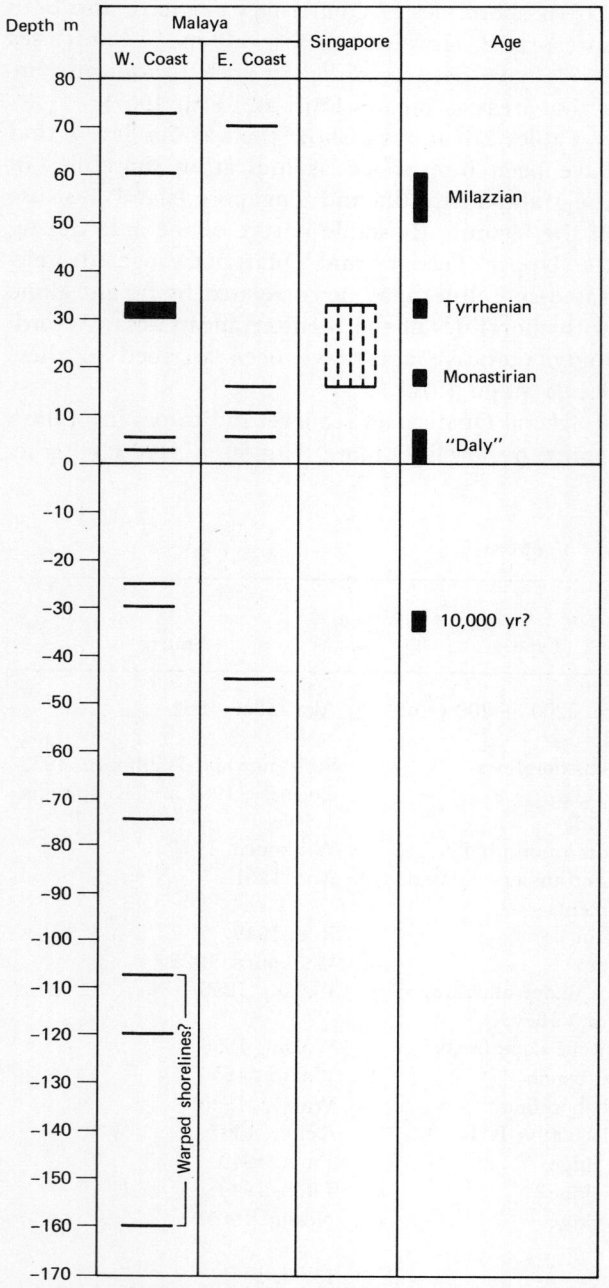

or 100 m (330 ft) lower than the present one. In Johore and Singapore, Old Alluvium (see Chapter 6) ranges from 45 m (150 ft) below to 69 m (226 ft) above the present sea level and is supposedly early Pleistocene or late Pliocene in age (Burton 1964). Alexander (1950) identified in these deposits shallow-marine as well as fluviatile features and postulated that they were deposited in an estuarine environment.

Two theories have been advanced to explain the vertical distribution of the sediments. Alexander (1950) assumed a constant sea level and a subsiding base upon which the sediments were laid down. Burton (1964) related the present base of the Old Alluvium with the average depth of the Sunda Shelf, which has been proved to represent a drowned peneplain 40 to 50 m (132 to 165 ft) below the present sea level (Molengraaff and Weber 1919). Zeuner (1959) wrote that the sea level progressively subsided about 100 m (330 ft) during the Quaternary, upon which were superimposed a number of fluctuations that were due to glacial and interglacial periods. If this is so, the Sunda peneplain must have had a total subsidence of 150 m (500 ft) during the Quaternary. To account for the top limit of the Old Alluvium at about 70 m (230 ft), Burton suggested an interglacial maximum sea level at about 75 m (250 ft). Such a sea level would also explain similar deposits in Perak and Selangor, respectively, the Boulder Beds and the Old Alluvium of the Kinta Valley (Walker 1956), and the Boulder Beds on top of the coal measures of Batu Arang (Roe 1953).

Additional indications for Quaternary downwarping or faulting along the west coast of Mayala consist of the thick alluvium of the Sungei Bernam (at least 161 m, 537 ft), and Sitiawan (108 m, 360 ft), both in Perak (Willbourn 1923, 1941). These localities are less than 10 m above sea level.

As elsewhere in the Indonesian Archipelago, three of the marine Daly sea levels that range from about 2 m (6 ft) below to 5 to 6 m (16 to 20 ft) above the present sea level have also been reported from Malayan coasts. By comparison with radiometrically dated levels elsewhere, the present writer has assigned ages to the Daly levels ranging from 6000 to 7000 yr B.P. for the 5 to 6 m level to less than 3500 yr B.P. for the lowest elevated level at 0.5 to 1 m (Tjia 1970b).

Some folding and faulting in Quaternary deposits have been reported for certain restricted localities. The Boulder Beds of Batu Arang, Selangor, have minor faults, near which the sedimentary layers are inclined as much as 40° (Roe 1953). The Old Alluvium in Johore and Singapore shows local, tight folds and even vertical beds; in Johore Bahru an outcrop has a fault of approximately 5 m (16 ft) throw that is accompanied by drag folds (Burton 1964). These very local and minor deformations are most likely related to slumping or collapse of basal support, especially if the sediments are resting on a calcareous substratum.

CHAPTER 3

Lower Paleozoic

C. R. JONES

The Lower Paleozoic rocks include a sedimentary cycle that was initiated in the Late Cambrian and can be traced through a series of well-defined geosynclinal deposits to the early Devonian. Associated with these sediments are acid pyroclastic rocks and an ophiolitic suite of igneous rocks. The geosynclinal conditions were arrested temporarily in early Devonian time by orogenic uplift in the northwest part of Malaya and an easterly migration of the depositional areas elsewhere.

The events of the Lower Paleozoic in Malaya constitute a well-defined episode in the development and orogenic modification of a geosynclinal belt that extended more than 3000 km (1800 miles) from Burma and Yunnan in the north to the Thai–Malay Peninsula in the south. The area of deposition was established in Cambrian time; it underwent structural modification in mid-Paleozoic time, was enlarged in the late Paleozoic, and finally rose to form the Thai–Malayan orogene in early Mesozoic times.

DISTRIBUTION AND CLASSIFICATION

The Lower Paleozoic rocks are confined to the western half of the Malay Peninsula (Figure 3.1), where they attain their greatest distribution along the east and west flanks of the Main Range Granite.[1]

[1] The Taku Schist of north Kelantan has been considered as Lower Paleozoic (Paton 1966), since it is overlain unconformably by probable Upper Paleozoic. However, since its age is unproven, it is here excluded from the Lower Paleozoic and is described in Chapter 9.

On the east side of the granite they form a narrow outcrop averaging some 11 km (7 miles) wide which extends southward for more than 300 km (185 miles) through the west parts of the states of Kelantan and Pahang into east Negri Sembilan and Malacca. The outcrop can be traced northward up the west side of the Main Range through west Negri Sembilan and east Selangor into Perak. In central and northern Perak the valleys, which lie between the Main Range and the subsidiary granite ranges to the west, are floored by these rocks. They also compose extensive tracts of country in the states of Kedah and Perlis and the Langkawi Islands, in the northwest part of the peninsula (Figure 3.4), and can be traced northward to the Satun and Songkhla provinces of peninsular Thailand.

The Lower Paleozoic rocks are commonly deformed to varying degrees, so that much of the fossil content has either been destroyed or is so poorly preserved that it is unrecognizable. Only in certain regions is the fossil record sufficiently complete to allow a detailed chronology to be worked out. In the northwest part of the peninsula it is possible to distinguish between rocks of the Cambrian, Ordovician, and Silurian systems, but in most other areas the Lower Paleozoic sequences cannot be so differentiated (Figure 3.1).

As strata of Lower Paleozoic age were encountered and studied over increasingly larger areas of the peninsula, they were found to fit into a definite pattern of sedimentary organization, that left little doubt about their origin in the component deposi-

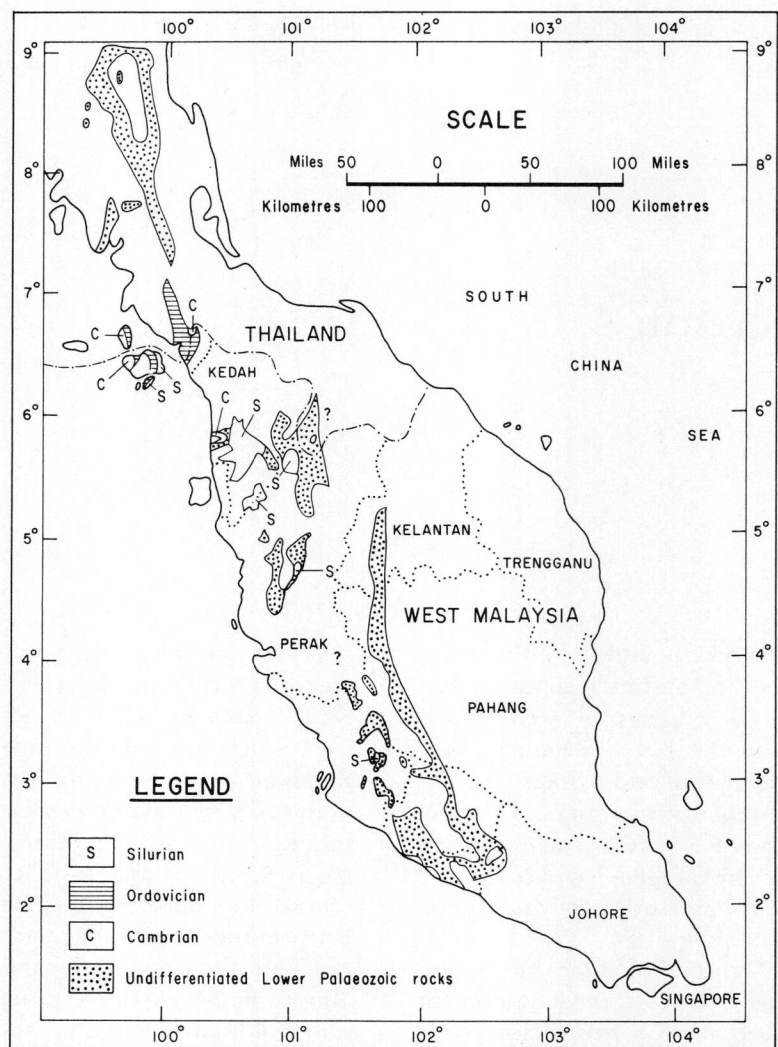

FIGURE 3.1 *Distribution of Lower Paleozoic rocks in the Malay Peninsula. Based on data of the Geological Survey of Malaysia.*

tional zones of a developing geosyncline. In the extreme northwest part of the country, in the Langkawi Islands, and in Perlis, the succession consists of thick deltaic deposits and shelly limestone—typically the products of shallow-water deposition—laid down on the continental shelf at no great distance from the foreland. To the east in Kedah and parts of Perak, the sequence is quite different; it comprises carbonaceous, siliceous, and pyritic fine-grained sediments; mostly fissile shale, siltstone, and cherty rocks, which contain an abundance of planktonic organisms but are almost devoid of benthonic fossils. These rocks obviously originated in deeper water and accumulated in a peculiar environment that was shielded from the freely circulating waters of the open ocean. Such rocks are the product of the miogeosynclinal basin.

In east Kedah and north and central Perak, the sequence again differs. The euxinic-type sediments of the miogeosyncline are again represented, but they are interbedded and interdigitated with beds of coarse sandstone and lenses of limestone. The succession is a rapidly alternating sequence of shallow-water and deeper water sediments, which could have originated only on a submarine ridge that was alternately raised and depressed. Breaks in the faunal sequence and evidence of intraformational erosion indicate that, at times, the ridge was exposed above sea level to form a chain of islands. This intrageosynclinal ridge may be termed a geanticline.

Eastward again, in east Perak, east Selangor, Negri Sembilan, and Malacca, and beyond the Main Range Granite in west Pahang and south Kelantan, the succession comprises normal deep-water sediments, followed by a sequence of poorly graded and rapidly accumulated sediments that are associated with contemporaneous and penecontemporaneous volcanic, pyroclastic, and ophiolitic rocks. In its early stages the environment of deposition was a deep-water basin, which later became subjected to rapid infilling. It continually subsided and eventually formed the deep crustal roots or tectogene of the geosyncline, wherein mobilization developed. The environment was typically that of the eugeosyncline.

The internal configuration of the geosyncline during early Paleozoic times is now known with a fair degree of accuracy (Figures 3.2 and 3.3). However, we still do not know whether the geosyncline was of the epicontinental type, situated along the east margin of the Gondwanaland continent or whether it was intercratonic, lying between Gondwanaland on the west and Cathaysia on the east. Certain evidence from the Cambrian succession of the Langkawi Islands suggests that the depositional belt might have developed by outward migration of a preexisting marginal geosyncline, sited along the edge of Gondwanaland; and the tendency of the intrageosynclinal basins to move eastward throughout the Paleozoic era, lends support to this idea.

On the other hand, various authors, notably Grabau (1924) have established the existence of a continental block, Cathaysia, lying to the east of the geosyncline and complementary in position to Gondwanaland on the west. This type of relationship is futher considered in the section devoted to the paleogeography and geological history of the Lower Paleozoic.

The intrageosynclinal environments of deposition provide a natural and convenient basis on which to classify and describe the Lower Paleozoic strata and associated igneous rocks. Table 3.1 summarizes the different sedimentary formations and igneous rocks that have been recognized within this broad environmental classification. The stratigraphical nomenclature employed herein is also introduced in this table. Figure 3.2 shows the distribution of the various facies in relation to the granites of late Paleozoic to Mesozoic ages.

ROCKS OF THE MIOGEOSYNCLINE

The sediments laid down in the Lower Paleozoic miogeosyncline comprise two main depositional types: those of the basin and those of the shelf. There occurs a certain amount of mixing of the facies along the westerly limb of the basin, where the rocks of the shelf give way to those of the basin. On the east side of the basin, the strata pass laterally into the mixed facies of the geanticinal ridge. The basin deposits are the most widespread, covering large areas of central and south Kedah (Figure 3.4). They form an essentially euxinic sequence of slowly deposited carbonaceous shale, mudstone, and siliceous rocks, with minor developments of more sandy beds. Metamorphosed equivalents of these rocks are recognized in Selangor. The rocks of the shelf region outcrop widely in the extreme northwest part of the Malay Peninsula in the Langkawi Islands, Perlis, and west Kedah. They include a thick sequence of deltaic sandy beds and shelly limestone. Shelf limestone also occurs in central Perak and in central Selangor.

The rocks of the miogeosynclinal zone have not been subjected to such strong tectonic deformation as those of the more internal zones of the depositional belt, with the result that the fossil record is more complete. The stratigraphy is therefore more clearly defined here than is the case elsewhere. Consequently, it has been possible to erect a number of rock stratigraphic units, with fairly well-defined time limits, to cover the depositional sequence in the various parts of the miogeosyncline.

In the discussion that follows, the rocks of the miogeosyncline are described under the three main headings of shelf deposits, basin deposits, and mixed facies.

Shelf Deposits

The rocks of the shelf area include a basal deltaic facies and overlying shelly limestone. The whole succession is well exposed along the fine coastal sections of the Langkawi Islands and the nearby Thai island of Terutau. A similar sequence occurs in Perlis, but the deltaic rocks are brought to the surface in only a small area. Deltaic sandstone, possibly the diachronous equivalent of the rocks in the Langkawi Islands, occurs in the Gunong Jerai area of west Kedah, and around Grik in north Perak. In Kedah, the overlying shelf limestone is poorly developed or missing, and in north Perak the sand-

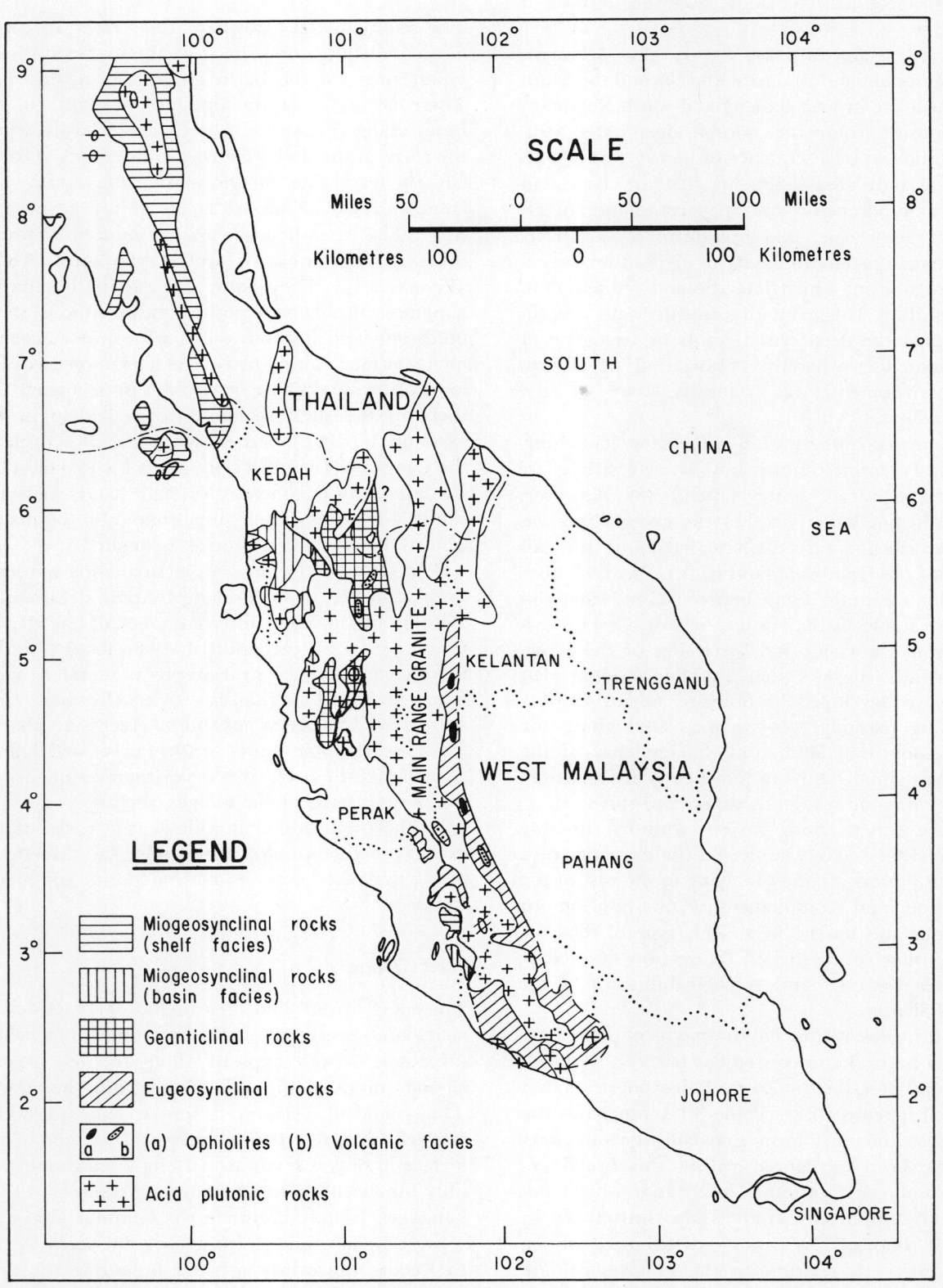

FIGURE 3.2 *Lithofacies map of the Lower Paleozoic rocks of the Malay Peninsula and their relation to the late Paleozoic and Mesozoic granite. Based on data of the Geological Survey of Malaysia.*

TABLE 3.1 Schematic classification and correlation chart of Malayan Paleozoic formations

	Northwest Malaya				Central and South Malaya	
	Mainly Shelf Facies	Miogeosynclinal Rocks — Mainly Basin Facies	Geanticlinal Rocks		Eugeosynclinal Rocks	
			Mixed Facies	Mixed Facies	Basin Facies	
Age	Langkawi–Perlis	Kedah	Perak–Selangor	Kedah–Perak	Pahang–Negri Sembilan	Negri Sembilan–Malacca
LOWER DEVONIAN	eroded surface				**Ophiolite** — hypabyssal basic and ultrabasic intrusive rocks often strongly serpentinized.	
LUDLOW	**Setul Formation** — Upper Detrital Member; Upper Setul limestone, ca. 120 m shelly limestone	Mahang Formation — unknown thickness of black carbonaceous and siliceous graptolitic argillite, siltstone, cherty rocks, and occasional lighter colored arenite of euxinic aspect	Kuala Lumpur Limestone (1830 m) and Chemor limestone — thickly bedded shelly limestone laid down after structural modification of the geosyncline in the late Lower Paleozoic	**Baling group** — Bendang Riang formation, ca. 1200 m carbonaceous shale, phyllite, and hornfels, with lenses of limestone and subsidiary quartzite and schist	Older Arenaceous Series (Foothills Formation) — 1550 m flyschlike sequence of conglomerate, sandstone, shale, schist, and phyllite marked by rapid lateral and vertical facies changes	**Bentong Group** — Arenite and psammitic schist sandstone, siltstone, quartz-biotite-garnet schist, and gneiss
WENLOCK			Hawthornden Schist — 920 m carbonaceous schist and phyllite, probably miogeosynclinal			
LLANDOVERY	Lower Detrital Member; Lower Setul limestone — ca. 1200 m well-bedded, dark grey, crystalline shelly limestone	Sungei Patani limestone and Pulau Bidan limestone	Dinding Schist — 3400 m quartz-mica and amphibole schist and quartzite, possibly eugeosynclinal	Grik siltstone — carbonaceous siltstone locally developed	Volcanic Facies — Lawin tuff and other pyroclastic rocks, composed of lenses of bedded rhyolite and rhyodacite crystal tuff	Argillite and pelitic schist — mudstone, siltstone, phyllite, and quartz-mica and amphibole schist
ORDOVICIAN					Papulut quartzite — ca. 1500 m arenite with subsidiary shale, limestone, conglomerate, and mylonite	Schist Series — 1550 m schist, phyllite, slate, and hornfels, including some amphibole schist
UPPER CAMBRIAN	Machinchang Formation — 1980 m conglomerate, sandstone, siltstone, and flaggy mudstone of deltaic aspect	Jerai Formation — 1425 m of arenite, schist, phyllite, and hornfels				

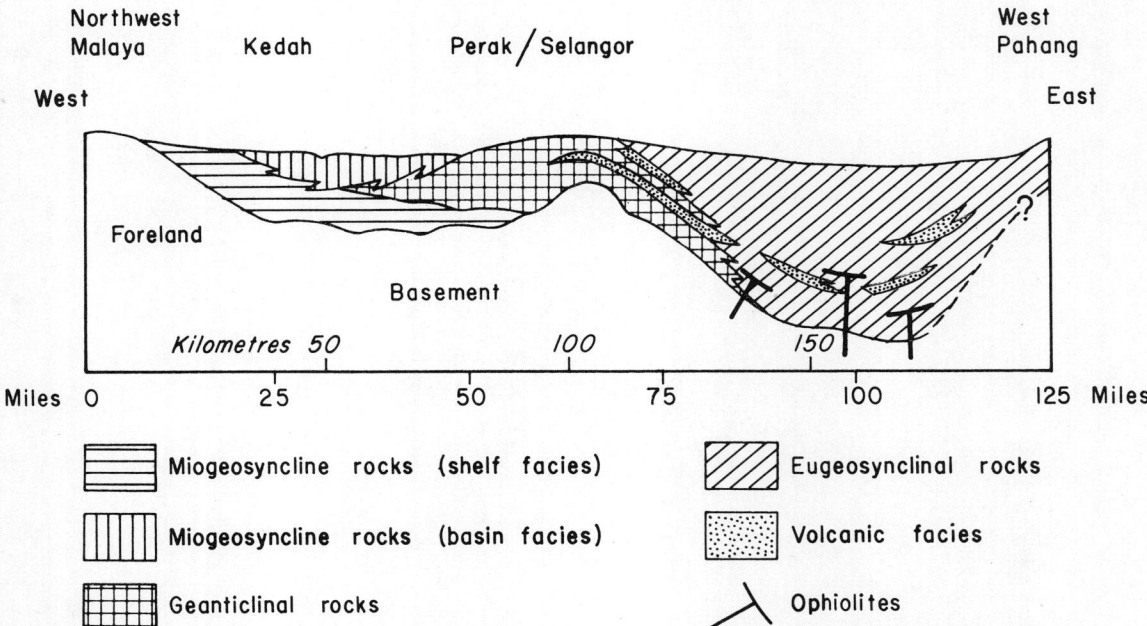

FIGURE 3.3 *Idealized section through the Malayan Lower Paleozoic geosyncline. Vertical scale is arbitrary.*

stone is overlain by the mixed facies of the geanticlinal ridge. In late Lower Paleozoic time, structural modifications to the geosyncline occurred; this resulted in an easterly migration of the component depositional belts, with the result that shelf limestone of Upper Silurian age is found in central Perak, where it overlies geanticlinal rocks, and in Selangor, where it occurs above basin deposits of miogeosynclinal and possibly eugeosynclinal affinities.

The type area for the shelf deposits is the Langkawi Islands, where Jones (1961) recognized the following generalized conformable succession:

2. Shelf limestone of a general Ordovician to Silurian age, with two detrital developments in the upper part of the succession. This unit is named the Setul Formation.

1. Current-bedded sandstone and subgreywacke, with siltstone, mudstone, and flaggy shale of Upper Cambrian age. This unit is named the Machinchang Formation.

DELTAIC FACIES

Machinchang Formation. The strata of the Machinchang Formation are exposed at the surface in the core of a large anticlinal fold which has affected the Lower Paleozoic rocks of Langkawi and Pulau Terutau. The rocks form the spectacular pinnacled ridge of the Machinchang Hills, which rise precipitously to 750 m (2450 ft) above sea level in northwest Langkawi. The rocks can be traced northward to the Thai Island of Terutau, where they develop superb scarp and dip-slope topography (Plate 3). The strata are the oldest known sedimentary deposits in the Malay Peninsula and, in fact, represent the only indisputable rocks of Cambrian age so far discovered in the part of Asia lying south of the Himalayas and North Vietnam. Certain possible correlatives of the Machinchang Formation occur in Burma and elsewhere in Thailand (other than Pulau Terutau), but their isochronous relationships are far from conclusively demonstrated.

The Machinchang Formation outcrops in the west of the Langkawi Islands. It comprises 2000 m (6560 ft) of sandstone, quartzite, and subsidiary beds of conglomerate, grit, siltstone, flaggy shale, and mudstone. The following succession is found on the north coast of Pulau Langkawi, between Telok Kubang Badak and Tanjong Chinchin.

4. Basal Ordovician limestone of the Setul Formation.

3. 30 m (100 ft) conformable passage beds, comprising grey to brown quartzite, sandstone, and fissile shale, with a discrete 16 m band of silicified limestone. Fossils are rare and include casts of orthid brachiopods and fragmentary saukid trilobite impressions.

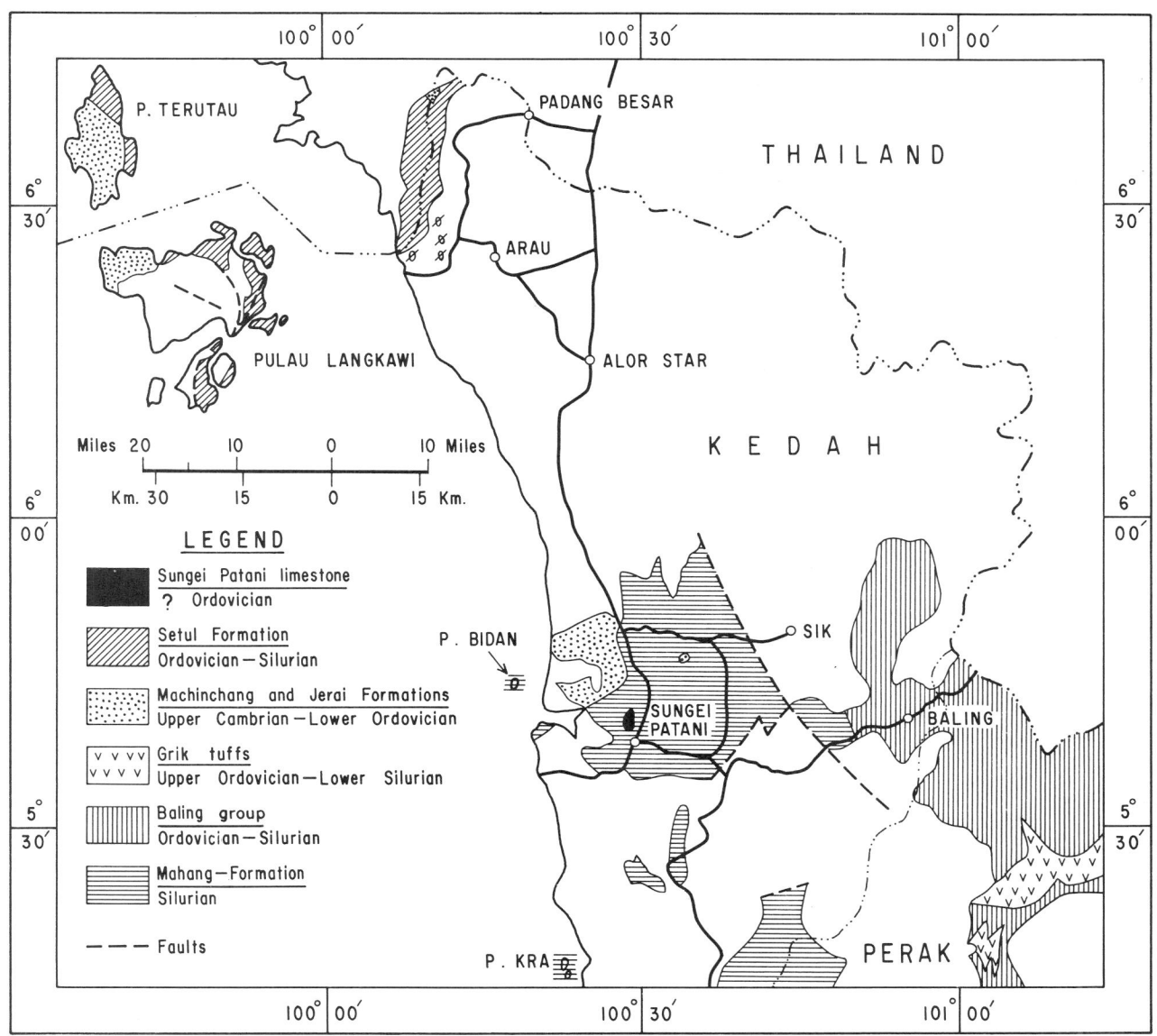

FIGURE 3.4 *Distribution of the Lower Paleozoic formations in northwest Malaya. Based on data of the Geological Survey of Malaysia.*

2. Some 900 m (2950 ft) thickly bedded and conspicuously current-bedded, grey, white, and purple sandstone and feldspathic grit with minor developments of pebbly sandstone, conglomerate, and brown mudstone. Fossils are confined to the upper beds and include fragmentary saukid trilobites.

1. About 1080 m (3500 ft) fairly thickly bedded grey subgreywacke, siltstone and flaggy mudstone, and shale. Somewhat phyllitic in places, and devoid of fossils.

The origin of the upper part of the Machinchang Formation is characteristically shallow water and deltaic. The beds composing the lower part of the succession are of a less varied lithology; their grain size is finer, and they have an appreciable content of detrital muscovite. These deltaic beds at the base of the Lower Paleozoic succession in Langkawi occur also to the east in Kedah and Perak, and they suggest a diachronous phase of coarse detrital deposition as the early Paleozoic seas progressively transgressed the basement.

In lithology the rocks vary in composition from subgreywacke to protoquartzite. Feldspathic varieties, including true arkose, are locally common.

PLATE 3 *Quartzite of the Upper Cambrian Machinchang Formation outcropping on the east shore of Pulau Terutau and forming the hills in the background. Photograph C. R. Jones.*

The feldspar-bearing rocks are usually of a speckled appearance on account of the white grains of altered feldspar which they contain. Red and purple banded sandstone is common and may contain up to 40% ferruginous matrix. The quartz grains are usually angular and poorly sorted. Bands of grit, and conglomerate composed of rounded pebbles of white quartz and grey quartzite bound together in a coarse arenaceous matrix, occur at certain levels in the upper part of the succession. Red, purple, and grey shale and mudstone are comparatively rare, except for a thick development of reddish grey mudstone that occurs near the top of the succession on the south coast of Pulau Terutau. Micaceous siltstone, and dark grey flaggy shale with muscovite and chlorite, are prominent in the lower part of the succession. Rocks thought to be tuffaceous, and containing small fragments of andesitic material, have been recorded from one locality on the north coast of Langkawi. The band of fine-grained stylolitic and silicified limestone, which occurs near the top of the Machinchang Formation on the north coast of Pulau Langkawi, is lithologically similar to the basal strata of the Setul Formation.

Late Cambrian trilobites and brachiopods are present in the upper part of the Machinchang Formation. In the Langkawi Islands, organic remains are rare and are limited to *Eoorthis* sp., fragmentary saukid trilobites, and some *incertae sedis*. However, on Pulau Terutau, a rich fauna has been collected from several horizons. This was described by Kobayashi (1957) and includes the brachiopod *Apheoorthis* (?) sp. and the trilobites *Pagodia thaiensis* Kobayashi, *Thailandium solum* Kobayashi, *"Eosaukia" buravasi* Kobayashi, *Saukiella terutaoensis* Kobayashi, and *Coreanocephalus planulatus* Kobayashi (Plate 5).

The late Cambrian age of the fauna is based on the presence of the genera *Pagodia*, *Saukiella*, and *Coreanocephalus*. However, the recent discovery of the trilobite genus *Asaphus*, in the topmost beds of the succession on Pulau Terutau (Kobayashi, in correspondence), indicates that the Machinchang Formation includes strata of early Ordovician age. The lower and middle parts of the sequence are unfossiliferous. However, on the grounds of the considerable thickness of the formation, it is probable that the lowest strata are of an appreciably earlier Cambrian age.

In north Perlis, 57 km (35 miles) east of Langkawi, a small area of unfossiliferous quartzite has been found lying conformably beneath Ordovician limestone (Jones in prep). A fine-grained quartzite has also been recorded from a borehole at a depth of 18 m (59 ft) under the coastal alluvium near Kuala Perlis. In both these areas, the rocks occur within the core of anticlinal folds and are correlated with the Machinchang Formation on a basis of their lithological similarity and stratigraphical disposition.

Jerai Formation. At a distance of 80 km (50 miles) south of Perlis, in central Kedah, there occurs a sequence of unfossiliferous detrital rocks, termed by Bradford (1965) the Jerai Formation. These beds have been correlated with the Machinchang Formation on structural and lithological grounds. The strata form a tract of mountainous country in the core of the Gunong Jerai dome. The rocks form the lower part of a conformable succession, and the upper beds include shelf limestone recorded from boreholes in the Sungei Patani area to the south of Gunong Jerai, as well as Lower Silurian graptolitic argillite that outcrops some distance to the east of the dome.

The strata composing the Jerai Formation are variously affected by regional metamorphism. The sequence has been divided into two members:

2. 525 m (1720 ft) quartz arenite with occasional felspathic grit.
1. 900 m (2950 ft) quartz-mica schist, and phyllite, with occasional amphibole and garnet-bearing schist.

Apart from their metamorphosed condition, the rocks composing the Jerai Formation show a strong lithological resemblance to those of the Machinchang Formation. Instances of graded bedding and current bedding are present among the arenite, but they are not noticeable features, having probably lost their identity through metamorphism. The stratigraphy of the two formations also shows much in common.

Correlatives in Thailand. Fossiliferous Cambrian strata, homologous to the Machinchang Formation, are known with certainty in Thailand only on Pulau Terutau. Brown *et al.* (1951) regarded the Phuket Group of Phuket Island and the nearby part of Peninsular Thailand as of possible Cambrian age, but these strata are now correlated with the Upper Paleozoic Singa Formation of the Langkawi Islands.

SHELF LIMESTONE

Setul Formation. In the Langkawi Islands, the late Cambrian Machinchang Formation is conformably overlain by thick shelly limestone of Ordovician to Silurian age. The limestone, together with beds of detrital sediment present in the upper part of the succession, was named the Setul Formation by Jones (1961) from the Setul Boundary Range of the Perlis frontier, where their identity and age were first established. Detrital beds have not been distinguished with certainty in the succession in Perlis; in Langkawi, however, two well-defined sequences of quartzite, siltstone, carbonaceous argillite, and chert of appreciable thickness are interbedded with the limestone. They mark the encroachment of the deeper water conditions of the miogeosynclinal basin onto the shelf. The detrital bands were formally named the Lower and Upper Detrital Members by Jones (1968), and they are considered further under the section on miogeosynclinal basin deposits.

The Setul Formation is exposed along the eastern seaboard of the Langkawi Islands. The rocks are disposed about the easterly dipping flank of a major anticlinal arch, but the structure is distorted by the doming effect of the central granite mass, which has been intruded into the limb of the anticline. Major thrusting, faulting, and folding have greatly complicated the continuity of strata. Fossils have helped greatly in defining the general chronological sequence, but in places metamorphism has completely destroyed all evidence of original organic content.

The sequence is most fully represented and the strata are least affected by metamorphism in northeast Langkawi in the vicinity of Selat Pulau Peluru and Pulau Langgun. Only the basal beds of the formation, which lie inland, are poorly exposed. Nevertheless, this part of the succession is well exposed further west along the north coast near Kuala Kubang Badak, where a conformable passage between the Machinchang Formation and the Setul limestone can be seen. The type area for the Setul Formation is in the northeast part of Langkawi Island, where it comprises four main divisions as follows:

Upper Detrital Member: ?140 m (460 ft) of grey quartzite and subgreywacke, brown and grey siltstone, and black carbonaceous and red shale, with graptolites and tentaculites of Lower Devonian age.

Upper Setul limestone: ca. 120 m (395 ft) of banded, light grey, stylolitic limestone containing occasional fossils.

Lower Detrital Member: 27 m (89 ft) of black carbonaceous siltstone, fissile and flaggy shale, and cherty beds containing graptolites, trilobites, and other shelly fossils of Lower Silurian (Llandovery) age.

Lower Setul limestone

3. 150 m (492 ft) of banded, well-bedded stylolitic limestone, with a conspicuous discrete pink limestone, containing trilobite and conodonts of Lower Silurian age.

2. 1020 m (3347 ft) of thickly bedded, dark grey, finely crystalline limestone with rich shelly horizons containing coelenterates, brachiopods, gastropods, cephalopods, cystids, and conodonts of the late Lower, Middle, and Upper Ordovician age. The rock is variably dolomitic and of a characteristic banded appearance, with strongly developed stylolitic partings and other irregular segregations.

1. 90 m (295 ft) of thickly bedded basal grey limestone. The rock is finely crystalline, conspicuously siliceous, and strongly indurated. Silicified remains of Lower Ordovician sponges and cephalopods occasionally occur.

The limestone of the Setul Formation is herein referred to informally as the Setul limestone. The Lower Detrital Member in the Langkawi Islands does, however, divide the limestone sequence into two parts, which can be conveniently termed the lower Setul limestone and the upper Setul limestone. Owing to great variations in thickness of the upper Setul limestone and to the absence of detrital developments in Perlis, this nomenclature has not been formalized.

The Setul limestone is a hard, brittle, dark-colored, thickly bedded, finely crystalline rock. Magnesia is usually present in small proportions, although strongly dolomitic rocks are known in certain areas. Noncarbonate impurities include silt, clay, ironoxide, and finely dispersed carbonaceous matter. Table 3.2 gives some chemical analyses of the Setul limestone. Occasionally, the rock is strongly silicified. Oolitic limestone showing current bedding has been recorded, but this feature is uncommon. It is possible that more of the rock was originally oolitic, but diagenetic and metamorphic changes may have destroyed these structures.

Within granite aureoles, the limestone has been recrystallized to white, grey, green, or pinkish marble that may contain tremolite, diopside, wollastonite, garnet, and clinozoisite. Along the contact zones, dense, dark-colored diopside-garnet and diopside-tremolite-wollastonite skarn rocks have developed. The rock usually shows a characteristic banded appearance that is due to the presence of stylolites (Figure 3.5), and other more irregular partings, which developed as a result of diagenetic segregation of the original fine clastic material from the pure carbonate (Hutchison 1963). Other irregular partings are due to the localized development of dolomite.

The Setul limestone gives rise to a spectacular karst topography. Along the eastern seaboard of the

TABLE 3.2 *Chemical analyses (weight percent) of Lower Paleozoic limestones*

	Setul Limestone			Kuala Lumpur Limestone			Baling Limestone
Specimen*	1	2	3	4	5	6	7
SiO_2	8.85	6.19	0.11	1.93	0.03	0.68	0.55
Al_2O_3	0.67	1.05	{0.11}	0.62	0.02	0.42	0.22
FeO, Fe_2O_3			{0.75}				
MgO	2.35	1.23	17.57	0.91	0.72	18.30	0.22
CaO	47.70	50.38	34.72	53.45	55.81	34.40	55.53
CO_2	38.64	40.52	45.88	42.72	43.28	46.12	43.20
Total	98.21	99.37	99.14	99.63	99.86	99.92	99.72

* Specimens 1–7 as follows:
  1. Dolomitic limestone, Bukit Lagi Quarry, Perlis, (081046, p. 77).
  2. Magnesian limestone, Bukit Wang Mu Quarry, Perlis, (081048, p. 77).
  3. Dolomite, Bukit Lagi, Perlis, (082004, p. 87).
  4. Limestone, Pong Onn, Ampang, Selangor, (081020, p. 71).
  5. Limestone, Huin Fatt, Salak South, Selangor, (081022, p. 73).
  6. Dolomite, Batu Caves, Selangor, (082003, p. 87).
  7. Limestone, P. W. D. Quarry, Baling Town, Kedah, (081026, p. 73).

Numbers and page references in parentheses refer to Alexander et al. (1964).

FIGURE 3.5 *Marmorized and stylolitic Setul limestone, east of Kuah, Pulau Langkawi. Photograph C. S. Hutchison.*

Langkawi Islands the limestone forms hills that rise precipitously to heights of 450 m (1470 ft). Cliffs and exposures of bare rock are common, and the vegetation is normally thin and scrubby due to the lack of soil (Plate 4). In west Perlis, the Setul limestone builds the Setul Boundary Range, which presents a continuous high barrier of cliffs and deeply dissected terrain, extending for over 30 km (19 miles) in Perlis and continuing into Thailand for many tens of kilometers.

The upper Setul limestone varies considerably in thickness. Thus on Pulau Langgun it measures about 120 m. On Pulau Timun, 14 km (9 miles) to the south, it thins to 4 m of white marble; and southwest again by 10 km (6 miles) on Pulau Tuba, it thickens to 90 m of limestone and marble. The smaller thickness in the southeast corresponds to a thickening of the Upper Detrital Member. This is probably because Pulau Timun is situated closer to the miogeosynclinal trough, with the result that the shelf limestone thins away and is replaced by the basin deposits in this direction. It would seem that the final infilling of the miogeosynclinal trough with coarse clastic sediment spread westward to cover the shelf limestone, and this phase is represented in the Langkawi area by the Upper Detrital Member.

The Setul limestone fauna in the Langkawi Islands is fairly rich. Kobayashi (1958, 1959), Igo and Koike (1966), and Yochelson and Jones (1968) have carried out studies on part of the faunal content, but much work remains to be done in this field. The macrofauna is essentially a gastropod–cephalopod association, but brachiopods, bryozoans, cystids, trilobites, and other fossils occur. Jones et al. (1966) mentioned that the basal limestone contains poorly preserved cephalopods including two species belonging to *Robsonoceras* or its allies, indicating a Lower Ordovician age.

Kobayashi (1959) described the molluscan fauna contained in the middle part of the lower Setul limestone. This includes the gastropods *Hormotoma* (?) sp., *Helicotoma jonesi* Kobayashi, *Helicotoma* (?) *costata* Kobayashi, *Palaeomphalus giganteus* Kobayashi, *Lesueurilla zonata* Kobayashi, *Malayaspira rugosa* Kobayashi, and the cephalopods *Endoceras* (?) sp., *Ormoceras langkawiense* Kobayashi, *Discoceras* (*Hardmanoceras?*) *chrysanthimum* Kobayashi, and *D.* (*H?*) *laeviventrum* Kobayashi (Plate 5). The fauna is similar to the Neichiashan fauna of central China and is approximately Llandeilian in age.

A common fossil is a horn-shaped gastropod operculum showing a fibrous internal structure. This was described under name *Teiichispira kobayashi* by Yochelson and Jones (1968), and it indicates a late Lower Ordovician age. The brachiopods (which include orthid and strophomenid genera) and the trilobites have not yet been investigated.

Numerous Ordovician and Silurian conodonts were described by Igo and Koike (1966) from the lower and upper Setul limestone. The genera recognized in the lower Setul limestone include *Acodus, Acontiodus, Drepanodus, Oistodus, Scolopodus,* and *Ulrichodina.* The following conodont zones were erected to cover the succession between the Middle Ordovician and the base of the Lower Detrital Member on Pulau Langgun, which is a little way above the base of the Silurian:

3. Acodus mutatus–Acontiodus hamari Zone.
2. Acodus similaris–Drepanodus altipes Zone.
1. Scolopodus staufferi–S. giganteus Zone.

The upper Setul limestone on Pulau Langgun contains the conodont genera *Acodus, Acontiodus, Amorphognathus, Bryantodus, Cordyolus, Distamodus, Drepanodus, Kockerella, Ozarkodina, Panderodus, Plectospathodus, Trichonodella,* and others.

A further conodont zone, the Panderodus unicostatus Zone, was established to cover this part of the succession. The upper Setul limestone also contains shelly fossils, among which can be distinguished hyolithids, nautiloids, and trilobites. Its

PLATE 4 *Thick-bedded limestone of the Setul Formation on the northeast coast of Pulau Langkawi. Photograph C. R. Jones.*

age, based on the graptolites in the contiguous detrital strata, is late Llandovery to Ludlow.

In Perlis the Setul Formation is entirely of limestone. Fossil evidence shows that only Ordovician rocks are represented, and structural observations suggest that the upper part of the formation is unconformably overstepped in the east by Upper Paleozoic strata.

Fossils occur in the limestone at a number of places. Kobayashi (1958, 1959) described the following gastropod–cephalopod association from Bukit Lagi quarry near Kangar and the Thye San mine in north Perlis: *Malayaspira* sp., *Lytospira rectangularis* Kobayashi, *Stereoplasmoceras* (?) sp., *Ormoceras langkawiensis* Kobayashi, *Armenoceras chediforme* Kobayashi, *Actinoceras perlisense* Kobayashi, and *Actinoceras* sp. The nautiloid-bearing rock can be correlated with the Toufangian Limestone of eastern Asia and the Black River–Trenton Limestone of North America, which are both of Middle Ordovician age.

SHELF LIMESTONE ELSEWHERE IN MALAYA. Some 80 km (50 miles) south of the Langkawi–Perlis region, thick limestone was reported by Jones (1968) from boreholes sunk in the Sungei Patani area of west-central Kedah. The rock is similar in appearance and composition to the Setul limestone, but no fossils were recovered. The beds lie under shales which, a little to the east, contain Lower Silurian graptolites and which overlie the Jerai Formation. The Sungei Patani limestone is therefore probably equivalent to part of the Setul Formation.

According to Bradford (1965), thin beds of limestone also outcrop on Pulau Bidan off the Kedah coast near the dome of Gunong Jerai. The rock is a grey limestone with carbonaceous and pyritic inclusions. It contains some poorly preserved straight nautiloids and crinoid stems. No precise date can be allotted, but the presence of nautiloids and the structural position of the beds suggest that the beds can be correlated, in part, with the Setul limestone.

Other minor calcareous developments were noted by Bradford (1965) in the Sungei Patani area, by Burton (1967b) in central Kedah, and by Jones (in prep.) in north Kedah. Either these occurrences correspond to the transient extension of shelf conditions into the basin, or they represent beds that lie within the zone of transition between the shelf facies to the west and the basin facies to the east.

Lower Paleozoic limestone also occurs in east Kedah, north and central Perak, and central Selangor. In east Kedah and north Perak, it appears as thin lenses in a mixed facies, which was laid down in the shallow oscillating seas of the geanticlinal ridge. Limestone formations in central Perak and in central Selangor are much thicker and represent sustained calcareous deposition. The strata are of shelf aspect and are generally younger than the limestone of northwest Malaya. In central Perak, the rocks contain Silurian and Devonian fossils and overlie Lower Silurian graptolitic mudstone, which is interbedded with thin limestone beds. In Selangor, a rich shelly fauna of Upper Silurian age has been recovered from the limestone that overlies unfossiliferous carbonaceous phyllite and schist.

Shelf conditions could have developed in late Lower Paleozoic time in these two areas in two different ways. They may have become established in the shallow waters of the infilled miogeosynclinal trough, or they may have arisen by structural modification and easterly migration of the component depositional basins of the geosyncline. To some extent these processes are interrelated, but there is evidence to suggest that the easterly movement of the geosynclinal belt was partly independent of the rate of development of intrageosynclinal deposition. Thus in Selangor the late Lower Paleozoic shelf limestone overlies basin deposits of miogeosynclinal aspect, whereas in Perak the limestone lies above a mixed facies of interbedded limestone, carbonaceous argillite, and subsidiary arenite, laid down in the environment of the geanticlinal ridge.

Gobbett (1964) named the Upper Silurian shelf limestone of central Selangor the Kuala Lumpur Limestone. The mid-Paleozoic shelf limestone of central Perak is herein referred to informally as the Chemor limestone, from the village of that name.

The Chemor Limestone. The prominent feature of the Kanthan hills, which lie near Chemor at the north end of the Kinta Valley in central Perak, is formed of hard, grey, crystalline limestone. Alexander and Müller (1963) reported Lower and Upper Devonian conodonts from these rocks.

Kuala Lumpur Limestone. Gobbett (1964) described 1830 m (7000 ft) of Upper Silurian marble from the Kuala Lumpur Valley of central Selangor. The strata outcrop only at the north end of the valley, where they form the spectacular Batu Caves. The marble, however, is widespread as bedrock in the many alluvial tin mines scattered throughout the valley (Figure 3.6). The beds overlie a thick miogeosynclinal sequence of carbonaceous schist and phyllite, which in turn pass down into schist of eugeosynclinal aspect. The marble is a finely crystalline grey to cream, thickly bedded, variably dolomitic rock. Banded marble, saccharoidal dolomite, and pure calcitic limestone also occur. Chemical analyses are given in Table 3.2. Stylolitic segregation of noncarbonate impurities is common, and oolitic limestone has been recorded from one locality. The rock is richly fossiliferous at certain horizons and contains corals, brachiopods, gastropods, and crinoids, which have been described by Thomas (1963) and Boucot et al. (1966).

Among the fauna are distinguished the corals *Ketaphyllum* aff. *turbinatum* (Linnaeus), *Heliolites* aff. *barrandei* var. *spongodes* Lindstrom, *Favosites* sp., *Thecia swinderniana* (Goldfuss), and *Halysites* sp.; the brachiopod genera *Dalmanella, Capellinella, Cymbidium,* "*Conchidium*", *Atrypella,* and *Delthyris*; and the gastropods *Poleumita* cf. *discors* (Sowerby), *Poleumita scamnata* Clarke and Ruedemann, *Euomphalus* (*Philoxene*) sp., and *Loxonoma* sp. (Plate 5). The brachiopod element indicates a Ludlow age for the limestone.

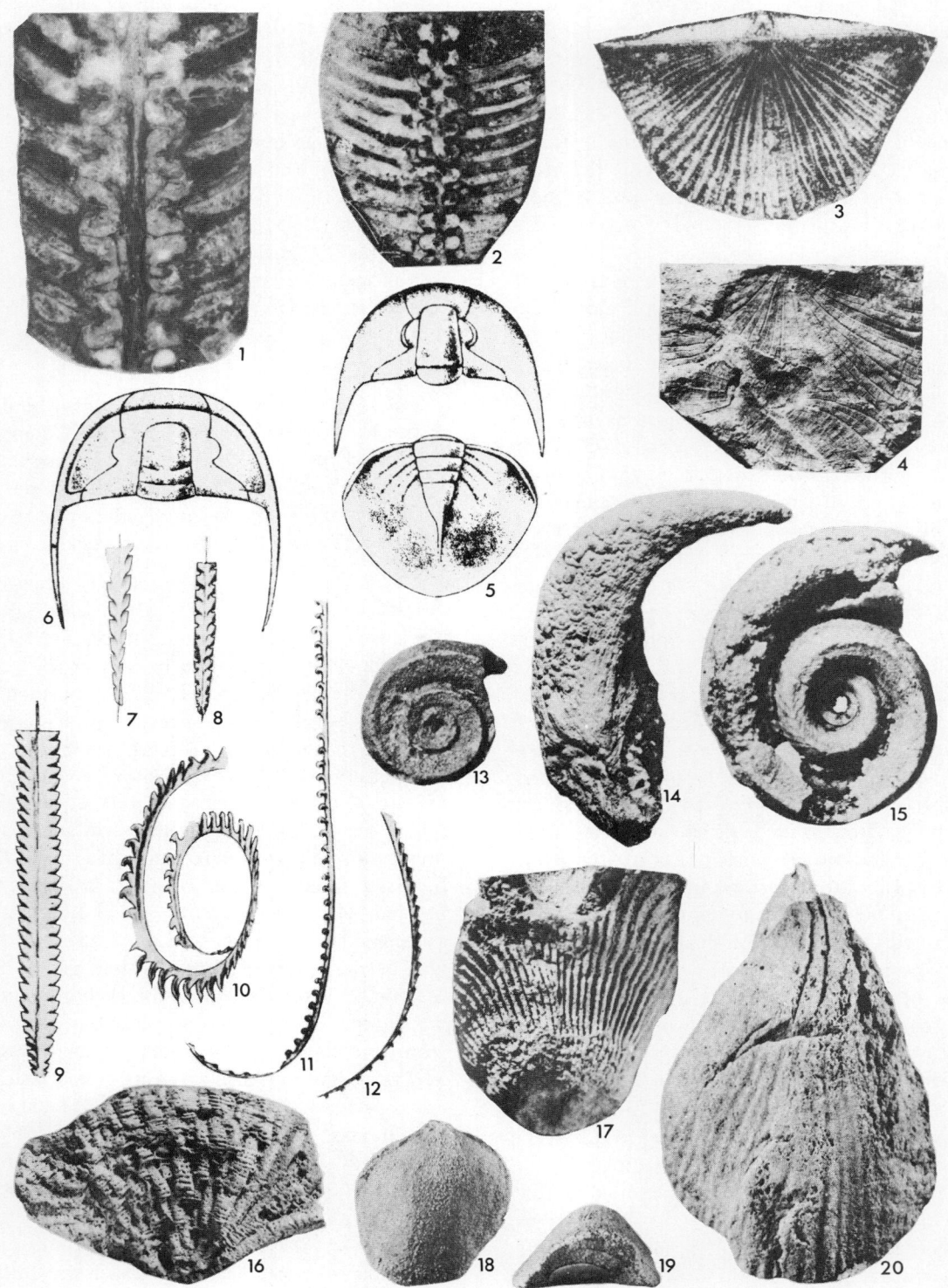

PLATE 5 *Lower Paleozoic fossils.* *1.* Actinoceras *sp., Upper Ordovician Setul Formation, Pulau Langgun, Langkawi,* ×1. *2.* Ormoceras langkawiense *Kobayashi, Ordovician Setul Formation, Pulau Tembus Dendang, Langkawi,* ×2. *3.* Cyrtonotella thailandica *Hamada, Ordovician Setul Formation, Satun, Thailand,* ×3. *4.* Rafesquina komalarjuni *Hamada, Ordovician Setul Formation, Satun, Thailand,* ×3. *5.* Coreanocephalus planulatus *Kobayashi, Upper Cambrian Machinchang Formation, Pulau Terutau, Thailand,* ×1. *6.* Thailandium solum *Kobayashi, Upper Cambrian Machinchang Formation, Pulau Terutau, Thailand,* ×1.

FIGURE 3.6 *Kuala Lumpur Limestone exposed in an open-cast tin mine at Segambut, Kuala Lumpur. Removal of the alluvium has revealed a dissected pavement formed of steeply dipping recrystallized limestone. Photograph D. J. Gobbett.*

It is probable that both the limestone forming the rock pinnacle of Bukit Takun (Figure 3.7), 22 km (14 miles) north of Kuala Lumpur and the bedrock in the nearby tin mines are of the same age as the Kuala Lumpur Limestone.

Limestone in West Pahang. The stratigraphical position of sporadic lenses of unfossiliferous limestone, which lie immediately to the east of the outcrop of the Lower to Middle Paleozoic eugeosynclinal rocks of the Bentong Group in west Pahang and east Negri Sembilan, is uncertain. Alexander (1968) considered that they formed part of an Upper Paleozoic sequence that lay unconformably on the eugeosynclinal rocks. On the other hand, Ja'afar (in prep.) regarded them as part of the Bentong Group, and thus they might well represent the shallow-water deposits of the eastern margin of the geosyncline, assuming an intracratonic setting. The limestone outcrops in isolated patches, and the largest of these forms the hill of Chintamani, southeast of Bentong. The rock which varies from a light-colored coarsely crystalline marble to an almost black, finely crystalline limestone, is sparingly dolomitic.

CORRELATIVES IN THAILAND AND BURMA. The Setul limestone of the Langkawi–Perlis region can be traced northward to Pulau Terutau and the Satun and Songkhla Provinces of peninsular Thailand. Its most northerly outcrop in the Thai Peninsula is near Surat Thani, 270 km (168 miles) north of the Perlis frontier. In Thailand, the unit was described by Brown et al. (1951) under the name Thung Song limestone. It is identical in lithology and faunal content to the Setul limestone. Near Thung Song, the limestone is interbedded with Lower Silurian graptolitic argillite and Lower Devonian

---

7. *Dimorphograptus malayensis* **Jones**, *Lower Silurian Setul Formation, Pulau Langgun, Langkawi,* ×2. 8. *Climacograptus scalaris* **(Hisinger)**, *Lower Silurian Setul Formation, Pulau Langgun, Langkawi,* ×2. 9. *Diplograptus modestus* **Lapworth**, *Lower Silurian Setul Formation, Pulau Langgun, Langkawi,* ×2. 10. *Spirograptus spiralis* **(Perner)**, *Lower Silurian Mahang Formation, south Kedah,* ×2. 11. *Streptograptus nodifer* **(Törnquist)**, *Lower Silurian Mahang Formation, south Kedah,* ×2. 12. *Pernerograptus revolutus* **(Törnquist)**, *Lower Silurian Mahang Formation, south Kedah,* ×2. 13. *Helicotoma jonesi* **Kobayashi**, *Ordovician Setul Formation, Pulau Langgun, Langkawi,* ×1.5. 14. *Lytospira rectangularis* **Kobayashi**, *Ordovician Setul Formation, Kangar, Perlis,* ×1. 15. *Malayaspira rugosa* **Kobayashi**, *Ordovician Setul Formation, Pulau Langgun, Langkawi,* ×1.5. 16. *Favosites* sp., *Upper Silurian Kuala Lumpur Limestone, Kuala Lumpur,* ×1. 17. *Ketophyllum aff. turbinatum* **(Linn.)**, *Upper Silurian Kuala Lumpur Limestone, Kuala Lumpur,* ×1.5. 18–19. *Atrypella* sp., *Upper Silurian Kuala Lumpur Limestone, Kuala Lumpur,* ×1.5. 20. *Conchidium cf. triangulum* **Khodalevich**, *Upper Silurian Kuala Lumpur Limestone, Kuala Lumpur,* ×1.

FIGURE 3.7 *Bukit Anak Takun formed of Kuala Lumpur Limestone, Templer Park, Kanching, Selangor. Photograph D. J. Gobbett.*

tentaculites beds, so that the position here is identical to that represented by the Setul Formation in the Langkawi Islands.

In the Shan States of east Burma, a geosynclinal sequence of Lower Paleozoic rocks similar to those occurring in the Malay Peninsula has been recognized (Pascoe 1959). Although it is certain that the sediments of the two regions were laid down in parts of the same depositional basin, we lack detailed information and thus the precise stratigraphical interrelationships between the two areas are not always clear. It would seem that much of the Ordovician and Silurian succession of the Shan States is composed of a mixed facies of interbedded argillaceous, arenaceous, and calcareous strata, similar to that found in the Baling group of Malaya, which originated along the line of the geanticlinal ridge. Thin limestone beds occur in the Ordovician succession of the Naungkangyi Stage and the Pindaya Formation, and in the Upper Silurian to Lower Devonian Zebingyi Series.

The only thick limestone of shelf aspect, which has so far been described, forms the upper part of the Mawsoon Series and the Orthoceras Beds in the Southern Shan States. The strata are considered to be lateral equivalents of the Pindaya Formation and the Upper Naungkangyi Stage of the Northern Shan States. The limestone forms great escarpments similar to the Setul Boundary Range of Perlis. The fauna includes the gastropods *Lophospira* sp. and *Helicotoma* sp. and the nautiloids *Actinoceras* sp., *Ormoceras* sp., and *Endoceras* sp., thus indicating a Middle to Upper Ordovician age. The similarities in thickness, lithology, faunal content, and topographical expression indicate that the limestones of the Mawsoon Series and the Orthoceras Beds are homologous with the Ordovician limestones of the Setul Formation in Malaya and the Thung Song limestone in peninsular Thailand.

Basin Deposits

The shelf limestone of the Langkawi–Perlis region passes southward and eastward, through transitional beds, into the strata laid down in the deeper waters of the miogeosynclinal basin. This depositional environment was sited over central and south Kedah, and extended for some distance into Penang State and the northern part of Perak. The basin was shielded from the open sea to the east by a geanticlinal upwarp, and this proved to be an effective barrier during much of Ordovician and Silurian time. The environment was one of stagnant conditions; ocean currents were unable to penetrate the barrier; faunal access was prevented except for plankton, which drifted in by way of the shallow waters covering the oceanic ridge; and the anaerobic conditions of the deeper waters could not sustain benthonic life. Little coarse clastic sediment found its way into the basin. During late Cambrian time, the foreland was probably reduced to a peneplain, and it subsequently provided only a modicum of fine sediment, which was deposited contemporaneously with Ordovician and Silurian shelf limestone, while some found its way further east into the basin.

Much of the miogeosynclinal basin sequence corresponds to the euxinic facies of Pettijohn (1957). The succession consists of slowly accumulated sediment, of silt and mud grade, with an abundance of finely disseminated carbon probably derived from phytoplankton. The silica content is fairly high. The source of this silica may be contemporaneous volcanic activity, the center of which was sited a little to the east of the geanticlinal ridge. On the other hand, Burton (1967b) suggested tropical climatic conditions as an adjunct to the presence of high concentrations of silica in the restricted seas of the basin.

A fauna composed mainly of planktonic organisms, including abundant graptolites, occurs; but there are significant breaks in the faunal sequence. Such gaps probably correspond to periods when the geanticlinal barrier rose above sea level, preventing ingress of plankton from the open sea. Occasional benthonic organisms have been found, often in association with the plankton; these may have been transported into the depths of the basin from the shallow waters of the geanticlinal ridge through such agencies as undertow or the slumping of fossiliferous sediment down the slopes of the submarine ridge. In other instances animals normally associated with a benthonic mode of existence may have become adapted to a floating mode of life. This is suggested by the presence of trilobites of the family Cyclopigidae, which have strongly swollen glabellae.

Occasional beds of arenite are interdigitated with the carbonaceous argillite, mainly along the eastern margin of the geanticlinal basin. This material may have been derived by slumping or turbidity flow down the slopes of the geanticlinal ridge. Arenite becomes increasingly conspicuous in the upper part of the succession and spreads onto the shelf, where it overlies Upper Silurian limestone. These changes in sedimentation are thought to be related to the active rise of the geanticlinal ridge. This supplied an abundance of coarse detritus with which the miogeosynclinal basin was eventually filled.

The sediments of the miogeosynclinal basin were first investigated in the Mahang area of South Kedah by Courtier (in prep.). Burton (1967b,c, 1972) described the northerly continuation of the beds in central Kedah and formalized the name Mahang Formation, which Courtier had previously given to the succession. The Mahang Formation now embraces the lower part of what Bradford (in prep.) termed the Sungei Patani formation in west Kedah. This unit was defined as a predominantly argillaceous sequence, with minor arenaceous and calcareous developments, of Ordovician to Carboniferous age.

Bradford had insufficient evidence to differentiate between a lower unit, now known to be equivalent to the Mahang Formation, and an unconformably overlying upper unit of Middle Devonian to Carboniferous age. Although both these formations are composed essentially of argillaceous sediments, some lithological distinction can be made between them. Thus in the Sungei Patani area the Mahang element is composed of shale that is predominantly red at the surface due to its high content of ferric oxide; however, it reveals its true black carbonaceous character at depth. Further north, where strata of the younger formation outcrop, the shale is distinguished by its intrinsic pale grey color and the absence of siliceous and flaggy constituents.

Miogeosynclinal basin deposits are also recognized in the Langkawi Islands, where they occur as thin beds of detrital sediment interbedded with shelf limestone, and in Selangor, where metamorphosed rocks of euxinic affinities underlie Upper Silurian Shelf limestone.

MAHANG FORMATION. More than 1000 km^2 (386 mile^2) of central and south Kedah is covered by the Mahang Formation. The rocks appear in three main outcrops. By far the largest is that which covers the country between the towns of Sungei Patani, Bedong, and Jeniang, in central Kedah.

A small faulted block, originally part of this outcrop, occurs around Bukit Batu Bertelor to the west of Baling. The strata reappear some 32 km (20 miles) away, in the Mahang area of south Kedah, and from this area they can be traced in an unbroken outcrop to beyond Selama in north Perak. West of the Kulim granite, some small isolated outcrops occur in neighboring Penang State.

Burton (1967b) divided the Mahang Formation into four lithological types:

1. A dominant argillaceous facies.
2. A subsidiary arenaceous facies.
3. A minor siliceous facies that grades imperceptibly into the argillaceous.
4. A restricted calcareous facies.

The argillaceous facies comprises carbonaceous shale and somewhat siliceous mudstone, siltstone, flaggy shale, and metaargillite. The rocks are usually hard and black due to the appreciable amount of carbon and an abnormally high silica content. They are frequently reduced on weathering to soft, pale grey or reddish products, in which the carbon has been leached away and residual iron has become concentrated under humid tropical surface conditions. The shale may pass in certain places into dark cherty rocks and carbonaceous radiolarite with more than 90% silica. Lenses of coarser sediment, constituting the arenaceous facies, have been observed in the eastern and southwestern parts of the larger outcrop. They are made up predominantly of pale-colored protoquartzite and subgreywacke and minor developments of orthoquartzite and greywacke.

The calcareous facies comprises several small bodies of calc-silicate rock and limestone, which are found only in the marginal areas of the outcrop of the Mahang Formation, where the basin strata pass laterally into the shelf facies on the west and the mixed facies of the geanticlinal ridge on the east. To these four facies must be added a very minor occurrence of volcanic rocks. Courtier (in prep.) recorded isolated instances of fine vitric and crystal-vitric tuff in the southerly outcrop near Mahang. Volcanic rock is elsewhere absent from the miogeosynclinal sequence. However, thick deposits of rhyolite tuff occur only 30 km (19 miles) to the east of the Mahang area, in the vicinity of Grik in north Perak, where they form part of the sequence laid down on the east side of the geanticlinal ridge. It is not improbable that a minor amount of pyroclastic material found its way into the eastern periphery of the miogeosynclinal basin.

The strata of the Mahang Formation produce a low undulating landscape with rounded hummocky hills divided by broad areas of flat swampy ground. Occasionally, well-defined hilly areas and ridges of greater relief rise to heights of between 150m (492 ft) and 350 m (1225 ft) above sea level. These features coincide with the outcrop of the more arenaceous or cherty rocks or where the strata have been more strongly metamorphosed than elsewhere.

The macrofauna of the Mahang Formation is dominantly planktonic, being composed of Silurian graptolites and an early Devonian graptolite-tentaculite association. Small numbers of brachiopods, lamellibranchs, crinoids, and trilobites have also been reported by Burton (1967b). Abundant microfossils include radiolaria, foraminifera, and hystrichospheres.

The graptolites and the dacryoconarid tentaculites are the most reliable stratigraphical indicators. Graptolites are well distributed through the succession and indicate the existence of conspicuous faunal gaps. No Ordovician graptolites have been discovered, and certain late Llandovery, early Wenlock, and Ludlow forms are absent. These gaps in the faunal record appear to correspond to interruptions to the free access of plankton from the open ocean that were caused when the geanticlinal barrier temporarily rose above sea level.

The graptolite fauna is poorly to moderately well preserved. The majority of localities have yielded graptolites of Llandovery age, which include *Climacograptus scalaris* (Hisinger), *Climacograptus rectangularis* (McCoy), *Glyptograptus lunshanensis* Hsü, *Diplograptus thuringiacus* Eisel., *Petalolithus tenius* (Barrande), *Pristiograptus gregarious* (Lapworth), *Pristiograptus regularis* (Lapworth), *Pristiograptus nudus* (Lapworth), *Monograptus marri* Perner, *Monograptus sedgwicki* (Portlock), *Streptograptus lobiferus* (McCoy), *Pernerograptus revolutus austerus* (Törnquist), *Spirograptus minor* (Bouček), *Spirograptus spiralis contortus* (Perner), *Demirastrites convolutus* (Hisinger), *Demirastrites decipiens* (Törnquist), *Rastrites hybridus* Lapworth, *Rastrites maximus* Carruthers, and *Cyrtograptus lapworthi* Tullberg (Plate 5).

The graptolite faunas may be compared to those of the Pristiograptus cyphus Zone to the zones of *Rastrites linnaei* and *Spirograptus minor*. No faunas representing higher Llandovery zones, except possibly the zone of *Spirograptus spiralis* and *Stomatograptus grandis*, have been observed.

A few localities yielding Upper Wenlockian graptolite faunas have been found. The species include *Pristiograptus meneghini* (Gortani), *Monoclimacis flumendosae* (Gortani), *Monograptus* aff. *flemingii* (Salter), *Monograptus* aff. *flexilis* Elles, and *Cyrtograptus lundgreni* Tullberg. The associations indicate the uppermost Wenlockian zones of *Monograptus flexilis* and *Cyrtograptus lundgreni*.

No graptolites of Ludlow age have yet been discovered in the Mahang Formation, but this does not imply a lacuna in sedimentation during the Ludlow. Jaeger (1959, 1962) demonstrated that graptolites of the *Monograptus hercynicus* group were confined almost entirely to the Gedinnian and the early part of Siegenian Stages of the Lower Devonian; these faunas have been recorded in association with dacryoconarid tentaculites at six localities. In most cases the graptolites are very poorly preserved. The crowds of tentaculites, which are frequently intimately associated with the graptolites on the same bedding plane, may have provided the graptolite rhabdosome with an uneven surface to rest upon at the time of deposition, causing the delicate skeleton to fracture. This theory is supported by the observation that, the graptolites are much better preserved when there are few or no tentaculites present. Some well-preserved fossils allowed comparison with the following described species: *Monograptus hemiodon* Jaeger, *Monograptus thomasi* Jaeger, and *Monograptus yukonensis* Jackson and Lenz (Jones 1967a). These species suggest the zones of *Monograptus praehercynicus* and *Monograptus hercynicus* in the upper Gedinnian and lower Siegenian stages.

A large number of localities yielding tentaculites have been found in the Mahang Formation. According to Burton (1967b), the organisms are small, thin-walled, conical shells less than 10 mm in length, and their impressions occur crowded together in localized beds. Two main types have been distinguished; these include some with transverse ribs and others with smooth walls. Bouček (*in* Burton 1967b) stated that the transverse-ribbed type includes a species close to *Nowakia acuaria* (Richter) of middle Lower Devonian age, and another form with fewer rings is similar to some Eifelian or Givetian species of *Nowakia*. The smooth-walled shells comprise large forms belonging to the group of *Styliolina fissurella* (Hall), which is characteristic of the Eifelian and Givetian in Europe, and smaller impressions that may be related to other styliolinids known to occur with *Nowakia acuaria* in Europe.

Other fossils recorded by Burton (1967b) from the Mahang Formation include radiolaria, hystrichospheres of the Silurian genus *Dictyotidium*, thin-shelled brachiopods and lamellibranchs, Ordovician trilobites of the family Cyclopygidae, and crinoids.

Summarizing the evidence afforded by the paleontological record, the age of the Mahang Formation can be placed between the limits of Ordovician and Lower Devonian time. The tentaculites corroborate the early Lower Devonian range of the *Monograptus hercynicus*-type graptolites; but since they may include younger forms, the uppermost strata may be of a somewhat later Devonian age. The base of the Mahang Formation has not been observed, and the absence of a diagnostic fauna in the lower beds does not allow a more precise dating than is afforded by the existing Ordovician cyclopygids.

DETRITAL MEMBERS OF THE SETUL FORMATION. Carbonaceous siltstone and cherty rock containing graptolites and subordinate arenite and tentaculites shale, all identical to the strata present in the Mahang Formation, occur interbedded with shelf limestone in the upper part of the Setul Formation of the Langkawi Islands. Two discrete developments, the Lower and Upper Detrital Members (Jones 1968), represent, respectively, a transient incursion of the basin facies on to the shelf in the Lower Silurian and the final phase of infilling of the miogeosyncline in the Early Devonian.

The Lower Detrital Member outcrops on Pulau Langgun and on Pulau Tanjong Dendang in the northeast of the Langkawi Islands, on Pulau Timun in the east, and on Pulau Tuba and nearby islands in the south of the group. The band varies in thickness from 22 m on Pulau Timum to 25 m on Pulau Tuba and Pulau Langgun. It is best exposed on Pulau Langgun where the strata are least affected by metamorphism. It includes black carbonaceous siltstone, flaggy shale, and chert (Figure 3.8). On Pulau Tuba the beds have been metamorphosed to spotted slate and phyllite, quartz-biotite schist, and cordierite-bearing schist.

The Lower Detrital Member comprises units 2 to 7 in the following succession, which is present on Pulau Langgun:

Unit	Thickness (m)	Description
8	—	Bedded Setul limestone
7	4.8	Black, carbonaceous siltstone and fissile shale with abundant graptolites
6	9.5	Closely bedded, black carbonaceous and siliceous shale and chert without macrofossils
5	5.5	Black carbonaceous and cherty mudstone with occasional very fragmentary graptolites
4	3.6	Closely bedded, carbonaceous quartzite and siltstone, veined by quartz stringers
3	0.4	Thickly bedded, carbonaceous siltstone and gritty subgreywacke with shelly fossils
2	1.1	Well-bedded, carbonaceous siltstone with abundant graptolites
1	—	Bedded Setul limestone

The strata are richly fossiliferous at certain levels: the fauna includes shelly fossils and graptolites.

Unit 2 contains abundant *Glyptograptus persculptus* (Salter) and occasional *Diplograptus modestus*, but no monograptids. The level belongs to the Glyptograptus persculptus Zone at the base of the Silurian.

The shelly fauna from unit 3, which was described by Kobayashi et al. (1964), includes the forms *Megalomphala?* sp., *Lophospira* sp., *Dalmanitina malayensis* Kobayashi and Hamada, and *Stenopareia* sp. *Dalmanitina malayensis* is closely allied to other species of *Dalmanitina* from Europe and Asia, which occur at a similar level just above the base of the Silurian.

Unit 7 is richly graptolitic. The bottom 0.8 m of this unit contains *Climacograptus scalaris normalis* Elles and Wood, *Orthograptus vesiculosus* (Nicholson), *Dimorphograptus malayensis* C. R. Jones, and *Pristiograptus cyphus* (Lapworth). The association is indicative of the Lower Llandovery Orthograptus vesiculosus and Pristiograptus cyphus Zones. The overlying 1.2 m has yielded an assemblage from the Middle Llandovery Pristiograptus Zone. Among the species present are: *Climacograptus scalaris* (Hisinger), *Climacograptus yangtzeensis* Hsu, *Rhaphidograptus törnquisti* (Elles and Wood), *Pristiograp-*

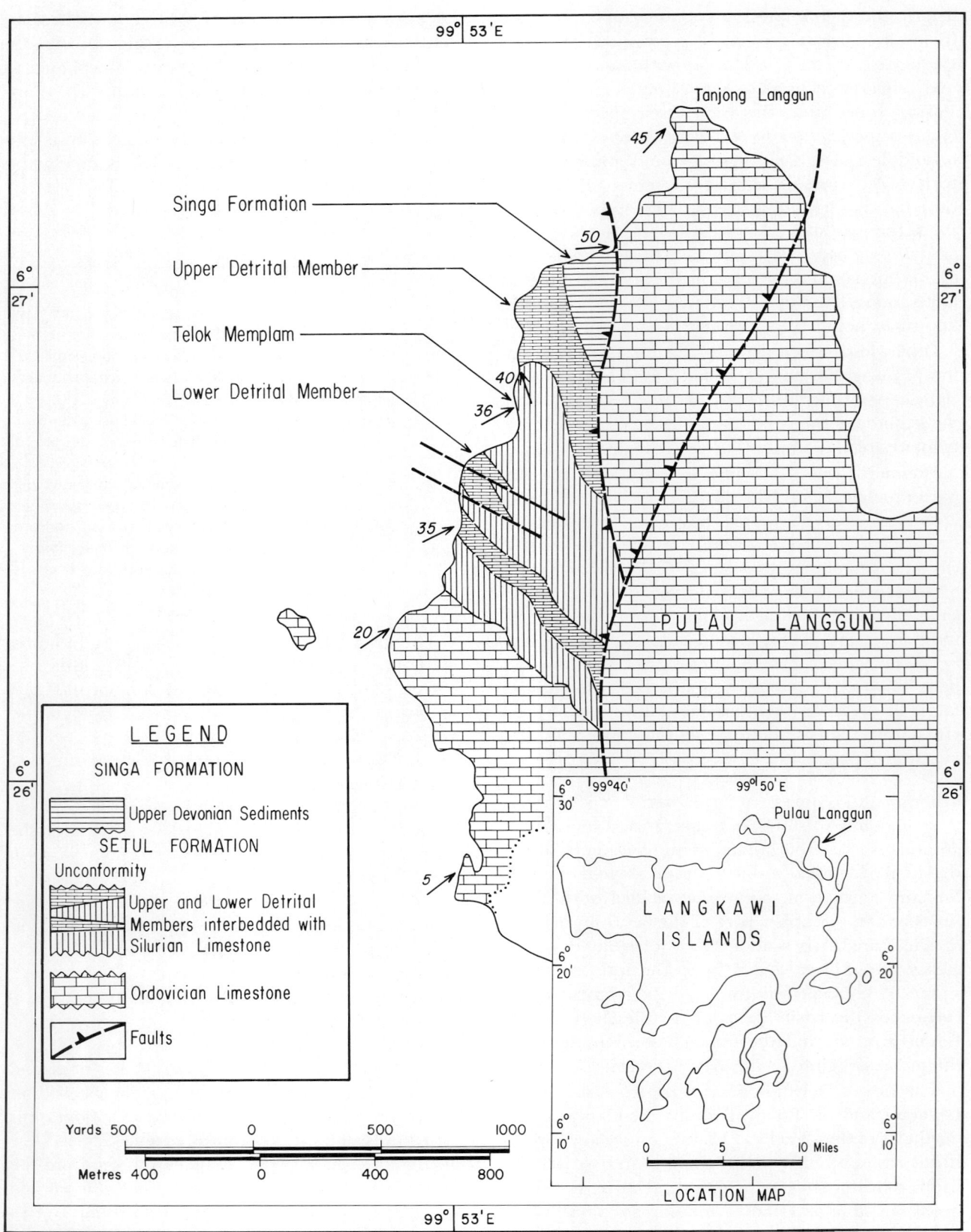

FIGURE 3.8 *Geological sketch map of the northwest coast of Pulau Langgun, Langkawi Islands.*

tus concinnus (Lapworth), *Pristiograptus sandersoni* (Lapworth), *Pristiograptus argutus* (Lapworth), and *Pernerograptus revolutus* (Kurch) (Plate 5). The topmost 2.7 m of strata contains *Petalolithus tenuis* (Barrande), *Retiolites (Pseudoretiolites) perlatus* (Nicholson), *Pristiograptus regularis* (Törnquist), *Monograptus sedgwicki* (Portlock), *Monograptus distans* (Portlock), and *Demirastrites convolutus* (Hisinger). The fauna belongs to the Middle Llandovery Demirastrites convolutus Zone, but the abundance of *Monograptus sedgwicki* in the top few centimeters and the disappearance of *Demirastrites convolutus* indicate the basal Upper Llandovery zone of *Monograptus sedgwicki*.

Above the Lower Detrital Member on Pulau Langgun, there occurs 120 m of Llandovery and Upper Silurian limestone. These beds pass conformably up into the Upper Detrital Member, which is thicker than the other clastic beds and possesses different lithology. The member comprises units 2 and 3 in the following succession:

Unit	Thickness (m)	Description
4		Fossiliferous Upper Devonian conglomeratic mudstone forming the basal beds of the Singa Formation
		unconformity
3	ca. 120	Strongly contorted brown and grey shale, siltstone, subgreywacke, and orthoquartzite
2	21	Fossiliferous dark grey and red bedded mudstone and shale
1		Bedded Setul limestone

It is impossible to work out the details of the succession in the upper part of the sequence because of intense and complicated folding and faulting. The basal beds are crowded with dacryonconarid tentaculites, with which are associated lamellibranchs and occasional graptolites of the *Monograptus hercynicus* group. The graptolites include *Monograptus* cf. *uniformis* Pribyl and *Monograptus langgunensis* C. R. Jones. The latter form is closely related to the late Silurian *Monograptus formosus* Bouček, and *Monograptus uniformis* ranges from the uppermost levels of the Silurian into the lower part of the Gedinnian. The age of the beds, therefore, is latest Silurian or earliest Devonian. According to Bouček (personal communication), the tentaculites resemble *Tentaculites matlockianus* Chapman and *Styliolina fissurella* (Hall), which suggest a somewhat later Devonian age; but these could represent deformed species of *Paranowakia*, which occur in the lower levels of the Devonian in Europe.

The overlying silty and sandy beds are unfossiliferous, but they are unconformably overlain by fossil-bearing conglomeratic mudstone of Upper Devonian age. It is concluded therefore that the age of the Upper Detrital Member is Lower Devonian.

The Upper Detrital Member thickens as it is traced southward. On Pulau Timun the thickness exceeds 200 m (655 ft), and on Pulau Tuba it is estimated to be 250 m (820 ft) thick. The succession is similar to that on Pulau Langgun, but nearly everywhere the rocks are strongly deformed and metamorphosed to metaquartzite, schist, phyllite, slate, and spotted hornfels.

HAWTHORNDEN SCHIST. The Hawthornden Schist comprises 900 m (3000 ft) of unfossiliferous carbonaceous schist and phyllite, which outcrop in the Kuala Lumpur area of Selangor; its affinities are not known with certainty. The unit overlies a thick sequence of psammitic and pelitic schist of eugeosynclinal aspect and passes conformably up into the thick Upper Silurian Kuala Lumpur Limestone. The Hawthornden Schist would therefore appear to be Lower Silurian in age, and it may represent the southerly continuation of the Mahang basin of Kedah.

However, there is little evidence to support this contention, since the Lower Paleozoic succession in the intervening parts of Perak is of the mixed facies laid down along the line of the geanticlinal ridge, and if the deposits of the Kedah miogeosynclinal basin were to be projected southward, the rocks would be expected to outcrop well to the west of the Kuala Lumpur area. This offset, however, may be due to major wrench faulting of a much younger age (Shu 1969). It is thought more likely that the schist was laid down during a transient phase of miogeosynclinal conditions, since the zones of geosynclinal deposition migrated progressively eastward.

CORRELATIVES IN THAILAND AND BURMA. Disturbed phyllitic carbonaceous shale with *Climacograptus* and *Diplograptus*, interbedded with Silurian shelf limestone, was reported by Kobayashi (1964, p. 6) from Ban Na, in peninsular Thailand, 280 km (174 miles) north of the Malayan frontier. The fauna, which includes *Dimorphograptus malayensis* C. R. Jones, is similar to that found in part of the Lower

Detrital Member of the Setul Formation in the Langkawi Islands and represents the Lower Llandovery Pristiograptus cyphus Zone. According to Burton, the sequence contains further bands of carbonaceous shale, and the top is marked by a development of tentaculites shale and arenite. The conditions would therefore appear to be identical to those obtaining in the Langkawi Islands during the Silurian and early Devonian times, when miogeosynclinal basin deposition spread temporarily to the shelf.

Brown et al. (1951) reported the occurrence of tentaculites shale at Na Suan, 240 km (149 miles) northwest of Bangkok. The strata form part of a mid-Paleozoic unit termed the Kanchanaburi Series. It is not know whether the beds at this locality are from a sustained argillaceous sequence representing basin conditions or from a mixed facies similar to the Baling group in Malaya, for tentaculites beds are known among both.

Apart from the thick shelf limestone of the Mawsoon Series and the overlying Orthoceras Beds, much of the Ordovician sequence of the Shan States region of Burma is of a mixed facies. No rocks of sustained euxinic aspect have been reported. Graptolite and tentaculites-bearing carbonaceous shale and siltstone have, on the other hand, a fairly wide distribution in the Silurian succession of the Northern and Southern Shan States (Pascoe 1959). In the Northern Shan States graptolites, representative of the Russian Lower and Middle Llandovery zones of *Orthograptus vesiculosus* (?), *Pristiograptus cyphus*, and *Demirastrites triangulatus*, occur in the Panghsa-pye Series. *Cyrtograptus* sp. was also recorded, proving the presence of the Late Llandoverian or Wenlockian material. Arenite of shallow-water origin is also present in the succession. As in the case of the Mahang Formation, these features may have originated from elsewhere by slumping or turbidity flow into the miogeosynclinal basin, or they may indicate that the depositional conditions of the Panghsa-pye Series were not consistently of the basin type.

The equivalents of the Panghsa-pye Series in the Southern Shan States contain a graptolite fauna representative of the Russian Llandovery zones of *Orthograptus vesiculosus* (?), *Pristiograptus cyphus*, *Demirastrites triangulatus*, *Demirastrites convolutus*, and *Monograptus sedgwicki*. Shallow-water sandy beds seem to be missing, and euxinic deposition was sustained over a longer interval of Lower Silurian time, so that true basin conditions appear to be represented. The Zebingyi Series of the Northern Shan States (and its equivalent in the Southern Shan States) lies some distance above the Lower Silurian graptolite-bearing strata. The beds are of limited thickness and contain tentaculites and graptolite shale, associated with shelly limestone.

The graptolites *Monograptus* cf. *riccartonensis* Lapworth, *Monograptus dubius* (Suess), and *Monoclimacis vomerina* (Nicholson) were recorded by Reed (1906), but the validity of the determinations is suspect. The graptolites occur associated with tentaculites and representative specimens in the collections of the Sedgwick Museum (Cambridge, England) are undoubtedly monograptids belonging to the early Devonian *Monograptus hercynicus* group. Therefore, the Zebingyi beds are regarded as the equivalents of the strata occurring at the top of the Mahang Formation, the Setul Formation, and the Baling group in Malaya, which contain the graptolite—tentaculite association.

ROCKS OF THE GEANTICLINE

Lying between the Lower Paleozoic miogeosynclinal sequence of northwest Malaya and the eugeosynclinal deposits of central Malaya, there occur rocks of a mixed facies, which were laid down under the generally shallow and continuously changing conditions of the geanticlinal ridge. The strata comprise an alternating and interdigitating sequence of lenticular quartzite, subgreywacke, shale, mudstone, calcareous shale, and thin limestone. Occasionally the limestone developments attain an appreciable thickness. Thick lenses of bedded rhyolite tuff are associated with the sequence along the eastern side of the geanticlinal zone. The pyroclastic rocks originated and spread onto the geanticlinal ridge from a volcanic arc sited near to the western edge of the eugeosyncline.

Rocks of the mixed sedimentary facies have been recorded from the Baling area of east Kedah by Burton (1967c, p. 172, 1972), where they were informally referred to as the "Baling Formation (or Group)." They also form extensive areas in the Grik district of north Perak, where they have been investigated by Jones (1970b). In the latter area the strata aggregate at least 2700 m (8910 ft) in thickness, and a number of component formations have been mapped. It is therefore considered that a group status for the Baling unit is warranted. From Grik, the beds can be traced intermittently for more than 80 km (50 miles) southward to the Kuala Kangsar area and the Kinta

Valley of central Perak. They may also exist in south Perak and Selangor. The rocks are nearly everywhere strongly folded and metamorphosed and therefore rarely fossiliferous. However, sufficient paleontological data have been obtained to prove an Ordovician to early Devonian age for the group.

Baling Group in East Kedah

The strata of the Baling group in east Kedah were laid down on the west or outer slopes of the geanticline; consequently they include a fair proportion of argillaceous rocks which can be related to the transition zone between the miogeosynclinal basin and the geanticlinal ridge. No volcanic rocks have been recorded. The unit is defined as an interdigitating arenaceous, argillaceous, and calcareous sequence, often carbonaceous and always metamorphosed.

The Baling group outcrops over a wide area around Baling town. The rocks are strongly folded and widely intruded by granite; in the southwest they are faulted against the Triassic Semanggol Formation. Further large outcrops probably occur in the Weng Valley and in the headwaters of the Sungei Muda to the north. The stratigraphical succession is difficult to determine owing to the rapid lateral and vertical facies changes which affect the sequence. The unit has been classified on a lithological basis into argillaceous, arenaceous, limestone, and calc-silicate facies. The argillaceous facies is the most widespread. It is composed of black indurated and carbonaceous shale, spotted shale, phyllite, metamorphosed siliceous shale, siltstone, chert, and quartz-biotite hornfels.

The next most important lithological group is the arenaceous facies. This comprises grey to pale-colored protoquartzite and some subgreywacke, which may be occasionally carbonaceous. Also included is calcareous quartzite containing diopside, plagioclase, sphene, hornblende, and epidote. The limestone beds, which normally only occupy small parts of the sequence, consist of pure white or light grey massive or thickly bedded marble in the thicker developments (see Table 3.2 for chemical analysis), and dark grey or black carbonaceous and argillaceous marble where the beds are thinner. Quartz, tremolite, and wollastonite have been reported from some marble occurrences.

Occasionally, limestone lenses form a thick part of the sequence; as an example, consider the beds that form the spectacular hill of Bukit Baling, which rises to a height of 600 m (1968 ft) a short distance to the north of Baling town. More than 1500 m (4920 ft) of strata has been measured in this outcrop. Elsewhere, the limestone facies is represented by beds only a few millimeters in thickness. The calc-silicate facies is of considerable importance. It comprises hard, brittle, often banded, green, grey, white and yellow, silt-grade hornfels variously composed of quartz, calcite, garnet, wollastonite, plagioclase, clinozoisite, epidote, hornblende, biotite, tremolite, microcline, zoisite, and sphene. The banding is due to slight differences in original composition of the component layers of sediment.

The generally metamorphosed condition of the sediments in the Baling group makes topographical expression in land of moderate relief. The indurated argillaceous rocks, hornfels, and arenaceous beds form fairly steep isolated hills, which rise to heights exceeding 300 m (985 ft) and are separated by low, wide, and swampy valleys. Where the limestone was sufficiently thick, it eroded to form spectacular precipitous hills which display well-developed karst topography.

No fossils have been recovered from the Baling group in east Kedah. However, not far to the west of Baling town, tentaculites and graptolites that have been found in carbonaceous argillites have been classed with the Mahang Formation. There is little doubt that these beds pass laterally eastward into strata of the Baling group. The graptolite fauna from one locality is indicative of the Middle Llandovery Pristiograptus gregarius or Demirastrites convolutus Zones. The tentaculites are similar to the forms from the Mahang Formation and point to an early Devonian age.

Baling Group in North Perak

Rocks of the Baling group cover wide areas of north Perak in the neighborhood of the towns of Grik and Kroh. They outcrop to form an enclave of sediment between the granite batholiths of the Bintang and Main ranges and are folded into a major synclinal structure. In respect of their great lithological heterogeneity and rapid vertical and lateral facies changes, they are closely comparable to their equivalents in east Kedah. Metamorphism is less widespread, but the strata are considerably deformed by regional shearing, with the result that phyllite and rocks showing varying degrees of schistosity are common. As in east Kedah, arenaceous, argillaceous, and calcareous facies can be distinguished, but the structure in north Perak allows the identification of a stratigraphic sequence. The following component formations were recognized by Jones (1972):

3. *Bendang Riang formation:* mainly carbonaceous shale and phyllite with numerous small lenses of limestone and some beds of quartzite and schist.

2. *Grik siltstone:* essentially carbonaceous siltstone.

1. *Papulut quartzite:* predominantly arenite with minor lenses of shale, limestone, and schistose subgreywacke.

The Papulut quartzite constitutes a well-defined unit reaching 900 m (2970 ft) in thickness. The strata outcrop along the limbs of the synclinal trough adjacent to the granite and also extend over wide areas of the Papulut Forest south of Grik, where the sedimentary cover is shallow. The rocks form some of the most rugged terrain in north Perak, with numerous craggy escarpments and hills of which Gunong Kendrong, rising to a height of 1000 m (3280 ft), is the most spectacular example (Plate 6). The strata consist of light brown, thickly bedded protoquartzite and subgreywacke and subsidiary lithic greywacke, and conglomerate with chert and quartzite pebbles. The rocks are fairly well sorted and may contain appreciable feldspar. Current bedding is occasionally present, but little grading is apparent. The quartzite is of shallow-water deposition and, by virtue of its low position in the Lower Paleozoic succession, it may possibly be equivalent to the Machinchang Formation of northwest Malaya. However, no fossils have been found to support this contention.

The Grik siltstone, which is locally developed only to the northwest of Grik town, can be regarded as a thick transition sequence of carbonaceous siltstone, lying between the Papulut quartzite and the argillite of the Bendang Riang formation.

The Bendang Riang formaion is composed predominantly of black carbonaceous shale, mudstone, siliceous shale, and phyllite. Thin lenses of limestone and arenite are also present. Metamorphism has, in places, transformed the sediment to quartz-biotite and quartz-chlorite-sericite hornfels and schist. In some areas the sequence contains lenses of rhyolite tuff. The limestone strata have a very restricted vertical and lateral distribution. Only in one area are they thick enough to form hills developing the characteristic features of karst topography. The limestone is light to dark grey, massive to thickly bedded, containing large quantities of noncarbonate impurity. It is frequently platy and banded owing to the segregation of the argillaceous impurities from the carbonate. Calc-silicate hornfels

PLATE 6 *View west across the Grik area toward Gunong Kendrong, north Perak. The country is formed of shales and quartzites of the Lower Paleozoic Baling group. Photograph C. R. Jones.*

composed of diopside, wollastonite, zoisite, clinozoisite, garnet, and coarsely crystalline marble are found in the vicinity of the granite margins.

Fossil remains are not common in the Baling group of north Perak. Some limestone shows evidence of organic content, but as with the argillite, metamorphism has mostly destroyed the fossils. However, in the valley of the Sungei Rui, between Grik and Kroh, there occurs shale that is less deformed than is usual and contains reasonably well-preserved graptolites and other fossils. The graptolites include *Climacograptus scalaris* (Hisinger), *Stomatograptus* cf. *grandis* (Suess), *Monograptus praecedens* Bouček, *Streptograptus nodifer* (Törnquist), *Spirograptus grobsdorfiensis* (Hemmann), and *Spirograptus minor* (Bouček). The faunal assemblages indicate the Upper Llandovery zones of *Rastrites linnaei* and *Spirograptus minor*, *Streptograptus crispus*, *Monoclimacis griestoniensis*, *Spirograptus spiralis*, and *Stomatograptus grandis*.

An unusual fauna of mixed planktonic–benthonic character was obtained from red mudstone near Kampong Pahit in the Rui Valley. The most abundant fossils are Lower Devonian tentaculites, identified by Bouček (in correspondence) as *Nowakia* cf. *acuaria* Richter and *Styliolina* sp. Associated with these are shelly fossils, which include the brachiopod genera *Plectodonta* and *Echinocoelia* (Hamada 1969b) and trilobites. The trilobites were considered by Kobayashi (in correspondence) to belong to the genera *Trinodus* and *Selenoharpes* of Ordovician age, but the brachiopods suggest a Lower Devonian age for the mudstone. It seems possible that the fauna represents a *remanié* assemblage of Ordovician shelly fossils redistributed in Lower Devonian times and laid down with contemporaneous brachiopods and tentaculites.

Ordovician trilobites have been collected from another locality south of Sik, Kedah. These are poorly preserved but from among them Kobayashi and Hamada (1970) identified *Isotelus*, *Ogygiocaris?*, and the cyclopygid *Microparia*.

Other Formations

Rocks of the mixed facies of east Kedah and north Perak appear to form an almost continuous belt of country, broken only by outcrops of granite, extending southward along the Perak River valley to beyond Kuala Kangsar in central Perak. In the vicinity of Lenggong, limestone lenses form a group of prominent hills and assume some importance. The rock is a coarsely crystalline marble, often containing tremolite and flakes of graphite. Chondrodite-bearing hornfels is also common. The belt of sediments continues southward hemmed in on either side by the steep granite ranges of Kledang to the east and Bintang–Gunong Bubu to the west, until it disappears beneath the coastal alluvium in the vicinity of Parit.

The sequence around Kuala Kangsar consists of unfossiliferous carbonaceous shale and subgreywacke, which form a low undulating terraine. Thick developments of limestone, strongly metamorphosed and intruded by aplite dykes, account for the precipitous hill of Gunong Pondok at the head of a sedimentary enclave, running westward from Kuala Kangsar into the Bintang–Gunong Bubu granite. The northerly continuation of this limestone beyond the granite outcrop can be found at Batu Kurau near Taiping, where an isolated hill rises above the alluvial plain. Limestone of probable Lower Paleozoic age has also been recorded from the bedrock in alluvial tin mines around Selama, north of Taiping. West of this area, the Lower Paleozoic rocks are either faulted against or unconformably overlain by the Triassic Semanggol Formation.

Lower Paleozoic rocks of the geanticlinal mixed facies are also present in the bedrock of the Kinta Valley of central Perak. The strata were investigated, although their true indentity was not recognized, in the Sungei Siput area mapped by Savage (1937) and in the Ipoh and Kampar areas surveyed by Ingham and Bradford (1960). These authors regarded the sedimentary sequence of interbedded limestone, shale, schist, and quartzite as belonging to two units of probable Upper Paleozoic and Triassic age. The much longer range of the stratigraphical succession in the Kinta Valley has been proved subsequently by the recognition of fossiliferous limestone of Silurian, Devonian, Carboniferous, and Permian ages.

The Lower Paleozoic sequence of the Kinta Valley, which has been determined mainly from temporary exposures in tin mines around Sungei Siput and Chemor north of Ipoh, comprises interbedded limestone and carbonaceous shale, mudstone, phyllite, and schist, together with subsidiary arenite that underlie the Chemor Limestone. The limestone is frequently a mottled light to dark grey, saccharoidal dolomite. It occasionally contains fossils such as corals belonging to the Silurian genera *Halysites*, *Heliolites*, and *Favosites*, which have been described by Thomas and Scrutton (1969). Graptolites have been recovered from shale at a number of places. At one locality the fauna contains *Diplograptus* cf. *magnus* H. Lapworth, which suggests that the level was that of the Middle Llandovery Pristiograptus gregarius Zone. Calcareous shale at another locality yielded poorly preserved Lower Devonian tentaculites of the genus *Nowakia*. Strata of the mixed facies pass upward into thick limestone of Upper Silurian and Devonian age, which are described in the section on shelf limestone.

Interbedded limestone and shale are known from farther south in Perak and north Selangor. The limestone has yielded Devonian, Carboniferous, and Permian fossils. No Lower Paleozoic beds have been proved, but it is very probable that such rocks exist.

Correlatives in Thailand and Burma

Rocks of the mixed geanticlinal facies cannot be distinguished with certainty in Thailand and Burma. Much of the Burmese Lower Paleozoic succession is of a mixed facies, apparently similar to that composing the Baling group of Malaya; but until more is known about the paleogeography and lithofacies of the Thai and Burmese sequences, we cannot relate the rocks to their intrageosynclinal environment of deposition.

The Kanchanburi Series of Thailand, described by Brown et al. (1951), is of Silurian to Carboniferous age. It is composed of arenaceous and argillaceous strata metamorphosed in places to slate, phyllite, and quartzite. Tentaculites shale, similar in faunal content to beds described from Malaya, occurs near Thung Song in peninsular Thailand and at Na Suan northwest of Bangkok. The absence of any limestone tends to suggest that the strata were laid down in the deeper conditions of an intrageosynclinal basin, rather than in the shallower waters of a geanticline.

Pascoe (1959) wrote that the Naungkangyi Series in the Northern Shan States of Burma and its equivalent the Pindaya Beds, in the Southern Shan States, comprised various argillites, sandy marls, and lenses of crystalline limestone. The rocks contain a rich shelly fauna of Middle to Upper Ordovician age. The overlying Panghsa-pye Series, Namhsin Series and Zebingyi Series of the Northern Shan States are composed, respectively, of graptolite-bearing carbonaceous shale of Llandovery age; sandstone, conglomerate, and marl with a shelly fauna of Wenlock and Ludlow age; and limestone and interbedded shale with tentaculites and graptolites of an early Devonian age. In the Southern Shan States, much of the sequence is identical to the mixed facies of the Northern Shan States.

ROCKS OF THE EUGEOSYNCLINE

Along the east side or internal zone of the Malayan Lower Paleozoic geosyncline, there occurs a thick sequence of schist and flysch-type detrital sediments which accumulated in the eugeosynclinal basin. Typically associated with these rocks are pyroclastic rocks and basic and ultrabasic intrusive rocks. The sediments comprise conglomerate, quartzite, sandstone, argillite, and cherty rocks, which commonly show rapid lateral and vertical facies variation. These overlie a schist formation which appears to be absent from the sequence in certain regions. Interbedded with the sediments, and more conspicuous on the geanticlinal flank of the basin, are rhyolitic and rhyodacitic crystal tuffs, which mark a period of contemporaneous explosive volcanic activity. The intrusive rocks comprise an ophiolitic suite of peridotite, pyroxenite, and dolerite, often strongly serpentinized, emplaced during a phase of mid-Paleozoic basic and ultrabasic igneous activity.

The eugeosynclinal rocks outcrop along the flanks of the Main Range Granite and also form enclaves and roof pendants within the granite. The main outcrop forms a continuous band averaging 12 km (7.5 miles) wide, extending in a north-south direction for 350 km (217 miles) along the east side of the Main Range Granite from Ulu Nenggiri in west Kelantan, through west Pahang and east Negri Sembilan, to Malacca. In Malacca the outcrop circumscribes the granite and can be recognized strik-north along the west margin of the granite into Negri Sembilan and Selangor. The eugeosynclinal facies may also be represented in Perak, although its identity there has not been conclusively demonstrated.

The Lower to mid-Paleozoic eugeosynclinal sediments on the east side of the Main Range were among some of the first rocks to be mapped in detail by the Geological Survey of West Malaysia. However, owing to lack of fossil evidence, their age and affinities were not established for many years. Scrivenor (1911), using lithological comparison, referred much of the sequence to the Triassic succession. He considered the strata to be the equivalents of fossiliferous Triassic beds, which outcrop widely farther to the east in Pahang, from which they were separated by an outcrop of Permian and Carboniferous calcareous rocks.

On structural grounds, Richardson (1946) and Alexander (*in* Ingham 1949, p. 29) were convinced that the rocks formed a distinct unit of considerably greater age. Richardson (1946, p. 218) applied the name "Foothills Formation" to the upper more arenaceous part of the succession and considered its age to be early Carboniferous or older because the strata underlay the Permo-Carboniferous calcareous rocks.

Alexander (*in* Ingham 1949) referred these rocks to a new unit named the Older Arenaceous Series, to distinguish them from the Triassic Upper Arenaceous Series, and gave them a tentative Lower Carboniferous age. Alexander (1959) subsequently changed this nomenclature to the Bentong Group, which was allotted a Carboniferous age. This unit did not include the underlying schist, which was erroneously correlated with the early Carboniferous Kuantan Group of east Pahang.

Only within recent years has evidence been produced to reveal the true identity and age of these sedimentary rocks (Ja'afar in prep., Jones 1967b). In the Karak area of south Pahang, Ja'afar (in prep.) recognized rocks equivalent to the Foothills Formation and the Older Arenaceous Series, and he named them the Karak Formation. Shale near the top of the unit yielded diagnostic fossils which indicate that the bulk of the succession is Lower Paleozoic in age.

The sedimentary rocks of the eugeosyncline are now fairly well known on the east side of the Main Range in Pahang. On the west, in Malacca, south Negri Sembilan, and Selangor, information is more limited. The type area lies in the vicinity of Bentong and Karak in west Pahang. In the following account Alexander's (1959) name, Bentong Group, is employed to include both the lower unit of schist and the overlying flysch-type sequence. Ja'afar (in prep.), working in south Pahang, was unable to distinguish the component formations present in the Bentong Group farther north, and described the whole sequence under the name Karak Formation.

Bentong Group in West Pahang

Richardson (1939, 1950) was the first geologist to tackle the detailed mapping of the eugeosynclinal rocks in northwest Pahang. The rocks were classed with the Triassic Upper Arenaceous Series and described as being composed of shale, quartzite, and conglomerate, all rich in muscovite and metamorphosed in places to phyllite and schist. The local presence of actinolite schist was noted. No attempt was made at stratigraphical analysis of the sequence. As noted previously, Richardson (1946) subsequently considered the rocks to belong to a distinct unit of early Carboniferous age or older.

Alexander (1968) worked immediately to the south in the neighborhood of Bentong, where the Bentong Group comprises a lower unit, which he termed the Schist Series, and an upper unit, the Older Arenaceous Series. The Schist Series was described as a strongly folded and foliated sequence at least 1500 m thick, composed of quartz and graphite schist with some phyllite, slate, and hornfels, and subsidiary garnet and amphibole schist and schistose grit. The amphibole schist is composed predominantly of actinolite, with minor quantities of quartz, chlorite, and epidote. The origin of these components has been discussed by several authors, including Richardson (1947a), Roe (1953), and Alexander (1968), and has been variously ascribed to such processes as the regional metamorphism of sills and dykes of basic and ultrabasic igneous rock, or the alteration of impure limestone and calcareous sediment. The former theory is perhaps the most acceptable, since hypabyssal intrusions of basic and ultrabasic igneous rock are known to be associated with the eugeosynclinal sediments, whereas unmetamorphosed calcareous rocks are virtually unknown. Ja'afar (in prep.) investigated the question from a geochemical aspect and demonstrated that the composition of the amphibole schist most closely resembles dolerite.

The conformably overlying Older Arenaceous Series is composed of over 2100 m (6890 ft) of strata of varied lithology and flysch-type affinities. The following four main lithological types were distinguished:

1. Conglomerate and grit.
2. Micaceous quartzite and subgreywacke.
3. Shale, phyllite, and schist, locally carbonaceous.
4. Carbonaceous cherty rock.

The whole sequence is characterized by poor sorting and rapid lateral and vertical facies changes, with the result that persistent and mappable units are for the most part lacking. No definite stratigraphical order of deposition was distinguished except that the conglomerate and grit lenses were said to occur near the top, the chert in the lower half, and the more arenaceous strata in the upper half of the sequence. Tuffaceous sediments are common in those areas where pyroclastic rocks are developed. Clastic fragments of rhyolite have also been recorded from the coarse grades of sediment.

The conglomerate and grit often form lenses several tens of meters thick near the top of the Older Arenaceous Series. These rocks form prominent ridges and hills rising to heights of 600 m (1968 ft) in the vicinity of Bentong. The conglomerate is poorly sorted, often of intraformational character, and it shows slumping and current bedding. It consists of semirounded phenoclasts of quartzite, chert, and occasional white quartz, usually several and sometimes many centimeters in diameter, set in sandy or shaly matrix. A good exposure of sheared conglomerate, in which elongate phenoclasts with their long axes set parallel to the bedding, occurs in a road-cutting on the Kuantan Road half a mile east of Karak (Figure 3.9). The arenaceous rocks comprise protoquartzite and subgreywacke. They contain muscovite and feldspar and much clastic material of intraformational origin, such as grains of rhyolite, shale, carbonaceous quartzite, and chert. Shale and flaggy beds are fairly common. They are often carbonaceous and siliceous and are locally metamorphosed to slate, phyllite, and schist. The chert is carbonaceous closely bedded rock. It grades into siliceous shale with increasing argillaceous content and is interbedded with carbonaceous phyllite. The origin of chert beds is discussed in the section on the miogeosynclinal basin deposits.

In south Pahang, the sediments of the Bentong Group are predominantly argillaceous and commonly metamorphosed—the Karak Formation Ja'afar (in prep.). The strata are described as tightly folded, and they comprise at least 4800 m (15,840 ft) of schist and hornfels interbedded with chert, quartzite, shale, phyllite, conglomerate, intraformational breccia, and rhyolite tuff. Certain lenticular masses of unfossiliferous limestone, which occur above the detrital rocks and which were alluded to in the section on shelf limestone, are included with the eugeosynclinal sequence by Ja'afar (in prep.). On the other hand, Alexander (1968) and Richardson (1946) regarded the limestone as belonging to a younger sequence of deposition. Richardson afforded evidence for an unconformity between the two successions. Combined with occurrence of much coarse detrital sediments at the top of the flysch sequence, the unconformity strongly suggests a major break below the limestone.

As in the area to the north, no strict stratigraphical order of deposition was distinguished in the Karak Formation. The unit was divided into five lithological members. Apart from the limestone mentioned earlier, these include:

1. A schist member.
2. A chert member.
3. A conglomerate member.
4. A pyroclastic member.

FIGURE 3.9 *Sheared intraformational conglomerate in the Bentong Group near Karak, Pahang. Photograph C. R. Jones.*

The grade of metamorphism increases conspicuously from east to west toward the granite. The schist member includes graphitic, quartz-muscovite, andalusite, chlorite, and tourmaline schist; amphibole schist composed of epidote, actinolite, tremolite, and hornblende; and less metamorphosed pelitic rocks including slate, phyllite, and shale. The chert member comprises grey and black radiolarite and radiolarian graphitic schist. The conglomerate member is composed of lenticular bodies of conglomerate and chert-breccia, associated with beds of sandstone and shale, and is confined mainly to the eastern edge of the outcrop in the upper part of the succession. The pyroclastic rocks include rare occurrences of rhyolite tuff and tuffaceous sandstone.

Fossils, sufficiently diagnostic to prove the undoubted Lower to mid-Paleozoic age of the Bentong Group and Karak Formation, have been discovered only in recent years. The first organisms found were poorly preserved orthoceratids in siltstone along the Bentong to Bilut Valley Road. Subsequently, a mixed planktonic–benthonic fauna was collected from bleached shale occurring near the top of the Karak Formation in a road-cutting on Tuan Estate south of Karak, Pahang (Jones 1967b).

The fauna described by Jones (1970a) includes the sponges *Hydrodictya cylix* Hall and Clarke and *Lyrodictya?* sp., the brachiopod *Orbiculoidea sinensis* Mansuy, the phyllocarid *Ceratiocaris* sp., and the graptolites *Monograptus* cf. *praehercynicus* Jaeger and *Linograptus* aff. *posthumus* (Richter). The graptolite association indicates an early Lower Devonian age. *Monograptus praehercynicus* is confined to the Monograptus praehercynicus Zone in central Europe, which is correlated with the upper part of the Gedinnian stage of the Rhenish succession. The age of this fauna proves that the bulk of the eugeosynclinal sequence is Silurian and possibly older.

Other Formations

The outcrop of the Bentong Group can be traced southward through east Negri Sembilan to Malacca. In the Ulu Triang area, carbonaceous mudstone has yielded *Monograptus hercynicus* Perner. This graptolite occurs in the upper layers of the Monograptus praehercynicus Zone and in the Monograptus hercynicus Zone of central Europe; these levels are correlated with the upper part of the Gedinnian Stage and the lower part of the Siegennian Stage of the Rhenish succession. In Malacca and the neighboring part of south Negri Sembilan, Rishworth (in prep.) distinguished two major stratigraphic units, one of argillite and pelitic schist and the other comprising arenite and psammitic schist. Amphibolite also occurs. Superficially these formations have a close correspondence, in lithology and relative age, to the Schist Series and Older Arenaceous Series of the Bentong Group in Pahang. However, Rishworth distinguished structural differences between the two units, which led him to propose that the psammitic sequence was considerably younger and quite distinct from the pelitic rocks. He tentatively correlated the arenite and psammitic schist with the Triassic formations and proposed a Middle to Late Paleozoic age for the argillite and pelitic schist.

However, it is contended by the writer that the two units can be justifiably correlated with the component formations of the Bentong Group. The argument is based on the apparent continuity of outcrop of the two sequences along the margin of the Main Range Granite and the stratigraphical and lithological similarity between the two successions. The differences in strike between the two formations in Malacca, noted by Rishworth (in prep.), may also exist in west Pahang. On the other hand, it is possible that the structural differences in Malacca and Negri Sembilan may be related to the shape of the southern extremity of the Main Range Granite batholith.

The argillite formation comprises fine-grained silty beds with minor sandy, shaly, and marly bands, along with very subordinate arenaceous beds with occasional chert and conglomerate. The rocks are invariably altered to quartz, quartz-chlorite-sericite, biotite, and muscovite bearing schist and phyllite. Two areas of amphibole schist composed of actinolite, hornblende, plagioclase, and quartz occur. The arenite formation consists of quartz sandstone and siltstone, metamorphosed to schistose and semischistose sericitic and chloritic metaquartzite, quartz, quartz-biotite and silliminite, and garnet-bearing schist and gneiss. There also occurs occasional quartz-tremolite-zoisite hornfels, which suggests that the sedimentary sequence contained some slightly calcareous beds.

Lower Paleozoic rocks exhibiting eugeosynclinal affinities have been recorded from the west side of the Main Range Granite in central and north Selangor. In both these areas the rocks closely resemble the components of the Bentong Group on the east side of the Main Range. The Dinding Schist, described by Gobbett (1965a) from the

Kuala Lumpur area, consists of 3400 m (11,200 ft) of quartz-mica schist, quartzite and subsidiary actinolite, diopside and epidote schist, and schistose conglomerate. The quartzite forms a discrete member, 1770 m (5840 ft) thick, in the middle part of the succession. In north Selangor, Roe (1951, 1953) classified as Triassic a sequence of quartz-mica, graphitic, and amphibole schist intruded by basic dykes. In view of the recent advances made in the elucidation of the geology of central Selangor, these rocks can now safely be reassigned to the Lower Paleozoic eugeosynclinal sequence.

The eugeosynclinal rocks of north Selangor appear to extend northward into south Perak, but in this region the geology has not been investigated in detail. East of the Lower Paleozoic geanticlinal mixed facies of north Perak, in the upper reaches of the Perak River, there occurs a sequence of schist (including amphibole schist and argillite) and arenite intruded by an ophiolitic suite of igneous rocks. These rocks are identical to the eugeosynclinal sequence of Pahang; but again, the region has not been surveyed in detail.

Volcanic Facies

TUFF. Acid pyroclastic rocks occur in the eugeosynclinal succession on both sides of the Main Range Granite. They are particularly conspicuous along the geanticlinal flank in north Perak, where they form large tracts of country in the Grik and Sungei Siput areas. The rocks consist essentially of rhyolite crystal tuff that has been emitted under subaerial conditions, probably from a series of vents disposed about an island arc. The distribution pattern and relative thickness of the tuff indicates that the arc probably lay along the geanticlinal margin of the eugeosyncline.

The pyroclastic strata occupy a large area of country to the east and south of Grik, where they form a characteristic hummocky topography. They are termed the Lawin tuff, from the locality where the facies is particularly well developed. The rocks occur interbedded with, and pass imperceptibly into, quartize, shale, and thin limestone of the Baling group. They are probably of Ordovician and Lower Silurian age.

Pyroclastic rocks similar to the Lawin tuff occur in the eugeosynclinal succession east of the Main Range in south Pahang. The strata are more attenuated and occur interstratified with schist, shale, and conglomerate of the Bentong Group. At one locality, beds of tuff occur below fossiliferous early Lower Devonian shale, which indicates that the volcanic activity persisted into the late Silurian or early Devonian.

These volcanic rocks are described in Chapter seven.

OPHIOLITES. Associated with the eugeosynclinal strata are small bodies of basic and ultrabasic igneous rock, which frequently are so strongly altered by dynamothermal metamorphism that their original character is difficult to ascertain. The rocks take the form of dykes, sills, and other small bodies and represent a phase of basic intrusive activity peculiar to the internal zones of the geosyncline. The bodies are essentially intrusive, but there is a possibility that certain layers of amphibole schist may represent strongly altered submarine basalt flows.

The ophiolites vary considerably in composition. They include pyroxenite and peridotite, which are nearly always strongly altered to actinolite schist and serpentinite. The serpentinite is perhaps the most characteristic representative of the suite, although its distribution is very local. In a brief reference to serpentinite in Pahang, Negri Sembilan, and north Perak, Scrivenor (1931, p. 51) included it in his broad classification of "granite and associated rocks," thus implying that it originated during the phase of synorogenic plutonism. Willbourn (1933, 1934) described serpentinite and layered rocks composed of bands of serpentine alternating with hornblende schist, garnet-pyroxene, and diallage-enstatite rocks, from the Lipis District of Pahang. Richardson (1939) reported the occurrence of serpentinite from the Raub area of Pahang, in which olivine and pyroxene were completely replaced by serpentine minerals. Accessory minerals identified in these rocks include magnetite, chromite, pyrite, pyrrhotite, tremolite, chlorite, and magnesite. It was concluded from the mineralogy that the serpentinite was formed by the regional metamorphism of ultrabasic rocks such as pyroxenite, saxonite, and peridotite. An occurrence of peridotite passing into fully serpentinized rock has since been recorded from south Pahang by Ja'afar (in prep.).

Richardson (1939, 1950) and Alexander (1968) mapped amphibole schist forming isolated outcrops in the eugeosynclinal succession of west Pahang. The rocks are fine grained and strongly foliated. They range in composition from epidote-chlorite-actinolite schist to quartz-chlorite-actinolite schist. Their origin was ascribed to dynamothermal metamorphism of either basic igneous rock or impure calcareous sediment. The theory is probably inaccurate, however, because calcareous strata are absent from those areas where the strata are less strongly metamorphosed. Ja'afar (in prep.) demonstrated that the composition of the amphibole schist most closely resembles dolerite.

Ophiolites appear to be rarer on the west side of the Main Range Granite. Roe (1951, 1953) described the occurrence of dolerite dykes in north Selangor, and Gobbett (1965a) referred to the local occurrence of beds of actinolite schist in the Dinding Schist of central Selangor. A small serpentinized body of ultrabasic rock has also been described by Jones (1970b) from the Grik area of north Perak.

The ophiolites seem to be absent from the upper part of the eugeosynclinal succession and this suggests that the basic and ultrabasic igneous activity was penecontemporaneous with sedimentation.

PALEOGEOGRAPHY AND GEOLOGICAL HISTORY

The paleogeography of the Malayan region in Lower Paleozoic times was discussed briefly by Jones (1961) and Hutchison (1961). Burton (1967a) dealt in greater detail with the distribution of land and sea as it affected deposition within the miogeosynclinal basin of the Lower Paleozoic geosyncline. On a regional scale the evolution of the areas of land and sea has been the subject of discussions by Grabau (1924), Sun (1946), Lu (1950), Pascoe (1959), and Nikiferova and Obut (1965).

Cambrian

Grabau (1925) demonstrated that in early Cambrian time the greater oceanic areas of the world were confined to three main basins, each characterized by an indigenous trilobite fauna. These were defined as the Atlantic Province with its *Holmia* fauna, the Pacific Province, in which *Holmia* was replaced by *Olenellus,* and the Australo-Asian Province, characterized by *Redlichia*. In southern Asia the existence of two major landmasses was postulated. Gondwanaland covered what is now peninsular India, and Cathaysia lay in the area at present occupied by southeast China, Indochina, and the Southeast Asian Archipelago. Between these continents a wide gulf of the Redlichia Sea was depicted extending northward through the present Bay of Bengal and Assam toward a farther landmass termed Tibetia. Immediately south of this land the gulf divided into two arms—one stretching westward along the line of the Himalayas as far as the Punjab and Tien Shan, the other extending northward to cover parts of south and central China and Mongolia. In Middle Cambrian times there was evidence of worldwide marine transgressions. This resulted in the expansion of the northeasterly arm of the Redlichia Gulf, which eventually joined the Pacific Province, allowing intermingling of their respective faunas.

The earliest evidence of this geographical pattern in the Malayan region lies with the Upper Cambrian Machinchang Formation and its possible equivalents in northwest Malaya and peninsular Thailand. The rocks include deltaic sandstone probably laid down in the shallow waters of the gulf that separated Gondwanaland from Cathaysia (Figures 3.10 and 3.11). The exact configuration of the depositional area is uncertain however.

In Ordovician and Silurian time the existence of a foreland lying immediately west of the Langkawi region is satisfactorily established. The coastline was well-defined in this direction and corresponded to the east margin of Gondwanaland, but whether it existed in this position during Upper Cambrian time is a matter for conjecture.

The sandstone in the lower part of the Machinchang Formation comprises flaggy argillite and subgreywacke which indicate flysch conditions. These strata may well have originated in a pre-existing geosynclinal basin, much of which lay to the west of the Langkawi area in Middle Cambrian and earlier times. This contention is supported first by the existence of the overlying sandstone of deltaic aspect, which appears to correspond to the molasse facies, and second by the observed progressive easterly migration of the geosynclinal isopic zones throughout the Paleozoic.

The acceptance of this interpretation would mean that during Cambrian time the coastline of Gondwanaland was positioned well to the west of the Langkawi area, somewhere within the region now occupied by the Andaman Sea. The theory would also suggest that the Malayan Lower Paleozoic geosyncline belonged to the epicontinental pattern, being sited along the eastern margin of Gondwanaland with an ocean basin lying beyond.

However, the available evidence better accords with an intercratonic setting. Current bedding in the deltaic sandstone of the Machinchang Formation of the Langkawi Islands indicates a source from a nearby land area to the north. The existence of such a source, combined with the absence of

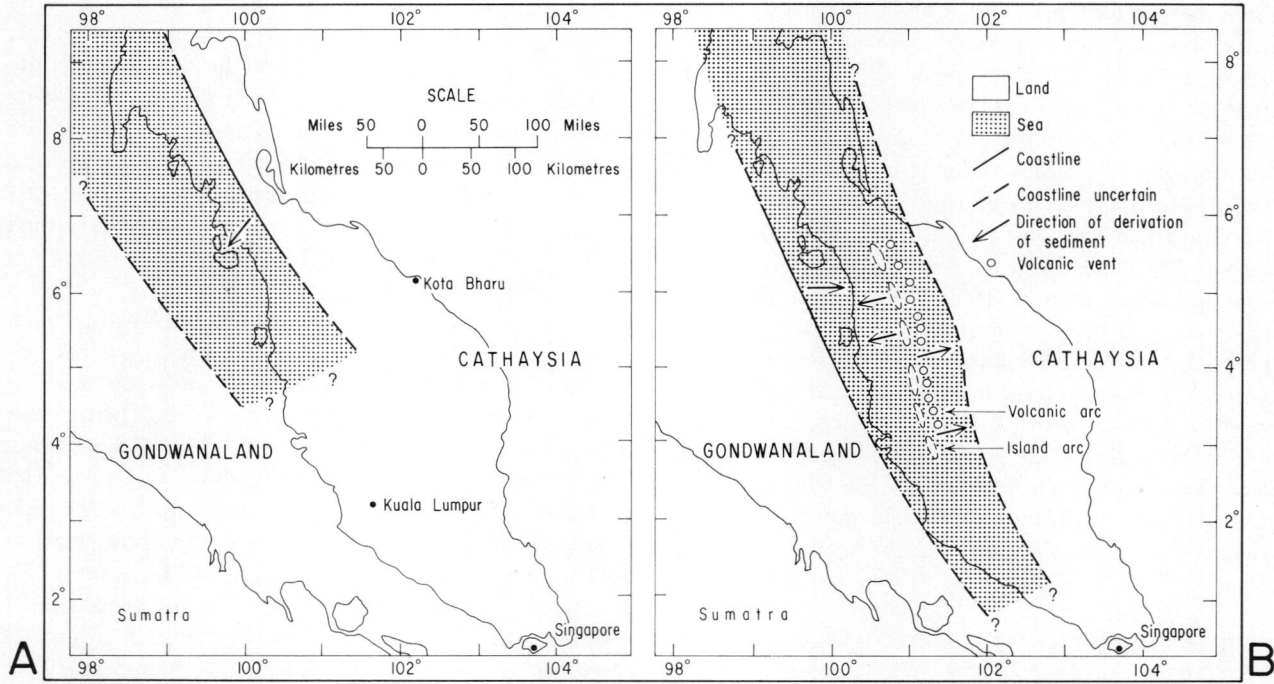

FIGURE 3.10 *Paleogeographic maps of the Malay Peninsula during the Lower Paleozoic:* (a) *During late Cambrian;* (b) *during Ordovician and Silurian times.*

Lower Paleozoic rocks from the east half of Malaya and over most of Indonesia, supports the existence of land in this direction and agrees with Grabau's concept of the continent of Cathaysia covering this region in Lower Paleozoic time.

Hutchison (1961) has proposed that the Taku Schist of north Kelantan represented part of the Precambrian crystalline basement. His arguments were based on the high metamorphic grade of the rocks (almandine amphibolite facies) and the apparent alignment of the outcrop with the Precambrian Mogok Gneiss of Upper Burma, and the occurrence of probable basement rocks in Sarawak. To accept this idea would be to agree with the theory of the intercratonic setting of the geosyncline, allowing the siting of the western coastline of Cathaysia between north Kelantan and north Perak throughout the Lower Paleozoic. However, subsequent work in south Kelantan suggests that the Taku Schist may be the metamorphosed and partially migmatized equivalent of Paleozoic strata.

The paleogeography at the end of the Cambrian was that of a narrow seaway covering northwest Malaya and extending northward through west peninsular Thailand and Burma to connect with the Himalayan and Chinese troughs. In the Malayan region the available evidence suggests that the depositional basin was a restricted shallow embayment, occupying an irregularity in the west coast of Cathaysia and connected westward to more extensive oceanic waters that ran south to Australia. The paleogeographical model for the Late Cambrian (Figure 3.10) has been constructed in accordance with an intercratonic pattern. The southerly confines of the depositional area which are uncertain, have been drawn to agree best with the available evidence. The Machinchang Formation of the Langkawi Islands and Pulau Terutau are the only rocks of definite Upper Cambrian age; probable correlatives occur in the Jerai Formation of central Kedah and the Papulut quartzite of north Perak, but it is possible that the eugeosynclinal Bentong Group of central Malaya also includes isochronous equivalents.

The link between the Malayan region and Australia is substantiated by the occurrence of *Pagodia* and other late Cambrian trilobite genera, which are present in the Machinchang Formation, in rocks forming the Macdonell Ranges of the Northern Territory of Australia (Opik 1956; Casey and Gilbert-Tomlinson 1956). The connection may have been by a seaway sited to the west of Sumatra or by way of Yunnan, South China, and the Pacific basin. The other marine connections between the

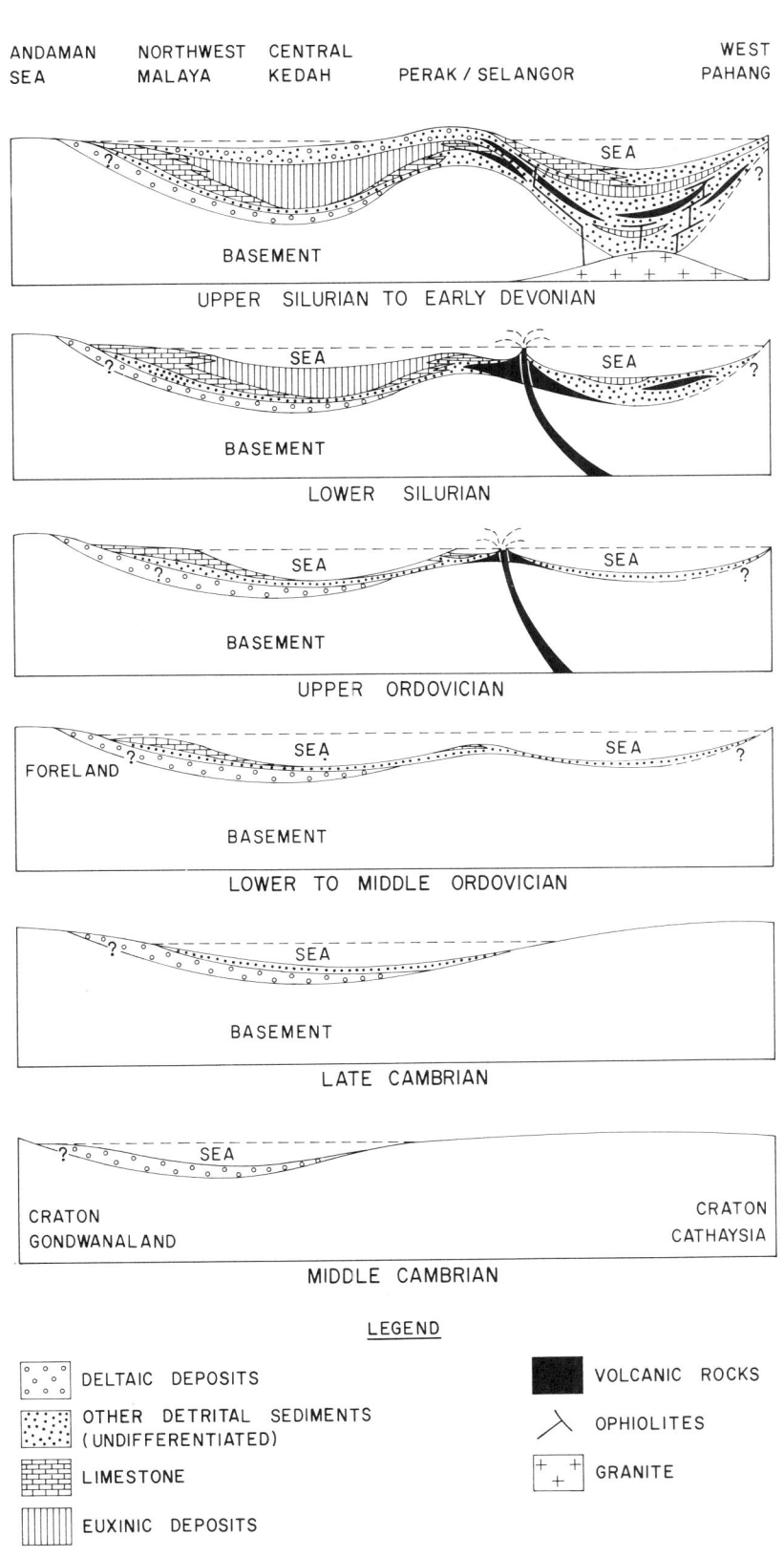

FIGURE 3.11 *Schematic sections tracing the postulated evolution of the Lower Paleozoic phase of the Malayan geosyncline.*

Malayan Region and the Australo-Asian Province of Cambrian deposition, as propounded by Grabau, are well demonstrated by the distribution of the saukid and other trilobite faunas. Saukids were reported from Yunnan by Sun (1946) and from as far west along the Himalayan trough as Iran by King (1937). According to Kobayashi (1957), the trilobites *Pagodia* and *Coreanocephalus* are important genera among the Upper Cambrian faunas of northwest China and Korea, thus proving the continuity of the marine link in this direction.

Ordovician

The Bawdwin Volcanics of the Burmese Lower Paleozoic succession suggest that geosynclinal conditions had become well established by the end of the Cambrian period. In Malaya, however, the Upper Cambrian and the lower part of the Ordovician appear to be confined to the orthoquartzite-carbonate facies of Pettijohn (1957), deposited in waters of shallow to moderate depth. It was not until the upper part of the Ordovician that the depositional zone became mobilized, with effective downwarping to form miogeosynclinal and eugeosynclinal troughs, and a complementary intrageosynclinal upwarp or geanticline (Figure 3.11). The paleogeography and intrageosynclinal evolution now became well defined (Figure 3.10). The western margin of the sea was aligned approximately north-south, immediately to the west of peninsular Thailand; the Langkawi Islands and Kedah were aligned with the eastern shores, situated some distance to the east of the present Main Range Granite.

Although this pattern is well defined in north Malaya, longitudinally it is less clear. Basin deposits of probable Ordovician age are known from as far south as Malacca. Whether the marine environment extended beyond this area is a matter for conjecture. Lower Paleozoic rocks are unknown from south Malaya and much of Indonesia. Moreover, there is a close relationship in outcrop patterns between the Lower Paleozoic sediments and the Main Range Granite and batholiths to the west, which may have originated in the tectogenes of the Lower Paleozoic geosyncline (Figure 3.2). These circumstances tend to indicate that the belt of deposition ended at no great distance to the south of Malacca.

The geological history of much of the Ordovician and Silurian times in Malaya relates to the series of isopic deposits laid down in the component zones of deposition within the geosyncline. The miogeosynclinal trough was defined by the shallow-water shelf limestone of the Setul Formation and the deeper water euxinic facies of the Mahang Formation, lying on the west side of the depositional zone. The shelf extended southward from the Thung Song region of peninsular Thailand, over the Langkawi–Perlis area, and beyond the west coast of Kedah. This gave way eastward to the Mahang basin of central and south Kedah, which extended southward into north Perak and possibly Selangor, where the Hawthornden Schist represents a metamorphosed sequence of deposits of euxinic aspect.

Sited to the east of the miogeosyncline was the geanticlinal upwarp along which the mixed facies of alternating and interdigitating arenaceous, argillaceous, and calcareous rocks of the Baling group were laid down. This zone covered east Kedah and the Grik area of north Perak, and extended southward into central Perak and possibly north Selangor. Eastward again was the eugeosynclinal basin in which argillaceous and overlying flysch-type rocks of the Bentong Group were deposited. This zone was situated along the line of the present Main Range, and the deposits can be recognized flanking the outcrop of the Main Range Granite from north Perak and west Kelantan in the north, through Selangor and west Pahang, to Negri Sembilan and Malacca in the south.

Lying within the eugeosyncline, near to the geanticlinal margin, was a volcanic arc that persisted throughout Upper Ordovician and Silurian time. This line of volcanic vents provided large amounts of pyroclastic material that became interbedded with the deposits of the eugeosyncline and the east flank of the geanticline. Burton (1967c, p. 185) suggested that volcanic pulsations, in phase with orogenic movements, were responsible for the abundant silica intermittently supplied to the seawater, which resulted in the deposition of chert and radiolarite in the intrageosynclinal basins.

In Ordovician times only a small amount of fine sediment found its way into the geosynclinal basins. In the late Ordovician and Silurian, however, the geanticline became a controlling factor in the pattern of deposition. At first it rose to form a submarine ridge, which effectively cut off the miogeosynclinal basin from the open sea to the east, causing stagnant conditions to develop. This is shown by the Mahang Formation with its predominant euxinic facies, rich plankton and phytoplanton, and impoverished benthonic faunas.

Later, in Silurian time, orogenic pulses caused the geanticline to rise periodically above the waters to form an island arc. This formed a temporary barrier between the miogeosyncline and the open sea, preventing access of plankton to the miogeosynclinal basin and so causing breaks in the sequence of graptolitic faunas. The oscillating conditions of the geanticlinal ridge are made apparent by the great variety and rapid lateral and vertical facies changes that characterize the sequence laid down along this zone. Burton (1967c, p. 185) considered that the orogenic movements were responsible for the progressive translation, by turbidity flow and slumping, of some of the coarser clastic sediments laid down on the geanticlinal flanks to the depths of the miogeosynclinal trough, where they became interbedded with euxinic sediments.

Silurian and Early Devonian

Evidence of a more universal marine transgression in Lower Silurian time is indicated by the extension of the basin facies on to the shelf. This is demonstrated in the Langkawi Islands by the occurrence of Middle to Upper Llandovery graptolitic siltstone which is found interbedded with the shelf limestone.

The worldwide Lower Silurian transgression considerably affected the paleogeography of Asia. During Ordovician time the distribution of land and sea had been modified by marine connections between the Australo-Asian Province and the Pacific and Atlantic Provinces. This is demonstrated by the close association of the gastropod–nautiloid faunas of the Setul Formation with the Black River–Trenton Groups of North America and the Vaginatenkalk of the Baltic states of Europe. However, the localized aspect of the early Ordovician graptolite faunas indicate that the major oceanic areas of Asia still remained very restricted.

Only in the Lower Silurian epoch do the graptolite faunas show a cosmopolitan character. Nikiferova and Obut (1965) have arrived at a fairly detailed paleogeographical model for much of Asia during Silurian time. This shows numerous geosynclinal basins covering large areas of Asiatic Russia and China. As in Cambrian time, the Malayan geosynclinal gulf connected northward with the Shan States and the Yunnan area, beyond which it split into three arms. To the west lay the Himalayan trough, which connected with the Ural geosyncline by way of the Tien Shan Sea of Uzbekistan. Northward lay a wide strait between the Tarim and Chinese continents. This extended over southwest China and branched into a number of geosynclinal gulfs, the largest of which was the Mongolian Sea. To the northeast was a long narrow seaway covering eastern China and North Korea and stretching to the northern end of the Japanese islands.

In Upper Silurian and Lower Devonian time, the orogenic pulses in the Malayan region became more pronounced. These led to major changes in the geography and depositional pattern. The geanticline rose to form a line of mountainous islands which provided an abundant supply of coarse clastic material to both the miogeosynclinal and eugeosynclinal troughs. The basins were gradually filled with a flysch-type sequence of arenite, greywacke, and intraformational conglomerate. In the miogeosyncline the euxinic argillite passes up into a sequence of arenite near the top of the Mahang Formation. The flysch facies spread westward on to the shelf, where the strata of the Upper Detrital Member covered the shelf limestone. The presence of tentaculite–graptolite plankton in the flysch sequence indicates that access with the open sea, by way of interisland straits, was maintained into the early Devonian epoch. A shelf formed along the east side of the island-arc on which the thick shelly Kuala Lumpur Limestone and Chemor limestone accumulated. Further east, in the eugeosynclinal trough, arenite and intraformational conglomerate of the Older Arenaceous Series of west Pahang, was laid down, along with the arenite and psammitic schist of Malacca and Negri Sembilan.

Infilling of the miogeosynclinal basin was probably complete by the end of the Lower Devonian. Subsequent tectonic adjustment led to folding of the rocks and the rise of land in the northwest part of the Malay Peninsula. Elsewhere, the marine environments remained as shallow seas in which shelly limestone was deposited throughout the Devonian. In the Upper Devonian, the reoccurrence of widespread transgressions resulted in the submergence of northwest Malaya, where clastic sediments were laid down unconformably on the Lower Paleozoic to early Devonian beds, and the sea spread eastward to cover much of the rest of Malaya. These marine incursions heralded a further phase of geosynclinal conditions that were finally brought to a close by the Malayan orogeny in the Mesozoic. The events of the Lower Paleozoic and early Devonian can therefore be regarded as comprising a fairly well-defined episode in the evolution and easterly migration of the Malayan geosyncline.

Mid-Paleozoic Igneous Activity

The question of whether plutonism occurred during the mid-Paleozoic orogeny has not been conclusively answered. The existence of a well-defined volcanic facies, and an ophiolitic suite of minor basic and ultrabasic intrusion, proves that mobilization and magmatic differentiation was undoubtedly achieved, at least in the Lower Paleozoic eugeosynclinal tectogene. The volcanic rocks were erupted in the late Ordovician and Silurian from a series of vents sited along the geanticlinal limb of the eugeosyncline. The ophiolites were intruded into the strata of the eugeosynclinal sequence, probably penecontemporaneously with deposition during the late Lower Paleozoic and early Devonian.

There is some evidence that granite emplacement accompanied the orogenic movements in the northwest part of Malaya. Koopmans (1965, p. 511) described boudinaged granite sills in the Upper Detrital Member of the Setul Formation of the Langkawi Islands, for which he afforded evidence of synkinematic intrusion during the first generation folding. Granite pebbles and boulders have been recorded from the Upper Devonian to Lower Permian Singa Formation of the Langkawi Islands. Their origin may lie with synorogenic granite intruded in Devonian times, but weathering of the Precambrian granites of Indosinia might also have been the cause. However, recent radiometric evidence has shown that the granites of Indosinia, which have long been regarded as Precambrian, are no older than Upper Permian in age, at least in part (Izokh et al., 1964; Lasserre et al. 1968).

Jones (1968) was of the opinion that the most important evidence in support of plutonism in the middle Paleozoic period is the close relation in outcrop which exists between the Main Range Granite and the eugeosynclinal sequence of the Lower Paleozoic geosyncline. It was said that the batholith was complex in structure and included several phases of intrusions, but there was no evidence to suggest that it was substantially older than any of the other Mesozoic tin-bearing granites of the Malay Peninsula. Several radiometric age determinations on the granite were available to prove that its emplacement was synchronous with the main Malayan orogeny. However, there is a strong possibility that both Main Range Granite and the batholiths lying to the west originated in the tectogene roots of the Lower Paleozoic geosyncline, although they were not elevated to high crustal levels until the main mountain-building episode of the Mesozoic.

CHAPTER 4

Upper Paleozoic

D. J. GOBBETT

Marine Upper Paleozoic rocks were recognized early in the geological exploration of the Malay Peninsula. Cliff-bounded limestone hills form prominent features in many parts of the northern half of Malaya and, in the northwest and in Pahang, some of the limestone yielded Permian and Carboniferous fossils. For many years it was assumed that all the limestone formations and associated shale were of Carboniferous and Permian age. In Ulu Pahang they were termed the Raub Series and the associated volcanic rocks were designated the Pahang Volcanic Series (Scrivenor 1911, 1931). Richardson (1939) substituted the term "Calcareous Formation" for "Raub Series," but later (Richardson 1950) changed this to "Calcareous Series." The latter name was accepted and applied generally to an upper division of the Malayan Carboniferous and Permian rocks by Alexander (1956). Alexander also distinguished a lower division, the Older Arenaceous Series, which comprise clastic formations underlying the Calcareous Series in the Langkawi Islands and in west Pahang. Later, in a general review of the Pre-Tertiary succession in Malaya, Alexander (1959) put forward the name Raub Group for the Calcareous Series and distinguished the Lower Carboniferous of east Pahang as the Kuantan Group.

More extensive mapping and fossil collecting, undertaken in the last decade mainly by the Geological Survey of Malaya, has shown that the Calcareous Series is a rock facies ranging in age from lower Ordovician to Upper Triassic. The Older Arenaceous Series is mainly Lower Paleozoic, but in southwest Pahang the upper part continues into the Lower Devonian (Jones 1967b). In Perlis and Langkawi some of the limestone is Lower Paleozoic and some is Upper Paleozoic (Jones 1968); an unconformity has been demonstrated between the two, most of the Devonian being missing. The limestone of the Kinta Valley, Perak, which was previously thought to be Carboniferous and Permian, is now known to be in part Lower Paleozoic (Ingham and Bradford 1960). The Viséan fauna from the Sungei Lembing, described by Muir-Wood (1948), has been recognized in central Pahang, and Carboniferous and Permian limestone is now known from Kelantan, Trengganu, Pahang, and Johore (Jones et al. 1966).

The Upper Paleozoic outcrops now recognized follow the approximate north-south regional strike of the Malay Peninsula (Figure 4.1). Facies changes occur across the strike, and four north-south zones may be distinguished, each with a distinctive sequence and rock assemblage. In general, knowledge of the Upper Paleozoic decreases from west to east in these zones, which, may be summarized as follows:

Zone 1: Northwest Malaya. In Langkawi, Perlis, and northwest Kedah the greater part of the Devonian is missing and the uppermost Devonian rests unconformably on the Lower Devonian or older rocks. The Carboniferous is represented by a clastic sequence that probably persisted into the Permian in Kedah but was replaced in Langkawi and Perlis by a shallow-water organic limestone facies in Mid-

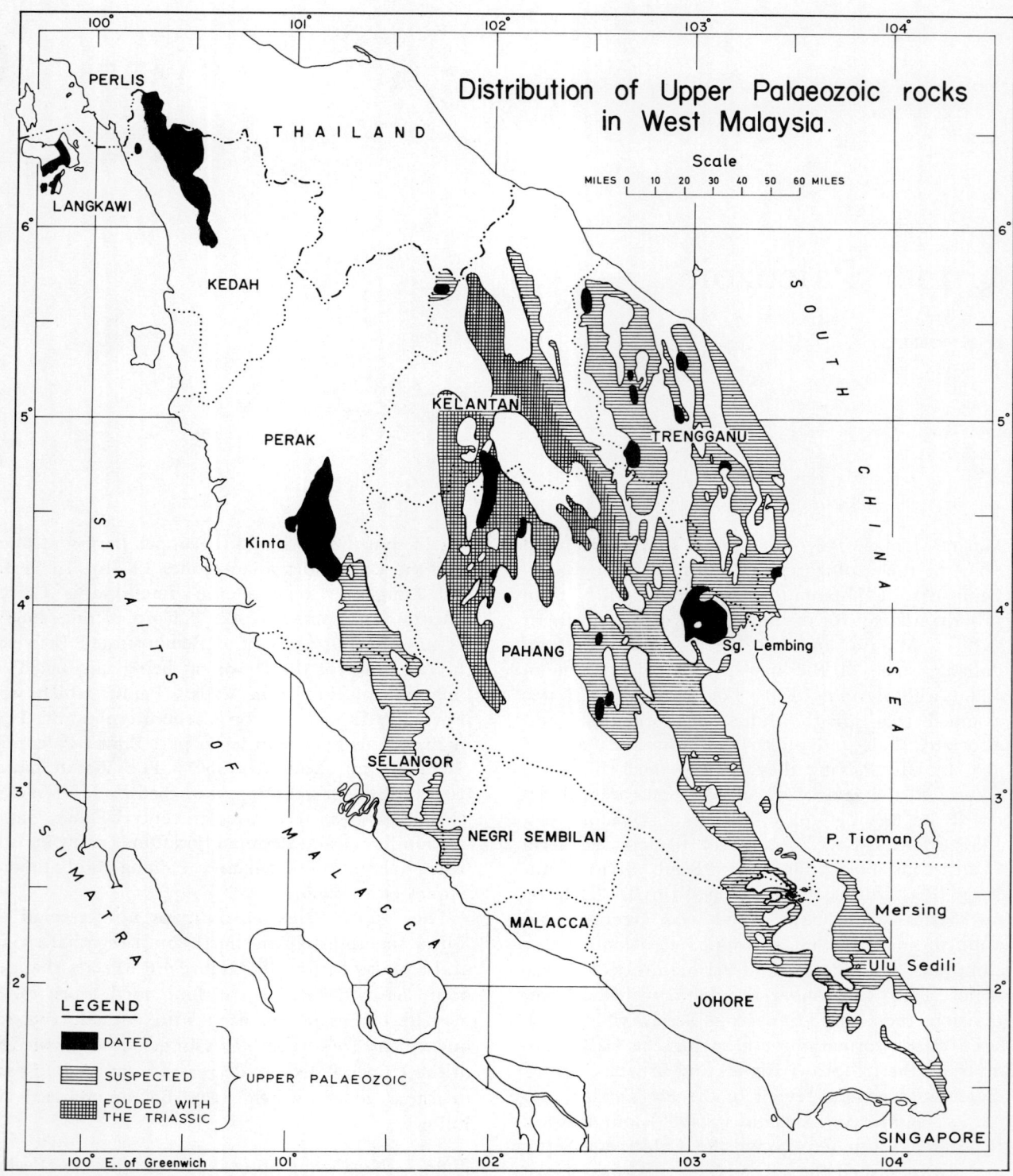

FIGURE 4.1 *Distribution of Upper Paleozoic rocks in West Malaysia. Based partly on data of the Geological Survey of Malaysia and on fossil occurrences (Jones et al. 1966).*

dle Permian time. During the Mesozoic, the sequence was flexure folded about north-south axes, but was not metamorphosed. In Langkawi it was overthrust by the Lower Paleozoic, probably in the late Permian.

Zone 2: Kinta Valley, Perak. In the Kinta Valley, shallow-water organic limestone, with thin shale intercalations, forms an apparently continuous sequence from the Lower Devonian to the Middle Permian; higher beds probably having been removed by erosion (Suntharalingam 1968). The limestone is extensively recrystallized and more strongly folded and faulted than the Upper Paleozoic of zone 1.

Zone 3: Central Malaya. Viséan to Upper Permian limestone and shale, associated with volcanic rocks, outcrop in north and central Pahang and are folded together with the overlying Triassic rocks. Most of the metamorphic rocks of central and south Kelantan are probably Upper Paleozoic.

Zone 4: East Malaya. The older sedimentary rocks of Trengganu, east Pahang and east Johore are thought to be largely of Upper Paleozoic age, although they have not been studied very closely and are regionally metamorphosed over much of the outcrop. Fossiliferous Viséan and Lower Permian limestone and shale, sometimes with volcanic rocks, have been recorded from widely scattered localities within this zone. Some shale, associated with sandstone, has yielded plant fossils of Upper Carboniferous age.

The Upper Paleozoic of zones 1 and 2 suggest that the miogeosynclinal conditions prevalent in the Lower Paleozoic persisted, with an interruption in the northwest caused by Devonian movements. However, zones 3 and 4, although containing abundant volcanic rocks, are not characterized by sediments of typical eugeosynclinal type. Thus the major divisions of the Malayan geosyncline, described in the preceding chapter, cannot be similarly defined when considering the Upper Paleozoic.

In addition to these zones of known Upper Paleozoic outcrop, an area of clastic sediments in Selangor known as the Kenny Hill Formation is probably of Upper Paleozoic age. It is described separately at the end of the chapter.

NORTHWEST MALAYA

The Upper Paleozoic of northwest Malaya (Figure 4.2) is best known in the Langkawi Islands, where almost continuous coastal exposures are to be found. There, two formations are recognized. The Singa Formation, a clastic sequence of Carboniferous and Lower Permian age, is succeeded by the Chuping Formation, a massive limestone that is probably entirely Permian (Jones in prep.). The Chuping Formation outcrops also in Perlis and north Kedah. The Singa Formation extends into north Perlis, but to the south it passes laterally into a more sandy, thicker bedded facies, the Kubang Pasu Formation (Jones in prep.). The probable relationships of these formations appear diagrammatically in Figure 4.3.

Singa Formation

In the Langkawi Islands the Singa Formation is typically developed in the southwest on Pulau Singa Besar and neighboring islands (Figure 4.4), where the easterly dip of the beds averages 12 to 15°. It is uncomplicated by isoclinal folding or faulting and indicates a thickness of about 1500 m (4900 ft). The base of the formation is not exposed in this area; probably, however, it is included in a section exposed on the northwest shore of Pulau Langgun (Figure 3.8) where quartz sandstone and massive red mudstone overlie unconformably rocks of Lower Devonian age, which form the youngest beds of the Setul Formation (Jones in prep.). In places, the red mudstone is sandy and contains pebbles of quartzite and vein quartz. It has a fauna of molluscs, trilobites, brachiopods, and ostracods, probably of uppermost Devonian age (Jones et al. 1966, Hamada 1968, 1969a). Higher beds are faulted out, but a similar fossiliferous red mudstone is known near the base of the Kubang Pasu Formation in Perlis.

In southwest Langkawi, the red sandstone and mudstone of Pulau Rebak Besar and Rebak Kechil contain elements of this fauna (T. E. Yancey, personal communication). These rocks may be best regarded as a separate formation (Gobbett, 1972). They appear to directly overlie the Upper Cambrian Machinchang Formation. Although the contact is below sea level, it may be regarded as an unconformable one; the Setul Formation being cut out and overstepped by the Singa Formation. The evidence for Devonian folding, before the deposition of the Singa Formation, has been discussed by Koopmans (1965).

The Singa Formation comprises a thin-bedded sequence of rapidly alternating black mudstone, silty shale, siltstone, and lithic and quartz sand-

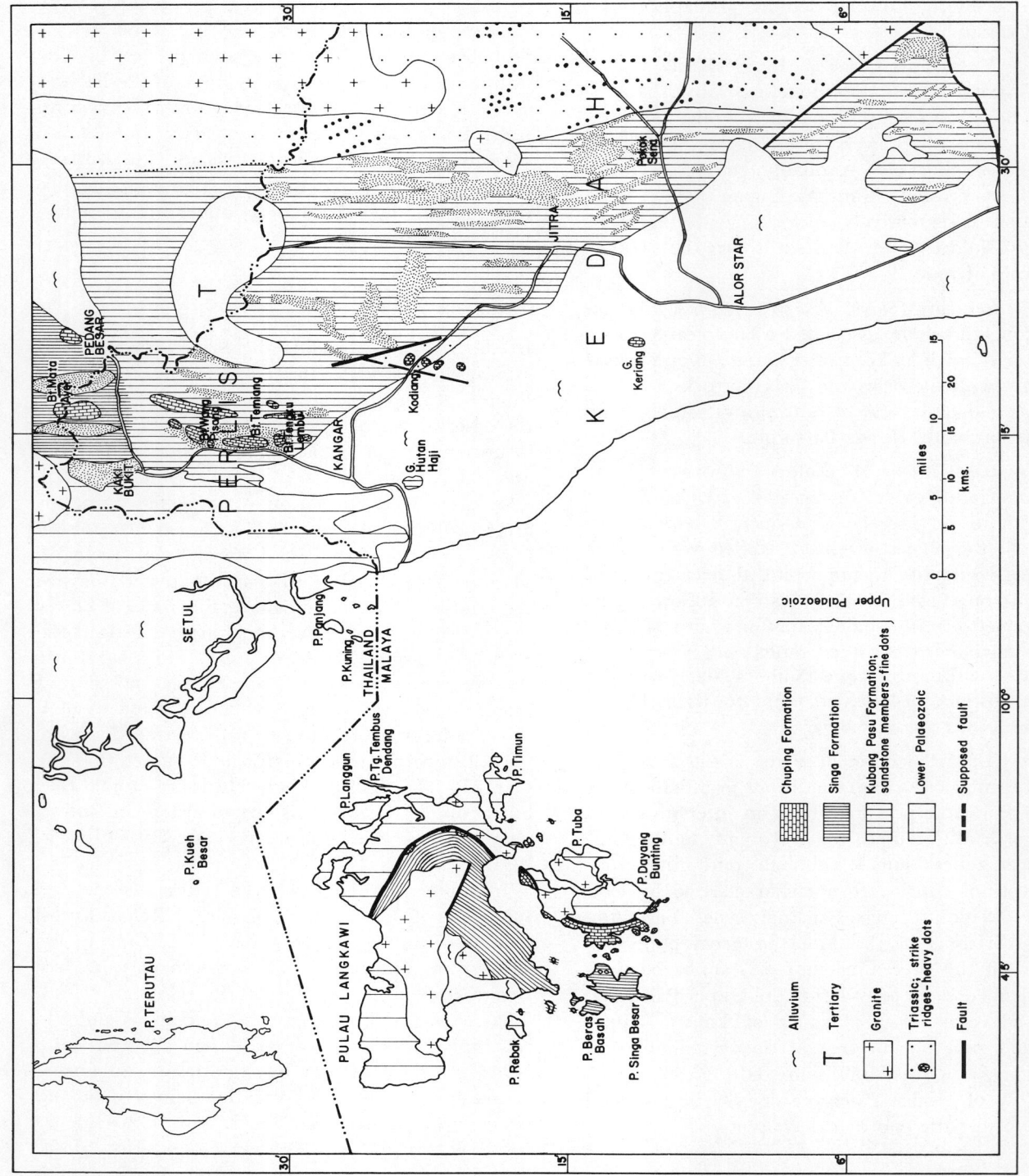

FIGURE 4.2 *Geological map of northwest Malaya emphasizing the Upper Paleozoic formations. Based partly on publications and data of the Geological Survey of Malaysia and on interpretation of aerial photographs.*

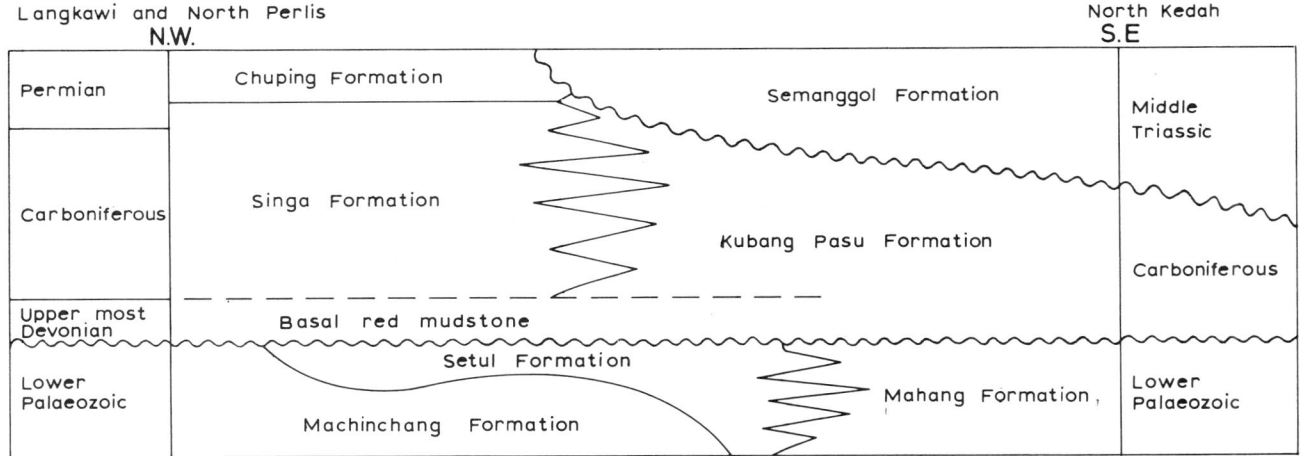

FIGURE 4.3 *Diagram of the inferred relationships of Upper Paleozoic formations in northwest Malaya; no scale.*

stone (Figure 4.5a). The black mudstone is calcareous and its weathered surface has a characteristic fretted appearance. The coarser quartz sandstone is also frequently calcareous. Flaggy siltstone characterizes the lower beds and forms the cliffed western shores of Pulau Tepor, Pulau Beras Basah, and Pulau Singa Besar. The middle part of the formation contains some horizons of friable yellow sandstone which outcrop on Pulau Singa Besar and Pulau Ular. Much of the upper part is composed of mudstone with numerous small, cross-laminated silt lenses, deformed by subsequent compaction and sometimes probably by slip-bedding before compaction. The sandstone and siltstone are commonly cross-laminated on a small scale and have been locally deformed into recumbent folds by submarine sliding and slumping (Figure 4.5b). The top of the formation becomes increasingly calcareous, and on Pulau Singa Kechil it is seen to pass upward without any apparent stratigraphic break into the Chuping Formation.

A particular feature of the Singa Formation is the presence of scattered pebbles, forming local pebble spreads throughout the succession; however, these spreads occur most commonly in the black mudstone. The pebbles vary from angular to rounded and from a few millimeters to about 20 cm in diameter. They are of local origin and can be closely compared with lithologies of the Lower Paleozoic rocks of Langkawi, including quartzite, arkose, limestone, and dolomite, as well as to lithologies characteristic of the Singa Formation itself, calcareous sandstone, siltstone, and silty mudstone. Pebbles of granite and porphyry, which are less common, attain a diameter of 20 cm; this size suggests that they are of local provenance. However, granite older than the Singa Formation is not proved *in situ* in the Malay Peninsula, apart from small sills intruded into the Lower Paleozoic of Pulau Tuba and deformed during the Devonian (Koopmans 1965).

In places, trace fossils are abundant. Raised markings on bedding planes include forms resembling *Planolites, Helminthopsis,* and "*Eophyton.*" Structures disposed vertically to be bedding are straight sand-filled tubes, about 5 mm in diameter and up to 0.5 m long; there are also concentric cones, 50 to 80 mm in diameter and 25 to 30 mm high, positioned with the apex downward and resembling *Conostichus*. The absence of body fossils throughout much of the Singa Formation makes it difficult to date.

Apart from the fossiliferous red mudstone on Pulau Langgun, which dates the base of the formation as uppermost Devonian, only a few poorly preserved solitary rugose corals, large crinoid stems, and internal moulds of a brachiopod (?*Composita* sp.) have been found near the top of the formation. These fossils do not serve to diagnose its age. However, the passage of the Singa Formation into the Chuping Formation, which contains Middle Permian fossils near its base, allows the top of the Singa Formation to be correlated with the Lower Permian. On Pulau Kueh Besar, a Thai island north of Langkawi, siltstone similar to that of the Singa Formation contains *Marginirugus, Worthenia,* and *Dielasma* of Lower Carboniferous age (Jones in prep.).

In Perlis, the Singa Formation has been mapped by Jones (in prep.) north of the Kaki Bukit to

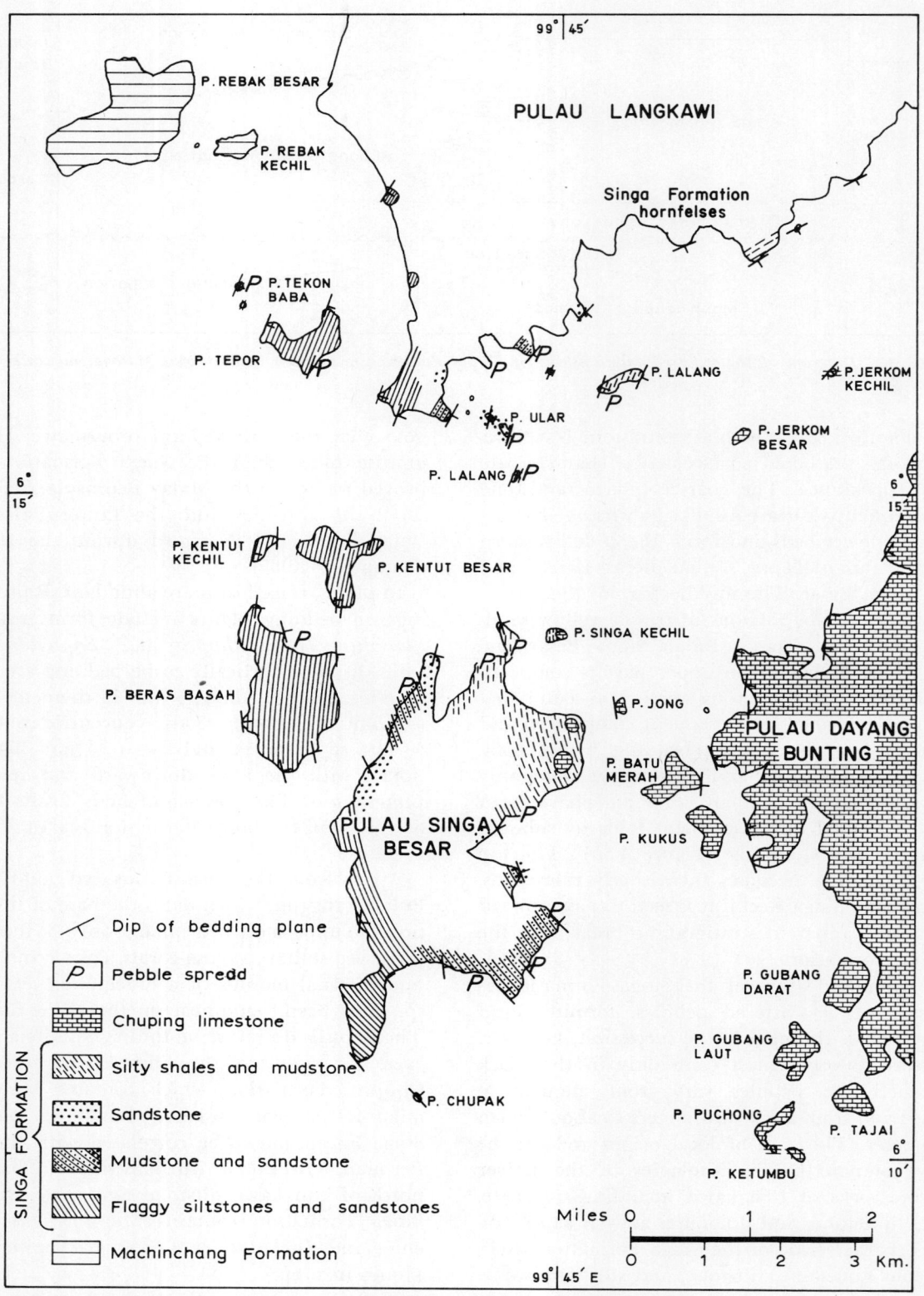

FIGURE 4.4 *Geological sketch map of southwest Langkawi, showing details of the Singa Formation. Pulau Rebak Besar and Rebak Kechil we incorrectly shown as formed of Machinchang Formation.*

FIGURE 4.5 *Singa Formation, Langkawi Islands:* (a) *Finely interbedded siltstone and mudstone showing load structures, southwest Pulau Langkawi. Photograph D. J. Gobbett.* (b) *Slump folds in mudstone, north point of Pulau Singa Besar. Photograph P. H. Stauffer.*

Padang Besar road, where it differs from the type development in the presence of massive quartz sandstone members. A section through the Singa Formation of north Perlis is shown in Figure 4.6. The Singa Formation overlies limestone of the Setul Formation with a slight angular unconformity in the Sungei Benut Valley, north of Kaki Bukit. The lower part of the sequence consists of dark colored mudstone, siltstone, and lithic sandstone. The Sungei Chuchoh valley is underlain by dark, grey muddy siltstone, folded isoclinally and containing trace fossils. Thick-bedded quartzite and arkose form the prominent north-south ridge of Gua Keh and reappear on the east limb of the Mata Ayer syncline to form high ground along the Thai border. The core of the syncline is occupied by interbedded lithic sandstone, siltstone, and shale, becoming increasingly calcareous toward the top, where they pass up into the base of the Chuping Formation. The upper part of the succession is fossiliferous and contains *Spirifer* and other unidentified forms (Jones in prep.). In central and south Perlis, the rocks underlying the Chuping Formation become more arenaceous and pass laterally into the Kubang Pasu Formation.

The flyschlike nature of the Singa Formation suggests that the original sediments were deposited in a rapidly varying environment not far from a shore line that had considerable relief. Devonian folding was probably accompanied by local uplift, and the rapid erosion of islands so formed. The

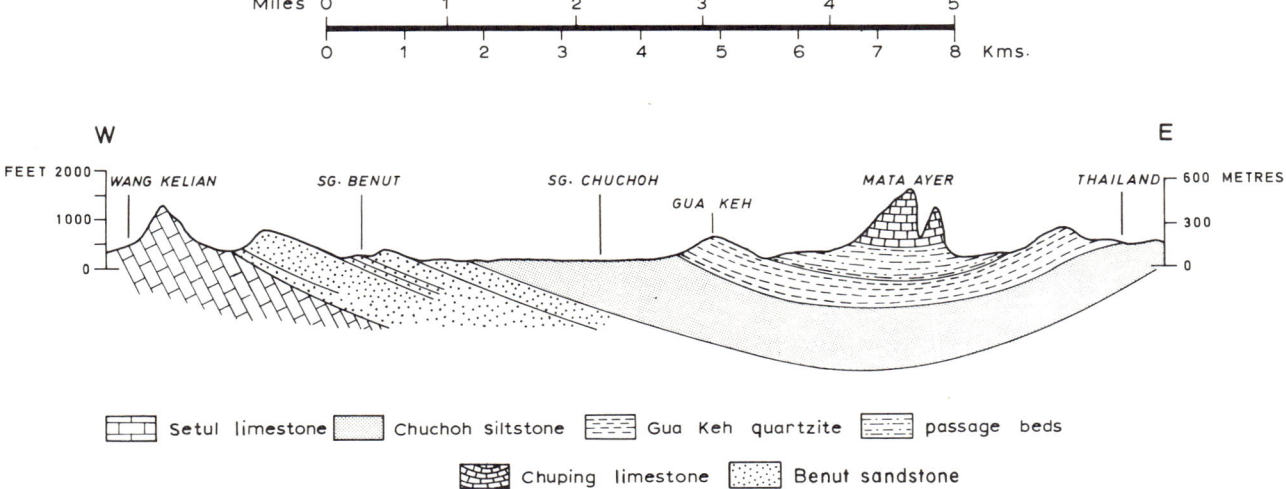

FIGURE 4.6 *East-west section through north Perlis showing the Singa Formation succession.*

poorly sorted nature of the sediments indicates rapid deposition, and the thin-bedded sequence implies rapidly varying conditions that could have been caused by tectonic instability. The slump bedding and pebble spreads are further evidence for this. However, from their relations to the mudstone matrix, the pebbles appear to have been dropped into the mud from above rather than rolled down a slope. They may have been carried in floating masses of vegetation; but it is unlikely that they were ice rafted, since the fauna of the Malayan Permian period is essentially Tethyan and indicates low latitudes.

Kubang Pasu Formation

The Kubang Pasu Formation outcrops in central and east Perlis and northwest Kedah. The Kedah outcrop was named the Kampong Sena Formation by Burton. It differs from the Singa Formation in being largely composed of thick-bedded quartz and felspathic sandstone (grey, red or purple) and interbedded with subordinate varicolored mudstone. The sequence is folded with a general north-south strike. The intensity of folding probably increases in the east, where isoclinal folds with easterly dipping axial planes cause the sandstone members to outcrop as a series of parallel ridges. Northeast of Alor Star the strike swings to a south-southeast direction (Figure 4.2).

The base of the Kubang Pasu Formation is nowhere exposed, but it is assumed to rest unconformably on the Setul Formation. At Gunong Hutan Haji in south Perlis, massive grey and red mudstone contains a fauna similar to that of the basal mudstone of the Singa Formation exposed on Pulau Langgun (Hamada 1969a) and contains the trilobite *Cyrtosymbole* (Kobayashi and Hamada 1966) (Plate 7). Similar strata form the Thai islands of Panjang and Kuning. Mudstone in the main outcrop of the formation in Kedah is locally fossiliferous. *Posidonia* sp. of Carboniferous type have been recorded from several localities, and a Lower Carboniferous fauna, including poorly preserved goniatites and chonetid brachiopods, is known from Pokok Sena (Jones et al. 1966). Chert lenses occur in the Kubang Pasu Formation in central Perlis, and from one of these the Moscovian fusulinid *Fusulina konnoi* was recorded by Scrivenor (1926).

The upper part of the formation is seen in central Perlis where pale-colored quartz sandstone, often calcareous, passes up into the base of the Chuping Formation. The passage beds have a Lower Permian fauna. They are well exposed at Bukit Temiang (Figure 4.7), where a massive bed of sandy limestone crowded with *Cancrinella* cf. *cancrini* (Vern) underlies calcareous siltstone and shale with a rich fauna of algae, fenestellid polyzoa, brachiopods, and molluscs (Jones et al. 1966). From a similar horizon in north Perlis, Newton (1926) described a Lower Permian fauna including *Schwagerina* cf. *granum-avenae* and *Pseudodoliolina*.

Chuping Formation

The Chuping Formation consists of massive, generally pale-colored, finely crystalline, pure calcitic limestone. Chemical analyses are given in Table 4.1. In Perlis the formation outcrops as groups of precipitous towerlike hills (*kegelkarst*) (Plate 1), lying in the cores of synclines and forming typical old-age karst topography. In Langkawi the outcrop is more continuous, and in the western part of Pulau Dayang Bunting it forms irregular jagged ridges surrounding deep, cliffed dolines. The top of the formation, which has been eroded away in Perlis, is overthrust by Lower Paleozoic limstone in Langkawi. The base is transitional to the Singa and Kubang Pasu Formations. Jones (in prep.) included the passage beds within the Chuping Formation, but here they are considered as part of these underlying formations. The bulk of the Chuping Formation is an unbedded and unfossiliferous limestone (Plate 8). However, the basal part is normally composed of well-bedded dark grey limestone; commonly there are chert nodules in layers parallel to the bedding, and locally the shelly fauna is abundant.

In Perlis the base is exposed below the vertical cliffs of Bukit Tengku Lembu, Bukit Wang Pisang, and Bukit Temiang (Figure 4.7). The lowest part of the sequence is of fine-grained black shelly limestone with nodules of black chert. Fossils, which are difficult to extract from this rock, are known to include *Sinopora dendroides* (Yoh), *Bellerophon*, *Euomphalus*, and small biconvex brachiopods resembling *Composita*. The overlying thick-bedded grey limestone is poorly fossiliferous but contains *Marginifera* and *Hamletella* on Bukit Tengku Lembu. The main part of the cliff is formed of massive pale grey limestone that seems to be unbedded, although the surface is frequently obscured by a dripstone coating. In the Langkawi Islands the base of the Chuping Formation is seen on

PLATE 7 *Devonian and Carboniferous fossils. 1–2.* Stringocephalus perakensis *Gobbett, Middle Devonian Kinta Limestone, Kampar, Perak, ×1. 3.* Cyrtosymbole perlisensis *Kobayashi and Hamada, basal Carboniferous Kubang Pasu Formation, Hutan Haji, Perlis ×2. 4.* Murchisonia *sp., Middle Devonian Kinta Limestone, Kampar, Perak, ×1. 5.* Straparollus *sp., Carboniferous Kinta Limestone, Kampar, Perak, ×1. 6.* Amygdalophyllum *sp., Lower Carboniferous, Sungei Lembing, Pahang, ×3. 7.* Spirifer scrivenori *Muir-Wood, Lower Carboniferous, Sungei Lembing, Pahang, ×1. 8.* Linoproductus kokdscharensis *(Gröber), Lower Carboniferous, Sungei Lembing, Pahang, ×1. 9.* Punctospirifer pahangensis *Muir-Wood, Lower Carboniferous, Sungei Lembing, Pahang, ×1. 10–11.* Pugnax asiaticus *Muir-Wood, Lower Carboniferous, Sungei Lembing, Pahang, ×1. 12.* Phricodothyris *sp., Lower Carboniferous, Sungei Lembing, Pahang, ×1.5. 13.* Posidonia *sp., basal Carboniferous Kubng Pasu Formation, Hutan Haji, Perlis, ×1.*

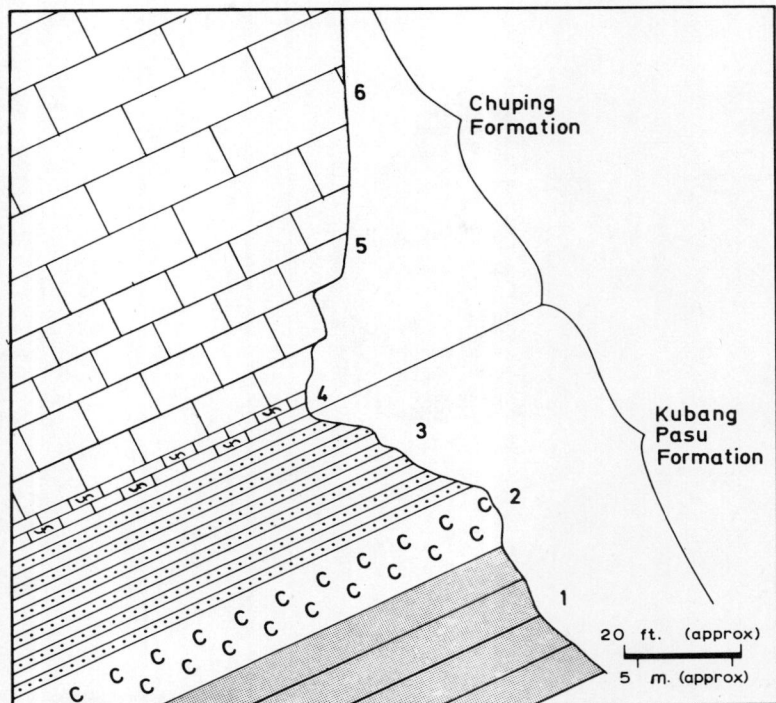

FIGURE 4.7 *Diagrammatic section through the junction of the Kubang Pasu and Chuping Formations on Bukit Temiang, Perlis, viewed from the north.*

6. *Pale grey massive fine grained recrystallized limestone.*
5. *Dark grey recrystallized limestone.*
4. *Black shelly recrystallized limestone with chert.* Sinopora, Bellerophon.
3. *Grey-green calcareous flaggy siltstone and shale.* Cypricardella, Edmondia, Bellerophon, Derbyia, Dielasma, Fenestella *and* Algae.
2. *Impure Cancrinella limestone.*
1. *Brown quartz sandstone.*

Pulau Singa Besar and Pulau Singa Kechil (Figure 4.4). Silty shale at the top of the Singa Formation become calcareous and passes up through 1 or 2 m of impure flaggy limestone into a cliff-forming, well-bedded, grey limestone with tabular masses of white chert up to 15 cm thick, which lie parallel to the bedding. This rock is typically developed on Pulau Jong (Plate 9) where, however, the base is not seen. In places it contains abundant fossils, including the corals *Sinophyllum, Wentzelella, Michelina,* and *Sinopora dendroides* (Yoh), the brachiopods *Derbyia* and *Reticulatia,* and polyzoa (Sakagami 1963).

In Langkawi the Chuping Formation has a minimum estimated thickness of 600 m (Jones in prep.). The upper part of the formation, although unfossiliferous, may be of Upper Permian age. In north Kedah, limestone hills near Kodiang appear similar to those of Perlis. However, the Kodiang limestone contains Upper Triassic conodonts (Ishii and Nogami 1966). We must now ask whether it is continuous with the Chuping Formation, which becomes younger to the south, or whether it is a distinct formation. At Kodiang the pale grey recrystallized limestone is frequently crinoidal. It is thin-bedded, contains chert nodules, and passes down into thin interbedded chert and siltstone. Thus it resembles the basal part of the Chuping Formation in being transitional below into noncalcareous beds, although the detailed lithology is different. The outcrop of this limestone is bounded to the east by a fault (Jones in prep.), and at present it is best considered as a faulted outlier of a distinct Triassic formation.

Summary

Evidence for Devonian folding in northwest Malaya was presented by Koopmans (1965). This deformation was most intense in the southeast of the

TABLE 4.1 *Chemical analysis of Upper Paleozoic limestones (weight percent)*

Specimen	Chuping Formation			Kinta Limestone				Kuantan Limestone
	1	2	3	4	5	6	7	8
SiO_2	—	0.29	0.05	4.78	0.43	0.33	0.23	1.32
Al_2O_3	0.04 ⎫	0.23	— ⎫	1.02	0.62	— ⎫	0.38	— ⎫
FeO, Fe_2O_3	0.19 ⎭		0.04 ⎭			0.21 ⎭		0.59 ⎭
MgO	0.22	0.35	0.30	1.75	7.37	0.65	19.58	0.99
CaO	55.73	55.56	55.58	50.88	47.09	55.36	32.54	54.68
CO_2	43.12	43.32	—	41.12	44.72	—	47.16	42.16
Total	99.30	99.75	55.97	99.55	100.23	56.55	99.89	99.74

Specimens 1–8 as follows:
1. Limestone, Bukit Chuping, Perlis, (081069, p. 81).
2. Limestone, Gunong Keriang, Kedah, (081033, p. 75).
3. Limestone, Gunong Keriang, Kedah, (081084, p. 85).
4. Magnesian limestone, Kanthan Quarry, Perak, (081044, p. 77).
5. Dolomitic limestone, Temoh, Perak, (081058, p. 79).
6. Limestone, Chemor, Perak, (081060, p. 79).
7. Dolomite, Chemor, Perak, (082005, p. 87).
8. Limestone, Bukit Charas, Kuantan, Pahang, (081055, p. 79).

Numbers and page references in parentheses refer to Alexander et al. (1964).

Langkawi Islands, where it was preceded by the intrusion of granitic sills into the Lower Paleozoic sediments. It can be dated as Post-Gedinnian on fossil evidence (Jones 1968). Uplift and erosion followed before late Devonian time, when marine sedimentation was resumed.

However, uplift and erosion continued in adjacent areas during the Carboniferous and early Permian. In the southeast, immature sandstone and mudstone were formed (Kubang Pasu Formation), and in the northwest, flysch-type sediments, probably derived from mountainous islands, were deposited in relatively deep water (Singa Formation). There was a complete absence of volcanicity during this and the subsequent period.

In the Lower Permian the supply of clastic sediment was reduced either because of cessation of uplift or because the surrounding areas were actually submerged. Deposition of pure limestone followed (Chuping Formation). This was probably originally mainly a biogenic limestone, but it was largely recrystallized later in its history.

Intrusions of adamellite in Langkawi were probably emplaced at the end of the Permian or in the Lower Triassic. A radiometric date of 242 ± 10 m.y. was obtained from the granite of Gunong Raya (Snelling 1965). Similar rock invaded the thrust plane along which the Lower Paleozoic had overridden the Permian. Therefore the thrust movement is likely to be of Upper Permian age, and it probably terminated the deposition of the Chuping Formation at least locally. In Kedah the Lower Triassic is absent, and upper Middle Triassic rocks rest unconformably on the Paleozoic (Figure 4.3). Thus it is probable that widespread earth movements occurred in late Permian and early Triassic times in northwest Malaya.

KINTA VALLEY

Although the geology of the Kinta Valley has been studied and discussed for about 80 years, little detailed work has been done on the Prequaternary sedimentary rocks. The valley is floored by limestone, which also forms prominent cliffed hills similar to those of Perlis. The limestone forming the hills is invariably marmorized and it is often disharmonically flow-folded; consequently it lacks fossils. The limestone beneath the Quaternary alluvium is intermittently and temporarily exposed in the workings of hydraulic open-cast tin mines. In places it is less recrystallized and contains Upper Paleozoic fossils, although until recent years few fossils had been discovered.

PLATE 8 *Bukit Ketri, Perlis. Photograph D. J. Gobbett.*

The most recent compilation on Kinta geology is by Ingham and Bradford (1960), who summarized the information available up to about 1953. These authors included all the Prequaternary sedimentary rocks in the Calcareous Series. This consists mainly of pale-colored, pure crystalline limestone (the calcareous facies; for chemical analyses, see Table 4.1) interbedded with a variety of mainly pelitic schists (the argillaceous facies). The schist is in places tourmalinized and includes, west of Batu Gajah, horizons of tourmaline-corundum rock originally described from loose boulders by Scrivenor (1910). The age of the Calcareous Series was thought to be Carboniferous. This conclusion was based on slender fossil evidence and on general lithological resemblance to crystalline hill-forming limestone found in Pahang, which was known to be Viséan. The discovery in 1956 of Lower Paleozoic fossils in the Calcareous Series northwest of Chemor was footnoted but not discussed (Ingham and Bradford 1960).

In recent years many new fossil localities have been discovered in the limestone, mainly by the Geological Survey. These have yielded molluscan, brachiopod, and coral faunas of Ordovician, Silurian, Devonian, Carboniferous, and Permian age. A detailed study of the stratigraphy in the southern part of the valley, west of Kampar, has shown an apparently continuous Devonian to Permian succession of limestone (Suntharalingam 1968). Present knowledge of the Prequaternary geology of the Kinta Valley is summarized in Figure 4.8. The information is still extremely sketchy, since only very localized areas of the limestone floor of the valley are temporarily exposed by tin mining. The limestone hills appear to be formed of more highly

PLATE 9 *Pulau Jong, Langkawi Islands, formed of horizontally bedded limestone of the Chuping Formation. Photograph C. R. Jones.*

metamorphosed limestone; this material is often flow folded, contains no fossils, and is partly obscured by dripstone. The schist is poorly exposed on low hills and occasionally in tin mines. It is normally weathered to a high degree, and its original stratification and structure are often obliterated by collapse and brecciation consequent on the solution of the limestone beneath.

Before the stratigraphy and geological history of the Kinta Valley can be elucidated, it is necessary to consider its structure. The strike of the Paleozoic rocks is in general north to north-northwest, although local variations abound. In many areas the dips are steep and the form of the limestone hills, as seen in vertical aerial photographs, frequently suggests near-vertical bedding. The distribution of diagnostic fossils suggests that the youngest beds lie to the southwest and the oldest to the northeast. This is a gross simplification, however for the sequence is certainly strongly folded and faulted.

The granite masses bordering the valley are concordant to the sedimentary rocks, but the nature of the contact is uncertain. The straight scarp of the Kledang Range, about 26 km (16 miles) long, strongly suggests that the west side of the valley is bounded by a major fault. Several less continuous faults can be postulated for the east side. If these faults exist, the Kinta Valley can be regarded as a graben, as proposed by Scrivenor (1913a). Ingham and Bradford (1960) argued that the concordant nature of the granite and the complexity in detail of its margin are evidence for a normal intrusive contact. If this is so, the Kinta Valley can be regarded as a synclinorium bordered by granite-cored anticlinoria and plunging south-southwestward.

A major structural feature, not only of Kinta but of Malaya as a whole, has been overlooked by previous authors. This is a system of approximately nothwest-southeast lineations along which strike-slip movement has occurred, usually in a sinistral sense. Aerial photographs clearly indicate numerous northwest-southeast lineations, expressed as negative features, in the granite highlands on either side of the valley. In places the linear features are occupied by quartz dykes. The outcrop pattern suggests that many are strike-slip faults and must affect the sedimentary rocks flooring the valley. They would also appear to disrupt a normal fault along the Kledang scarp (Gobbett 1971). Evidence for some vertical movement along these lineations is provided by the peculiar reversal of drainage at the northern end of the valley. The watershed crosses the valley about 6 km (4 miles) north of Chemor (Figure 4.8).

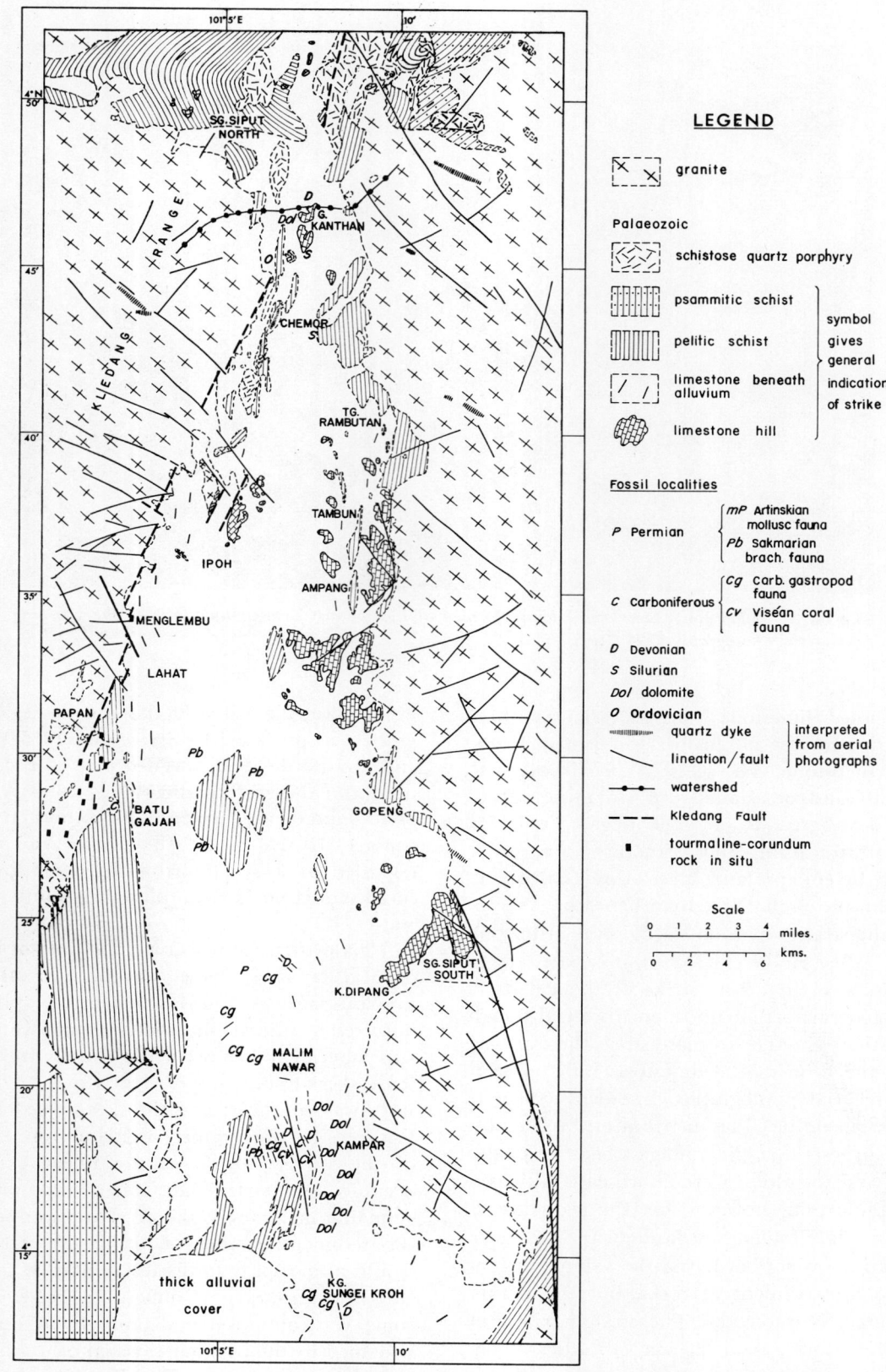

FIGURE 4.8 *Geological sketch map of the Kinta Valley. Based on Ingham and Bradford (1960), data of the Geological Survey of Malaysia, Suntharalingam (1968), and aerial photograph interpretation.*

Thus the structure of the Paleozoic rocks is probably due to folding along north-south axes, normal faulting along the strike, and oblique strike-slip faulting, which together would produce an intricate pattern of small structural units. The sequence of limestone and interbedded schist has a similar lithology throughout the valley, and no extensive marker horizon has been recognized. However, Suntharalingam (1968) defined local lithological divisions and, fortunately, was able to date these by fossils. It seems likely that much of the limestone and schist is of Devonian, Carboniferous, and Permian age, but folded and faulted inliers of Lower Paleozoic limestone and schist occur, particularly in the north. The following discussion of the Upper Paleozoic is limited to areas in which the age of the rock is established.

Devonian

Devonian fossils were first recognized in Malaya from Gunong Kanthan (Alexander and Müller 1963). These were conodonts, further studied by Müller (in correspondence). The limestone forming Gunong Kanthan is a thin-bedded, fine-grained, black, carbonaceous, calcitic rock, with some argillaceous beds. It is traversed by white calcite veins and contains no macrofossils. Its thickness was estimated as at least 330 m (1080 ft) by Savage (1937). The conodonts include Lower, Middle, and Upper Devonian species, and there seems to be a continuous Devonian sequence. Immediately west of the hill a mottled dolomite (leopardstone) is exposed in tin mines. It is interbedded with cream-colored dolomite which has yielded Silurian tabulate corals (Jones et al. 1968, Thomas and Scrutton 1969).

The other known Devonian localities lie at the southern end of the Kinta Valley. West of Kampar, the Thye On beds (Table 4.2) are of massive pure, calcitic limestone, grey in color but weathering to form a chalky white surface patina. Richly fossiliferous, they have numerous colonies of tabulate corals, stromatoporoids, abundant murchisoniid gastropods, onocerid nautilioids, and the brachiopod *Stringocephalus perakensis* (Gobbett 1966) (Plate 7). The Thye On beds are thus Givetian in age. They overlie massive cream-colored dolomite (Kim Loong no. 1 beds) with poorly preserved tabulate corals of *Thamnopora* type, amplexiform corals, murchisoniid gastropods, and amboceliid brachiopods. This dolomite, which is conformable with the Thye On beds, is probably of Lower Devonian age and perhaps older. According to Suntharalingam (1968), its thickness exceeds 1300 m

TABLE 4.2 *Upper Paleozoic sequence west of Kampar, Kinta Valley, Perak. After Suntharalingam (1968).*

H. S. Lee beds	(Top not seen) Biohermal limestone	20 m	LOWER PERMIAN
	Bioclastic and fusuline limestone	20 m	
Nam Loong beds	Impure carbonaceous brachiopod-polyzoan limestone	60 m	
	Crinoidal limestone	100 m	
Kim Loong No. 3 beds	Pyritiferous black shale and argillaceous sandstone	100 m	?UPPER CARBONIFEROUS
Kuan On beds	Thin-bedded grey recrystallized calcitic limestone with dolomitic beds, interbedded with thin calcareous and carbonaceous shales	500 m ?	CARBONIFEROUS
			LOWER CARBONIFEROUS
Thye On beds	Massive grey recrystallized calcitic limestone	150 m	? UPPER DEVONIAN
			MIDDLE DEVONIAN
Kim Loong No. 1 beds	Mainly pure cream-colored dolomite (Base not seen)	600 m ?	LOWER DEVONIAN ?SILURIAN

(4265 ft), but the succession is probably repeated by strike faults.

Near Kampong Sungei Kroh, 10 km (6 miles) south-southeast of the Thye On mine, limestone probably of Devonian age is again exposed. It is a well-bedded grey, chalky weathering, calcitic limestone with abundant stromatoporoids, tabulate corals, and amplexiform corals. Molluscan fossils are poorly preserved; *Stringocephalus* seems to be absent.

Doubtless other outcrops of Devonian limestone exist in the Kinta Valley but, in the absence of fossils, these cannot be recognized as such.

Carboniferous

Limestone of Carboniferous age is probably extensively exposed in the southern half of the Kinta Valley. It is best known west of Kampar (Kuan On beds, Table 4.2). The Kuan On beds differ from the older limestone in being thinner bedded and showing a greater variation in lithology. Some beds are dolomitic, but since these are interbedded with calcitic limestone, dolomitization presumably occurred soon after deposition. Although the limestone is generally recrystallized, some horizons are still recognizably oolitic. Thin interbedded calcareous and carbonaceous shale is usually highly weathered and is unfossiliferous. Fossils, which are common in the limestone, are usually ill preserved. The fauna is in many places dominated by gastropods. The majority of these belong to the superfamilies Pleurotomariacea, Murchisoniacea, Neritacea, and Platyceratacea; and since many forms belong to long-ranging genera, they are not very accurate age determinants. Some beds are crowded with elongate colonies of an organism resembling the ?stromatoporoid *Amphipora*. Bivalves showing an affinity to *Schizodus* are locally common; cephalopods are uncommon and seem to be limited to poorly preserved nautiloids. The common preservation of the aragonitic gastropod shells suggests that the Carboniferous sea was warm. Rapid burial and lack of turbidity would also tend to inhibit the solution of aragonite (Jefferies 1962).

Brachiopods and corals have not been discovered associated with this molluscan fauna. However, the Viséan coral *Siphonophyllia* has been recognized from two localities. West of Kampar, *Siphonophyllia* cf. *cylindrica* occurs in thick-bedded sandy limestone, lying stratigraphically between the Devonian and Carboniferous gastropod-bearing limestone; near Batu Gajah, *Siphonophyllia* aff. *gigantea* and *Zaphrentites*? have been recorded (Jones et al. 1966). Limestone with a fauna similar to that of the Kuan On beds is known from near Kampong Sungei Kroh and northwest of Malim Nawar.

Lying above the gastropod limestone west of Kampar is a series of pyritiferous black phyllitic shale and poorly sorted argillaceous sandstone and siltstone, the Kim Loong No. 3 beds (Table 4.2). Although unfossiliferous, this material is assumed to be Upper Carboniferous, since it underlies the Lower Permian Nam Loong beds. Other outcrops of pelitic schist shown on the map (Figure 4.8) may be of Carboniferous age, but their stratigraphical relations are obscure.

Permian

Lower Permian limestone is best known from two tin mines west of Kampar, where the limestone is richly fossiliferous. The Nam Loong beds (Table 4.2) are formed of thick-bedded, light grey to black carbonaceous and argillaceous limestone. They weather to give a porous and friable rock, easily broken with the fingers. Strike faults cut the exposures and obscure the stratigraphic relations of the various lithologies. However, crinoidal limestone containing crinoid cups as well as columnals, and including small ?dielasmid brachiopods probably lies near the base of the Nam Loong beds. It is succeeded by marly limestone with scattered productid brachiopods, particularly of the genus *Waagenoconcha,* and bellerophontid and pleurotamarian gastropods. Included in this part of the sequence are one or more massive biostromal limestone masses crowded with brachiopods and fenestellid polyzoa, to the exclusion of other fossil groups. The brachiopod fauna is Lower Permian (Sakmarian) and includes the genera *Derbyia, Linoproductus, Cancrinella, Waagenoconcha, Spirifer, Stenocisma,* and *Cleiothyridina*. The top of the Nam Loong beds consists of black argillaceous limestone with bellerophontid gastropods.

An abrupt change in lithology marks the base of the overlying H. S. Lee beds, but the contact may be faulted. The lower part of the H. S. Lee beds is formed of interbedded limestone and dolomite, which pass up into a massive grey bioclastic limestone. This material is composed almost entirely of the worn tests of *Pseudofusulina krafti*, but in places it also includes inorganic limestone clasts. The Pseudofusulina limestone is followed by grey massive limestone formed largely of comminuted shells and complete shells of small gastropods.

The upper part of the H. S. Lee beds is a biohermal limestone with a rich fauna dominated by gastropods (Plate 10), many of which are forms unknown elsewhere (Batten 1972, in correspond-

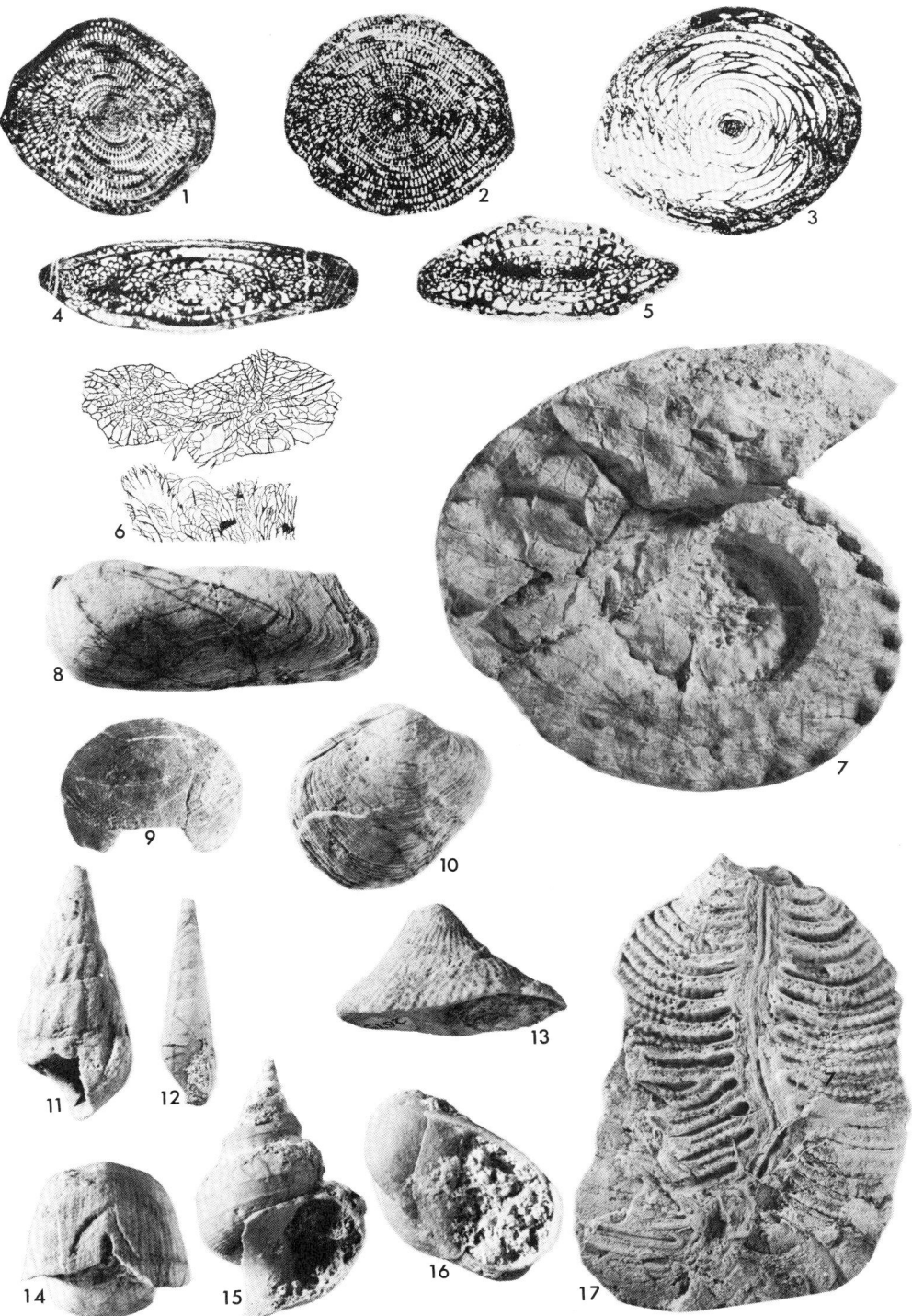

PLATE 10 *Permian fossils.* *1.* **Neoschwagerina cheni** Sheng, Middle Permian, Jengka Pass, Pahang, ×5. *2.* **Yabeina asiatica** Ishii, Middle Permian, Jengka Pass, Pahang, ×5. *3.* **Verbeekina verbeeki** (Geinitz), Middle Permian, Kampong Awah, Pahang, ×5. *4.* **Pseudofusulina gobbetti** Igo, Middle Permian, Jengka Pass, Pahang, ×5. *5.* **Schwagerina** *cf.* **gumbeli** Dunbar and Skinner, Middle Permian, Jengka Pass, Pahang, ×5. *6.* **Wentzelella malayensis** Igo, Permian, Ulu Sungei Atok, Pahang, ×2. *7–16. Middle Permian Kinta Limestone, Kampar, Perak.* *7.* **Metacoceras** *sp.*, ×1. *8.* **Parallelodon** *sp.*, ×1.5. *9.* **Lepidopleurus** *sp.*, ×1. *10.* **Schizodus** *sp.*, ×1.5. *11.* **Pseudozygopleura** *sp.*, ×2. *12.* **Meekospira** *sp.*, ×1. *13.* **Sallya** *sp.*, ×1. *14.* **Buchania** *sp.*, ×12. *15.* **Eotomaridae** gen. and sp. nov., ×1. *16.* **Planospirina** *sp.*, ×2. *17.* **Leptodus** *sp.*, Middle Permian, Jengka Pass, Pahang, ×1.

ence). It also contains dasycladacean algae (Elliot 1968), the fusuline *Misellina claudiae,* ostracodes, cidaroid spines and plates, waagenophyllid corals, chiton valves, the bivalves *Schizodus, Parallelodon,* and a large problematical bivalve, the nautiloid *Metacoceras,* and Middle Permian goniatites. Brachiopods are relatively rare. The limestone weathers to a friable chalky rock from which the fossils are easily extracted. The large problematic bivalve is found concentrated in a shell bed of up to 4 m thickness. The H. S. Lee beds can be correlated with the Pseudofusulina ambigua Zone of the Japanese Permian, equivalent to the lower part of the Leonardian of North America.

East of Batu Gajah, tin mining has exposed bioclastic limestone that is probably of Lower Permian age. It includes crinoidal and fenestellid limestone with spiriferid and productoid brachiopods with a general similarity to the Nam Loong beds.

Summary

The Upper Paleozoic rocks of the Kinta Valley differ markedly from those of nothwest Malaya. In Kinta there is no evidence of Devonian earth movements. On the contrary, deposition of limestone appears to have been almost continuous from Silurian until Middle Permian time. It can be postulated that the site of the Kinta Valley was a stable shelf during the Upper Paleozoic. From time to time, the calcareous sediments accumulating on this shelf were interrupted by argillaceous deposits invading it from the more rapidly sinking adjacent areas. The absence of coarse-grained clastic sediments indicates either that the area lay some way from land or that the neighboring land was of low relief. A hypothetical cross section of conditions in the Lower Permian is presented in Figure 4.9.

The youngest dated Paleozoic sedimentary rocks in Kinta are Leonardian. A potassium-argon radiometric date of 232 ± 10 m.y. from the granite of Kuala Dipang indicates intrusion at the end of the Permian (Snelling 1966). This was preceded by deformation of the limestone, which was intensive in many places in the Upper Permian. Triassic sediments have not been recognized in the Kinta Valley. They may have been present once but later were eroded away with most of the other sedimentary cover of the intrusive granite masses. The Paleozoic of the Kinta Valley seems to have been preserved in a down-faulted area within the granite.

CENTRAL MALAYA

Less is known of the Upper Paleozoic in the central part of Malaya than of that in the west; the present knowledge is summarized in Figure 4.10. Upper Paleozoic fossils have been collected from a number of isolated localities, usually from limestone or shale associated with pyroclastic rocks. Triassic fossils have also been found frequently, adjacent to the Upper Paleozoic ones and in a similar rock facies, the Gua Musang formation. The rocks have been

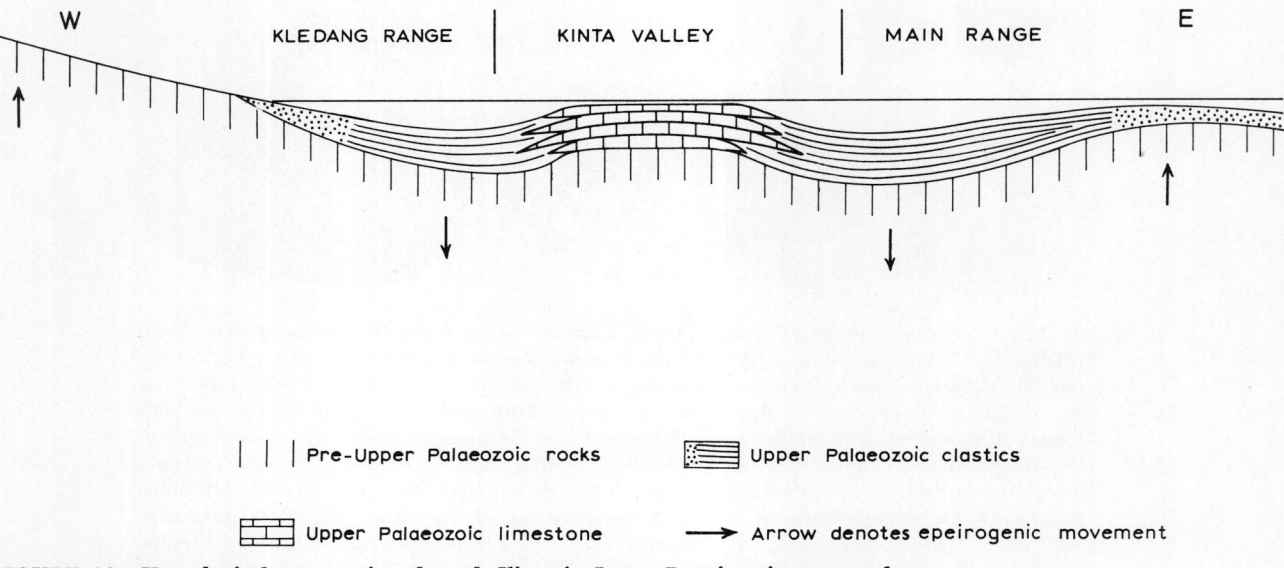

FIGURE 4.9 *Hypothetical cross section through Kinta in Lower Permian time; no scale.*

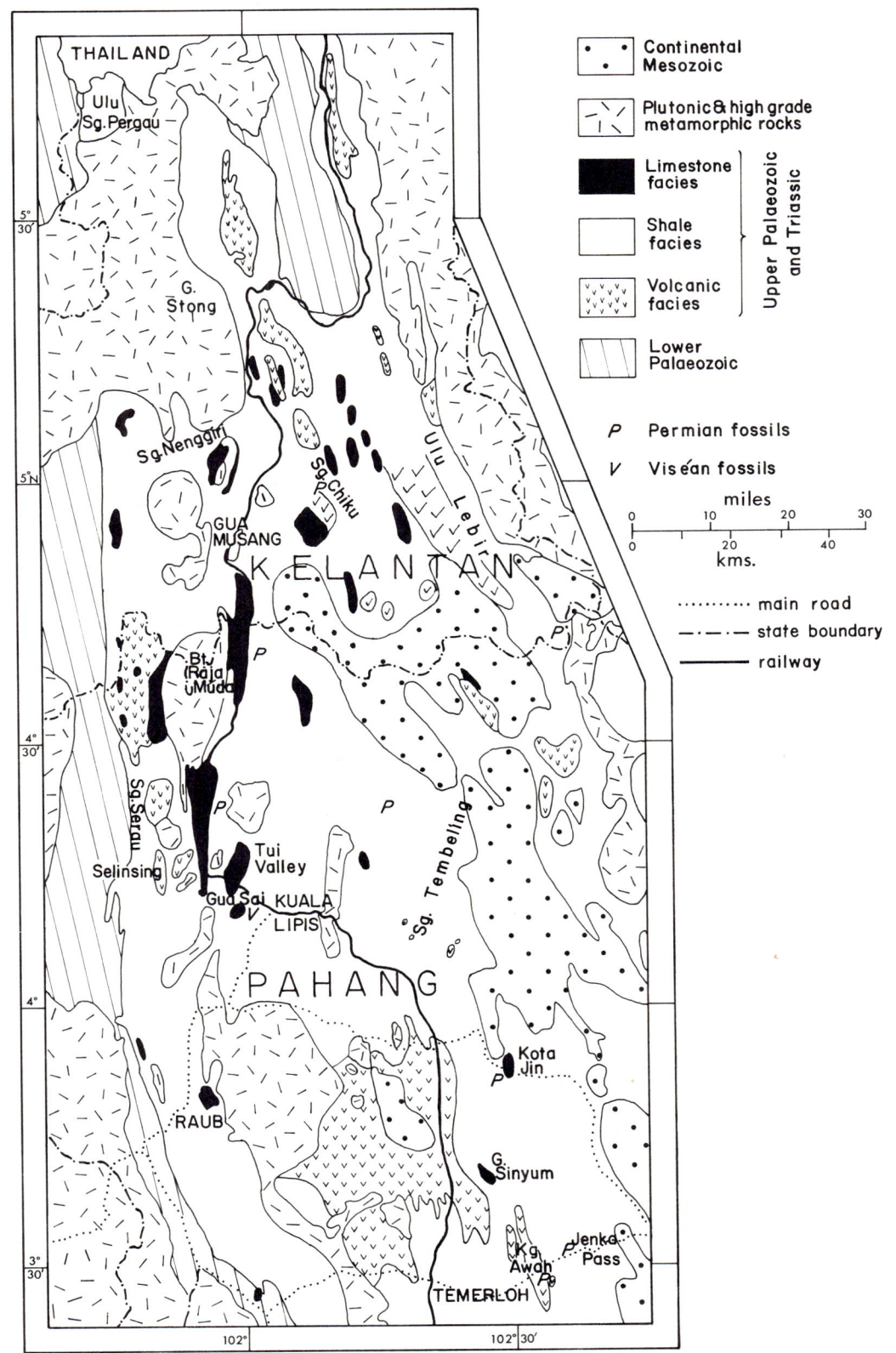

FIGURE 4.10 *Geological sketch map of central Malaya. Based on data of the Geological Survey of Malaysia and Koopmans (1968).*

intensively folded and faulted and, because of the general lack of continuous exposures and marker horizons, it has so far proved impossible to map individual strata or groups of strata or even to delineate the Upper Paleozoic from the Triassic part of the sequence. The regional dip is toward the east; the beds are folded isoclinally and overturned toward the west.

In south and central Pahang, shale with Middle and Upper Triassic fossils predominates, and Upper Paleozoic fossils are known only from small inliers east of Temerloh. To the north, limestone and pyroclastic rock become more important, and a number of late Permian and early Triassic faunas (Hada 1966, Tamura 1968) are known Viséan limestone outcrops near Kuala Lipis. The fossil distribution suggests that the folds culminate to the north.

North Pahang and South Kelantan

In North Pahang, Richardson (1947b, 1950) traced the distribution of rock facies and also constructed a vertical sequence based on the observation that the rocks dip and apparently young toward the east. No fossils had been discovered in the area when Richardson published the results of his mapping. He admitted that the sequence was not necessarily a stratigraphic one because of the probability of repetition of the beds by folding and faulting. Lateral facies variation was demonstrably rapid and would obscure such repetition. Fossil discoveries make it apparent that widely separated stratigraphic horizons are repeated across the strike, and it can be assumed that beds, ranging in age from Viséan to Upper Triassic (Carnian), have been isoclinally folded and faulted. Figure 4.11 represents diagrammatically the kind of structure envisaged, but it probably bears no relation in detail to the actual structure.

Richardson's (1947b) sequence of eight lithological subdivisions may be regarded as a structural sequence but not a stratigraphic one. This structural sequence, somewhat modified, is given in Table 4.3, which demonstrates the difficulty of separating the Upper Paleozoic from Triassic rocks of shale facies (Daonella biofacies of Jones et al. 1966) in this part of central Malaya.

Richardson (1947b) divided the sedimentary rocks of north Pahang into three main facies (calcareous, argillaceous, and volcanic) and two mixed "subfacies" (limestone-shale and shale-rhyolite tuff). It is convenient to describe the rocks under these headings, and the following account is based on Richardson's work.

CALCAREOUS FACIES. The calcareous facies consists largely of pale-colored calcitic limestone which is hard, nonporous, brittle, and splintery. Grey and black varieties contain small amounts of carbonaceous, argillaceous, and pyroclastic impurities. The limestone has been recrystallized into a fine-grained calcite mosaic, in places porcellaneous; original sedimentary structures and organic remains have in most instances been entirely obliterated by

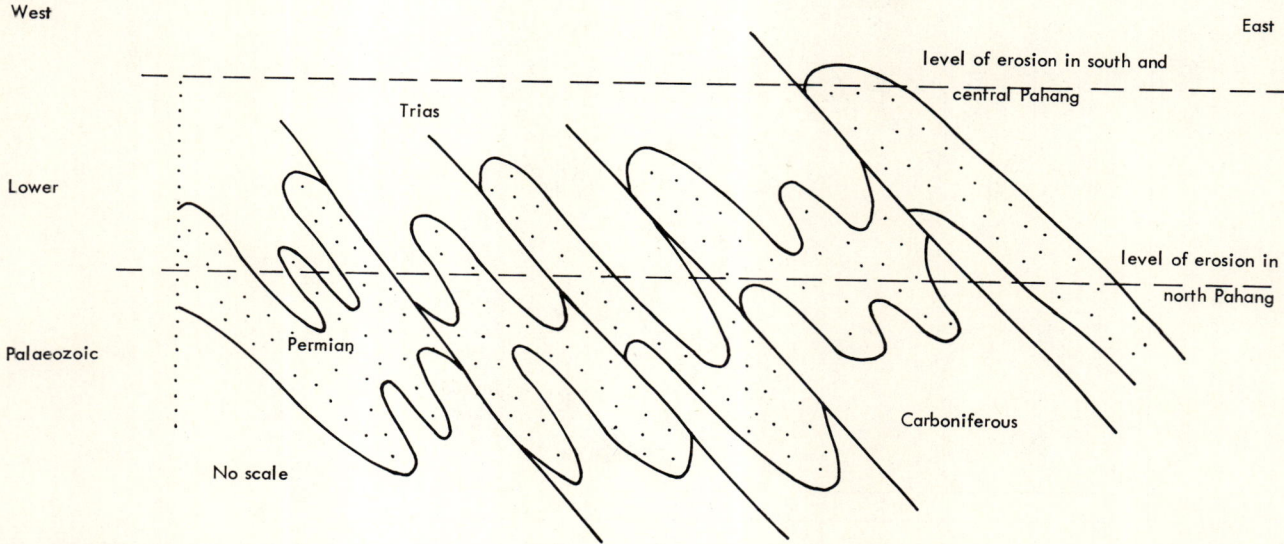

FIGURE 4.11 *Diagram showing the probable kind of structure of the Upper Paleozoic and Triassic rocks in central Malaya.*

TABLE 4.3 *Upper Paleozoic and Triassic ("Calcareous Series" and "Pahang Volcanic Series") of north Pahang, Based on Richardson (1947b) and Jones et al. (1966). Numbers refer to a structural sequence, 1 being the lowest rocks and 8 the highest rocks.*

← WEST · EAST →

	1	2	3	4	5	6	7	8
NORTH ↑	RHYOLITE TUFF minor shale (Triassic) limestone and mixed facies ----(transition to)---- SHALE (Permian) minor rhyolite tuff and limestone ----(transition to)---- Limestone, mixed facies, tuff ----(transition to)---- SHALE, minor mixed facies, limestone and rhyolite tuff ----(transition to)---- SHALE, mixed facies and limestone	Interlaminated rhyolite tuff and shale of the SERAU VALLEY	LIMESTONE (Middle Permian and Scythian) Shale (Middle Permian, Upper Permian, Scythian, Anisian), TUFF and agglomerate ? mainly shale	RHYOLITE—ANDESITE AGGLOMERATE and TUFF, shales and minor thin limestones	Massively bedded mudstones and siltstones of the TANUM VALLEY	Rhyolite–andesite agglomerate and tuff, and shale	Limestones of the TUI VALLEY (Viséan)	Shales (Middle Permian and Middle Triassic), andesite tuffs and agglomerates of KECHAU and ULU TANUM VALLEYS
↓ SOUTH								

this recrystallization. However, the bedding is often preserved as alternating grey and white laminae. Silicification and dolomitization have occurred locally but are generally unimportant. These processes have occasionally preserved original structures, particularly oolites and cross-lamination. Thin beds of shale, tuff, and sandstone, and nodules of chert, are found infrequently within the limestone.

The most important occurrence of this facies is in division 3 of the structural sequence (Table 4.3), which flanks the granite mass centered around Bukit Raja Muda and extends northward to Gua Musang. The limestone outcrops as lenses elongated parallel to the strike and gives rise to cliff-bound ridges and lines of isolated towerlike hills. *Verbeekina* sp. and other poorly preserved fusulines from Gua Musang show that some of the limestone is Permian. However, another similar limestone mass near Gua Musang contains Lower Triassic (Scythian) ammonoids (Hada 1966).

Massive limestone, with traces of an original oolitic structure, forms the hills of Gua Sae and Gua Bama, west-northwest of Kuala Lipis. It contains Viséan fossils and is here tentatively correlated with the limestone of the Tui valley, placed in division 7 of the structural sequence (Table 4.3).

SHALE FACIES. Shale, mudstone, and shaly siltstone outcrop much more extensively than limestone, and it can be assumed that these materials form the bulk of the sequence in the south of the area. The rocks are often calcareous, but the sequence rarely contains limestone beds. The shale is thin bedded, laminated, and highly fissile. It is commonly carbonaceous, and individual laminae may be relatively light or dark colored, depending on their carbon content. Occasionally shale and mudstone are found associated with dark-colored bedded chert. The sequence may also contain beds of argillaceous sandstone composed of fine to medium grained angular quartz in a matrix of limonitic or carbonaceous clay.

Fossils from this facies range in age from Middle Permian to Middle Triassic. Those of Permian age are mainly productoid and spiriferoid brachiopods but include *Leptodus* and *Uncinunellina*, lophophyllidiid corals, fenestellid polyzoa, and crinoid ossicles often mixed with bivalve molluscs of Triassic type.

VOLCANIC FACIES. The bulk of the volcanic facies consists of fine to medium grained pyroclastic rocks. Rhyolite tuffs outcrop extensively in the west of the area. They are mainly mixed grey crystal-lithic tuffs, massive to fine bedded, and well jointed. They sometimes occur interlaminated with, and color banded by, dark grey tuffaceous shale and black carbonaceous shale. Green tuffs of trachytic and andesitic composition become important in the east of the area, high in the structural sequence. Agglomerate, present at a few localities, is mainly rhyolitic and sometimes contains limestone clasts. Some of the coarser agglomerate may have been deposited subaerially, but usually the pyroclastic rocks are closely associated with marine sediments and are locally fossiliferous. The fossils are generally fragmentary and include crinoid ossicles, polyzoa, and brachiopods of Upper Paleozoic aspect.

Lavas are very subordinate. Flow-banded spherulitic rhyolite trachyte, trachyandesite, and andesite lavas have been identified. Since they are associated with shale and water-deposited tuff, these were probably extruded on the sea floor.

Acid rhyolitic pyroclastic rocks are most abundant in the north and west, and more basic types are characteristic of the southeastern part of the area. Richardson considered that the sequence shown in Table 4.3, which is herein considered a structural sequence, was in the main a stratigraphic one. He thus concluded that acid volcanic activity preceded the production of more basic material. It is clear that one center of acid volcanicity lay to the northwest and one of andesitic type to the southeast, but the age relationship between the two cannot be worked out at present because of the structural complexity discussed previously.

SUBFACIES. A mixed limestone and shale subfacies is restricted to the southwest around the Selensing gold mine and in the Serau and Satak valleys. Limestone beds up to 7.5 cm thick are separated by thin shale partings. This rock is transitional between limestone and shale and do not appear to be very important.

Another transitional rock type, described by Richardson (1947), consists of thin (2.5 to 5 cm) beds of pale grey rhyolite tuff interlaminated with darker shale, giving a prominently color-banded rock.

North Kelantan

The Upper Paleozoic and Triassics rocks of south Kelantan strike northward toward the metamorphic complex of Gunong Stong. The shale and volcanic facies both outcrop extensively, and metasediments in the Nenggiri Valley include a thick marble formation. These sediments and metasediments are probably at least partly of Upper Paleozoic age. Plant remains, probably of Lower Permian age, have been described from the Sungei Chiku in central Kelantan (Edwards 1926), and in the extreme northwest of Kelantan (Ulu Sungei Pergau), Permian fossils have been recorded from a black limestone (Jones et al. 1966).

Central Pahang

The sediments of north Pahang strike south-southeastward into central Pahang, where, however, the Upper Paleozoic is largely buried beneath the Triassic. Small inliers of Permian, mainly of calcareous or mixed facies, outcrop in the eastern part of the Temerloh district.

A road cutting at the Jengka Pass about 24 km (15 miles) east of Temerloh on the Maran road, exposes steeply dipping, isoclinally folded and faulted beds of thin-bedded bituminous bioclastic limestone, black shale, and tuffaceous sandstone (Figure 4.12). The limestone contains a rich fossil assemblage of algae and foraminifera of Middle to Upper Permian age (Verbeekina–Neoschwagerina Zone) (Plate 10), and a coral–brachiopod fauna including the genera *Wentzelella, Michelinia, Leptodus, Spiriferina, Horridonia, Linoproductus,* and *Derbyia*. The black shale contains crinoid ossicles, marginiferid and other brachiopods, polyzoa, and plant fragments; the brachiopod *Proboliolina* is abundant in one of the sandstone beds. Owing to the high degree of deformation and the absence of criteria to indicate which way the beds young, the stratigraphic succession at the Jenka Pass is not known. These Permian rocks are overlain unconformably by late Triassic or early Jurassic sandstone (Ichikawa et al. 1966) (Figure 4.12).

Algal–foraminiferal limestone with a fauna identical to that of the Jenka Pass is exposed in the Sungei Jengka and as blocks in an andesitic agglomerate at Kampong Awah, a few kilometers to the west. At Kampong Awah, limestone and andesite lava have an intimate relationship. Small fragments of limestone and individual fossils occur in the lava, and blebs of lava appear to be contained within the limestone. Thus it would seem that the lava was extruded onto the sea bed during the deposition of the limestone. After consolidation, the mixed lava and limestone were brecciated by explosive volcanic acitivity or by submarine faulting and slumping.

Summary

Viséan and Lower to Upper Permian rocks have been dated by fossils in central Malaya. This Upper Paleozoic sequence is characterized by apparently conformable relations with the overlying Triassic, by contemporaneous volcanicity, and by rapid facies variation. The environment seems to have been

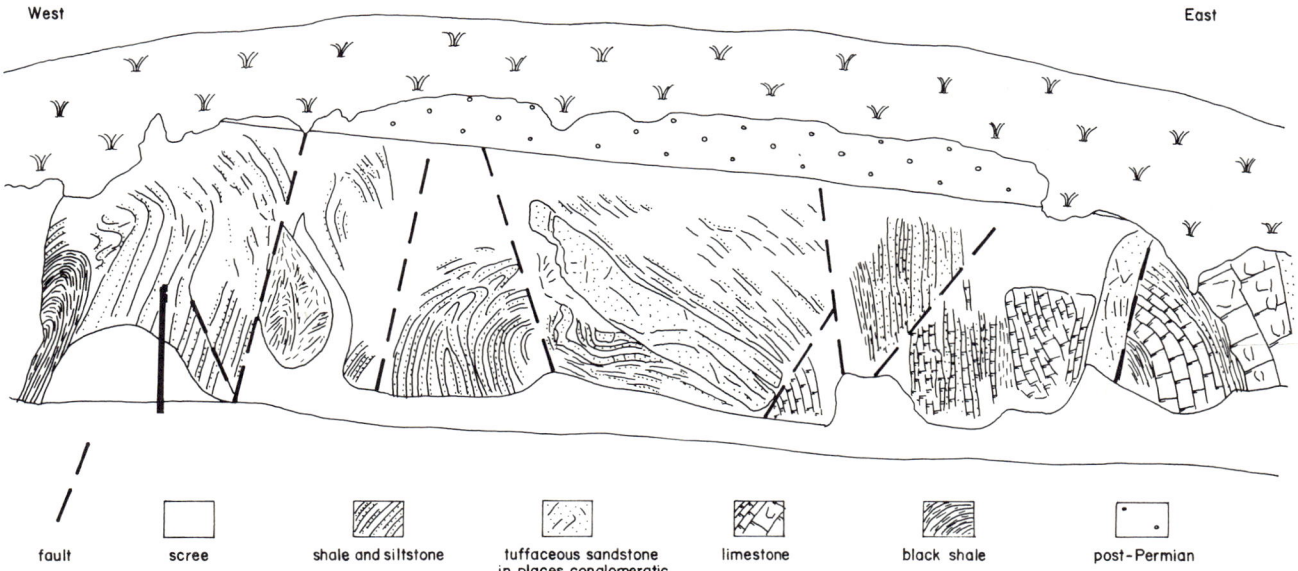

FIGURE 4.12 *Section exposed on the north side of the road at the Jengka Pass, Pahang. Drawn from photographs and field sketches. Length of section is approximately 50 m (165 ft); telegraph pole at the west and indicates the scale.*

marine throughout, and coarse-grained clastic rocks are rare. Limestone which is frequently closely associated with pyroclastic rocks, may have been restricted to shallower parts of the sea in the vicinity of volcanic islands and sea mounts. Laminated and bituminous shale is common, suggesting the frequent absence of a bottom fauna and the presence of reducing conditions on the sea bed, possibly in relatively deeper water.

In north Pahang and south Kelantan, marine sedimentation may have been more or less continuous across the Paleozoic–Mesozoic boundary. In view of this important possibility, further studies on the stratigraphy and paleontology of this area are most desirable.

EAST MALAYA

For the present purpose, east Malaya is defined as Trengganu and the eastern thirds of Pahang and Johore (Figure 4.13). It is separated from the Upper Paleozoic rocks of central Malaya by a zone largely composed of continental Mesozoic rocks and granite.

The geology of Trengganu is probably more poorly known than that of any other part of Malaya. Much of the strata has been regionally metamorphosed to a higher degree than in Pahang, and considerable parts of the state are occupied by large granite masses concordant to the regional structure. In general lithology and structure, the sediments and metasediments resemble those of central Malaya. Pyroclastic rocks of rhyolitic to andesitic composition are widespread and locally important in bulk. The limestone facies is more restricted. Massive hill-forming limestone is present only at Bukit Biwah in Trengganu and in two other areas west and northwest of Kuantan. As in central Malaya, structural complexity has confounded the stratigraphy, and rocks of similar facies but different ages are easily confused. There is no paleontological evidence of marine Triassic rocks in East Malaya, and it is likely that the majority of the older, more or less metamorphosed marine sediments and associated volcanic rocks are of Upper Paleozoic age. The area northwest of Kuantan is relatively well known and is discussed first.

Kuantan Area

The Upper Paleozoic rocks around Sungei Lembing (Figure 4.13) have been described by Fitch (1952), and the fossils they contain were monographed by Muir-Wood (1948). Fitch divided his "Calcareous Series" [later named the Kuantan Group by Alexander (1959)] into calcareous and argillaceous facies but was not able to define their stratigraphic relations. He deduced a general dip toward the east, indicating that the beds youngened in that direction.

The calcareous facies consists of massive, recrystallized, hill-forming limestone (for chemical analyses see Table 4.1) which forms lens-shaped outcrops striking north-northeast; Bukit Charas and Bukit Sagu are the largest. The coral-brachiopod fauna (Plate 7) of this limestone is Viséan, probably late Viséan (Dibunophyllum Zone) (Muir-Wood 1948). However, Igo and Koibe (1968) described Lower Namurian conodonts associated with the brachiopod fauna from Bukit Charas.

The dominant argillaceous facies comprises thin-bedded shale and siltstone. These are sometimes finely laminated but frequently contain traces of burrows and other signs of organic activity, ripple marks, plant debris, and occasionally the brachiopod *Lingula*. These characters suggest shallow-water deposition. The fauna, dominated by brachiopods and polyzoa, differs from that of the limestone but is also Viséan. Abundant plant remains are usually too poorly preserved to be identified; however, *Lepidodendron* and *Stigmaria* have been recorded. The shale and siltstone are commonly tuffaceous, and Fitch (1952) mapped one lens of purple-weathering acid tuff.

Shale in the Sungei Luit, some 48 km (30 miles) southwest of Sungei Lembing, contains a poorly preserved fauna which, however, resembles that of the Viséan Kuantan Group. It includes *Schizophoria*, indeterminate productoid and spiriferoid brachiopods, and phillipsiid trilobite pygidia (Jones et al. 1966). Cuttings along the main road from Kuantan westward to the Sungei Lepar have exposed steeply dipping mudstone with slump structures (Figure 4.14). This is in places cut by granite dykes and resembles the mudstone of the Kuantan Group, although it has not yielded any fossil remains.

Sandstone, grit, conglomerate, and black carbonaceous shale lying to the east of the Viséan outcrop were grouped by Fitch (1952) into an "Arenaceous Series," lying unconformably on the Viséan and possibly of Triassic age. Field evidence for an unconformity is obscure, and there is no reason to suppose these sediments are Triassic. The sediments contain poorly preserved plant remains of Upper Paleozoic type (Fitch 1952) and strike north toward an area of Carboniferous rocks in Trengganu. The

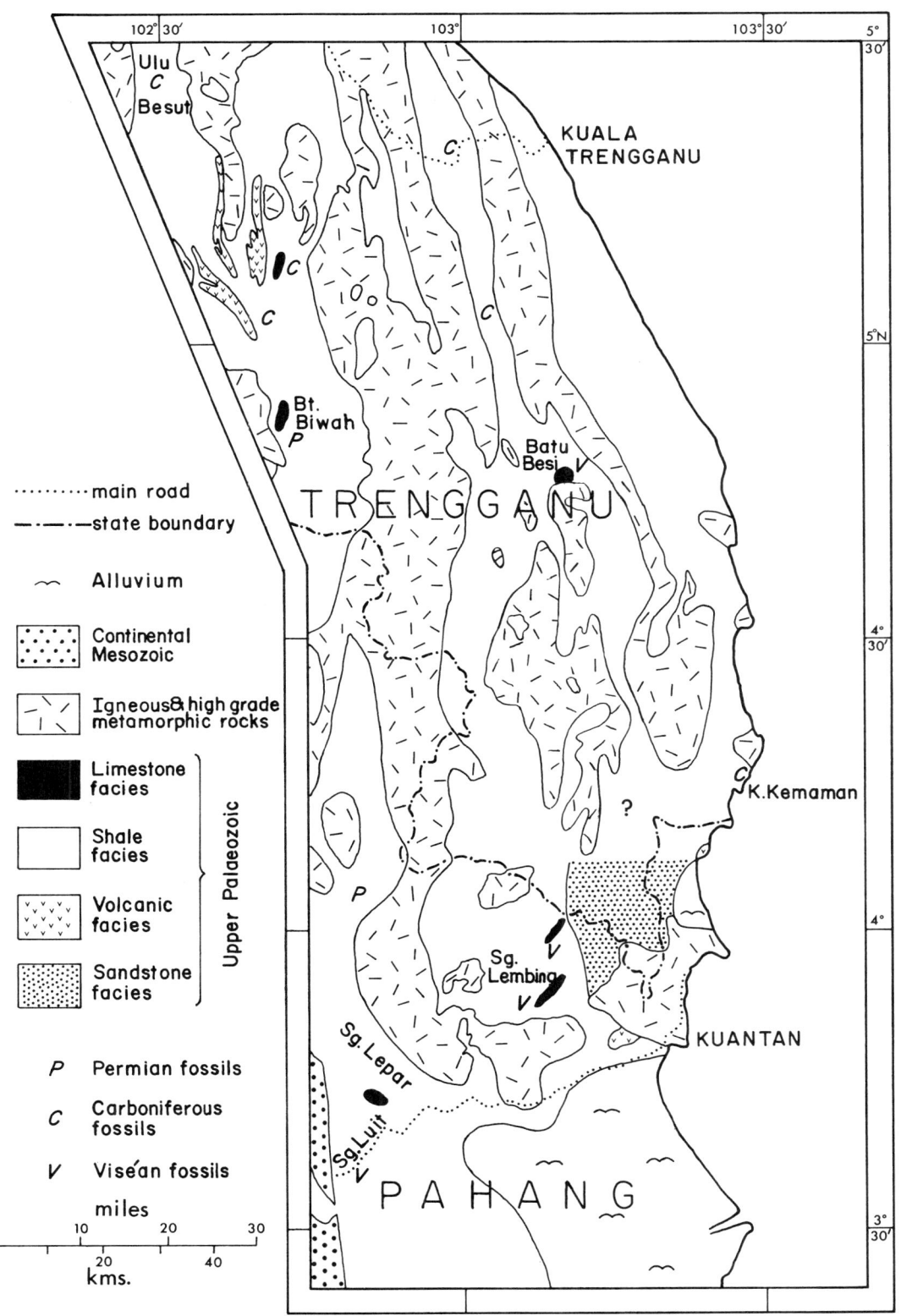

FIGURE 4.13 *Geological sketch map of east Malaya. Based on data of the Geological Survey of Malaysia.*

FIGURE 4.14 *Slump folds in Carboniferous(?) sediments. View south, Temerloh–Kuantan road near Kertam, about 55 km (35 miles) west of Kuantan. Photograph P. H. Stauffer.*

marked change in lithology of the so-called Arenaceous Series suggests a break in sedimentation after the deposition of the Viséan period and uplift of an adjacent area to supply the coarser detrital material. On these grounds, an unconformable relationship to the Viséan could be postulated. However, the junction of the Viséan "Calcareous Series" and the "Arenaceous Series" may be a tectonic one.

The "Arenaceous Series" is locally associated with rhyolite tuff, volcanic breccia, and subordinate lava.

Trengganu

Facies mapping of the northern part of Trengganu has been carried out by the Geological Survey of Malaya. MacDonald (1968) suggested that Lower to Middle Carboniferous shale was overlain by a formation that was mainly sandstone, which in turn was overlain by Permian limestone and shale. However, there is insufficient knowledge of the structure and field relations of these rocks to enable us to formulate a reliable stratigraphy. The facies present include shale, frequently black and fissile, occasional small limestone lenses, sandstone, pyroclastic rocks, and subordinate lavas.

Carboniferous fossils have been recorded from several scattered localities, mainly in the northern part of the state. Lenses of recrystallized limestone have yielded the corals *Lithostrotion* and *Amygdalophyllum* and small spiriferid brachiopods. *Girtyoceras*?, chonetid brachiopods and bivalve molluscs, *Lepidodendron* fragments and fronds of *Rhodea*?, *Sphenopteridium*?, *Neuropteris*, and cf. *Sphenophyllum* have been recorded from carbonaceous shale (Jones et al. 1966). These fossil occurrences show that rocks of a Viséan to Westphalian age are present in the area.

Fossiliferous Permian rocks occur near the junction of the Kelantan–Trengganu–Pahang state boundaries but are unknown elsewhere in Trengganu. In the Sungei Biwah and at Bukit Biwah dark-colored fine-grained foraminiferal limestone dips eastward. It contains *Parafusulina* and waagenophyllid corals and is thus dated Lower Permian. In the extreme southeast of Kelantan, sandy bioclastic limestone contains fusuline foraminifera characteristic of the Middle Permian (Jones et al. 1966) and, from shale in northwest Pahang, Igo (1964) identified a shelly fauna which he regarded as Middle Permian.

The Upper Paleozoic sediments of Trengganu resemble those from elsewhere in east and central Malaya in being associated with volcanic rocks. MacDonald (1968) noted that in northeast Malaya volcanic rocks become more basic than rhyolite to the east; in north Trengganu andesitic rocks are dominant, although volcanic rocks as a whole decrease from west to east.

An arenaceous facies recognized by MacDonald (1968) is best developed in north Trengganu. It includes quartz, lithic and tuffaceous sandstone, arkose, and conglomerate. It is doubtful to what extent this facies is of Upper Paleozoic age: some of the conglomerate most probably belong to the continental Mesozoic. If MacDonald's stratigraphy proves to be correct and the arenaceous rocks do lie stratigraphically between the Middle Carboniferous and the Permian, they may be the lateral equivalents of Fitch's "Arenaceous Series" in the Kuantan area.

Southeast Pahang and Johore

In southeast Pahang and Johore, the Upper Paleozoic is probably represented by the steeply dipping sandstone and shale sequence in the Mersing District and possibly by the metasediments of east Johore; both are well exposed intermittently along the coast but poorly known inland. No fossils have been obtained from these rocks. They are overlain unconformably by the continental Mesozoic. Rhyolite lava, older than the granite and cut by andesite sills, is present on Pulau Tioman and Pulau Ujol. Rhyolite and andesite lavas also occur on the mainland, where they lie beneath the continental Mesozoic rocks and are thus presumbly of Upper Paleozoic or Triassic age.

In Ulu Sedili, east Johore, poorly preserved Lower Permian fusulinids have been collected from a limestone lens. The associated clastic sediments are presumably also Lower Permian. The overlying Linggiu Formation contains a rich Upper Permian flora (Kon'no et al. 1970).

Summary

Although most of the "pregranite" bedded rocks of east Malaya are probably Upper Paleozoic, apart from those in the Sungei Lembing area they have received little attention and are poorly understood. They resemble the Upper Paleozoic of central Malaya, particularly in their volcanic content. The oldest dated rocks are Viséan, the youngest Middle Permian, and the Upper Permian appears to be missing. The Upper Carboniferous may be represented by a coarser clastic facies denoting uplift in neighboring areas (? to the east) at about this time. Uplift of east Malaya itself may have occurred in the upper part of the Permian and continued into the Triassic.

Although these events are conjectural, they may be considered as associated with the concordant granite intrusions which were emplaced on a large scale. A lower Permian potassium-argon radiometric date of 260 ± 10 m.y. has been obtained from the granite of Ulu Sungei Kemaman, an Upper Permian date (240 ± 8 m.y.) from granite at Sungei Lembing, and a mid-Triassic date (215 ± 8 m.y.) from the granite of Bukit Ubi, near Kauntan (Snelling 1965).

KENNY HILL FORMATION

P. H. STAUFFER

The name "Kenny Hill formation" has been used for a number of years for a sequence of clastic sedimentary rocks occurring in the Kuala Lumpur area, although the unit has never really been properly defined. In the first published geologic account of the region, Willbourn (1922a) grouped all the clastic rocks into one "series" thought to be younger than the limestone of the Kuala Lumpur area, which was then supposed to be Carboniferous. Later it was shown (Yin in prep.; Gobbett 1965a) that the clastic rocks fall into two groups, one underlying and the other overlying the limestone, which is now known to be at least in part Silurian (Gobbett 1965a). The younger clastic sequence has been called the "Kenny Hill formation" after exposures in the Kenny Hill residential area of Kuala Lumpur town (Yin in prep.).

Occurrence

The known occurrence of the Kenny Hill formation exists as a broad synclinal belt, generally 7 to 10 km (5 to 6 miles) wide, running from Kuala Lumpur town southward through the suburbs of Petaling and Petaling Jaya and farther to the south for at least 30 km (19 miles) (Figure 4.15). Further extensions to the south and west have been suggested (Yin in prep.) but remain to be proven. At its northern end the outcrop is terminated abruptly along a possible fault of west-northwest–east-southeast orientation. The western and eastern margins of the outcrop belt are in part intrusive contacts against granitic rocks and in part depositional (?) contacts with older sedimentary rocks, especially the Silurian carbonates called the Kuala Lumpur Limestone (Gobbett 1965a).

Lithology

The Kenny Hill formation consists of a monotonous sequence of interbedded shales, mudstones, and sandstones. A single report of a conglomerate bed from a borehole (Yin in prep.) has not been corroborated. The shales occur in thin (a few centimeters) to thick beds (1 m or more) and are often finely laminated. Beds of mudstone and sandstone are thicker, commonly one to several meters thick. Beds are laterally persistent in the main for all lithologies, although lenses of mudstone occur in

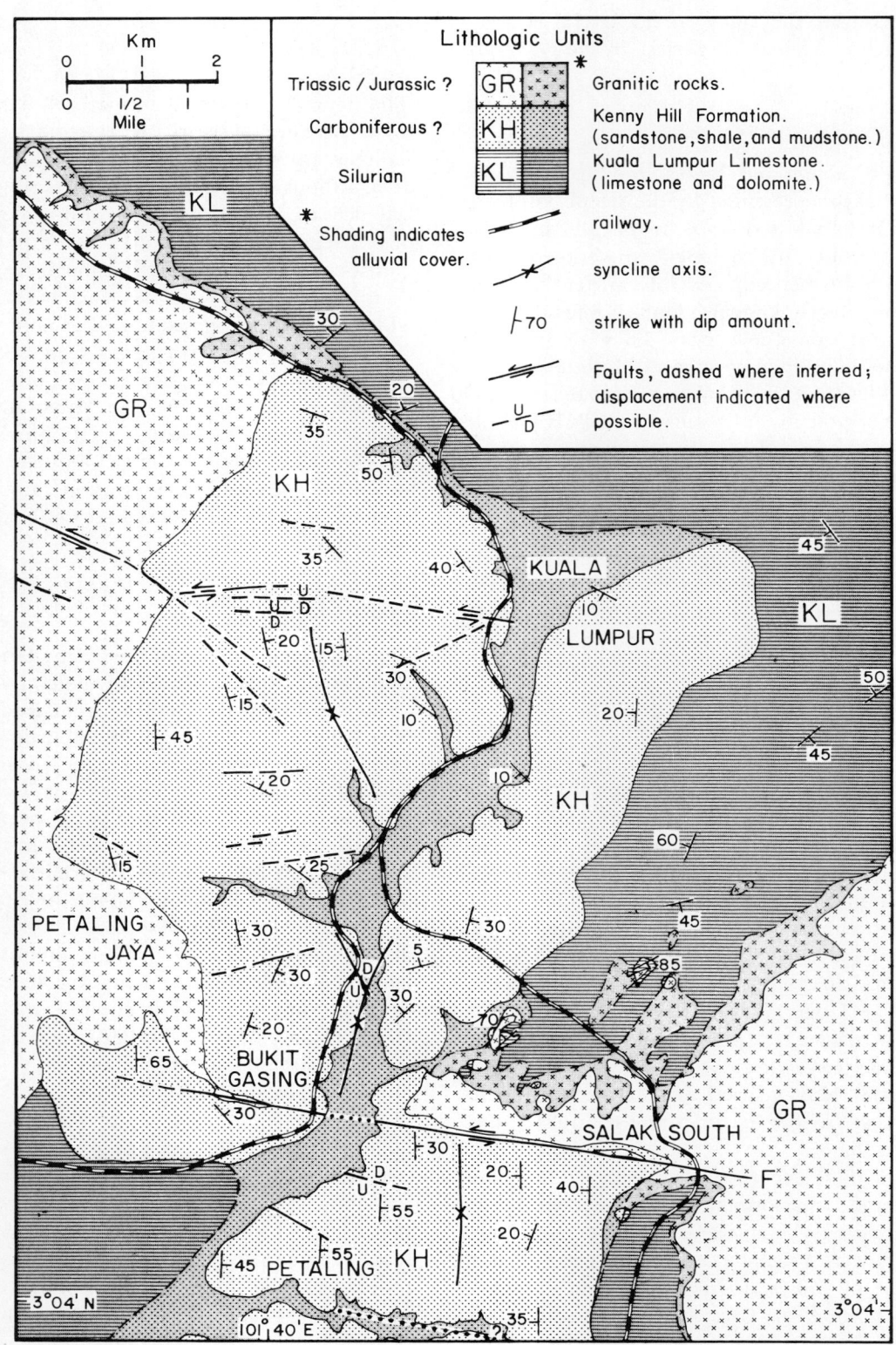

FIGURE 4.15 *Geological sketch map of the Kenny Hill formation in the Kuala Lumpur area. Based mainly on Choy (1970) and Yeap (1970).*

some thick sandstones. The sandstone is mostly fine grained to medium grained (median diameter 0.125 to 0.5 mm).

Both fine and coarse lithologies are generally very much weathered—the lutites to a plastic mass of clay, the sandstones to a friable aggregate of grains. Some of the coarser sandstone beds are more resistant and remain hard and coherent in fairly shallow cuts. In thin section, these hard sandstones are seen to consist generally of quartz grains, mostly monocrystalline with slightly undulose extinction, together with some chert grains and rarer weathered feldspars. The texture is an interlocking mosaic implying considerable postdepositional changes. Indicated changes include much pressure solution of quartz, some overgrowths on quartz grains, and recrystallization of chert and fine-grained quartz matrix, causing blurred and serrated edges on the larger quartz grains. Iron oxide films are present between grains and along the solution surfaces. Secondary and metamorphic minerals include pyrite cubes and crystals of chlorite and mica in the matrix. Heavy minerals separated from crushed specimens are mainly well-rounded pink zircons, with some tourmaline and opaque minerals. The petrography is given by Choy (1970) and Yeap (1970).

The shales and mudstones are phyllitic, commonly showing foliation and incipient cleavage, generally at an angle to the bedding. The lutites are often light in color in weathered outcrops, but deeper cuttings show that most were originally dark and carbonaceous rocks.

Primary structures in the Kenny Hill formation comprise a distinctive assemblage of graded bedding, soft-sediment deformation, and organic tubes and burrows. In most outcrops, however, the only conspicuous structure is the planar and regular bedding surfaces, an effect heightened by the common presence of shear surfaces subparallel to bedding. Some sandstone beds are graded, and in a few outcrops in the Petaling area rhythmic sequences of uniformly well-graded standstone beds are visible, about 20 to 50 cm thick and separated by thinner shales. Yeap (1970) reports one occurrence of medium-scale cross-bedding in sandstone in the very basal part of the unit, only 20 m from outcrop of the underlying limestone.

Pebble-size clasts of grey clay are seen in some of the sandstone beds, stretched and flattened along bedding to form thin, sharp-edged lenses (Figure 4.16a). Within the sandstones, discontinuous and subparallel laminae are seen, giving an impression

FIGURE 4.16 *Kenny Hill formation, Petaling, Selangor. Photographs P. H. Stauffer.* (a) *Sheared and streaked-out lenses of dark mudstone in sandstone; note ripped up mudstone remnant at top;* (b) *sandstone and dark mudstone showing soft sediment deformation and organic tubes and burrows (undeformed); note tube to left of pencil near point transecting lithologic boundary.*

of "flaser bedding." Locally, very tight recumbent slump folds are visible in the lutites, and some slump or rip-up structures in which both lutite and sandstone beds have been partly disintegrated (Figure 4.16b). These structures are often strongly planar and stretched out parallel to bedding, and the presence of some undeformed organic burrows transecting the structures (Figure 4.16b) proves that they formed in the still soft sediment and are therefore penecontemporaneous with deposition.

Organic tubes and burrows have been observed in Kenny Hill sediments in a number of outcrops. These structures are vertical to horizontal, usually cylindrical, and range from 2 to 20 mm in diameter, with most being 4 to 5 mm. They occur abundantly in occasional beds, generally finely laminated clay lutites, in which the lamination has been partly destroyed by the bioturbation. They also occur in some slumped sequences involving sandstones. Since they are visible only because the tube filling is a different lithology from the host rock, they may also occur undetected in massive beds.

Structure

Dips in the Kenny Hill formation are moderate, commonly less than 30°, and their outline a broad syncline with an approximately north-south axis (Yin in prep.). The western limb is somewhat steeper, with dips of 50 and even 70°. Yin (in prep.) feels that repetition of section by strike faults or unrecognized folds must be present, because thickness calculations were based on the assumption of a simple syncline yield rather large values. He estimated the actual thickness at only about 300 m (1000 ft). Choy (1970) and Yeap (1970), however, are convinced from detailed mapping that no major fault repetition is present. Based on attitudes in the syncline, they estimate the exposed thickness of the unit as 1200 to 1500 m.

The gentle folding in the Kenny Hill is in marked contrast to the structure in the underlying Lower Paleozoic rocks. These are intensely and tightly folded, in part along almost east-west axes (Gobbett 1965a, p. 69; Yeap, 1970).

The Kenny Hill formation is cut by numerous small and some larger faults. Prominent are vertical (wrench) faults at orientations of 100 to 110°, 135°, and 65°. The first two seem to be sinistral, but the last may be dextral. Faults of moderate to steep dip and having a variety of other strikes also occur. Some of the wrench faults have crush zones up to several meters wide; most of the dipping faults show only very minor crush zones. Numerous small shears, often subhorizontal and subparallel to bedding, are present, but are difficult to detect except where they cross a lithologic boundary at an angle so large that they are not mistaken for lenticular bedding. These shears also show almost no crushing, but appear to have offsets ranging up to several meters.

Joints in the Kenny Hill sediments form a definite pattern, dominated by two near-vertical sets arranged approximately north-northeast–south-southwest and east-west (Choy 1970, Yeap 1970).

Fossils and Age

No datable fossils have yet been found in the Kenny Hill formation. Poorly preserved impressions and carbonaceous imprints, presumably of plant fragments, are locally common in the mudstones and shales. Recent discovery of abundant trace fossils (the tubes and burrows mentioned previously and shown in Figure 4.16b) and a few very poorly preserved body fossils (Yeap 1970) offers the hope of an eventual dating based on paleontology. The body fossils include probable pelecypods and brachiopods and some long, branching, segmented tubes of unknown affinities.

The age of the Kenny Hill formation can be fixed at present only by its relations to other rock units. It appears to be younger than the Kuala Lumpur Limestone (Gobbett 1965a), which has yielded Silurian fossils. The Kenny Hill has a simpler and much different structural pattern from the limestone and other older sediments, and it overlies them along a discordant contact which is probably an angular unconformity (Yin in prep.). The contact itself is never well exposed, but where it has been penetrated in borings there is often a disturbed and brecciated zone in the Kenny Hill sediments adjoining it. This may represent collapse consequent upon solution of the limestone by ground water. It also suggests, however, the possibility that the contact may be a low-angle thrust fault more or less along bedding in the Kenny Hill. Yin (in prep.) has already noted that the contact has some of the characters of a "decollement." Recent workers, nonetheless, concur (Choy 1970; Yeap 1970) in calling the contact an unconformity.

No younger sedimentary rocks (other than Quaternary alluvium) are seen in contact with the Kenny Hill formation. The formation abuts against granitic rocks in the Damansara area along an intrusive contact (Choy 1970), is cut by granitic dykes, and by quartz dykes at Bukit Gasing in Petaling Jaya and at Salak South (Yeap 1970). Yeap has also documented intrusive relations of granite and Kenny Hill formation in mines north of the Sungei Besi, and xenoliths of the sedimentary rocks were noted in the marginal parts of the granite. J. B. Scrivenor, in his field notes for 1904 (referred to by Yin in prep.), mentioned tin mineralization in the Kenny Hill sediments at Pantai, just southwest of Kuala Lumpur. The ore was mainly in three lodes striking 104°, 110°, and 130°, but it also occurred disseminated in the sandstones and shales. The major episode of granite emplacement in this part of Malaya is thought to have occurred in early- to mid-Mesozoic. (Potassium-argon dating of samples from Ulu Langat, about 15 km (10 miles) southeast of Kuala Lumpur, indicates a minimum age of lowermost Jurassic for the granite there—J. D. Bignell personal communication.) The tin mineralization probably occurred shorty thereafter.

Hence the age of the Kenny Hill formation can be bracketed as younger than the Kuala Lumpur Limestone and older than the granitic rocks and the tin lodes. This leaves a range of possibility from Silurian to mid-Mesozoic. In northwestern Malaya there is an unconformity between dominantly calcareous rocks of Ordovician to lowermost Devonian age (the Setul Formation) and clastic sediments of late Devonian to early Permian age (Singa and Kubang Pasu Formations). The latter sediments are in many ways similar to the Kenny Hill formation, and it seems most reasonable at our present state of knowledge to regard the Kenny Hill formation as Upper Paleozoic, probably mainly Carboniferous, and the approximate correlative of the Singa and Kubang Pasu Formations.

ORIGIN OF SEDIMENTS. The sediments of the Kenny Hill formation were probably deposited in marine waters not far from an eroding landmass, which itself probably consisted mainly of sedimentary rocks of moderate or low relief. The depositional area was likely on a slope or unstable shelf, with a moderate but not large depth of water.

The evidence for these inferences are as follows: Instability is indicated by the deformational structures and evidence of resedimentation (Figure 4.16). The possible presence of brachiopods in a very sparse fauna suggests marine waters that were not extremely shallow (or at least not particularly favorable to preservation). The lack of very coarse material in the sands and their high mineralogical maturity indicates low relief and advanced chemical weathering in the source area, or extended transport and reworking, or both. The high chemical maturity, and especially the dominance of well-rounded zircons in the heavy mineral suite, suggest derivation from older sediments. The abundance of carbonaceous fragments, and the abundance of sand, together with the lack of flysch-type sedimentary structures, argue against a deep-water turbidite-pelagite origin. Lack or extreme rarity of cross-bedding, channeling, and other traction-current structures, on the other hand, signify that an origin of fluvial or inner deltaic type is unlikely. The most probable sites for deposition of the Kenny Hill sediments appear to be the outer portions of a delta or shelf, or the upper portion of a submarine slope.

The directions of transport and paleogeographic relations of the Kenny Hill sediments are unknown, but Yeap's (1970) reported cross-bedding in the base of the unit gives transport directions to the south or west.

CORRELATION AND SYNTHESIS

Each of the four zones just considered had a distinct Upper Paleozoic history. This is outlined in Table 4.4. Granite intrusion toward the end of Permian time may have been broadly contemporaneous in three of the zones, but the sedimentary formations cannot be traced across zone boundaries. Some correlation can be made by way of the standard chronostratigraphic scale by using fossils; of these fusulinids and goniatites (unfortunately rare) are the most reliable. Condonts may provide a useful dating tool in the future. Many of the strata are unfossiliferous and, since in the structurally complex central and east zones a stratigraphy has yet to be worked out, correlation can only be based on a very broad comparsion.

Correlation with Neighboring Countries

The Upper Paleozoic of Malaya may be further understood by reference to its extension along the strike to the north and south, as far as this can be recognized. Upper Paleozoic rocks outcrop widely in peninsular Thailand, Burma and Sumatra, and in more restricted areas in west Sarawak (Figure 4.17).

Representatives of the Upper Paleozoic age of northwest Malaya cannot be traced to the south. However, they can be clearly traced to the north and correlated with rocks in peninsular Thailand which form the upper part of the Phuket Group (Mitchell et al. 1971), the Kanchanaburi Series, and the Rat Buri limestone. The lithology of the Phuket Group closely resembles that of the Singa Formation. The characteristic pebble spreads, black mudstone alternating with flaggy siltstone, and common trace fossils are all present at the type locality on Phuket Island, 240 km (149 miles) north of the Langkawi Islands. Fossils recorded from the Phuket Group by Young and Jantaranipa (1970) include two trilobite pygidia comparable to *Cyrtosymbole perlisensis*, originally described from the basal part of the Kubang Pasu Formation in Perlis. Thus the Phuket Group is clearly the lateral continuation of the Singa Formation and the Kubang Pasu Formation and

TABLE 4.4 *Comparison of the four zones of the Malayan Upper Paleozoic*

	Northwest Malaya	Kinta Valley	Central Malaya	East Malaya
PERMIAN	Thrusting to the west (Kisap). Granite intrusion	Granite intrusion	Shale–pyroclastic–limestone sequence (Viséan to Upper Permian) showing rapid facies variation, at least in some areas, continuing into the Lower Triassic	Granite intrusion
	Limestone deposition (Chuping formation)	More or less continuous (Silurian to Permian) limestone deposition (Kinta Limestone) on shoal area flanked by a "sediment trap" receiving mainly fine-grained clastic material		Shale, sandstone, with subordinate pyroclastic rocks, and limestone of Viséan and Lower Permian age. Upper Carboniferous is represented by shale with Westphalian plants and ? sandstone and grit.
CARBONIFEROUS	----transition to---- Flysch deposition (Singa Formation) in the northwest; and deposits of immature sandstone and mudstone (Kubang Pasu Formation) in the southeast		Plutonic activity during this period not proved.	
DEVONIAN	Period of folding, low-grade regional metamorphism, and some plutonic activity		No evidence	No evidence
	Fine-grained clastic rocks forming the top of the Setul Formation			

similarly underlies limestone of Permian age. Farther north along the strike, the Mergui Series of Peninsular Burma may be in part equivalent to the Singa Formation and the Phuket Group. However, the Mergui Series has a more varied lithology, which includes pyroclastic rocks (Brown and Heron 1923), and it is separated from the overlying Permian Moulmein limestone by a distinct formation of sandstone and shale.

The Kanchanaburi Series has been regarded as late Silurian to Carboniferous. The lower part includes shale with monograptids and tentaculitids and can be correlated with the uppermost part (Lower Devonian) of the Setul Formation of northwest Malaya. The unconformity, which represents most of the Devonian period in northwest Malaya, has not been proved in Thailand; but the next youngest fossiliferous beds, lithologically similar to the Kubang Pasu Formation, contain a Lower Carboniferous fauna in Patalung (Reed 1920).

The Rat Buri limestone, usually assumed to be "Permo-Carboniferous," has yielded to date only Permian fossils. Its age probably varies in different parts of Thailand. In peninsular Thailand it can be correlated with the Chuping Formation. On the island of Ko Muk, west of Trang, the Rat Buri limestone is well-bedded and grey; it has chert nodules and contains an Artinskian fauna of polyzoa, brachiopods, and corals (Sakagami 1966b). It overlies black shale and sandstone, within which lies a thin conglomeratic limestone containing derived Lower Carboniferous brachiopods (Hamada 1960).

Extensive limestone outcrops in west-central and north Thailand are also referred to the Rat Buri limestone. Lower and Middle Permian fossils are known from the limestone in the north. The Moulmein limestone of peninsular Burma is similar and contains Lower to Middle Permian fossils in Tenasserim (Chhibber 1934). The widespread plateau limestone of the Shan states is partly Devonian and partly Permian. The character of the Carboniferous is obscure.

The Upper Paleozoic limestone of the Kinta Valley has a unique character and represents a localized facies that is not seen elsewhere. However,

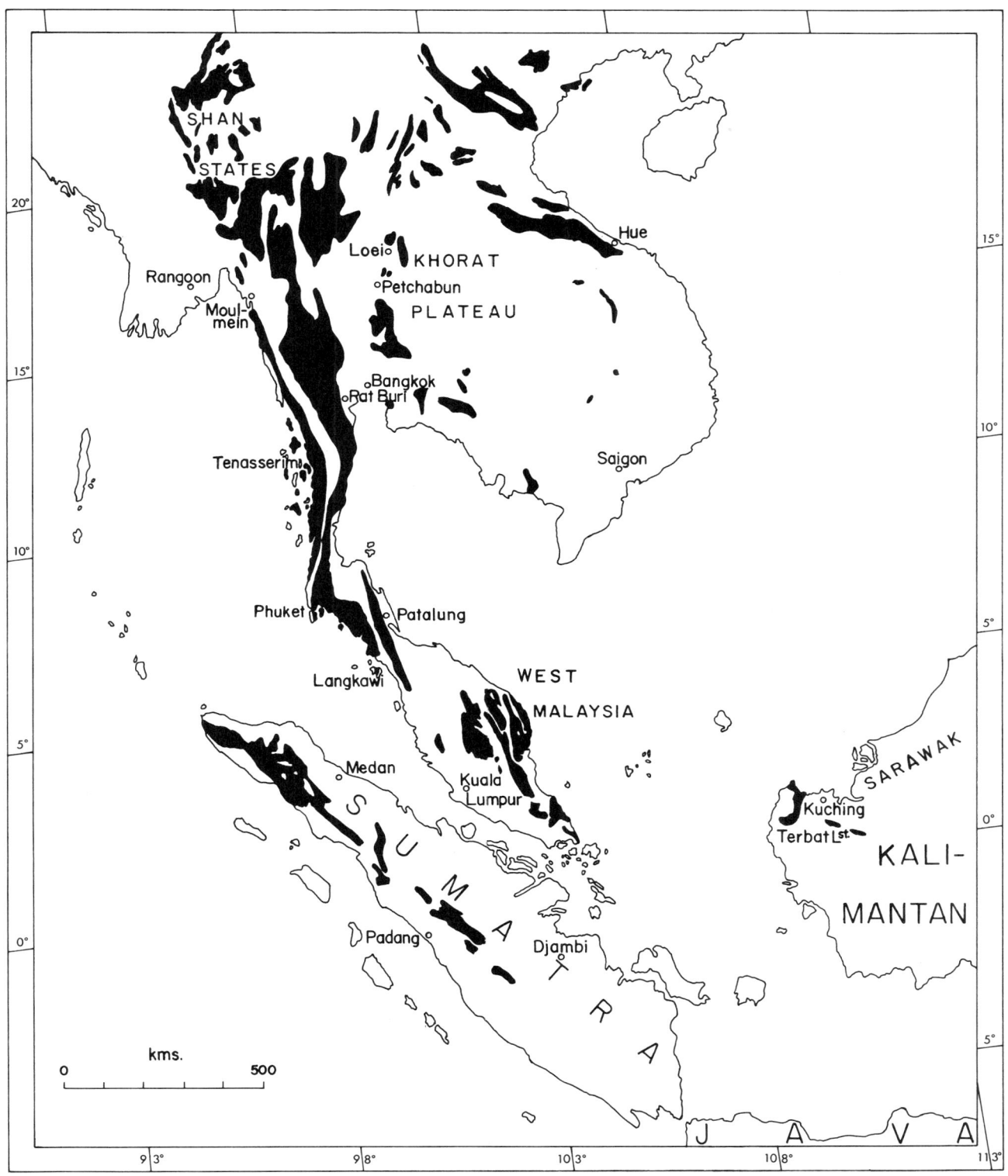

FIGURE 4.17 *Upper Paleozoic outcrops in Southeast Asia. Based on data extracted from published sources.*

some of the characters of the Upper Paleozoic of central and east Malaya can be recognized in the Upper Paleozoic of central and northeast Thailand and of Sumatra. Near Rat Buri, southeast of Bangkok, the Kanchanaburi Series includes tuffaceous sediments of Lower Carboniferous age (Sakagami 1966a) and the Rat Buri limestone is Artinskian (Sakagami 1968) and Guadalupian (Waterhouse and Piyasin 1970).

Along the west flank of the Khorat Plateau, the Rat Buri limestone commonly includes clastic beds. The fusuline limestone facies ranges from Lower to Middle Permian (Pseudoschwagerina to Neoschwagerina—Verbeekina Zones) (Pitakpaivan 1965). In the Loei area a sequence from the Lower Carboniferous to the Middle Permian has been tentatively worked out (Kobayaski 1964). The lowest beds (?Devonian) are volcanic. The Carboniferous beds are mainly clastic and the Permian mainly of limestone facies (Rat Buri limestone). The Upper Permian, however, consists of sandstone and shale with a Cathaysian flora (Asama et al. 1968); this is also present at Petchabun (Kon'no 1963, Asama 1966).

To the south of Malaya the Upper Paleozoic of the "Djambi Nappe" in Sumatra (Bemmelen 1949) consists of gently folded, unmetamorphosed sediments and volcanic rocks of dacitic to andesitic composition. The sediments include fusuline limestone, shale and sandstone that are sometimes tuffaceous, polymict conglomerate, and thin coals; all are probably of Permian age. The oldest fusuline limestone horizon contains *Pseudoschwagerina* and is overlain by coal and rootlet beds with an autochthonous Cathaysian flora older than that from Thailand. Bemmelen interpreted these sediments as transported tectonically and originally deposited farther to the northeast.

The phyllite and slate with occasional sandstone and fusuline limestone lenses, which form large areas of the Sumatra Highlands farther to the southwest, are poorly known but appear to have little in common with the Malayan Upper Paleozoic. In west Sarawak the Terbat Formation, consisting of cherty fusuline limestone and shale, ranges in age from Moscovian to Lower Permian (Wilford 1965); it shows no resemblance to any of the Upper Paleozoic formations of the Malay Peninsula.

Geological History

In summary, the Upper Paleozoic history of the Malay Peninsula may be outlined as follows. At the beginning of the Devonian age, sedimentation probably followed the same basic pattern it had displayed during the Lower Paleozoic (Chapter 3). In northwest Malaya, and possibly also in central Malaya, a major break occurred toward the end of Lower Devonian time. In Langkawi and Perlis, folding caused a cleavage and boundinage to develop in the affected rocks, and minor granitic intrusion occurred in the Middle or Upper Devonian. However, no evidence of these earth movements can be traced in the Kinta Valley, where Middle Devonian limestone appears to be conformable with both Lower and Upper Devonian limestone. It seems that a stable shelf occupied the Kinta Valley area throughout the Upper Paleozoic. This was probably surrounded by deeper water, which received clastic sediments derived from the erosion of the folded and uplifted areas.

The Devonian of central and east Malaya is unknown except for early Devonian strata forming the top of an essentially Lower Paleozoic geosynclinal sequence in west Pahang.

In the northwest, the uplifted area, produced by the folding, was eroded to produce clastic material for Carboniferous sediments which were deposited unconformably on the older rocks. This sedimentation, which was discussed earlier, appears to have been relatively constant throughout the Carboniferous and Lower Permian.

The Lower Carboniferous of central and east Malaya, where known, is composed of shale and limestone, associated with volcanic tuffs. The fossil-bearing beds are Viséan. The absence of coarse clastic material suggests that, if any uplift did occur in the Devonian, was local and was soon destroyed by erosion in the Upper Devonian or Tournaisian.

In central Malaya, the Lower Carboniferous pattern of sedimentation and paleogeography probably continued with little change into the Upper Carboniferous, although there is no fossil evidence of this period. However, in east Malaya a change in sedimentation (probably accompanied by gentle earth movements) took place after the deposition of the Viséan limestone and shale. This change was brought about by the uplift of a land mass to the east, which then supplied clastic material of sand and grit grade and plant remains of Upper Carboniferous European type.

The land mass can be identified with the Indosinian land that first arose from the Middle Carboniferous earth movements of Indochina. Uplift in the west central part of Malaya, which provided sediment for the Kenny Hill formation of Selangor, may have occurred at about the same time. The influence of Indosinia waned during the Permian period when limestone and shale deposition was again dominant in east Malaya, at least until Middle Permian time. A sea with volcanic islands occupied central Malaya during the Permian, and in it were laid down shale and pyroclastic rocks; limestone was formed locally around the volcanic islands and shoals. Occasional plant remains drifted into the area, either from the east or from the south of the Malay Peninsula, where at times the sea regressed and terrestrial conditions prevailed.

In northwest Malaya, the Middle Permian is represented by pure limestone, indicating the absence of further uplift and the presence of a widespread shallow sea, extending far to the north in Thailand.

The late Permian marked an important stage in the history of the Malay Peninsula, as indeed it did on a worldwide scale. Widespread uplift took place, accompanied by granite intrusion in the northwest, in the Kinta Valley, and in the east. The Triassic is unconformable on the Paleozoic in the former area and absent in the latter two areas. In some parts of central Malaya, however, marine conditions continued apparently unbroken, and without much facies change, through the Upper Permian and into the Lower Triassic.

The most striking aspect of the Malayan Upper Paleozoic is that the narrow north-south trending zones, which are in close proximity to each other, should have such contrasts in their history.

CHAPTER 5

Mesozoic

C. K. BURTON

Dedicated to the memory of my friends and co-workers in Malaya,

Simon MacDonald
Russel Patton
Donald Rishworth

The Mesozoic Era witnessed the main geological events that have given the Malay Peninsula its present form and constitution. When the era opened, the region that extends from west Yunnan, through east Burma and west Thailand to Malaya, had already experienced a long geosynclinal history, commencing as early as the Ordovician, time. The peninsula had been differentiated into a miogeosynclinal zone, occupying most of peninsular Thailand and the western third of Malaya, and a eugeosynclinal zone covering axial and eastern Malaya and the immediately adjacent part of south Thailand (Figure 5.1). Between the two lay a more or less persistent geanticline, or miogeosynclinal ridge in the terminology of Aubouin (1965); in the extreme northwest there are indications of a shelf area, and a second upwarp or eugeosynclinal ridge may have existed near to the present east coast of Malaya.

The final stages of the geosynclinal phase occurred in the Triassic period, ushering in the culminating orogenic revolution of the region—the Thai–Malayan Orogeny—which was accompanied by granite emplacement on a vast scale. Like the Rocky Mountain Orogeny of western North America (Gilluly 1963), this orogeny seems to have had a revolutionary phase that lasted for a long time (early Triassic to early Cretaceous) and to have comprised several pulses of diastrophism and of plutonism. When the orogeny had passed, the Malay Peninsula had been transformed from a mobile to a stable region, henceforth forming a portion of the Sunda Shield.

During the Mesozoic age there accumulated considerable masses of late to postorogenic sedimentary rocks, derived from the newly generated mountains. There was also substantial faulting both of a dip-slip (block) and a strike-slip (wrench) nature.

LOWER TRIASSIC

Occurrence and Nomenclature

In West Malaysia, Lower Triassic sedimentary rocks are known with certainty in only one restricted area in the northern part of the country. Here the dominantly calcareous sediments of Permian age extend up into the Lower Triassic and are preserved in a tectonic downwarp lying between the migmatite complex of Gunong Stong in Kelantan and the granite batholith of Gunong Benom in Pahang. The rocks occur in a synclinorium that extends eastward from the line of Gunong Stong–Gunong Benom to the Boundary Range Granite, which occupies the boundary between Kelantan and Trengganu. To the southeast, the Lower Triassic passes beneath younger rocks.

In south Kelantan, these Permian and Triassic sedimentary rocks have been designated by Yin (*in* Kobayashi et al. 1967) as the Gua Musang Formation (Figure 5.2). The same formation is represented by a sequence of low-grade metasediments and metapyroclastic rocks in Ulu Nenggiri and also between the Gunong Stong Complex and the Taku Schist.

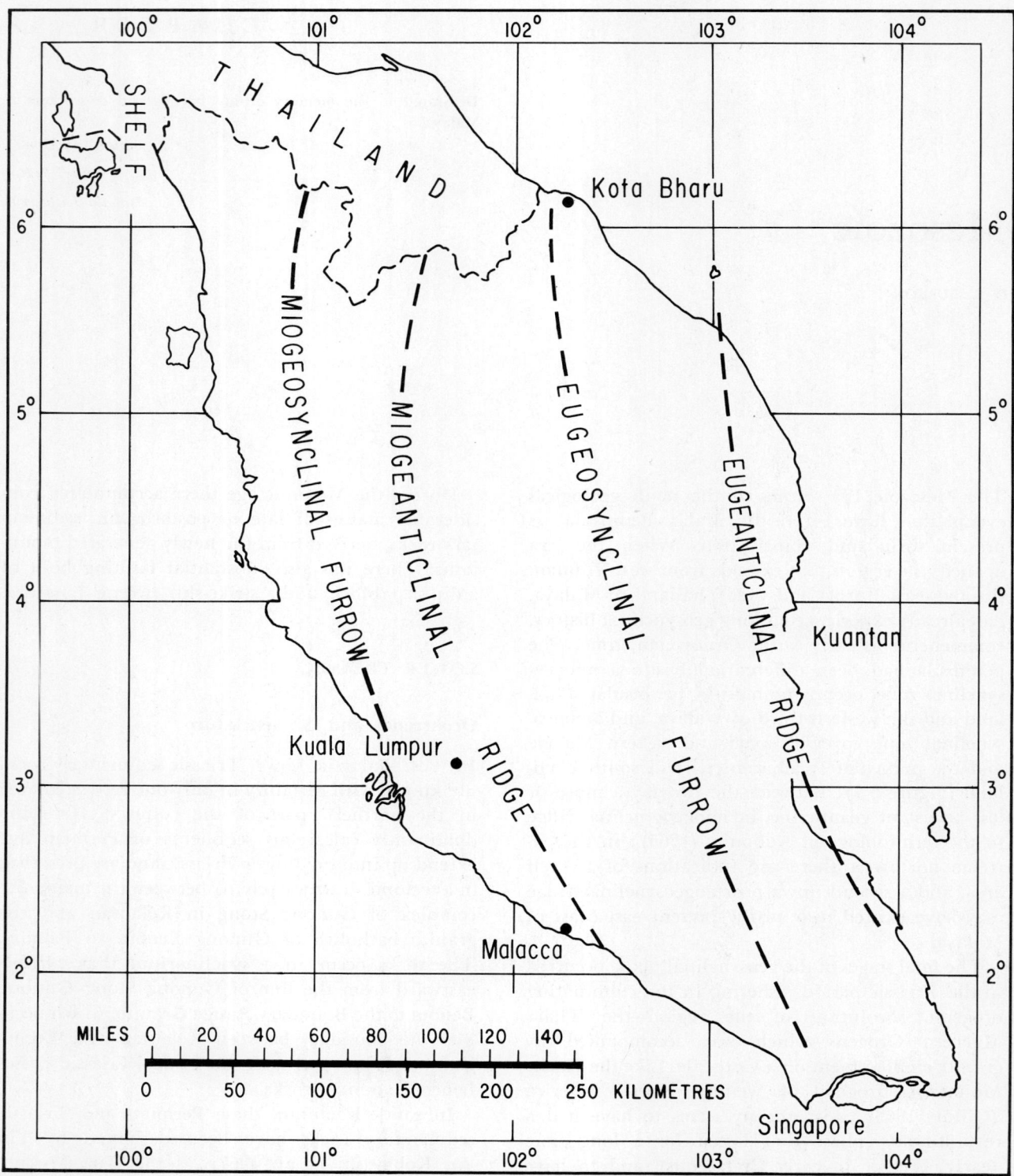

FIGURE 5.1 *Geosynclinal organization in Malaya from the Ordovician to the Upper Triassic.*

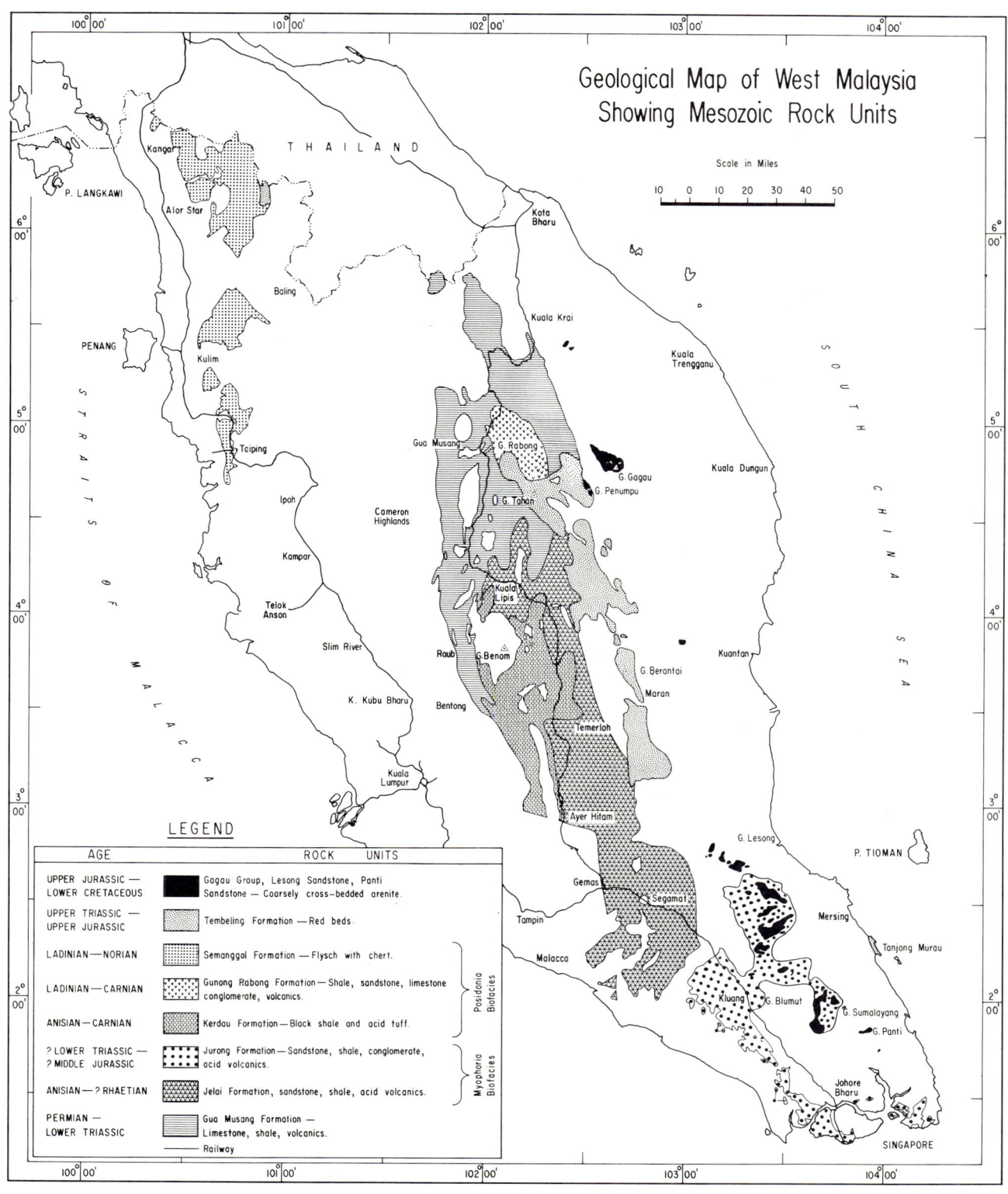

FIGURE 5.2 *Distribution of Mesozoic rocks in the Malay Peninsula. Based partly on data of the Geological Survey of Malaysia.*

PLATE 11 *Triassic fossils.* *1.* Daonella pahangensis *Kobayashi, Ladinian, Temerloh, Pahang,* ×1. *2.* Claraia griesbachi concentrica *(Yabe), Scythian, Merapoh, Pahang,* ×1. *3.* Costatoria pahangensis *Kobayashi and Tamura, Carnian, Temerloh, Pahang,* ×1. *4.* Costatoria *cf.* myophoria *(Boettiger), Upper Triassic, Singapore,* ×1. *5.* Costatoria singapurensis *Kobayashi and Tamura, Upper Triassic, Singapore,* ×1. *6.* Costatoria chegarperahensis *Kobayshi and Tamura, Anisian, Chegar Perah, Pahang,* ×1. *7.* Costatoria quinquicostata *Kobayashi and Tamura, Ladinian, Chegar Perah, Pahang,* ×1. *8.* Myophoria? newtoni *Kobayashi and Tamura, Carnian, Temerloh, Pahang,* ×1. *9.* Halobia charlyana *Mojsisovics, Carnian, Nami, Kedah,* ×1.5. *10.* Pteroperna malayensis *Newton, Carnian, Kuala Lipis, Pahang,* ×1. *11.* Paratrachyceras regoledanus *(Mojs), Carnian, Johore,* ×1. *12.* Ptychites *cf.* rectan-

On approximately the same strike, in the vicinity of Johore Bahru at the southern extremity of the Malay Peninsula is a thick sequence of rhyodacitic volcanic rocks, referred to as the Gunong Pulai volcanic member of the Jurong formation (Burton in prep.). These rocks are invaded by an adamellite for which radiometric evidence suggests a Lower Triassic age. Because of the chemical and mineralogical similarity between the adamellite and the rhyodacitic volcanic rocks, it is thought that the Gunong Pulai member does not long predate the adamellite and therefore may also be of Lower Triassic age. Similar, possibly equivalent rhyodacitic rocks are widespread in Johore State.

Gua Musang formation

The designation of the Gua Masang formation is taken from a prominent limestone hill in south Kelantan which has also given its name to a small town on the East Coast Railway, 13 km (8 miles) from the Pahang border. This stratigraphic name is at present informal.

It is evident that the Gua Musang formation extends well beyond the south Kelantan area. It extends southward at least as far as Kuala Lipis, whence it may continue to the Temerloh–Maran district of central Pahang—a linear extent of 193 km (120 miles). Northward for some 32 km (20 miles) from Gua Musang, this rock unit becomes progressively more metamorphic in the direction of the Stong Migmatite Complex. It appears to the east and north of Gunong Stong as phyllite and metavolcanic rock near Dabong and as partial calcareous metasediments around Kampong Belimbing near the Thai border (Figure 5.2). It is not yet certain what proportion of the Gua Musang formation is of Triassic age, and the bulk of this rock unit may be Middle and Upper Permian.

In the type area, however, Yin (1965a,b) made important discoveries of Lower Triassic ammonoids and pelecypods. The Scythian bivalve genus *Claraia* was found in shale interbedded with limestone some 12 km (7.5 miles) south of Gua Musang (Ichikawa and Yin 1966, Tamura 1968). Ammonoids were collected from limestone in a railway cutting at Gua Panjang, 11 km (7 miles) to the northeast of the Scythian lamellibranch locality (Fig. 1 *in* Ishii et al. 1966). These fossils were examined by Hada (1966), who identified the following species (Plate 11):

Owenites koeneni Hyatt and Smith
Owenites carpenteri Smith
Arctoceras sp. cf. *A. blomstrandi* (Lindstrom)
Paranannites aspensis Hyatt and Smith
Prosphingites austini Hyatt and Smith
Pseudosageceras multilobatum Noetling.

Hada stated that this fauna clearly indicated the mid-Scythian zone of *Meekoceras gracilitatis*. From the same locality (locality F Ph 2 *in* Igo et al. 1965), Sato reported the ammonoids *Prenkites?* sp. and *Isculites* of upper Scythian or Anisian age. Numerous conodonts found here suggest an Anisian age (Igo et al. 1965, pp. 8–9). The same authors noted that limestone 0.8 km (0.5 mile) farther east, which yielded schwagerinids resembling *Parafusulina*, neoschwagerinids, and possibly *Verbeekina*, is evidently Middle Permian. Since there is no indication of any major stratigraphic break between the two localities, it was assumed that sedimentation was continuous from the Middle Permian to the Anisian.

Further evidence that the Gua Musang formation may extend up into the Anisian comes from Chegar Perah, some 40 km (25 miles) south of Gua Panjang, where Sato recognized the following ammonites (Jones et al. 1966, p. 343):

Frecherites sp.
Paraceratites or *Frecherites?* sp.
Paraceratites sp.
Paraceratites cf. *trinodosus*.

Beyond this locality, in the Sungei Perenggan near Kuala Lipis, shale contains fossils said by Cox (*in* Service 1949) to resemble the Lower Triassic *Pseudomonotis* (*Claraia*) *aurita* Hauer (Plate 11). This event may represent another development of the Lower Triassic portion of the Gua Musang formation. No Triassic faunas are known in this formation farther to the south, which seems to be entirely of Permian age there. A detailed description of part of the formation is contained in Richardson's (1950) account of the geology of north Pahang, but some Lower Carboniferous strata of a different rock unit are included with the present Gua Musang formation in Richardson's

---

gularis *Kraus, Anisian, Kuala Lipis, Pahang,* ×1. *13.* **Paranannites aspensis** *Hyatt and Smith, Scythian, Gua Musang, Kelantan,* ×1. *14.* Owenites carpenteri *Smith, Scythian, Gua Musang, Kelantan,* ×1. *15.* Paraceratites trinodosus *(Mojs), Anisian, Kuala Lipis, Pahang,* ×1.

Calcareous Series (Calcareous Formation on his map).

In the Gua Musang–Merapoh areas of south Kelantan and north Pahang, limestone constitutes up to 80% of the Gua Musang formation (Yin in correspondence) and builds prominent, steep-sided hills up to 455 m (1460 ft) high. To the north and south, however, the proportion of limestone rapidly decreases and becomes subordinate to shale. The limestone is variably dolomitic and partly siliceous, sometimes pale in color, sometimes carbonaceous, sometimes banded in pale and dark hues. Some chemical analyses of the limestone are presented in Table 5.1. Typically it is well bedded and apparently a calcilutite or microsparite (recrystallized micrite), although more coarsely crystalline rock and shelly limestone are not infrequent (Richardson 1950, MacDonald 1968).

The Gua Musang formation as a whole is predominantly argillaceous, but the argillaceous rocks are often poorly exposed because their resistance to weathering is relatively low compared with that of the cliff-forming limestone. The argillite is often black, not infrequently calcareous, and often laminated. Arenaceous rocks are of minor importance.

Volcanic material is commonly interbedded with the argillaceous rocks, less commonly with the limestone. From Richardson's (1950) account it seems that the volcanic rocks of the Gua Musang formation are mainly tuffs of trachytic, trachyandesitic, and andesitic composition. Rhyolitic volcanic rocks also occur.

Yin (1965a), described complex folding, which made it difficult to erect subdivisions in the Gua Musang formation. Nevertheless, Yin (1965b) was able to give the following details of the succession:

Age	Description
Middle Triassic (Anisian)	Predominantly limestone with shale and volcanic beds contain ammonoids and conodonts.
Lower Triassic (Scythian)	Predominantly argillaceous limestone with shale and volcanic beds containing *Claraia*.
Upper Permian—	Shale and siltstone with strong development of tuff, lava, and lesser agglomerate. Isolated limestone lenses containing *Leptodus* shelly fauna.
Middle Permian—	Predominantly limestone (with fusulinids) and minor shale beds.

The Anisian part of this succession is here excluded from the Gua Musang formation because it is typically associated with the *Myophoria* biofacies (see below).

Ichikawa and Yin (1966, p. 101) remarked on the strong deformation of the Lower Triassic strata

TABLE 5.1 *Chemical Analyses of Triassic limestone (weight percent)*

Specimen	Gua Musang formation			Kodiang limestone	
	1	2	3	4	5
SiO_2	13.80+	0.12+	0.44+	2.60+	tr
Al_2O_3	1.16	tr	tr	0.86	0.43
FeO, Fe_2O_3					0.11
MgO	0.55	0.27	2.03	4.99	0.75
CaO	46.40	55.60	53.12	48.40	55.20
CO_2	37.01	43.94	43.90	42.80	43.56
Total	98.92	99.93	99.49	99.65	100.05

+ insoluble residue

Specimens 1–5 as follows:
1. Carbonaceous limestone, Sungei Merapoh, Pahang, (081076, p. 83).
2. Limestone, Merapoh, Pahang, (081013, p. 71).
3. Magnesian limestone, Sungei Merapoh, Pahang, (081006, p. 69).
4. Dolomitic limestone, Kodiang Quarry, Kedah, (081032, p. 75).
5. Limestone, Bukit Kechil, Kodiang, Kedah, (081070, p. 81).

Numbers and page references in parentheses refer to Alexander et al. (1964).

near Gua Musang. Northward the metamorphic grade increases, and 32 km (20 miles) north of Gua Musang, the rocks are intruded by the Stong Migmatite Complex.

Gunong Pulai member

In mapping the Jurong formation of south Johore, Burton (in prep.) recognized a lower, volcanic Gunong Pulai member and an upper, clastic Bukit Resam member. The former is clearly intruded by hornblende-bearing adamellite, which radiometric evidence suggests to be of early Triassic age. Similar volcanic rocks have been seen at various places throughout Johore State, but we do not know to what extent these can be assigned to the Lower Triassic, since almost identical volcanic rocks appear also to postdate the granite. On the island of Nanas in the Johore Strait, such rocks carry granite clasts that are indistinguishable from the granite outcropping in south Johore. Moreover, similar volcanic rocks in the Kerdau formation of central Pahang are almost certainly of Middle Triassic age.

The principal rock type of the Gunong Pulai member is a tuff of rhyodacitic composition. Agglomerate, tuff, and lava of dacitic to andesitic character are also represented, and minor amounts of clastic material are intercalated.

The rhyodacite tuff is essentially a crystal-lithic tuff. When fresh it is a dark green or greenish-grey material mottled with coarse white feldspars, glass-clear quartz, and variously colored lithic fragments. The feldspar comprises both andesine and orthoclase and, in addition to quartz, epidote and hornblende are common. Flinter (personal communication) has suggested that this rock is welded tuff, at least in part, since a flow structure is evident and the crystals (particularly those of quartz) are frequently corroded.

The presence of hornblende links these rhyodacitic volcanic rocks with the hornblende-bearing adamellite that intrudes them. The feldspars are also of similar composition in the two rock types. This is particularly evident in the Ulu Endau area of north Johore. Details of the chemistry of these rocks are given in Table 5.2. In view of the chemical and mineralogical similarity of the rhyodacite tuff and the hornblende-bearing adamellite, the two rocks are possibly not far removed in time. Moreover, there is no recognizable break in the field between the preadamellite Gunong Pulai member and the apparently postadamellite Upper Triassic

TABLE 5.2 *Comparison of chemical analyses* and norms of the granodiorite (GS 25929) and the rhyodacite (GS 25930) from Ulu Endau, Johore.*

	Analysis weight %			Niggli molecular norm	
	Granodiorite	Rhyodacite		Granodiorite	Rhyodacite
SiO_2	71.7	72.3	Quartz	32.27	31.33
Al_2O_3	14.8	13.2	Corundum	4.02	1.38
TiO_2	0.31	0.16	Orthoclase	25.20	28.18
Fe_2O_3	0.60	1.33	Plagioclase	33.48	33.26
FeO	2.78	2.48	Ab_xAn_{100-x}	$Ab_{87}An_{13}$	$Ab_{82}An_{18}$
MnO	0.07	0.07	Magnetite	0.65	1.43
MgO	Trace	0.37	Ilmenite	0.44	0.23
CaO	1.01	1.28	Apatite	0.15	0.06
Na_2O	3.14	2.96	Calcite	0.10	0.18
K_2O	4.14	4.63	Hypersthene	3.67	3.94
H_2O+	1.34	0.79	$En_x Fs_{100-x}$	En_0Fs_{100}	$En_{27}Fs_{73}$
H_2O-	0.26	0.35	Total salic	94.98	94.15
P_2O_5	0.07	0.03	Total femic	5.02	5.85
CO_2	0.04	0.07	Quartz + albite + orthoclase	86.52	86.89
Total	100.26	100.02	QTZ:AB:OR	37:34:29	36:32:32

* Both analyses are of composite specimens. The analyses were performed by Leong Pak Cheong (Alexander et al. 1964 pp. 22–23). The Niggli norms were computed on the University of Malaya IBM 1130 computer using the program of Hutchison and Jeacocke (1971).

Bukit Resam member. In addition, volcanic rocks similar to the Gunong Pulai members seem to form part of the Bukit Resam clastic member. Hence a Lower Triassic age is suggested for the Gunong Pulai volcanic member.

Synthesis

During the Permian age, much of Southeast Asia was covered by an ocean in which carbonate deposition was prominent. The deposition was apparently centered in Thailand, extending therefrom into eastern Burma, southern China, Indochina, and the northwest quadrant of Malaya. In Malaya this carbonate facies extended across the miogeosyncline in the western part of the peninsula to the eugeosyncline in the medium part, although the eugeosynclinal zone can still be distinguished by the occurrence of volcanic rocks intercalated with the limestone in Kelantan and Pahang. Farther east, in Trengganu, and east Johore, the place of limestone is largely taken by black shale containing some plant remains, evidently indicating shoaling of the Permian sea.

Paton (1960) noted that the Permian strata of Trengganu become more coarsely clastic upward. Thus since there is no evidence of Triassic rocks in Trengganu, east Pahang, nor east Johore, and since part of the Trengganu granite has been dated as Permian, Paton deduced a Permian uplift for this part of the country. This possibly marks the site of a geanticlinal ridge (Figure 5.1). Further evidence is afforded by the difference of the granites. In the region of the proposed geanticlinal ridge, the large batholith of the Kapal and Maras–Jerong granite (MacDonald 1968) is more acid in nature and is strongly associated with tin mineralization, whereas the batholiths of the Boundary Range and the Kemachang granite (MacDonald 1968), which occupy the region of the proposed eugeosynclinal furrow, are less acid and are not associated with tin mineralization.

Conditions established in the Permian period appear to have continued into early Triassic time, at least in the eugeosyncline; no Lower Triassic rocks, either sedimentary or igneous, are known from the miogeosyncline. In south Kelantan and northwest Pahang, massive limestone, apparently built largely of micrite, continues from the Permian into the Triassic period and suggests relative tectonic stability (the Gua Musang formation). Volcanic activity occurred from time to time, however, as indicated by volcanic layers intercalated in the shales.

The general paucity of volcanic material within the limestone itself suggests that volcanicity inhibited limestone formation. There is some evidence to suggest that the limestone sea of the Lower Triassic shoaled toward the east over what may be termed the east Malayan geanticline.

From the radiometric age determinations presently available (Snelling 1965, and personal communication; Anon. 1966) it seems that there was considerable plutonic igneous activity in the Lower Triassic or late Upper Paleozoic in the Malay Peninsula. Evidently the extensive granite batholiths that lie to the east of the Main Range—the Boundary Range Granite and the granites of south Johore and Singapore—were emplaced in the region of the eugeosynclinal furrow at this time, as were the batholiths of the Lawit granite, Kapal granite, and the Maras–Jerong granite, in the region of the eugeanticlinal ridge. Some granite may also have been intruded in the region of the miogeanticlinal ridge at this time.

Potassium–argon radiometric determinations indicate that the Taku Schist underwent dynamothermal metamorphism to the almandine-amphibolite facies at this time in deeply buried levels within the eugeosynclinal furrow. Some migmatization occurred with the Taku Schist, forming localized interfoliated granite, which has also yielded a lower Triassic radiometric age. In the north of the schist outcrop is a large body of essentially gneissic granite, the Kemachang granite (MacDonald 1968), which is intimately associated with the Taku Schist. Thus there appears to have been a distinct metamorphic episode in the Lower Triassic age, characterized by migmatization as well as granite intrusion. The occurrence of a foliated granite at Kupang, near Baling in Kedah, may also indicate that Lower Triassic plutonism occurred within the miogeosyncline.

The Stong Migmatite Complex represents deep-seated conditions similar to those in which the Taku Schist was metamorphosed. No radiometric date is available, but it would be reasonable to correlate the Stong Complex and the Taku Schist both to the same tectonic level and to the same age of metamorphism. The extensive migmatite terrane of the Stong Complex (MacDonald 1968), considered with the common occurrence of ptymatic folding, indicates that orogeny accompanied the late Upper Paleozoic to Lower Triassic plutonic activity.

The Lower Triassic therefore represents a period of orogenic activity, and it is not then surprising that there is a general absence of Lower Triassic strata in the Malay Peninsula. Apart from the Gua Musang formation, Lower Triassic beds are missing in Malaya; indeed in the miogeosyncline much of the Permian, all the Scythian, and possibly all the Anisian are absent also. The occurrence of Lower Triassic sedimentary rocks may be summarized as follows: unknown in Thailand; recorded at one locality only in Burma (Pascoe 1959, p. 902); unproven in west Yunnan (Pascoe 1959, p. 905); of restricted occurrence in Indochina (Saurin 1956); unestablished in Bangka, Borneo (Van Bemmelen 1949, Marks 1956) and in Billiton (Adam 1960); and noted at only one place in Sumatra (Marks 1956, p. 197).

It must be clearly stated, however, that the Lower Triassic strata that are known in Malaya do not reveal any lithofacies indications of contemporary or impending orogeny. Thus, in the vicinity of Gua Musang, Kelantan, Igo et al. (1965, p. 7) declare that

> No evidence of any major stratigraphic break is noted and it is therefore assumed that sedimentation went on more or less continuously from the late Middle Permian to the Anisian.

Similarly, in south Johore, no proof of unconformity was found between the Gunong Pulai volcanic member and the Bukit Resam clastic member of the Jurong formation, although the former is intruded by granite and the latter appears to postdate in granite (Burton, in prep.). Conceivably, only these strata, which were little affected by the Lower Triassic earth movements, were preserved. Far-reaching changes seem to have been effected at this time for, apart from the emplacement of vast quantities of granite, the Malayan area seems to have reverted from the shelf like condition of Permian and earliest Triassic time to a more diversified geosynclinal condition in the Middle and Upper Triassic age.

MIDDLE AND UPPER TRIASSIC

Occurrence and Nomenclature

Middle and Upper Triassic sedimentary rocks are widespread in axial Malaya; they extend in a broad tract of country, between the Main Range and the Kelantan—Trengganu Boundary Range, from Kelantan through Pahang State. Southward extensions of sedimentary belt are to be found in Johore and in Singapore. A second zone of Middle and Upper Triassic strata lies in northwest Malaya in the states of Kedah and Perak. The Main Range Granite, which is situated between these two tracts, appears generally to have been emplaced in Middle to Upper Triassic time.

The Triassic rocks to the east of the Main Range were placed by Alexander (1959) into his "Lipis Group," which he described briefly as ". . . a characteristically argillo-arenaceous sequence . . . including an argillo-calcareous formation of late Triassic age." This terminology was merely cursory, since no attempt was made to define the extent of this rock unit and no mention was made of component formations. Nevertheless, the name is retained here to cover a number of clastic formations in axial Malaya which succeed the calcareous Gua Musang formation and its equivalents.

Consequent upon the discovery of numerous fossils of the pelecypod *Myophoria*, the strata at Kuala Lipis were named "Myophorian Sandstone" by Newton (1900). In 1911 Scrivenor allotted most of the Lipis Group to the Raub Series of Carboniferous to Permian age, whereas the Myophorian Sandstone was placed in the Gondwana rocks dated as Upper Triassic to ?Middle Jurassic. A similar convention was adopted on the geological map of Malaya included in Scrivenor (1931), where much of the Lipis Group was allotted to the Carboniferous (and Permian?) and the remainder was assigned to the Triassic (and Permian?).

For the next quarter century it became the practice in the Malayan geological Survey ". . . where no palaeontological evidence can be found, to assume . . . that the calcareous beds are Carboniferous and Permocarboniferous, and that the arenaceous beds are Triassic. However, argillaceous beds may belong to either system" (Ingham 1938, p. 9). Thus most of this group, with its low content of calcareous rocks, was put into the Triassic and a smaller portion was placed in the Carboniferous and Permian on Ingham's geological map of Malaya (Ingham 1948).

When it became apparent that there were two separate arenaceous developments in Malaya, it was to the Younger Arenaceous Series or the Newer Arenaceous Series that the rocks of the Lipis Group were referred (Richardson 1950). Jones et al. (1966, p. 335) recognized that two major biofacies are represented in much of the Triassic of Malaya:

> One consists of a shallow-water lamellibranch fauna typified by *Myophoria*. The other is a lamellibranch fauna almost entirely confined to the genera *Daonella*, *Halobia* and *Posidonia*.

These biofacies approximate to the earlier lithological grouping of Arenaceous Series (*Myophoria*) and Calcareous Series (*Daonella*), although calcareous rocks are actually of very restricted occurrence in the latter. In this chapter, a lithostratigraphic subdivision is preferred. Within the Lipis Group we recognize the Jelai formation in Pahang and Kelantan and the Jurong formation in South Malaya, both of *Myophoria* biofacies, and also the Kerdau formation in Pahang and the Gunong Rabong formation in Kelantan, both of *Daonella* biofacies (Table 5.3).

Axial Malayan Province

JELAI FORMATION. In accordance with modern stratigraphic usage, Newton's (1900) Myophorian Sandstone is here renamed (informally) the Jelai formation and redefined in lithostratigraphic terms; the characteristic thick-shelled pelecypods are regarded merely as physical criteria (Article 4c of the *American Code of Stratigraphic nomenclature*, 1961). The name is taken from the Sungei Jelai, which flows through the town of Kuala Lipis.

The outcrop distribution of the Jelai formation is shown in Figure 5.2.

Published descriptions of the Jelai formation refer only to the fossil localities, and it is therefore difficult to form a balanced lithological picture. The formation is probably mainly argillaceous, the arenaceous component being accentuated in outcrop by its superior resistance to erosion. It comprises shale and mudstone, generally of a dark color and sometimes sandy. Intercalated with the argillaceous rocks are beds of immature sandstone of varying grain size. This sandstone is often termed "argillaceous" (e.g., Service 1949, p. 19); presumably, the description refers to weathered rocks wherein the labile constituents have been converted to clay minerals. Minor limestone is reported (Procter, in correspondence), and pyroclastic material is widespread; andesite occurs at Jerantut (Edwards 1933).

Collections of shelly faunas have been made from the shale, mudstone, and sandstone (Plate 11). Examination of these suggests that deposition may have begun as early as the Anisian in north Pahang whence Tokuyama (1961) described two collections containing *Myophoria goldfussi lipisensis* ssp. nov. and *Neoschizodus laevigatus elongatus* Philippi.

On the strength of these Middle Triassic fossils, Tokuyama erected a separate "Lower Myophorian Sandstone." The localities from which this fauna was taken were numbered 189 (Ulu Sungei Kenong, northeast of Kuala Lipis) and 195 [5 km (3 miles) south of Kuala Lipis] in the comprehensive list of Malayan fossils compiled by Jones et al. (1966). The authors quoted Tokuyama as assigning an Anisian age to the above-mentioned fossils, although he actually inclined more toward a Ladinian age. However, Cox (*in* Service 1949, p. 19) regarded locality 195 as Upper Triassic, whereas a collection 4 km (2.5 miles) away from locality 189 was considered to be Upper Triassic (Cox, *in* Jones et al. 1966, pp. 336–369).

Kobayashi et al. (1966, p. 100) indicated, however, that part of the formation must be Anisian, their opinion was based on the recent discovery of *Paraceratites* cf. *trinodosus* farther north at Chegar Perah. Two other fossil collections from the Jelai formation are thought to be Middle Triassic, both located in the southern outcrop, in the Triang area. From one of these, Howarth (*in* Jones et al. 1966, p. 343) recognized the ammonite genus *Analcites*, together with plant remains; at another, Cox identified *Posidonia* sp.?, *Parallelodon* sp., *Cucullaea* sp., small turriculate gastropods, and conchostracan remains (*in* Jones et al. 1966, p. 324).

Although most of the fossils in the Jelai formation have been assigned to the Upper Triassic (complete faunal lists are given in Jones et al. 1966, pp. 336–368), it is doubtful whether the Rhaetian stage is present. Newton (1900) referred the first collection from the Kuala Lipis area to the Rhaetian because of the occurrence of *Chlamys valoniensis* Defrance, a characteristic Rhaetian species in Europe. Newton later identified another form which he referred to *Pleurophorus elongatus*? with *Modiolopsis gonoides* Healey—originally described from the Rhaetian Napeng Beds of Burma. Weir (1925) maintained that the age of this horizon was Prerhaetian because of the absence of *Avicula contorta* and the uncertainty of the identification of *Modiolopsis gonoides*.

Subsequently Cox (in correspondence 1940) described a *Palaeoneilo* from the Jelai formation near Kuala Lipis that was comparable with *Palaeoneilo curvirostris* Healy from the Burmese Rhaetian. A *Modiolus*, also comparable to a Burmese Rhaetian form, was identified by R. P. S. Jefferies (in correspondence 1960), who further reported that a collection from a locality east of Temerloh included the species *Entolium quotidianum, Palaeocardita singularis,* and *Nucula peelii*, which are again found in the Napeng Beds, together with *Nucula perlonga*, which is known in the Rhaetian of Tonkin. Recently however, Kobayashi and Tamura (1968a)

TABLE 5.3 *Correlation of the Mesozoic Formations.*

Period	Stage	Northwest Malaya	North — Axial Malaya — South	South Malaya
CRETACEOUS	Upper		Gagau Group	Tebak formation
	Lower			
JURASSIC	Upper	Saiong beds	Tembeling Formation	
	Middle			
	Lower			
TRIASSIC	Rhaetian	Semanggol formation (Conglomerate member, Rhythmite member, Chert member)	Murau Conglomerate	Murau Conglomerate
	Norian		Gunong Rabong Formation	Pasir Panjang member
	Carnian	Kodiang limestone		Jurong formation
	Ladinian		Jelai formation	Bukit Resam member
	Anisian			
	Scythian	Upper Paleozoic	Gua Musang formation	Gunong Pulai member
			Paleozoic	Upper Paleozoic

revised the myophoriids and concluded that those from the Jelai formation were of Middle Triassic age.

The moderately coarse clastic nature of the Jelai formation, together with its abundant thick-shelled benthonic fauna, indicates a shallow-water environment. Fossil plants from a locality halfway between Temerloh and Triang suggest that this shallow water was fairly near to the shore, although current-oriented shells of *Neoschizodus laevigatus* at Ulu Kenong, northeast of Kuala Lipis, were considered by Tokuyama (1961) to indicate neritic rather than littoral or paralic conditions. Tokuyama's conclusion is perhaps supported by the apparent absence of rudaceous rocks.

Some volcanic rocks are interbedded in the Jelai formation, and to the south of Kuala Lipis the Jelai formation appears to be interdigitated with black shale of the Kerdau formation (see below). It seems, therefore, that the Jelai formation was deposited in shallow offshore water, possibly over limited shoals or bars, within the eugeosynclinal basin or basins of central Malaya.

The fauna of the Jelai formation has been compared with that of the Alps, particularly with that of the St. Cassian beds at the top of the Ladinian (Weir 1925, Cox 1936, Cox *in* Savage 1950). The supposed Rhaetian element of the fauna has been likened to that of the Napeng Beds of Burma (Newton 1923, Cox 1936, Jefferies in correspondence) and of Tonkin (Jefferies in correspondence). Other components of the Jelai fauna have been said to resemble fossils from Yunnan (Tokuyama 1961, Jefferies in correspondence), Transjordan (Cox 1936, Tokuyama 1961) and Indochina (Cox 1936, Jefferies in correspondence), implying continuous, well-aerated shallow-water connections along the Tethys and some of its Asian ramifications in Middle and Upper Triassic time.

JURONG FORMATION. Because of poor exposures, no suitable type section has been found for a sequence of clastic sediments and volcanic rocks that occurs in south Johore (Burton, in prep.). These rocks are better exposed farther southeast (in the western part of Singapore), where a number of coastal sections and artificial cuttings provide more information relating to their lithology, structure, and age than is available in south Johore. In particular, several fossil discoveries have been made near Jurong in southwest Singapore, and it is thought that this vicinity is best suited to furnish a type locality for these rocks, which were therefore referred to informally as the Jurong formation.

In south Johore, it was possible to divide the Jurong formation into a lower Gunong Pulai member and an upper Bukit Resam member. There is no evident unconformity between the two members. However, the Gunong Pulai member is clearly intruded by granite that is evidently of Lower Triassic age and, although the field relationships of the Bukit Resam member with the granite are obscure, on Singapore Island this member contains an Upper Triassic pelecypod fauna. It thus appears that the Bukit Resam member is quite discrete from the Gunong Pulai member and is separated from it by a major plutonic episode.

Alexander (1950) divided the sedimentary rocks of Singapore into three units. Unfortunately, the boundaries of their outcrops were not shown on her map, and correlation with the observations of other workers is difficult. The oldest of her divisions is a strongly contorted mica schist which is here excluded from the Bukit Resam member because of its high degree of metamorphism; it is apparently separated by an angular unconformity from the next division. This is the argillaceous series, composed mainly of shale with beds of sandstone of various grades, tuff, and lava (including spilitic pillow lava) associated with chert. Succeeding this is the arenaceous series composed of sandstone and conglomerate with shale, some coal seams, and containing numerous plant remains.

Chin Fatt (1965) recognized two rock units in the Singapore sedimentary rocks. The lower unit (which, unfortunately he also termed the Jurong formation) was underlain by volcanic breccia (cf. the Gunong Pulai member). It consisted of thinly interbedded immature sandstone and mudstone, both dominantly grey-black and apparently lacking conglomerate.

Above this was the Pasir Panjang formation, comprising rhythmic alternations of conglomerate-sandstone-mudstone, characterized by a predominant red color. Chin Fatt's (1965) lower division seems to correspond approximately to Alexander's (1950) argillaceous series, and this in turn corresponds to the Bukit Resam member of Johore. Similarly, the Pasir Panjang formation appears to be more or less the equivalent of Alexander's (1950) arenaceous series, which is possibly only moderately developed or missing in south Johore.

Since there is no record of a mappable stratigraphic break between these two rock units (Alexander 1950) and since, according to Chin Fatt, both contain Upper Triassic fossils, they are here included together in one formation, the upper unit being separately identified as the Pasir Panjang member of the Jurong formation. At the same time it is recognized that in the future it may prove necessary to follow Chin Fatt and to treat the Pasir Panjang member as a separate formation.

The name of the Bukit Resam member is taken from the hill called Bukit Resam in south Johore. This rock unit covers much of the western part of Singapore and extends northwestward through south Johore to north Johore, whence it may pass into the Jelai formation in south Pahang.

BUKIT RESAM MEMBER. The Bukit Resam member of the Jurong formation consists of frequent alternations of shale and sandstone, with minor siltstone and conglomerate with a little tuff and minor lava. Individual beds range in thickness from a few millimeters to a few tens of meters, but they are usually thin. Rapid variation in the thickness of strata is not unusual, and some beds are demonstrably lenticular.

Shale is the most common rock type, normally carbonaceous and of a dark grey color; 6.2% carbon is reported in one instance (Chin Fatt 1965). Some purple and red shale seen in south Johore may represent a limited development of the Pasir Panjang member. In some shale exposures, rounded bodies of chert of pebble to boulder grade are seen. These are probably autochthonous concretions rather than large clastic grains, since an occurrence in Singapore was found to be associated with silicified fossil gastropods. There is a graduation from shale through silty shale into siltstone. One very thinly bedded variety of siltstone comprises beds a few millimeters to a few centimeters thick, in alternating reddish-purple and pale yellow-grey colors. It has yielded fossil plants near the village of Lima Kedai in south Johore. It is conceivable, however, that this rock should also belong to the Pasir Panjang member, which is dominantly red, and in which silty beds have proved to contain plant remains on Singapore Island (Alexander 1950, p. 12).

The sandstone of the Bukit Resam member is strongly weathered in outcrop, and the labile constituents are broken down to produce the argillaceous sandstone of Newton (1923). It is clearly immature in character, and in Johore it apparently varies within the range protoquartzite-subgreywacke-greywacke. In Singapore, Chin Fatt (1965) reported that the sandstone is generally grey to black and poorly sorted, with 17 to 18% of clay matrix. It carries fragments of chert and metamorphic rock. This rock includes fossils (*Gonodon* sp. and *Posidonia* sp.), whereas adjacent lutites seem to be barren of organic remains. Bukit Resam sandstone, in Johore as well as in Singapore, shows small-scale cross-bedding.

Conglomerate is reported to be absent from the Bukit Resam member in Singapore (Chin Fatt 1965). In south Johore, the occurrence of conglomerate is only limited, but again it is conceivable that those conglomerate occurrences actually represent a small development of the Pasir Panjang member.

To the northwest of Johore Bahru, the clastic rocks are intercalated with tuff, generally of acidic composition; to the south, pyroclastic grains are often admixed in the detrital material. Alexander (1950, p. 12) mentioned three occurrences of volcanic rock in Singapore, including spilitic pillow lava associated with chert and some tuff. An andesitic breccia was recorded by Chin Fatt at the base of the Bukit Resam member, and boulders of spilite were seen at a higher level.

PASIR PANJANG MEMBER. The Pasir Panjang member of the Jurong formation is named from a locality on the south coast of Singapore. It is also seen farther north, where it occurs together with the Bukit Resam member in the Jurong area. As noted earlier, occasional red and purple shale, very thinly bedded red and yellow siltstone, and scattered beds of conglomerate in south Johore may also belong to this rock unit. The Pasir Panjang member consists of conglomerate, sandstone, and lutite; these materials are repeatedly intercalated, possibly in a rhythmic manner, and are dominantly red.

The conglomerate appears to be polymict, containing phenoclasts of sandstone, metaquartzite, mudstone, and rhyolitic volcanic rock. Alexander (1950, p. 12) found pebbles of micaceous schist and noted that these are identical in appearance to the mica schist that occurs *in situ* elsewhere in Singapore. Scrivenor (1931, p. 49) observed pebbles of schorl rock in conglomerate on the offshore island of Belakang Mati, and he noted that sheared granite forms phenoclasts in similar rock on the nearby Indonesian island of Sambo.

Sandstone in this member has been described as "muddy fine to medium sandstones or siliceous immature subgreywacke" (Chin Fatt 1965). It contains chert, metamorphic rock fragments, and biotite. The lutite comprises shale, mudstone, and some siltstone, often of reddish-brown color. Alexander (1950, p. 12) also reported ". . . occasional tiny seams of coal" in this member, and Scrivenor (1926) found a 6-in. seam of coal at Mount Guthrie, Singapore.

A number of fossil localities have been found on Singapore Island, but it is not always certain to which member of the Jurong formation they belong. As noted previously, Chin Fatt found *Posidonia* and *Gonodon* in sandstone beds in the Bukit Resam member. He also discovered an abundant thick-shelled pelecypod fauna, characterized by *Myophoria*, in a black mudstone band at the top of this rock unit. Chin Fatt also claimed that an earlier collection of similar biofacies from Morse Road, Mount Faber, lay in the outcrop of the Pasir Panjang member, but there is doubt about this.

There seems to be some structural dislocation here for, as Scrivenor (1931, pp. 67, 106–107) pointed out, this Upper Triassic locality is apparently structurally above the Middle Jurassic locality at Mount Guthrie. According to Newton (1923), the Mount Faber collection included *Promathildia colon*, *Lopha* cf. *montis-caprilis*, *Chlamys valoniensis*, *Spondylus dubiosus*, *Prospondylus comtus*, *Gervillia scrivenori* (sp. nov.), *Cassianella* cf. *tenuistria*, *Palaeocardita* cf. *crenata*, *Myophoria ornata*, *Myophoria* cf. *goldfussi*, *Myophoria bittneri*, *Modiolopsis gonoides*, and *Spiriferina* cf. *fragilis*. Newton (1923) thought that these may be considered Upper Triassic or Rhaetic. The myophoriids were revised by Kobayashi and Tamura (1968a), who concluded that they were Carnian or Norian in age.

F. E. S. Alexander also made some fossil collections in Singapore which were examined by Cox (written communication 1952). These identifications are omitted from the otherwise complete survey of the fossil record of Malaya and Singapore made by Jones et al. (1966). The most prolific of these localities is near Huat Choe village in the Jurong area. From the dark grey shale exposed at the fossil locality, it seems likely that this belongs to the Bukit Resam member. Here Cox reported: two species of "*Arca*," *Modiolus* cf. *nachamensis*, three species of "*Gervillia*," *Cassianella* sp., two species of "*Pteria*," *Posidonia* sp., *three species of "Pecten,"* three species of "*Lima*," *Myophoria* cf. *goldfussi*, *Myophoria inaequicostata*, *Trigonia* cf. *zlambachensis*, *Myophoriopsis* cf. *carinata*, *Myophoriopsis* sp., *Trigonodus* sp., *Pachycardia?* sp., *Myoconcha* sp., *Lucina* sp., *Prolaria* sp., *Pleuromya* sp., and *Cuspidaria* sp. Cox concludes that the age of this assemblage is Upper Triassic, as shown by the occurrence of the two species of *Myophoria*. Another locality in Pulau Ayer Chawan yielded *Estheria mangalensis* Jones, which indicates an Upper Triassic age.

Working 80 km (50 miles) along the strike to the northwest, immediately south of Yong Peng in central Johore, Burton discovered ammonites and pelecypods in rocks similar in aspect to the Bukit Resam member. According to Sato (1963) the ammonites were Carnian in age, comprising:

Trachyceras (Paratrachyceras) cf. *regelodanum*,
Trachyceratidae gen. st. sp. indet.
Orestites sp.

The foregoing fossil determinations indicate a Late Triassic age for much of the Bukit Resam member. Less chronologic information is available concerning the Pasir Panjang member. Alexander (1950 p. 13) and Chin Fatt (1965) agree that a fossil locality at Mount Guthrie, found by Scrivenor and Hanitsch in 1905, is in the Pasir Panjang member. This area included the following forms (Newton 1906):

Cucullaea scrivenori sp. nov.
Gervillia hanitschi sp. nov.
Volsella cf. *compressa*.
Astarte scrivenori sp. nov.
Astarte guthriensis sp. nov.
Goniomya scrivenori sp. nov.
Goniomya singaporensis sp. nov.
Podozamites cf. *lanceolatus*.

Newton considered that these were possibly of Middle Jurassic age, at about the horizon of the Bajocian or Inferior Oolite. This dating has never been confirmed, and geologists in Malaya have been inclined to doubt its validity (e.g., Scrivenor 1908), but as Jones et al. (1966, p. 346) have pointed out, the Mount Guthrie fossils are quite different from any of the Triassic fossils so far described from Malaya. Kobayashi and Tamura (1968b, p. 142) suggested that the fauna was probably Lower Jurassic.

Further fossil plant collections from the Pasir Panjang member await expert examination.

The repeated intercalation of lutite and immature arenite in the Bukit Resam member, together with the general absence of fossils in the lutite and the fine cross-bedding in the arenite, link much of this member with flysch, altogether there is insufficient evidence to assign it positively to this facies. In particular there is no proof of turbidity currents. The occurrence in the sandy horizons of *Posidonia*, believed to have been nektoplanktonic in habit (Jefferies and Minton 1965), is also reminiscent of the flysch of the Semanggol Formation in the northwest and, to a lesser extent, the Gunong Rabong formation in the north and part of the Kerdau formation in central Malaya. Alignment of these shells and their concave-downward arrangement suggested a neritic environment to Chin Fatt (1965). Conditions were probably very different when the interlayered lutite was formed. The dominance of black lutite seems to indicate restricted circulation and toxic bottom waters hostile to organic life, and indeed, life was confined to the upper waters.

Most of the fossils in the Bukit Resam member, however, are of shallow-water benthonic forms. Chin Fatt recorded these at the very top of the member, where they may point to a change in conditions; but there is no evidence that all the occurrences here of the *Myophoria* biofacies have this stratigraphic position. Although this fauna may be somewhat reworked (Chin Fatt found it to be randomly oriented in black mudstone), it does indicate shallow-water, well-aerated conditions in some part of the Bukit Resam depositional province.

Newton (1923) noted the similarity, both in lithology and paleontology of the Morse Road, Mount Faber, locality with the "Myophorian Sandstone" of Pahang (the Jelai formation) more than 320 km (200 miles) to the north. Newton also pointed out the strong affinities of the fossils to the St. Cassian beds of the Austrian Tyrol.

The Pasir Panjang member clearly represents a different set of conditions. Of the fossil locality at Mount Guthrie, Newton (1906, p. 487) remarked:

The association of land and marine organisms would at once suggest an estuarine or lagoon origin for the beds containing them, more especially as the mollusca belong to genera of families which may be regarded as of shallow-water habitat, whilst the plant remains might be accounted for by the close proximity of land or the transporting agency of river action.

Further evidence of shallow water is provided by Scrivenor's (1926) discovery of a (15-cm) seam of coal near this locality. Alexander (1950) also reported occasional tiny seams of coal in this member and further noted that some of the conglomerates on Pulau Senang looked like beach deposits.

The incidence of cross-bedded sandstone, polymict conglomerate, plant fossils, and coal beds, togther with the possibility of beach deposits and of original red coloration, all combine to suggest that the Pasir Panjang member is of late-tectonic molasse facies (Pettijohn 1957, pp. 618–622).

Alexander (1950, p. 12) concluded from variation in pebble size that the Pasir Panjang conglomerate was derived from land lying to the south or southeast. From van Bemmelen's (1949, pp. 304, 308–309, 314) account, however, it is evident that similar rocks also occur to the south in the Riouw–Lingga Archipelago. It is of interest to note Scrivenor's (1931) report of pebbles of schorl rock in Bukit Resam conglomerate on Singapore and of sheared granite on the Indonesian island of Sambo in the Riouw Archipelago. They indicate that the Pasir Panjang member postdates a phase of granite emplacement, which would be in accord with its molasse character and with the widespread occurrence of late Upper Paleozoic to Lower Triassic granite in the neighborhood.

KERDAU FORMATION. The Kerdau formation is named from a village on the railway line 11 km (7 miles) north of Mentakab on the Temerloh district of central Pahang. A fossil locality (locality 6 in Kobayashi 1963a) in the Sungei Cheriau to the west of Kerdau station (3°35′N; 102°22′E) might possibly provide a type section. Since this section has not yet been described, the status of the rock unit must remain informal; however, extensive field work has been carried out in the area by Ja'afar bin Ahmad.

Presumably because it includes some limestone, much shale, and relatively little arenaceous rock, the present Kerdau formation was included by Scrivenor (1911a) with the limestone and shale of his Raub Series, of Carboniferous to Permian age. Similarly, on the 1930 and 1948 editions of the geological maps of Malaya (Scrivenor 1931, Ingham 1948), the Kerdau formation outcrop is allotted to the Carboniferous (and Permian) systems, which were then thought to be represented in Malaya only by essentially calcareous and argillaceous strata. Consequent upon a number of recent fossil determinations (Kobayashi 1963a), the type area of the Kerdau formation is correctly shown as Triassic on the 1963 map (Alexander 1965), but the western part, to the east of Karak, is shown as Permian.

The Kerdau formation outcrops along a north-south belt on the west side of the Pahang River (Figure 5.2). It occupies a zone at least 80 km (50 miles) long and 32 km (20 miles) wide. Its eastern margin is unknown and may possibly lie to the east of the river. To the west it is bordered partly by strata of the Foothills Formation (Richardson 1946) and partly by the Benom granite. To the north and south its outcrop is delimited by rocks of the Jelai formation, of which it is a lateral equivalent; and in the north it appears to interdigitate with rocks of the Jelai formation.

Although the present Kerdau formation has previously been grouped with calcareous rock sequences, it contains only a very limited development of limestone. The formation is predominantly of black carbonaceous shale. Normally interbedded with the shale is rhyolitic tuff, which increases in prominence northward, where it becomes locally dominant. The shale is frequently tuffaceous. Minor arenaceous bands occur, which are often also dark in color. These rocks are immature in character and include greywacke. In some cases (as in the southeast of the outcrop) the arenite bands are closely spaced, giving the rock the character of flysch. Rare beds of dacitic tuff, rhyolitic lava, and conglomerate have also been recorded (Ja'afar bin Ahmad in prep.). The limestone seems to be confined to three localities in the western part of the outcrop (Ja'afar bin Ahmad 1965a), although it is tempting to equate this occurrence with the similarly lenticular limestone of unknown relationship which is found at Gunong Sinyum and Gunong Sah some 48 km (30 miles) away and a little to the east of the most easterly known occurrence of the formation.

Courtier, Yin, Burton, and Procter collected numerous thin-shelled lamellibranchs (Plate 11), some ammonites, and crinoids from the Kerdau formation (localities 211 and 218 *in* Jones et al. 1966, p. 341). These fossils were examined by Kobayashi and Sato (Kobayashi 1963a), who reported several species of the bivalve *Daonella*, and the ammonoids *Paraceratites*, of *trinodosus* type, *Arpadites* cf. *cinensis*, and *Arpadites* sp. ex. gr. *A. szaboi*.

They concluded that the fauna represented a time range from Anisian to uppermost Ladinian or lowest Carnian. A nearby locality (224) further yielded the Anisian forms *Paraceratites trinodosus* and *Acrochordiceras* sp.

Ja'afar bin Ahmad (1965a,b, in prep.) has made numerous additional collections farther to the west, all with a similar aspect. Some of these have also been examined by Kobayashi (Kobayashi 1963a, Jones et al. 1966, p. 341—localities 210, 225, 227, 228) and several species of *Daonella* have been recognized. *Posidonia* cf. *japonica* also occurs, and the presence of *Halobia* (reported in Ja'afar bin Ahmad 1965b) may signify a higher stratigraphic level than has been recognized to the east. Indeterminate plant remains were also found at one point.

It seems that the nearby fossil locality at the 91st milestone on the road from Karak to Mentakab (locality 226 *in* Jones et al. 1966) should also be allotted to the Kerdau formation because of the presence of *Halobia* and *Daonella* (Cox written communication 1958). Plant remains were also found here, and Ball (written communication 1958) recorded what he believed were poorly preserved conchostracan remains. In a similar fauna in northwest Malaya, fossils regarded as conchostracan (Jones 1905) were later shown to be small halobiids (Newton 1925, Kobayashi et al. 1966).

To the north are three more fossil localities that are adjacent to or within the northern Jelai formation outcrop, although they appear to represent outposts of the Kerdau formation. The Kerdau fauna can be recognized (Cox *in* Service 1949, p. 19) in argillaceous sandstone in the Sungei Som (locality 201), comprising *Posidonia* cf. *wengensis* and *Daonella* or *Halobia* sp.

Kobayashi has pointed out that a further fossil locality (no. 199) at Lubok Sukom near Kuala Tembeling resembles the Kerdau formation in lithology and carries similar fossils, including *Posidonia* and *Arpadites* (?) of probable Ladinic age (Jones et al. 1966, p. 329). Comparable black shale with siltstone in the Sungei Tua (locality 197) contains the Anisian cephalopods *Paraceratites trinodosus*, *Sturia sansovini*, *Acrochordiceras* sp., and *Ptychites* sp. (Kummel 1960).

Thus the faunal evidence shows that the Kerdau formation is on the whole Anisian to Ladinian in age, possibly extending into the Carnian. Kerdau rocks, however, are said to be metamorphosed near to the margin of the Benom granite (Ja'afar bin Ahmad 1965a, in prep.).

Like the fauna of the Jelai and Jurong formations, the Kerdau fauna has widespread Tethyan affinities. The various species of *Daonella* in the formation have been compared with Tethyan forms in the Alps, the Himalayas, Yunnan, Kueichou, Indochina, and Timor (Kobayashi 1963a, p. 105). The ammonites are also said to be of types widely distributed in the Tethys and the general circum-Pacific region (Kummel 1960). They compare particularly with Alpine forms (Sato 1963).

It appears that the ammonites were the first (Anisian) immigrants to the Kerdau province. This early arrival may have resulted from their free-swimming habit in life or from the capacity of their shells to float for great distances after death, as in the case of the modern *Nautilus* (Hamada 1964, Toriyama et al. 1965).

The ammonites were later joined by a pelecypod fauna characterized by *Daonella*, which was suggested by Jefferies and Minton (1965) as nekto-planktonic in habit. Such a mode of life seems likely in the present case because of the thin-shelled nature of the fauna and because, as implied by the large-scale development of black shale, the bottom water was apparently toxic. On the other hand, Kobayashi and Tokuyama (Kobayashi et al. 1966) wrote that in Malaya and Japan ". . . *Halobia, Daonella* and *Posidonia* are all muddy bottom lovers." Although the molluscs are considered here to be nektoplanktonic, there is some indication of benthonic life in the Kerdau formation.

Kobayashi (1963a) was uncertain of the true nature of the "crinoid stems" that occur here. They correspond, as Kobayashi observed, to similar fossils associated with a similar fauna in similar black shale in the Middle to Upper Triassic Semanggol Formation of northwest Malaya. The latter have subsequently been identified as *Serpulites* (Kobayashi et al. 1966, pp. 104, 119–120). Since they are tubular, it would seem that these worms were benthonic forms. Howell (1957, pp. 812–813) quotes three authors (Parks, Richter, and Ruedemann) to show that some marine worms, particularly *Serpulites*, were able to live in poorly aerated water, and sometimes on and in black muds rich in hydrogen sulfide.

The Kerdau fauna—mainly nektoplanktonic, but displaying a few forms that evidently are particularly tolerant of deoxygenated conditions—might point to deep water, as might the black shale lithology. However, marine worms seem to be virtually exclusively shallow-water dwellers (Howell 1957, p. 805), whereas plant remains imply a near-shore situation, and ripple marks usually suggest shallow water. Furthermore, the rhyolitic lava of the Kerdau formation is sparsely vesicular and has metamorphosed the subjacent sediments. Apparently, then, the lava is subaerial (Ja'afar bin Ahmad in prep.); hence local shoaling of the Kerdau depositional zone occurred.

Outcrops of strata of the shallow-water Jelai formation, deposited under aerated conditions as witnessed by the prolific bottom fauna, occur to the north and south. It may be that these strata represent contemporary shallows or bars that caused stagnation in the intervening Kerdau area and, conceivably, also trapped any terrigenous sediment moving along the length of the geosyncline. In the consequently restricted Kerdau basin, decay of organic matter was largely inhibited; black shale therefore slowly accumulated, and pulses of volcanicity produced tuff and rare lava, possibly associated with local uplift followed by erosion, which resulted in minor developments of coarse clastic sediments. The Kerdau formation locally assumes the character of flysch when immature arenite is closely intercalated with black shale.

GUNONG RABONG FORMATION. Some 145 km (90 miles) to the north of the Kerdau formation outcrop is another development of strata with a similar thin-shelled pelecypod fauna and a strong argillaceous development. This rock unit was first recognized by Yin (1965a,b, *in* Kobayashi et al. 1966) and termed the Gunong Rabong formation after the 1530-m (5049-ft) mountain of that name, which lies 11 km (7 miles) east-southeast of Gua Musang. Good exposures are available on the mountain for selection of a type section.

This rock unit outcrops in south Kelantan to the east of the Middle Permian to Lower Triassic Gua Musang formation. Its boundaries have not been clarified and its full extent is therefore not yet known. Rocks assigned to the Jelai formation occur at Ulu Lakit, 35 km (22 miles) to the northeast of Gunong Rabong (locality 162, *in* Jones et al. 1966). It may therefore be that the Gunong Rabong formation occupies a north-south belt some 24 to 32 km (15 to 20 miles) wide (Figure 5.2). Yin (1956b) believed that the sediments in the Gunong Rabong range extended southward into the Gunong Tahan range, but Koopmans (1966, 1968) demonstrated that the Tahan range is comprised of the Tembeling Formation, which is distinguished from the Gunong Rabong formation by generally coarser clastic character, strong original red coloration, limited development of volcanic rocks, the presence of coal shale, and a sparse but discrete fossil content. The Tembeling Formation may be, in part, the lateral equivalent of the Gunong Rabong formation and the Jelai formation; on the whole, however, it appears to overlie them.

A full lithological description of the Gunong Rabong formation is not yet available. Yin (1965a) noted that in the south this rock unit consists of a thick nonfossiliferous succession of shale, siltstone, immature quartzite, subgreywacke, and some greywacke and conglomerate of the flysch type. Northward, the conglomerate bands disappear and volcanic rocks become more conspicuous. Limestone also occurs and sometimes forms bodies of considerable thickness. Following further field work, and assisted by four fossil determinations, Yin (1965b) was able to divide the Gunong Rabong formation into two members, as yet unnamed:

2. Flysch-type deposits of protoquartzite-shale-subgreywacke.
1. Sandstone and shale with volcanic bands, some conglomerate and isolated developments of limestone.

Yin (*in* Kobayashi et al. 1966) says that the formation is generally only moderately affected by subsequent regional metamorphism and plutonism, although the fossils are somewhat deformed; indeed, one specimen was reduced in height by 50% (Kobayashi et al. 1966, p. 109). This was thought to be an expression of a regional northward increase in intensity of metamorphism. This increase of metamorphism probably is a result of accentuated uplift in north Malaya, exposing to denudation rocks that have been buried at deeper tectonic levels than elsewhere to the south.

Five fossil collections from the Gunong Rabong formation have been studied. Casts from the Sungei Kelubi (locality 165 *in* Jones et al. 1966) were determined by Gerth as possibly *Daonella* (Scrivenor 1931, p. 65). Nearby, on the Sungei Lebak, Yin found (his locality FPhN 4) more fossils, identified by Kobayashi (in correspondence) as *Posidonia* sp. and *Paratrachyceras*? sp., evidently of Carnian age. Then 5 km (3 miles) to the east in the Sungei Chiku (locality FPhN 5) Yin found *Daonella*, apparently of Ladinian age. These two discoveries have been included in one locality (164) in Jones et al. (1966, p. 340). Subsequently two more of Yin's collections were examined by Kobayashi (written communciations) and were found to yield *Posidonia* cf. *kedahensis*.

From the paleontological determinations, it appears that the Gunong Rabong formation is of Ladinian to Carnian age and therefore contemporaneous with much of the Jelai formation, outcrops of which occur to the south (Kuala Lipis) and to the east (Ulu Lakit). The Gunong Rabong formation fossils present a very different aspect from the stout-shelled shallow-water pelecypods of the Jelai formation, but since the descriptions of both rock units are incomplete it is difficult to compare the lithofacies. Yin, who stated that the Gunong Rabong is of flysch facies, has not yet detailed any convincing evidence of this, and his reference to a protoquartzite-shale-subgreywacke association was not indicative of flysch. However, personal discussions have indicated a number of similarities between the Gunong Rabong formation and the Ladinian to ?Norian flysch of the Semanggol formation of northwest Malaya. It appears that amid the shallow-water condition that supported a thick-shelled pelecypod benthonic fauna and was widespread in central Malaya in Middle to Upper Triassic time, another separate basin or subbasin existed in south Kelantan, from which benthos was evidently excluded. A nektoplanktonic pelecypod faunule lived here, as in the Kerdau formation, although the sediments deposited in this region differ from those of the Kerdau formation in having a much higher proportion of coarse clastic material.

West Malayan Province

Another important, although smaller, development of Middle to Upper Triassic sedimentary rocks is found in western Malaya. These strata are generally less metamorphosed than those to the east and lack volcanic intercalations. The most important rock unit in this region is the essentially clastic Semanggol formation, which is largely of flysch facies. In the far northwest this is succeeded by a limited occurrence of limestone, referred to here as the Kodiang limestone.

SEMANGGOL FORMATION. The Semanggol formation is disposed in three outcrops increasing in size from south to north and lying, respectively, in northwest Perak, in south Kedah, and in central to north Kedah, continuing into Thailand, where its extent has not been defined (Figure 5.2).

Because of widespread wrench faulting (Burton 1965, 1967d) outcrops of the Semanggol formation have, in part, rectilinear and sharply angular boundaries. This faulting seems to have brought about the separation of the north and central outcrops, and possibly the southern outcrop was also adjacent to these prior to faulting.

The Semanggol formation is generally bounded on the east by the Main Range Granite batholith and on the west by Paleozoic sedimentary rocks and the Kulim granite. It is clearly intruded by granite, but owing to the strong wrench faulting and only moderately good exposure, its stratigraphic relationships with other rock units are obscure. Its base has not been seen and its top is eroded. To the north of the Sungei Muda in northeast Kedah, the Semanggol formation is uncomformably overlain by the Saiong beds, a sequence of red conglomerate and shale resembling the Jurassic Tembeling Formation (Ong 1968).

The name of this rock unit was introduced by Alexander (1959), who said that his Lipis Group of other parts of Malaya was represented on the west coast by the "Semanggol Formation of Middle Triassic age." He gave no further details on this occasion, nor when he mentioned the name again (Alexander 1962). Courtier (in prep.), who assumed (no doubt correctly) that Alexander referred to the Triassic rocks of Gunong (Putus) Semanggol in north Perak, used the name informally for rocks of similar age and facies in nearby south Kedah. The name was perpetuated for extensions of the same rocks to the north (Burton, unpublished). The name was subsequently used in a publication (Burton 1965), and by the time that a discrete account of this rock unit was written (Burton *in* Kobayashi et al. 1966), Semanggol formation had virtually been established by usage.

The Gunong (Putus) Semanggol locality does not yield exposures suitable for a type section. It is therefore proposed that the type section be located at Bukit Merah, some 8 km (5 miles) to the north along the strike, where good exposures of closely interbedded arenite and argillite with occasional conglomerate bands are seen.

Scrivenor (1911) initially included the rocks of Gunong Semanggol in his Triassic to Jurassic Gondwana rocks, believing that they constituted an extension of the Gondwana System of India. Willbourn (1926) later allotted the Kedah outcrops of the Semanggol formation to a group of quartize and shale with chert, which he also considered to be Triassic to Jurassic in age. When the unconfirmed Jurassic element of the dating had been eliminated, Willbourn's grouping was incorporated on the 1930 (Scrivenor 1931) and 1948 (Ingham 1948) geological maps of Malaya. Recent work has shown that much of this so-called Triassic is actually part of the Silurian–Devonian Mahang Formation and of the mainly Carboniferous Kampong Sena Formation (now equated with the Kubang Pasu Formation—see Chapter 4). These errors were partly amended on the 1963 map (Alexander 1965), but a new flaw was introduced in that part of the Semanggol outcrop was wrongly excised from the Triassic and allotted to the Permian System.

The Semanggol formation is affected by repeated and rather tight folding and the succession is consequently difficult to elucidate. The predominance of eastward or southeastward dips, however, points to stratigraphic younging in that direction, and Kobayashi (1963b, p. 117) has furnished some paleontological support for this supposition. It has been shown (Burton *in* Kobayashi et al. 1966, p. 101) that although the formation is characterized by alternations of rather thin strata of varying lithology, in general three members can be recognized from west to east—and presumably from base to top if the formation is correctly deduced to young toward the east:

3. *Conglomerate Member*—interbedded arenite and lutite with important conglomerate horizons. More arenite than in the other two members. Apparently no chert.

2. *Rhythmite Member*—alternating thin to very thin arenite and lutite beds. Few conglomerate horizons. Apparently no chert.

1. *Chert Member*—rather thinly interbedded lutite and arenite with important chert horizons. No conglomerate.

It appears that the chert is strictly confined to the western (lower?) part of the outcrop, but otherwise there seems to be a gradation from one member to another. All three members are represented in the large central-north Kedah outcrop; only the first two are known in south Kedah, and possibly only the second type occurs in north Perak.

Ong (1968) subdivided the Semanggol formation of northeast Kedah into a lower unit of dark grey mudstone which he called the Lang mudstone, and an upper conformable rhythmite sequence, 1400 m (4590 ft) thick; this second unit which he called the Gubir beds, is formed of sandstone and mudstone with some conglomerate horizons.

Chert Member. The chert member occupies most of the south Kedah outcrop of the Semanggol formation and the westernmost part of the central-north Kedah outcrop. Although constituting only about 10 to 15% of the chert member, chert dominates the outcrop by virtue of its resistance to erosion. It comprises a number of elongate bodies which, at their maximum development, are several miles long and at least 600 m (1970 ft) thick. The chert may be white, cream yellow, pale grey, dark grey, black, green, pink, red, or purple; paler hues are more common, expressive of the purity of the rock (one specimen was assayed at 98.2% silica). Infrequently carbonaceous chert is encountered; one or two specimens studded with rhombohedra of a carbonate mineral have also been seen. The Semanggol chert is generally well bedded with individual beds ranging in thickness from several millimeters to several centimeters (Figure 5.3).

The chert is often more or less recrystallized. Small white spots apparent in some hand-specimens prove to be radiolarian tests infilled with spherulitic chalcedony or microcrystalline quartz. On occasion these are very abundant in the more argillaceous varieties, and in some cases the internal structure of the tests can be seen. Willbourn (1926, p. 312) reported that sponge spicules were often very common. Siliceous spicular bodies seen by the present writer, however, have been regarded as radiolarian spines because they are normally associated with radiolarian tests. Other microfossils (?foraminifera) also occur. Closely packed lamellibranchs have been

FIGURE 5.3 *Bedded chert of the Semanggol formation, Tawar, Kedah. Photograph C. K. Burton.*

found at a few localities in the chert. Details of these are given below.

Lutite and arenite are more abundant than chert. Exposures of arenaceous rocks are more abundant than those of argillaceous rocks, but this is thought to be a consequence of the intense tropical weathering to which these rocks have been exposed. Such weathering would no doubt cause exposures of lutite to be preferentially disintegrated, eroded, and concealed relative to those of arenite.

The principal type of argillaceous rock in the chert member is a black, carbonaceous shale or mudstone. Beds are commonly several centimeters to 1 or 2 meters thick, usually with thin silty or sandy partings at intervals of a few millimeters to a few centimeters. Lamellibranch impressions, together with some serpulids and cephalopods, quite often occur along these partings. Siltstone is not uncommon, and it is frequently an ultrafine greywacke. Where closely associated with chert, the lutite is sometimes siliceous and exhibits continuous gradation through porcellanite and argillaceous chert to chert proper.

The arenite of the chert member is of greywacke, subgreywacke, and protoquartzite. Greywacke is common and ranges beyond the sand grade, from silt into granules or even pebbles, although no actual conglomerate has been seen. Granite fragments have been found at three points in the greywacke of the chert member. Graded bedding is sometimes seen, but it is often obscured by weathering and only becomes apparent on microscopic examination of a series of specimens in stratigraphic order.

On a large scale, the chert member is characterized by alternations of coarse and fine grained beds. This is repeated on a smaller scale, particularly in the siltstones. In many of these there occurs a fine lamination with individual layers a fraction of a millimeter to 1 or 2 mm thick. These laminations may be regular and parallel to the bedding (the *horizontal laminations* of McBride 1962). Frequently, however, they show other characteristics such as pinching and swelling, which may lead to disruption of the bedding and development of a series of disconnected lenses.

Another feature often seen is convolute lamination. In cross section this ranges from gentle undulations to tightly contorted folds, which may also develop into disrupted bedding. These convolutions have a limited (few centimeters) vertical extent and are underlain and overlain by uncontorted layers. The fine laminations also sometimes show small-scale cross-bedding, the sets of beds involved being normally only a few millimeters to a few centimeters thick. Small pebbles of siltstone and shale are sometimes seen in arenite. Subcylindrical zones, a few millimeters in diameter and a few centimeters in length, are also present. In these the bedding is slightly disturbed. They lie at all attitudes and their cross sections resemble clasts of the surrounding rock which have been slightly transported.

Rhythmite Member. A rhythmite facies is represented in all the three main outcrops of the Semanggol formation. It constitutes possibly the whole of the most southerly outcrop and the southeastern part of the central one, reaching its maximum development in the central strip of the northern area. This facies consists of a closely alternating succession of arenite and lutite in beds that are usually a few centimeters thick. Sporadic thicker beds of either rock type also occur, and although the two are typically in approximately equal amounts, there is a complete range from shale with a few thin sandy layers to arenite with infrequent shale partings. Occasionally conglomerate also occurs. The bedding in this facies is commonly uniform and well defined, although sometimes it is diffuse.

The principal arenaceous rock is greywacke, although subgreywacke is not uncommon. K. Y. Foo (in correspondence) remarked on the high micaceous content of the arenite and the lutite in the Taiping area, and Ong (1968) reported contents of up to 8% sericite and about 1% biotite and muscovite in northeast Kedah. A minority of arenite beds attain a thickness of 1 m and some reach 10 m. The thicker beds include lenticles and partings of shale. Graded bedding is also often exhibited by the arenite and thicker strata may include two or three graded units. Cross-lamination is also of frequent occurrence and has been seen within beds as thin as 1 cm. Cross-bedded units are never more than a few centimeters in thickness. Typically the arenite strata in the rhythmite belt have a sharp and uneven base. Often this represents a microunconformity, for the irregularities are paralleled by minor folding in the underlying shale, whereas the bedding in the succeeding arenite is undisturbed, indicating that the arenite is filling undulations on the lutite surface. In some cases the irregularities on the bottom of the arenite constitute sole marks, including flute casts, groove casts, and cigar-shaped projections, usually closely spaced and either round or flattened in section. Similar bodies sometimes occur in the midst of lutite, apparently quite discrete from any arenaceous bed. Ripple marks with a wavelength of 20 to 30 cm and a height of 3 to 6 cm occur infrequently (Ong 1968).

Other structures exhibited by the greywacke and subgreywacke include convolute lamination, prolapsed bedding, and sedimentary pullaparts. All these can be seen in the excellent exposures at Bukit Merah in north Perak (Figure 5.4).

FIGURE 5.4 *Rhythmite member of the Semanggol formation at Bukit Merah, Perak. Photographs C. K. Burton. (a) Laminated and cross-bedded arenite overlain by lutite beds showing pinching and swelling; (b) convolute lamination; (c) prolapsed bedding (slumping), slump movement from left to right.*

The argillaceous component of the rhythmite is normally a black shale or mudstone similar to that of the chert belt. It carries some pyrite and sometimes intercalated laminations of arenite or siltstone, especially in the thicker beds. Occasionally the beds attain a thickness of 10 m. Fossils are abundant in the lutite, especially in what appears to be the higher part of the rhythmite belt, often but not always lying on the sandy or silty partings. In addition to the cigar-shaped bodies of arenite in lutite, which lie parallel to the bedding, smaller cylindrical arenite bodies occur, frequently cutting across the bedding planes. Possibly these represent worm burrows infilled with sand (Ong 1968).

The conglomerate that is interbedded with the rhythmite is generally lenticular and of rather limited extent. Foo (in correspondence) recorded an exceptional development 20 m thick near Gunong Semanggol. These rocks are polymict ultracoarse greywackes, typified by chert phenoclasts of granule to cobble grade. Normally the chert of the clasts does not correspond to the pale varieties typical of the Semanggol formation; rather, it tends to be a dark-colored type, such as occurs in the nearby Silurian to Devonian Mahang Formation. Other phenoclasts are of quartz, metaquartzite, sandstone, and shale. Foo (in correspondence) noted that in the Taiping area the phenoclasts are often angular.

Conglomerate Member. The mean grain size of the arenite becomes coarser to the east, and the rhythmite member passes into the conglomerate member, occupying the Thai border strip of the most northerly Semanggol outcrop. Information regarding this member is still rather scanty because it lies in deep jungle country that has only been covered by reconnaissance survey. In northeast Kedah this facies occurs in the Bukit Pakir Terbang range, which is built of conglomerate for at least 22 km (14 miles). Willbourn (1926, p. 313–314) believed that this range of conglomerate continued farther south beyond the Bukit Jelutong granite batholith. It now seems, however, that Paleozoic rocks occur to the south of the granite, and it may be that the Pakir Terbang range is offset by a right lateral wrench fault to Bukit Dada Ayam in central Kedah.

Nevertheless, the age of the conglomerate belt is in doubt, for Ong (1968) showed that the Semanggol formation rhythmite rocks to the north of the Sungei Muda are overlain unconformably by a sequence of red conglomerate and mudstone of typical continental aspect. Foo correlated this sequence with the Tembeling Formation of Jurassic age and named it the Saiong beds. Conglomerate constitutes only a small, but impressive, proportion of the facies. The conglomerate has more quartz and quartzite phenoclasts and fewer of chert than does conglomerate interbedded with the rhythmite. Arenite, which plays a more important role in this facies, is both thicker and more frequent than in the rhythmite and chert members. Its thickness is normally to be reckoned in decimeters and it exhibits excellent graded bedding. Black shale resembling that elsewhere in the formation also occurs in the conglomerate member.

Paleontology. Some 170 occurrences of macrofossils have been found in the Semanggol formation. These frequently lie on partings in the shale of the chert member and more commonly in the rhythmite member. Three localities have been recorded in chert and one in siliceous shale. Fossils have not so far been found within the conglomerate member, where field investigations have only been of a reconnaissance nature. Thirty-seven fossil collections have been studied in detail [Jones 1905, Newton 1925, Kobayashi 1963b, Kobayashi et al. 1966 (which includes a list of 36 localities), and Ichikawa in correspondence]. Much material still awaits expert examination, but the salient features of the fauna are known.

The most striking feature of the fauna is the abundance of individuals (particularly in view of the lithofacies) and the paucity of genera. *Posidonia* is by far the most numerous and most widespread genus. Of the 37 collections studied to date, *Posidonia* (7 species) appears in 23 (Kobayashi et al. 1966, p. 112) and this genus has been found in a large number of other localities. *Halobia* (10 species) is recorded from 15 of the collections and is known at more than 20 other places (Plate 11), whereas *Daonella* (8 species) is reported in 14 of the examined collections and has also been seen elsewhere, although less frequently than *Halobia*.

Tubular bodies, initially considered to be crinoids (Newton 1925) and now regarded as *Serpulites* s. l. sp. (Kobayashi et al. 1966), have been noted in eight of the documented collections and are very widespread. Ammonites have been found at nine places. Of the lamellibranchs *Pecten* (Kobayashi 1963b), *Gervillia?* (Kobayashi et al. 1966), and *Costatoria?* (Ichikawa in correspondence), one instance each is known. A number of poorly preserved plants and some medusoid impressions completes the list of Semanggol fossils.

A considerable minority of the lamellibranch shells are of minute forms, 1 or 2 mm in diameter. Most of them have been referred to the genus *Posidonia*. These diminutive forms were discussed by Kobayashi and Tokuyama (Kobayashi et al. 1966, p. 108), who could not decide whether they are immature forms, dwarfs, or adults of a very small species. Since the forms have now been found associated with shells of all intermediate sizes up to that of the large adults, it seems likely that these minute valves represent nepionic individuals.

The large volume of shale is almost uniformly black, which seems to indicate that the muddy bottom of the Semanggol Sea could not support the life of the myriads of bivalves found in the Semanggol formation. The thin-shelled nature of this fauna suggests a nektoplanktonic existence. However, the occurrence of the tubiculous *Serpulites* together with possible worm casts in the Semanggol lutite might be held to show that the muddy sea floor was not inimical to organic life.

The lamellibranchs supply us with some chronological information concerning the Semanggol formation. The Upper Ladinian form *Daonella indica* was found at two localities in the chert belt, and *Daonella* cf. *pichleri* collected nearby also seems to point to an Upper Ladinian age (Kobayashi et al. 1966, p. 104). *Daonella indica* is also found in the rhythmite belt at Ulu Pedu, but *Halobia*, essentially of Carnian age, is abundant in this facies. In the west (near the top?) was found the species *Halobia aotii*, which may be uppermost Carnian to lower Norian (Kobayashi 1963b, pp. 116–117).

Both paleontological and structural evidence suggest that the chert member is the oldest member of the Semanggol formation, although it is possible that the boundary with the rhythmite member may be diachronous and that to some extent chert may pass laterally, as well as vertically, into rhythmite.

The origin of the Semanggol chert is somewhat uncertain. Its association with siliceous arenite and argillite might suggest a replacement origin, as might the occurrence of patches rich in sericite in one example. The presence of radiolaria appears to indicate a primary rather than replacement origin, although not all authorities would agree. Cohen et al. (1963), for instance, ascribed bedded chert with radiolaria to silicification of greywacke; the absence of radiolaria in the associated greywacke is explained by postulating solution of the radiolaria in interstitial waters in that part of the rock which did not undergo silicification. However, the considerable thickness and extent of the chert bands in the Semanggol formation, together with their regularity of form and chemical purity, contradicts a replacement origin. Nevertheless, replacement may have occurred peripherally. The concentration of radiolarian tests is not usually high, and thus it seems that this chert does not represent a siliceous biolith and that it is, most likely, a chemical precipitate.

The source of the silica remains unexplained. Cohen et al. (1963 p. 147) stated that there is a worldwide correlation between the occurrence of chert in greywacke sequences and volcanic activity. Although no volcanic rocks are known from the Semanggol formation, the chert is at least approximately contemporaneous with considerable volcanicity in other parts of the Malayan region. In particular, the upper part of the Kerdau formation in central Pahang has a strong pyroclastic component and contains the same fossils as, and seems to correspond in age to, the chert member (Kobayashi 1963a). The silica may therefore have originated in ash falls into the Semanggol basin originating from contemporaneous volcanic activity in other parts of the country. For this silica to have accumulated as chert, a natural barrier to retain it seems to be required, as well as an absence of clastic sediment.

A barrier to the east of the Semanggol basin is perhaps indicated by the perfect mutual separation of the volcanic facies in axial Malaya and the chert facies in northwest Malaya. The rising geanticlinal barrier might eventually have sealed off the Semanggol basin from the source of silica, thereby offering an explanation of why the Semanggol chert ceases abruptly upward. The same geanticlinal bar may have further restricted the circulation of the water within the depositional basin and thus promoted generation of the black shale, which is the dominant lithology in the Semanggol formation.

The repeated alternation of arenaceous and argillaceous clastic rocks, the abundance of greywacke, the occurrence of chert and the absence of limestone, together with the presence of recurrent graded bedding, sole marks, convolute lamination, and other evidence of soft-sediment deformation, all demonstrate that the Semanggol formation is a flysch facies as defined by Sujkowski (1957) and Pettijohn (1957 pp. 615–618). Conceivably the enhanced importance of arenite and rudite in the upper part of the succession implies a transition from flysch to molasse.

At present, information on contemporary sediment transport is too sparse to permit the construction of any detailed paleocurrent or paleoslope models. Sole marks in the northern outcrop of the Semanggol formation suggest southward movement, whereas cross-bedding in the central and southern outcrops, and slump features in the southern one, indicate transport to the north. This conflicting evidence may result from the reworking of the arenite by deep marine currents, as postulated by Hsu (1964). In any case the directions indicated seem to lie along the (miogeosynclinal) basin axis.

We must look to other criteria in any attempt to reconstruct the paleogeography. Most modern theories would derive the greywacke from the geanticline, but the rocks themselves show evidence of a source area to the west rather than to the east. Thus the rhythmite facies becomes coarser westward and, in particular, there seems to be more conglomerate around Bukit Merah and Gunong Semanggol than elsewhere. The conglomerate in the areas named contains many angular phenoclasts. Similarly, dark-colored chert phenoclasts, typical of the Semanggol conglomerate, seem to be derived from the Mahang Formation, whose outcrop is mainly situated to the west of the Semanggol outcrop. In addition, the only two known speciments of shallow-water benthonic lamellibranchs from the Semanggol have both been found on the west side of the outcrop (*Gervillia*? at Bukit Merah and *Costatoria*? in north Kedah). To the northwest lies the contemporaneous Kodiang limestone, which may represent a shelf facies, possibly peripheral to an emergent landmass.

Possibly, then, the postulated geanticline was not an important source of sediment throughout much of Semanggol time, although finally it may have yielded substantial contributions to the conglomerate belt, which is best developed in the east.

KODIANG LIMESTONE. The name "Kodiang limestone" was introduced by Jones et al. 1966) to designate carbonate rocks in the extreme northwest of Malaya, whose Triassic age had been demonstrated by Ishii and Nogami (1966). This rock unit was earlier included in the Carboniferous (and Permian?) by Willbourn (1926), Scrivenor (1931), and Ingham (1948) and was shown as Permian by Alexander (1965) and Burton (1965).

The Kodiang limestone outcrops in a series of hills aligned in a narrow belt 13 km (8 miles) long and trending SSW-NNE, in the vicinity of Kodiang railway station in north Kedah (Plate 12). The hill known as Bukit Kechil seems to constitute a suitable type locality with good exposures of limestone, underlain by mudstone; it has yielded algae, foraminifera, bivalves, and diagnostic Triassic conodonts (Ishii and Nogami 1966). Two conodont faunules have been described: that from Bukit Kechil is Late Triassic, and that from Bukit Kalong is Middle to Late Triassic.

According to Ishii and Nogami, this formation is made up essentially of white intrasparitic limestone together with some micritic limestone, chert and mudstone. Two chemical analyses of the limestone are given in Table 5.1.

Apart from the occurrence of small amounts of chert and mudstone, there is little evidence of a transition between the Kodiang limestone and the contemporary Semanggol formation, which succeeds it within 24 km (15 miles) to the southeast. The occurrence of intrasparite shows that some contemporary erosion and redeposition of limestone took place and the micrite indicates periods of quiescence. It is not considered that this sequence is a calcareous equivalent of the Semanggol flysch, however, which would probably be represented by intramicrite rather than intrasparite. It appears that

PLATE 12 *Bukit Kodiang, Kedah, formed of Triassic limestone. Photograph C. R. Jones.*

the environment of deposition was quite different in the Kodiang area, and possibly the shelf conditions evidenced here in Lower to Middle Paleozoic time (Burton 1966, 1967c) still persisted, or were regained in the Middle and Upper Triassic.

Middle to Upper Triassic Granite

Between the two belts of Triassic sedimentary rocks in northwestern and axial Malaya is a linear expanse of mountainous granite country, largely corresponding to the Main Range, and to the north the Kledang and Bintang Ranges, which branch off from the Main Range.

Rocks of the Semanggol formation appear to be metamorphosed by granite in the Bintang Range. However, recent radiometric determinations (Snelling 1965, Anon. 1966, Snelling et al. 1968) suggest that the bulk of the granite in the Main Range, the Kledang Range, and the Bintang Range is of Middle to Upper Triassic age and hence is generally contemporaneous with the deposition of the Semanggol formation rocks. Nevertheless, some of the granite may predate and some may postdate the formation (Table 5.3).

A full description of this granite and its radiometric dating is deferred until Chapter 8.

Synthesis

It appears that early Triassic diastrophism in the Malayan region resulted in the erosion of most of the Lower Triassic and some of the Permian strata. Geosynclinal conditions were renewed in peninsular Malaya with connections to Burma, the Himalayas, Jordan, and the Alps to the west, and with Indochina and Indonesia to the east, as deduced by the faunal evidence. The eugeosynclinal basin of central Malaya probably was the first to be reestablished, since Anisian sedimentary rocks are recorded from that region whereas the oldest strata known from the reestablished miogeosyncline of west Malaya are of Ladinian age (Table 5.3). It also seems that the miogeanticlinal ridge was upraised anew between the mio- and eugeosyncline, since there are quite marked differences in the rock sequences deposited in the two belts.

The Middle to Upper Triassic rocks of the eugeosyncline comprise mainly clastic sediments with important volcanic members, predominantly of rhyolitic composition; only modest amounts of limestone and chert are present. The Anisian strata are disposed in the middle of the eugeosynclinal furrow in Pahang. They lie on the east side of the Gunong Benom range about a line parallel to the NNW-SSE regional strike and approximately in line with the Scythian rocks, which occur farther to the north in Kelantan. The regional plunge is to the south along the fold axis, and this may be why there was little or no break between the Scythian and Anisian in the axial part of the eugeosyncline. Conceivably this zone was depressed rather than uplifted during the Lower Triassic orogenesis, and the nearby Gunong Benom granite batholith was emplaced near to the core of the synclinal downwarp.

The Middle to Upper Triassic eugeosynclinal strata exhibit two distinct developments corresponding, respectively, to the shelly and graptolitic facies of classic Lower Paleozoic geosynclines. The shelly facies is here composed of sandstone and shale with a rich thick-shelled benthonic fauna characterized by the lamellibranch *Myophoria*. The Jelai formation of Pahang and Kelantan and most of the Bukit Resam member of the Jurong Formation in south Malaya represent this facies. The other facies is composed of black shale with thin immature arenite beds, containing a nektoplanktonic fauna consisting essentially of *Posidonia, Halobia,* and *Daonella,* and some ammonoids. This facies is represented by the Kerdau formation of Pahang, the Gunong Rabong formation of Kelantan, and the lower part of the Bukit Resam member in Singapore. Because of their habit, individuals of the *Daonella* fauna may appear amid the *Myophoria* biofacies, but the two faunas are usually separate.

The *Myophoria* facies, with its rich thick-shelled benthonic fauna and generally clastic character, is clearly of shallow-water origin. The *Daonella* facies, on the other hand, might be thought to represent a deep-water environment because of its abundant black shale and frequent flyschlike aspect (Dzulynski and Walton 1965, p. 6). However, the occurrence of *Serpulites*, plant remains, ripple marks, and subaerial lava in the Kerdau formation, and the incidence of current-oriented fossils in the Bukit Resam member, all seem to indicate that the *Daonella* facies was also generated in shallow-water conditions. The restricted or absent benthos and the high carbon and pyrite content of the black lutite do not necessarily indicate a deep-water origin but suggest the presence of deoxygenated bottom waters, which inhibited both life on the sea bed and the decay of organic matter that accumulated there.

The two facies of the eugeosynclinal area are not arranged in longitudinal belts, but the *Daonella*-bearing black shale seems to constitute relatively limited zones in a regional development of Myophorian sandstone. Possibly the black shale outcrops represent portions of the geosyncline wherein the circulation was restricted by bordering shoals either of *Myophoria* facies sediments or of volcanic origin.

The *Daonella* facies also occurs, as the Semanggol formation, in the miogeosyncline, but it begins later, in the Ladinian. This formation lacks volcanic rocks and contains in its lower horizons a strong development of bedded chert that is remarkably pure and pale in color. Possibly contemporaneous volcanic activity in the eugeosyncline furnished silica that entered the miogeosynclinal basin either by way of marine currents or by direct infalls of ash. The silica may have been retained by the geanticlinal bar, enabling it to accumulate as chert. Conceivably the same barrier caused restriction of circulation of the waters in the Semanggol basin, thereby promoting the formation of black shale.

Into this environment were introduced also a number of coarse clastic intercalations, typically of greywacke composition but with some subgreywacke and protoquartzite. By analogy with similar coarse clastic rocks higher in the succession, these appear to be the products of turbidity flows.

The postulated geanticline seems to have been a positively rising feature that eventually reached such a height that the basin was excluded from the source of the silica, causing chert deposition to cease abruptly near the beginning of the Carnian. Black shale now became the sole autochthonous deposit, indicating extensive stagnation of bottom waters that was largely or wholly inimical to life and that, moreover, inhibited the decay of organic matter. In the upper waters, however, thrived a nektoplanktonic *Daonella* fauna. Minor developments of dark-colored mudstone in the Kodiang limestone may indicate that at times these conditions had some effect in the shelf area to the west, although there is no trace there of the Semanggol fauna.

During the Carnian period, arenaceous incursions into the basin increased in number and frequency. Detailed strutures show clearly that these are turbidites and, together with the intercalated shale, they constitute a typical synorogenic flysch deposit. Rather surprisingly, there is some evidence that the turbidite sandstone was derived from the west, where a foreland may have been situated, rather than from the supposed geanticline to the east. The numerous turbidity flows were probably triggered off at relatively short intervals by repeated earthquakes, or seaquakes, as envisaged by Wood and Smith (1959). Subsequently, however, the geanticline seems to have been uplifted to such an extent that it became an important source of sediment, yielding coarse detritus to the east side of the basin. This stage occurred during the emplacement of the Main Range Granite batholith, which eventually came to occupy the site of the geanticline and part of the miogeosyncline.

This orogeny and its attendant plutonism marked the termination of the geosynclinal stage in Malaya and caused the stabilization or cratonization of the Malayan orogen.

JURASSIC AND CRETACEOUS

Middle Jurassic fossils were reported from Singapore by Newton (1906). This dating was unconfirmed by other evidence, however, and thus it has been largely ignored and the fossiliferous rock has been allotted to the Triassic period, which is widely represented in the neighborhood (Scrivenor 1931, p. 107, Ingham 1948). It was believed that the interval between the Triassic and the Tertiary corresponded to a gap in the sedimentary record in Malaya, and it was supposed by many authors that much of the Malayan granite might be Cretaceous in age (Ingham 1938, Roe 1951, 1953, Ingham and Bradford 1960).

It was not until 1959 that Paton (1959) was able to establish conclusively the occurrence of Post-Triassic Mesozoic sediments in Malaya. These are referred to as the Gagau Group and comprise mainly postorogenic nonmarine sandstone which outcrops in the Kelantan–Trengganu–Pahang borderland. Subsequently, Survey geologists found similar sandstones in south Johore, in northeast Johore, and in southeast Pahang. Koopmans (1966, 1968) described a late to postorogenic, but Pre-Gagau, molasse facies of Upper Triassic to Jurassic age from north Pahang. He called this the Tembeling Formation. Possibly equivalent rocks are known in Johore State, and the long-alleged Middle Jurassic age of some of the strata in Singapore now appears more plausible, since these are of similar facies to the Tembeling Formation.

Recently obtained radiometric age determinations (Snelling 1965, Anon. 1966) have indicated that the bulk of the Malayan granites are of Triassic age or older. However a concentration of potassium-argon determinations around the Jurassic–Cretaceous boundary (Hutchison 1968b, Snelling et al. 1968) indicates that an important event occurred approximately contemporaneously with the beginning of deposition of the Gagau Group. There is no definite indication that this event was associated with magmatism, because the rocks that have yielded potassium-argon dates in this concentration have given older ages by the rubidium-strontium method. This implies that the event being dated by the potassium-argon method is tectonic rather than magmatic (Snelling et al. 1968). Extensive wrench faulting probably took place at this time; it is thought that these displacements may be related to continental drifting.

Epeirogenic movements also occurred in Jurassic to Cretaceous time. Initially the floor of the Tembeling basin sank as sediments accumulated, but it later rose to uplift the Tembeling Formation prior to the succeeding Gagau Group deposition. The sediments of the Gagau Group were similarly deposited in subsiding basins, and it appears that sedimentation to these basins was terminated also by uplift. The uplift shows a pronounced southward tilt and probably was of an epeirogenic character.

Tembeling Formation

OCCURRENCE AND NOMENCLATURE. A thick sequence of coarse arenaceous-rudaceous rocks alternating with argillaceous strata in the Tembeling River basin of Pahang was originally referred to by Scrivenor (1907) as the Tembeling Series. Later Scrivenor (1911) came to believe that these rocks, together with other clastic sequences in Malaya, were an extension of the Upper Gondwanas of India of Upper Triassic to ?Middle Jurassic age; he therefore termed these Gondwana rocks. This name was abandoned later (Scrivenor 1931), when the present Tembeling Formation was allotted to the Triassic of Malaya, a precedent which was followed on the 1948 geological map of Malaya (Ingham 1948). These rocks were subsequently included in the Triassic Younger Arenaceous Series or Newer Arenaceous Series of Richardson (1950) and in the Triassic Lipis Group of Alexander (1959).

On the 1963 geological map of Malaya (Alexander 1965), only a minor portion of this rock unit was assigned to the Triassic, the bulk being allotted to the Late Carboniferous and to the Permian, whereas a small area in north Pahang was placed in the Jurassic to Cretaceous, following Rishworth (in prep. B), who included part of the present Tembeling Formation in the Gagau Group. Ichikawa et al. (1966), describing a small section of the Tembeling Formation, referred to it informally as the Jengka Pass formation. Scrivenor's original name was resuscitated by Koopmans (1966) as the Tembeling rocks, subsequently formalized as the Tembeling Formation (Koopmans 1968).

On the basis of its distinctive fluviatile-deltaic-lacustrine origin and late to postorogenic molasse facies, the Tembeling Formation was excluded from the Lipis group by Koopmans (1968). A type section (Figure 5.5) was designated in the Sungei Tekai upstream for a distance of about 12 km (9 miles) from its confluence with the Sungei Tembeling. This rock unit was said to constitute a belt trending north-northwest which averaged 40 km (26 miles) in width and which was at least 180 km (112 miles) long. It occupies the Tahan Range or Tahan Coulisse (Scrivenor 1907), which culminates in the highest peak in Malaya: Gunong Tahan, 2192 m (7186 ft) above sea level. The outcrop extends northward to the vicinity of Gua Musang in south Kelantan, and Koopmans (in correspondence) considered that it continued onto Gunong Rabong and the part of the country that had been included by Yin (1965a,b) in his Gunong Rabong formation (q. v.). Southward, the Tembeling Formation extends beyond the Pahang River to Bukit Chermingat. A further outcrop extends southeastward from the mountain Tungku in west central Pahang.

Koopmans (1968) also considered that the formation recurred farther afield in northeast Johore and pointed to similar rocks in Singapore (the Pasir Panjang member of the Jurong formation) and in the nearby Lingga Archipelago. Similar molasse facies rocks in northeast Kedah (Figure 5.5) have also been ascribed to the Tembeling Formation (Ong 1968). The outcrop extends northward into Thailand.

LITHOLOGY. Koopmans (1968) described the Tembeling Formation as a succession of conglomerate, felspathic and quartzitic sandstone, and grey and reddish shale and mudstone. The base is generally formed of a thick, bedded conglomerate, often red-purple and polymict but tending to become white and oligomict in some places. The phenoclasts are not well rounded and they include rare examples of granite. Surprisingly, Koopmans (1968) selected Tanjong Murau in northeast Johore as the type locality of the basal conglomerate. This is 273 km (170 miles) from the Sungei Tekai type section, and a complete absence of chronological information makes it impossible to establish that this locality is equivalent to the basal conglomerate of the main Tembeling outcrop, which is that displayed along the road from Maran

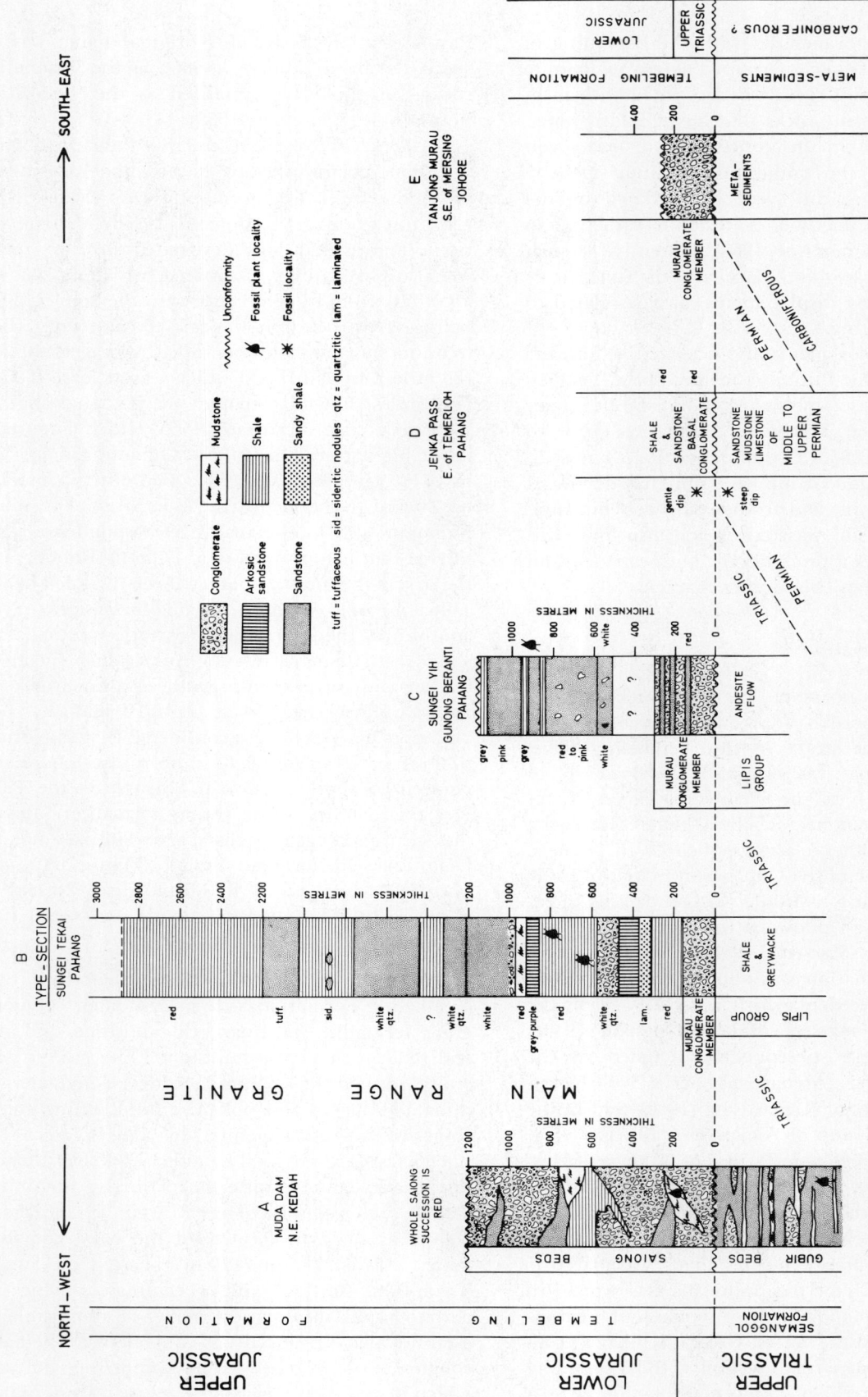

FIGURE 5.5 *Stratigraphic sections in the Tembeling Formation:* (a) *Northeast Kedah after Ong (1968);* (b) *type section, Sungei Tekai after Koopmans (1968);* (c) *Sungei Yih after Mohammad bin Ayob (1968);* (d) *Jengka Pass after Ichikawa et al. (1966);* (e) *east Johore after Koopmans (1968).*

FIGURE 5.6 *Bedded Murau Conglomerate exposed on the road between Maran and Kuantan. Photograph P. H. Stauffer.*

to Kuantan (Figure 5.6). Hence the Murau conglomerate must be excluded from the Tembeling Formation proper, although there may be equivalence in terms of age, facies, or both. Similar conglomerate outcrops on the north Trengganu coast at Kuala Keluong and Pulau Rhu. Mohammad (1968) has used the name "Murau Conglomerate Member" for a red polymict conglomerate at least 300 m thick, which outcrops in the Sungei Yih between 1.2 and 2.4 km (0.75 and 1.5 miles) east of the road from Maran to Jerantut in the Gunong Berantai area of Pahang. It forms the oldest rock of the Tembeling Formation in this region.

Other conglomerate horizons occur in the Tembeling Formation, generally rather low in the succession. They are intercalated with felspathic sandstone, mudstone, and a few lentils of coal shale. Higher up, the rudaceous and labile components decrease and quartz sandstone becomes well developed (Plate 13). The top third of the formation is comprised largely of thick homogeneous red shale and mudstone. Koopmans (1968) records a layer of tuffaceous sandstone high in the type section, and Ichikawa et al. (1966, p. 124) noted frequent intercalation of tuffaceous layers in the vicinity of the Jengka Pass, 72 km (45 miles) to the south.

Tembeling Formation sandstone shows abundant cross-bedding (Figure 5.7), load-structure, occasional ripple marks, and a few layers containing reworked shale pellets. Plant fossils were discovered in grey mudstone, but no organic remains have been seen in the red beds. At the Jengka Pass the formation is gently dipping (Ichikawa et al. 1966), but on the regional scale flexure folding proves to be the dominant structural form, major folds having a wavelength of about 8 km (5 miles) (Koopmans 1968). Some cases of localized overfolding have been observed west of Maran.

Although the top of this rock unit is not seen in the type section, Koopmans (1968) estimated that the Tembeling Formation there is 3000 m (10,000 ft) thick. It seems to thin to the west—although again the top is missing. Shale and greywacke of Anisian age are said to underlie the Tembeling Formation near the type section. Elsewhere in this area the formation rests upon various rock types thought to belong to the Middle or Upper Triassic. In places, this

FIGURE 5.7 *Cross-bedded fluvial sandstone in the lower part of the Tembeling Formation, Sungei Jempol, Pahang. Photograph P. H. Stauffer.*

PLATE 13 *Easterly dipping cross-bedded, fluvial sandstone in the lower part of the Tembeling Formation, Sungei Jempol, Pahang. Photograph P. H. Stauffer.*

contact seems to be conformable, and Koopmans (1968 p. 39) wrote that in the west these ". . . continental deltaic rocks are partly grading into . . . rocks of a marine facies belonging to the upper part of the Lipis group."

However, because of the variety of rocks underlying the Tembeling Formation and because of the lithological character of the basal conglomerate, Koopmans suggested an unconformable contact in the Tahan Range. Clearly there is considerable overlap, for in the Jengka Pass, to the west of the Tahan Range, the Tembeling Formation is unconformable on Permian strata (Ichikawa et al. 1966), and at Tanjong Murau the Murau conglomerate overlies metasediments, probably of Carboniferous age.

The top of the Tembeling Formation has been recognized only at Gunong Penumpu on the Kelantan–Pahang border, some 50 km (35 miles) north of the type section. Here Rishworth (in prep. B) recorded rocks of the Gagau Group, but part of this succession was placed in the Tembeling Formation by Koopmans (1968), who said that the Gagau Group here lay unconformably on the Tembeling Formation.

In the basin of the Sungei Jeram in southeast Pahang, north of Bukit Chermingat, Burton found a very similar succession of coarse, gently dipping sandstone and conglomerate apparently succeeded by a strong development of red lutite. Farther south, the Pasir Panjang member of the Jurong formation in Singapore can be correlated with the Tembeling with regard to the following features: age and the presence of polymict conglomerate—including granite clasts, red coloration (including red shales), plant remains, and coal seams. Like the Tembeling Formation, this rock unit seems to be closely related to the underlying rocks.

In northwest Malaya, some 290 km (180 miles) northwest of the Tembeling type section, Koopmans (in correspondence) and Ong (1968) have reported another succession comparable in lithology and stratigraphic position to the Tembeling Formation. Unconformably overlying the Ladinian–Carnian Semanggol formation around Bukit Saiong, near the Thai border in northeast Kedah, is a sequence of red-purple conglomerate and red mudstone that appears to be of molasse facies. These rocks have previously been included, erroneously it now seems, in the conglomerate member of the Semanggol formation outcrop, just as Yin (1965a,b) seems to have included part of the Tembeling succession in the contemporaneous Gunong Rabong formation.

PALEONTOLOGY AND AGE. At Gunong Penumpu the Gagau Group, which overlies the Tembeling Formation, is represented by the Lotong Sandstone of Upper Jurassic to Lower Cretaceous age (Rishworth in prep. B). This dating sets an upper limit to the age of the Tembeling Formation. At the base of the Tembeling Formation in the Jengka Pass, a small collection of fossils was made and described by Ichikawa et al. (1966). The collection comprised *Aequipecten* sp., nuculanid?, *Isocrinus* sp., brachiopoda, and the plants *Sagenopteris* sp. and *Equisetites* sp., which were identified by E. Kon'no. These fossils indicate a Mesozoic age and in particular the *Aequipecten* suggests late Middle or Upper Triassic.

A number of plants, spores, and pollen grains have also been obtained in the middle part of the rock unit, near Maran, Pahang. Serra (1968) described *Ptilophyllum* sp., *Zamites* sp., and *Klukia?* sp. and stated that these showed affinities to Rhaetic and Liassic species. B. E. Morgan (*in* Koopmans 1968, p. 39) recorded the palynomorphs *Classopolis classoides* Pflug and *Circulina* sp., both known from the Jurassic. However, Smiley (1970a,b) lists, from two collections obtained near Maran, 18 species of plants. Of these, 12 are identical with species described by Kon'no from the Gagau flora and recognized as late Jurassic or early Cretaceous (see p. 132). Thus the Tembeling Formation may range in age from Upper Triassic to Lower Cretaceous.

SUMMARY. The presence of pectinids, crinoids, and brachiopods at the Jengka Pass shows that the basal part of the Tembeling Formation was deposited under marine conditions (Ichikawa et al. 1966, p. 126). This environment seems to represent the last remnant of the extensive seas that covered much of the Malayan region in the Middle to Late Triassic and were largely expelled by the Upper Triassic orogenic revolution. Apart from these basal layers, the Tembeling Formation seems to be of nonmarine origin.

Scrivenor (1908) regarded the succession as being of estuarine origin. Koopmans (1968) believed that the Tembeling Formation was laid down in a fluviatile-deltaic-lacustrine environment. The crossbedding of the arenaceous rocks in the Tembeling Formation is thought to suggest deltaic conditions. Interbedded grey to red shale is considered to be a lacustrine deposit formed in a relatively warm and dry climate, and carbonaceous rocks, including coal shale, seem to indicate local swamps. Sustained warm conditions are evidenced by the red beds in the top part of the formation.

Evidently coarse clastic material, derived from the newly upraised (Upper Triassic) mountains, rapidly infilled the remnant seas, which were thereby converted into a paralic type of environment. As the uplands were worn down by erosion, progressively finer sediments were produced, although the high feldspar content of the Tembeling Formation arenite indicates that deposition was still rather rapid and implies that granite occurred in the hinterland.

There is some uncertainty regarding the location of the source area. Insofar as the Tembeling Formation may, in part pass west into the upper beds of the marine Upper Triassic, an easterly source seems to be indicated. The Murau conglomerate of Johore is clearly transported from the east (Koopmans 1968), and Mohammad (1968) reported that crossbedding in the lower sandstone member and imbrication in his "Murau Conglomerate Member" indicate a current direction from the east, whereas in the upper sandstone member of the Berantai area the current direction is generally from the north. In Singapore, Alexander (1950 p. 12) suggested derivation from the southeast or south for the rock unit referred to here as the Pasir Panjang member of the Jurong formation. However, different environmental conditions may have applied in the case of these south Malayan sequences, which are from 270 to 350 km (170 to 220 miles) distant from the main Tembeling outcrop and which have not been established as isochronous with the Tembeling Formation. It is theorized that an eastward source may have made only minor contribution because immediately to the east lies the Gagau Group which seems to be derived from the west. The Gagau may in part be the equivalent of the Tembeling or it may succeed the latter closely in time. Moreover, in the vicinity of the type section, the Tembeling Formation appears to be conformable on the Triassic, but to the west, in the Tahan Range, the junction was inferred to be unconformable. This leads us to consider a source to the west rather than to the east, since an unconformity is more likely to occur near to the margin rather than in the center of a depositional basin.

The high labile composition and coarsely clastic nature of the lower part of the Tembeling Formation indicates rapid accumulation in shallow water, and the thick red shale and mudstone of the top part are considered to be lacustrine deposits, probably laid down in water of no great depth. Shallow-water conditions throughout a stratigraphic thickness of some 3000 m (10,000 ft) indicates considerable subsidence, which appears to have been rather rapid, at least in the early stages.

Gagau Group

OCCURRENCE AND NOMENCLATURE. In northeastern Malaya, in the common borderlands of Kelantan, Trengganu, and Pahang, occurs a sequence of gently dipping, incompletely consolidated sandstone and conglomerate with subsidiary lutite and volcanic rocks. These occupy an elevated position—up to 1380 m (4525 ft) above sea level—and they rest unconformably upon presumed Lower Triassic Boundary Range granite and ?Permian country rocks (Plate 14). Rishworth (in prep. B) referred to this sequence as the Gagau Group. In the type area extending mainly westward and northward from Gunong Gagau, this rock unit outcrops over 108 km² (42 miles²) (Figure 5.2).

Comparable rocks cap the ridge that extends from Gunong Penumpu to Gunong Perlis, some 13 km (8 miles) southwest of the Gagau upland. Small outliers of similar rocks occur on the Kelantan–Trengganu border, south of Bukit Temiang and 83 km (52 miles) west of Sungei Lembing, Pahang (Koopmans in correspondence).

In 1952 Paton first discovered and realized the distinct identity of these rocks in the Gagau area, and later he issued a preliminary note concerning them (Paton 1959). This sequence was then named the Gagau formation by Alexander (1959) and the same term was used by Koopmans (1966), who provided some additional information obtained by the examination of aerial photographs. Alexander (1962) proposed to name the basal part of the Gagau succession the "Paton conglomerate." Although the outstanding pioneer work of the late J. R. Paton is readily acknowledged, it is not in

PLATE 14 *View east across the Gunong Gagau plateau formed of flat-lying sandstone of the Gagau Group. Photograph D. E. H. Rishworth.*

accord with modern practice to name a stratigraphic unit after an individual. Thus Rishworth (in prep. B) subsequently renamed the Paton conglomerate as the Badong Conglomerate. He further concluded that the Gagau formation should be upgraded to group status, embodying a lower Badong Conglomerate and an upper Lotong Sandstone, and he accorded formation status to both of these units. Rishworth's nomenclature, being based on detailed field examination, is here adopted, although homotaxial rock sequences in the southeast Pahang and northeast Johore areas, some 255 and 350 km (160 and 220 miles) distant, respectively, are here treated as discrete lithostratigraphic units.

On the 1930 and 1948 geological maps of Malaya (Scrivenor 1931, Ingham 1948), the outcrop of the Gagau Group is depicted partly as granite and partly as Triassic sandstone and shale; but on the 1963 map (Alexander 1965) it was more correctly shown (after Rishworth, in prep. B) as Jurassic to Cretaceous.

Since Paton's reconnaissance survey, a number of geologists have worked in the Gagau area. Foremost among them is Rishworth, who also made an exhaustive compilation (Rishworth, in prep. B) of the available data relating to the Gagau Group. It is from this work that much of the following description is taken.

LITHOLOGY. Rishworth (in prep. B) stated that the Gagau Group is distinguished from the underlying rocks by its consistently gentle dips. The Gagau outcrop forms a structural basin in which Burton has measured 126 dip angles that range from 2 to 28° and average 11.4°. Rishworth also noted that there was a notable lack of crumpling, shearing, and cleavage in nearly all the Gagau rocks. The harder beds were often jointed, sometimes closely, but joint patterns were generally regular and the effects of strong or protracted tectonic disturbances were absent except in beds adjacent to faults, due to local gravity sliding and slumping. By contrast, most exposures of the pre-Gagau basement contained beds that were strongly fractured or cleaved and sometimes boundinaged. The Gagau Group contained weakly cemented beds of conglomerate and siltstone; some of the quartz sandstone was very hard due to secondary cementation, but the constituent grains were not strained and rarely had sutured contacts.

Quartz veins have been seen in the Gagau rocks, but we know that these are by no means indicative of postdepositional igneous or hydrothermal activity, since F.E.S. Alexander (1959) found that quartz veins may be formed as a by-product of weathering under tropical conditions.

Rishworth (in prep. B) further observed that the Gagau rocks were either strongly colored, purplish, red, or chocolate, or were very pale, near-white, or sometimes pale green. Black or dark grey strata were thin and rare. Some pre-Gagau rocks were red or white, but most were grey to black or, where pale, were interbedded with dark beds; pale green beds were uncommon.

The base of the Gagau Group lies on a surface of considerable relief, ranging in altitude from 183 m (600 ft) to 1010 m (3300 ft) above sea level.

Badong Conglomerate. The lower component of the Gagau Group, the Badong Conglomerate, is named from the Sungei Badong because it was in the headwaters of the eastern branch of this river that a type section was selected. The formation is found on the western and southern flanks of the Gagau upland, where it forms a dissected escarpment composed of vertical cliffs and steep, razorback ridges. The Badong Conglomerate is up to 460 m (1530 ft) thick and may formerly have attained a thickness of about 600 m (1990 ft), but it thins rapidly and dies out to the east and north. Its base, however, rests on a distinctly uneven surface and often constitutes a striking angular unconformity (Plate 15). It is composed of polymict conglomerate with subordinate lithic sandstone and minor siltstone and shale. A few sections, particularly at and near its base, contain trachytic lava flows and some possible intrusions. The formation is typically red or reddish brown; the finer grained rock types being more strongly colored.

At the base the conglomerate is typically coarse boulder and cobble beds, but in accord with the crude overall grading that exists within the Gagau Group, it is somewhat finer higher up. The rocks are characterized by angular to subrounded lithic phenoclasts of chert, metaargillite, quartz, metaquartzite, and rhyolite, with a high proportion of less durable material such as sandstone, siltstone, shale, feldspar, and biotite. These coarse grains are randomly oriented in a sandy to silty matrix of similar composition. Arenite in the Badong Conglomerate likewise has a constitution similar to the rudite; the cement may be ferruginous, and impregnation and replacement by calcite has been noted. The few argillaceous rocks that occur are strongly colored (although apparently never black), and sometimes are micaceous.

In the Badong Conglomerate, the stratification may be indistinct, but often cross-bedding is well

PLATE 15 *Badong Conglomerate lying unconformably on steeply dipping Permian? shale and quartzite, Gagau plateau. Photograph D. E. H. Rishworth.*

developed in units up to several meters thick. Graded bedding is reported, and in one instance inverse grading was seen (Rishworth, in prep. B). Possible ripple marks were also found.

The Badong Conglomerate passes up without any marked stratigraphic break into the Lotong Sandstone, named after the Sungei Lotong, which flows from the Gagau upland to join the Sungei Sat. Rishworth (in prep. B) designated a type section in the upper reaches of the Sungei Lotong, south of Gunong Badong. The formation is exposed in benches and high cliffs fringing the inner basin of the Gagau upland and in the valley of the Sungei Pertang, which drains this basin.

Lotong Sandstone. The Lotong Sandstone is an irregular elongate lenticular body of variable thickness. In the north and east it often overlaps the Badong Conglomerate onto the irregular surface of the basal Gagau unconformity. It seems certain that some of the lower strata recorded in, for example, the type section are absent elsewhere, and it is possible that in part these beds may be the lateral equivalents of some or all of the wedge-shaped Badong Conglomerate. An indeterminable thickness of the upper beds has been removed by erosion, doubtless at a relatively rapid rate in this elevated situation and tropical climate. Rishworth (in prep. B) estimated that the present minimum thickness ranges from 300 to 730 m (985 to 2400 ft). The top of the formation is the present erosion surface. No overlying sequences have been found in the Gagau type area or where homotaxial rocks occur elsewhere.

The Lotong Sandstone is distinguished from the Badong Conglomerate by its clearly defined stratification, by a general absence of conglomerate (although some horizons do occur), by an overall finer grain size and higher textural and compositional maturity. The characteristic beds of this formation are composed of white orthoquartzite, protoquartzite, and subarkose, which are best developed near the top. Lower down, the arenite is interbedded with red, purple, and green siltstone with some tuff, lava, mudstone, and conglomerate. The lowest beds, overlying the Badong Conglomerate, consist mainly of red siltstone. Rishworth (in prep. B) believes that the depositional history of the Lotong Sandstone includes several distinct subcycles.

Orthoquartzite is the most common type, containing occasional fragments of chert, metaquartzite, feldspar, mica, and argillite. With an increase in the lithic grains, this rock passes into protoquartzite, and with more feldspar (which may reach 20%) it grades into subarkose. Some arenite corresponds in composition to subgreywacke. Some contains rare fragments of granite; some is clearly tuffaceous. A calcite cement is sporadically developed in the arenaceous rocks.

Some siltstone may also be tuffaceous. It may be red, grey, green, or mottled, and it is usually interbedded with subordinate bands of sandstone. One example of micaceous quartz siltstone was found to contain small fossil lamellibranchs. The mudstone is dark grey, dark brown, or green and may contain plant fossils. Rishworth (in prep. B) described a pale green bentonitic mudstone consisting of montmorillonite, mica, and minor quartz; he noted that the bentonitic character of this rock suggested a volcanic origin.

Conglomerate is rather restricted in the Lotong Sandstone and generally comprises paraconglomerate whose phenoclasts may be sparsely distributed in a sandy matrix. The clasts are of well-rounded pebbles, cobbles, and boulders of chert, quartz, and metaquartzite. Large clasts of rhyolite and tuff at two horizons are similar in composition to volcanic rocks intercalated in the formation. One occurrence of conglomerate was found to carry limestone pebbles. In this rock and in some other conglomerate, veins and patches of secondary calcite occur.

At a few localities, veins and lenses of coal up to 10 cm thick occur within the Lotong Sandstone. Lava of generally trachytic composition underlies the formation at some places, and interbedded within the sequence are lavas of rhyolitic, dacitic, trachytic, trachyandesitic, andesitic, and basaltic composition and some rhyolitic tuff and agglomerate. The trachyte beneath the Lotong Sandstone may be the equivalent of the similar rock below the Badong Conglomerate, evidently outpoured in the erosion interval predating the Gagau Group.

Cross-bedding on large and small scale alike is well exhibited in the Lotong Sandstone. The large units are reported to be wedge shaped and the smaller ones lenticular. The cross-bedding has not been systematically studied, but in one section, transport from the west is indicated (Drummond written communication). Graded bedding is said to be quite common, and some units are graded from conglomerate at the base to fine sandstone at the top. The grading may be repeated through a number of successive units. Rishworth (in prep. B) also noted two other types of rhythmic sequences; these consist, respectively, of massive siltstone frequently intercalated with thin beds of sandstone and of thick strata of coarse sandstone alternating with thin flaggy beds of fine sandstone. One bedding plane appeared to contain fossil suncracks infilled by sediment.

The morphological expression of the Lotong Sandstone consists of gentle dipslopes, inclined toward the interior of the Gagau upland. The outcrop is delimited on the outer sides by a high scarp face formed in the upper sandstone beds. The foot of this escarpment is buried in a thick talus slope deposited at the inner margin of a broad structural bench up to 3 km (2 miles) wide, underlain by the siltstone, subordinate sandstone, and volcanic beds that constitute the middle and lower units of the Lotong Sandstone. In some localities the arenaceous or volcanic strata form restricted horizons that cap small waterfalls and cliffs; but generally the much softer siltstone forms a slope, declining gently down to another marked break in the profile at the boundary between the Lotong Sandstone and the underlying Badong Conglomerate.

PALEONTOLOGY AND AGE. No organic remains have been reported in the Badong Conglomerate, but fos-

PLATE 16 *Gagau flora. 1. Pinna of* Gleichenites pantiensis *Kon'no, Tebak Formation, Gunong Panti, Johor, ×1. 2. Pinna of* Gleichenites gagauensis *Kon'no, Gagau Group, Gunong Gagau, Kelantan, ×1. 3. Pinna of* Gleichenites stenopinnula *Kon'no, Gagau Group, Gunong Gagau, Kelantan, ×1. 4.* Frenelopsis malaiana *Kon'no, Gagau Group, Guong Gagau, Kelantan. 5.* Ptilophyllum *cf.* pterophylloides *(Yokoyama), Tebak formation, Ulu Endau, Johore, ×1.5.*

sils have been collected at five places in the Lotong Sandstone. Four of these collections comprise plant remains and one consists of lamellibranchs. The plants were examined by Kon'no (1966, 1968), who reported the following: Equisetales: *Equisetites burchardti;* Filicales: *Gleichenoides gagauensis* (nov.), *Gleichenoides stenopinnula* (nov.), *Gleichenoides serratus* (nov.); Cycadales: *Otozamites gagauensis* (nov.), and *Pelourdea* cf. *megaphylla* (Phillips); Coniferales: *Cupressinocladus acuminifolia* (nov.), *Frenelopsis malaiana* (nov.), *Sphenolepis* cf. *kurriana, Nageiopsis?* spp., and *Conites spinulosus* (nov.); Incertae sedis: *Carpolithes* sp. (Plate 16).

This assemblage indicates an Upper Jurassic to Lower Cretaceous age, most probably early Lower Cretaceous, and some of the forms present are said to bear a close resemblance to the Wealden flora of Europe.

The associated lamellibranchs are difficult to

identify because they are poorly preserved. They are said to be freshwater forms which look similar to *Sphaerium* and *Nakamuranaia* (Jones et al, 1966).

SUMMARY. The western border of the Gagau outcrop is approximately rectilinear, parallel to, and some 3 km (2 miles) east of, what is thought to be a major fault along the valley of the Sungei Lebir (Burton 1967e). It thus seems that the margin of the Gagau outcrop may be bounded by a fault which is related to that in the Sungei Lebir. We then must consider the age relationship of the Gagau Group to this faulting.

The Lebir fault possibly originated as a wrench or tear fault, but as shown earlier, strong post-Lower Triassic vertical displacement must have occurred along this fault to bring, in the adjacent part of Kelantan State, catazonal Taku Schist of Lower Triassic metamorphism and related gneissose Kemachang Granite, on the west side of the fault, into juxtaposition with much less metamorphosed sedimentary rocks and mesozonal Boundary Range Granite on the east. This granite was evidently uplifted well above contemporary base level in the Lower Triassic to Upper Jurassic, for its surface underneath the Gagau rocks shows considerable relief.

To explain how such morphology came to form the floor of a depositional basin, it is suggested that the vertical movements about the Lebir fault immediately predated the Gagau Group deposition and that this rock unit was consequently generated in a down-faulted basin, delimited to the west by a strong eastward-facing fault scarp. This reconstruction would accord with the rapid deposition evidenced by the coarse Badong Conglomerate, its prismatic or wedge shape, and its situation on the south and west flanks of the Gagau Group outcrop as an apron or basin-margin accumulation derived from the upthrown side of the Lebir fault. The northeastern boundary of the Gagau outcrop is also faulted (Koopmans 1966).

A phase of volcanicity preceded deposition of the Gagau Group. Since the disposition of trachytic extrusions formed at this time is evidently controlled by the pre-Gagau surface (Rishworth, in prep. B) and since they appear to be related to similar volcanic rocks intercalated higher in the succession, it would seem that the initial volcanic phase did not long predate the Gagau clastic strata. Possibly this early volcanicity was related to the faulting, which appears to have constructed the environment of deposition.

Sedimentation began with the Badong Conglomerate, whose dominant coarse grain and coarse cross-bedding, with graded bedding and possible ripple marks, indicate deposition in shallow water. The characteristic red coloration of this formation seems to indicate warm nonmarine conditions, and this interpretation is supported by the apparent absence of fossils. The generally coarse, polymict nature of this rock unit, which has numerous angular and subangular clasts, implies short transport and rapid accumulation.

Succeeding the Badong Conglomerate is the Lotong Sandstone. In part the Lotong Sandstone may be the lateral equivalent of the Badong Conglomerate. But elsewhere, although there is no sharp break between the two formations, an unconformable relationship is indicated, since the Lotong Sandstone is reported to overlap the Badong Conglomerate onto the basal Gagau unconformity. The Lotong Sandstone depositional province was more extensive than that of the Badong Conglomerate, and Lotong Sandstone strata are finer grained, better stratified, and notably more mature. Nevertheless, it seems that conditions had not changed radically in Lotong Sandstone times. The graded bedding and the cross-bedding indicate subaqueous deposition, and the large scale of the latter (together with the occurrence of fossil plants, lenses of coal, and possible suncracks) implies that the water was shallow and near shore. The fossil plant material and freshwater lamellibranchs, and the red color of the lower strata, indicate warm continental conditions. The higher maturity may suggest that the source area was now less elevated and that conditions were more stable.

The significance of orthoquartzite in the Lotong Sandstone may be misinterpreted, however, for sediments deposited under tropical conditions probably attain a high maturity index more rapidly than is generally believed. Furthermore, the occurrence of granite and limestone clasts in the Lotong Sandstone shows that deposition was still rapid, whereas the occurrence of what appears to be intraformational conglomerate suggests intermittent uplift. Indeed, Rishworth (in prep. B) considers that the formation comprises several subcycles of deposition, possibly consequent upon repeated uprise of the source area. Volcanic intercalations in the Lotong Sandstone, including some of basaltic character, may be related to these postulated phases of uplift. The carbonaceous argillite and coal beds accumulated in swamps that may have resulted from interruption of the surface drainage either by the same tectonic movements or by flows of lava.

It must be emphasized, however, that the overall decrease in grain size and increase in maturity from base to top of the Gagau Group represents a coherent sedimentary cycle, evidently resulting from the continuous degradation of the source area following its initial uplift. The subsequent vertical movements seem therefore to have been merely subsidiary in nature, producing only minor breaks and rather indistinct subcycles of sedimentation both in the Badong Conglomerate and in the Lotong Sandstone formations.

Since the Gagau Group is thought to comprise at least 900 to 1400 m (2950 to 4600 ft) of essentially shallow-water sediment, it is evident that subsidence of the basin floor took place as deposition proceeded. The coarse red Badong Conglomerate is a very-shallow-water deposit, whereas the generally paler colored and finer Lotong Sandstone was probably laid down in rather deeper water, suggesting that on the whole, deposition failed to keep pace with subsidence. The coarse grain and high labile content of the Gagau Group demonstrate that sediment was supplied to the basin and buried at a rather fast rate. It is therefore inferred that subsidence of the Gagau basin floor was also rapid.

The Badong Conglomerate is evidently a red bed association of Newark type (Pettijohn 1957, pp. 625–630), as demonstrated by its coarse-grained, coarsely cross-bedded, strongly colored rocks, which are almost wholly clastic and which thin rapidly away from a postulated contemporary fault scarp. Pettijohn (1957) observes that the Newark-type association is a kind of oxidized molasse. That this interpretation applies also to the Badong Conglomerate is supported by the associated volcanicity. Molasse is characteristically a late-geosynclinal sequence, and Aubouin (1965 p. 97) has noted that this tectonic stage is commonly signalized by volcanic rocks of an andesitic character (including trachyandesite). The bulk of the volcanics in the Gagau Group are of trachytic and trachyandesitic composition.

The basal part of the Lotong Sandstone represents depositional conditions similar to those evidenced by the Badong Conglomerate, but the indications of continued contemporary vertical movements and associated basaltic volcanic rocks also link the Lotong Sandstone with the somewhat later postgeosynclinal phase of Aubouin (1965).

Since the Gagau Group furnishes evidence of contemporary strong negative as well as positive vertical movements, but shows little folding even though it overlies folded and intruded rocks, it appears that the group was generated at the conclusion of a tectonic cycle, identified by Cady (1950) as the "Stage of final differential uplift and local subsidence." We know that deposition of the Gagau Group was followed by further uplift because the group now occurs at elevations up to 1384 m (4525 ft) above sea level.

Tebak formation

OCCURRENCE AND NOMENCLATURE. Generally flat-lying sandstone forms scattered outcrops in east Johore and southeast Pahang. In places it occupies low-lying ground, but typically sandstone outcrops as upland mesas and cuestas, which attain a height of 834 m (2751 ft) on Gunong Beremban and which are evidently remnant from a formerly continuous and extensive body of strata (Plate 17).

This formation was unrecognized or was confused with the Triassic on the geological maps of Malaya compiled by Scrivenor (1931), Ingham (1948), and Alexander (1965). However, in the last ten years its outcrop has been mapped in the Gunong Panti area by P. V. O. Drummond, who referred to it informally as the Panti sandstone formation; in the Sungei Tebak area by S. S. Rajah; and in the Ulu Endau area by C. R. Jones and C. K. Burton. Other outcrops, delimited by the examination of aerial photographs, are shown on the geological map included in this volume. Rishworth (in prep. B) referred to these rocks as the "Lesong sequence" and correlated them with the Lotong Sandstone of the Gagau Group. In a short review, Rajah (1969) proposed the informal name "Tebak formation" to include all the outcrops in Johore. This name is used here to include also outcrops in southeast Pahang.

The Tebak formation overlies older granite and Upper Paleozoic rocks with a marked angular unconformity. In most outcrops the beds are near horizontal, but locally they dip up to 35° (Rajah 1969). Because of the strong relief of the underlying surface, its thickness varies considerably, although it can generally be assessed rather easily owing to the gentle dip. The maximum thickness in the Sungei Tebak area was estimated by Rajah (1969) to be at least 300 m (985 ft). The upper limit of the formation is everywhere the present erosion surface and, apart from recent alluvium and colluvium, no younger rocks have been found resting on it.

LITHOLOGY. The Tebak formation is mainly composed of flaggy or massively bedded quartz

PLATE 17 *View north over south Pahang to Gunong Lesòng showing flat-topped hills capped by sandstone of the Tebak formation. Photograph C. R. Jones.*

sandstone, frequently cross-bedded. There are minor amounts of felspathic sandstone, micaceous siltstone, grey to purple mudstone, and grit; conglomerate occurs in the basal beds. At Bukit Peradong, 10 km (6 miles) southeast of Gunong Lesong, Dawson (in correspondence) recorded that an eroded undermass of granite, shale, and quartzite is overlain by cross-laminated arenite. It comprises a lower, purple-red, partly conglomeratic unit, succeeded by a white unit containing subordinate carbonaceous shale. At the junction with its basement is a sheet of porphyritic rhyodacite, some 150 m (490 ft) thick.

On Gunong Panti, Kee (1966) stated that at some places the basal part of the formation consisted of up to 15 m (49 ft) of dark grey sandy mudstone interbedded with dark grey shale and a few beds of sandstone. Some of this mudstone contained numerous plant fossils. At some localities a basal clay containing coarse quartz grains closely resembled weathered granite and was underlain by an iron pan 15 cm thick. Elsewhere, although there was no true basal conglomerate, near the bottom of the Tebak formation lay pockets and layers of conglomerate and conglomeratic sandstone. These contained angular to rounded clasts of shale and quartz, with feldspar, metasiltstone, metaquartzite and rare granite, quartz porphyry, and possibly volcanic rocks. Other small bodies of conglomerate occurred sporadically at higher stratigraphic levels.

The sandstone that constitutes the bulk of the Tebak formation on Gunong Panti varies in composition from orthoquartzite to protoquartzite and subarkose. It ranges in grain size from coarse to fine, and the grains vary in shape from subangular to subrounded. About 80% of them are quartz (Kee 1966); chert, muscovite, biotite, acid volcanic rock, and feldspar are also present. Feldspar is sometimes present in important amounts (Drummond, in correspondence). A siliceous cement binds the rock together, and the secondary quartz overgrowths, together with modification of grain shape by secondary pressure solution, produce a pattern of interlocking grain margins, giving the rock the superficial aspect of a metaquartzite.

The beds show internal laminations and frequently cross-laminations, and according to Kee the cross-bedding occurs in sets of 0.15 to 0.9 m, the individual foresets being 0.5 to 1.25 cm thick. Drummond (in correspondence) states that the sequence comprises cross-bedded units about 3 m thick. On occasion the sandstone beds are separated by layers of grey shale, often between 8 and 30 cm thick and sometimes micaceous and silty.

Megaripples were reported in sandstone by Drummond (in correspondence), and Kee (1966) observed an occurrence of graded bedding at the base of the formation. Drummond also noted worm casts in shale, possible load-cast structures at various places, and a possible mollusc track in siltstone. Most investigators have remarked the presence of lutite clasts in the sandstone. These are of various shapes and include slabs up to 18 cm across, sometimes broken and contorted. In one exposure on Gunong Panti, Drummond (in correspondence) found small patches of coal, 15 cm wide and 5 to 8 cm thick, which gave the impression of extending into the body of the rock for a considerable distance. He suggested that these were carbonized tree trunks. Jones (in correspondence) found a small coal seam near to the summit of Gunong Panti West, and farther to the west Rajah also reported thin coal seams.

PALEONTOLOGY AND AGE. Plant remains are common in the Tebak formation but are frequently too poorly preserved to be identified. However, Kon'no (1966) described *Ptilophyllum* cf. *pterophylloides* (Yokohama) from Ulu Endau and the following plants from Gunong Panti (Plate 16): Filicales: *Gleichenoides pantiensis* Kon'no (nov.). Coniferales: *Frenelopsis malaiana* Kon'no (nov.), *Frenelopsis malaiana parvifolia* Kon'no (nov.). Incertae sedis: *Carpolithes* spp.

He considered the range of this flora to be Upper Jurassic to Lower Cretaceous, more probably early Lower Cretaceous.

The Palaeontology Laboratory of Brunei Shell Petroleum Ltd. (*in* Kee 1966) found sporomorphs common in lutite clasts from the Tebak formation at Gunong Panti. These included *Classopollis* sp. (cf. *Classopollis classoides*) and *Aequitradites* sp. (cf. *Aequitradites* cf. *verrucosus*). The former ranges from Upper Triassic to Upper Cretaceous, but the latter is thought to be confined to the Lower Cretaceous.

The Tebak formation closely resembles the Lotong Sandstone of the Gagau Group in its orthoquartzite-protoquartzite-subarkose composition, with conglomerate layers, subordinate dark-colored lutite with plant material, and minor occurrences of coal. Both formations are strongly and thickly cross-bedded, both rest on a very uneven surface of hornblende-bearing granite and metasediments of Upper Paleozoic age, and both formations have a similar structural form and topographic expression.

In Ulu Endau, the basal conglomerate of the Tebak formation is similar to the Badong Conglomerate, and volcanic rocks occur low in the sequence both on Gunong Gagau and in Ulu Endau. In Johore the basal conglomerate is more restricted and volcanic rocks appear to be absent.

Thus it seems probable that the Gagau Group and the Tebak formation originally formed a continuous body of strata.

SUMMARY. Rather frequent coarse pebble beds, in combination with sandstone showing much large-scale cross-bedding and some ripple marks, indicates that the Tebak formation was generated in a shallow-water environment. The occurrence of coal and of plant remains (in interbedded argillaceous rocks) supports this conclusion, as do worm burrows in shale and possible mollusc tracks in siltstone. Slightly curled lutite clasts suggest exposure to the atmosphere of newly deposited sediments (Kee 1966). Since some of these clasts are 18 cm long, they probably have not traveled far; and it is implied that shallow water was not far distant from their present location.

It is not yet possible to state with certainty whether the indicated shallow water was fresh or marine, although Kee (1966) has pointed out that the absence of marine fossils is significant, and he considers that the leaves and slender stems of gymnosperms which he found would be unlikely to reach the sea. It further seems likely that the possible carbonized tree trunks recorded by Drummond (written communication) indicate conditions were nonmarine or transitional between nonmarine and marine. The occurrence of planar cross-bedding, with long foresets and partings of plant material, was thought by Kee (1966) to indicate a fluvial environment, and he regarded the evidence of subaerial exposure shown by the lutite clasts as furnishing in agreement with this supposition. It still seems possible, however, that these rocks were deposited in either estuarine or deltaic conditions.

No systematic survey has been made of the cross-bedding, although one good example described by Kee (1966) indicates an eastward transport, and another related feature seems to show current flow to the south. Although scanty, these paleocurrent indications are in accord with Drummond's suggestion (in correspondence) that the quartz and feldspar in the Tebak formation is possibly derived from granite, for large expanses of granite lie to the north and west.

That deposition of this rock unit was preceded by a protracted erosion interval is implied by the strong relief of the surface underlying the Tebak formation and by the occurrence in the Sungei Tebak area of underlying weathered granite and gravelly clay (which seems to be a highly weathered granite) with an iron pan at the base of the formation near Gunong Panti. In the case of the Gagau Group it is considered that the change from a similar erosional province to a depositional province was due to downfaulting. There is no known evidence of comparable faulting in the Panti area, however, and the lower Badong Conglomerate of the Gagau Group, evidently a fanglomerate derived from a fault scarp, is represented by only a very limited and discontinuous development of rudaceous rocks at Gunong Panti.

Possibly the operative force here was downwarping and/or upwarping in the source area. The sedimentary cycle was initiated by a polymict conglomerate that accumulated rapidly, as indicated by the angular nature of some of the clasts and the occurrence of granite pebbles among them. This detritus infilled some of the hollows on the earlier topography, but soon there supervened swampy conditions of vastly reduced sediment supply and of limited aeration in which dark grey shale and mudstone with much plant material was formed.

The sediment imported to this depositional basin became coarser in grain once more when the main body of the sandstone was then laid down. Scattered occurrences of coal are thought to indicate occasional recurrence of local swampy conditions. The amount of feldspar in the sandstone is variable but generally rather limited. If, as Drummond believed, the feldspar is of granitic origin, then it seems either that quartzite also contributed large amounts of coarse quartz, or that exposed granite was some distance from the site of deposition so that much of the feldspar was destroyed in transport. Granite outcrops nearby, but it is also widespread over the possible source area, so that distant granite might have produced the sediment.

A number of outliers of the Tebak formation show that it was formerly considerably more extensive than its present outcrop. It is also possible that this rock unit was originally appreciably thicker, for the evidence of pressure solution at Gunong Panti probably requires a greater superincumbent load than the 46 to 168 m (152 to 551 ft) which now occurs there. As noted concerning the Gagau Group, the exposed situation and rather poor degree of consolidation of this rock unit would render it very susceptible to rapid erosion by tropical rains.

Synthesis

From Ordovician to late Triassic time, the Malayan region was occupied by an orthogeosyncline, with a miogeosyncline located along the western coastal strip and a eugeosyncline situated in axial and eastern Malaya. This pattern was terminated by an orogenic revolution in Upper Triassic time when massive uplift occurred. A huge volume of granite was emplaced to form the Main Range in Middle to Upper Triassic in the vicinity of the miogeosyncline–eugeosyncline border zone. This region, which seems to have acted as a geanticline during at least part of the geosynclinal stage, became a mountain range at the end of the Triassic and eventually constituted the "backbone" of Malaya, known as the Main Range.

Although the geosynclinal sea was on the whole driven out by this orogeny, for a short time a remnant sea persisted in central Malaya, and deposition of the Upper Triassic to Jurassic Tembeling Formation commenced here. In part this may be conformable on the late geosynclinal marine sediments of the Lipis Group (Middle to Upper Triassic) and in part the lateral equivalent thereof, but it is essentially distinct both structurally (being less folded and normally with an unconformable base) and lithologically (constituting terrestrial oxidized molasse).

It is interesting to note that in the same area where the Tembeling Formation may be conformable on the Lipis Group, sedimentation may also have proceeded continuously through the Lower Triassic orogenic phase.

On the west side of the Main Range, a similar red molasse facies (Ong 1968) unconformably overlies Karnian (miogeosynclinal) flysch. This suggests that the newly uplifted Main Range yielded oxidized molasse to either side. East of the Tembeling outcrop lies the Gagau Group, which comprises two formations—a lower Badong Conglomerate and an upper Lotong Sandstone. It is open to question whether the former is to be equated with the Tembeling Formation. The lower Badong Conglomerate is similarly of oxidized molasse facies but seems to pass conformably up into the (Upper Jurassic to Lower Cretaceous) Lotong Sandstone, whereas elsewhere the Tembeling Formation was strongly uplifted prior to the deposition of the Lotong Sandstone.

To the south, Koopmans (1968) has correlated the Murau conglomerate of northeast Johore with the conglomeratic lower part of the Tembeling Formation. In Singapore, the Pasir Panjang member of the Jurong formation is also similar in facies to the Tembeling Formation and is likewise closely associated with Upper Triassic marine eugeosynclinal strata. Farther south, on Pulau Bintan and the nearby Indonesian islands, the gently dipping Bintan formation (Jongmans 1951) may be equivalent. This rock unit, which is composed of arkosic sandstone and clay shale with coal layers, was formerly included in the Tertiary "Plateau Sandstone" (van Bemmelen 1949, pp. 309–310); however, plant fossils show that it is Upper Triassic to Lower Jurassic in age, most probably Rhaetic (Jongmans 1951). From van Bemmelen's account (1949, pp. 304, 308–309, 314) it is clear that similar rocks occur elsewhere in the Riouw and Lingga archipelagoes.

To the north of Malaya, near to the mouth of the Chumphon River in peninsular Thailand, is a sequence of coarsely cross-bedded red clastic strata. In this vicinity the evidently Jurassic lamellibranch *Eomiodon chumphonensis* has been described (Hayami 1960) and it seems that these strata may be the equivalents of the Tembeling Formation, 790 km (480 miles) to the south. Red beds of molasse aspect also occur near Thung Song and Rat Buri in south Thailand. In the northern part of the country, between Lampang and Chiengrai, the geosynclinal Upper Triassic shows a similar development to that in axial Malaya and is again closely associated with postorogenic red beds. Furthermore, the latter seem to be the equivalents of part of the lithologically similar uppermost Triassic to Cretaceous Khorat Group of east Thailand.

Some chronological data are available in the Khorat Group and the following components are evidently correlative with the Tembeling Formation in both age and facies: The Nam Phong Formation of Rhaetic–Lias age; the Phu Kadung Formation, which is possibly Jurassic; the Phra Wihan Formation, which is evidently Jurassic; and the Sao Khua Formation, which is also Jurassic (Lamoreaux et al. 1958, Ward and Bunnag 1964, Iwai et al. 1966).

These strata extend from Thailand into Indochina. Ward and Bunnag (1964) correlated them with the Middle Indosinian ("Lower Redbeds") of Upper Triassic (Carnic–Noric) age and with the lowest part of the Upper Indosinian ("Upper Redbeds" or "Upper Sandstone") which extends from Rhaetic to Upper Cretaceous. Borax and Stewart (1966) agreed with these correlations. It is evident then that the Tembeling Formation of Malaya forms part of a regional development of oxidized molasse facies genetically associated with the culminating Late Triassic orogeny in the Yunnan–Thai–Malayan geosyncline.

The basal part of the Tembeling succession, and possibly of equivalent rock units, was indeed deposited in marine conditions remnant from the geosynclinal stage, but coarse clastic sediments, yielded from the newly formed mountains, quickly converted the Tembeling depositional province into a paralic type of environment, where the bulk of the formation was accumulated in rivers, lakes, deltas, and swamps.

The main body of the Tembeling Formation was possibly derived from the west from the Main Range, and observations on the cross-bedding of the rocks at Chumphon in south Thailand show that these also had a westerly source. It seems, however, that contemporary rocks in Johore and Singapore were derived from a preexisting elevated area lying to the east of southern Malaya, and an easterly source may also have made a minor contribution to the Tembeling Formation.

The generally shallow-water character of the Tembeling Formation throughout its stratigraphic thickness of 3000 m (9850 ft) indicates syndepositional depression of the basin floor. The coarse and high-labile character of the lower strata suggests that deposition, and consequently subsidence, was initially quite rapid but slowed down as the red argillite of the top third of the formation was accumulated. Subsequently, depression of the Tembeling basin gave way to uplift. Tembeling rocks were evidently deposited near to sea level, but now they build Gunong Tahan, 2192 m (7186 ft) in height, the highest peak in Malaya. It is not contended that the whole of this elevation occurred directly after deposition, but since the highest elevation of the Tembeling Formation is 812 m (2661 ft) higher than the highest elevation of the top of the succeeding Lotong Sandstone, it seems that the immediately post-Tembeling elevation was of this order at least. As uplift occurred, folds of long wavelength were also imposed upon the strata.

Equivalents of the Tembeling Formation in northeast Johore, Singapore, Indonesia, and south Thailand show no indication of having experienced strong post-Tembeling uplift, since they all now occur within a few hundred feet of sea level at the most. Since the Badong Conglomerate of the Gagau Group, located close to the Tembeling outcrop, was also not significantly affected by these movements, it might be inferred that it is in fact post-Tembeling in age.

Before any conclusions are made regarding these age relationships, however, it must be taken into consideration that the two rock units are separated by the Lebir fault, which seems to be a major structural feature in northeast Malaya (Burton 1967e). The Tembeling Formation lies on the upthrown side of this fault and the Gagau Group on the downthrown side. Indeed, it is this downthrow which seems to have transformed a previous erosional province into the Gagau depositional basin.

It may then be that either vertical movement on the fault predates the Tembeling Formation and that the Badong Conglomerate is of Tembeling age and the post-Tembeling movement was confined to the west of the fault; alternatively, the main displacement on the fault may have been syngenetic with the post-Tembeling uplift and the Badong Conglomerate, banked up against the fault scarp, may therefore be later than the Tembeling Formation.

The post-Tembeling movements were succeeded by another depositional phase over much of eastern Malaya and, as suggested previously, possibly it was the same movements which constructed the new sedimentary province. Nearly everywhere sedimentation now took place upon an irregular surface, which shows every indication of having experienced a prolonged erosion interval. In only one place are these new sediments known to rest on the Tembeling Formation (at Gunong Penumpu where the contact is unconformable).

These post-Tembeling rocks typically comprise coarsely cross-bedded, pale-colored, nonmarine orthoquartzite–protoquartzite–subarkose, with minor conglomerate and carbonaceous lutite and a very small amount of coal. At least part of this facies is of Upper Jurassic to Lower Cretaceous age, as indicated by plant remains at several places. In the Gagau basin it is represented by the Lotong Sandstone, preceded by the Badong Conglomerate, and it occurs in south Malaya as the Tebak Formation.

Rocks that have been compared with the Lower Tertiary "Plateau Sandstone" on the Indonesian islands south of Singapore may also be contemporaneous with the Lotong Sandstone. On grounds of lithology and topographic expression, Rishworth (in prep. B) has correlated the Lotong Sandstone with the Phu Phan Formation of the Khorat Group in east Thailand. The two rock units also seem to be more or less equivalent in age. Ward and Bunnag (1964) assigned the Phu Phan Formation to the Middle to Upper Jurassic, and Borax and Stewart (1966) allotted it to the top of the Middle Jurassic. The formation was further correlated by these authors with part of the Upper Indosinian of Laos. Like the Tembeling Formation, therefore, the Lotong Sandstone and the Tebak Formation of Malaya form part of a regional lithofacies development.

In Malaya, this lithofacies is distributed approximately along the strike of the former eugeosyncline. From the northernmost outlier of the Lotong Sandstone to the southernmost outcrop of the Tebak Formation is a distance of 448 km (280 miles). Within this belt there are gaps of 58, 83, 128, and 171 km (36, 52, 80, and 107 miles) wherein rocks of this association have not been found. Possibly therefore, rather than one continuous basin, there may have been at least two depositional basins, one near Gunong Gagau (apparently fault determined) and one in the Panti–Lesong area but they may have been interconnected.

Throughout the depositional province, the succession was initiated by conglomerate, rapidly deposited and partly infilling the hollows on the uneven floor. This rock is much more strongly developed in the Gagau area (Badong Conglomerate) than elsewhere, possibly because of the proximity of a contemporary fault scarp. These coarse basal rocks were succeeded by the main sandstone, with minor lutite, evidently laid down in fluviatile, lacustrine, and paludal conditions. Transport from the west is suggested by the very limited paleocurrent data available.

Rishworth (in prep. B) believes that the Gagau Group comprises several subcycles of deposition which may have resulted from repeated uplift of the source area. This is perhaps supported by the occurrence of basalt amid the volcanic rocks intercalated in the Gagau Group, for basaltic volcanicity seems to be normally associated with vertical movement (Joplin 1960, Aubouin 1965, p. 100). The rather coarse grain and immature nature of the sandstones shows that sedimentation was fairly rapid, and their thickness implies that the sedimentation was accompanied by sympathetic subsidence.

Clearly, conditions of deposition were similar to those of the Tembeling Formation, but the Lotong Sandstone and related rocks are less oxidized and more mature. The former characteristic may be explained by deeper water in the depositional zone; and the higher maturity index may result from lower relief of the source area, a more humid climate, longer transport of the sediment, a more mature source (the Tembeling Formation itself?), reworking within the basin of deposition, or a combination of these factors.

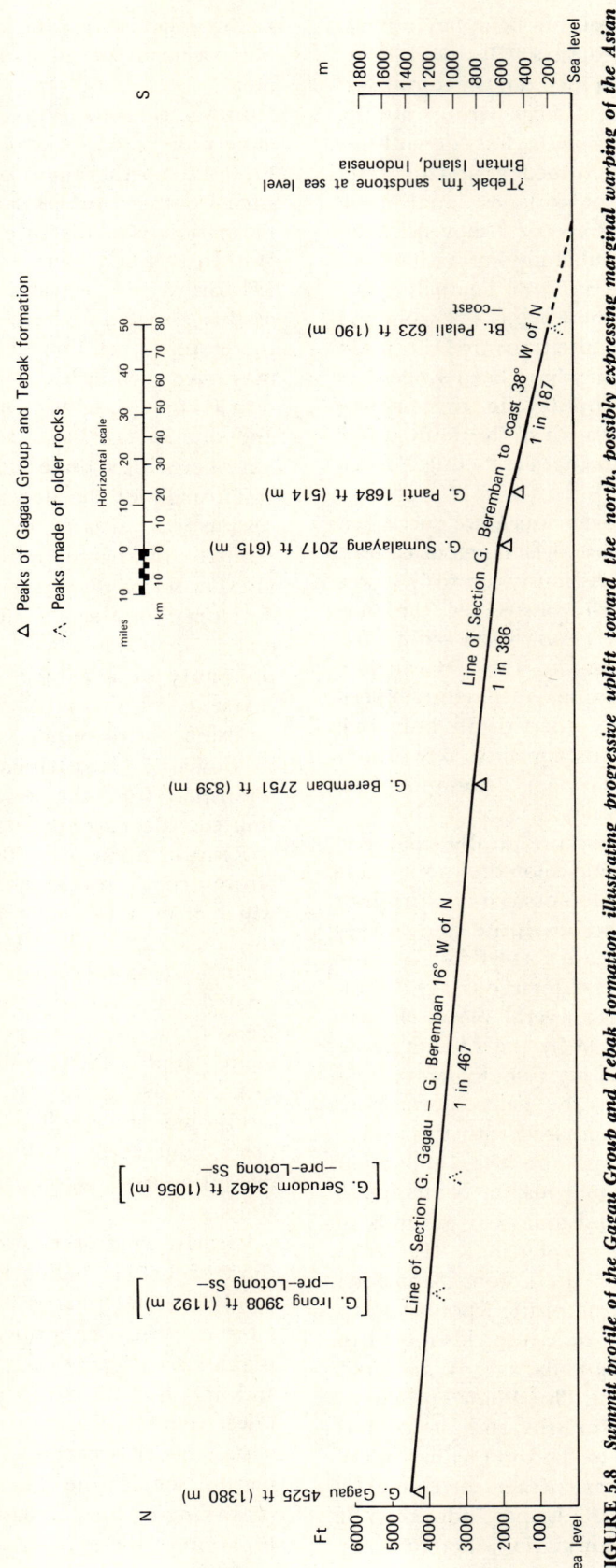

FIGURE 5.8 *Summit profile of the Gagau Group and Tebak formation illustrating progressive uplift toward the north, possibly expressing marginal warping of the Asian continent.*

Potassium-argon radiometric age determinations indicate a tectonic event about 135 m.y. ago, approximately contemporaneous with the deposition of the Lotong Sandstone. However, two specimens from a granite mass in Thailand have both given whole rock ages of 110 m.y. This age is significantly younger than the Upper Jurassic potassium-argon ages determined on Malayan rocks. The extent of these Middle Cretaceous granites elsewhere, in Malaya and Indonesia, is not known (Snelling et al. 1968).

The 135 m.y. concentration of potassium-argon radiometric dates may well indicate the age of wrench faulting, which has strongly displaced the Triassic Main Range Granite and the Semanggol formation sedimentary rocks in northwest Malaya (Burton 1965). A potassium-argon date of 150± 8 m.y. (Anon. 1966) on the Kupang granite may also be related to the faulting. The faults are found throughout the Malay peninsula; the main ones are aligned northwest-southeast and show sinistral displacements. A minor set runs northeast-southwest and exhibits dextral shifts. The cumulative effect of these movements would be north-south extension of the peninsula accompanied by progressive eastward displacement toward the south (i.e., westward-convex curvature). It has been suggested that this phenomenon is an expression of continental drift of the peninsula eastward from Gondwanaland (Burton 1967a).

Deposition of the Gagau Group and the Tebak formation was succeeded by another phase of strong uplift so that these rock formations now stand well above sea level. The age of the elevation is uncertain, but the strata are known to be Upper Jurassic to Lower Cretaceous and all proven Tertiary sediments in both Malaya and Thailand are quite distinct both in situation and in facies, having accumulated in small usually intermontane basins. This uplift could be correlated with the Late Cretaceous (70 m.y.) concentration of radiometric ages (Snelling et al. 1968). It was accompanied by minor folding, so that the main outcrop of the Gagau Group now constitutes a structural basin and the Tebak formation is gently undulating in form. Elevation is the dominant feature, however, and there is a pronounced southward tilt. Uplift evidently attained its acme where the Lotong Sandstone reaches an altitude of 1380 m (4525 ft) at Gunong Gagau.

At Gunong Beremban, 251 km (156 miles) to the south-southeast, the top of the Tebak formation is at 839 m (2751 ft). It reaches 615 m (2017 ft) on Gunong Sumalayang (some 87 km (54 miles) to the southeast, and 18 km (11 miles) away on the same bearing it attains only 514 m (1684 ft) above sea level on Gunong Panti West.

These points lie close to a straight line passing through the Gunong Gagau and Sumalayang summits with an overall gradient of 1 in 453 (Figure 5.8). In detail, however, the line joining the peaks of Gunong Beremban is inclined at a slightly steeper angle, and the short sector from Gunong Sumalayang to Gunong Panti West is considerably steeper, with a gradient of 1 in 174. If this line be extended it is found to pass very near to the top of Bukit Pelali, 190 m (623 ft), in south Johore and to plunge into the sea off the northern shore of Bintan island, 24 km (15 miles) southeast of Johore. Farther north, a number of other peaks in Malaya are close to this summit profile, and it also seems significant that rocks of the equivalent Khorat Group of Thailand reach an elevation of 1316 m (4318 ft), at Phu Kadung, very similar to that of 1380 m (4528 ft) at Gunong Gagau.

In effect, because of the stronger uplift of north Malaya relative to the south, it is only in the northern part of the country that rocks of the catazone reach the surface (Taku Schist, Stong Complex, Kemahang granite). Thus the northward increase in the grade of metamorphism exhibited by Malayan Triassic rocks (noted by Kobayashi et al. 1966, p. 111) may be illusory, and explicable in terms of greater uplift and deeper penetration by erosion in the north.

Since the imperfectly consolidated Lotong Sandstone and its equivalents are relatively susceptible to tropical weathering, it might be argued that the summit profile that has been constructed is unlikely to be related to a preuplift topographic surface. However, various eminences made of other rock types lying close to the profile attest to its validity.

It further appears that present-day erosion of the Lotong association is essentially a process of scarp retreat and that vertical degradation by attendant and other processes is at a minimum. The top of the Lotong Sandstone and Tebak formation may hence represent a depositional surface. Where the same plane is formed of older rocks, it is clearly erosional in origin, and the feature thus seems to be an inclined gradation plane, comprised of both depositional and erosional elements. Since this surface appears to continue beneath the Java Sea, it may represent the northward extension of the Sunda peneplain (Molengraaff 1921, Brouwer 1925). The increasing tilt of the plane to the south seems to be an expression of marginal downwarping of the Asian continent.

CHAPTER 6

Cenozoic

P. H. STAUFFER

Throughout Cenozoic time, the southern part of the Malay Peninsula was largely or entirely emergent. It was also relatively stable tectonically, activity being confined to epeirogenic uplift and tilting, some fault movements, and some local gentle downwarps. There is no evidence of any major marine transgression nor of the formation of any major terrestrial basins during the Cenozoic. The known Cenozoic deposits are all superficial and relatively thin and, except near the coastal margins, all appear to be terrestrial.

Known or suspected Cenozoic stratified rocks and sediments include the following:

1. A series of small lacustrine–paludal filled basins between the Main Range and the west coast. Five such basins have been recorded, some of them having yielded usable lignitic coal, and others may be suspected. Dated by plant fossils as late Cenozoic, probably late Miocene or Pliocene.

2. Local surficial basalt flows, near Kuantan in Pahang and at Segamat in Johore. The material, was formerly thought to be early Tertiary, but recent dating of the Kuantan basalt by the potassium-argon method indicates they may be early Quaternary.

3. Widespread deposits of weathered gravel, sand, and clay, with peat and partly lignitized plant fragments up to tree size, and locally including a basal boulder gravel. These rocks and sediments occur mainly under the coastal plains and the broad lower courses of major stream valleys, but they also form low dissected hills up to about 75 m (250 ft) elevation. Most of these deposits are clearly fluvial or colluvial in origin, but some are enigmatic. Dated by ^{14}C as older than 39,000 B.P. and dated locally by vertebrate bones as Pleistocene, they are suspected to be mainly early and middle Pleistocene and are referred to herein as Old Alluvium.

4. Terrace deposits of weathered gravel and finer material bordering some of the larger rivers are undated but presumably Pleistocene.

5. A locally conspicuous rhyolite ash deposit, overlain only by modern soil. At some localities it has apparently been reworked by streams. Undated, but suspected to be from a Pleistocene eruption in Sumatra.

6. Extensive modern deposits, especially on the broad lowland coastal plains and along river courses, and including only slightly weathered or unweathered gravel, sand, clay, and peat inland, are herein referred to as Young Alluvium. They include multiple beach ridges and marine and lagoonal clays near the coasts. Locally dated as Holocene but may include some late Pleistocene.

Also present are various residual products, including laterite, iron pans, bauxite, as well as ordinary soil and colluvial deposits, which must have developed during Cenozoic time. This chapter, however, does not deal with these varieties in detail.

TERTIARY SEDIMENTARY ROCKS

The known Tertiary sedimentary deposits of West Malaysia occur in a series of small basins between the Straits of Malacca and the mountains of the interior. All are low lying, being barely above the level of the alluvial plains, and except where mining operations for coal have been carried out, the deposits are very poorly exposed. These basins in West Malaysia appear to be the southward extension of the series of Tertiary basins in peninsular Thailand, where the deposits have been called the Krabi series (Brown et al. 1951, p. 40).

The sediments in the West Malaysian basins consist of partly consolidated gravel and sand; soft shale, often carbonaceous; seams of low-grade coal (lignite); and rare calcareous shale and limestone. Thicknesses are of the order of a few hundred feet. The sediments are mainly, and probably entirely, continental, lacustrine, paludal, and fluvial. They are generally almost flat-lying, but dips of 30° to 40° occur, and a synclinal or basin structure has been demonstrated for some of the Tertiary areas. All the Tertiary occurrences lie unconformably on much older, Paleozoic to early Mesozoic rocks that are generally folded and partly metamorphosed. Information about the Tertiary deposits has most recently been briefly summarized by Renwick and Rishworth (1966).

The five areas of known Tertiary deposits in West Malaysia (Figure 6.1) are, from north to south:

1. Bukit Arang–Betong.
2. Enggor.
3. Batu Arang.
4. Kepong.
5. Kluang–Niyor

Bukit Arang–Betong

The existence of coal-bearing strata in Perlis near the Thai border was first reported by Scrivenor (1913b). Thin coal seams seen in pits near Bukit Arang led to the name "Bukit Arang Coal Beds" and to suggested correlation with the coal-bearing beds at Batu Arang in Selangor. Intermittent later investigations, mainly concerned with the possibility of commerical coal, were climaxed by deep exploratory drilling in 1941 (Alexander 1947). More recently, the area has been studied in some detail in the process of preparing a memoir on the geology of Perlis, north Kedah, and the Langkawi Islands (Jones, in prep.). The following account is taken largely from these sources.

EXTENT AND THICKNESS. The Tertiary deposits underlie an area of rolling land mostly at about 60 m (200 ft.) above sea level, but rising to 122 m (400 ft). Natural exposures are almost nonexistent, and so the extent of the deposits is known only where drilling or pitting has been carried out. The original discovery was near the border in Perlis. A series of test pits dug in 1958 showed that this area of deposits extended 7 or 8 km (4 to 5 miles) to the southwest into Perlis (C. R. Jones, quoted in Renwick and Rishworth 1966, p. 29). In addition, another area of similar deposits has been found on the Kedah–Thai border from 7 to 13 km (4 to 8 miles) southeast of Bukit Arang. Both areas seem to widen toward Thailand, and since coal has been reported from southwest of Sadao across the border, it is likely that both are parts of one larger basin that lies mainly within Thailand. The area within Kedah and Perlis has been estimated at about 40 km^2 (15 mile^2) (Jones in prep., pp. 283–284).

The exact thickness of the deposits is not known, but it is more than 180 m (600 ft). A drill hole at Bukit Arang on the Perlis–Thai border had reached that depth and was still in soft sediments in late 1941, when work was terminated by the war. The underlying rocks at the edge of the Tertiary area are clastic sedimentary rocks of Carboniferous age (Jones in prep.).

LITHOLOGY. Bore holes in the Bukit Arang beds have shown a variety of sediments of a fluvio-lacustrine type. Most abundant are plastic clays of varied colors, but also present are sandy clay, loose sand, coarse feldspathic sand, gravel, and poorly sorted boulder gravels. Thin seams of soft lignitic coal occur but are not abundant. Most are only a few centimeters thick. Some of the clays are chocolate-brown in color, and some may be slightly bituminous. Oil shale has been reported on the Thai side of the border.

Pits dug into the deposits show that beds near the surface are lenticular and the deposits are in general poorly bedded and poorly sorted, with gravel lenses. The large clasts appear waterworn and well rounded, and they range up to 30 cm in diameter. Most are of quartzite and quartz, with some granite and chert. Small amounts of cassiterite occur in the sands.

The very coarse sediments appear to be on top, since Jones (in prep.) remarked that high ground is underlain by gravels and low areas by clays, and the report on the deep drilling (Alexander 1947) did not mention boulders or difficulties in drilling such as might have been expected if a boulder deposit had been encountered.

FOSSILS AND AGE. No fossils have been reported from the deposits in the Bukit Arang area. The

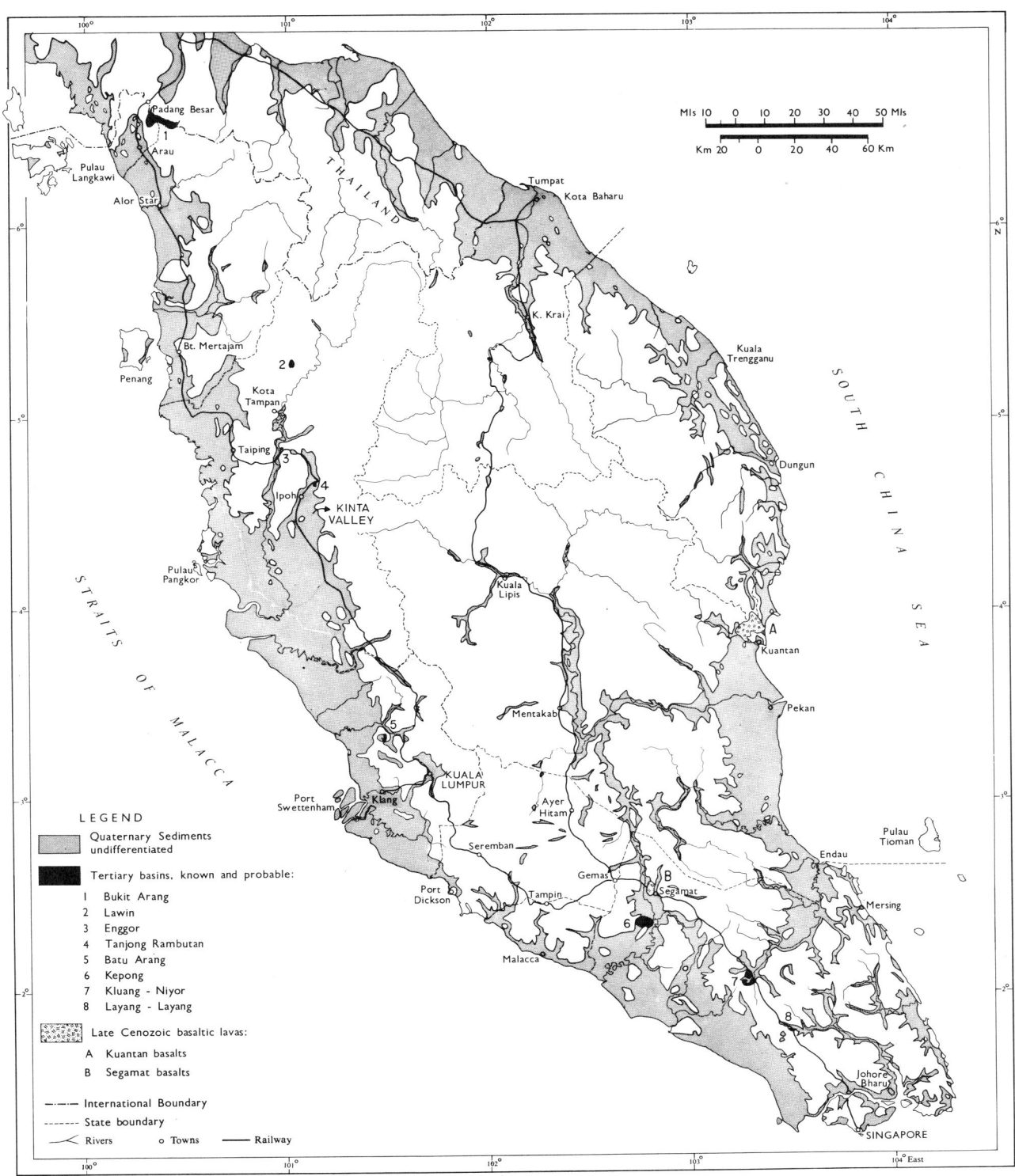

FIGURE 6.1 *Distribution of Cenozoic sedimentary rocks and basaltic lavas in the Malay Peninsula. Boundary of Quaternary sediments based on geomorphic interpretation of 1:63,360 topographic maps supplemented by published data and field checks. The "boulder beds," the Old Alluvium, and all the river terrace deposits are assumed to be Quaternary. Data for southern Thailand very approximate.*

strong similarity—in general character, in apparent origin, and in moisture content of the coals—to the better dated deposits at Batu Arang (see below) suggests a late Tertiary age.

Enggor

A small area of poorly consolidated coal-bearing strata was reported in 1917 (Scrivenor 1917) on the south slope of the Sungei Enggor, a tiny tributary of the Sungei Perak in central Perak, lying at elevations of 46 to 76 m (150 to 250 ft) above sea level. The area was prospected and drilled, and until 1928 mining operations to recover the coal were carried out, both underground and in surface pits.

EXTENT AND THICKNESS. The area is very small. Scrivenor (1917) estimated the area of coal-bearing rocks at about 8.5 hectares (21 acres), and F. T. Ingham's map (reproduced in Renwick and Rishworth 1966, after p. 29) shows the total area of Tertiary sediments as about 20 hectares (50 acres). Despite considerable prospecting in surrounding areas, no major extensions have been found.

Several shafts and bore holes were put down through the sediments at Enggor. The original one in 1916 reached a depth of 40 m (131 ft) (Scrivenor 1931, p. 116), and later bores penetrated as much as 62 m (205 ft) of sediments (Renwick and Rishworth 1966, p. 36). The Tertiary beds rest on partly metamorphosed calcareous and argillaceous sedimentary rocks of probable Paleozoic age.

LITHOLOGY. The sediments in the Enggor basin are mostly poorly consolidated clays and sands, with some lignite seams and calcareous shales. The 1916 shaft revealed two seams of coal 1.2 m thick (4 ft) and bottomed in calcareous shale. The coal is apparently lenticular, for other shafts and bores showed only scattered thinner coal seams, or no coal at all but up to 27 m (90 ft) of calcareous shale (Scrivenor 1931, p. 116).

FOSSILS AND AGE. No fossils other than fragmentary plant remains have been reported from the sediments at Enggor. Their similarity to the deposits at Batu Arang indicates that a late Tertiary age is likely.

Batu Arang

The most intensely studied and best known area of Tertiary sediments in West Malaysia occurs at Batu Arang in Selangor. This occurrence was discovered in 1908 by a local Malay man panning for tin ore in the Rantau Panjang Forest Reserve at about 21 m (70 ft) above sea level (Scrivenor 1931, p. 114). The occurrence was referred to in the early literature by the names Rantau Panjang or Selangor Coal Measures. Upon investigation, this area proved to have two thick [up to 14 m (45 ft)] and many thinner seams of lignitic coal, easily workable because tilting and erosion had brought them to the surface and they were covered only by soil.

Mining of the coal began in 1915 and continued until 1960, when the Malayan Collieries operation at Batu Arang finally ceased. The old mine pits are now flooded. Because of studies spurred by the economic interest, and thanks to the excellent exposures in the mines and several exploratory boreholes, the features of this Tertiary basin are known in considerable detail. The geology was described by Roe (1953) and later by Law (1961); the history of prospecting and mining is detailed in Renwick and Rishworth (1966). The following account is mainly after Roe's memoir and Law's thesis.

EXTENT AND THICKNESS. The sediments at Batu Arang form a roughly triangular basin encompassing an area of about 15 km^2 (6 mile^2) (Figure 6.2). They lie with marked unconformity on much older and steeply dipping metasediments, mainly quartzites. Detailed surface mapping has delimited the boundaries with reasonable precision, save on the southwest side, where Law's map (1961) is better than Roe's (1953). But on the western third of the perimeter, only an upper unit of boulder beds, probably Quaternary in age, is present. The actual area of Tertiary sediments is less, approximately 13 km^2 (5 mile^2).

The present thickness of the sedimentary fill at Batu Arang varies with the position in the basin, because the eroded top has only relatively small relief, whereas the base of the sequence shows marked basinal structure. A maximum thickness in excess of 427 m (1400 ft) is reached in the central part of the basin, as proved by diamond drill borings (Figures 6.2 and 6.3). However, most of this great thickness is in the upper, probably Quaternary unit of boulder beds, which are considered later.

The Tertiary sediments (the so-called coal measures) have a maximum recorded thickness of 265 m (869 ft) in diamond drill hole 4. Where the beds crop out, along the eastern and northern sides of the basin, the sequence is 183 to 244 m (600 to 800 ft)

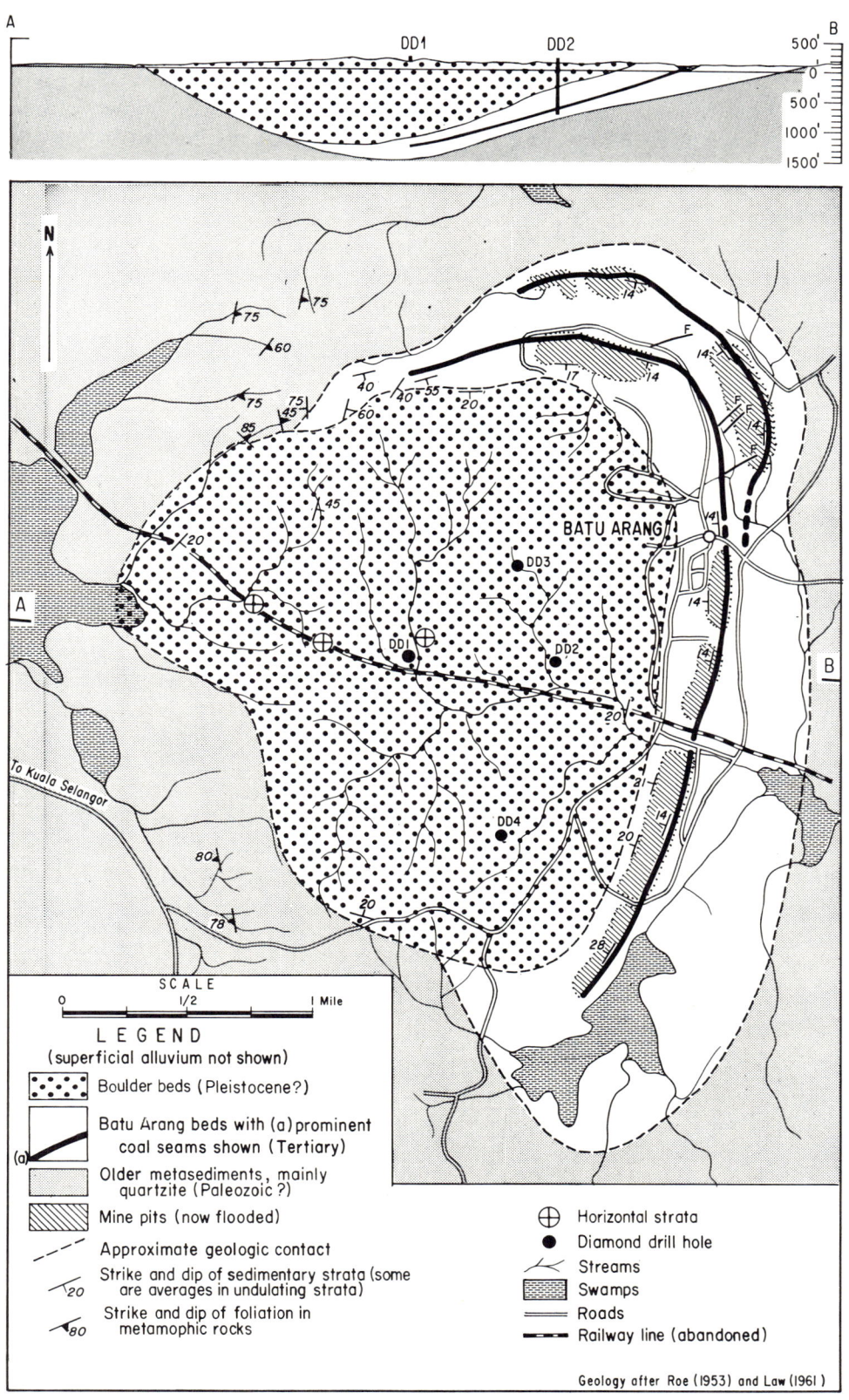

FIGURE 6.2 *Geological map of the Tertiary basin at Batu Arang, Selangor. Mainly after Roe (1953) and Law (1961) with some additions and corrections.*

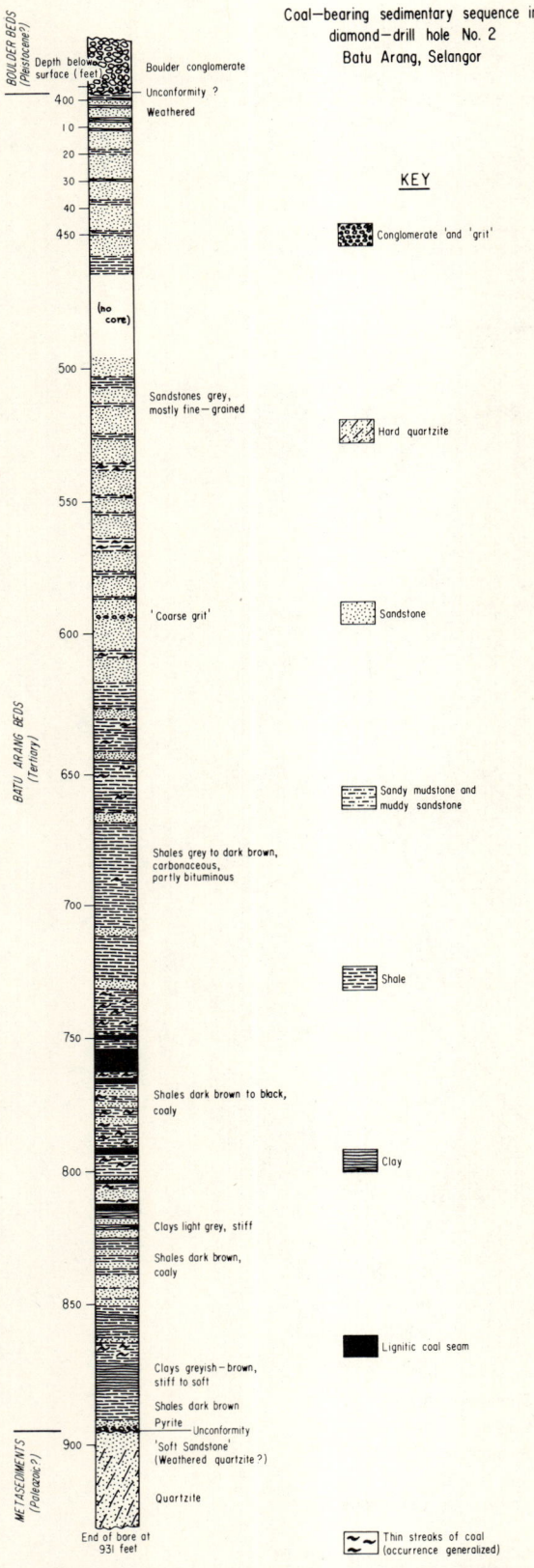

FIGURE 6.3 *Columnar section through the Tertiary sediments at Batu Arang, Selangor, based on a drill log given in Roe (1953, pp. 61–64). No thickness corrections have been made to take account of the dip (up to about 15°).*

thick, calculating from dips and map widths of outcrop. Drill hole results show that the sequence thins westward, to less than 91 m (300 ft) at the center of the basin (drill hole 1). Farther west it wedges out completely, since along the western side of the basin the overlying boulder beds lie directly on the older bedrock. This thinning appears to represent mainly an original variation in deposited thickness, since identifiable subportions of the section also thin westward. Some of the variation in thickness may also be due to erosion prior to deposition of the apparently unconformably overlying boulder beds.

LITHOLOGY. The coal-bearing sequence of sediments at Batu Arang consists mainly of clay and shale, commonly carbonaceous, and fine-grained sandstone. Also included are beds and lenses of coarser sandstone and conglomerate, as well as the lignite seams themselves. The only detailed account of the petrology of the deposits is that given by Law (1961), although their general character had already been described by Scrivenor (1911), McCall (1922), and others.

A somewhat detailed section through the deposits, derived from the log for diamond drill-hole 2 (Roe 1953, pp. 61–64) is given as Figure 6.3. No corrections have been made for the dip of the beds, here about 15°.

At the base of the section, just above the weathered top of the underlying quartzite, is a very thin conglomerate of quartzite fragments with secondary pyrite, followed by dark brown shale with pyrite in the lower part. Above this lie about 45 m (148 ft) of interbedded shale, clay, sandstone, and lignite, the lateral equivalent of the section containing the mined seams to the east of this drill hole.

The fine-grained sediments range from stiff, structureless clay to well-laminated and fissile shale. The color ranges from rare light grey in some stiff clays, through dark greyish brown, to blackish brown and black. Carbonaceous matter is abundant, both in fine disseminated form, which lends the color to the rock, and as discrete plant fragments, including some fairly well preserved leaves. The shale is often silty or even sandy and grades into clayey sandstone in places. In this section the shale is seen to consist mainly of clay minerals, chlorite, and sericite arranged in a planar fabric, with some grains of subangular quartz and zircon and numerous specks of opaque carbonaceous matter (Law 1961, p. 22).

The interbedded sandstone is mostly fine-grained but ranges up to coarse grained and even pebbly. The color ranges from white through shades of brown, depending on the amount of carbonaceous matter. Most of the organic matter in the sandstone is in the form of discrete wavy laminae consisting of matted plant fragments and coaly material. Small-scale cross-bedding is very common in the sandstone; the larger variety is not recorded. Sieve analyses reveal that the sandstone is often well sorted, although clayey and very coarse to pebbly sandstone is more poorly sorted (Law 1961). Pebbly beds sometimes contain rounded clay pebbles.

The sandstone consists mainly of subangular grains of quartz in a varying amount of clayey matrix, which contains some rounded grains of rutile and zircon. Heavy minerals separated from the disaggregated sandstones include monazite, rutile, zircon, apatite, tourmaline, and iron oxides (Law 1961, pp. 20–23).

The coal or lignite occurs mostly as thin laminae or "streaks" in dark shales, but it also builds thicker layers, including in this section (Figure 6.3) some five that are more than 30 cm thick. The thickest one here is 2.4 m (8 ft), but in the mined area to the east thicknesses several times that are reached.

The coal, which appears to be intermediate between high-grade lignite and subbituminous coal, is a hard, black rock with a resinous luster and a tendency toward conchoidal fracture. It ranges from dull to shiny and contains laminae of very shiny, hard, and brittle jet-black material. Some softer, granular dull coal also occurs.

The Batu Arang coal burns with a long, smoky flame and is noncoking. Analyses (Scrivenor 1931, p. 115, Roe 1953, p. 134, Law 1961, p. 25) reveal an ash content of 2 to 12%, a total moisture content of 14 to 21%, volatiles amounting to about 35%, and fixed carbon ranging from 35 to 45%.

Although some of the coal seams are underlain by clays that might be considered "seat-earths," others are underlain by sandstone, and lenses of sandstone and even conglomerate occur within the seams themselves. Erosional channels filled with sands ("washouts") cut the coal seams in places, and Law (1961, p. 21) showed that the sands filling one such channel are well sorted. Such evidence indicates that the coal is in the main transported organic material washed in by water, and not the *in situ* remains of vegetation. This was already pointed out by W. R. Jones in 1912 (Renwick and Rishworth 1966, p. 64).

Above the coal-bearing portion of the section is a thick sequence of grey to brown shale, becoming increasingly interspersed with sandstone beds toward the top. Thin streaks of coal occur here, as well as disseminated carbonaceous matter, but no thick coal layers are found. Some of the shales are slightly bituminous.

FOSSILS AND AGE. Fossils found within the coal-bearing sediments at Batu Arang include plant remains and some gastropods.

Fossil leaves and other plant fragments were studied by H. N. Ridley, then Director of the Botanical Gardens in Singapore, soon after the sediments were discovered. His conclusions, which were reported in Scrivenor (1911b) and Scrivenor and Jones (1919), have been referred to by Scrivenor (1931), Roe (1953), and Renwick and Rishworth (1966). Ridley noted the following plants:

Lauraceae—*Litsea*
        Tetradenia (*Neolitsea*) or *Lindera*
        Laurinea
Annonaceae—*Polyalthia*
Ericaceae—*Vaccinium scortechinii?*
Euphorbiaceae—*Macaranga*
Marattiaceae—*Angiopteris erecta*
Myrtaceae—*Eugenia*

This flora was commented on by Ridley as follows (quoted in Scrivenor 1931, p. 115):

The fossil leaves were a good deal decayed before they were covered by sand, and in most cases the nerves are indistinct. They are nearly all coriaceous leaves, and suggest an altitude or a drier climate than we have here now, but possibly only the zerophytic leaves have stood preservation. The petioles are missing in all. . . . *Tetradenia* (*Neolitsea*) or *Lindera* . . . occur now on the seashore or on hills at high altitudes. . . . The collection suggests a drier climate than we have at present, but of this one cannot be sure. They are, on the whole, of a modern type—i.e. the kind of plants living at the present day.

This assemblage indicates a young age—Late Tertiary or younger—for the Batu Arang beds. It also suggests either a drier climate than at present, as Ridley noted, or a partly upland source for the transported plant material.

Law (1961) figured a well-preserved leaf identified as *Monocarpia marginalis* (Annonaceae) but reported again that most leaves were too decomposed at burial to be identified.

Pollen analyses have been carried out by Law (1961) and by scientists of the Royal Dutch Shell Group at the Hague (Roe 1953, p. 54). Few definite conclusions could be drawn from these studies because of the lack of reference literature on this part of the world, but the prevalence of modern types at least confirmed the age of the deposits as Tertiary or younger.

Scrivenor (1931) reported poorly preserved shells of gastropods at Batu Arang and suggested that they were *Helix*. Partly crushed remains of thin-shelled gastropods are locally common in the shales. L. R. Cox more recently identified these as *Viviparus*, a long-ranging fresh water form (Renwick and Rishworth 1966, p. 69).

Scrivenor (1931) suggested that the Batu Arang coals were probably Miocene in age, since their moisture content is similar to that of Miocene lignitic coals in Indonesia. Evidence from fossils has merely endorsed the supposition of a Neogene age, but the consistent modernity of the plants makes a Pliocene age seem perhaps more likely than a Miocene one.

Kepong

In 1934, during the sinking of a well in a low-lying area near Kampong Bukit Kepong in Johore, bituminous shales similar to those at Batu Arang were accidentally discovered. The area was subsequently prospected intermittently for the possibility of commercial coal deposits, but the results did not justify actual mining. The history of investigations at Kepong (or Durian Chondong, after another nearby kampong) is detailed in Renwick and Rishworth (1966). The following account is very largely taken from their work.

EXTENT AND THICKNESS. The sediments at Kepong lie under a cover of soil and alluvium in an area where the ground surface is only about 15 m (50 ft) above sea level. Hence they are not naturally exposed, and their extent is only known very approximately from the prospecting pits and bores that have been put down. Renwick and Rishworth (1966, p. 43) report that prospecting in the 1950s covered more than 6000 hectares (15,000 acres), but they give no indication of the area over which the sediments were encountered in the pits. Willbourn estimated the sediments to extend more than 8 km (5 miles) in an east-west direction (Renwick and Rishworth 1966, p. 40).

Malayan Collieries Ltd. sank a number of deep bores in the Kepong area in 1936. The deepest of these reached 194 m (638 ft), and at this depth it passed into volcanic rock. E. S. Willbourn, after a visit to the site, concluded that this volcanic rock and other rock encountered at depths of 120 to 150 m (400 to 500 ft) were interbedded with the shales and, therefore, "that the bore had not yet reached Pre-Tertiary rocks, and that the full thickness of the coal-bearing strata had not been penetrated" (Renwick and Rishworth 1966, p. 41) Hence the total thickness appears to be more than 194 m (638 ft) but remains unknown. The lithologies underlying the Tertiary sediments also remain unknown. Burton (1964, p. 35) reported that the Tertiary beds are unconformably overlain by alluvium, which he considered to be probably early Quaternary in age, containing fragments of coal.

From the level of one volcanic layer in several bores, Willbourn estimated a dip of 1 in 9 (about 6°) to the east-southeast (Renwick and Rishworth 1966, p. 42). C. K. Burton, however, reported dips of 20 to 25° to the west for the sediments exposed in prospect pits (Renwick and Rishworth 1966, p. 44).

LITHOLOGY. The sediments of the Kepong area appear to be essentially similar to the other Tertiary sequences in West Malaysia. At shallow depths the rocks are mostly of carbonaceous shale and clay containing plant fragments, streaks of lignitic coal, and occasional thicker seams of coal. Some of the shale is bituminous, whereas some is calcareous, and Willbourn reported one bed of "hard, grey, impure limestone" (Renwick and Rishworth 1966, p. 40). The coal seams reach reported thicknesses up to about 6 m (20 ft) as shown in rather shallow prospect pits, but they are very lenticular and such thicknesses do not persist laterally.

C. K. Burton reported that the sediments revealed in pits were generally well laminated and that thin coal seams were interbedded with shale and some sandstone. The coal appeared to be transported material, and it was often argillaceus and impure, burning only poorly and with much smoke (Renwick and Rishworth 1966, pp. 44-45).

At greater depths in the deep bores the sediments include more coarse clastic beds and possible volcanic ash beds. Willbourn reported much sand with pebbles and boulders of tuff, shale, and lava at depths between 110 and 170 m (360 and 550 ft) in one bore (Willbourn 1937, Renwick and Rishworth 1966, p. 42). He appears to have regarded some of the volcanic rocks as genuine ash beds contemporaneous with the sediments, although he also implied that some were clasts derived from the erosion of older rocks (Renwick and Rishworth 1966, pp. 41-42).

FOSSILS AND AGE. Willbourn (1937) noted that the carbonaceous sediments in the Kepong area resembled those at Batu Arang and Enggor and suggested they were probably the same age.

Samples of grey marly shale with fossil shells, from depths of 27 to 44 m (88 to 145 ft) in the bores, were sent to L. R. Cox at the British Museum. The shells were identified as belonging to a new species of the freshwater gastropod genus *Viviparus* (which Cox named *Viviparus willbourni*), except for one damaged specimen of a ribbed gastropod tentatively referred to the genus *Thiara*. The presence of *Viviparus*, also known from Neogene freshwater formations elsewhere on mainland Southeast Asia, proves that the deposits are nonmarine and most probably lacustrine. The genus *Viviparus* ranges from Jurassic to Holocene, but in view of the rather good state of preservation, Cox considered it unlikely that the Kepong specimens could be older than Late Tertiary (Cox 1937).

Kluang–Niyor

J. B. Scrivenor in 1910 noted rocks in a railway cutting near Niyor, in Johore, which he thought might be Tertiary, and he recommended prospecting for coal in the area. Such prospecting was done intermittently until the 1930s, but results were uniformly discouraging. The record of activity and the results of some bores and pits were published by Renwick and Rishworth (1966).

EXTENT AND THICKNESS. Kluang–Niyor, like the Kepong area, is a low-lying area mostly between 30 and 75 m (100 and 250 ft) above sea level and with almost no natural exposures; hence the extent of the presumed Tertiary sediments is poorly known. Their reported occurrences range from Niyor south to Kluang [a distance of about 8 km (5 miles)] and from Kluang to the west and southwest about 9.5 km (6 miles). West of Kluang the general dip is to the south, and it is possible that the sediments form a basin several kilometers across.

The thickness of the sediments is unknown but appears to be at least 60 m (200 ft) in places. A bore at Niyor Railway Station was halted at 67 m (219 ft) while still in clays. However, some other bores located "along strike" from known occurrences have passed rather quickly into older bedrock (granites and volcanic rocks). The structure of the deposit is evidently not a simple basin.

Dips of about 15° to the south are reported from near Kluang, and here the presumed Tertiary beds are unconformably overlain by clayey alluvium of probable Pleistocene age (Burton 1964, p. 35).

LITHOLOGY. Most of the sediments reported from below the alluvium in the Kluang–Niyor area are grey to black shales and clays. One occurrence near Niyor was described by F. T. Ingham as being of compact grey shale with numerous concretions or nodules (Renwick and Rishworth 1966, p. 47). The bore at Niyor that reached 67 m passed through clay. Elsewhere, carbonaceous shale with plant material was reported, sometimes slightly bituminous.

A few seams of lignitic coal have been discovered in the sediments. One such seam, found in a well on an estate northwest of Kluang, was claimed to be more than 9 m (30 ft) thick but the quality of the coal was poor. This was underlain by clay and sandstone. J. B. Scrivenor found that it had 18.7% moisture and more than 20% ash (Renwick and Rishworth 1966, p. 50).

FOSSILS AND AGE. Other than unidentified plant fragments, no fossils have been reported from the sediments at Kluang–Niyor.

The sediments are older than the alluvium that unconformably overlies them, and they are generally similar to the beds at Batu Arang and in the other Tertiary basins in West Malaysia. Hence a late Tertiary age also seems likely here. Burton (1964), p. 35) mentioned the possibility of an early Quaternary age, basing his hypothesis on the modernity of appearance of the plant fossils as judged by D. Walker.

Other Areas

TANJONG RAMBUTAN. About 1.5 km (1 mile) south of Tanjong Rambutan in the northern Kinta Valley, an occurrence of dipping beds of shale, sand, and lignite, unconformably capped by later deposits, has been reported. Renwick and Rishworth (1966, p. 19) considered this to be part of the Quaternary alluvium, but a late Tertiary age would seem more likely. Scrivenor (Scrivenor and Jones 1919, p. 83) noted that the lignite here, of all the occurrences in the "alluvium," is "the most compact and homogeneous . . . that I have seen" and "the only seam . . . likely to be useful as fuel." Samples were sent to Shell for pollen analysis, and the results indicated a Miocene age or possibly younger (N. S. Haile personal communication).

LAWIN. A small area of probable Tertiary sediments has been described by C. R. Jones (1970b) in the Grik area of northern Perak. This occurrence is shown on Alexander's (1965) geologic map of Malaya. Jones reported that the sediments occur in a roughly circular area about 2.5 km (1.5 miles) in diameter, with natural exposures along streams, including the Sungei Lawin, a tributary of the Sungei Perak. The exposures are of semiconsolidated, poorly sorted feldspathic sand and gravel with some boulder beds. The beds dip 20° to 45° toward the center of the basin, suggesting a possible thickness of as much as 300 m (1000 ft). Lensing and cross-bedding indicate a fluvial origin, but no lignite has been reported. At the margin of the basin the sediments rest unconformably on granite, quartzite, and tuff. No fossils have been reported, but Jones judged the age as probably late Tertiary to Quaternary, based on the lithology and structure of the sequence in comparison with the known Tertiary basins.

LAYANG LAYANG. Burton (1964, p. 35) reported the occurrence of grey to grey-brown partly consolidated clays underneath Quaternary "older alluvium" near Layang Layang on the railway line south of Kluang in Johore. He noted that these beds might be late Tertiary by analogy with the stratigraphy at Kluang–Niyor and Kepong. But he also wrote that the well-preserved plant remains in these beds were judged to be probably Quaternary by D. Walker.

It is more than likely that the known areas of Tertiary sediments in West Malaysia do not represent the full extent of such occurrences. The ones that are known are generally low lying and poorly exposed, and in some cases completely covered by later alluvium and thick soils, so that discovery is only possible by digging a well or boring. Other occurrences of Tertiary beds may and probably do lie hidden under the alluvium. In addition, some of the "old alluvium" in, for instance, the Kinta Valley, may be late Tertiary in age, rather than Pleistocene as presently supposed. In all these deposits, including the known Tertiary basins, the scarcity of fossils, especially marine fossils, makes precise dating very difficult.

It should be noted here that the beds at Chegar Perah in north Pahang, listed as a Tertiary fossil locality in Jones et al. (1966, p. 358), are now best regarded as being of Quaternary age. This judgment is based on the very young appearance of the deposits in the field and the nondiagnostic character of the rather poorly preserved plant material (T. Kobayashi personal communication reporting work of H. Tsuyama).

Summary and Discussion

In looking at the known Tertiary sedimentary deposits of West Malaysia as a group, several features stand out.

First, all the known occurrences are on the western side of the peninsula, and most are 15 to 50 km (10 to 30 miles) from the coast on the Malacca Straits. The two Johore occurrences are a little farther inland and closer to the center of the peninsula, which here is much narrower than it is farther north.

Second, there is no evidence of marine deposits in any of the known basins. The assumption sometimes made that the sediments are partly estuarine remains unsupported, and the fossils that have been identified are true freshwater forms. The deposits would appear to be fully continental. They have the sedimentologic character of lacustrine, paludal, and fluvial sediments.

Third, all the Tertiary basins appear to have undergone only mild structural deformation. Dips are generally 20° or less; only rarely, as in the northwest part of Batu Arang, are steeper dips reported.

Thus the Tertiary consists of a chain of continental basins in a zone running generally on the west side of the peninsula. One may ask two questions here: Are these known areas of Tertiary sediments merely the scattered erosional remnants of a formerly continuous sheet of deposits? and What produced the basins in which they were deposited?

Tertiary sediments have certainly been more extensive than they are now, for in places we see the dipping beds truncated at the surface or along an unconformity. But it may be doubtful if they ever formed a continuous formation. The nature of the deposits—lacustrine, paludal, and fluvial—suggests discrete basins of sedimentation. If they had constituted the landward edge of a swampy coastal plain wedge, we might expect to find the brackish water or marine facies preserved somewhere, but this is not the case.

The mild deformation of the basins might suggest that the occurrences we see are merely the results of accidental preservation on the sites of downwarps. But the rapid lateral wedging of the deposits, clearly visible at Batu Arang, argues that local downwarp was also occurring at the time of deposition and in fact controlled deposition. There is no evidence of significant folding tectonics in West Malaysia since late Tertiary, although there is evidence of significant fault movements. Enggor, Batu Arang, and Kepong are all located on the trends of major structural lineaments, and at Batu Arang the overlying pile of Quaternary(?) boulder beds seem clearly related to major faulting. We may therefore postulate that the scattered Tertiary basins were produced by late Tertiary structural adjustments mainly involving faulting and that, although other basins that have been eroded away or lie yet undiscovered may have formed, they probably were always a series of discrete and isolated downwarps, possibly far from the coast but probably low lying.

Our knowledge that the known basins are limited to the western part of the Peninsula may reflect concentration of the tectonic activity to that region —radiometric potassium argon dating has indicated a strong Upper Cretaceous to Lower Tertiary tectonic event along this belt of country—or possibly continuance of preferential uplift of the east side of the Peninsula during late Tertiary or early Quaternary, resulting in removal of deposits by erosion. Alternatively, this limitation may reflect greater deformation and hence exposure of Tertiary sediments in the western areas, or more intense prospecting in the western part, or even pure coincidence, since the number of basins is rather small.

QUATERNARY DEPOSITS

Apart from soils and residual deposits (which are considered briefly below), the Quaternary of the Malay Peninsula includes extensive deposits of unconsolidated to semiconsolidated gravel, sand, mud, and clay, occupying the coastal lowlands and the floors of some inland valleys, locally also occurring on terraces or as erosional remnants of higher level deposits (Figure 6.1). All this has traditionally been referred to as "the alluvium," although it includes colluvial and residual materials and near the coasts it is demonstrably in part marine. A marine origin for some of the older inland deposits cannot yet be ruled out.

Because of the valuable concentrations of cassiterite in the "alluvium," much interest was focused on these deposits in the early part of the twentieth century. Mine exposures still provide the best access to the deposits, and in general they are very poorly known in areas without mining activity.

Much of the early work in the "alluvium" (summarized in Scrivenor 1931) unfortunately involved several long and exhausting controversies concerning, among other things, a hypothesis of glacial origin for some "boulder beds," the possibility of a former sea level about 75 m (250 ft) higher than at present, and the distinction between weathered bedrock and young unconsolidated deposits. Perhaps because of weariness induced by these controversies, interest in the "alluvium" subsided for two decades and, apart from the work of Walker (1956, 1962), did not revive until the late 1960s.

Recent studies (Ayob 1965, 1970; Sivam 1968, 1969; Newell 1971) have begun to put our knowledge of the Quaternary sediments on a more solid scientific basis, but all the fundamental questions about their nature and origin have not yet been resolved.

In addition to these sediments, and the residual deposits, the Quaternary of West Malaysia includes a few occurrences of rhyolitic volcanic ash and two occurrences of basaltic lavas that were formerly thought to be Tertiary but now appear to be possibly early Quaternary (Figure 6.1). These volanic deposits are dealt with first.

Volcanic Deposits

BASALTIC FLOWS. Fitch (1952) described flat-lying basalts that form the ground surface over an area of about 150 km² around Kuantan town in east Pahang (see Figures 6.1, 7.6). These basalts are very fresh in foreshore exposures and are clearly extrusive flows. They appear to retain a constructional topographic form, and through their gentle upper surface, higher hills of granite project. The highest point on the basalt surface is near its center, at Bukit Tinggi (138 m; 454 ft). The basalt is compact, microcrystalline, vesicular olivine basalt, with a partly glassy groundmass and locally containing nepheline. Columnar jointing is common.

Fitch (1952) thought that these basalts were associated with dolerite dykes, which intrude Lower Carboniferous rocks in the same region, and suggested for all these basic rocks a tentative Tertiary age. Recent radiometric dating by the potassium-argon method has given a minimum age of 1.7 m. y. B.P. for the basalt from the Kuantan area (J. D. Bignell personal communication). If this date is subsequently corroborated, it will indicate that the flows are early Quaternary in age.

In the Segamat area of north Johore (see Figure 6.1), mildly deformed basic extrusive and hypabyssal rocks have been described by Grubb (1965), who suggested they were contemporaneous with the Kuantan basalts. Grubb identified three individual flows, from 20 to 45 m (166 to 148 ft) thick, with a poorly vesicular basal part sometimes including fragments of the frothy top of the previous flow, and a more vesicular and glassy upper portion, especially near the top. The flows dip about 20° in a gentle syncline and show postconsolidation faulting, slickensides, and hydrothermal alteration. Vesicles are commonly filled by zeolites and metallic sulfides.

RHYOLITIC ASH. A superficial and presumably very young deposit of poorly to moderately consolidated rhyolitic ash has been reported from a few sites in West Malaysia, mostly in the Perak River valley and in western Pahang.

This ash was first reported by Scrivenor (1930, 1931) from Tanjong Perak Estate on the Perak River north of Kuala Kangsar. The ash was exposed on the bank of the river by erosion during the flood of 1926 to 1927, and it was seen to overlie gravelly river deposits, which in turn lay on granitic bedrock. Ash as thick as 9 m (29 ft) was seen in places.

Siliceous "spicules" and "spherasters" found in the ash were at first identified as coming from marine organisms, leading to the inference Scrivenor 1930, 1931) of a former sea level 70 m or more above the present one. Oakley (1940) suggested that the siliceous bodies, apart from some freshwater diatoms, were inorganic in origin; after considerable controversy in the literature, his views were accepted by Scrivenor, who acknowledged (1946, 1949) that the rhyolite ash provided no evidence on former sea levels.

The ash apparently occurs widely in the Perak River valley. Willbourn (1938) described occurrences at Kota Tampan Estate, upstream from Tanjong Perak (see Figure 6.1). Here the ash overlies gravel beds that contain both water-rounded pebbles and angular (colluvial?) fragments of quartz, quartzite, and other hard rocks (Walker, 1962). Some of these pebbles and fragments, which are chipped in a manner that suggests deliberate working, have been described as paleolithic ("Tampanian") tools (Collings 1938, Sieveking 1962). The tools occur only in the gravel layer, where they are diffusely scattered among natural fragments of the same size grade, ranging in shape through all degrees of angularity.

Similar rhyolitic ash has also been found in parts of western Pahang. Richardson (1939) reports up to nearly 5 m (16 ft) of ash overlying alluvium at Kampong Dong in the Raub area. At the Sungei Gok in the same area, 38 cm of ash are overlain by about 1.8 m of soil and alluvium at an elevation of 150 m (500 ft). Alexander (1968) reported that the ash occurs patchily in the Bentong area, at elevations below about 45 m (150 ft), between Triang and Mengkarak.

Character of the Ash. The material is generally a loosely consolidated, friable, and porous white rock, coherent enough to be cut into blocks in the Perak River occurrences. Scrivenor (1930, 1931) reported that the ash consisted mainly of comminuted pumice, locally up to 1 mm in size, but mostly smaller. The Pahang occurrences are described as loosely consolidated fine ashy material, ranging locally up to medium sand size (Alexander 1968). No larger pieces of volcanic material have been reported.

Examination of moderately consolidated ash from the Perak River terrace at and near Kota Tampan shows that it consists almost entirely of glassy volcanic material (Stauffer 1970). Some of this is in blocky bits of milky-white pumice, with flattened vesicles and flow structure; part of it is in clear transparent plates and fragments of glass. The latter often look like "spicules" or branched rods in thin section, but close examination reveals that in every case the plates are flat, curved, or box-corner fragments of thin vesicle walls.

Some of the glass is in the form of "spheraster"-like bubbles, but usually these are attached to a platy fragment. Also present are crystals of quartz, biotite, muscovite, and rare feldspar, all quite fresh. The coherence of the rock is largely due to very numerous small patches of secondary chalcedony cement. There is no evidence in these samples of admixture with ordinary river sediment.

Sieve analyses of the Kota Tampan deposit give a median diameter near the sand–silt boundary and indicate that the material is moderately to poorly sorted; elemental analysis shows its composition to be very comparable to that of an ordinary granite (Stauffer 1970 and unpublished work).

Origin of the Ash. Since the Malay Peninsula has no known Cenozoic sources for rhyolitic material, the ash must be blown by the wind from a nearby volcanic area. The likely source is clearly Sumatra. This layer of ash extending into central Malaya must have been produced by a very large eruptive event, whose deposits should be very prominent in Sumatra. Scrivenor (1931) proposed that the catastrophic prehistoric eruption that produced the caldera of Lake Toba in northern Sumatra, and whose deposits form thick rhyolite tuffs there, was the source of the Malayan ash.

The ash must certainly have been carried by wind to Malaya. But in most reported occurrences the material appears to have been washed and redeposited by running water. Richardson (1939, p. 78) showed that the ash in the Raub area contains the same suite of heavy minerals as the associated alluvium, and in places it includes seams and lenses of alluvial clay, sand, and gravel. The thicknesses reported (9 m from the Perak River valley, 5 m from Pahang) are excessive for a direct-fall ash deposit from volcanoes 320 km (200 miles) away. Thus we have the implication that the material has been washed together by streams and hill wash. The Kota Tampan occurrences, which show no evidence of mixing with ordinary river deposits, are found on a terrace above the river; they are only 1 m or less thick and may represent a relatively undisturbed original deposit. If so, its thickness is still large and we can posit a volcanic event of the first magnitude, whose traces might well be discoverable elsewhere in Malaya.

Stratigraphic Position and Age. At every locality from which it has been reported, the rhyolite ash is a partly or poorly consolidated, porous, superficial deposit, overlain only by soil and locally by colluvium or "hill wash." At Kota Tampan it overlies gravel containing supposed paleolithic tools. It seems safe to infer that the ash is no older than middle Pleistocene.

On the other hand, its very patchy and scattered preservation (although it may be more commonly present, unnoticed, in the subsoil) suggests that sufficient time has passed for most of it to be washed away. The "dunes" of ash reported by Walker (1962) from Kota Tampan may be locally cemented patches preserved as erosional remnants of a formerly continuous deposit that has been partly reworked by the river on the lower part of the terrace there. This terrace is, in turn, most probably Pleistocene in age. A Holocene age for the ash is conceivable, but a late or even middle Pleistocene age is perhaps more likely.

Since the ash has never been reported from mine exposures of alluvium, it is possible that it is younger than most of the inland alluvial deposits. However, weathering and solution of delicate glass shards may be rapid in saturated alluvium. The ash is not ordinarily looked for and is difficult to recognize without microscopic examination.

Soils and Residual Deposits

SOILS. Processes of soil formation in the Malay Peninsula reflect the humid tropical climate of the area. Weathering is deep, and intense chemical decomposition generally dominates physical disintegration. Abundant rainfall, with only a poorly developed dry season in most of the area, effects profound leaching of the regolith, producing deep, rather sterile profiles dominated by quartz and clay minerals and including greater or lesser amounts of ferruginous and aluminous residual materials.

The depth of weathered material can be very great (Figure 6.4). Thicknesses in excess of 30 m are not uncommon. On slopes the regolith tends to

FIGURE 6.4 *Soil and weathering profile developed on granite. Road-cut a few kilometers south of Georgetown, Penang. Cut face is about 8 m high. In the dark (red) lower part, the original plutonic texture is still preserved. Photograph P. H. Stauffer.*

move downhill, both in slow creep and in sudden landslides, especially after heavy rains. Such transported colluvial material, which is extremely common on lower slopes and near valley sides, forms a significant part of the "alluvium" near the base of hills and mountains.

In areas of massive rock, especially granitic igneous rock, weathering produces rounded core boulders of unweathered rock that may be "floating" in a thick layer of otherwise completely weathered material (Figure 6.5). When such material slides or slumps into a gully or small valley, the weathered loose part tends to wash out, leaving a jumbled pile of boulders. Such boulder piles choke many of the smaller stream valleys in the mountains and make travel difficult. Scrivenor (1931, p. 124) mentions the common occurrence of such masses, called *gugup* in Malay. Core boulders create a further difficulty in prospecting and site investigation: sometimes it is difficult to tell, in borings, whether one has really hit solid bedrock or whether the drill is on a core boulder.

The soils of the peninsula have been studied and mapped, mainly from an agricultural point of view, by the Malaysian government's Soil Survey, which has issued a series of regional studies including maps. More detailed treatment of the soils is outside the scope of this chapter and the reader is referred to the Soil Survey's publications.

FIGURE 6.5 *Very coarse colluvium exposed in the J.K.R. quarry at Kuala Dipang at the base of Bujang Malaka, a granite mountain at the southeastern edge of the Kinta Valley. Large core boulders of fresh granite in a matrix of decomposed granite. Photograph P. H. Stauffer.*

RESIDUAL DEPOSITS

Laterite. The residual ferruginous material in the soil is frequently in the form of lateritic pellets or nodules; locally it has developed into true laterite. The presence of laterite in Malaya was noted and discussed by Scrivenor (1931), who pointed out that some of the Malayan material had the characteristic (cited by Buchanan in the original laterite of India) of hardening upon drying. Indeed, Malayan laterite has been used as building stone in the Malacca area. The types of lateritic soils in Malaya have been discussed by Panton (1956).

The Malaysian Soil Survey has estimated that soil associations that include laterite and lateritic soils cover nearly 6% of the land area of Malaya (quoted in Eyles 1970), but fully developed laterite is much less extensive. The physiographic distribution of laterite in Malaya and its significance have been discussed by Eyles (1970). Laterite occurs mainly in the southern and western portions of the peninsula and is limited to low elevations, where it occupies a variety of physical settings, including summits of hills up to 100 m elevation. The laterite is best developed in areas where the bedrock is argillaceous rather than arenaceous or granitic.

Eyles (1970) regards the hilltop laterites as "fossilized" occurrences in which the laterite, originally formed in valley-bottom or foot-slope sites, has rendered the ground more resistant to erosion and hence has caused a reversal of relief upon uplift of the area or lowering of base level. M. L. Leamy (referred to in Eyles 1970) suggested that the hilltop laterites might be remnants of a Cretaceous peneplain. The present elevations of continental late Mesozoic deposits (see Chapter 5) show that northern and eastern Malaya have been preferentially uplifted by as much as 1000 m relative to the western and southern portions. This could account for the preservation of hilltop laterites mainly in the south and west of the peninsula, and their possible removal by erosion in the uplifted areas.

It has also been suggested by Eyles (1967) that the hilltop laterites might be related to a postulated early Quaternary higher relative sea level (about 70 m above the present one), but Haile (in press) has shown that evidence for this former sea level is very inconclusive.

Bauxite. In certain areas near the southern tip of the peninsula, lateritization has resulted in the formation of bauxite. In southeastern Johore, the bauxite is commercially mined and has been described in detail by Grubb (1966, 1968). Bauxite has formed here mainly on acid volcanic rocks, and the bedrock composition has been an important factor in its formation. But low-grade iron-rich bauxite (or aluminous laterite) has formed also on schist and granitic rocks in the same area of Johore, and acid volcanic rocks farther north have not developed bauxite. Suitable climatic conditions, prolonged stability of the area, and low rates of denudation have all contributed to the formation of the bauxite. This combination of circumstances serves to reemphasize that the southern tip of the peninsula has remained near sea level for a long time, whereas areas to the north have suffered uplift and erosion.

The mineable bauxite is almost exclusively at elevations above 15 m and below 70 m (between the 50 ft and 200 ft contours) in an area of low rolling hills, where few summits exceed the higher figure. It has been suggested (Tjia 1969) that the altitudinal distribution of bauxite in Johore and other land areas of the Sunda region is related to a Neogene erosion surface or peneplane graded to a base level about 50 m above present sea level, but this interpretation has been disputed (Wilford 1969).

The ore zone in the Johore bauxite is commonly 2 to 8 m thick and varies from mottled and pisolitic to massive and concretionary. Texture and composition also vary, but normally a matrix of finely granular gibbsite containing different proportions of limonite, hematite, chamosite, and kaolinite forms the actual ore (Grubb 1968, p. 51).

A few occurrences of bauxite in the same area are as accumulations of pebbles in the lowland alluvium. These are invariably close to residual bauxite on the slopes above, and the latter has clearly been the source. These alluvial bauxites, which occur exclusively in the valley bottoms, are thought to be very young, probably Holocene in age.

Fluvial Terrace Deposits

Fluvial terraces are well developed as geomorphic features along several of the larger rivers in West Malaysia. They are especially notable along the Pahang River and some of its larger tributaries, and along the Perak River.

Only in a few areas has the presence of actual river deposits on the terraces been verified, although it is likely that these are generally present where the terrace is well defined.

Along the Sungei Lipis (a tributary of the Pahang) south of Kuala Lipis town, horizontally bedded, unconsolidated river gravel and sand can be seen in road cuttings at the dissected edge of a well-developed low terrace (on which the Kuala Lipis airfield is sited).

River deposits including rounded gravel have been observed in the broad terrace of the Perak River north of Kuala Kangsar. At Kota Tampan, site of both the Quaternary rhyolite ash in its purest known occurrence and the "Tampanian" tool finds, the material underlying the superficial ash is a gravelly layer in which stream-worn pebbles are mixed with angular, probably colluvial quartz fragments and soil. It is in this mixture that the tools have been discovered.

Courtier (1962) noted fluvial terraces at elevations of 30 m (100 ft) and 45 m (150 ft) above present sea level in inland valleys in Province Wellesley, opposite Penang Island. The lower of these two levels seemed to be correlatable with a poorly preserved terrace fringing the coastal plain in this same region. The latter terrace contains sandy deposits with rounded quartz clasts, abundant mica, and fresh feldspar, which Courtier interpreted as fluvial sediments derived from granite.

The Malaysian Soil Survey recognizes a category of soil called "terrace soils," and these have been mapped in some areas (Gopinathan 1968). Their distribution gives at least some indication of the extent of fluvial terrace deposits.

No detailed studies of the fluvial terraces, their heights, ages, and origin have been made, but such investigations should be rewarding.

Inland Valley Fill

A number of inland valleys and lowlands have a filling or cover of unconsolidated sediments. These sediments include the very rich tin placers that have played so important a role in Malaya's prosperity.

The tin placer deposits are especially abundant near contacts between calcareous bedrock and granitic intrusions, and they are preserved generally on the limestone, which has commonly been eroded to a karst plain, with isolated hills rising above it (see Plate 1). The limestone suffers solution, both in the initial erosion down to a bedrock flat and also after burial under alluvium. The latter type of solution causes subsidence and collapse in the alluvium above the dissolving rock (Figure 6.15) and leaves a variety of insoluble residues, as well. As a result, the composition and especially the structure of the unconsolidated deposits are complex and often enigmatic in areas of limestone bedrock. Yet since these areas have been more exposed by mining and are better known, the following account is largely based on them. Valley fill in areas of arenaceous, argillaceous, or other noncalcareous bedrock is stratigraphically and structurally simpler but is less well known. Two of the important tin-producing areas, the Kinta Valley in Perak and the Kuala Lumpur lowlands in Selangor, have provided most of the detailed data.

BOULDER BEDS. At a number of localities, the base of the unconsolidated sediments is a deposit of very coarse gravel. Early work in the Kinta Valley found such deposits of "boulder beds" and "boulder clays" along both margins of the valley, near the base of the hills. These were at first regarded by Scrivenor (*in* Scrivenor and Jones 1919) as being Permian and Carboniferous glacial deposits ("Gondwanas"), but later (Scrivenor 1931, 1949) as Post-Mesozoic and probably Pleistocene sediments.

W. R. Jones (1917) showed that some of the clays and boulder clays were eluvial and traceable into unweathered schistose or phyllitic bedrock. This is the origin of the "Tekka Clays" (Rastall 1927) and the "Western Boulder Clays" (Willbourn 1936a). Some of the other occurrences of boulder beds in Kinta are probably colluvial, especially those just at the bases of hills. But some also appear to be genuine water-washed boulder gravels.

The "Gopeng Beds" (Scrivenor 1912) of the southeastern part of the Kinta Valley were largely mined away in the early decades of the twentieth century. Scrivenor originally thought them to be glacial in origin and Permo-Carboniferous in age, the occasional presence of large tree trunks being ascribed to recent collapse into forgotten excavations, but he later (1931) abandoned the glacial theory. Rastall (1927) and Jones (1917) interpreted the Gopeng Beds as high-level alluvium. The lithology of the deposits was mainly sandy clay, with scattered larger clasts weighing up to 50 lb. The clay was generally pale, but locally stained red, and it showed distinct bedding in most places.

The boulder beds under discussion here are distinguished from younger bouldery alluvium by their state of weathering. In these old boulder beds, granitic clasts, which commonly make up the majority of the larger clasts, are weathered more or less completely (Figure 6.6b). Scrivenor (1912) noted that the boulders and pebbles in the Gopeng Beds were generally so weathered that they fractured through as easily as the clay matrix and hence rarely stood out in relief. Most of such clasts in the boulder beds are now just masses of clay with scattered quartz grains and are generally similar in character to the matrix, which is normally a clayey sand of granitic derivation. The outlines of the large clasts are visible on a cut surface because of differences in texture and color and the presence of veins and other structures. The boulders and cobbles are rounded to subangular. In the matrix there sometimes occur rounded pebbles and cobbles of other lithologies, some of which may be less weathered, (e.g., vein quartz, schist, or quartzite).

FIGURE 6.6 *Boulder beds (early Quaternary):* (a) *At Batu Arang, Selangor, near top of sequence. Photograph P. H. Stauffer;* (b) *at Kanching, Selangor. Pervasively weathered granite clasts deformed by slumping and minor faulting (probably caused by subsidence over limestone removed by solution). Photograph S. P. K. Loganathan.*

Such basal deposits of weathered granitic boulders have been seen in mines in many smaller valleys, for instance, east of Gunong Tempurong at the south end of the Kinta Valley and in the Kanching Valley at Templer Park, north of Kuala Lumpur (Figure 6.6b).

These deposits presumably represent the initial phase of valley filling following a period of vigorous downcutting and excavation. Although they may not be the same age everywhere, they are most likely to be mainly or entirely Pleistocene. The deposits are probably contemporaneous with, and the lateral equivalents of parts of the "Old Alluvium," and they are best regarded as merely a local facies of that unit.

A possibly correlative and related deposit, pointed out as such by Walker (1956, p. 31), is the thick wedge of boulder beds at Batu Arang, mentioned earlier (p. 146) as overlying the Tertiary coal-bearing sediments. The boulder beds here are not buried by any later deposits, and the main lithology in the clasts is a very resistant quartzite derived from the local Pre-Tertiary bedrock. Hence these deposits appear fresher and "younger" than the granite-derived boulder beds and show little weathering of the clasts (Figure 6.6a). Nonetheless, their age is almost certainly early Quaternary: a maximum age is given because they overlie, with probable unconformity, late Tertiary lake beds; a minimum age is implied because the boulder beds are themselves deformed structurally, with dips up to 60° locally. An additional argument for a definitely Pre-Holocene age is the absence of any conspicuous source for the erosion of such a great mass of boulders and other quartzite debris (more than 300 m thick in the center of the basis). This source was most likely a steep fault-produced slope, and erosion sufficient to render it inconspicuous could not have taken place without considerable elapsed time.

Age. Although nothing very definite can be said about the age of these deposits of "boulder beds," circumstantial evidence suggests they are in the main early Quaternary. Wherever they are found, they form the basal part of the unconsolidated sediments and so probably represent the first stage of valley filling. At Batu Arang, the boulder beds unconformably overlie strata of possible Pliocene, or at most Miocene, age, and are themselves deformed and almost certainly Pre-Holocene and therefore Pleistocene. The "Old Alluvium" of Kinta and other areas (see below) is closely associated with the boulder beds but never underlies them. In the main, the Old Alluvium is probably lower to middle Pleistocene, and hence the boulder beds are most likely also lower to middle Pleistocene. They could, however, be older in some places.

OLD ALLUVIUM. Above the scattered occurrences of boulder beds and above the bedrock in many other areas is a complex of unconsolidated sediments that form the remainder and bulk of the "alluvium" in the Kinta Valley and other inland valleys. This complex includes gravel, sand, silt, and clay in all possible mixtures, together with peaty sediments, peat, and accumulations of partly lignitized wood and logs.

The sequence at any one locality is often complicated, slumped, and difficult to decipher (see Figures 6.7 and 6.10), and the deposits are extremely lenticular, so that detailed correlations between localities has not proved possible.

Stratigraphic subdivision of these deposits into an older set and a younger set was suggested long ago (Rastall 1927), and some distinction of this sort has been adhered to by nearly all subsequent workers. Although the distinction is sometimes difficult to make, and the age of the break may be different in different areas, it does seem that the division is a real one and consequently it is used here. I use the terms "Old Alluvium" and "Young Alluvium," which have been used by Walker (1956), Ingham and Bradford (1960), and Sivam (1969). The Old Alluvium as used here is essentially synonymous with the "high-level alluvium" of Rastall (1927) and others, and with the "Older Alluvium" of Burton (1964).

FIGURE 6.7 *View of open-cast tin mine near Bidor, Perak, showing typical exposures in the Old Alluvium. Note the abundance of wood fragments and white clay-rich sediments. Photograph P. H. Stauffer.*

Rastall (1927) proposed the presence in Malaya of an older "high-level alluvium," which he thought represented not only an earlier time but also a higher relative stand of sea level. Included in this "high-level alluvium" were the "Gopeng Beds," which he recognized as unconsolidated Cenozoic sediments and which were preserved in low hills up to an elevation of 80 m (260 ft) above present sea level. Also included were the deposits in southern Johore and on Singapore Island, which occur up to about 70 m (230 ft) above sea level and are dissected into low hills.

Kinta Valley. Walker (1956) proposed a classification of the alluvial deposits in the Kinta Valley into four types:

1. The Boulder Beds.
2. The Old Alluvium.
3. The Young Alluvium.
4. Organic Mud and Peat.

Although the peaty deposits occur as lenses and layers within both Old and Young Alluvium, the other three types are at least semistratigraphic, the Boulder Beds and Old Alluvium together being approximately equivalent to Rastall's "high-level alluvium."

Walker described the Old Alluvium (1956, p. 25) as consisting mainly of grey to brown sandy clay, with frequent intercalated layers of sand and gravel. The gravel contained clasts up to 15 cm long, mostly of "friable quartz" (?) but also of granite, schist, and hornfels. The bedding in these layers is locally tilted and distorted where solution of underlying limestone has caused subsidence. Walker suggested that much of the clay in the Old Alluvium might have been deposited as sand grains (of feldspar?) and weathered to clay *in situ*. But he does not explicitly state what criteria distinguish the Old Alluvium from the Young Alluvium, save that the Young Alluvium is mainly sand and gravel according to his description.

Walker pointed out that the Old Alluvium constitutes most of the valley fill in the Kinta Valley and occurs up to elevations of 70 m (230 ft) above sea level. Like Rastall (1927), he ascribed this to a former higher stand of sea level and, in spite of having attributed much of the clay to *in situ* weathering, he concluded from the fine grain size of the deposits that they were laid down in "deep, quiet water" in a lake, sea, or large estuary. This conclusion has been challenged by recent workers (Sivam 1969, Newell 1971), who cite evidence for fluvial origin.

Ingham and Bradford (1960), in their comprehensive memoir on the Kinta mining district, gave a brief discussion of the general character of the unconsolidated deposits, using the subdivisions that Walker established. They emphasized that both the "boulder beds" and the Old Alluvium were rich in clay, noting that this characteristic sometimes renders hydraulic mining difficult. They also mentioned the very common slumped and distorted nature of these deposits where underlying limestone bedrock has been removed by solution. This was previously pointed out also by Scrivenor (Scrivenor and Jones 1919, p. 55; Scrivenor 1931, p. 125) who observed that a "tin-bearing bed, originally horizontal, may lie vertically against the side of a solution cavity in the limestone and produce very erroneous ideas in prospecting" and that the "beds next the limestone are generally very stiff clay owing to an admixture of residual material from the limestone" (see Figure 6.15).

Some of the carbonaceous material in the Old Alluvium apparently became mobile at some stage and was redeposited as "coal veins," which are often vertical, reaching thicknesses of about 10 cm and cutting both the alluvium and weathered bedrock. Fermor (1939) and Richardson (1941) described these veins as consisting mainly of a vitrain-type material and ascribe their origin to "gelation" of colloids from wood pieces contained in the alluvium. Fitch (1952, p. 46) noted similar features in the Kuantan area of Pahang.

The thickness of the alluvium in the Kinta Valley averages less than 10 m north of Ipoh and increases somewhat irregularly southward, averaging about 20 m in the area of Kampar. Deeper troughs also occur along granite–limestone contacts, and low hills of schist bedrock and steep limestone hills rise above the alluvium in places.

Near the sides of the valley, the Old Alluvium often has the character of "granite wash"—angular grains of quartz mixed with kaolin and superficially similar to weathered granite. The two materials can be distinguished by the occasional presence of bedding structures in the former or of quartz veins or dykes in the latter. In the central portions of the valley, the Old Alluvium consists of more clearly bedded sand, clay, and peat, sometimes showing cross-bedding.

Most of the Old Alluvium consists mineralogically of quartz and clay minerals, with some mica. In addition, Ingham and Bradford (1960, p. 81) listed the following heavy minerals from the deposits in the Kinta Valley.

Abundant: ilmenite, zircon, tourmaline.
Common: monazite, cassiterite.

Usually Present: rutile, leucoxene, topaz, anatase, magnetite, limonite, hematite, corundum.

Rare: garnet, pleonaste, andalusite, brookite, tremolite, fluorite.

They also cited the following minerals as having very limited distribution: arsenopyrite, galena, native copper, gold, scheelite, cerussite, cobaltite, xenotime, pyroxene, amphibole, zoisite, yttrotungstite, stolzite, and columbite. Secondary pyrite, marcasite(?), and siderite are locally abundant.

Recently, detailed studies of aspects of the sedimentology of the alluvial deposits have been made in the northern Kinta Valley (Sivam 1969) and the southern Kinta Valley (Newell 1971). These studies have provided abundant evidence of the fluvial origin of most of the deposits and should serve to lay to rest the common earlier speculations that the abundance of clay indicated a marine origin for the Kinta "alluvium."

Sivam (1969) found that Walker's (1956) distinction between Old Alluvium and Young Alluvium could be applied in the field and that it represents a significant stratigraphic division marked by an unconformity. The characteristics of the Old Alluvium that differ from those of the Young Alluvium are (Sivam 1969, pp. 31–35) especially the greater degree of weathering, thicker soil profile, partial consolidation, and abundance of slumping and disturbance (see Figures 6.8, 6.9).

The Old Alluvium in north Kinta consists of clay-rich and locally peaty deposits of mud, sand,

FIGURE 6.8 *Old Alluvium from a mine west of Gunong Rapat, Kinta Valley, Perak, showing the effects of slumping. The light-colored round fragments are clay. Photograph P. H. Stauffer.*

and gravel, pervasively weathered so that clasts of granite and schist can be crushed between the fingers and do not protrude in general from the surface of a cut or slumped bank (Figure 6.9). Where the Old Alluvium forms the original ground surface, a soil 3 to 5 m thick developed on it (Sivam 1969). The Old Alluvium is semiconsolidated, because of its high clay content, it is plastic when wet, although rather hard when dry. Red or yellow parallel weathering bands from iron-staining ("Liesegang rings") are common, and framboidal

FIGURE 6.9 *Weathered gravel of the Old Alluvium, exposed in a mine near Tanjong Rambutan, Kinta Valley, Perak, showing advanced weathering of most lithologies other than quartz. Photograph P. H. Stauffer.*

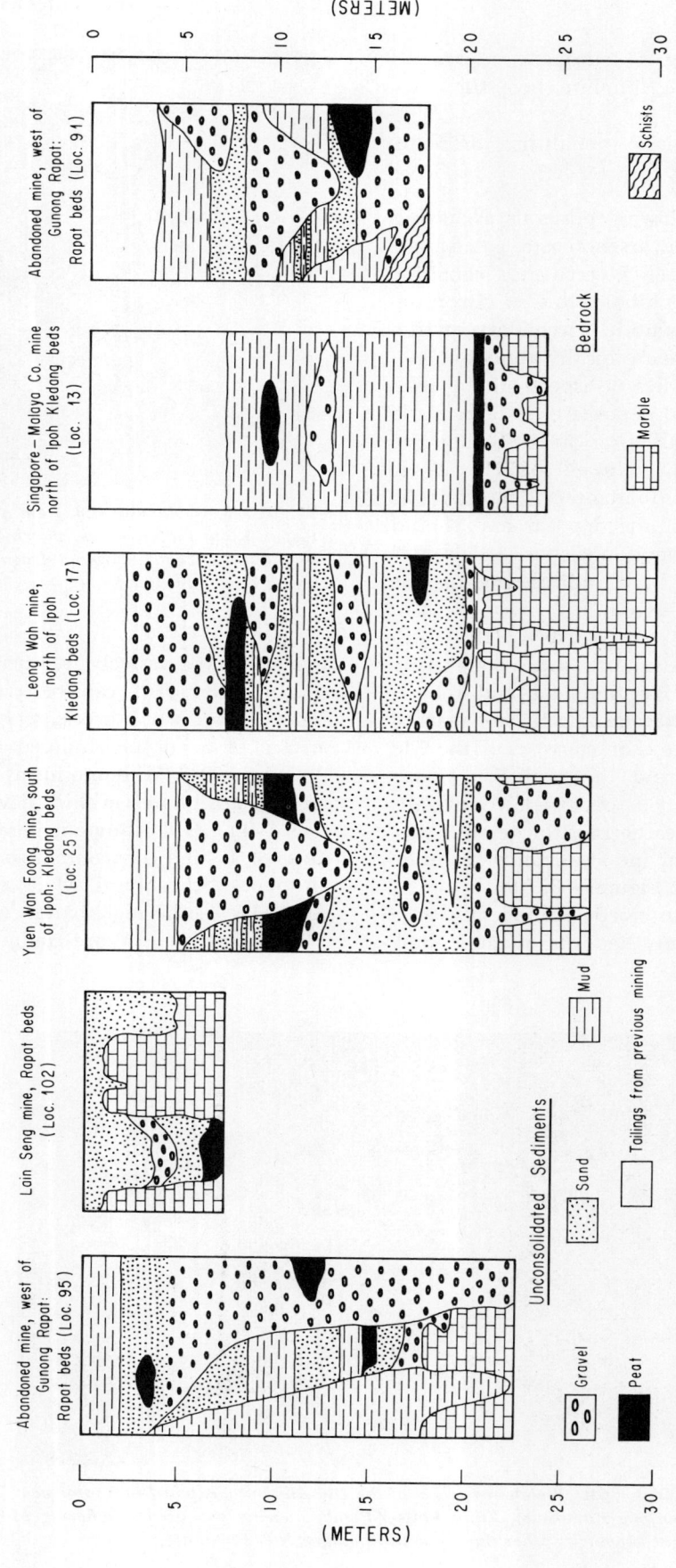

FIGURE 6.10 *Diagrammatic stratigraphic sections of Old Alluvium in the Kinta Valley, Perak. After Sivam (1969), to which the locality numbers refer.*

pyrite has been reported from carbonaceous sediments in the Old Alluvium (Sivam 1969).

Size analyses of 30 samples of Old Alluvium (Sivam 1969) showed that the median size ranged from coarse gravel to fine silt (pure clay samples were not analyzed); values of graphic standard deviation σ_I (size parameters from Folk and Ward 1957) were generally greater than 2.0 and commonly in excess of 3.0, showing the poor sorting of the deposits. Nearly all samples were fine skewed to strongly fine skewed. In analyses of eight samples of "granite wash" from the southern Kinta Valley, which he regards as belonging to the Old Alluvium, Newell (1971) found the median size in the coarse-to-fine sand range, sorting a little better than in Sivam's samples (σ_I from 1.0 to 2.0); nearly all samples were fine skewed.

Newell also studied the distribution of heavy minerals, and he remarked that this "granite wash" contains some of the highest values he found in his area. He also noted an apparent tendency for higher values of heavy mineral content to be associated with coarser grain size and relatively low to negative values of skewness (i.e., tending toward being coarse skewed).

In its detailed stratigraphy the Old Alluvium of the Kinta Valley is very complex and consists of numerous highly lenticular and discontinuous beds and irregular masses of different lithologies. The great variability of the stratigraphy is well revealed by the diagramatic columns of Sivam (Figure 6.10).

This extreme lateral variability of the Old Alluvium is a typical feature of fluviatile sediments. Other, additional characteristics of the deposits leave little doubt that they are mainly river sediments. In a broad sense, a fluvial environment is indicated by the peat and peaty clay, which occur as local and discontinous beds and lenses, and the occasional piles of branches and wood fragments (Figure 6.11) and even isolated large logs that are found buried and partly lignitized in the deposits. The deposits also show some primary sedimentary structures of the types expected in fluvial sediments: larger-scale cross-bedding (Figure 6.12); graded sequences; mud clasts (Figure 6.8), sometimes armored; and lenticular bedding (Sivam 1969, Newell 1971).

The Kinta Valley Old Alluvium can be interpreted as consisting of channel or substratum deposits and overbank or topstratum deposits (Sivam 1969). The channel deposits are coarse-grained sand and gravel, generally found at the base of the local sequence and less commonly as lenses higher up. They are sometimes elongate in the direction of transport and the primary structures, listed earlier, are mostly seen in them. It is also from these channel deposits that most of the tin is recovered.

The overbank deposits, which form the bulk of the Old Alluvium, are generally finer grained and range from laminated fine sand to clay. They occur not only above the channel deposits, but also at the base or, quite commonly, they compose the entire section. This abundance of overbank deposits, which suggests that deposition of the Old

FIGURE 6.11 *Section in Old Alluvium at a mine near Tanjong Rambutan, Kinta Valley, Perak. At the base is sticky, white, sandy clay, followed by (dark) peat and wood, then sand and silt, and finally soil. Limestone bedrock is exposed elsewhere in the mine. Photograph P. H. Stauffer.*

FIGURE 6.12 *Bedding structures in Old Alluvium near Kampar, Kinta Valley, Perak. The rock in both exposures is "granite wash" with large weathered feldspars (white) and angular quartz grains in a clay matrix. From Newell (1971). (a) Cross-bedding; (b) small scour channel.*

Alluvium took place during rather rapid infilling of the valley floor, is in contrast to the picture seen in the Young Alluvium (see below).

From studies of transport direction indicated by primary structures and grain-size trends in gravel, particularly maximum size, Sivam (1969) has been able to reconstruct in outline the paleocurrent or paleodrainage pattern in the Old Alluvium of the north Kinta Valley. The pattern that emerges is one in which the main streams drained off the Kledang Range on the west and the Main Range on the east and turned southward within the valley. One important drainage path was from north of Gunong Rapat (southeast of Ipoh) toward Batu Gajah, where no major stream flows today. This former drainage path is marked by a high proportion of channel deposits and consequently contains some very rich tin placers.

The bedrock surface below the alluvium is a surface of erosion, which may be rather flat or may, especially on limestone, have considerable local relief. Solution of the limestone, which apparently has taken place at least partly underneath a cover of loose sediments, produces a characteristic pinnacled topography ("microkarst" Figure 6.13) and has led to considerable slumping of the overlying sediments. In places the Old Alluvium forms a horizontally bedded filling between pinnacles of limestone, showing that solution occurred in part during subaerial exposure (Figure 6.14). It is not

FIGURE 6.13 *Surface of limestone bedrock near Batu Gajah, Kinta Valley, Perak, showing "microkarst" topography with pinnacles, fluting, and notches. Photograph P. H. Stauffer.*

FIGURE 6.14 *Old Alluvium horizontally bedded within a solution cavity in limestone bedrock. Batu Gajah, Kinta Valley, Perak. Photograph P. H. Stauffer*

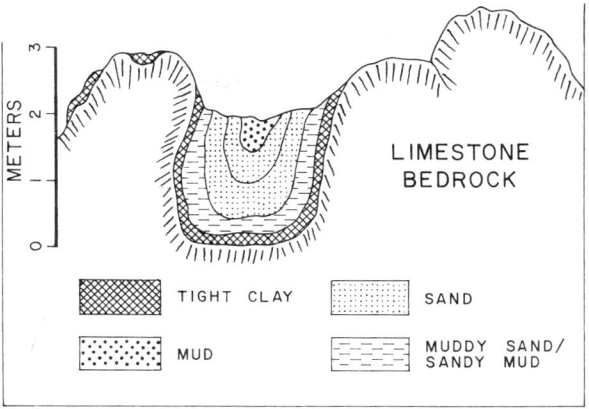

FIGURE 6.15 *Diagrammatic section of Old Alluvium filling "pothole" between pinnacles of limestone bedrock, Sungei Besi, Selangor. Tight clay next to the bedrock surface is mainly residual from solution, and the conformity of the strata to the bedrock surface shows that solution occurred after deposition of the alluvium. After Mohammed bin Ayob (1970).*

uncommon, however to find the bedding layers in the Old Alluvium steeply dipping or vertical as they acommodate to the limestone surface that is evolved by solution (Figure 6.15).

Although most of the Kinta Valley deposits are clearly of fluvial origin, and no real evidence of any marine deposition has been found, some of the "alluvium," especially massive clays and very badly slumped areas, is enigmatic in character. Near the valley sides, colluvial deposits and probably also landslide debris play a significant role in the Old Alluvium. Residual deposits, including weathered and slumped argillaceous bedrock and residues from limestone solution, are probably widespread and not limited to areas where it has been possible to demonstrate such an origin for the "alluvium."

Kuala Lumpur Area. Around Kuala Lumpur is a fairly large alluviated lowland, and in its unconsolidated deposits has been developed a tin-placer district perhaps second only to Kinta. The alluvial areas here are generally restricted to calcareous bedrock, and the other lithologies still form hills which surround the lowland.

The thickness of the alluvium varies from a few meters to a few tens of meters in most places; it reaches greater thicknesses in the bedrock "troughs" developed along contacts between calcareous rocks and granites. Such is the case at the famous Sungei Besi Tin Mines south of Kuala Lumpur, where thicknesses greater than 50 m (165 ft) are known and the bedrock surface is in places distinctly below present sea level (Mohammed Ayob 1970).

The sediments at Sungei Besi have been studied by Mohammed Ayob (1965, 1970), who correlated them with the Old Alluvium in the Kinta Valley on the basis of stratigraphic and petrologic similarity. The sediments consist of partly consolidated muddy sand and sandy mud, with some peat, clay, and gravel. Although generally well stratified, the material is commonly mildly to strongly disturbed by slumping. Clear examples of stratified sediments "draped" into depressions between limestone pinnacles (Figure 6.15) not only explain the cause of the slumping but also prove that the solution process has occurred in these cases under a cover of sediment.

The sand in the sediments at Sungei Besi has a median size mostly in the fine-to-medium sand range and is poorly to very poorly sorted. In composition it is mainly of quartz, with up to 10% of muscovite and varying amounts of tourmaline, pyrite, cassiterite, zircon, chlorite, and iron oxides.

Organic material occurs as muddy and sandy peat and partly lignitized plant remains. The latter include leaves, stems, twigs, bark, and tree trunks commonly to 30 cm and occasionally to 1.5 m in diameter. Some of these are found upright in apparent growth position.

The organic material and the sedimentary character of the deposits clearly indicate a fluvial origin, as pointed out by Mohammed Ayob (1970).

Johore and Singapore. Unconsolidated deposits distinctly older than the modern river and coastal sediments have been known on Singapore Island and in the Johore Bahru area for some time (Scrivenor 1924). Rastall (1927) included them in his "high-level alluvium," and Burton (1964) called them the "Older Alluvium" (Figure 6.16). There is general agreement that they are related to the Old Alluvium of the Kinta Valley and other inland areas, and they are here considered to be part of the Old Alluvium.

The Old Alluvium of Johore and Singapore was described by Burton (1964). The deposits occur as dissected low hills reaching elevations of about 70 m (230 ft) above sea level, and they also underlie some lowland areas. Burton (1964, p. 36) recorded bouldery material on several higher hills in Johore, which reach elevations of 138 m (458 ft), and he thought that this may be remnants of an older similar deposit. On Singapore Island the base of the deposit is locally as low as 45 m below sea level, and total thicknesses of about 60 m are known.

The deposits in Johore and Singapore are generally semiconsolidated and considerably weathered coarse sand, sandy clay, and gravel. The sand is generally poorly sorted (P. V. O. Drummond personal communication). The large clasts seldom exceed 30 cm in diameter and are well rounded, whereas sand grains are often angular. Commonly the composition is granitic-feldspathic, reflecting the granitic bedrock that often underlies the deposits, and such granite-derived sediment ("granite wash") is sometimes hard to distinguish from weathered granite. Scrivenor (1931) who noted the abundance of incompletely weathered orthoclase and microline (sometimes forming one-third of the total material) in sand of the Old Alluvium in Singapore, suggested that this might indicate formation in a colder climate than at present. Besides granitic rocks, pebbles and boulders of acid dyke rocks, vein quartz, quartz-tourmaline rock, various clastic sedimentary rocks, hornfels, and acid volcanic and pyroclastic rocks occur. The tuffs and volcanic rocks sometimes predominate in pebble beds in the Johore Bahru area. Heavy minerals recorded include magnetite, ilmenite, zircon, tourmaline, cassiterite, monazite, topaz, staurolite, pyrite, anatase, and sphalerite (Burton 1964).

Bedding is often massive in the coarser deposits (Figure 6.16), but thinly interbedded sequences

FIGURE 6.16 *Rudaceous Old Alluvium near Bedak, Singapore. The crude bedding is subhorizontal. This alluvium is largely derived from the underlying granite. The outcrop has now been removed for a building site. Photograph C. S. Hutchison*

also occur. Graded bedding, lenticular bedding, and cross-bedding have been noted. Dips up to 20° occur and Burton reports small tight folds locally. Drummond has noted the common occurrence of lenses, channeling, and larger cross-bedding, suggestive of fluvial deposition in the deposits near Johore Bahru (P. V. O. Drummond personal communication).

These deposits of Old Alluvium cover most of the eastern third of Singapore Island and much of the area around Johore Bahru to about 15 km (10 miles) inland. They also underlie an unknown area in the valley of the Sungei Johore and are reported at scattered inland sites farther north in Johore. Burton (1964, p. 34) proposed deposition in three separate basins but noted that the evidence was inconclusive.

Age and Origin of the Old Alluvium. Direct evidence of age in the Old Alluvium (fossils and radiometric ages), although scanty and generally nondiagnostic, are consistent with a dominantly early to middle Pleistocene age. Various lines of reasoning based on the physical character and distribution of the deposits tend to support that approximate age range.

Fossils other than plants are not common in the Old Alluvium. An elephant tooth from a depth of 3.6 m in alluvium at Salak, a few kilometers northwest of the Kinta Valley's north end, was identified as being of *Palaeoloxodon namadicus*, an extinct and characteristically Middle Pleistocene species (Hooijer 1963a). A collection of vertebrate bone fragments collected by Bill Bush from a mine near Batu Gajah in the southern Kinta Valley yielded identifications of a rhinoceros, a suid, deer (probably *Hydropotes* and possibly *Muntiacus*), turtle shells, and the spine of a catfish (D. A. Hooijer personal communication). Hooijer regarded the age of this collection as probably Pleistocene.

Savage (1937) and Richardson (1939) reported finds of elephant teeth in alluvium in north Kinta and near Raub, Pahang, respectively. The type of alluvium is not known in either case, but the Raub tooth was doubtfully referred to the extinct *Palaeoloxodon namadicus*.

Burton (1964, p. 39) reported rare shell fragments, echinoid spines, and doubtful otoliths from samples of Old Alluvium in Johore or Singapore, but he did not give the location or elevation of the site of these marine fossils.

The most abundant fossils in the Old Alluvium are plant remains ranging from fine peat and pollen to large stems and branches, and including leaves and lumps of *damar* (a resin from forest trees), as well as wood. Pollen from the deposits at Sungei Besi, near Kuala Lumpur, yielded a trisaccate grain of *Podocarpus imbricatus* type, known in Boreno only from late Pliocene and younger (Haile and Mohammed Ayob 1968).

Several radiocarbon determinations have been made on wood and peat from the Old Alluvium. Those in Kinta (Sivam 1968), which include one from the same site near Batu Gajah that yielded the vertebrate bones, proved to be beyond the range of the method (ages given as > 39,900 years B.P.). Of three determinations of material from Sungei Besi (Haile and Ayob 1968), two were similarly beyond the range of the method (ages given as > 41,200 and > 41,500 years B.P.), but a third sample yielded an age of 36,420 (+1255, −1085) years B.P. This sample was of woody stems, some apparently in upright growth position, from a level about 13 m above sea level, within the upper half and probably the upper third of the alluvial section present. This dating would make the age at that point late Pleistocene.

Indirect evidence of the age of the Old Alluvium is provided by the physical character and position of the deposits.

The high degree of weathering of the Old Alluvium, the high content of clay (which is quite likely in part the result of weathering), the common slumping and disturbance caused by solution of underlying limestone bedrock, and the occurrence of the deposits as low dissected hills in Kinta and on Singapore Island all indicate the passage of some time since the Old Alluvium was formed. These characteristics also serve to differentiate the Old Alluvium from the more recent fluvial and coastal deposits. Younger unconsolidated deposits overlie the Old Alluvium unconformably (see below).

Further indirect evidence is provided by the altitudinal distribution of the deposits. It is clear from the abundant evidence cited previously that most of the Old Alluvium in the Kinta Valley, the Kuala Lumpur area, and the Johore Bahru–Singapore area had a fluvial origin and that no marine sediments have been proven in the deposits. It is nonetheless possible that the Old Alluvium is related in its origin to a higher sea level than at present, which led to aggradation of river valleys. As noted earlier, occurrence of Old Alluvium in various areas are found up to elevations of about 70 to 75 m (230 to 250 ft) above present sea level. Haile (in press) has pointed out that we cannot simply conclude from this that sea level was 70 or 75 m higher than present, as has been commonly done in the literature. Fluvial sediments can be expected to be deposited at varying elevations above sea level. The deposits above 75 m in Johore, which Burton (1964) found troublesome and suggested were remnants of an even older deposit (presumably related to an even higher former sea level), are easily explained as having formed in the valleys of higher tributaries in the river system, as suggested by P. V. O. Drummond (personal communication).

But the widespread occurrence of the Old Alluvium significantly above present sea level, particularly near the present coast in Johore and Singapore, does suggest the former existence of a coastal or piedmont plain graded to a base level somewhat above present sea level. P. V. O. Drummond (personal communication) has interpreted an exposure in the Old Alluvium 7 km (4.5 miles) from Johore Bahru on the Kota Tinggi road as representing a possibly estuarine mud-flat environment, with numerous pebble-filled small channels in mottled clay. The elevation of this exposure is about 20 m (about 75 ft).

The radiocarbon dates prove that the Old Alluvium is Pre-Holocene. In the Pleistocene, sea level fluctuated by more than 100 m below its present position. During glacial periods, when sea level was low, rivers would have been able to erode and clean out their valleys down to elevations well below present sea level. As sea level rose at the start of each interglacial, the rivers would tend to aggrade rapidly up to a position in equilibrium with the new higher sea level. It seems probable that most of the Old Alluvium was deposited during one or more of these periods of rapid aggradation.

Evidence of rapid aggradation in the Kinta Valley is given by the low proportion of channel deposits in the Old Alluvium (Sivam 1969) and is consistent with this supposition. The Old Alluvium might thus have been deposited mainly in relatively brief periods at the end of the Gunz, the Mindel, and the Riss glaciations. Although we should expect the deposits of these separate periods to form distinct units, each unconformable on the older ones, the slumped and weathered condition of the Old Alluvium makes it impossible to sort these out, until better datings can be obtained. If the higher parts of the Old Alluvium do represent sea levels above the present one, then early Pleistocene is a likely age, since there is some evidence for such levels then (Zeuner 1959). The sediments dated at 36,420 (+ 1255, −1085) years B.P. at Sungei Besi (Haile and Mohammed Ayob 1968) may represent aggradation during an interstadial within the Würm glacial period.

We can state a maximum age for the Old Alluvium when we consider that it is unconformable on the underlying rocks, which locally include late Tertiary fluvial–lacustrine deposits. In central Johore it has been found difficult to distinguish between Old Alluvium and late Tertiary deposits (Burton 1964, p. 35). The sparse fossil data indicating a Pleistocene age, cited earlier, do not preclude the possibility that parts of the Old Alluvium are late Tertiary in age.

YOUNG ALLUVIUM. Locally overlying the Old Alluvium and filling channels and depressions in it, as well as overlying bedrock at many other places, are unconsolidated deposits of sand and gravel, with some peat and clay. Walker (1956) called these deposits "Young Alluvium."

In the Kinta Valley the deposits are easily distinguished from the Old Alluvium by a number of criteria (Walker 1956, Sivam 1969):

1. The clasts in the Young Alluvium are unweathered or only slightly weathered, and soils developed on Young Alluvium are commonly less than 2 m thick. Wood fragments in the Young Alluvium are only slightly lignitized. Banded iron-staining ("Liesegang rings") has not been reported.
2. The Young Alluvium is unconsolidated. Bedding structures are generally well preserved, and signs of slumping or disturbance of the beds are extremely rare, even where the deposits lie on limestone bedrock.
3. North of Batu Gajah, the Young Alluvium is never more than 15 m thick, and it occurs in linear belts flanking present-day major streams, to which it is clearly related. Its upper surface is flat and at the level of the general valley floor, never forming dissected hills.
4. Where both occur in the same exposure, Young Alluvium always overlies Old Alluvium, and the contact relationship is unconformable (Figure 6.17).

The stratigraphic section in the Young Alluvium at any particular locality commonly exhibits a simple graded sequence from gravel at the base, through sand to finer material at the top (Figure 6.18). The proportion of coarse material—gravel and sand—is markedly higher than in the Old Alluvium.

The texture of the Young Alluvium is loose and porous. Studies of grain size distribution in the

FIGURE 6.17 *Unconformity between Young Alluvium and Old Alluvium, Teblas Mine, near Tanjong Rambutan, Kinta Valley, Perak. Basal gravel of Young Alluvium overlies tight sandy clay of Old Alluvium. Photograph P. H. Stauffer.*

FIGURE 6.18 *Section in Young Alluvium at Teblas Mine, near Tanjong Rambutan, Kinta Valley, Perak. Note the fining upward from basal gravel to silt, sand, and the large logs trapped in the sediment. Sandy clay of the Old Alluvium is exposed in the bottom of the mine, right. Photograph P. H. Stauffer.*

sediments of the northern Kinta Valley (Sivam 1969) show median size values ranging up to 15 mm and values of sorting (σ_I) from 0.8 to 3.5 but mainly from 1.0 to 3.0 (poorly to very poorly sorted). The sediments are fine-skewed ($Sk_I > 0$). Newell (1971) analyzed sands from sediments in the southern Kinta Valley, which he equated with the Young Alluvium of Walker (1956). These sands show generally poor to very poor sorting (σ_I from 1.0 to 2.6).

Most of the samples were fine skewed or nearly symmetrical, but a minority were coarse skewed. Heavy mineral content was rather uniformly low ($< 0.5\%$), with a maximum of 1.26%. This agrees with the observation of Sivam (1969) that the *kaksa* (miners' term for tin-rich zone) is generally in the Old Alluvium. The richness of the older deposits may reflect repeated erosion and redeposition during successive periods of low and high sea level in the Pleistocene. The heavy cassiterite could not wash away, but it might become increasingly concentrated.

Sivam (1969) made grain counts of heavy minerals from five samples of Young Alluvium at the Teblas Mine (Figure 6.18) near Tanjong Rambutan, northeast of Ipoh. The results were fairly consistent and showed that about half the heavy grains are tourmaline and ilmenite, and most of the rest consist of zircon, cassiterite, and mica.

Age and Origin of the Young Alluvium. Primary sedimentary structures of a fluvial character are abundant in the Young Alluvium (Sivam 1969, Newell 1971). These include lenticular bedding and channels, trough cross-bedding and (less commonly) tabular cross-bedding, and clast imbrication in the gravel. Transport directions inferred from directional structures in the northern Kinta Valley (Sivam 1969) are dominantly southward and closely follow modern drainage trends. There can be no doubt that the Young Alluvium is a fluvial deposit related to the existing river system. Indeed, it is impossible to rigidly separate the Young Alluvium from the currently forming river deposits, since they are part of the same continuous series.

The minimum age of the Young Alluvium is therefore the present day. The maximum age is indicated by its universally unconformable relationship over the Old Alluvium and its comparatively unweathered, undisturbed, and unconsolidated condition. The unconformable relation to the Old Alluvium has been shown in the Kinta Valley (Sivam 1969; see Figure 6.17) but has in general not been documented elsewhere because the distinction between the two is not always made. Burton (1964, p. 37), however, mentioned the occurrence of younger alluvium incorporating boulders of the underlying probable Old Alluvium in a tin mine at Kajang, 24 km (15 miles) south of Kuala Lumpur, and it is likely that an unconformable relationship between the two is the general rule.

Fossils found in the Young Alluvium are uniformly of modern varieties. The occasional teeth of modern elephant reported (Jones et al. 1966) are probably from Young Alluvium. A radiocarbon age determination from the north Kinta Valley gave an age of 3070 ± 100 years B.P. (Sivam 1968). Another two determinations, however, made on pieces of wood and logs from the basal portion of the Young Alluvium where it overlies Old Alluvium in the same area, gave the unexpected results that the age was beyond the range of the method (age given as > 39,000 years B.P.; see Sivam 1969). This result is anomalous, since the containing sediments have a very young aspect, and hence the possibility that these pieces of wood were exhumed from older deposits and then reburied in more recent sediments must be considered likely.

The sharp demarcation of the Young Alluvium from the underlying Old Alluvium implies that the former represents a period of sedimentation following an intervening period of erosion. Since the Young Alluvium appears to be continuous with the modern river deposits, it is suggested that deposition has taken place during the postglacial period of sea level stand at or near the present level (i.e., that the Young Alluvium is Holocene). Apart from the two anomalous radiocarbon ages mentioned, all the evidence seems to be consistent with this view.

Sivam (1969) accepted the "anomalous" radiocarbon ages as valid indicators that parts of the Young Alluvium date from the last interglacial period. This interpretation is not impossible, but if correct it implies that the Old Alluvium is mainly early Pleistocene or even older. The character of the Young Alluvium—its coarseness and the dominance of channel deposits—is typical for a river that is not aggrading (Sivam 1969). If rapid valley-filling occurred at the end of a glacial period, then lateral migration of the rivers during the long stillstand of sea level in the subsequent interglacial period should produce such a deposit as the Young Alluvium, a thin layer of channel deposits unconformable on the underlying material. If part of the Young Alluvium represents such a capping dating from the last interglacial (Riss–Würm), then it is surprising that it is not affected by slumping caused by solution of underlying limestone, which should be most active during glacial periods, when the water table was probably lowered by downcutting of the streams and groundwater movement was favored by the increased local relief.

The Young Alluvium is limited to positions flanking the present streams. Then if part of it is of Riss–Würm age, the drainage pattern has been fixed, since at least Riss times, and the general valley fill of Old Alluvium would have to be pre-Riss. The rather sharp difference between Young Alluvium and Old Alluvium in terms of weathering and slump disturbance makes an age contrast of Mindel–Riss versus Riss–Würm seem inadequate, and if the Young Alluvium is really in part pre-Würm, it seems likely that the Old Alluvium is then mainly pre-Mindel (i.e., early Pleistocene or older). The parts of the Old Alluvium that appear to be related to a sea level significantly above the present one could be regarded as lending support to this interpretation, although of course a mainly early Pleistocene age for the Old Alluvium would also be consistent with a purely Holocene age for the Young Alluvium.

Further work is clearly needed on these problems, which have important implications for Quaternary history in the Malayan region.

CAVE DEPOSITS. Caves are common in the limestone hills of West Malaysia. These hills are best developed in the northwest corner of the country (Perlis, northern Kedah, and the Langkawi Islands) and in the Kinta Valley in Perak, but hills also occur at a number of other localities from Kuala Lumpur northward. Within the caves have been found a variety of deposits, including secondary calcareous coatings and cave formations (such as stalactites and stalagmites); limestone breccia formed by collapse of portions of the roof; cave earths, and guano deposits; and ordinary river alluvium washed in by streams using the caverns as part of their course to the sea.

These deposits are, of course, insignificant in volume and extent when compared with other sorts of Quaternary sediments, but they are of considerable importance for several reasons. Some of the caves were inhabited in prehistoric times, and because of better conditions for preservation in caves than outside, much of the archaeological information on the Malay Peninsula has come from cave deposits. These deposits have also given us the best Quaternary fossil material, both human and animal. Where bats inhabit the caves, their guano has been eagerly sought as fertilizer, and guano mining has unfortunately damaged or destroyed a number of potential archeological sites. Finally, in a few places, tin-bearing alluvium is found within caves and is mined by placer methods.

Archaeology is dealt with briefly in a later section, and I wish to elaborate here on only two aspects of the cave deposits.

Fossils in Cave Deposits. Several caves in West Malaysia have yielded mammalian bones. Hooijer (1963a) described a collection found at a depth of 9 m in tin-bearing deposits in a limestone cave near Ipoh in the Kinta Valley. Two teeth in this collection were identified as belonging to *Duboisia santeng* (Dubois), an extinct antelope previously known only from the Middle Pleistocene of Java. Also present was a radius of a *Hippopotamus* sp., indicating at least Pre-Holocene age. Associated with these were fragments that could be either Pleistocene or Holocene—parts of *Rhinoceros sondaicus* (Desmarest), *Sus* sp., *Cervus* sp. (? Rusa) *Bibos* c.q. *Bubalus* sp., and a vertebra of a small carnivore were included. If the collection is of one age, then it would seem to represent a fauna of at most Middle Pleistocene age.

Hooijer (1963b) identified teeth from Gua Cha rock shelter in Kelantan as being of *Rhinoceros sondaicus* (Desmarest), the Javan rhinoceros, known to have survived in Malaya till at least 1932, when the last known specimen was shot at Telok Anson (Medway 1969, p. 103). These teeth at Gua Cha were from a mesolithic horizon in the deposits.

Tin-Bearing Alluvium in Caves. In a few caves, streams have deposited ordinary alluvial sand, gravel, and mud washed in from outside. The most spectacular example is the system of caves near Kaki Bukit in Perlis. These caves and their deposits have been described by Jones (1965, in prep.). The caves are developed in the Ordovician–Silurian Setul Limestone of the Setul Boundary Range. The largest are entered from Wang Tangga, a valley about 1.5 km (1 mile) long, of the "karst window" type, completely enclosed by limestone hills but connected to the lowlands outside along a subterranean stream channel. The alluvium in these caves varies from reddish clay and semiconsolidated sand and gravel to hard, calcareously cemented sandstone and conglomerate. The sand is coarse and feldspathic, and the gravel contains well-rounded pebbles of quartzite, limestone, granite, and vein quartz, in addition to polished pellets of iron oxides. Cassiterite and monazite occur, the former resulting in the deposits being mined to more than 1 km (0.6 mile) horizontally and more than 100 m (330 ft) vertically from the cave entrance, down to a depth of 50 m (170 ft) below sea level (Jones in prep.).

Jones (1965, in prep.) stated that the deposits are usually well stratified and exhibit cross-bedding, grading, and poor sorting. There can be little doubt that they were laid down by a stream flowing through the cave, although no stream flows there today. The age of these deposits is unknown, but it is likely that they are Pleistocene.

Coastal Plains

Fringing much of the coastline of the Malay Peninsula is a coastal plain of very low relief and standing at most a few meters above sea level (Plate 2, Figure 6.1). The width of this plain varies greatly, from zero where bedrock hills form the actual coast, to 20 km (12 miles) or more along fairly large stretches of both east and west coasts, and reaching a maximum of about 60 km (37 miles) around the mouth of the Perak River.

The coastal plain represents the top surface of a wedge of unconsolidated sediments; the wedge generally thickens seaward, reaching thicknesses greater than 100 m at various points along the coast. The deeper parts of the wedge are known only from scattered boreholes, but they appear to be a complex intertonguing of marine and nonmarine strata. The surface of the coastal plain is made up of a fairly ordered horizontal succession of beach, lagoon, swamp, and river deposits, which in many areas record a marked progradation of the shore since the postglacial rise in sea level.

THICKNESS OF SEDIMENTS. The coastal plain at its inner edge is continuous with the surface of the alluvium in most of the inland valleys, such as the Kinta Valley. Even the Kuala Lumpur lowland, almost completely surrounded by hills, is connected with the coastal plain of Selangor via alluviated pathways. The depth of alluvium in the Kinta Valley is known to increase gradually southward (i.e., downstream and toward the sea). South of Kampar it becomes generally greater than 20 m and then, soon, too deep for economic working of open-cast mines (the tin-rich zones are commonly close above the bedrock surface). Beyond this our information comes only from scattered deep boreholes, some sunk for prospecting and others for groundwater or engineering purposes.

Boreholes giving the depth of unconsolidated deposits near the coast have been reported in a number of publications. They are here briefly reviewed, proceeding counterclockwise around the coast from Perlis. All the bores were sited on the coastal plain at elevations generally less than 10 m.

Jones (in prep.) gave data for 28 boreholes sunk for groundwater investigations in the coastal plain of Perlis and north Kedah. Several short seismic traverses in the same area show that the bedrock surface is generally a gently undulating one less than 30 m (100 ft) below sea level, with some steep slopes, hills, and channels, a few going below −30 m. The thickness of unconsolidated sediments encountered in the boreholes was about 18 to 22 m (59 to 62 ft) near the coast and somewhat less inland—15 to 18 m (48 to 59 ft) in the Kodiang area, 7 to 9 m (23 to 30 ft) in the Kangar area. The greatest thickness, 32 m (105 ft), was encountered at Arau, 16 km (10 miles) from the coast. The bedrock surface here is 27 m (88 ft) below sea level. This may be a channel or may reflect solution of the limestone bedrock.

A borehole sunk in 1912 at Prai, on the mainland opposite Penang Island, penetrated 84 m (159 ft) of sediments (Scrivenor 1931). A bore sunk in 1960 at nearby Butterworth reached hard bedrock at a depth of 137 m (449 ft), but the thickness of weathered bedrock traversed is not known (Bradford 1965).

Scrivenor (1931) reported bores at Bagan Datok at the mouth of the Perak River which reached depths of 84 m without hitting bedrock. Another bore on the lower Bernam River to the south penetrated weathered quartzite bedrock at a depth of 135 m (443 ft). A third bore near the mouth of the Selangor River farther south passed through 67 m (220 ft) of sediments. Scrivenor and Jones (1919) wrote of a boring on Carey's Island near Port Swettenham which penetrated 50 m (165 ft) of soft sediment and did not reach bedrock.

On Singapore Island, unconsolidated sediments have been proven to extend to at least 45 m (150 ft) below sea level (Burton 1964).

On the east coast of the peninsula, only a few deep coastal borings are reported from the Kuantan area (Fitch 1952) and near Kota Bharu (MacDonald 1968). Bores in and near Kuantan town reached granite bedrock at depths around 31 m (102 ft), but prospecting bores near the Sungei Soi, southwest of Kuantan, penetrated as much as 51 m (167 ft) of loose sediments. The borings near Kota Bharu proved that the thickness of unconsolidated sediments exceeds 30 m (98 ft) there.

SUBSURFACE DEPOSITS. No detailed study has ever ben made of samples from these deep boreholes, but the drilling logs provide some indication of the nature of the deposits encountered. These consist dominantly of clay and sandy clay, with thinner intercalated layers and lenses of sand, peat, and rare gravel. Reports of shells or decayed vegetation are not uncommon. Jones (in prep.) listed heavy minerals washed from samples in several boreholes in Perlis. The suite is dominated by limonite, tourmaline, zircon, hematite, pyrite, and ilmenite, but siderite, topaz, rutile, monazite, cassiterite, corundum, garnet, thorite, and xenotime, are also included.

Evidences of both marine and terrestrial deposits exist in the borehole records. Unbroken valves of marine molluscs were found in grey-green clay about 9 m (30 ft) below sea level in a bore on the coast at Kuala Perlis, and green clay with shell fragments also occurred at depths of 9 to 12 m

(30 to 40 ft) below sea level in two bores 10 km (6 miles) inland in the same region (Jones in prep.). Shell remains were reported at depths of 18 m (60 ft) below sea level in a bore-hole on the lower Bernam River (Scrivenor 1931). Haile (in press) also found marine shells at 10 m (33 ft) below sea level in a bore on the Selangor coastal plain. In addition, the tight blue, grey, or green clay frequently noted in the drilling records may be marine in origin.

Evidence of continental fluvial and swamp deposits below present sea level is provided mainly by the records of peat, wood, and oxidized iron in the boreholes. Layers of peat have been reported at elevations of −12 m (−40 ft) in North Kedah (Jones in prep.) and −23 to −36 m (−77 to −119 ft) on the lower Bernam River (Scrivenor 1931). Fragments of peat, pieces of wood, and "decayed vegetation" are also reported from depths of 18 and 37 m (59 and 122 ft) below sea level on the lower Bernam River (Scrivenor 1931) and about 12 m (40 ft) below sea level at Kuantan (Fitch 1952). Scrivenor (1931, p. 128) mentioned "laterite earth" at a depth of 16 m (55 ft) in the bore at Bagan Datok, and Jones (in 1972) noted "laterite pebbles" at about 9 to 14 m (30 to 45 ft) below sea level at Kuala Perlis.

This intermixing of marine and nonmarine deposits in the unconsolidated sediments underlying the coastal plains of the Malay Peninsula no doubt reflects the major fluctuations in sea level caused by glaciation and deglaciation in the Pleistocene time. It is in addition possible and perhaps probable that some Prequaternary sediments are present in this section as well. The Tertiary sedimentary section known in Sumatra and in offshore areas may locally extend under the coastal plain of the Malay Peninsula. Detailed studies of borehole samples are now in progress, and the results should eventually do much to clarify the late Cenozoic geology of the peninsula; in particular, they may shed considerable light on the probable age of the Old Alluvium encountered in the inland valleys.

SURFICIAL DEPOSITS. In contrast to the subsurface deposits, the surficial deposits of the coastal plains are readily accessible for study, and their geomorphic expression still in many cases reveals their origin. These sediments consist of present and former beach sand, lagoonal mud, marine clay, and, especially toward the inner edge, river alluvium and colluvium. They demonstrate progradation amounting to many kilometers along parts of the coast, and provide evidence of a slight drop in relative sea level during the Holocene.

Jones (in prep.) has described the coastal plain in Kedah and Perlis, which stands about 2 m (7 ft) above sea level at the coast and about 6 m (20 ft) above sea level at its inland edge. Much of the plain is composed at the surface of bluish or greenish clay with marine shells. Former beaches are found far inland. On the west side of Gunong Keriang, about 10 km (6 miles) from the coast near Alor Star, is a beautifully preserved shoreline (previously noted by Scrivenor 1931), with an erosive notch cut in the limestone bedrock. These are borings of *Lithophagus*, oyster shells still attached to the rock, and an accumulation of marine shells at the base of the cliff. Similar shell accumulations occur at the bases of other hills on the Kedah–Perlis coastal plain. The usual forms present are *Arca, Meretrix, Ostrea, Cyrena,* and *Telescopium*. One such deposit, at the base of Bukit Chuping in Perlis, more than 16 km (10 miles) from the coast, has been analyzed by the radiocarbon method; it yielded an age of 5200 ± 200 years B.P. (Jones in prep., Alexander 1962).

BEACH RIDGES. Inland ridges composed of white sand, known locally as *permatang* and representing former beaches, are a common feature of many parts of the coastal plain in Malaya. They have been noted by Scrivenor and Jones (1919) in southern Perak, by Scrivenor (1931) in Kedah and near the Perak–Selangor boundary, by Fitch (1952) in the Kuantan area of Pahang, by Courtier (1962) in Province Wellesley, by Bradford (1965) in southern Kedah, by MacDonald (1968) in Kelantan and Trengganu, and by Jones (in prep.) in Perlis, Kedah, and on Pulau Langkawi. Nossin (1964a, 1965b) has made detailed studies of the ridges on the east coast.

A description of the beach ridges in the Kuantan area of Pahang has been given by Fitch (1952). The ridges are long, linear or curved features that tend to occur in nested sets paralleling the present coastline. They are composed mainly of white beach sand with some shell fragments, and they support a very limited flora. Between the ridges are swales representing former lagoons. These are quite distinct, and in many cases still occupied by swamps or lakes, when they occur near the coast; farther inland they become indistinct and support dense vegetation.

Fitch divided the inland beach ridges into two series. The older series occurs between 1.6 and 6.4 km (1 and 4 miles) from the present shore and stands at up to 11 m (36 ft) above sea level; the younger series occurs up to 4 km (2.5 miles) from the present shore and stands at up to 4.5 m (15 ft) above sea level. Fitch interpreted the levels of these two series as evidence of two successive lowerings of relative sea level: from 11 m (36 ft) down to 4.5 m (15 ft) and from 4.5 m down to the present beach level. Since, however, it has been found by Nossin (1964a) that the presently forming ridge on the modern beach is generally 3 to 4.5 m (10 to 15 ft) above sea level, we cannot consider that the second series of ridges of Fitch prove any change in relative sea level. Fitch is, however, supported by Courtier (1962), who stated that the first *permatang* in parts of Province Wellesley, which occurs about 365 m inland, is 1.5 m (5 ft) higher than the top of the present beach.

There can be no doubt, however, about the higher series of Fitch's beach ridges representing a drop in relative sea level amounting to at least 6 m (21 ft). Jones (in prep.) documented three series of beach ridges on the west side of Pulau Langkawi, at elevations of about 9, 6, and 3 m (30, 20, and 10 ft). At least the highest of these can also be taken to indicate some relative drop in sea level. Other evidence of a comparatively recent drop of relative sea level in West Malaysia amounting to perhaps 6 m has been summarized by Haile (in press).

The advance of the coastline by as much as 24 km (15 miles) that has been recorded in these successive beach ridges (Nossin 1965a) and by the marine clay now found many kilometers inland clearly took place mainly if not entirely during the Holocene, and continues at present. Working out the chronology of these coastal changes using radiocarbon dating offers a fruitful field of research. In addition, blocks of pumice occur in some of the beach ridges, and it might be possible to relate these eventually to particular Indonesian or Philippine eruptions.

During the early part of this process of progradation, relative sea level was slightly higher (by 6 or 7 m) than at present. The radiocarbon date from Bukit Chuping (5200 ± 200 years B.P.) indicates that the process of horizontal progradation that produced the present coastal plain began with the postglacial rise in sea level to approximately its present level. The slight drop in relative sea level since then is most likely a eustatic change, since its evidence is so widespread and uniform in the Malay Peninsula, which is supposedly a relatively stable landmass.

But isostatic or epeirogenic rise of the land cannot be ruled out as a possible cause.

It is possible also that some of the highest, oldest, and most poorly preserved beach ridges may be relics of a pre-Würm interglacial high stand of sea level. But I regard this as unlikely because erosion during low glacial stands of sea level could effectively destroy such unconsolidated deposits.

Modern coastal sediments and coastal processes have been studied by Hill (1966), Nossin (1961, 1962, 1964a, 1964b, 1965b), and Tjia (1970a). Much of their work is geomorphological in nature and outside the scope of this chapter.

Tektites

An interesting feature of the Quaternary geology of countries in Southeast Asia is the presence of tektites—curious meteoritic bodies of siliceous glass, whose shapes are usually solids of revolution and whose surfaces show a variety of sculpturing thought to be at least partly caused by aerodynamic effects encountered during high-velocity entry in the earth's atmosphere. The Malay Peninsula lies within the great tektite-strewn field stretching from Tasmania to southern China (Barnes 1963) and between the area of occurrence of "Indochinites" on the Southeast Asian mainland and Indonesian tektite localities in the Natuna Archipelago. It is not known to which, if either, of these groups the Malayan tektites belong.

Tektites from West Malaysia were listed, described, or figured in Scrivenor (1909, 1931), Tilley (1922), Prior (1927), Lacroix (1932), Willbourn (1936a), Roe (1953), Baker (1963), Barnes (1963), and Hosking and Stauffer (1970). Barnes 1963) listed five localities in Malaya: Kuantan, Seremban, Gemas, west-central Kelantan, and north Pahang. Tektites from the Kuantan and Seremban areas were reported as early as 1909 (Scrivenor 1909).

A tektite weighing 464 g, presented to the British Museum by J. B. Scrivenor, is apparently still the heaviest known from Malaya (Baker 1963). Although this specimen and another weighting 316 g are usually listed as having come from Kelantan, their origin is in fact obscure, since they were found unlabeled in a drawer at the Raffles Museum in Singapore (Scrivenor 1931, p. 181).

Malayan tektites were described by Scrivenor (1931) as being spherical or oval in form and having a finely sculptured glossy jet-black surface. They are generally discovered when they are washed out of the alluvium in open-cast tin mines. In the Gambang Valley of Pahang the miners say the tektites come from just on top of the bedrock. The alluvium there is only about 5 m thick and rather fine grained, with scattered residual quartz pebbles lying on the bedrock at its base. Quite probably, then, the tektites here are reworked and possibly much older than the alluvium covering them.

Since it seems very probable that all Southeast Asian and Australian tektites fell in a single event that occurred about 700,000 years ago (although some authors regard Australian tektites as much younger; see Barnes 1967), finds with precisely known discovery points could serve as valuable stratigraphic markers in the Quaternary deposits. It is hoped that care will be taken to watch for such finds in the future.

Archaeology

One aspect of the Quaternary which is of great interest to man is the record of his own emergence as a species and of the development of culture. It is therefore not out of place to make at least brief mention of the archaeology of the Malay Peninsula. This account is based mainly on the fuller summary given by Tweedie (1965).

The Paleolithic is represented in Malaya only by the "Tampanian" tool industry reported from Kota Tampan, Perak, which was discussed in an earlier section. No skeletal material from this period has been found in Malaya, but since *Homo erectus* (*Pithecanthropus*) lived both in Java and on mainland Asia, such discoveries on this natural migration route must be considered a possibility in the future.

The mesolithic is represented by abundant stone tools, most commonly hand-axes, ascribed to the Hoabinhian culture common to all mainland Southeast Asia. The people of this culture lived in caves, and the stone tools are often found there in midden deposits, which also contain abundant evidence of the food they used: shells of the river snail *Thiara variabilis*, bones of fish, turtles, pig, deer, bear, monkeys, and rodents. Human bones are also commonly found; these are similar to the modern Melanesian or Papuan type. Their presence in the middens suggests either deliberate burial or cannibalism. Tweedie (1965) suggests an age range of 9000 to 4000 years B.P. for the mesolithic in Malaya, based on the Post-Pleistocene aspect of the associated animal remains and the supposed arrival of neolithic culture by the latter date. Recent work in Thailand (Gormann 1969) indicates that the Hoabinhian culture there dates back to at least 9000 years B.P. and by that time already included domesticated plants.

The neolithic of Malaya is represented by carefully shaped and smoothed stone axe-heads and adze-heads, stone bracelets, and pottery of great variety and artistry of form. Especially common among the stone implements are very skillfully made adze-heads up to 45 cm (18 in) long; some are beaked, some beveled, and some shouldered. The neolithic people probably lived in houses. but caves were used occasionally for shelter and for burial of the dead. Tweedie (1965) gives a detailed account of the neolithic midden in Gua Cha in Kelantan.

Stone implements have frequently been reported from open-cast mines in various parts of Malaya (Scrivenor 1931, Richardson 1939, 1950). Scrivenor (1931) reported one stone tool that was brought up by a bucket-dredge from a bed of peat 15 m (50 ft) below the surface at Batu Gajah.

Scattered relics of the bronze and iron ages have been found in Malaya, especially the delicate and strangely curved iron blades called *tulang mawas* ("ape-bones") by the Malays. The possible use of these odd and awkward-looking blades is unknown.

Indian colonization of northwestern Malaya from about the fourth to the twelfth century A.D. has left abundant evidence, including many Hindu temples, especially on the slopes of Gunong Jerai (Kedah Peak), which then may have stood on the end of a peninsula. The stone foundations of some of these temples have been restored.

A wooden boat recovered from the mouth of the Sungei Pontian in southern Pahang in the 1920s contained a great deal of pottery, including apparently a glazed stoneware jar that was probably of early Sung (about tenth century A.D.) origin.

Numerous relics of more recent centuries have, of course, also been found. Among the more interesting to the geologist and miner are the timbers encountered in open-cast mines which give evidence of early endeavors to mine the tin placers by sinking shafts (the so-called *lombong siam* or Siamese mine).

CONCLUDING REMARKS

In Cenozoic time, the Malay Peninsula was apparently stable tectonically and most of it was emergent continuously. There is evidence of considerable Post-Cretaceous uplift, especially in the northern part. No Tertiary deposits older than the late Miocene or Pliocene lake beds are known at all, and it seems likely that the entire peninsula was then dry land and formed a portion of emergent Sundaland. Minor basic volcanism occurred sometime in late Tertiary or early Quarternary. By late Miocene or Pliocene the lowland areas were near sea level, and local subsidence resulted in the formation and preservation of lacustrine and other terrestrial deposits, but no evidence of actual marine transgression has been found.

During the Quaternary, a relative subsidence of at least the margins of the peninsula occurred, placing the bedrock surface at the coasts, as much as 100 m or more below sea level. In spite of this, there is evidence of a drop in relative sea level since early Quaternary (?) time.

The glacio-eustatic fluctuations of sea level have undoubtedly been the fundamental control on events in the peninsula during the Quaternary. Erosion during low stands of sea level and deposition during high stands have shaped much of the bedrock surface and produced the complex of sediments that fills the valleys and mantles the lowlands. The evidence for Quaternary sea levels in the Sunda region have recently been compiled by Tjia (1970b), and the evidence for West Malaysia and nearby areas has been critically reviewed by Haile (in press). The relation of these sea level changes to the sequence and distribution of sediments is a story that has yet to be worked out in detail.

CHAPTER 7

Volcanic Activity

C. S. HUTCHISON

All Malayan volcanic and pyroclastic rocks have previously been referred to as the Pahang Volcanic Series because they are best developed in Pahang and were thought to be restricted to Carboniferous, Permian, and Triassic time. In the original classification, Scrivenor (1931) also included a large number of occurrences of hypabyssal rocks that he thought were pregranite in age. However, he did admit the possibility that not all volcanic and pyroclastic rocks in Malaya belonged to this "important manifestation of late Palaeozoic and early Mesozoic volcanic activity." Willbourn (1917) used the same classification, and for many years the name Pahang Volcanic Series came into general use to include all volcanic, pyroclastic, and some related subvolcanic igneous rocks.

Richardson (1950) discussed the classification of the Malayan volcanic rocks, and he proposed that the term Pahang Volcanic Series be limited to include only rocks of undoubted volcanic origin, contemporaneous with the interstratified sedimentary rocks.

More recent field investigations have shown that contemporaneous volcanic and pyroclastic rocks are much more extensive geographically than within the state of Pahang; occurrences have been described from Trengganu and Kelantan to the north, from Johore and Negri Sembilan to the south, and from Perak, Kedah, and Selangor in the west; the main bulk, however, is undoubtedly centered in Pahang. Contemporaneous volcanic and pyroclastic rocks have been found interstratified with Lower Paleozoic, Upper Paleozoic, and Mesozoic sedimentary strata, and postorogenic basalt flows of Tertiary or Quaternary age occur in the Kuantan area of Pahang (Fitch 1952) and in the Segamat area of Johore (Grubb 1965).

The term Pahang Volcanic Series is therefore misleading and has clearly outlived its usefulness.

Volcanic rocks varying from rhyolite to andesite are widely distributed throughout the Malay Peninsula, but they are much less common than pyroclastic rocks of corresponding composition. The majority of the rocks are rhyolitic; andesitic and trachytic rocks are subsidiary. Basalt flows are locally important, but restricted in occurrence (Figure 7.1 and Table 7.1).

The volcanic and pyroclastic rocks are described according to area of occurrence; where stratigraphic age is known, this category is also used. In addition, some problematical rocks of doubtful volcanic (or plutonic?) origin, occurring within and to the east of the Main Range, are described in this chapter because they have definite volcanic affinities.

LOWER PALEOZOIC VOLCANISM

Rhyolitic pyroclastic rocks form large tracts of country in the Grik and Sungei Siput areas of Perak. The Grik occurrence has been referred to as a sheared quartz porphyry (Scrivenor 1931) and the Sungei Siput occurrence as a foliated, phyllitic, and sheared pregranite quartz-porphyry (Savage 1937). A similar quartz-porphyry occurs in the Jerai

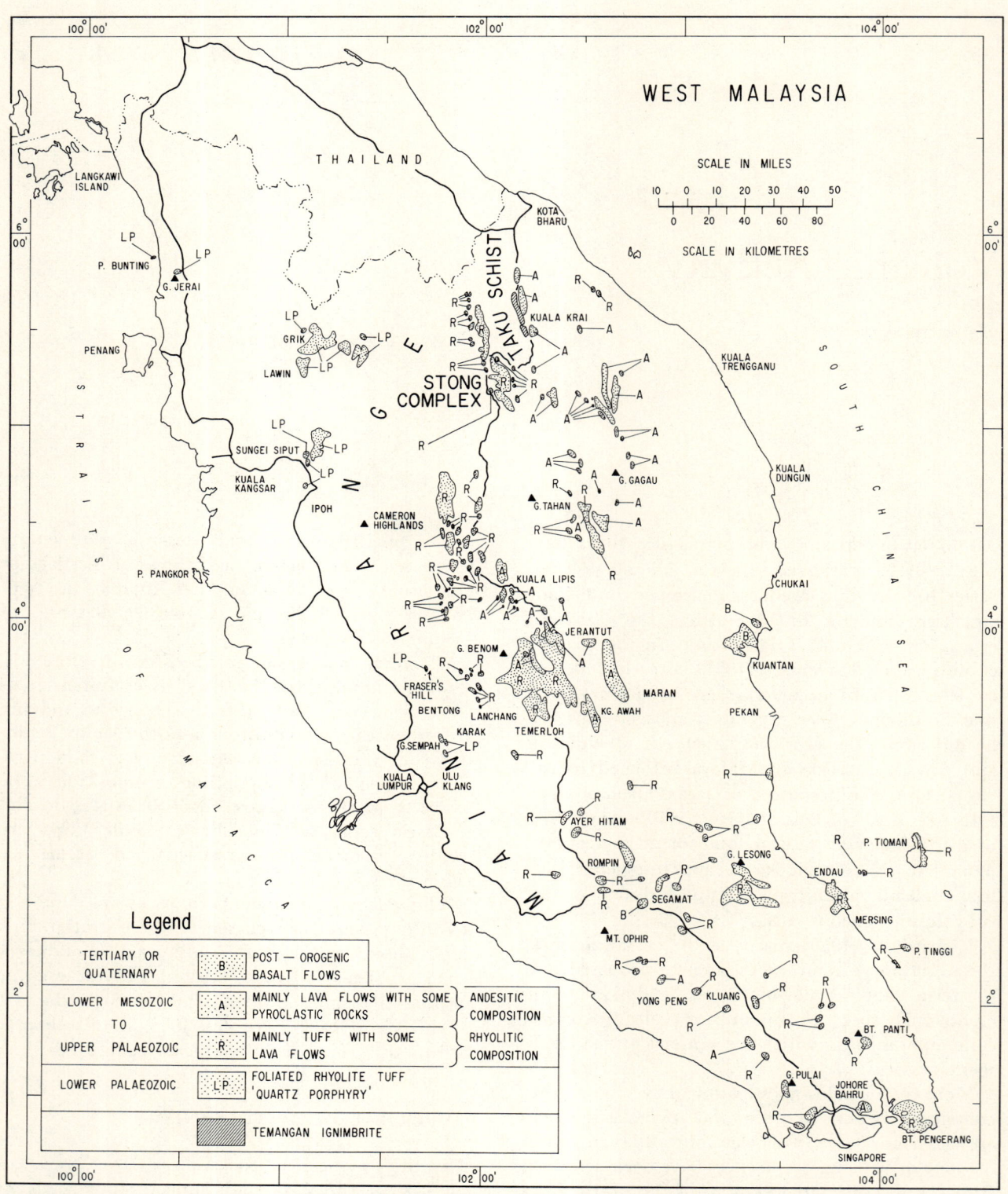

FIGURE 7.1 *Sketch map of the Malay Peninsula showing the major occurrences of volcanic and pyroclastic rocks. Based mainly on Alexander (1965) and other publications of the Geological Survey of Malaysia.*

TABLE 7.1 *Schematic correlation chart of volcanic and pyroclastic activity in the Malay Peninsula*

		Kedah	Perak	Selangor	Main Range	West ←——— Kelantan and Trengganu ———→ East ←——— Pahang ———→ ←—— Johore ——→		
	QUATERNARY		Rhyolite ash				Segmat postorogenic Basalt	Kuantan Basalt
MESOZOIC	TERTIARY							
MESOZOIC	CRETACEOUS					Minor occurrences of intercalated flows and tuff of rhyolite, trachyte, andesite, and basalt in the Gagau Group		
MESOZOIC	JURASSIC	No known occurrences of volcanic rocks					Gunong Pulai Volcanic Member of the Jurong Formation: mainly rhyodacite tuff	
MESOZOIC	TRIASSIC						? ?	
UPPER PALEOZOIC	PERMIAN					Very abundant occurrences: The Pahang Volcanic Series: predominantly of tuff, with less frequent lava and agglomerate. Predominantly of rhyolitic composition, andesite is less common, and trachyte is infrequent. Andesite agglomerate at Kampong Awah is dated as Permian.		
UPPER PALEOZOIC	CARBONIFEROUS							
UPPER PALEOZOIC	DEVONIAN				Minor occurrences: Andesite agglomerate and quartz porphyry of uncertain genesis in the neighborhood of Ginting Sempah			
LOWER PALEOZOIC	SILURIAN			Minor occurrences: Foliated rhyolite tuff in the Hawthornden Schist				
LOWER PALEOZOIC	ORDOVICIAN	Minor occurrences: Foliated rhyolite tuff in the Jerai Formation	Abundant occurrences: Foliated rhyolite tuff in the Grik (lawin tuff) and the Sungei Siput and Kinta Valley areas					
LOWER PALEOZOIC	CAMBRIAN					No evidence		

Formation on Gunong Jerai in Kedah. It was described by Bradford (1965), who also discussed its origin. Farther south and east, in the Ulu Klang area near Kuala Lumpur, Yin (in prep.) and Willbourn (1917) have described possible volcanic rocks interfoliated with the Lower Paleozoic Dinding Schist. Minor occurrences of metavolcanic rocks occur in the Foothills Formation east of the Main Range, where they are interfoliated with schist and metaconglomerate.

Abundant occurrences of rocks that have been referred to as "quartz porphyry" and "microgranite" outcrop within and near a large elongate roof pendant of metasedimentary rocks, which lies within the Main Range along the boundary between Selangor and Pahang in the vicinity of Ginting Sempah. These rocks are now thought to be best regarded as metavolcanic or metapyroclastic and not, as Scrivenor (1931) and Alexander (1968) suggested, to represent marginal phases of the Main Range Granite.

The term quartz porphyry has been used in Malaya to signify a rock containing phenocrysts of quartz and alkali feldspar, with or without mica, in a microcrystalline or cryptocrystalline groundmass. If the phenocrysts are abundant, the rock is called granite porphyry; if they are absent it may be called microgranite, but felsite, an uncommon term in Malayan geological literature, could equally well be used. In the vast majority of cases, the term quartz porphyry has been applied purely as a descriptive name to signify the appearance of the rock in the hand specimen.

Subsequent misinterpretation has occurred because the term porphyry is generally applied today to mean a rock of hypabyssal intrusive genesis. However, it should be remembered that the term quartz porphyry was once commonly applied in Europe to all Pre-Tertiary extrusive equivalents of granite, whereas rhyolite was applied to those of all Tertiary or Post-Tertiary age. It is therefore advised that in reading Malayan literature the terms quartz porphyry, granite porphyry, and porphyry be understood purely in the petrographic sense with no genetic implication. Many may be more appropriately renamed rhyolite or rhyolitic tuff in our present usage.

The various occurrences of volcanic and pyroclastic rocks, which are thought to be of Lower Paleozoic age, are described in geographic order from west to east (Table 7.1).

West Kedah

A small occurrence of quartz porphyry occurs in the Jerai Formation 3 km (1.75 miles) northeast of the summit of Gunong Jerai (Bradford 1965). The rock is continuous and conformable with the metasediments of the Jerai Formation and is not intrusive. Bradford (1965) ascribed its present similarity to quartz porphyry to metamorphism and suggested that the rock may have been a tuff or volcanic rock that has been "porphyrized." The rock is described as a quartz-feldspathic hornfels or semischist. It contains varying proportions of porphyroblasts (or relic phenocrysts) of quartz, microcline, orthoclase, plagioclase, perthite, and biotite in several combinations. The groundmass is predominantly of quartz and feldspar, with subsidiary muscovite, biotite, chlorite, tourmaline, epidote, clinozoisite, apatite, magnetite, hematite, and pyrite.

Bradford (1965) suggested that the relic phenocrysts may indicate that the rock originally was a grit or conglomerate, but it seems more likely that they represent the original crystals of a crystal tuff or the original phenocrysts of a volcanic rock. Undoubtedly much of the present mineralogy has resulted from dynamothermal metamorphism, because this rock is enclosed by an envelope of schist and metaquartzite, which commonly contains epidote and garnet. Metavolcanic rock also outcrops around the reservoir at Malaysia Waterfall, above the main Gurun quarry on Gunong Jerai. This rock has been mapped erroneously as granite; actually, it is a fine-grained porphyry and is interfoliated with the Jerai Formation metaquartzite, into which it grades imperceptibly.

A comparable quartz porphyry occupies most of Pulau Bunting, some 16 km (10 miles) to the northwest of the Gunong Jerai occurrence. This has been described by Chung (1959) as being finer grained and having a more homogeneous groundmass than the occurrence on Gunong Jerai. It is also more uniform over a wider area. It contains phenocrysts of colorless to smoky quartz. The groundmass is of quartz, muscovite, and biotite. Like the Gunong Jerai porphyry, it is associated with metasediments resembling gneisses, which contain bands of biotite. Epidote, garnet, magnetite, and andalusite have been identified. Chung (1959) regarded the porphyry as being a hypabyssal intrusion into the sedimentary strata because of the metamorphic condition of the sedimentary strata along its contact in northeast Pulau Bunting. However, the metamorphism of the Gunong Jerai region is dynamothermal and perhaps attains almandine amphibolite facies, so that a porphyry intrusion on Pulau Bunting would be unlikely to have caused the metamorphism in the surrounding rocks. It seems more reasonable to regard the Pulau Bunting porphyry as similar to that in the Gunong Jerai area.

The age of the Gunong Jerai and Pulau Bunting porphyries must be taken as equivalent to that of the enclosing Jerai Formation, which is regarded as Upper Cambrian to Lower Ordovician.

Upper Perak—Grik Area

Scrivenor (1931) noted that quartz porphyry was quarried at Lawin and outcrops nearer Grik, in Upper Perak, where it forms long rapids in the Sungei Perak and Sungei Temengor (Figure 7.1).

He also noted that some of it is sheared. Jones (1970b) referred to this rock as the Lawin tuff and he demonstrated that it is of pyroclastic origin (Figure 7.2a). It is a foliated rhyolitic and rhyodacitic tuff, interfoliated with Ordovician to Lower Silurian metaquartzite, shale, and limestone beds of the Baling group. The bodies are lenticular and form several horizons within the succession. They often pass laterally into quartzite in an imperceptible manner. The tuff contains occasional lapilli and bombs of scoriaceous rhyolite. No undoubted lava

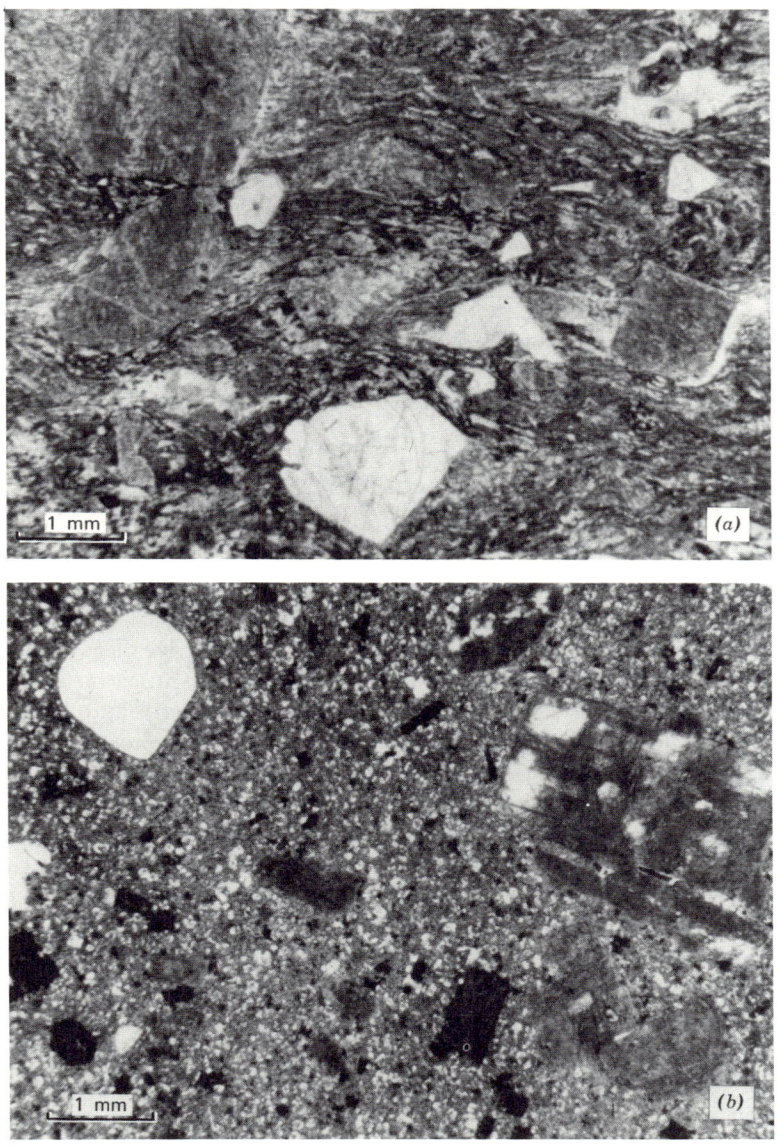

FIGURE 7.2 Photomicrographs (C. S. Hutchinson) of rhyolitic rock in ordinary light; scale as indicated: (a) Sheared rhylotic tuff of the Lower Paleozoic Lawin Tuff, near Lawin, Grik, Perak, specimen UM 6735; (b) porphyritic rhyolite of Triassic age, Yew Ling Quarry, Lanchang, Pahang. specimen UM 5925.

flows have been recorded. The crystal tuff contains fragments and crystals of quartz, K-feldspar and plagioclase. The matrix is fine grained or cryptocrystalline and is composed of quartz, mica, chlorite, and iron oxide (Jones 1970b).

Central Perak—Northeast and South of Sungei Siput North

The quartz porphyry in central Perak is medium grey, tinged with green, but occasionally mauve in color. The groundmass is fine grained to cryptocrystalline and contains larger rounded crystals of quartz and angular crystals of feldspar, crystals of plagioclase zoned from andesine to albite, sometimes perthite; biotite is locally abundant. The rock is invariably strongly foliated (Savage 1937); some varieties are extremely phyllitic, and in some sheared varieties the biotite is chloritized and epidote is common.

At Jalong Tinggi Estate, the schistose rhyolitic rock is interfoliated with limestone (Savage 1937), and at Kuala Tamang its pregranite age is demonstrated by an intrusive contact between the unfoliated granite and the foliated tuff. Savage (1937) thought that the tuff was a laccolithic intrusion of quartz porphyry, but in view of the concordance of of its foliation with that of the associated metasediments, it is better to regard it as a contemporaneous pyroclastic or volcanic rock.

Ingham and Bradford (1960) described occurrences of similar rock to the south of Sungei Siput, northeast and southeast of Chemor in the Kinta Valley, to the east of the road from Chemor to Jelapang. The main outcrop follows a north-south ridge. Scrivenor (1931) also mentioned an occurrence at Ampang in the Kinta Valley, where the rock resembles a foliated or sheared metaquartzite. At Ampang the tuff is associated with quartzite, into which it apparently grades imperceptibly.

Near Chemor, the tuff is usually greenish to greenish grey. The groundmass is very fine grained and is rich in pyrite. Rounded crystals of quartz are more common than angular crystals of feldspar. The rock is strongly foliated and is closely associated with schist (Ingham and Bradford 1960). Vein quartz cuts the tuff at the 10.5 milestone along the road from Chemor to Jelapang.

Ingham and Bradford (1960) wrote that the tuff is a pregranite intrusion of quartz porphyry that has been sheared before the intrusion of the unsheared Kledang Range granite. However, its close association with the schist and its invariably foliated nature suggest that it is better considered as a contemporaneous volcanic or pyroclastic rock within a sedimentary sequence that has been dynamothermally metamorphosed. These metasediments are known to be of Lower Paleozoic age (see Chapter 3).

Selangor

Several small occurrences of amphibole schist in the neighborhood of Rawang and Serendah, to the north of Kuala Lumpur, may represent metamorphosed basalt flows or dolerite dykes intruded into the Lower Paleozoic metasediments (Roe 1953). Exposures are generally small and discontinuous, and hence the field relationships between the amphibole schist and the country rocks are unknown. Three main areas of occurrence have been described by Roe (1953): to the east and south of Ulu Yam, in the Serendah area between the Sungei Choh and Kapar Bharu Estates, and in the Sungei Tua and Devon Estates. These rocks are described in Chapter 9.

Several occurrences of foliated, possibly metavolcanic, rocks associated with the Dinding Schist in the Ulu Klang district have been described by Willbourn (1922). For example, at Bukit Dinding the metaigneous rock is foliated together with the schist and is light violet-grey, containing bands of biotite and tourmaline, augen of quartz, and phenocrysts of feldspar. At the 7 milestone on the Ulu Klang road, a similar foliated rock contains bands rich in hornblende, others rich in augite, and others rich in sphene (Willbourn 1922).

Selangor–Pahang Boundary

Along the border between Selangor and Pahang, and for a few miles on either side, several occurrences of volcanic and pyroclastic rocks are associated with the metasediments of the large roof pendants that are present within the Main Range Granite. Immediately to the west of the Main Range these metasediments are of undoubted Lower Paleozoic age (Gobbett 1965a). The Lower Paleozoic Foothills Formation, which occurs immediately to the east of the Main Range, is of similar lithology, and the similarity of the metasediments in the roof pendants suggests also a Lower Paleozoic age. In addition to a few occurrences of definite volcanic and pyroclastic origin, several important occurrences of uncertain genesis are within and associated with these roof pendants. These are described here even though the present evidence conclusively proves neither a volcanic nor an intrusive origin. Their aphanitic nature does, however, indicate that they are either volcanic or subvolcanic.

At and around Ginting Semph, on the main road betwen Kuala Lumpur and Bentong where the road crosses the Selangor–Pahang boundary, are several occurrences of microgranite and quartz porphyry that have confounded geologists for a considerable time. Some regard these as local modifications of the Main Range Granite; others believe that they are pregranite volcanic or pyroclastic rocks. Scrivenor (1931) has discussed this problem and has listed the following occurrences.

On the Pahang side of Ginting Sempah, on the main road, is an outcrop about 4 km (2.5 miles) long of pyroxene-bearing grey microgranite, which superficially resembles the granite but lacks the coarse porphyritic texture that is typical of the Main Range Granite. On the Selangor side of the pass, a grey and purple quartz porphyry occurs which is thought to be older than the granite (Scrivenor 1931). However Scrivenor found a new quarry in which the quartz porphyry appeared to pass into the microgranite, "suggesting that the former may be a marginal modification of the granite bordering a large island of sedimentary rocks" (Scrivenor, 1931, p. 25).

Actually the large roof pendants of "sedimentary rocks" that occur in this vicinity within the Main Range Granite on the Selangor side of Ginting Sempah, have been dynamothermally metamorphosed to greenschist facies, and phyllite predominates. The same type of quartz porphyry is found at Gombak, near Kuala Lumpur, and near Ginting Bidai, to the south of Ginting Sempah. Scrivenor (1931) was uncertain if the pyroxene-microgranite, which occurs on the Selangor side and on the Pahang side of Ginting Sempah, is a phase of the Main Range Granite or predates the granite. Willbourn (1922) recorded the microgranite at the Ginting Sempah pass near the 20.5 milestone on the Pahang side of the pass in larger masses associated with boulders of quartz porphyry, as huge boulders in the district of Ginting Bidai, as small boulders in the Sungei Langat near Dusun Tua, and as a boulder at the foot of Bukit Lanjan near Sungei Buloh.

Willbourn (1922) described the microgranite as a greenish-grey rock containing phenocrysts of quartz; plagioclase (oligoclase to labradorite); brown, strongly pleochroic biotite; light-colored acmite, and perhaps also some orthopyroxene, apatite, magnetite, and a little pyrite. The phenocrysts, which seldom exceed 6 mm in length, are set in a groundmass of a clear quartz-feldspar intergrowth of average grain size 0.1 mm. The pyroxene phenocrysts are usually surrounded by a corona of pleochroic pale green mica. The phenocrysts are very angular and often broken.

Foliated hornblende-augite-schist, which may represent a metavolcanic rock, has been described by Willbourn (1922) from several localities in the Ginting Bidai region of Ulu Klang. The hornblende is fibrous and strongly pleochroic.

In his account of the Pahang Volcanic Series, Willbourn (1917) recorded boulders of dark-colored quartz porphyry scattered in many places on the Main Range. He specifically mentioned occurrences near the Ginting Sempah road at the 11.2, 17, 20.5 and 22.5 milestones and at several places going down the Pahang side of the pass. It must be noted however, that many parts of the road from Kuala Lumpur to Bentong have been straightened, and thus the foregoing milestone readings cannot be directly applied to the present road. Willbourn also recorded the porphyry along the old road from Ulu Klang to Ginting Bidai, where it occurs both as boulders and *in situ*, and also as a single boulder near the 8 milestone on the road from Kuala Lumpur to Rawang.

In hand-specimen, the porphyry shows readily distinguishable white phenocrysts of feldspar and quartz set in a dark purple or black compact, homogeneous groundmass. Along the new road northward from Ginting Sempah toward the Ginting Highlands, extensive road cuts in this porphyry reveal that the natural color is black and that the purple color is secondary but widely developed. The feldspar is mostly orthoclase, but plagioclase, strongly zoned from labradorite to oligoclase, is also present. The feldspar and quartz phenocrysts are angular and sometimes rounded (Willbourn 1917). A little uralitized augite is occasionally present, suggesting a connection with the microgranite that was described earlier (Willbourn 1917). Well-shaped crystals of biotite are common, often very much altered to chlorite, epidote, and iron oxide. Some of the porphyry contains acmite and altered cordierite (Willbourn 1922).

Very often the groundmass is foliated. Willbourn (1917, p. 457) interpreted this feature as follows:

> Very often [the biotite crystals] have been bent by shearing movements which have imparted a flow-structure to the groundmass, these same shearing movements in all probability being responsible for the angularity of the quartz and feldspar crystals.

On the other hand, a locality displaying a microspherulitic texture in the groundmass prompted Willbourn (1922) to say that possibly the rock might be a rhyolite, but that it was noteworthy that in some slides the phenocrysts were so irregular in outline that the rock showed some resemblance to an ash.

The groundmass of the porphyry is extremely fine grained, in part glassy (Haile personal communication) and resembles that of rhyolite. The occasional microspherulitic nature supports this interpretation. However, Willbourn (1917, pp. 457–458) seemed to be biased against a possible volcanic origin:

> Some specimens much resemble tuffs, but no fragments of lava have been observed in them, and their tufflike appearance is probably due to brecciation by shearing movements. . . . From the uniformity in the appearance of the quartz-porphyry both in the handspecimen and under the microscope, when examined as boulders in the different localities above described and *in situ,* it seems to be clear that the rock is an intrusion and not a lava, and this view is supported by the total absence of any appearance of stratification or variation when traced through the 100 feet or so of thickness of rock which is exposed at Jeram Gading.

Jeram Gading is in the Sungei Sleh (= Seli), a right tributary of the Sungei Klang in the region of the Klang Gates reservoir near Kuala Lumpur. It does not, however, seem that homogeneity is a good criterion for deducing an intrusive mode for this rock. The homogeneity, the angularity of the quartz and feldspar crystals, and the glassy nature of the groundmass could better be explained by considering the quartz-porphyry to be a welded tuff or ignimbrite. In the Sungei Sleh area, Willbourn (1917) described the rock as being associated with chert, grit, and quartzite. Scrivenor (1931) inclined to the view that the quartz porphyry was a marginal phase of the Main Range Granite, but he admitted that it might be volcanic.

The homogeneous nature of the quartz porphyry over distances of the order of 2 km (1 mile) of outcrop on the new Ginting Highlands road eliminates the possibility that this rock might be a marginal phase of the Main Range Granite. Its aphanitic to glassy groundmass, persistent over extensive areas, indicates that the rock is volcanic or subvolcanic.

Granite-porphyry also occurs in the tributaries of the Ulu Klang, particularly near Kuala Sleh and Ulu Langat (Alexander 1968), where orthoclase occurs in addition to quartz and plagioclase in the phenocrysts, and pleochroic pale green mica forms a corona round the crystals of pyroxene.

A volcanic breccia was found by Willbourn (1917) occurring as boulders near Kuala Sleh 21 km (13.5 miles) from Kuala Lumpur along the old road to Ginting Bidai. It is green and composed of angular fragments of quartz porphyry, quartzite, and small rounded pieces of homogeneous finegrained siliceous rock, which might be of sedimentary origin or of devitrified lava. Grains of quartz make up a considerable proportion of the rock; some are angular, others are rounded and corroded, and broken crystals of orthoclase and oligoclase-andesine plagioclase are common. There is very little fine-grained matrix, but among it can be noticed some grains of epidote (Willbourn 1917).

A recent study has been made of the rocks around Ginting Sempah by Haile (personal communication). prompted by the construction of a new mountain road from Ginting Sempah northward to the Ginting Highlands. Extensive outcrops of the quartz porphyry have been made in deep road cuts along this new road. Although no lapilli or volcanic inclusions have been found in the porphyry, the groundmass is often glassy, suggesting a volcanic origin. However, Haile believed that the pyroxene microgranite exposed on the main road from Ginting Sempah to Bentong is a high-level instrusive rock. He felt that in this region there may be a suite of volcanic and subvolcanic rocks, apparently all related to the Lower Paleozoic roof pendants in the Main Range Granite.

The origin of the Ginting Sempah porphyry and pyroxene microgranite is therefore uncertain, but it seems highly unlikely that these rocks are related to the Main Range Granite because of their radically different mineralogical and textural nature. They are more likely to be related to the roof pendants of greenschist facies metasediments of apparent Lower Paleozoic age which occur in this region within the Main Range Granite.

Near the 21 milstone along the Ginting Sempah road, Willbourn described a band of fine-grained, nonporphyritic, dark grey rock composed of oligoclass laths 0.2 mm long set in a groundmass of serpentinite or chlorite, quartz, ilmenite, and calcite. It is closely associated with foliated chert and feldspathic grit, and "its mode of occurrence suggests that the rock is a pillow-lava" (Willbourn, 1922 p. 56). At the 17.5 milestone, a similar rock occurs as a band 1.8 m (6 ft) wide in phyllite. It contains poorly shaped hornblende crystals in addition to the plagioclase laths.

Undoubted volcanic rock has been found elsewhere within these metasediment roof pendants in the granite. Roe (1951) described andesite and andesite porphyry in the Sungei Sempan basin in Pahang, 6.5 km (4 miles) west-northwest of Fraser's Hill. The andesite is exposed in a series of small waterfalls in the Sungei Merah. It has been metamorphosed, as have the associated sedimentary rocks, and the biotite and amphibole are considered to be of metamorphic genesis.

East and South of the Main Range

A few occurrences of amphibolite in the Schist Series of the Bentong Group may represent metamorphosed basalt flows or dolerite dykes (Alexander 1968). Most of the quartzite and metaconglomerate strata of the Foothills Formation contain a small proportion of volcanic fragments that are usually rhyolitic in composition (Alexander 1968). All these occurrences are too small to show on any of the 1:63,360 geological maps.

A band of volcanic conglomerate at Jeram Limau (4°33.5'N; 101°46'E) in the Sungei Serau and the Sungei Bisek has been described by Richardson (1950). It is medium grained, composed of rounded fragments of vein quartz, sheared quartz-sericite rock (altered rhyolite), calcareous spherulitic rhyolite, and fragments of quartz, orthoclase, and plagioclase. Limonite and pyrite are abundant.

The grit that is associated with chert in the east of Negri Sembilan may contain a certain amount of volcanic material. A deposit of volcanic tuff occurs at the 42.2 milestone on the Kuala Jelai road near the chert (Willbourn 1922).

Farther south, around Gemas in north Johore, tuff containing fragments of andesite, oligoclase to andesine plagioclase crystals, and numerous quartz grains, occurs interbedded with shale and is exposed in several railway cuttings (Willbourn 1922).

Serpentinite bodies in Negri Sembilan were thought by Willbourn (1922) to represent a serpentinized lava flow that was extruded after deposition of shale and before a quartzite sequence. These occur 0.4 km (0.2 mile) west of the Durian Tipus road and 3.2 km (2 miles) from the Pahang border, in the Sungei Pesasi a few meters north of the road from Jelebu to Pertang, and in several exposures near the road from Tampin to Kuala Pilah.

A sheared granophyre, of uncertain genesis, has been described by Scrivenor (1931) as occurring at the old Pasoh gold mine in Negri Sembilan. It contained pockets of gold-bearing pyrite.

Chemistry

A selection of available chemical analyses of the Lower Paleozoic volcanic, pyroclastic, and problematical rocks is given in Table 7.2, together with their Niggli molecular norms calculated by the computer method of Hutchison and Jeacocke (1971). From the normative compositions, we see that all these rocks are abundant in quartz and that all have K-feldspar to plagioclase ratios within the medium range of rhyodacite. The names in the table are as given by the field geologists, and it appears that rhyodacite would be a more appropriate group name. They all have a chemical composition equivalent to the major proportion of the Main Range Granite batholith, which is generally of adamellite. It is not therefore, surprising that they can be easily misinterpreted as fine grained marginal modifications of the granite in the absence of good exposures of their contacts with the country rocks or of any definite volcanic features.

UPPER PALEOZOIC TO LOWER MESOZOIC VOLCANISM

East of the Main Range, we know of extensive outcrops of pyroclastic and subsidiary volcanic rocks, mainly in the states of Pahang and Kelantan, with significant occurrences in Johore and lesser occurrences in Trengganu. The largest and best studied are in the area between Kuala Lipis and Gua Musang, where Richardson (1950) described outcrops totaling 310 km^2 (120 mile^2). Tuff, interbedded with shale and limestone, occupies lowland areas between the Sungei Jelai, the Main Range foothills, and the western margin of the highlands that extend northward from near Chegar Perah into south Kelantan. In Kelantan, similar rocks occur interbedded with metasediments in the region between the granite of the Kelantan–Trengganu Boundary Range and the Main Range Granite (MacDonald 1968). Volcanic and pyroclastic rocks are less abundant in Trengganu, where they are more localized geographically and stratigraphically. Occurrences of tuff are also known from the east coast around Kuantan (Fitch 1952).

In central Pahang, andesite agglomerate is well exposed in the neighborhood of Kampong Awah, where it has been extensively quarried for road metal, and to the north in the Jenka Triangle. Rhyolite forms a prominent ridge striking north-south near Lanchang, and the ridge is also quarried for road metal.

To the south, in Johore, rhyolite occurs on Pulau Tioman, where it is cut by andesite sills, and on Pulau Ujol and also on the mainland, where it is well exposed in the neighborhood of Endau to the north of Mersing, in the central area between Mersing and Kluang, around Gunong Pulai, and on

TABLE 7.2 *Chemical analyses and calculated Niggli molecular norms of selected volcanic, pyroclastic, and problematical igneous rocks* of Lower Paleozoic age*

	Weight Percentage					
	1	2	3	4	5	6
SiO_2	70.14	69.33	68.86	68.2	66.74	71.9
Al_2O_3	14.58	13.40	13.67	12.9	14.90	13.3
TiO_2	0.64	0.40	0.58	0.39	0.70	0.42
Fe_2O_3	0.64	2.91	0.30	0.92	0.47	0.97
FeO	2.97	1.56	3.77	4.01	3.77	3.33
MnO	0.11	0.05	0.09	0.19	0.07	0.04
MgO	0.71	0.86	1.63	2.02	1.76	1.21
CaO	2.68	2.42	2.40	4.10	2.93	0.07
BaO	n.d.	0.05	0.05	n.d.	0.06	n.d.
Na_2O	2.73	2.28	2.48	2.29	2.39	6.25
K_2O	4.39	5.29	4.94	3.37	4.71	0.22
H_2O+	0.24	0.16	0.13	0.24	1.12	1.98
H_2O-	0.46	1.02	0.95	1.31	0.20	0.12
P_2O_5	0.07	0.18	0.11	0.10	0.05	0.12
CO_2	n.d.	0.06	0.05	0.29	0.21	0.04
ZrO_2	n.d.	trace	nil	n.d.	nil	n.d.
F	n.d.	0.04	0.03	n.d.	0.01	n.d.
S	n.d.	nil	0.06	n.d.	0.08	n.d.
Cl	n.d.	0.03	0.01	n.d.	0.01	n.d.
Less O for F, Cl, S	nil	0.02	0.04	nil	0.04	nil
Total	100.36	100.02	100.07	100.33	100.14	99.97

	Niggli Molecular Norm Percentage					
	1	2	3	4	5	6
Quartz	27.43	28.94	24.54	27.76	22.99	29.05
Corundum	0.70	0.19	0.33	nil	1.27	3.06
Orthoclase	26.48	32.31	29.90	20.56	28.52	1.33
Plagioclase	38.13	31.64	33.81	36.70	35.18	57.38
Ab_xAn_{100-x}	$Ab_{66}An_{34}$	$Ab_{66}An_{34}$	$Ab_{67}An_{33}$	$Ab_{58}An_{42}$	$An_{62}An_{38}$	$Ab_{100}An_0$
Halite	n.d.	0.10	0.03	n.d.	0.03	n.d.
Total salic	92.75	93.17	88.62	85.02	87.99	90.81
Magnetite	0.68	3.00	0.32	0.99	0.50	1.04
Hematite	nil	0.09	nil	nil	nil	nil
Ilmenite	0.91	0.58	0.83	0.56	1.00	0.60
Apatite	0.15	0.44	0.27	0.22	0.12	n.d.
Fluorite	n.d.	0.11	0.09	n.d.	0.02	n.d.
Calcite	n.d.	0.16	0.13	0.76	0.54	0.10
Pyrite	n.d.	nil	0.16	n.d.	0.21	n.d.
Diopside	nil	nil	nil	2.39	nil	nil
	—	—	—	$Wo_{50}En_{26}Fs_{24}$	—	—
Hypersthene	5.51	2.45	9.59	10.07	9.60	7.45
	$En_{36}Fs_{64}$	$En_{100}Fs_0$	$En_{48}Fs_{52}$	$En_{51}Fs_{49}$	$En_{52}Fs_{48}$	$En_{46}Fs_{54}$
Total femic	7.25	6.83	11.38	14.98	12.01	9.19

* Specimens 1–6 as follows:

1. Pyroxene granodiorite porphyry (013002 p. 20), Kula Seli, Klang River, 12th mile, south of Ginting Sempah, Selangor.
2. Rhyolite tuff (quartz porphyry) (012018 p. 12). Sungei Naning, north of Ginting Sempah. Also analyzed for Cr_2O_3 nil, NiO nil. Specific gravity 2.68.

the south east extremity of the peninsula of Johore State. Minor lava flows occur on Singapore Island.

All the pyroclastic and volcanic rocks described in this section are thought to date from the Carboniferous, Permian, and Lower Triassic. In some cases these ages have been ascertained from numerous fossil localities in interbedded shale and limestone sequences. In many of these areas, however, as in Kelantan and Johore, the volcanic and pyroclastic rocks are associated with metasediments; accordingly, it is not possible to be precise about their age, but they are generally considered to be Upper Paleozoic. The members of this group of rocks, which has been classically referred to as the Pahang Volcanic Series, are typically products of contemporaneous explosive volcanicity that attained its maximum activity in the Permian and Lower Triassic.

Not all areas of volcanic and pyroclastic rocks are equally well known. Many outcrops shown on the geological map of Malaya (Alexander 1965) have not ben described. Others are well known through publications of the Geological Survey. The following account, based on these publications, is accordingly somewhat incomplete. The obvious course is to begin with the State of Pahang, where the rocks are best developed and best described, and then trace their lateral variations.

Central and North Pahang

The main area of occurrence of pyroclastic and volcanic rocks in Malaya is central and North Pahang. The rocks are interbedded mainly with shale, and less frequently with limestone of the Gua Musang formation, and confined therefore to the Permian and Lower Triassic, with the bulk in the Permian. The following account is taken from Richardson (1950), who proposed a succession summarized as follows:

Group 1. Rhyolite tuff is predominant in the north; interlaminated rhyolite tuff and shale occurs farther south; rhyolite tuff decreases southward toward Kuala Lipis.

Group 2. Rhyolite tuff, mostly of medium to coarse grain; some contains fragments of trachyte, trachyandesite, and andesite. This group is more basic than group 1 and, together with the rocks of group 3, forms a transitional series between early, dominantly rhyolitic tuff of group 1 and the later chloritic intermediate rocks of group 4.

Group 3. Tuff with some agglomerate, mainly of intermediate composition, but with some of rhyolitic composition. This group is more basic than group 2.

Group 4. Medium to coarse-grained tuff with subordinate agglomerate; colored green, grey-green, or green-black; containing chlorite; and of intermediate composition. The tuff contains abundant fragments of trachyte, trachyandesite, and andesite, and some rhyolite. In the Sungei Tui valley, for example, the tuff is predominantly andesitic. Rocks of this group reach their maximum development in the Sungei Kechau basin, to the north of Kuala Lipis.

From this succession, Richardson (1950, p. 40) assumed ". . . that the volcanoes first produced material dominantly of rhyolitic composition: later, material ranging in composition from trachyte to andesite was produced, and the supply became greater as time went on." The validity of this interpretation was discussed in Chapter 4.

The lavas are associated with neritic sediments. The extrusions are thought to be submarine, since no examples of scoriaceous texture have been found and the enclosing interstratified sediments show no signs of contact metamorphism (Richardson 1950).

PETROLOGY OF THE VOLCANIC ROCKS. Although widely distributed, lava is not abundant and is much less common than pyroclastic rocks. It ranges in composition from rhyolite to andesite.

Rhyolite. Thin bands of spherulitic rhyolite are intercalated with sedimentary rocks near Kampong Kermoi in the Sungei Jelai, in the Sungei Bisek, the Sungei Ingsor, the Sungei Tinggu, and in the tributaries of the Sungei Chiniau. Fragments of spherulitic rhyolite also occur in agglomerate at Foo Brothers Hydraulic Gold Mine (Richardson 1950). The outcrops of all these rocks are small, not more than a few meters wide, and they are interbedded with sedimentary rocks.

---

3. Pyroxene microgranite (012013 p. 12), near milestone 90, Bentong–Ginting Sempah road. Specific gravity 2.69.
4. Foliated rhyolite tuff (012024 p. 14), Sungei Nak Sap, west of Grik, Upper Perak.
5. Tuff or rhyolite (granite porphyry) (012017 p. 12). Vicinity of Ginting Bidai, between Ulu Klang and Ginting Sempah, on the Selangor–Pahang boundary. Specific gravity 2.66.
6. Granite porphyry (005005 p. 118). Sungei Piah, Lenggong area, Perak.

Numbers and page references are to Alexander et al., (1964).
n.d. = not determined.

The rhyolite often shows flow banding; the flow lines are clearly marked by laminae of chlorite, but sometimes the material is compact. The color varies from dark grey through green-grey to pale grey, and the color is occasionally banded.

The rhyolite is spherulitic, the spherules being ellipsoidal and filled with quartz-orthoclase-chlorite aggregates of approximately 0.35 mm diameter. Often the feldspar in the spherules is altered to sericite and epidote. The spherules occasionally exhibit a radial texture. The rhyolite contains phenocrysts of orthoclase, microcline, albite, and quartz, often slightly corroded (Richardson 1950). The rhyolite in the agglomerate contains three generations of phenocrysts; the first of quartz crystals and large prisms of orthoclase in excess of plagioclase; the second of quartz with medium-grained orthoclase and plagioclase in equal amounts; and the third of microphenocrysts of orthoclase and plagioclase, which are generally much corroded. The groundmass of all the rocks is very fine grained to cryptocrystalline and contains microlites of quartz, sericite, and chlorite with accessory magnetite; locally, it contains epidote.

Trachyte. Thin bands of trachyte are interstratified with shale in the Sungei Yu about 2 miles west of the railway line and in the tributary of the Sungei Chekua. In addition, fragments of trachyte are fairly common in the pyroclastic rocks. Trachyte can be either dark green or purple-grey and it contains phenocrysts of orthoclase and chlorite that have replaced mafic minerals, set in a base of orthoclase laths and chlorite. The Sungei Yu locality is characterized by conspicuous areas of calcite rimmed by chlorite (Richardson 1950).

Trachyandesite. Only two localities are recorded as containing trachyandesite. One is a small lenticular outcrop of massively jointed, speckled grey, porphyritic hornblende trachyandesite in the Sungei Serau at Lubok Liang near Kampong Liang (Richardson 1950). Both orthoclase and intermediate plagioclase are present as phenocrysts, commonly altered to calcite, epidote, and chlorite. Brown hornblende occurs as euhedral phenocrysts, some with inclusions of apatite. The groundmass is of orthoclase laths with parallel trachytic texture, together with glass, patches of chlorite and calcite, epidote, iron oxides, sericite, kaolin, and quartz. Ilmenite occurs as an accessory and is almost completely replaced by leucoxene. The other occurrence is in the Sungei Jelai near the Bukit Betong railway bridge. Here the rock is medium grained and greenish-grey, composed of phenocrysts of plagioclase and a few of orthoclase and quartz in a fine-grained groundmass of quartz sericite, and chlorite, containing abundant microlites of orthoclase. Patches of chlorite with iron oxides which are secondary after mafic minerals, and epidote, ilmenite, and magnetite are common accessories.

Fragments of trachyandesite similar to that just described are plentiful in calcareous agglomerate exposed at the 153.5 milestone on the railway line. The lava is medium grained and comprises phenocrysts of oligoclase-andesine in a groundmass of orthoclase laths, calcite, chlorite, and grains of magnetite.

Andesite. Massive fine-grained jointed andesite is exposed in the Sungei Kasai Kechil. It is traversed by veinlets of calcite and quartz. Elongate prisms of turbid feldspar and patches of chlorite after mafic minerals are set in a groundmass of chlorite. Pyrite is a common accessory. In addition, andesite is very common in agglomerate and tuff at the 153.5 milestone on the railway line, where it consists of laths of intermediate plagioclase in a groundmass of black glass. Fragments of augite andesite from a rhyolite-trachyte-andesite tuff in the Sungei Jelai at Kuala Medang comprise corroded phenocrysts of pale green, nonpleochroic augite and crystals of plagioclase in a cryptocrystalline, grey-brown groundmass (Richardson 1950).

PETROLOGY OF THE PYROCLASTIC ROCKS. Tuff is overwhelmingly preponderant and agglomerate is restricted to a few localities.

Tuff. The tuff is well jointed, hard, and tough when fresh, but soft, friable, and eventually in coherent when weathered. Some occurrences of tuff are massive, others thinly stratified, and many are finely laminated. The joint pattern is highly characteristic, consisting generally of three directions in addition to the more feeble bedding plane direction. The most important set of joints trends approximately at right angles to the strike, others cross the bedding at angles of 30° to 40° on either side of the strike direction. Excellent examples of well-developed joints occur in Jenut Batu Papan and in many places in the Sungei Serembun, the Sungei Lah, and the Sungei Jelai at Jeram Star.

The tuff ranges in grain size from fine to coarse, but it is mostly medium grained. Few tuffs are well graded. The constituent fragments are generally subangular to well rounded, but some are very angular. Feldspar crystals are more angular than quartz, and the coarse material is more angular than the finer. The volcanic constituents are both crystal and lithic, and generally the tuff is a mixed crystal-lithic type.

The tuff was deposited in water (Richardson 1950) and is without doubt contemporaneous with interbedded neritic sedimentary rocks. Good examples of interbedded sequences of tuff with shale and limestone occur at numerous localities, for example, in the following rivers: Jelai, Serumbun, Henderik, Kau, Malim, Chadu, Chiniau, and Bisek.

Rhyolite tuff is the most abundant variety; tuff containing fragments of trachyte and andesite is much less common. No single occurrence has been found of a tuff containing rock fragments exclusively of one volcanic rock—all occurrences are polygenetic.

The following is a complete list of the constituent materials in the pyroclastic rocks of north and central Pahang as given by Richardson (1950).

1. Volcanic constituents
  a. Crystal material—comminuted crystals and crystal dust
    Common: quartz, sericite, orthoclase, plagioclase (albite to calcic andesine).
    Rare: microcline, quartz-orthoclase spherulites, augite, hornblende.
    Doubtful: biotite, tourmaline, muscovite (they may have been derived from granite).
    Accessory: pyrite, ilmenite, magnetite, leucoxene, limonite.
    Others: calcite, chlorite (both anisotropic and isotropic varieties), epidote, picotite, magnesite.
  b. Lithic material—rock fragments, comminuted rock material, volcanic dust, and lapillae
    Acid: rhyolite (holohyaline, hypocrystalline, devitrified, spherulitic, flow-banded, cryptocrystalline, microcrystalline, fine-grained, medium-grained, feebly porphyritic, strongly porphyritic, chloritic, calcitic, and epidotic varieties).
    Intermediate: quartz-trachyte, trachyte (microcrystalline, fine-grained, nonporphyritic, porphyritic, and chloritic varieties), trachyandesite (medium-grained, porphyritic, and hornblendic varieties), andesite (hypocrystalline, medium-grained, porphyritic hornblendic, and augitic varieties).
    Basic: nil.
    Ultrabasic: sheared serpentinite (containing picotite, calcite, and magnesite).
2. Sedimentary constituents
  a. Torn from the walls of volcanoes during eruption: silicified limestone, carbonaceous limestone, calcareous chert, chloritic phyllite, shale, carbonaceous shale, probably biotic-quartz-hornfels.
  b. Introduced by rivers flowing from a land surface undergoing subaerial denudation: clay, silt, sericite, muscovite, quartz, limonitic and chloritic material, probably biotite, tourmaline, and some feldspar.
  c. From marine organisms: carbon dust (?), fragments of lamellibranch shells.
  d. Deposition from seawater: calcite, silica.
3. Constituents produced by metamorphism— Minerals produced by the reconstitution of clastic material due to thermal and dynamothermal metamorphism: sericite, muscovite, quartz, actinolite, hornblende, augite, biotite, epidote, graphite.

The various types of tuff have been described as follows by Richardson (1950).

Rhyolite Tuff. Typical rhyolite tuff is well exposed in many rivers in the area of Pahang between 4°18'N to 4°40'N: and 101°46'E to 101°59'E. In addition, there is an exposure at the Sungei Yu Halt, 176.5 milestone on the east coast railway. The tuff is typically grey or pale yellow-grey and may be speckled with white feldspar or clear quartz crystals. Usually it is compact, hard, massively jointed, and medium to coarse grained; sometimes, however, it is fine grained. The quartz crystals vary from angular to well rounded, and certain cases have been described as waterworn. Orthoclase is usually present, and plagioclase, of albite to oligoclase range, sometimes occurs. Some microcline may be present. Fragments of aphanitic and porphyritic rhyolite tuff usually appear in the rhyolite tuff. A few occurrences of trachyte and biotite-quartz hornfels fragments have been recorded. Tourmaline occurs sporadically.

The groundmass of the tuff normally ranges from fine to medium grain and contains biotite, chlorite, iron oxides, limonitic material, clay, epidote, and pyrite. Some occurrences have a carbonaceous base, and some are calcareous where interbedded with limestone.

An unusual tuff in the Sungei Bisek contains abundant quartz-orthoclase spherulites in a very-fine-grained aggregate of quartz and feldspar.

Tuffaceous Shale. Tuff, thinly interlaminated with shale and tuffaceous shale, occurs between 4°19' to 4°28.75'N and 101°50' to 101°53'E. The largest outcrop extends northward from Ulu Sungai Meledu into the headwaters of the Sungei Selor. Typically these rocks comprise alternating laminae of fine-, medium-, and coarse-grained rhyolite tuff and shale; the individual seams seldom exceed 3 mm in thickness. The sequence is strongly color banded, the tuff being almost white, pale grey, or medium grey, and the shale medium grey, dark grey, or black. The black bands are usually carbonaceous.

Shaley tuff is common throughout an area between 4°21' to 4°40'N and 101°46' to 101°50.5'N. It is also usually banded, but the rhyolite tuff layers contain abundant clay. In some cases the tuff includes fragments of black shale.

Trachyte-Rhyolite Tuff. Trachyte-rhyolite tuff is restricted in occurrence. A few specimens have been recorded within the region between 4°17.5' to 4°44.75'N and 101°48.5' to 101°59.5'E. The trachyte fragments are generally nonporphyritic; they range from microcrystalline to fine grained and are well flow banded. The tuff is always a crystal-lithic type, with fragments of quartz, orthoclase, and plagioclase, as well as of rhyolite and trachyte. The rhyolite fragments vary from feebly to strongly porphyritic and many contain interstitial glass in the matrix. Many occurrences of this tuff are rich in chlorite.

Andesite - Trachyte - Rhyolite Tuff. Richardson (1950) recorded andesite-trachyte-rhyolite tuff, a material of more intermediate composition, only from Kuala Medang in the Sungei Jelai. It is medium grained, grey, and calcareous. Crystal fragments include quartz, orthoclase, and plagioclase ranging from oligoclase to andesine. The lithic material comprises very-fine-grained rhyolite containing feldspar microlites in a glassy matrix, porphyritic rhyolite with conspicuous orthoclase crystals in a microcrystalline matrix, and trachyte composed of small ortholase prismatic crystals in a matrix of feldspar microlites and pale green glass. In addition, it contains numerous fragments of augite andesite. The augite, which is pale green and nonpleochroic, occurs as corroded phenocrysts. The andesite is composed of phenocrysts of augite and calcic andesine in a grey-brown matrix. The tuff matrix is of chips of quartz and feldspar mixed with volcanic dust, clay, calcite, and pyrite.

Tuffaceous Sandstone. In the Sungei Serumbun an unusual lithic sandstone occurs interbedded with typical grey rhyolitic tuff. It contains fragments of rhyolite and serpentinite and is dark grey to black with patches of green. Its composition is of fragments of quartz, chloritic and aphanitic rhyolite, and green fibrous sheared serpentinite that contains crystals of magnesite, pyrite, and picotite, and patches of chlorite. The matrix of the sandstone is of argillaceous material, containing a calcite mosaic, magnesite crystals, and carbon dust. The rock is considerably sheared.

Agglomerate. Agglomerate is of much more restricted occurrence than tuff. It has been found only at a few localities along the eastern margin of the granite body that extends from the Sungei Chiniau into Kelantan, along the railway line near Bukit Betong, in the Sungei Jelai, the Sungei Tui, and the Sungei Mesah valleys.

The most common type is rhyolite agglomerate, but more basic types containing fragments of trachyte, trachyandesite, and andesite also occur. The agglomerate may vary from grey and greenish-grey to purple and green. It contains fragments that may be as large as 15 cm in diameter. The fragments consist of a variety of materials, including white silicified limestone; chloritic phyllite; white, green, blue-green, and mottled rhyolite, which may be aphanitic, porphyritic, or spherulitic; feebly porphyritic andesite; quartz trachyte, dark green-grey trachyandesite; and quartz, orthoclase, and plagioclase fragments. They are embedded in a matrix that is often chlorite rich, although it may be calcareous or carbonaceous. Usually it contains some epidote and pyrite. The most accessible locality is at the 153.5 milestone on the railway line north of Bukit Betong.

North Kelantan and North Trengganu

Lava, tuff, and agglomerate of rhyolitic to andesitic composition outcrop extensively in the north Kelantan–north Trengganu area and undoubtedly represent an unbroken northward extension of the volcanicity described previously in north and central Pahang. The lava and pyroclastic rocks are interbedded with sedimentary rocks, and they occur as narrow bands and as extensive outcrops. All degrees of interbedding occur, and areas of volcanic and pyroclastic rock grade into areas of sedimentary rocks. However, the andesite–sedimentary rock boundary is usually more abrupt. This area, which has been mapped by J. R. Paton, J. Dawson, D. Slater, D. Santoh Singh, and S. MacDonald, was described by MacDonald (1968). The following description is taken from that publication.

The volcanic and pyroclastic rocks have been subdivided into two groups: (1) almost entirely of andesitic composition (2) predominantly of rhyolitic and trachytic composition, but with subordinate dacitic and andesitic components. Rapid variation, makes it impossible to subject the second group to further differentiation.

Volcanic rocks of boths groups occur most extensively in Kelantan between the Stong Migmatite Complex and the Boundary Range Granite. In the regions of Bertam, Kemubu, and Dabong on the east coast railway line, and to the north of Dabong, the rocks are predominantly of rhyolitic and trachytic tuff and lava with subordinate agglomerate. To the north and south of Kuala Krai, their character is more andesitic. In Trengganu, volcanic and pyroclastic rocks are less abundant, more localized, and confined to a few horizons; moreover, they are not thinly interbedded as in Kelantan. Rhyodacite occurs in north Trengganu, whereas to the south only andesite has ben recorded, occurring mainly in the neighborhood of Gunong Tembat, 58 km (36 miles) west of Kuala Trengganu. Here the andesite mass has been interpreted as having an anticlinal form (MacDonald 1968). Andesite occurs to the west of this locality in the Boundary Range Granite, but it is absent farther west in Kelantan, and there rhyolite and rhyolite tuff predominate. One of the most accessible andesite localities forms a hill between the railway line and the road from Kuala Krai to Kota Bharu.

The rhyolitic flows and pyroclastic rocks in north Kelantan, which are interbedded with metasediments of the Gua Musang formation, are therefore mainly of Permian to Lower Triassic age. They have been foliated and dynamothermally metamorphosed to greenschist facies.

In Trengganu, the andesitic rocks are associated with undifferentiated Upper Paleozoic sediments and metasediments, which may be predominantly of Permian age but possibly extend down to the Carboniferous. Hence in this region the contemporaneous volcanic activity appears to be limited to the range Carboniferous to Lower Triassic, with maximum activity in the Permian.

PETROLOGY OF THE VOLCANIC ROCKS. Lava is less abundant than pyroclastic rocks. Fine-grained rhyolite is difficult to distinguish from rhyolitic tuff because of dynamothermal metamorphism. However, andesite is easily distinguishable from andesitic tuff because it is more coarsely grained.

Andesite. Andesite is remarkably uniform throughout Kelantan and Trengganu and shows no essential difference from the same type of rock found in central and north Pahang (described previously). The major metamorphic effect on the andesite is the replacement of augite phenocrysts by amphibole.

The rock is typically porphyritic, usually green and purple, but some varieties are grey-green or black. Generally it is fine grained with abundant pyroxene phenocrysts, with or without feldspar phenocrysts. Rarely only feldspar phenocrysts are present. The groundmass is holocrystalline, composed of feldspar, pyroxene, and opaque minerals.

The most common feldspar is of andesine composition, but it may range almost to oligoclase, and rarely to labradorite. In unmetamorphosed andesite, the feldspar is fresh and free from inclusions, occasionally being slightly zoned. The phenocryst orientation is generally random, but some cases of flow-texture have been recorded. Broken and irregular feldspar phenocrysts occur infrequently. Orthoclase is generally absent. In metaandesite, the feldspar is considerably altered.

The pyroxene is usually augite, pale green, honey-colored, or colorless in thin section. It is usually euhedral, and commonly twinned. Pigeonite has been recorded in Kelantan. The pyroxene in the groundmass is generally granular. In Trengganu the augite of the andesite has 2V values ranging from $48.5°$ to $63°$, and $n\beta$ ranges from 1.681 to 1.683. The augite of the groundmass appears to be of similar composition to that of the phenocrysts.

The most abundant opaque mineral is magnetite, which usually occurs in the groundmass but may also occur as larger irregular blebs. Pyrite and pyrrhotite are common, especially in the andesite near Temangan north of Kuala Krai. Ilmenite has been recorded only rarely.

Hornblende is rare as a primary constituent. It is olive green to brown, and occurs together with augite. Most hornblende in andesite is of metamorphic origin, pseudomorphing augite. Other replacing minerals are calcite, chlorite, epidote, magnetite, and sericite. Both pyroxene and feldspar are replaced during metamorphism.

A rare occurrence of glass in andesite has been recorded. Quartz infrequently occurs as phenocrysts and as interstitial grains.

Vesicular andesite is rare, but a few localities in Kelantan and Trengganu have been described. The vesicles are infilled with chlorite, calcite, and quartz. At one locality in Trengganu, vesicles occur in an andesite that contains primary hornblende phenocrysts.

Rhyolite. Rhyolite is variable in color from grey to green and, rarely, red. It occurs as narrow bands interbedded with shale and tuff and also as thick massive well-jointed bodies. Generally it is porphyritic. The phenocrysts are of quartz, orthoclase, or sanidine, and (rarely) a ferromagnesian mineral which is usually biotite, set in a groundmass of variable texture. Many bodies of rhyolite have been metamorphosed.

Quartz is generally clear, usually as irregular fragments but sometimes as corroded and embayed phenocrysts. K-Feldspar is euhedral, occasionally corroded, and often twinned.

The groundmass is often microcrystalline, composed of a granular aggregate of quartz, feldspar, and mica. It may also be of glass with feldspar crystallites, or it may be microfelsitic; in rare instances it may contain spherulites. Flow texture is common, and the glass may contain perlitic and elongate vesicular structures.

Secondary minerals include calcite (very common), chlorite, sericite, epidote, and iron oxides. Siderite and tourmaline are seldom found.

Dacite. Dacite is of restricted distribution and has a mode of occurrence similar to that of rhyolite. It is grey to dark grey and occasionally greenish. Quartz, plagioclase, hornblende, and biotite occur as phenocrysts in a generally aphanitic groundmass, which often shows excellent flow structure and contains opaque dust. Secondary calcite is very common, and epidote, iron oxides, chlorite, and sericite are often present.

Trachyte. Although trachyte occurs very rarely, it has been recorded at a few localities, interbedded with rhyolite. This typically white, grey, or green rock is porphyritic; it has abundant crystals of orthoclase or sanidine, and rarely plagioclase, in a fine-grained groundmass that contains some glass. Iron oxides are common in the groundmass. Trachytic flow texture is occasionally well developed. Amphibole and biotite are common accessory minerals. Calcite, chlorite, pyrite, and epidote are common secondary minerals.

Intermediate Varieties. Trachyandesite and rhyodacite occur frequently in Kelantan and are generally closely associated with other lavas. Extensive outcrops of rhyodacite, grading into rhyolite, occur in north Trengganu. They are considerably altered.

PETROLOGY OF THE PYROCLASTIC ROCKS. Pyroclastic rocks are considerably more abundant than lava flows and are most strongly developed in the region between the Boundary Range Granite and the Stong Migmatite Complex (Figure 7.1). They have a wider geographic distribution than the lava and occur interbedded with pelitic sediments and metasediments all over north Kelantan. However, except for andesite tuff and agglomerate, which occurs in close association with andesite lava, pyroclastic rocks are very rare in Trengganu. Yet the abundant greywacke of north Trengganu may be composed to a considerable extent of pyroclastic crystal material.

There are pyroclastic equivalents of all the volcanic types previously described, but as in central Pahang, rhyolite tuff is by far the most common (Figure 7.3b). Andesite agglomerate and lapilli tuff is normally asociated with lava flows. However, rhyolite agglomerate is generally absent. The finer grained rhyolite tuff shows all degrees of admixture with sedimentary formations and may grade into shale, occasionally limestone, and quartzite. Most of the tuff occurrences are considerably metamorphosed, which makes interpretation difficult. The following lithological descriptions are taken from MacDonald (1968).

In the valley of the Sungei Lebir south of Kuala Krai are several tuff occurrences of predominantly andesitic composition. They vary in grain size from coarse agglomerate to volcanic dust and devitrified glass. The varieties seem to be haphazardly distributed. They are usually grey to dark grey and contain white crystal lapilli or small green lapilli of hydrated glass. The tuff may also be white, black, green, or brown; the color often depends on weathering. Crystal ash and glass are common in all tuffs.

Rounded bombs of andesite up to 30 cm in diameter, glass lapilli, fragments of feldspar, quartz, and pyroxene, are common volcanic materials in the tuff. In addition, fragments of tuff and of shale occur within the tuff. The larger of the rock fragments are generally rounded to subangular, but as size decreases, angularity increases.

The feldspathic crystal ash is of shattered crystals which retain some crystal faces. The composition is between oligoclase and andesine. Fragments of quartz crystals also occur. Generally ferromagnesian minerals are absent from the tuff. Glass in the form of lapilli or in the matrix usually forms 30% of any rock, but it is nearly always devitrified to a cryptocrystalline, colorless, brownish or opaque mass as seen in thin section. Some specimens are composed entirely of glass lapilli that have been flattened to lens shapes and often altered to fine-grained aggregates of epidote.

Glass also occurs in a variety of forms from hairlike wisps, for example, in limestone, to devitrified obsidian masses. Of 150 examined thin sections of tuff from the Lebir Valley, 55% contain quartz and no pyroxene, 27% contain plagioclase only, 15%

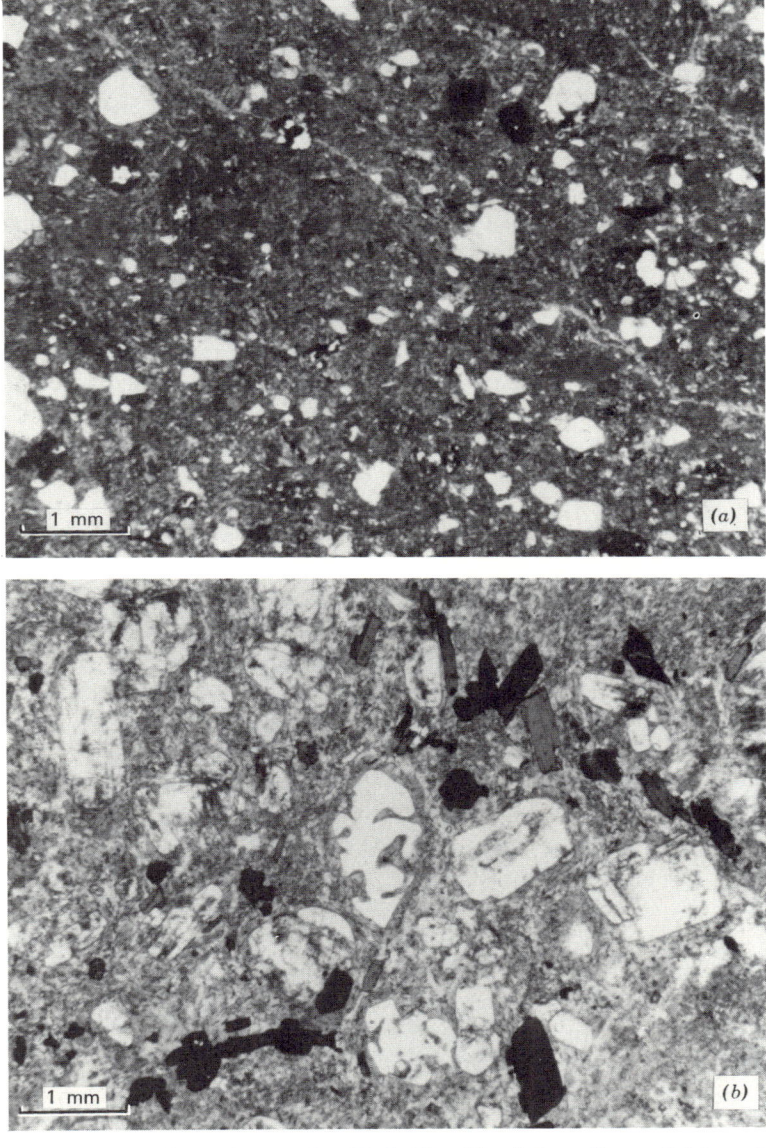

FIGURE 7.3 *Photomicrographs (C. S. Hutchison) in ordinary light; scale as indicated:* (a) *Temangan ignimbrite, Sungei Hau, Kelantan, specimen UM 3724;* (b) *acid tuff, Sungei Pergau near Kuala Yong, Kelantan, Specimen UM 6734.*

have pyroxene and no quartz, and only 3% have quartz together with pyroxene. All contain feldspar, and in the majority the matrix is andesitic. The lava in the neighborhood is entirely of andesite. Hence it is difficult to explain why most of the tuff contains abundant quartz and is deficient in pyroxene.

Slater (*in* MacDonald 1968) attempted to explain this apparent anomaly by suggesting that the frequency of occurrence of quartz, feldspar, and pyroxene in the tuff is related to the ability of intratelluric crystals in the andesitic magma to survive the shock of explosive activity—quartz is suggested to be most likely to survive and pyroxene, although most abundant in the magma, is the least likely to survive. The pyroxene crystals are therefore considered to disintegrate almost entirely and to form the groundmass of the tuff.

Alteration of the feldspar in the tuff is frequently to saussurite and calcite. Calcite may occur as euhedral crystals or as irregular masses. By this process, calcareous tuff is formed. Epidote occurs frequently and forms fine-grained pseudomorphs after feldspar.

Near Temangan in Kelantan, fine-grained agglomerate is associated with andesite. The angular rock fragments rarely exceed 5 cm in diameter and are of andesite.

In the Kelantan River, near Tanah Merah, a grey-green tuff contaning euhedral crystals of feldspar and pale yellow augite occurs. The pyroxene occurs as euhedral to anhedral crystals but commonly also as irregular fragments. The feldspar crystals are commonly altered. The groundmass is fine grained. Chlorite commonly forms pseudomorphs after pyroxene. The feldspar has altered to mica and chlorite. The lava fragments are of very-fine-grained andesite consisting of feldspar laths and microlites set in a groundmass of augite and epidote granules, with micaceous and chloritic material and a little glass. The feldspar is andesine. Some lava fragments have a trachytic texture. Magnetite also occurs, along with a little quartz, which results from devitrification of the glass.

Rhyolite, rhyodacite, and dacite tuff include vitric, crystal, and lithic varieties. They are light in color (often grey or green) and are thinly interbedded with sedimentary rocks or may form larger bodies. Welded rhyolite tuff is common. Most acid tuff is considerably altered; and particularly in the finer grained varieties, the processes of devitrification, silicification, and replacement of secondary minerals make the distinction between a tuff and a lava flow often impossible. Greenschist facies metamorphism of tuff and some thin lava flows often has produced sericite-chlorite schist and phyllite.

Plagioclase usually alters to sericite, chlorite, and epidote, although fresh unaltered orthoclase and plagioclase crystals are not uncommon. Quartz usually occurs as corroded crystals or as granular aggregates, often enclosed by chlorite patches; sometimes it is found intergrown with K-feldspar as a micrographic intergrowth. Iron oxides and apatite frequently occur in the acid tuff, and calcite is sometimes very common.

The calcite content varies a great deal both in tuff and lava, but it may form as much as 80% of dacitic tuff in Kelantan. Not all this calcite can be of secondary origin, and some must be sedimentary. Most of the epidote must be attributed to the greenschist facies of dynamothermal metamorphism, but some may result from saussuritization of the feldspar.

The Temangan Ignimbrite

In Kelantan an ignimbrite forms a prominent ridge about 24 km (15 miles) long and 0.8 km (0.5 mile) wide, trending approximately north-south. Good exposures occur on the Kelantan River northwest of Kuala Krai, on the Sungei Galas southwest of Kuala Krai, and south of Temangan on the road to the old iron mine. At the last locality, the contact with shale and sandstone is exposed. The ignimbrite trends north-south and has a high easterly dip, whereas the sediments dip at 35° in the direction of 155°. This evidence suggests an intrusive nature for the ignimbrite. The following description is after Aw (1967).

The ignimbrite is generally massive, but minor flow structure occurs locally. The rock body is homogeneous both laterally and vertically. Jointing is present but not prominent. The color varies from pink to dark brown in fresh specimens but is greenish grey in altered specimens. The rock is conspicuously porphyritic, with phenocrysts of quartz and feldspar in a felsitic matrix (Figure 7.3a). In some specimens, fragments of shale and black pumiceous material are present. More than 70% of the phenocrysts are of quartz, ranging from less than 0.1 to 2 mm in diameter and from euhedral to highly angular. Nearly all are corroded, and some phenocrysts contain small inclusions of glass.

Feldspar phenocrysts include plagioclase (An_{30}) and K-feldspar. They are often sieved through resorption and corrosion. K-Feldspar is generally altered, but the border zones are clear. Some crystals are fractured into separate parts. Green biotite occurs sparsely. The groundmass of the rock is of cryptocrystalline material and rare glass shards that show incipient devitrification. Northwest of Kuala Krai, on the Sungei Kelantan, observers have noted that the shards are tricuspate and their borders devitrified. In other specimens the shards show some crude alignment. The groundmass probably contains a high proportion of plagioclase, but it is too fine to identify. Euhedral crystals of magnetite are ubiquitous in the groundmass. There are also narrow wisps of chlorite, and small spherulites are present in some localities. The cryptocrystalline groundmass probably represents devitrified glass.

This rock body is not a true lava, nor a pyroclastic rock, nor a hypabyssal intrusion. Aw (1967) wrote that it is a "dyke" of ignimbrite that plugs an old feeder fissure, and he compared it with intrusive ignimbrites described by other authors. Its characteristics are those of an extrusive rock, but its shape and attitude to the sedimentary rocks suggest an intrusive nature. This rock has been described as a sillar, implying induration by recrystallization. Aw (1967) suggested that it was emplaced as a tuff flow.

Attention has been drawn to the unique nature of the linear outcrop of the Temangan ignimbrite by Burton (1967e). The extension to the south-southeast of the line of the ignimbrite "dyke" marks the eastern margin of the Boundary Range Granite for a length of 61 km (38 miles). At the same time it appears to be related to the adjacent long straight course of the Sungei Lebir, which H. D. Tjia (personal communication) has shown to be the site of a wrench fault. Therefore, instead of being contemporaneous with the Permian to Lower Triassic metasediments, the Temangan ignimbrite may be related to the later Lebir wrench fault.

East Pahang

In the neighborhood of Kuantan, tuff and rhyolite occur associated with metasediments of general Upper Paleozoic age, but in some localities there is good evidence to suggest a Lower Carboniferous (Viséan) age. Purple weathered tuff occurs in the Sungei Batu and the Sungei Reman near Bukit Sagu, associated with Lower Carboniferous shale and phyllite. Rhyolite occurs interbedded with psammitic rocks of presumed Upper Paleozoic age in and around Bukit Cherating, around Bukit Minyak and Bukit Gerai, at Bukit Rangin, in the Saravanamuthu Pillai Estate, and at a few isolated localities in the Paya Besar. These rocks were described by Fitch (1952).

TUFF. The tuff, which is always weathered in outcrop, is usually red to purple with white flecks of altered feldspar. It consists of angular to subangular quartz grains; some grains have crystal faces, and flakes of chlorite and grains of fine quartz aggregate and are set in a quartz-sericite matrix. Some of the quartz crystals are strained and the chlorite crystals bent. On advanced weathering the tuff becomes limonitic.

RHYOLITE. The rhyolite is buff colored when partly weathered and red or brown due to limonite when completely weathered. Quartz phenocrysts are easily visible in the hand specimen. Some occurrences are of pure rhyolite lava, others are of rhyolite containing pyroclastic crystal fragments and fragments of sedimentary rocks. Flow-banded rhyolite is rare, but it has been recorded at Bukit Setongkol near the road from Kuantan to Jerantut, near Kampong Pandan, and near Cherating. It contains corroded quartz crystals in a cryptocrystalline matrix of devitrified rhyolite glass.

More commonly, the rhyolite contains rock fragments and grades into tuffaceous rhyolite in which corroded quartz crystals, fragments of quartz crystals, aggregates of fine-grained quartz and, in some specimens, fragments of feldspar crystals, are set in a devitrified rhyolite glass matrix. Enclosed in this rock are rounded pebbles of quartzite and silicified shale up to 5 cm long. Such rocks occur at Tanjong Batu Pak Mok.

At Bukit Minyak, metamorphism of the rhyolite has produced muscovite in a sheared rhyolite matrix.

Fitch (1952) suggested that the volcanic and pyroclastic rocks in the Kuantan area are submarine but were extruded at or near a shoreline.

Scrivenor (1931) proposed that the rhyolitic rocks should be correlated with the quartz porphyry dykes which postdate the dolerite dykes at Bukit Ubi in Kuantan town. Fitch (1952), however, believed that his was a wrong correlation because some of the rhyolite bodies in the Kuantan area have been tilted and now dip at 30°, whereas he porphyry dykes at Bukit Ubi appear to have retained their steep dip. In addition, many rhyolite occurrences are sheared and metamorphosed and accordingly are quite unlike the completely unaltered quartz-porphyry dykes of Bukit Ubi.

The Jengka Triangle

The Jengka Triangle rural development scheme area is bounded by the main roads between the three towns of Jerantut, Temerloh, and Maran (Figure 7.4). The geology of this region has been mapped by Chong and Yong (1967).

Volcanic and pyroclastic rocks occur interbedded with a sedimentary sequence of sandstone, shale, conglomerate, and limestone. At Kampong Awah the limestone has been dated as Permian (see Chapter 4). It can therefore be assumed that, because the andesite is closely associated with the limestone, the volcanism in this region was Permian.

Coarse-grained andesite occurs on the eastern side of the Triangle and outcrops at the waterfall at the 85.5 milestone along the road from Jerantut to Maran. This andesite body extends southward to Ulu Jempol, and at its southern extent it outcrops in the Sentol quarry as andesite agglomerate at the 98.75 milestone.

Andesite tuff outcrops near the upper region of the Sungei Jengka and in its western tributaries. The tuff can be divided into andesite tuff, crystal-lithic tuff, and acidic tuff.

Porphyritic andesite and agglomeratic andesite are usually intimately associated. They are found at Kampong Awah and in the Sentol quarry.

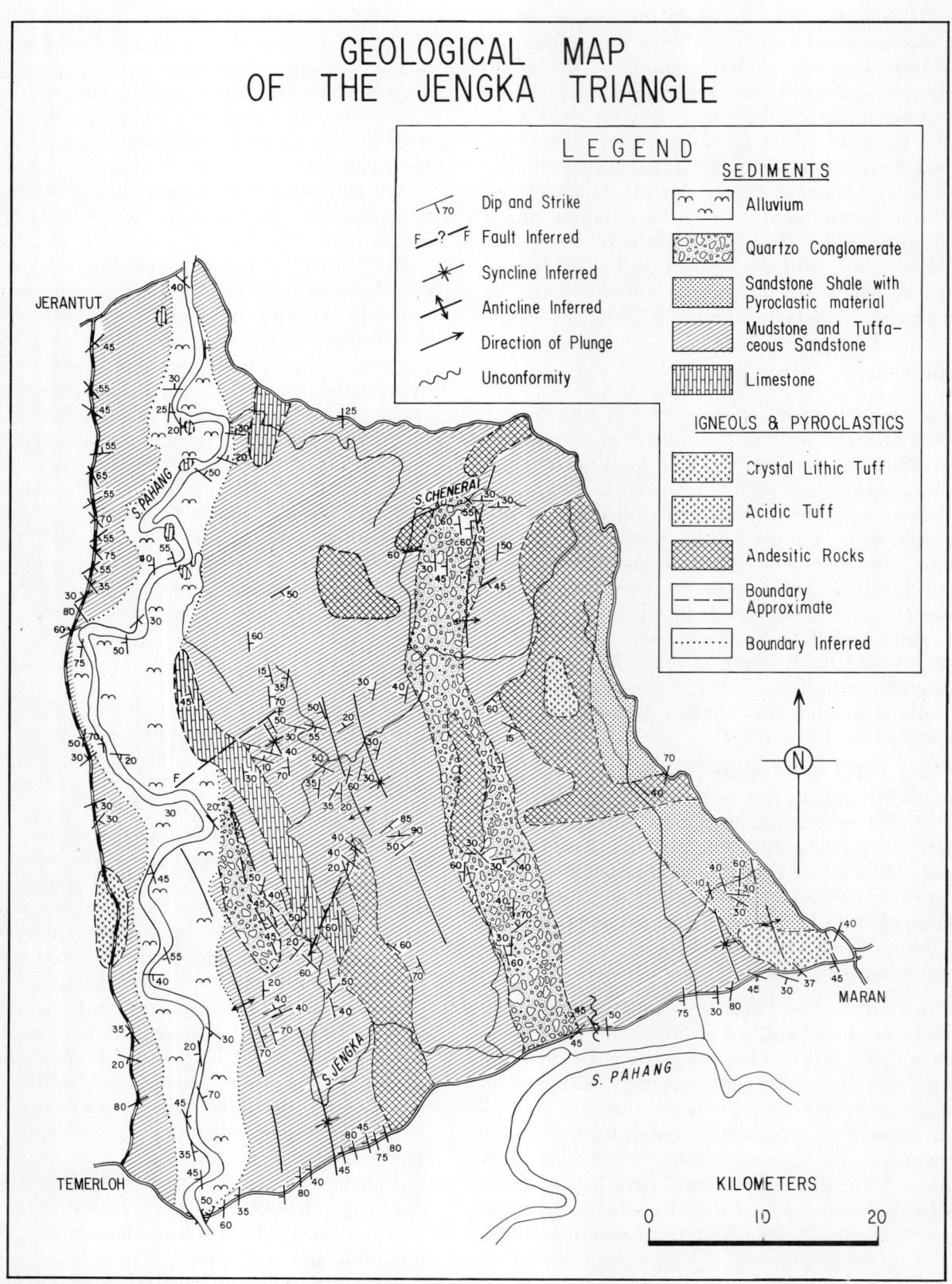

FIGURE 7.4 *Geological map of the Jengka triangle, Pahang. After Chong and Yong (1967).*

Crystal lithic tuff occurs at Ulu Jempol and at the 94.5 milestone on the road from Jerantut to Temerloh. Acidic tuff is exposed at milestones 118 and 119 along the same road.

The andesite agglomerate at Kampong Awah has been described by Wong (1960). There are two road metal quarries on the sides of Bukit Kepayang, but one is no longer used. Kampong Awah is about 16 km (10 miles) east of Temerloh along the main road to Maran. In the main quarry there is an intimate mixture of limestone and pyroclastic andesite, but no andesite lava flow.

The limestone is of two lithologies, light grey and black, and both are fossiliferous. The pyroclastic fragments range from 50 cm diameter to less than 1 mm. However, most fragments are larger than 32 mm; both angular and rounded occur, so the general term agglomerate is warranted. The fragments in the agglomerate are andesite, limestone, and more rarely gabbro, sandstone, and greywacke. Fragments and blocks of andesite occur in a limestone matrix, and fragments and blocks of andesite and limestone occur in an andesitic matrix. The relationships are very complex. It can be concluded that these complexities result from explosive submarine andesitic activity in a sea of limestone deposition. The feeder channels for the andesite eruption must have passed through a sedimentary sequence of limestone, sandstone, greywacke, and gabbro before reaching the sea floor. The gabbro inclusions are much more restricted than other lithologies and are found in the andesite only.

Wong (1960) has described the various components of the agglomerate as follows:

The andesite is porphyritic. Phenocrysts are of:
1. Augite—nonpleochroic, $2V = 57° + ve, \gamma \wedge c = 43°$, usually euhedral but frequently with corroded margins and alteration to chlorite;
2. Plagioclase—tabular crystals of andesine, usually almost completely saussuritized;
3. Hornblende—present only occasionally, pleochroic from greenish-yellow to brown, crystal outlines are usually embayed; some of the hornblende pseudomorphs augite.

The groundmass is generally composed of laths of plagioclase, granular augite, magnetite, pyrite, and brown or black glass. Pyrite also occurs in veins. Some zeolite veins also occur. Some specimens contain chlorite in the groundmass, together with secondary calcite. Biotite is sparsely present. Epidote occurs occasionally. The augite of the groundmass may be replaced by hornblende.

The agglomerate consists of various rock fragments in a groundmass composed of crystal fragments of augite and saussuritized plagioclase. But many agglomerate specimens have a calcareous groundmass of microcrystalline opaque calcite, containing andesitic crystalline fragments.

The contact between the andesite fragments in limestone groundmass or between the limestone fragments in the andesite groundmass is always sharp.

West Pahang

BENTONG AREA. Pyroclastic rocks occur commonly in an area east of the Sungei Bilut, where they are frequently interbedded with shale, sandstone, conglomerate, and limestone of presumed Permian to Lower Triassic age. They range in composition from rhyolitic through trachytic to trachyandesitic Alexander (1968). A few occurrences are andesitic.

The most common rock type is medium-grained rhyolite tuff, containing fragments of rhyolite, grains of quartz, and flakes of chlorite. More basic types are less common and are usually agglomeratic; they become present more frequently toward the east. The greater majority of the tuff occurrences are of admixtures of andesitic, rhyolitic, and sedimentary material, and they grade into tuffaceous sandstone and shale. A few rare occurrences of tuff containing granite, granophyre, and microsyenite fragments have been recorded. Some tuff occurrences are massively bedded, others are more commonly thinly stratified.

Some agglomerate occurs in the Klau Forest Reserve. It is poorly graded and the matrix is rich in volcanic glass.

Lava flows are virtually absent from the Bentong area; fragments of rock varying from rhyolite and andesite occur in the tuff and agglomerate.

Porphyritic rhyolite with prominent quartz phenocrysts and/or feldspar phenocrysts occurs in a few small, narrow outcrops. The groundmass is finely crystalline, but traces of spherulites have been seen and well-marked flow texture is often present. The quartz phenocrysts are usually corroded.

Trachyte and latite are rare. They are composed of orthoclase and oligoclase, with calcite and limonite and display a well-marked flow texture.

Dacite and andesite become more frequent toward the east. Andesite fragments in the tuff are dark purple or green-black. They are composed of porphyritic feldspar, sometimes quartz, augite, groundmass feldspar, an opaque iron oxide, and chlorite. The feldspar is strongly zoned from oligoclase to andesine and is sericitized. The augite is greenish and may be altered to chlorite. Epidote and calcite are common alteration products.

At Lanchang village, 32 km (20 miles) due east of Bentong on the main road from Karak to Temerloh, the road cuts through a prominent north-south trending mass of fine-grained rhyolite (Figure 7.2b). Outcrops can be seen on the main road about 1 km (0.75 mile) east of Lanchang, but excellent exposures are found in a road metal quarry just off to the south of the road, immediately east of the village. The rock is pale grey and resembles chert in the hand-specimen, with a characteristic conchoidal fracture.

Phenocrysts are few and of 3-mm size. They consist of angular but completely anhedral quartz, albite that is euhedral and tabular but corroded and strongly sericitized, and biotite as long tabular crystals, completely chloritized.

The groundmass is of a microcrystalline mosaic (grain size 0.1 mm) composed of the same minerals. There is a crude foliation, which is indicated by incomplete layers of the larger phenocrysts.

This rock is very homogeneous through its outcrop, and although shown by Alexander (1965) as "granite," consideration of its constant microcrystalline nature clearly indicates volcanic or subvolcanic origin. The concordance of the rock mass to the regional structure suggests that it may be a contemporaneous rhyolite.

RAUB AREA. Richardson (1939) has described two bodies of rock in the region to the north of Raub, Pahang, which may be interpreted as intrusive microgranite or extrusive rhyolite.

The Bukit Kajang Granite Porphyry. Granite porphyry forms the Bukit Kajang range, extending northward for nearly 11 km (7 miles) from 5 km (3 miles) due north of Raub toward Cheroh village. It has a width of about 2 km (1 mile) on average. It is a strongly jointed, compact, pale grey rock containing conspicuous phenocrysts of quartz and feldspar, commonly between 2 to 5 mm diameter. The rock groundmass is fine grained and commonly micrographic. Dark minerals are usually biotite and commonly tourmaline, but some varieties have no dark minerals. The phenocrysts are corroded. Feldspar is normally euthedral. Microcline and orthoclase are present. The plagioclase is albite, on average An_8. Tourmaline occurs as stellate clusters. Richardson (1939) suggested that this rock mass is related to the Main Range Granite, which lies some 5 km (3 miles) to the west. But there is no obvious connection, and the Bukit Kajang granite porphyry is separated by a belt of metaconglomerate and other metasediments that are not intruded by granite. This mass is concordant to the overall regional strike, and no evidence has been found to help determine if it is intrusive or volcanic contemporaneous with the enclosing sediments and metasediments. In view of the close association of this body with the Raub gold mineralization, it is important that the question of the genesis of the Bukit Kajang granite porphyry be kept open until definite evidence is produced to confirm its origin.

The Gali and Dong Granite and Quartz Porphyry. A large and irregularly shaped body of granite and quartz porphyry extends northward for about 11 km (7 miles) from Bukit Serdam, which lies to the east of the main road from Raub to Benta. Its maximum width is 1.5 km (1 mile), and like the Bukit Kajang body, it is conformable with the regional strike.

The Gali and Dong granite and quartz porphyry is exposed in the old quarry at Kampong Lepar, about 4.8 km (3 miles) north of Bukit Serdam. It is a compact, pale bluish-grey or purplish-grey, fine-grained rock containing phenocrysts of quartz, orthoclase, and plagioclase. The usual size of the phenocrysts is 2 to 3 mm. The rock is strongly jointed. Orthoclase phenocrysts are rounded, strongly corroded, and deeply embayed. Plagioclase, of albite composition, also occurs as corroded phenocrysts. Quartz is also somewhat corroded. Octahedra of magnetite occur. Muscovite is also present. The groundmass, which is cryptocrystalline, is composed of quartz, feldspar, some muscovite, ilmenite, and magnetite octahedra. A slight foliation is shown by the mica minerals. The contact zone of the porphyry with the metasediments is represented by a schist of sericite, feldspar, and quartz.

North of the Sungei Dong, much of the body is of quartz porphyry. It is a strongly foliated rock, containing corroded quartz crystals in a cryptocrystalline matrix in which the mica has a strong preferred orientation.

The Dong and Gali rock mass, with its cryptocrystalline matrix, corroded crystals, and foliated nature, seems to be of volcanic origin. Apparently Richardson (1939) was mistaken in interpreting it as intrusive, and he produced no evidence to support his view.

South Johore

Grubb (1968) has described lava and pyroclastic rocks in southeast Johore which are associated with graphitic and muscovite schist, assumed to be of Carboniferous? to Triassic? age. The following description is taken from his publication.

The lava flows vary from weakly pillowed andesite through intermediate types to normal rhyolite. Pyroclastic rocks are by far the more abundant and possess a similar range of composition. Grubb (1968) proposes an initial extrusion of andesitic lava followed by intermediate to acid flows, then by explosive pyroclastic deposition.

LAVA. Four distinct types of lava have been recognized. The first to be described is thought to be the oldest; the last the youngest.

Andesite. Andesite occurs at Tanjong Bulat, where it shows a pillow texture. The pillows vary from 15 to 30 cm in diameter. We can infer the age of the andesite because xenoliths in the neighboring granite, which intrudes rhyolite tuff, are predominantly of andesite, thus indicating that andesite may underlie and, hence, predate the rhyolitic rocks.

The andesite is dark grey and highly vesicular, in places scoriaceous. Vesicles, which may be as large as 10 cm in diameter, are generally infilled with albite, epidote, and chlorite. The rock consists of tabular andesine phenocrysts varying from 0.2 to 8 mm in length. They are zoned to rims of oligoclase, are well sericitized, and show good flow texture. The groundmass is of fine granular andesine laths and pyroxene granules in an altered glassy matrix. The glass has been devitrified to nearly isotropic chlorite, containing fine brown granules of sphene and garnet. The pyroxene in the groundmass is hypersthene ($2V = 70$ to $80°$, $-ve$) and augite, both now largely altered. The plagioclase laths, however, are relatively fresh and range from An_{36} to An_{40}. Quartz sometimes occurs in vesicles and as minute patches in the groundmass. Accessory magnetite and ilmenite also occur as minute skeletal crystals.

Andesite fragments occur as xenoliths in the micropegmatitic zone of granite at Tanjong Sepang. They have been largely assimilated by the granite. The original pyroxene crystals have been uralitized to form striking large crystals of dark green and blue pleochroic hornblende ($2V = 72°$ $-ve$) as well as small amounts of cummingtonite. The glass of the groundmass has been altered to brownish-green biotite, which in turn has been replaced by hornblende. Some albitization of the feldspar has occurred. The xenoliths also contain epidote.

Dark nonbanded rhyolite. Dark nonbanded rhyolite occurs only in a small area between Bukit Raja, Bukit Bopeng, and Bukit Tanah Merah, and outcrops on the coast at Tanjong Pasang in southeast Johore. It is overlain by high-grade bauxite and is probably the source of the ore. The rock is closely jointed, and no flow structure is seen. It consists of small phenocrysts of subhedral feldspar zoned from andesine to oligoclase, and of subhedral to anhedral quartz, all set in an extremely fine-grained matrix of quartz and feldspar. Microscopic autoxenoliths of rhyolite and some of ?andesite occur. Some of the rock is microbrecciated. The matrix is of oligoclase, sanidine, quartz, green hornblende, and epidote, with accessory magnetite and ilmenite now altered to leucoxene and rimmed by anatase. There is some mineralization of pyrite and chalcopyrite.

Banded porphyritic rhyolite. Banded porphyritic rhyolite is dark purple to brown and contains conspicuous sanidine, oligoclase, perthite, and quartz phenocrysts, along with numerous vesicles of light green chalcedony. The fine-grained matrix is predominantly of quartz, secondary mica, calcite, magnetite, sphene, and leucoxene. Flow banding is coarse but regular. Flows as thin as 30 cm show well-developed vesicular tops. Xenoliths are numerous, varying from 1 to 10 cm in diameter. Most of the xenoliths are of rhyolite. Alteration is advanced.

Banded felsitic rhyolite. The banded flow structure of felsitic rhyolite is highly contorted. It is purple to pink, with flow bands 1 to 3 cm thick, and it has a fine grained felsitic texture. This rock is considerably more basic than the normal rhyolite, possibly intermediate between rhyolite and andesite in chemical composition. It is weakly porphyritic. Oligoclase phenocrysts are almost completely sericitized, and quartz crystals are corroded. In addition, there are occasional clusters of epidote, porphyroblasts of hornblende, and skeletal magnetite crystals. The groundmass is of a mosaic of coarse poikilitic quartz grains enclosing minute laths of sanidine ($2V = 14°$), oligoclase (An_{18}, low-temperature form), and lesser orthoclase. Accessory minerals include magnetite, ilmenite, sphene, epidote, brownish chlorite, and zircon.

TUFF. The appearance, composition, and attitude of the tuff changes rapidly within very short distances. The tuff has been subdivided by Grubb (1968) into "ashy tuff" (< 3 mm grain size) and "agglomeratic tuff" (> 3 mm grain size). However, it is not easy to apply this subdivision strictly because there are all degrees of gradation and intermixing of the two types.

Ashy Tuff. More common than agglomeratic tuff, ashy tuff is mainly of a crystal-lithic type. Vitric fragments are common; some glass shards have been observed, but they are usually devitrified. On the whole, the lithic fragments consist of banded ashy material. The crystal fragments are of quartz, either as minute particles or as embayed subangular grains. The matrix is quartzo-feldspathic and is very uniform and extremely fine grained. Montmorillonite has replaced almost all the feldspar. Accessory minerals include abundant octahedra of magnetite, marginally altered to limonite, and lesser amounts of siderite, zircon, and chlorite.

Some tuff exposures are completely massive, others are well-laminated. Their colors range from black, brown, and purple, to almost white.

Agglomeratic tuff. Agglomeratic tuff is highly variable in texture and composition. The color ranges from almost white to dark grey or black. The fragment diameters are between 3 and 40 mm. The fragments themselves may be rounded, subangular, or angular, and they consist mainly of fine-grained ashy tuff, with subordinate amounts of rhyolitic and vitric materials. Crystal fragments are of quartz and oligoclase.

Chemically the tuff varies from rhyolitic to andesitic. The groundmass of the tuff at Bukit Bangsal Kunyit is a quartzofeldspathic mosaic of minute granules of bluish sodic amphibole and some patches of albite.

The tuff is generally masive in outcrop, or it may have a crude stratiform appearance. Intercalations of ashy tuff accentuate the bedding in places, but, much of the tuff has been laterized.

The island of Nanas in Johore Strait is of lava and tuff, ranging from rhyolite to andesite (Scrivenor 1931). These rocks contain secondary biotite and hornblende, which probably result from metamorphism. In one disused quarry on the island the tuff contains abundant fragments of biotite granite in which secondary hornblende and biotite are common.

Burton (in prep.) described many occurrences of volcanic rocks throughout Johore State. Most of these occurrences are possibly of Triassic age. He has classified them as belonging to the Gunong Pulai volcanic member of the Jurong formation. The principal rock type is a tuff of rhyodacite composition. Agglomerate, tuff, and lava of dacitic to andesitic character are also present in minor amounts. The rhyodacite tuff is of crystal-lithic type. It is dark green to greenish-grey when fresh and mottled with white feldspar phenocrysts. Clear quartz and many colored lithic fragments can also be seen. The feldspar is both of andesine and orthoclase. Quartz, epidote, and hornblende are common. Some occurrences appear to be ignimbrite, characterized by flow-texture and corrosion of the crystals, particularly those of quartz.

Central Johore

Rajah (1967) has described wide occurrences of volcanic rocks in central Johore near Gunong Blumut. He has tentatively subdivided them into the Sedili volcanic rocks and the Chemendong volcanic rocks.

THE SEDILI VOLCANIC ROCKS. The Sedili volcanic rocks occur in the Sungei Ulu Sedili–Sungei Tempenis area and around the Sungei Payong. They are predominantly of ignimbrite, with subordinate lava flows, lithic tuff, agglomeratic tuff, agglomerate, and volcanic ash. Their color is generally dark grey to greenish grey, with local variations. The tuff is largely of rhyodacitic composition, although locally it may be rhyolitic or andesitic. The welding of the tuff varies from poor to intense. Larger crystals of clear quartz and pink feldspar can be seen in the outcrops. The feldspar is oligoclase to andesine. Hornblende, epidote, and chlorite are common. Flow structure is usually seen, and the quartz crystals are usually corroded.

The lava flows are mainly rhyolitic and rhyodacitic. Some outcrops show folding. Restricted flows of andesite and dacite occur within the tuff bodies. The Sedili volcanic phase was therefore mainly of the explosive type, as witnessed by the preponderance of pyroclastic rocks.

CHEMENDONG VOLCANIC ROCKS. The Chemendong volcanic rocks include lava, tuff, and agglomerate. The lava varies from rhyolite to rhyodacite, and in some places quartz-andesite occurs. The lava is generally porphyritic and shows flow banding. The most common rock type is a greenish-grey rhyodacitic flow, although at some localities its place is taken by crystal welded tuff. The rock has well-formed quartz and feldspar phenocrysts in a fine-grained mosaic of quartz and feldspar, together with varying amounts of hornblende. Biotite, epidote, and chlorite frequently occur. Pyroxene is not abundant. Variable amounts of magnetite are present.

Rhyolite occurs mainly on the western parts of the northern foothills of Gunong Blumut to the south of the road from Kluang to Jemaluang and

Mersing. The color varies from grey through pale green to salmon pink.

Rajah (1967) deduced that the larger bodies of volcanic rocks are Triassic, although some may be Permian or possibly Carboniferous.

East Johore

J. H. Bean (in prep.) has mapped Pulau Tioman, which lies some 48 km (30 miles) off the east coast of Johore from Mersing. Apparently on this island most of the sedimentary and volcanic rocks have been dynamothermally metamorphosed and are also highly permeated by granite dykes and veins. Volcanic rocks appear mainly to be metarhyolites, with some metatrachyte and metaandesite. Tuff is also common. The rhyolite commonly shows flow banding. Metamorphism has caused the formation of abundant biotite, epidote, chlorite, and some garnet in the volcanic and pyroclastic rocks. Some rhyolite and tuff bodies contain lapilli varying from less than 3 cm to 60 cm in diameter. Metagglomerate also occurs. Because of the metamorphic condition of these rocks, they can generally be correlated with the metasediments of undifferentiated Upper Paleozoic age, which occupy most of the east coast of Malaya.

On the mainland adjacent to Pulau Tioman, between Endau and Mersing, are excellent outcrops of rhyolite tuff and flow-banded rhyolite. Similar rock is present in Ulu Endau (Figure 7.5).

Rhyolitic volcanic rocks, which are predominantly pyroclastic, outcrop in a narrow belt along the east coast of Malaya from the extreme south tip northwards into south Pahang. In the Mersing area, they have been named the Jasin Volcanics and ascribed a Triassic age by Chong et al. (1970). This age, rather younger than previously held, is in keeping with the spectacularly different deformation style of these volcanic rocks and the underlying Upper Paleozoic metasediments of presumed Permian and Carboniferous age. The metasediments are isoclinally folded and their foliation generally of near vertical dip, whereas the "Jasin Volcanics" always exhibit shallow to subhorizontal lithological layering. A major unconformity exists therefore between these two formations. The Upper Carboniferous east coast granitic episode must clearly be related to the pre-"Jasin Volcanics" deformation which uplifted the eastern coastal belt.

Chemistry

A selection of available chemical analyses of the Upper Paleozoic to Lower Mesozoic volcanic and pyroclastic rocks is given in Table 7.3, together with their Niggli molecular norms calculated by the computer method of Hutchison and Jeacocke (1971). From the normative compositions, the andesite demonstrates a considerable range—from saturated with normative quartz to undersaturated with normative olivine. Rocks that have been named trachyte differ from rhyolite only in a considerably lower content of quartz. Since the acid volcanic rocks are comparable in composition to the plutonic igneous rocks of Malaya, the lack of definite volcanic features has made their interpretation in some areas difficult if not impossible.

High corundum values in the norm indicate that some of the analyzed rhyolite specimens may not be wholly volcanic but partly tuffaceous.

MIDDLE TO UPPER TRIASSIC VOLCANISM

Although considerably less important than the Upper Paleozoic to Lower Triassic volcanism described previously, some volcanic activity persisted in axial and eastern Malaya. In the Jerantut area of central Pahang, some andesite is intercalated with sedimentary rocks of the Middle to Upper Triasic Jelai formation. Tuff and lava, some of which is spilitic, and andesite breccia occur in the Jurong formation in southwest Singapore, and rhyolite tuff occurs northwest of Johore Bahru.

Rhyolite tuff interbedded with black shale becomes prominent in the northern part of the Kerdau formation. The shale is often tuffaceous, and rare beds of dacitic tuff have been recorded. Rhyolite lava in the Middle of Upper Triassic Kerdau formation is sparsely vesicular.

In the Middle to Upper Triassic Gunong Rabong formation, volcanic rocks increase in bulk toward the north.

JURASSIC TO CRETACEOUS VOLCANISM

Volcanism persisted into the Jurassic and Cretaceous time in the axial belt of Malaya, but the phenomenon became more localized than previously.

True volcanic rocks appear to be absent from the Upper Triassic to Upper Jurassic Tembeling Formation, although grey micaceous tuffaceous sandstone has been recorded from restricted horizons.

In the Upper Jurassic to Lower Cretaceous Gagau Group, an important horizon of trachyte occurs low in the sequence, and some of the sandstone horizons are tuffaceous. Within the Lotong Sandstone are

TABLE 7.3 *Chemical analyses and calculated Niggli molecular norms of selected volcanic and pyroclastic rocks* of Upper Paleozoic to Lower Mesozoic age*

	Weight Percentage						
	1 Andesite	2 Andesite	3 Trachyte	4 Rhyodacite	5 Rhyolite	6 Tuff	7 Tuff
SiO_2	53.0	51.62	54.22	71.10	79.4	72.9	56.6
Al_2O_3	16.9	18.80	16.98	13.82	12.5	15.3	15.9
TiO_2	1.06	0.98	0.54	0.42	0.10	0.18	2.00
Fe_2O_3	2.09	1.43	1.79	nil	0.96	1.34	4.70
FeO	4.88	7.46	10.16	3.73	0.35	0.17	8.20
MnO	0.34	0.11	0.20	0.03	0.04	0.01	n.d.
MgO	5.34	4.10	2.41	1.99	0.26	0.10	2.66
CaO	5.32	8.34	1.74	2.50	0.11	0.09	6.00
Na_2O	3.94	2.66	4.57	2.09	0.24	3.24	2.30
K_2O	3.55	1.96	2.31	1.43	3.84	4.70	1.60
H_2O^+	2.59	2.15	3.71	2.30	1.98	1.58	n.d.
H_2O^-	0.30	0.24	0.57	0.13	0.63	0.33	n.d.
P_2O_5	0.55	0.40	0.37	0.08	trace	0.07	n.d.
CO_2	0.08	n.d.	0.19	0.15	0.03	0.07	n.d.
Total	99.94	100.25	99.94	99.86	100.44	100.08	99.96

	Niggli Molecular Norm Percentage				
	1	2	3	4	5
Quartz	nil	1.77	5.35	41.93	62.05
Corundum	nil	nil	5.80	5.57	9.11
Halite	nil	nil	0.07	0.07	nil
Orthoclase	21.35	11.91	14.35	8.88	24.21
Plagioclase	54.28	59.10	48.54	31.05	2.68
Ab_xAn_{100-x}	$Ab_{66}An_{34}$	$Ab_{42}An_{58}$	$Ab_{89}An_{11}$	$Ab_{63}An_{37}$	$Ab_{86}An_{14}$
Total salic	75.63	72.78	74.10	87.50	98.04
Cassiterite	n.d.	n.d.	n.d.	0.03	n.d.
Pyrite	n.d.	n.d.	0.14	nil	n.d.
Magnetite	2.22	1.54	1.97	nil	0.75
Hematite	nil	nil	nil	nil	0.22
Ilmenite	1.50	1.40	0.79	0.62	0.15
Apatite	1.17	0.86	0.81	0.18	trace
Calcite	0.21	n.d.	0.51	0.40	0.08
Diopside	3.54	4.27	nil	nil	nil
$Wo_{50}En_xFs_{50-x}$	$En_{37}Fs_{13}$	$En_{27}Fs_{23}$	—	—	—
Hypersthene	7.43	19.14	21.68	11.28	0.77
En_xFs_{100-x}	$En_{74}Fs_{26}$	$En_{55}Fs_{45}$	$En_{32}Fs_{68}$	$En_{51}Fs_{49}$	$En_{100}Fs_0$
Olivine	8.30	nil	nil	nil	nil
Fo_xFa_{100-x}	$Fo_{74}Fa_{26}$	—	—	—	—
Total femic	24.37	27.22	25.90	12.50	1.96

* Specimens 1–7 as follows:

1. Hornblende andesite (024021 p. 36). Anak Sungei Tembeling, near Kampong Peling, Pahang.
2. Porphyritic andesite (024009 p. 32). Sungei Trengganu (Lata Terap), Trengganu.
3. Trachyte (022004 p. 28). Second road cutting from the seaward end of the new road from Ulu Jurong village to Tanjong Kling village, Singapore. Also contains ZrO_2 trace, Cl 0.02, S 0.05, BaO 0.15 (less 0.04 in total for S and Cl).
4. Rhyodacite (013006 p. 20). Small tributary of the Sungei Chiniau, Pahang. Also analyzed for ZrO_2 nil, Cl 0.02, S nil, BaO trace, SnO_2 0.07. Specific gravity = 2.72.

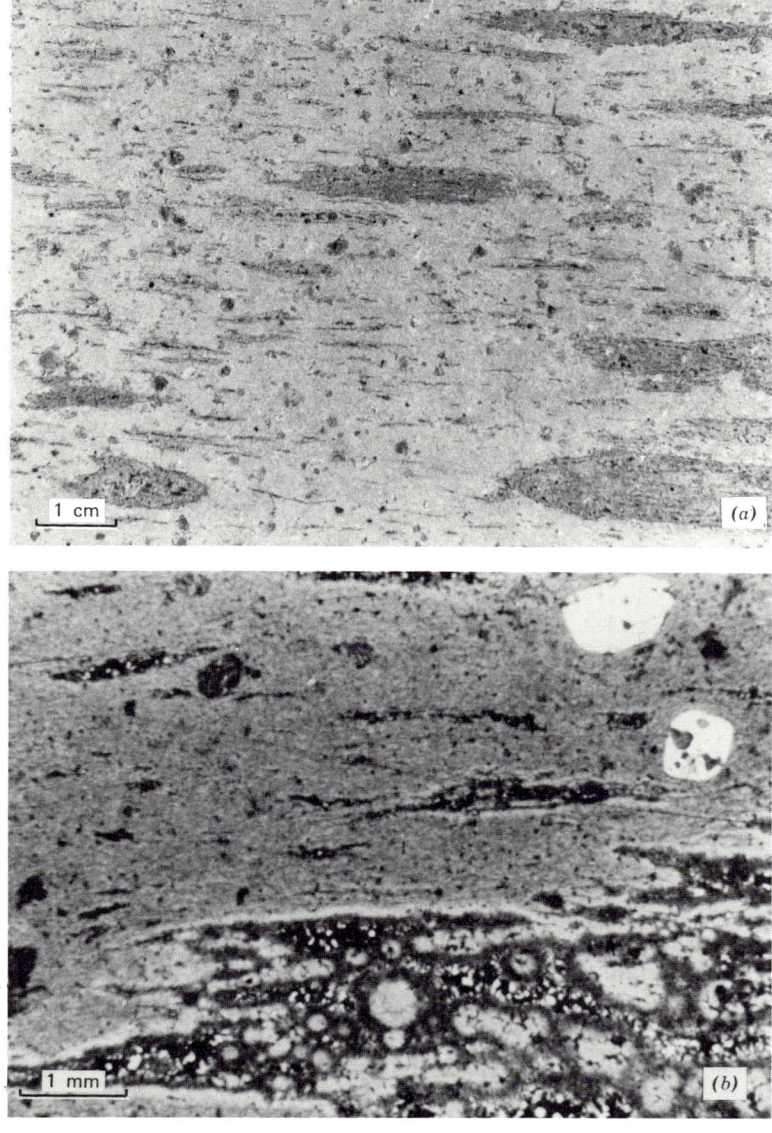

FIGURE 7.5 *Flow-banded rhyolite, Ulu Endau, Johore:* (a) *Photograph Jaafar bin Haji Abdullah;* (b) *photomicrograph (C. S. Hutchison) in ordinary light of specimen UM R3032.*

flows of rhyolite, dacite, trachyte, and andesite, as well as some basalt flows. The latter are apparently the earliest established basalt flows in the history of the Malay Peninsula. Rhyolite tuff and agglomerate also occur. Trachyte forms a prominent base to the Lotong Sandstone, and rhyodacite forms a similar prominent base to the Tebak formation in north Johore. Also within the Tebak formation, low in the sequence, occurs a porphyritic rhyodacite 150 m thick.

CENOZOIC VOLCANISM

Two main occurrences of postorogenic alkali basalt are known in Malaya—the larger in the neighborhood of Kuantan in Pahang (Fitch 1952), the

5. Rhyolite (012035 p. 16). Anak Sungei Pengau, Pahang.
6. Tuff (06300 p. 52). Sungei Sat, Pahang.
(References for 1–6 are to Alexander et al. (1964).
7. Agglomeratic tuff. Bukit Bangsal Kunyit, Southeast Johore. Grubb (1968, p. 23).

other in the neighborhood of Segamat in north Johore (Grubb 1965) (Figure 7.1).

There is no definite stratigraphic indication of the age of these rocks. Fitch (1952) suggested that the Kuantan basalt and associated dolerite dykes may be of Eocene age, similar to the basalt, dolerite, and diorite that occur in Java intercalated with Eocene beds (Willbourn 1922). The dolerite and basalt in the Kuantan area are cut by quartz porphyry dykes, which Fitch (1952) has suggested to be of Miocene age. A recent potassium-argon radiometric determination suggests that the Kuantan basalt may be Quaternary, with an age of less than 2 m.y. (J. D. Bignell in correspondence).

On the basis that both localities (Kuantan and Segamat) are of feldspathoid-bearing olivine basalt (basanite), Grubb (1965) has suggested that they are contemporaneous.

Kuantan

Basalt overlies and surrounds the granite hills north and northwest of Kuantan (Figure 7.6) and extends over Upper Paleozoic arenaceous sedimentary rocks with intercalated contemporaneous volcanic rocks. The highest point on the basalt outcrop is at Bukit Tinggi (136 m; 454 ft), almost at the center of the lava outcrop. The basalt outcrops at sea level on the coast at Batu Hitam, and although the maximum thickness of the flow is unknown, it is unlikely to exceed 136 m. Numerous dolerite dykes cut the underlying rocks, showing perhaps that the eruption was of the fissure type. Recently, however, one of these dykes was radiometrically dated at 110 m.y. (Bignell in correspondence) and is thus considerably older than the basalt. Fitch (1952) was unable to subdivide the Kuantan basalt into individual flows, as Grubb (1965) has done for Segamat.

After erosion of much of the Upper Paleozoic sedimentary cover and of some of the granite, basaltic magma rose through fissures, probably in late Tertiary or Quaternary time. Much of the magma flowed out over the land surface as basaltic lava, but some solidified as dolerite dykes before it reached the surface (Fitch 1952).

Many dolerite dykes probably acted as feeders to the lava, and these bodies occur in large numbers as intrusions through the granite. Excellent examples are exposed north of Gambang, at Bukit Ubi quarry in Kuantan town, and in good coastal sections at Tanjong Tembeling and Telok Chempedak, where the dykes exhibit glassy chilled margins.

PETROLOGY OF THE BASALT. Generally the rock is a compact, microcrystalline, vesicular olivine basalt, black to greenish-black, with a matte appearance on broken surfaces. Columnar jointing is common and is particularly well displayed on the Jeram Kuantan Estate.

The minerals determined in thin section are calcic plagioclase, colorless to purple augite, colorless olivine, titaniferous magnetite, and ilmenite. In some of the basalt, the occurrence of nepheline as small crystals with rectangular outline was proved by staining tests (Fitch 1952). The nepheline-bearing lava should more properly be referred to as nepheline basanite.

The texture of the basalt and alkali basalt is normally intergranular, with crystals of augite occupying the spaces between matted plagioclase laths. In some of the coarser grained rocks, the texture is subophitic. The phenocrysts, which are always euhedral with little or no corrosion, are predominantly of olivine and rarely of plagioclase and augite, which are mainly to be found in the groundmass. In a few cases, a brown or green glass matrix, containing crystallites of magnetite, has been described (Fitch 1952).

The basalt is commonly vesicular; the vesicles being normally spheroidal or irregular and some are lined or filled with zeolites or calcite. Some specimens are extremely rich in vesicles.

PETROLOGY OF THE DOLERITE. The dolerite is normally coarser grained than the basalt, and it is nonvesicular, although in some dykes there are bands of amygdales either near the wall or in the center. The amygdales are lined with chlorite or vermiculite and are filled with calcite and a little quartz (Fitch 1952). Ophitic textures are common. The minerals include titanaugite, calcic plagioclase, colorless olivine, accessory opaque iron oxides, some pyrite and chalcopyrite, and apatite, biotite, and quartz in small quantities, but these may be xenocrysts from the granite through which the dykes intrude (Fitch 1952). Some of the dolerite is rich in augite, others are rich in olivine. Phenocrysts of plagioclase up to 5 cm long are fairly common, especially in the dykes exposed along the coast. The cores of these phenocrysts are completely altered to chlorite, giving them a greenish color in hand-specimen, while the rims of the phenocrysts remain fresh. Feldspar laths in the groundmass show a parallel alignment around the phenocrysts. Euhedral crystals of augite up to 1 cm long occur in the dolerite at the Sungei Tulang and Anak Sungei

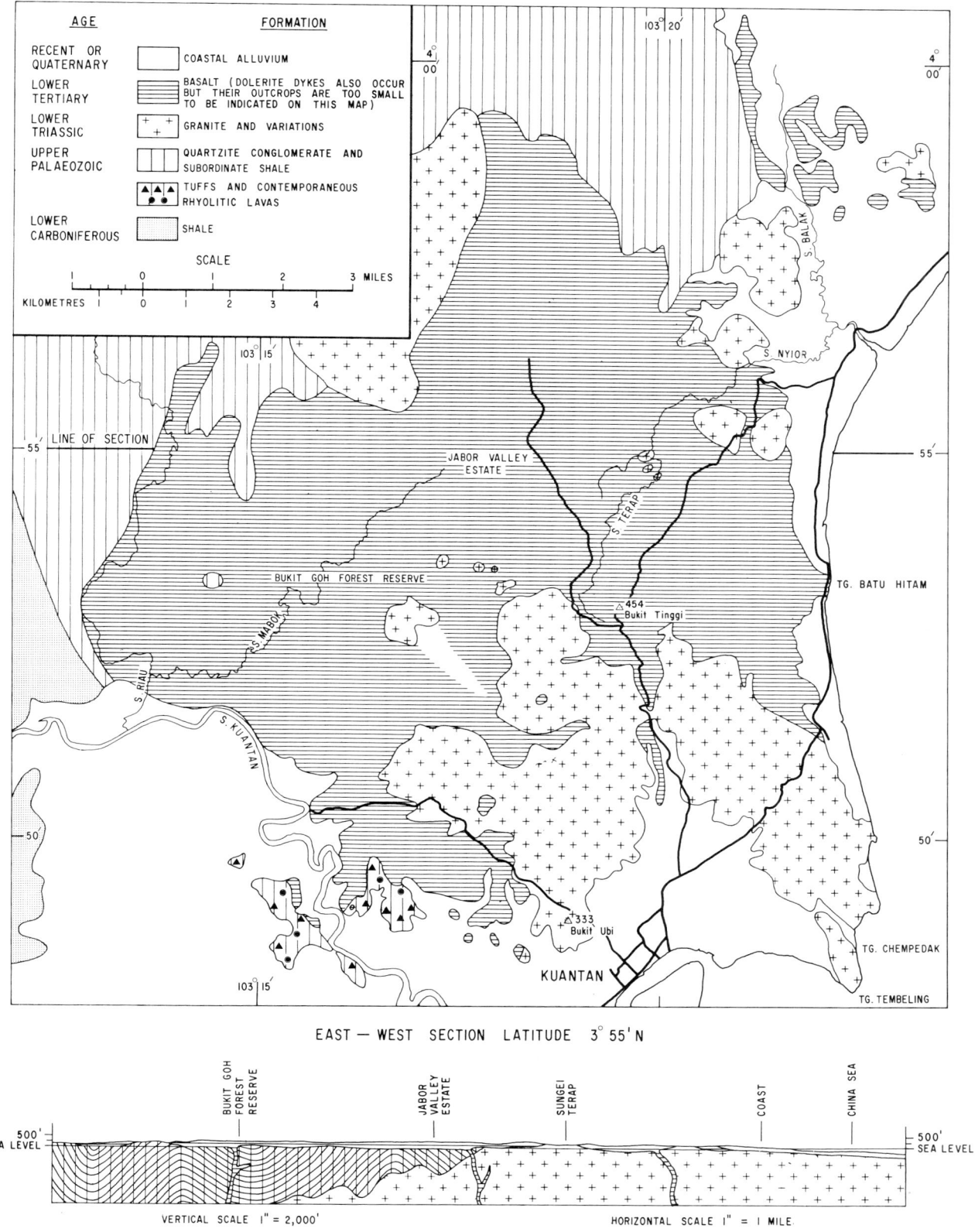

FIGURE 7.6 *Geological map and cross section of the Kuantan basalt and feeder dyes. After Fitch (1952).*

Belat. Olivine occurs invariably as phenocrysts. At Tanjong Tembeling, the dolerite dykes contain xenoliths of very altered rock.

The dolerite weathers more readily than the basalt to give a brown spheroidal crust, so characteristic of basic rocks.

Segamat

In the immediate neighborhood of Segamat, in north Johore, there occur three recognizable alkali basalt lava flows which dip 20° in a southwest direction. The three flows are of leucite tephrite, leucite basanite, and potassic ankaramite. The flows are all intruded by shoshonite dykes varying in width from 15 cm to 1 m and at Kampong Jabi, some 3 km (2 miles) north of Segamat, they are cut by a large, undifferentiated conformable sill of shonkinite, 30 m thick.

The individual flows vary from 18 to 42 m (60 to 140 ft) in thickness. They have a basal horizon of black nonvesicular or poorly vesicular basalt containing phenocrysts of augite. These basal horizons commonly contain rounded serpentinite xenoliths of from 1 to 6 cm in diameter. The lava flows are differentiated by gravity settling, and proceeding upward from the base, the flows become more vesicular and richer in feldspathoid and plagioclase phenocrysts, with correspondingly less olivine and pyroxene. The uppermost 3 m of each flow is characterized by glassy chilling, and the uppermost meter consists of a conspicuous red vesicular lateritic horizon. Frequently, fragments of this laterized surface of a preceding flow have been detached and incorporated within the succeeding flow, where they often show marginal darkening of their red color owing to contact baking. Hence the flows were subaerial to have produced the surface laterization, and the climate was distinctly tropical at the time of the eruptions. There must have been a considerable time lapse between each flow to allow for the laterization of the uppermost meter.

PETROLOGY OF THE BASALT. In the lower levels of the flows, the basalt is porphyritic and the groundmass either glassy or microcrystalline; only seldom are the component minerals recognizable. Phenocrysts of large zoned augite are abundant, and hypersthene, serpentinized olivine, and sericitized plagioclase less common. Vesicles range up to 2 mm in diameter. They were originally infilled mainly with calcite, rarely with natrolite; but their contents are now almost completely replaced by albite, epidote, riebeckite, caledonite, and chlorite (Grubb 1965).

Numerous small phenocrysts of leucite and hauyne have been pseudomorphed by oligoclase-andesine. The small black serpentinite xenoliths that are scattered throughout this horizon are exclusively of antigorite.

The pyroxene phenocrysts have the following optical properties: $\alpha = 1.687$, $\beta = 1.697$ to 1.698, $\gamma = 1.717$, $2V = 53$ to $60°$ $+ve$. The crystals are zoned with a slight outward increase of 2V and have a characteristic sharply defined marginal zone. Occasionally rims of hypersthene ($2V = 69°$ $-ve$) enclose cores of augite ($2V = 40°$ $+ve$). All phenocrysts are remarkably colorless in thin section, reflecting the low content of titanium dioxide in the basalt (Grubb 1965 and Table 7.4). Sodium cobaltinitrite staining has indicated that the groundmass is potassium rich (Grubb 1965), although all that can be identified in it are occasional crystals of augite, hypersthene, oligoclase, and apatite. There is also an abundant scattering of fine magnetite. The laterized upper portions of the flows are similar to the rest of the flow, but with less abundant mafic phenocrysts; and the ferric oxide, alumina, and alkali contents are higher than they are in the unlaterized flow.

PETROLOGY OF THE INTRUSIVE ROCKS. The shonkinite sill at Kampong Jabi differs from the lavas in having a coarse texture and in an apparent lack of magmatic differentiation. It contains phenocrysts of sericitized plagioclase (An_{33}) serpentinized olivine, zoned hypersthene ($2V = 63°$ to $69°$ $-ve$), and numerous zoned augite crystals ($2V = 48°$ to $52°$ $+ve$). The groundmass is of large poikilitic anorthclase with lesser oligoclase-andesine (An_{31}), occasional nepheline, augite ($2V = 52°$ $+ve$), hypersthene ($2V = 63°$ to $69°$ $-ve$), abundant biotite, numerous euhedral magnetite and infrequent apatite crystals. The sill is chilled against the lava flows and has a glassy matrix, now largely replaced by chlorite (Grubb 1965).

The shoshonite dykes differ from the lava and the sill in having a higher $Na_2O:K_2O$ ratio, caused perhaps by late-stage albitization. They contain sericitized andesine (An_{36} to An_{48}), augite ($2V = 60°$ to $68°$ $+ve$), and serpentinized olivine phenocrysts. The groundmass is of oligoclase-andesine laths (An_{30}), anorthoclase, a few augite crystals ($2V = 34°$ $+ve$), and accessory magnetite and apatite.

PETROGENESIS. Grubb (1965) considered that there was a common magma source for both the lava and hypabyssal rocks in the form of a subterranean basic body, perhaps a hybrid feldspathized ultrabasic complex similar to that described in the

Benta–Gunong Benom area (Richardson 1939). The presence of serpentinite xenoliths in the lava gives weight to this hypothesis. The main Segamat lava flows show mineralogical and chemical variations compatible with gravity differentiation. There is an upward increase of alumina, ferric oxide, and alkali and a corresponding fall in magnesium oxide and ferrous oxide. The differentiation can be explained by a gravity sinking of pyroxene and olivine phenocrysts accompanied by intense formation of vesicles toward the top of the flows, resulting in the upward concentration of plagioclase, leucite, and hauyne phenocrysts and vesicles. The bottom of the flows contains 30% (by volume) pyroxene, and 7% olivine phenocrysts, whereas the top has only 3% pyroxene and no olivine. On the other hand, the bottom contains 1% vesicles and feldspathoid phenocrysts and the top 16% (Grubb 1965).

It is the upper vesicular horizons that have been laterized by high temperature subaerial oxidation.

The shonkinite sill is undifferentiated because the prevailing hydrostatic pressure inhibits gaseous flotation, as in the flows.

The final solidification of the Segamat volcanic rocks was marked by the appearance of a weakly sodic residuum, which caused widespread autometasomatism throughout the entire volcanic region (Grubb 1965).

The fissure basalt eruptions of Kuantan and Segamat may well be related to intense and deep-seated strike-slip faulting that has caused a release of basic magma from the deper levels of the crust. The alkaline nature of the effused magma may result from contamination by the geosynclinal pile of sedimentary rocks with pyroclastic intercalations which the magma has encountered on its way to the surface.

Chemistry

A selection of available analyses of the Tertiary or Quaternary basaltic lava flows and associated feeder dolerite dykes is given in Table 7.4. The highly alkalic nature of the lava flows is shown by the high content of either normative orthoclase or nepheline; their generally undersaturated nature is indicated by the absence of quartz and the presence of normative olivine. The dykes are much more variable chemically and may either contain normative quartz or normative olivine.

Pleistocene Rhyolite Ash (see chapter 6)

Stratified rhyolite ash, up to 9 m thick, occurs on the right bank of the Perak River in the Tanjong Perak Estate opposite the mouth of the Sungei Plus in the Enggor District, Perak (Scrivenor 1930). It is of comminuted pumice, porous and friable, and contains diatoms, sponge spicules, and spherasters. At its lowest level it is 45 m (150 ft) above present sea level and is probably Upper Pleistocene in age (Scrivenor 1949). Deposits of similar rhyolite ash have also been recorded from Lenggon, Perak (Willbourn 1938); from the Raub area, Pahang (Richardson 1939); and from southwest Pahang (Willbourn 1940). These ash deposits lie at too great an altitude to have been deposited by a late Pleistocene or Recent sea, and they are assumed to be aeolian. They were probably derived from the major eruption that formed the caldera of Lake Toba in northern Sumatra. This caldera is surrounded by masses of rhyolite tuff. Blocks of rhyolite-pumice are frequently washed up on the east coast beaches of Malaya at the present day.

SUMMARY OF VOLCANIC ACTIVITY IN MALAYA

The geosynclinal history of the Malay Peninsula is characterized by contemporaneous explosive submarine volcanic activity from the Upper Cambrian to the Middle Triassic. Pyroclastic deposits are considerably more abundant than lava flows. Most of the tuff deposits are of a mixed character, both crystal and lithic, and contain volcanic fragments ranging from rhyolitic to andesitic composition. Pyroclastic rocks of a predominantly rhyolitic character are far more abundant than those of an andesitic character. Many of the rhyolitic rocks are intermediate between truly pyroclastic and truly volcanic and may be better regarded as ignimbrite.

The Lower Paleozoic activity is preponderantly rhyolitic and is represented by significant occurrences at Lawin in the Grik area, and to the north of the Sungei Siput in Perak. However, serpentinite and metabasite bodies within the Lower Paleozoic outcrop in Pahang and Negri Sembilan may represent basic lava flows. Andesitic rocks were very restricted and are found only at a few localities within roof pendants in the Main Range Granite. Quartz porphyry and microgranite occurrences in the same roof pendants could well be interpreted as Lower Paleozoic rhyolitic flows or ignimbrites.

Volcanic activity continued into the Upper Paleozoic and Lower Mesozoic and reached its paroxysmal stage in the Permian to Lower Triassic, when activity was centered in axial Malaya in the states of Kelantan, Pahang, and Johore. Although

TABLE 7.4 *Chemical analyses and calculated Niggli molecular norms of selected postorogenic basalt flows and related dolerite dykes**

	Weight Percentage					
	1	2	3	4	5	6
SiO_2	47.45	39.35	47.74	49.75	47.45	46.53
Al_2O_3	13.68	12.23	8.98	17.30	14.36	14.88
Fe_2O_3	3.45	6.58	5.56	8.38	2.82	2.93
FeO	8.03	7.01	5.48	2.20	12.14	10.10
MgO	7.48	9.82	10.59	1.95	5.06	6.97
CaO	8.00	11.64	11.96	7.40	7.94	8.82
Na_2O	3.59	2.84	1.31	2.29	2.65	2.62
K_2O	1.46	1.40	3.78	7.44	0.93	1.09
H_2O^+	3.11	4.85	2.38	2.53	2.33	2.46
H_2O^-	1.05	1.11	0.76	0.38	0.41	0.32
CO_2	0.12	0.13	nil	n.d.	0.65	0.72
TiO_2	1.85	2.14	0.97	0.82	2.48	2.18
P_2O_5	0.69	0.82	0.44	n.d.	0.60	0.43
MnO	0.12	0.13	0.18	n.d.	0.18	0.11
Cl	0.17	0.07	n.d.	n.d.	0.10	0.10
S	nil	nil	n.d.	n.d.	0.36	0.05
Cr_2O_3	0.02	0.02	n.d.	n.d.	0.01	0.02
NiO	0.01	0.01	n.d.	n.d.	tr	tr
BaO	0.03	0.05	n.d.	n.d.	0.02	0.02
Less O for Cl, S	0.04	0.02	—	—	0.16	0.04
Total	100.27	100.18	100.13	100.44	100.33	100.31

n.d. = not determined, tr = trace.

	Niggli Molecular Norms					
	1	2	3	4	5	6
Quartz	—	—	—	—	2.17	—
Orthoclase	8.95	8.78	23.11	45.77	5.74	6.66
Plagioclase	50.52	22.23	14.42	22.27	50.11	50.43
Ab_xAn_{100-x}	$Ab_{64}An_{36}$	$Ab_{20}An_{80}$	$Ab_{46}An_{54}$	$Ab_{30}An_{70}$	$Ab_{48}An_{52}$	$Ab_{47}An_{53}$
Nepheline	—	13.23	3.29	8.83	—	—
Halite	0.55	0.23	—	—	0.33	0.32
Total salic	59.83	44.46	40.82	76.87	58.36	57.41
Wollastonite	—	—	—	3.46	—	—
Magnetite	3.74	7.30	6.02	3.54	3.08	3.17
Chromite	0.02	0.02	—	—	0.01	0.02
Hematite	—	—	—	3.72	—	—
Ilmenite	2.68	3.16	1.40	1.19	3.61	3.14
Apatite	1.50	1.82	0.95	—	1.31	0.93
Calcite	0.32	0.35	—	—	1.72	1.88
Pyrite	—	—	—	—	0.98	0.13
Diopside	14.04	29.63	40.58	11.21	5.41	8.62
$Wo_{50}En_xFs_{50-x}$	$En_{36}Fs_{14}$	$En_{44}Fs_6$	$En_{45}Fs_5$	$En_{50}Fs_0$	$En_{26}Fs_{24}$	$En_{32}Fs_{18}$
Hypersthene	4.41	—	—	—	25.53	18.98
En_xFs_{100-x}	$En_{37}Fs_{27}$	—	—	—	$En_{52}Fs_{48}$	$En_{64}Fs_{36}$
Olivine	13.46	13.25	10.23	—	—	5.71
Fo_xFa_{100-x}	$Fo_{73}Fa_{27}$	$Fo_{89}Fa_{11}$	$Fo_{89}Fa_{11}$	—	—	$Fo_{64}Fa_{36}$
Total femic	40.17	55.54	59.18	23.13	41.64	42.59

rhyolitic pyroclastic rocks continued to predominate, rocks of andesitic composition became important in the late Upper Paleozoic and Triassic. These were characteristically lava flows and agglomerates. Nothing more basic than andesite was extruded during the Upper Paleozoic and early Mesozoic; during that time volcanism was widespread and of mixed rhyolitic and andesitic nature.

During the phase of molasse deposition in the Upper Triassic to Lower Jurassic, volcanicity seems to have waned, to be renewed in the Upper Jurassic as flows and pyroclastic intercalations in the Gagau Group postorogenic deposits. Since these beds are thought to be continental, the volcanism must have changed in character from submarine to subaerial in the interval. Although the composition of the volcanic rocks continued to be rhyolitic and andesitic, definite basalt flows now occurred for the first time in the history of Malaya. These basalt extrusions were of the fissure type and were undoubtedly connected with major wrench faults and epeirogenic uplift that controlled the deposition of the postorogenic sedimentary formations.

Basaltic activity continued after the uplift of Malaya and flows in the neighborhood of Kuantan and Segamat were apparently subaerial and are related to the present topography. Radiometric evidence suggests that they may be of Quaternary age, and they are undoubtedly related to the continuing major faulting that is characteristic of the subrecent history of the peninsula.

Volcanic ash from Indonesian eruptions occurs sporadically in Quaternary alluvial deposits of the peninsula, and pumice blocks continue to be washed up on the east coast beaches, presumably from present-day volcanic eruptions in Indonesia or the Philippines.

Figure 7.7, a plot of the variation of the cationic percentages of total iron, total alkalis, and magnesium for all available chemical analyses of Malayan volcanic rocks, shows that some rocks of basaltic nature are no more basic than others that have been petrographically named andesite. Several of the andesite specimens that have been analyzed are undersaturated and contain normative olivine. Hence it would be wrong to say that the andesite volcanism that characterized the early Mesozoic is vastly different chemically from the basaltic volcanism that represents the late Mesozoic and Cenozoic. The difference is petrographic rather than chemical and reflects the difference in geological setting of the extrusions. The petrographic difference is striking, however, from the submarine geosynclinal extrusions of andesite to the continental subaerial extrusions of basalt.

The triangular variation diagram (Figure 7.7) demonstrates that there exists in Malaya a complete and continuous range of volcanic compositions from rhyolite to basalt.

REGIONAL CORRELATION

Lower Paleozoic

BURMA. Some 2000 km (1240 miles) north of the Lawin tuff locality in north Malaya, in the neighborhood of Bawdwin in Tawng-peng State of the Northern Shan States of Burma, occurs an important series of volcanic and pyroclastic rocks of definite Lower Paleozoic and probable Cambrian age. These have been referred to as the Bawdwin Volcanic Series, and they are found along the boundary between the Chaung Magyi beds and the overlying fossiliferous Paleozoic formations (Pascoe 1959, p. 591). The Bawdwin Volcanic Series represents the only Lower Paleozoic volcanic and pyroclastic rocks in the Southeast Asian region that can be correlated with those in the Grik and Sungei Siput areas of Malaya.

---

* Specimens 1–6 as follows:

  1. Zeolite-filled vesicular olivine basalt. Rapids at road iunction to May and Jeram Kuantan Estates, in Semambu Estate, Kuantan, Pahang (Fitch 1952).
  2. Vesicular nepheline basalt (pyroxene, olivine, nepheline-bearing). Sungei Karang, below Bukit Jelatang, near Kuantan, Pahang (Fitch 1952).
  3. Potassic ankaramite lava at very bottom of main Segamat flow (contains black serpentinite xenoliths), Segamat, North Johore (Grubb 1965).
  4. Laterized leucite-tephrite from top of the main Segamat lava flow, Segamat, North Johore (Grubb 1965).
  5. Medium-grained dolerite (augite, plagioclase, chlorite, iron oxides) southwest side of Tanjong Tembeling, Kuantan, Pahang (Fitch 1952).
  6. Porphyritic dolerite (augite, plagioclase, biotite, iron oxides), near Kuala Sungei Chempedak, Kuantan, Pahang (Fitch 1952).

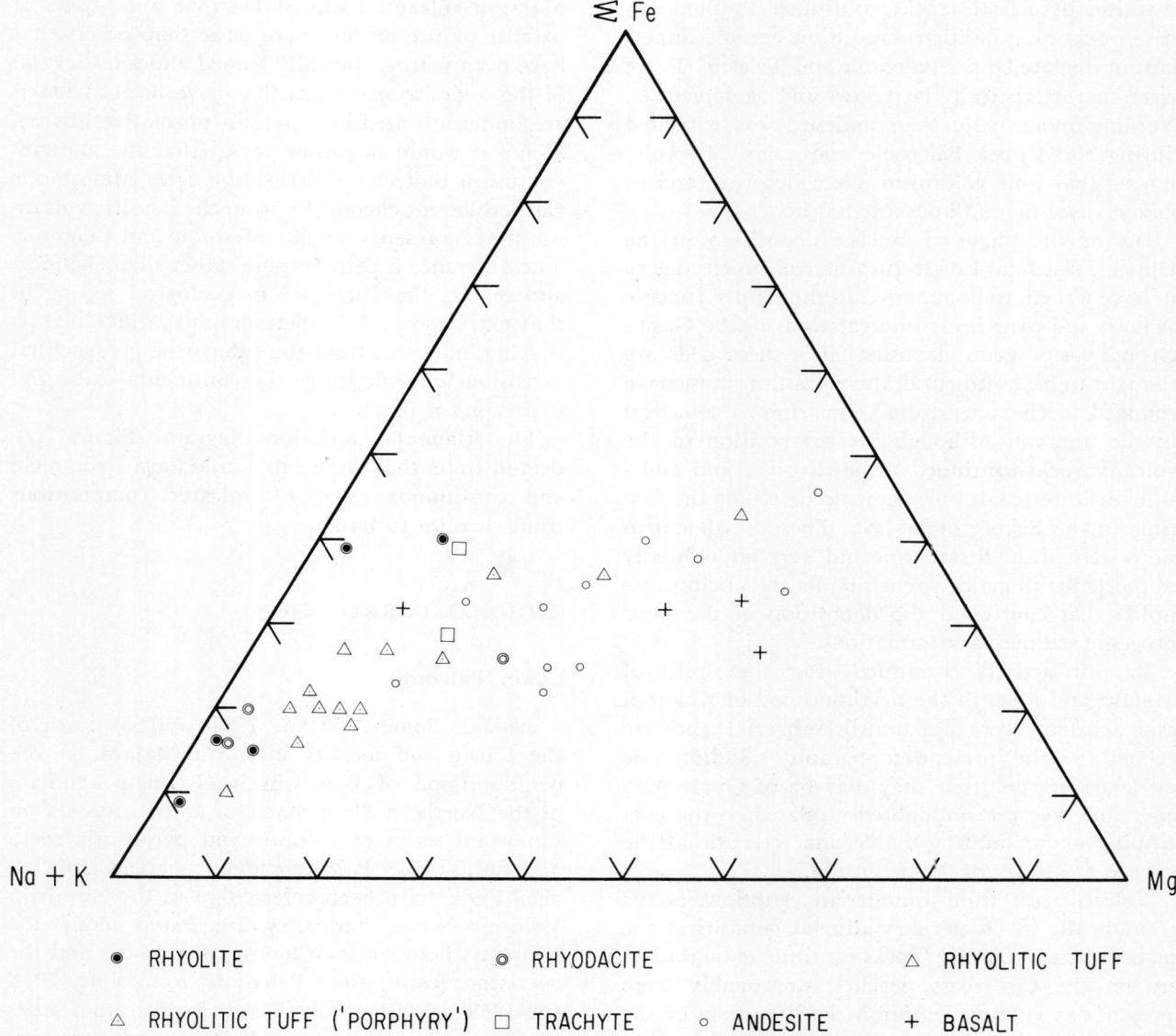

FIGURE 7.7 *Variation diagram of cationic percentages of total iron, alkalis, and magnesium for all available chemical analyses of Malayan volcanic rocks ranging in age from Lower Paleozoic to Cenozoic. Data computed from analyses listed in Alexander et al. (1964).*

The Series consists of tuff and ash beds interstratified with rhyolite flows. There are occasional bands of volcanic breccia and some quartz porphyries. Pink or brown rhyolite is subordinate to tuff, shows flow texture, contains corroded phenocrysts, and is occasionally spherulitic. Phenocrysts of quartz preponderate over feldspar and there is a characteristic absence of plagioclase and pyroxene. The tuff resembles the rhyolite in composition, and metamorphism makes it difficult to differentiate tuff from lava. As with the Lower Paleozoic tuff of Malaya, the Bawdwin tuff is metamorphosed, foliated, and often chloritized. Near Bawdwin, the rhyolite and the tuff have been extensively mined for silver and lead.

In the Southern Shan States, light-colored rhyolite of the Bawdwin suite occurs interbedded with limestone and shale of the Ordovician Pindaya beds (Pascoe 1959).

THAILAND. In the intervening region between Pindaya in the Southern Shan States of Burma and northern Malaya, no records of Lower Paleozoic volcanic rocks have been found in the literature, although occurrences are to be expected in peninsular Thailand along the same north-south trend.

INDOCHINA. East of Burma and Thailand, some metandesite occurrences in South Vietnam and eastern Cambodia are considered to be of Lower Paleozoic age (Fromaget 1941). They have been metamorphosed to amphibolite facies and consist of phenocrysts of plagioclase (An_{50}), devitrified glass inclusions, phenocrysts of hornblende which are often replaced by biotite, and some porphyroblasts of biotite, all set in a groundmass of andesine microlites, small crystals of biotite, and some magnetite.

INDONESIA. To the south of the Malay Peninsula no Lower Paleozoic rocks occur.

Upper Paleozoic to Lower Mesozoic

Important widespread volcanic and pyroclastic activity in the whole Southeast Asian region characterized the Upper Paleozoic to Lower Mesozoic. The volcanic rocks are characteristically rhyolitic and rhyodacitic but contain important andesite developments.

BURMA. Tuff and agglomerate are abundant in the Mergui Series, centered on Elphinstone and Maingay's Islands near Mergui in Tenasserim (Chhibber 1934) (Figure 7.8). The agglomerate includes fragments of pumice, devitrified glass, sedimentary rocks, feldspar porphyry, and rhyolite that is sometimes spherulitic. Associated with the agglomerate are rhyolite, microgranite porphyry, and tuff. The agglomerate is always of rhyolitic character. The centers of Upper Paleozoic volcanic activity appear to be the Mergui Islands, but volcanic tuff is widely distributed in the Mergui Series rocks in the regions of Tavoy and Mergui (Pascoe 1959). Near Tavoy, abundant agglomerate and greywacke horizons occur in the gorge of the Great Tenasserim River. They are composed of angular fragments of quartz, slate, quartzite, feldspar, and occasional rounded pieces of granite (average size 1 to 2 cm), all set in an argillite matrix. As with the Upper Paleozoic Singa Formation of northwest Malaya, this Pre-Mergui granite has never been found *in situ* (Pascoe 1959).

INDOCHINA. Volcanic rocks ranging in age from Upper Carboniferous to Upper Triassic (Lower Indosinias) are abundant in Vietnam, Cambodia, and Laos (Figure 7.8). They are of two general categories: an older group of lava flows of a more basic character, and a younger group of general dacitic character which includes some rhyolite. Although these rocks are similar petrographically to those of the same age in Malaya, in Indochina they are associated with continental sandstones, whereas in Malaya they are associated with marine deposits.

The basic rocks are generally considered to be Pre-Permian and are usually foliated. Northwest of the Indosinia mass, andesite is very abundant. It contains phenocrysts of plagioclase and hornblende, often altered to serpentine. The groundmass is glassy and contains microlites of andesine. In South Vietnam the andesite contains phenocrysts of andesine, augite that is often uralitized, and occasionally biotite. The groundmass is of microlites of andesine and small crystals of augite and sometimes hypersthene. Magnetite is always abundant. Several occurrences of andesitic tuff and agglomerate have also been recorded. In western Cambodia the most frequent type is a fine-grained clinopyroxene-bearing andesite, which grades into basalt in character. Some occurrences contain aegirine and nepheline.

The age of the dacitic group is thought to range from Permian to Upper Triassic. These rocks are composed essentially of a groundmass of small grains of quartz and feldspar containing crystals of all sizes of andesine, hornblende, and a few of magnetite. Biotite is locally abundant. The dacite flows are interbedded with tuff in South Vietnam. Northwest of Indosinia, the dacite ranges from rhyodacite to andesite in composition. Dacite in western Cambodia and in the Thailand region of Menam is characterized by the presence of pyroxene phenocrysts. The dacite is thought to be Post-Triassic in age (Fromaget 1941).

Rhyolite is abundant and widespread in Indochina. The age ranges from Permian to Lower Triassic. In the northwest of Indosinia, in the region of Pak Lay, numerous bands of rhyolite are interbedded with the sedimentary rocks. The groundmass is of cryptocrystalline quartz; the phenocrysts, in order of abundance, are of corroded quartz, orthoclase, plagioclase, and small grains of magnetite. In South Vietnam a large expanse of rhyolite occurs in the region of Nha Trang, where it is of remarkably constant texture and composition throughout. Some outcrops are spherulitic. In South Vietnam, the rhyolite flows postdate the dacite flows, and they all appear to be of Triassic age. In western Cambodia the flows are considered to be Post-Permian (Fromaget 1941).

THAILAND. In view of the great similarity of volcanic rocks in Indochina and in Malaya of Upper Paleozoic to Lower Mesozoic age, it is surprising that no mention of any volcanic occurrences of this age is to be found in the literature on the geology of Thailand. A glance at Figure 7.8 should

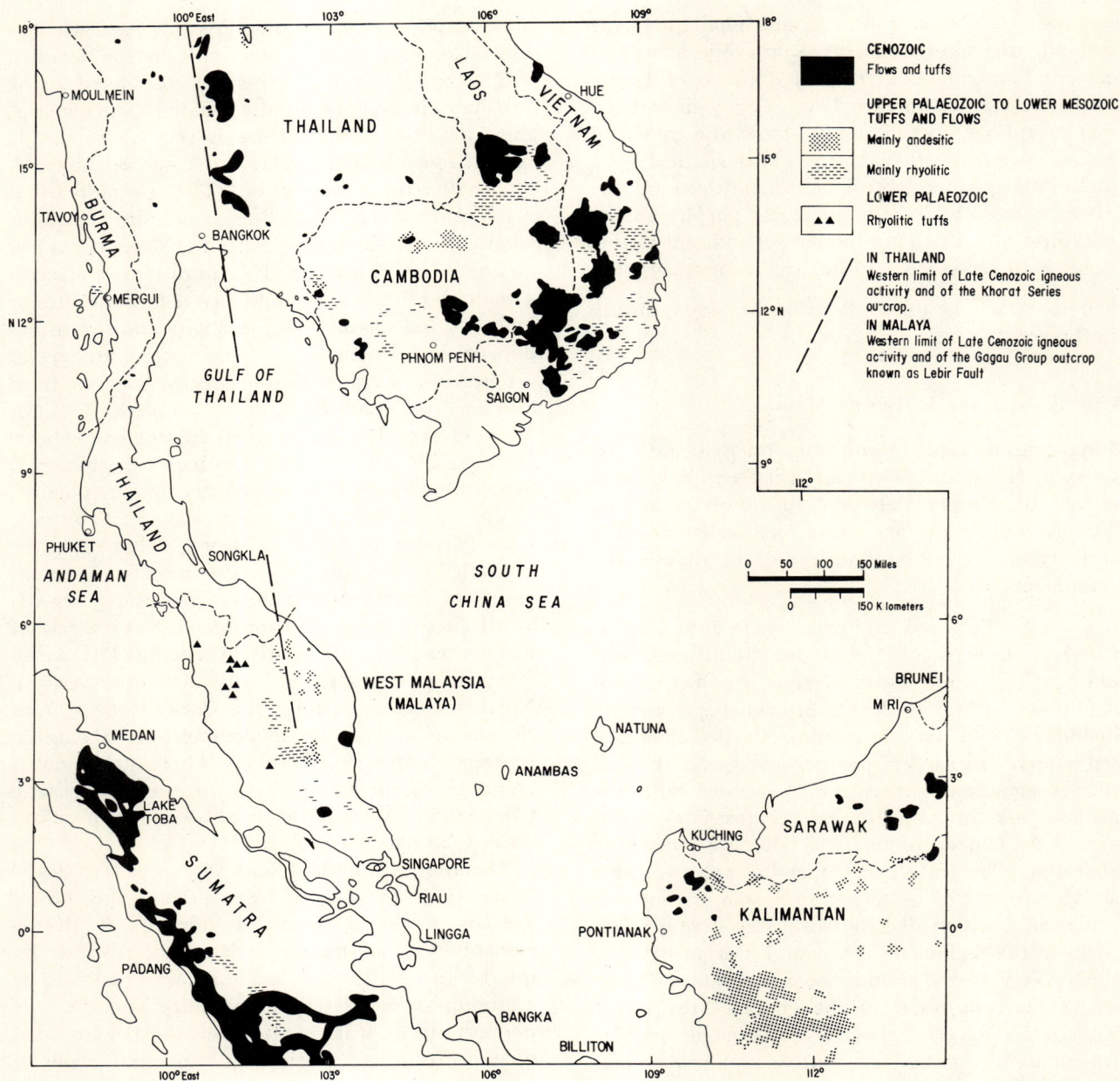

FIGURE 7.8 *Distribution of volcanic and pyroclastic rocks in southeast Asia. Based on Pascoe (1959), Brown et al. (1951), Klompé (1965), Klompé et al. (1961), Fromaget (1941), Alexander (1965), and Kirk (1968).*

indicate that perhaps the reason is not that these volcanic rocks do not exist, but that they have not been described.

SUMATRA. In the Padang Highlands of Sumatra, the Silungkang Volcanic Series of Permian and Carboniferous age has been described by Klompé et al. (1961). The volcanic series is of lava and tuff interbedded with a variety of sedimentary rocks. Hornblende dacitic tuff predominates over lava. The lava is of similar composition, composed of phenocrysts of hornblende and plagioclase in a fine-grained matrix. Some intercalated andesite tuff occurs in the overlying Triassic rocks.

In the Djambi region of west-central Sumatra, tuff and lava of dacitic and andesitic composition occur intercalated with the Permo-Carboniferous and Triassic sedimentary sequences. The andesite in this region is augite-bearing.

RIOUW–LINGGA ARCHIPELAGO. Permo-Carboniferous or Triassic volcanic rocks occur on several islands of the Riouw–Lingga Archipelago. They are considered by van Bemmelen (1949) to be comparable to similar rocks in Johore and Pahang to the north on the mainland of Malaya.

ANAMBAS. Andesite of Permo-Carboniferous age found in Anambas has been described by van Bemmelen (1949).

NATUNA. Tuff occurs in Natuna associated with the Permo-Carboniferous sedimentary rocks. Basic volcanic rocks are associated with a younger series of shale and conglomerate of probable Upper Triassic age (van Bemmelen 1949).

INDONESIAN BORNEO (KALIMANTAN). Intermediate to basic lava, tuff, and breccia occur intercalated with Permo-Carboniferous sedimentary rocks in west and central Borneo (Klompé et al. 1961). These are of a basalt and a basalt-andesite character and are associated with rocks of the Pulu Melaju Series (van Bemmelen 1949).

In the overlying Triassic rocks, acid to intermediate (quartz keratophyre) volcanic and pyroclastic rocks are common.

In the Serberuang area, submarine acid lava and tuff occurs associated with Cretaceous rocks.

MALAYSIAN BORNEO (SARAWAK). In west Sarawak, the Serian Volcanic Formation of Upper Triassic age has been described by Kirk (1968). The volcanic rocks are predominantly of andesitic composition, lava being more abundant than pyroclastic rocks. A series of rhyolite and trachyte has been named the Semabang Trachyte Member, which appears to be mainly younger than the andesitic rocks, although in part it is their lateral equivalent. Breccia and tuff occur but are subordinate to the lava.

SUMMARY. Throughout the Sundaland area, extending from Borneo in the south, through Sumatra, Malaysia, and Thailand, to Indochina and Burma in the north, a Permo-Carboniferous–Triassic period of strong volcanic activity resulted in the formation of a contemporaneously deposited sedimentary-volcanic sequence. In all areas the volcanic paroxysm probably took place in Permian to Lower Triassic time. The various volcanic products were supplied by volcanoes close to the coast of an extensive land mass or on islands (Klompé et al. 1961). These products, their distribution, and the facies of the contemporaneous sedimentary rocks suggest that at least part of the Sundaland area was during that period covered by a shallow sea in which an island archipelago existed. In Indochina, conditions appeared to have been mainly nonmarine.

Cenozoic

Postorogenic basaltic and other flows and tuffs are characteristic of the whole Southeast Asian region (Figure 7.8) in Cenozoic time.

INDOCHINA. Numerous extensive basalt flows occur in South Vietnam, Cambodia, and Laos. They are considered to be of Pliocene, Quaternary, and Recent age. The basalt in South Vietnam frequently contains pigeonite. Other mafic minerals are olivine, bronzite, and augite. The plagioclase may range from An_{20} to An_{70} in zoned crystals. In Cambodia the basalt is similar and is characterized by olivine, augite, pigeonite, and zoned plagioclase. Zeolitization is common in many localities. In western Cambodia, some of the basalt contains nepheline (Fromaget 1941) and in this respect is similar to basalt in Malaya.

THAILAND. Late Tertiary to Pleistocene basalt flows occur in several widely scattered areas in Thailand. In the Mae Moh basin the basalt lies directly on Tertiary beds of probable Pliocene age. Associated with the basalt flows are some andesite flows and dykes of diorite and quartz porphyry (Brown et al. 1951).

Many of the Thai late Tertiary and Quaternary rocks show a remarkable linear arrangement along the western margin of the Khorat Plateau. These rocks are thought to be related to deformations of the Tertiary alpine-himalayan orogeny, to faulting and uplift of the Khorat Plateau, and to the formation of the depression of the Gulf of Thailand (Klompé 1962). It is of interest that the same structural line (Figure 7.8), that marks the western limit of the Khorat plateau and the western extent of major basalt activity in Thailand, aligns with the Lebir fault in Malaya, which played an important part in controlling late Mesozoic sedimentation. This line also marks the western limit of basaltic activity in Malaya.

In addition to the basalt flows, late Tertiary to Pleistocene diorite and quartz diorite intrudes the Khorat Series. In Thailand Brown et al. (1951) recorded occurrences of andesite in small stocks, dykes and flows, cut by dykes and covered by flows of a younger rhyolite porphyry.

SUMATRA. Three cycles of Cenozoic volcanic activity can be distinguished in Sumatra (van Bemmelen 1949). First the so-called old-andesites were extruded from the Paleocene until the Miocene. These were followed by huge flows and deposits of rhyolitic tuff. The second phase, during the later Tertiary, was of acid flows and tuffs formed at Semangko, Ranau, and Toba. The acid Lampong Tuffs of southern Sumatra were also formed at this time, as also were the Bantam Tuffs of western Java. The third phase is represented by the young and still active volcanic cones of the Barisan Range of western Sumatra.

The volcanism described previously has no equivalent in Malaya except for the minor postbasalt quartz porphyry dykes of the Kuantan area. However, subrecent olivine basalt flows of Sukadana in the Lampong districts of Sumatra represent fissure eruptions on the "margins of the continental shield of the Sunda Land, comparable with other young basalt effusives along its borders" (van Bemmelen 1949, p. 230).

Other occurrences of young Quaternary olivine basalt in Indonesia were listed by van Bemmelen (1949) as at Midai, Niut, Murai, Beluh, and Karimondjawa.

INDONESIAN BORNEO (KALIMANTAN). Many extrusions of chiefly hornblende-hypersthene-andesite occurred in the Oligocene to Miocene in Kalimantan. They were also accompanied by more acid varieties.

In the Quaternary period, olivine basalt issued around the Niut stock of hornblende andesite, comparable to the basalt effusives of Sukadana in southern Sumatra (van Bemmelen 1949).

MALAYSIAN BORNEO (SARAWAK). Large volumes of dacite, andesite, and basalt were erupted in several areas within and adjacent to the Northwest Borneo Geosyncline (Kirk 1968). Most of the volcanic activity is clearly postorogenic. The dominant rock type is dacitic and is generally pyroclastic. The Hose Mountains represent a group of pyroclastic volcanic centers. The Usun Apau Plateau is similarly of dacitic pyroclastic rock. In the Nieuwenhuis Mountains, straddling the border between Sarawak and Kalimantan, andesite and basalt lavas and interbedded pyroclastic rocks were overlain by thick beds of andesitic breccia and tuff.

Younger basalt flows of Quaternary age occur to the south of the Usun Apau Plateau and in the Linau Balui Plateau. In western Sarawak there is a small outcrop of andesite lava at Semantan. The basalt has been described as containing olivine or hypersthene in addition to the clinopyroxene (Kirk 1968).

MIDAI. The small island of Midai at the southwest side of Natuna in the South China Sea is composed of a Quaternary basaltic volcano.

SUMMARY. Postorogenic olivine basalt flows occur in Malaya and all neighboring Sundaland countries. They appear to be Quaternary in age, although some may even be late Tertiary.

Cenozoic acid and intermediate volcanic activity is characteristic of countries to the south and west of Malaya, but completely absent from Malaya itself, except for the Quaternary deposits of rhyolite ash in Perak and other parts of Malaya which probably resulted from the widespread distribution of windborne ash from the eruption of Toba in Sumatra.

Basaltic flows become more abundant toward the north, and acidic and intermediate flows and pyroclastic deposits are more abundant toward the west and south. Present-day active volcanism is now confined to the Barisan Highlands of Sumatra (Figure 7.8).

CHAPTER 8

Plutonic Activity

C. S. HUTCHISON

Granite, adamellite, and granodiorite are by far the most extensively developed rocks of igneous aspect in the Thai–Malayan eroded mountain chain. Throughout this chapter all these rock variations are included under the term "granite," defined in the broad sense to include all holocrystalline coarse to medium grained rocks of plutonic aspect (but not necessarily of igneous origin), having a hypidiomorphic-granular texture and composed essentially of quartz, potash feldspar, and/or sodic plagioclase, and subordinate biotite, muscovite, and hornblende.

Throughout the Malay Peninsula, and especially in West Malaysia, the granite forms prominent mountain ranges which trend generally NNW-SSE, apparently occupying the cores of anticlinal structures, representing the upper parts of a huge batholith (Klompé 1962). Many of the ranges can be traced continuously northward into Thailand. The most spectacular of the granite ranges is the Main Range, and this granite is accordingly referred to as the Main Range Granite (Figure 8.1). This prominent granite range divides the densely populated west coast plains from the relatively unpopulated central and eastern parts of the peninsula. It can be traced northward from Malacca, where it is of low elevation, through the Kuala Lumpur area, to the Ipoh area, reaching its highest elevation in excess of 2100 m (7000 ft), and its maximum outcrop extent just north of the Cameron Highlands, thence northward into Thailand.

Other granite ranges, with similar trend but of less spectacular elevation and dimensions, outcrop along the eastern parts of the peninsula (Figure 8.1). In the south, in Johore and Singapore, the granite outcrops extensively, but with rather subdued topography. The granite extends southward to Singapore Island and to the Indonesian islands of Singkep, Bangka, and Billiton. Throughout the orogen, the granites are characterized by their association with tin mineralization.

So-called granite porphyry and quartz porphyry are of common occurrence in West Malaysia, but it is doubtful if these are of intrusive origin (see Chapter 7).

Compared with the granite, other plutonic igneous rocks are insignificant in the Malay Peninsula. In Pahang, there are several small ultrabasic bodies that are also elongated along the NNW-SSE regional strike. There are small basic bodies in Singapore and south Johore. Hypabyssal dyke intrusions are numerous and varied.

GRANITE

Following the general classification of Buddington (1959), there are three distinct tectonic levels of granite emplacement: the epizone, the mesozone, and the catazone. Granites of Malaysia can be ascribed to each of these three tectonic levels.

Epizone

The depth of emplacement of granite in the epizone category may vary from near surface to 5 km (3 miles). The plutons are always discord-

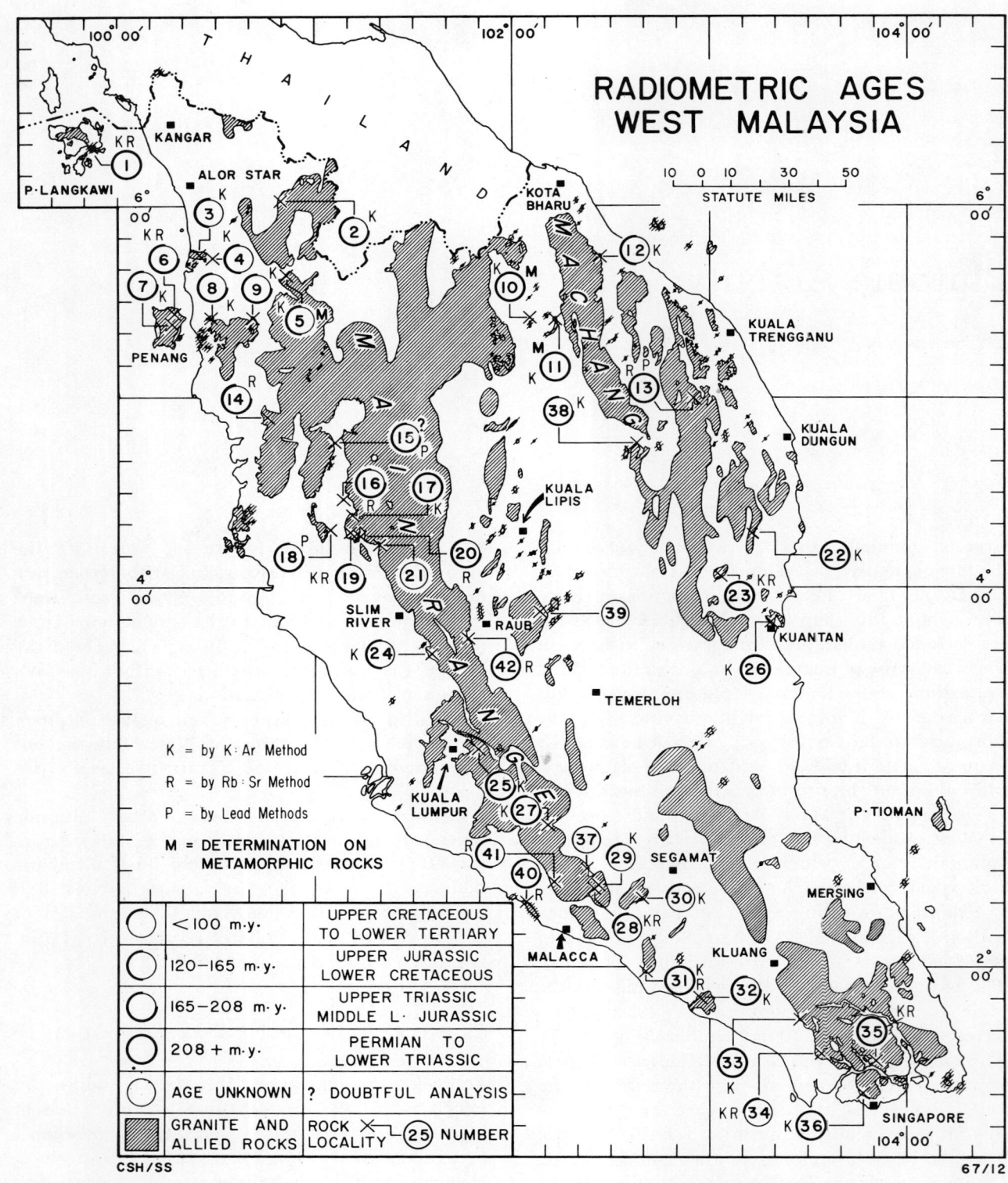

FIGURE 8.1 *Distribution of granite in the Malay Peninsula, after Alexander (1965). Numbered localities are of radiometric age determinations listed in Table 8.5.*

ant to the regional structure. Such intrusions are never associated with regional metamorphism, but they may cause contact metamorphism because they are strongly disharmonious with the strata they intrude (Buddington 1959). They are atectonic or post-tectonic.

Although such intrusions have not with certainty been established from field evidence in West Malaysia, several granites from the western margins of the Main Range Granite have yielded radiometric dates that indicate a late Cretaceous to Eocene age of intrusion (discussed later in this chapter). The main argument in favor of this conclusion is the concordance of the potassium-argon and the rubidium-strontium dates on single mineral and on whole rock analyses. Such concordance of dates indicates a fairly uncomplicated cooling history, characteristic of high-level, post-orogenic or atectonic intrusions (Armstrong 1966).

Further support for the existence of an Upper Cretaceous–Eocene granite in West Malaysia is the occurrence to the north in Thailand of granite of similar age (Brown et al. 1951). This is discussed more fully later.

At the present time, the only known occurrences of Late Cretaceous–Early Tertiary high-level post-orogenic granite emplacements are:

Bukit Senggeh granite, Malacca (locality 28, Figures 8.1 and 8.2)
Batang Malaka granite, Malacca (locality 29, Figures 8.1 and 8.2)
Mount Ophir (Gunong Ledang) granite, Johore (locality 30, Figures 8.1 and 8.2)
Gunong Pulai granite, Johore (locality 34, Figure 8.1)

The first three of these granites have been mapped by Paramananthan (1966) as distinct bodies of circular or elliptical outcrop (Figure 8.2), but they are likely to be of the same pluton. He was unable to find any relation in the field between these granites and the older Main Range Granite because of lack of outcrop.

The younger granites occur as small plutons adjacent to or satellitic to the Main Range Granite. But it must be stressed that these are the only four known occurrences at present, and there is no evidence to exclude the possibility of similar

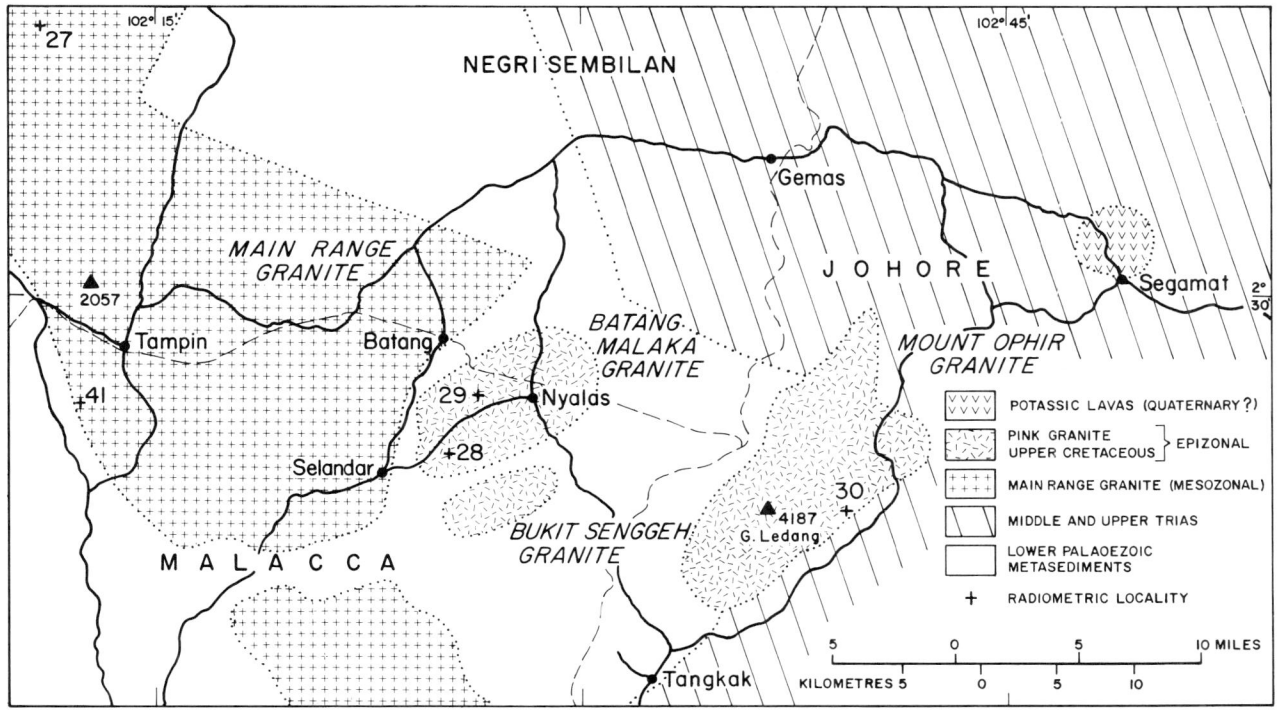

FIGURE 8.2 *Geology of the area around Mount Ophir (Gunong Ledang), Malacca, showing the relationship of the Upper Cretaceous pink epizonal granites to the southern margin of the older mesozonal Main Range Granite. After Paramananthan (1966).*

epizonal granites elsewhere, even on the eastern side of the peninsula. Indeed, Paton (1960) considered the granites of Trengganu to be Upper Cretaceous, but there is no radiometric evidence to support his view.

In the Kuantan and Gambang area of Pahang, several granite plutons of Upper Carboniferous (Hutchison and Snelling, 1971) and of Permian age must undoubtedly have been epizonal emplacements. Evidence for such a conclusion is two-fold: (a) the short time-lapse between the deposition of the country rocks (Permocarboniferous) and the intrusion, and (b) the often distinct and well-defined contact aureoles, outside of which the sedimentary rocks are fossiliferous and unmetamorphosed.

PETROGRAPHY. All the known occurrences of epizonal granites have a characteristic and distinctive petrography. They are characteristically pink in hand-specimen, the color being predominantly due to the alkali feldspars. They are medium to coarse grained and usually equigranular, although weakly porphyritic varieties occur locally with feldspar phenocrysts up to 1 cm long. There is distinct evidence of chilling of the intrusions, the interiors of the plutons being coarse grained, while the borders are usually of medium grained varieties (Paramananthan 1966). This is a characteristic of high-level intrusions. The high temperature mineralogy described below is restricted in Malaya to the late Cretaceous granites and all other granites studied so far contain highly triclinic K-feldspar.

The essential minerals are quartz, alkali feldspar, plagioclase, biotite, and magnetite. The quartz is usually subhedral and frequently forms myrmekitic growths along the microcline boundaries. The alkali feldspar is strongly perthitic orthoclase: X-ray diffraction studies now in progress indicate a high temperature structural state of zero triclinicity. The alkali feldspar crystals are characteristically pink, even in thin section, and are usually subhedral. Universal stage data indicate a range of 2V values, even in the same hand specimen, from approximately 58° to 85°. The plagioclase forms subhedral to euhedral crystals, white in hand-specimen and colorless in thin section. Twinning is common on carlsbad, albite, ala A and ala B laws and some zoning is seen. The average composition is oligoclase, An_{15}.

The accessory minerals are sericite, replacing the feldspars, apatite, zircon, and sphene. Examples of modal analyses of these granites are given in Table 8.1 (Nos. 1 and 8) and several have been included on the distribution diagram of modal analyses (Figure 8.5).

In composition, these rocks are adamellitic, but tend toward granite *sensu stricto*. Chemical analyses and Niggli norms are presented in Table 8.2.

Paramananthan (1966) has suggested that the high-level Upper Cretaceous–Lower Tertiary granite of Batang Malaka, Bukit Senggeh, and Mount Ophir are genetically related to the potassic lavas that occur only some 8 km (5 miles) to the east, at Segamat, Johore (Figure 8.2). Grubb (1965) correlated these potassic lavas with the basalts of Kuantan, Pahang, as being of the same postorogenic Late Cretaceous to Eocene episode of volcanic activity.

Mesozone

The depth of emplacement of granite in this category may vary from 5 to 11 km. The plutons are usually of batholitic size and are composite in character, being part concordant and part discordant to the regional structure. Argillaceous strata in the neighborhood of such granites are usually present as slate and phyllite (Buddington 1959). The granites themselves commonly exhibit planar foliation, representing the upward flow of the magma. Mesozone granites are usually late orogenic in the revolutionary phase of an orogenic development. Unlike the epizonal granites, these rocks are not completely disharmonious with the country rocks, which were generally maintained at the conditions of temperature and pressure of the greenschist facies, or even as high as the epidote amphibolite facies of regional metamorphism, while the batholiths were emplaced. Thus the plutons will be linked with regionally metamorphosed rocks more characteristically than with contact aureoles, although both types of metamorphism may be distinguished. However, it would be a mistake to ascribe all metamorphism of rocks associated with mesozonal granites to one type only.

Mesozoic late orogenic granites constitute the majority of granite bodies in the Malay Peninsula. The two major granite ranges belong to this category: the Main Range Granite (Scrivenor 1931), and the Boundary Range Granite (MacDonald 1968) which forms the prominent north-south-trending dividing range between the states of Kelantan and Trengganu in the eastern part of the peninsula. The southern extensions of these granites into Johore and Singapore are also of the mesozone. Included also are many smaller similarly elongated outcrops *en echelon* with the large massifs, which are thought to connect with them at depth.

These granite bodies form huge batholiths elongated parallel to the regional strike of the country rocks, but they distinctly cross-out the sedimentary formations. The bodies were probably emplaced in the mesozone during the late kinematic stage of the Mesozoic orogeny and are therefore not syntectonic (as described by previous authors), but rather late orogenic.

Syntectonic granites are generally completely concordant with the regional structure, and the bodies are sheetlike, not batholithic (Buddington 1959, Raguin 1965). Truly syntectonic granites do occur in West Malaysia, and these are described below as catazone granites. But the majority of the granites, which form the large subconcordant batholiths, must be considered as late orogenic. Characteristically they show no metamorphic foliation and may be considered as entirely of igneous origin.

These are the granites which are associated with tin mineralization throughout the Malay Peninsula and as far south as the Indonesian island of Billiton.

PETROGRAPHY. Grey porphyritic biotite granite, with large phenocrysts of microcline and plagioclase, is the commonest kind of granite exposed in the Main Range and the west of Malaya, and it can be considered to be the typical rock (Scrivenor 1931) (Figures 8.3 and 8.6a). In places the phenocrysts show a marked flow alignment (Figure 8.4). The following mineralogy is typical of the Main Range Granite.

1. *Phenocrysts.* The phenocrysts are of alkali feldspar, plagioclase feldspar, and quartz, commonly about 4 m long, they vary from 1 to 10 cm.

The alkali feldspar is commonly microline but occasionally orthoclase, although what has often

FIGURE 8.3 *Weathered porphyritic granite, Kampong Batu Hampar, Dindings, Perak. Photograph Jaafar bin Haji Abdullah.*

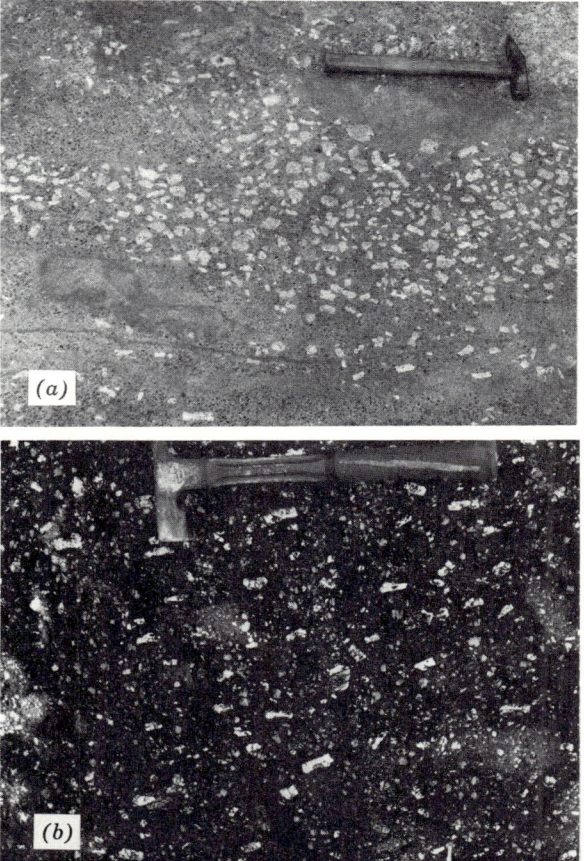

FIGURE 8.4 *Flow alignment of feldspar phenocrysts in porphyritic granite:* (a) *Sungei Lawin, Perak, Photograph C. R. Jones;* (b) *Kuah, Pulau Langkawi, Photograph C. S. Hutchison.*

been identified as orthoclase may be untwinned microcline. Large phenocrysts are characteristically microperthitic of the vein type. In some larger crystals, the perthitic nature can even be seen in the hand-specimen. Under cross polars, the veins are seen to be of sodic plagioclase, distinguished from the microcline host by the simple albite twinning and slightly different birefringence. When thin sections have been prepared for modal analyses by the two-color staining method of Bailey and Stevens (1960), the sodic plagioclase exsolution lamellae are stained pinkish, whereas the microcline or orthoclase host is yellow.

Many large alkali feldspar crystals are ornamented. Phenocrysts frequently contain one or more internal zones of small quartz and other crystals of the groundmass minerals. But the crystal has continued to grow in optical continuity beyond its original outline, so that a zone of groundmass material became enclosed within the final crystal, parallel to the final crystal outline.

The plagioclase feldspar is somewhat variable in composition, and commonly it is zoned. Composition may range from albite through oligoclase to andesine. It is often sericitized. Internal zones of small crystals of other minerals reveal that ornamentation is common in the phenocrysts, as it is with the alkali feldspars.

The quartz usually occurs as subhedral crystals, occasionally somewhat rounded and embayed (Hutchison 1964a).

2. *Groundmass.* The groundmass is commonly equigranular and of medium to fine grain. The main constituents are the same minerals as the phenocrysts; biotite, muscovite, hornblende, and tourmaline also occur, occasionally attaining larger crystal dimensions.

Biotite is the most abundant mafic mineral in Malayan granites. It is commonly fresh but may be chloritized in hydrothermal and shear zones. It is a typical biotite, brown and pleochroic with common zircon inclusions and pleochroic haloes.

Muscovite is generally rare as a primary mineral. It is to be found as a result of hydrothermal action, replacing the plagioclase and the biotite, and even the alkali feldspar to form a greisen. In such cases

TABLE 8.1 *Modal analyses of selected Malayan granites* in volume percent*

	1	2	3	4	5	6	7	8	9	10	11	12
Points counted	—	10,357	5676	6498	7108	8012	10,725	—	14,180	10,570	10,520	10,721
Quartz	41	38	36	35	34	32	31	25	29	24	22	20
K-Feldspar	30	27	33	14	21	33	37	40	26	28	34	26
Plagioclase	26	28	23	34	42	28	28	29	32	42	31	44
	An_{15}	An_{12}	An_{14}	An_{10}	An_{26}	An_{26}	An_{20}	An_{15}	An_{20}	An_{34}	An_{31}	An_{25}
Biotite	2.5	4	7	6	3	7	3	4	10	5	12	8
Apatite	—	3	0.2	—	—	0.1	—	—	0.2	0.1	0.1	0.1
Amphibole	—	—	—	—	—	—	—	—	—	—	0.3	2
Sphene + epidote	—	—	—	—	—	—	—	—	—	—	0.6	—
Opaque	0.5	0.3	—	0.2	—	0.2	0.1	2	—	0.1	—	0.1
Muscovite + sericite	—	3	0.6	11	—	0.4	1	—	3	1.4	—	1
Fluorite	—	—	0.1	—	—	—	—	—	—	—	—	—

* Specimens 1–12 as follows (Specimen numbers refer to the collections of the Department of Geology, University of Malaya, Kuala Lumpur):
 1. Pink granite (UM4387): Batang Malaka granite, 25 milestone Selandar–Nyalas road, Malacca. Radiometric locality 29.
 2. Porphyritic biotite adamellite (UM1584); New Jetty, Kuah, Langkawi.
 3. Biotite adamellite (UM881): Papan quarry, Perak.
 4. Greisenized granodiorite (UM1468): Ampang, Selangor.
 5. Granodiorite (UM5372): Batu Pahat Quarry, 80 milestone Batu Pahat–Pontian road, Johore. Radiometric locality 32.
 6. Porphyritic adamellite (UM1938), 32 milestone Bentong road, Pahang.
 7. Leuco-adamellite (UM1884): Bukit Buloh, J.K.R. quarry, Kelantan.
 8. Pink granite (UM4394): Mount Ophir quarry, 28.5 milestone Tangkak–Segamat road. Radiometric locality 30.
 9. Porphyritic adamellite (UM3963): Bukit Baloh, Sungei Linchong, Pahang. Radiometric locality 23.
 10. Porphyritic biotite adamellite (UM16): Lone Pine Hotel, Penang Island.
 11. Porphyritic adamellite (UM5371), Sungei Sok, Kedah. Radiometric locality 2.
 12. Pink hornblende adamellite (UM5306). Bukit Labohan, J.K.R. quarry, Trengganu.

it is commonly associated with tourmaline as a product of tourmaline greisenization (Hutchison and Leow 1963).

Hornblende is uncommon and occurs only sporadically as a primary constituent. It is more common in granites of the east coast. Hornblende or actinolite occurs as a result of contamination of the granite by partial assimilation of xenoliths. The hornblende may be redistributed in veins and dykes as hornblende-rich pegmatites, common in Singapore and Johore (Hutchison 1964a).

3. *Tourmaline.* This occurs sporadically as a result of greisenization.

4. *Accessory minerals.* The most common accessory minerals are apatite, magnetite, and fluorite. Less commonly, thin sections may show cassiterite, titanite, and allanite. Cassiterite is remarkably rare in Malayan granites despite the fact that they are in a tin province.

Several representative modal analyses of Main Range and other granites are given in Table 8.1, and the largest number of the specimens plotted on the distribution diagram (Figure 8.5) are of Main Range Granite. It can be seen from Figure 8.5 that the majority of the granites *sensu lato* have approximately equal amounts of alkali feldspar and plagioclase and hence would strictly be classified as adamellite. The small triangle *abc* within Figure 8.5 represents the limits of composition of granite as proposed by Tuttle and Bowen (1958), and very few rocks from West Malaysia fall without these limits.

The variability of the Main Range Granite has been well represented by Scrivenor (1931, p. 25) in his description of a section on the road from Kuala Kubu Bharu to Gap:

42½ milestone, coarse porphyritic granite and, close by, coarse nonporphyritic granite

45 ms dark porphyritic granite with much biotite

46¼ ms nonporphyritic granite

47 ms coarse nonporphyritic granite with tourmaline, biotite and muscovite

47 ms fine-grained nonporphyritic granite with tourmaline, biotite, muscovite, and fluorite

47¼ ms coarse nonporphyritic granite with biotite, muscovite, and fluorite

47½ ms microgranite with tourmaline veins, biotite, muscovite, and fluorite

48 ms coarse nonporphyritic and also porphyritic granite

49 ms granite porphyry with muscovite predominating over biotite

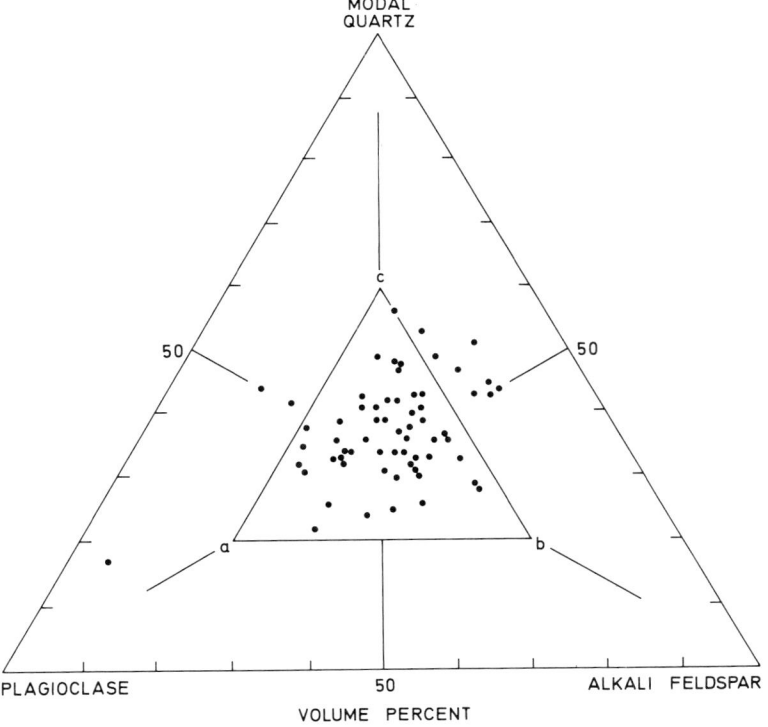

FIGURE 8.5 *Triangular distribution diagram showing the relative proportions of modal quartz, modal plagioclase, and modal K-feldspar for 63 typical Malayan granites.*

51 ms granite porphyry similar to that at the 49th mile, and also coarse porphyritic granite close by
Near 52nd milestone, granite porphyry
Near 53rd milestone, granite porphyry
53rd to 59th miles coarse porphyritic granite.

It must be realized that the road has been improved since Scrivenor's description, and the milestones are now not strictly applicable.

Unlike the Main Range Granite, the Boundary Range Granite is more variable in color, and pink varieties are common. Even in a single quarry face, both white or grey and pink varieties may coexist. But grey varieties are much more abundant than the pink varieties. The main rock types are: medium-grained microcline-microperthite granite, graphic granite, medium-grained and porphyritic granodiorite, tonalite, granodiorite porphyry, microgranite, aplite, pegmatite, and vein quartz (MacDonald 1968). In this granite, plagioclase is generally more abundant than alkali feldspar. In the porphyritic varieties, the large phenocrysts of pink microcline microperthite occasionally attain a length of 10 cm. Biotite is the most abundant mafic mineral, but hornblende occurs more frequently than in the Main Range Granite. Some of this hornblende is distinctly primary but much can be attributed to assimilation of xenoliths. Apparently the granites that contain hornblende do not contain cassiterite (MacDonald 1968).

In the southern extensions of the Boundary Range Granite, in the Kuantan area of Pahang, the prominent types are grey porphyritic and nonporphyritic biotite granite (Fitch 1952). Fitch (1952) noted that the relation between the porphyritic and nonporphyritic varieties was too haphazard to map separately. Phenocrysts of alkali feldspar are up to 6 cm long. In some cases the feldspars are pink. Quartz is so abundant that nonporphyritic leucogranite commonly occurs. In both granite and leucogranite, there is a common micrographic intergrowth of quartz and feldspar (Fitch 1952). Hornblende and muscovite are rare in this granite.

Farther south in Johore and Singapore, the granite is generally more abundant in plagioclase than in alkali feldspar. The plagioclase varies from albite to andesine, but is generally oligoclase. In Singapore Island, the alkali feldspar has been determined to be generally orthoclase microperthite (Hutchison 1964a). As elsewhere, biotite is the most abundant mafic mineral, although hornblende is locally present. The texture is extremely variable—from coarsely porphyritic with up to 65% of large phenocrysts set in a fine-grained groundmass, to medium to coarse grained nonporphyritic varieties (Hutchison 1964a).

In southeast Johore the granite grades from coarse grey slightly porphyritic to nonporphyritic granite, micropegmatite, and granophyre with graphic intergrowth of the minerals (Grubb 1968).

The granite of Gunong Benom (Figure 8.6b), in central Malaya, was described by Khoo 1968) as a coarse-grained nonporphyritic biotite granite.

Generally it may be said that the granites occurring east of the Main Range tend to be nonporphyritic, and variations in the granites of the east coast are only weakly porphyritic (in contrast to the very coarse porphyritic nature of the Main Range granite).

Exceptions do occur, however, and the granite at Bukit China, to the south of Mentakab in Pahang, is extremely coarsely porphyritic and is identical to the most coarsely porphyritic varieties of the Main Range.

This general difference between the Main Range granite (typically coarsely porphyritic) and the granites of central and east Malaya (typically equigranular) suggested to C. K. Burton (personal communication) that they might be of different ages, and the radiometric information (see later section) does seem to support this view.

SHEARING OF THE GRANITE. Zones confined generally to the marginal regions of the Main Range Granite have suffered shearing, which is often intense. One of the best examples of a granite showing cataclasis is in the Jabatan Kerja Raya (J.K.R.—Public Works Department) quarry 4.8 km (3 miles) east of Kuala Kubu Bharu, Selangor, along the Gap road. At this locality, the feldspar phenocrysts are commonly bent and broken. The quartz groundmass shows mortar structure, the cataclasis often being healed through recrystallization; but strong strain shadows can be seen under crosspolars. Albite twin lamellae of the plagioclase crystals are commonly bent, and biotite is completely chloritized and associated with epidote. The normal porphyritic texture is destroyed by the drawing out of the phenocrysts along the shear direction, and the groundmass, much of which is secondary, is redistributed and sometimes mylonitized. Slickensided mylonitic granite has also been described near Chenderiang in the Kinta Valley (Ingham and Bradford 1960).

CONTAMINATION OF THE GRANITE. Especially in the marginal zones, granites are commonly characterized by dark xenoliths. The shapes of the

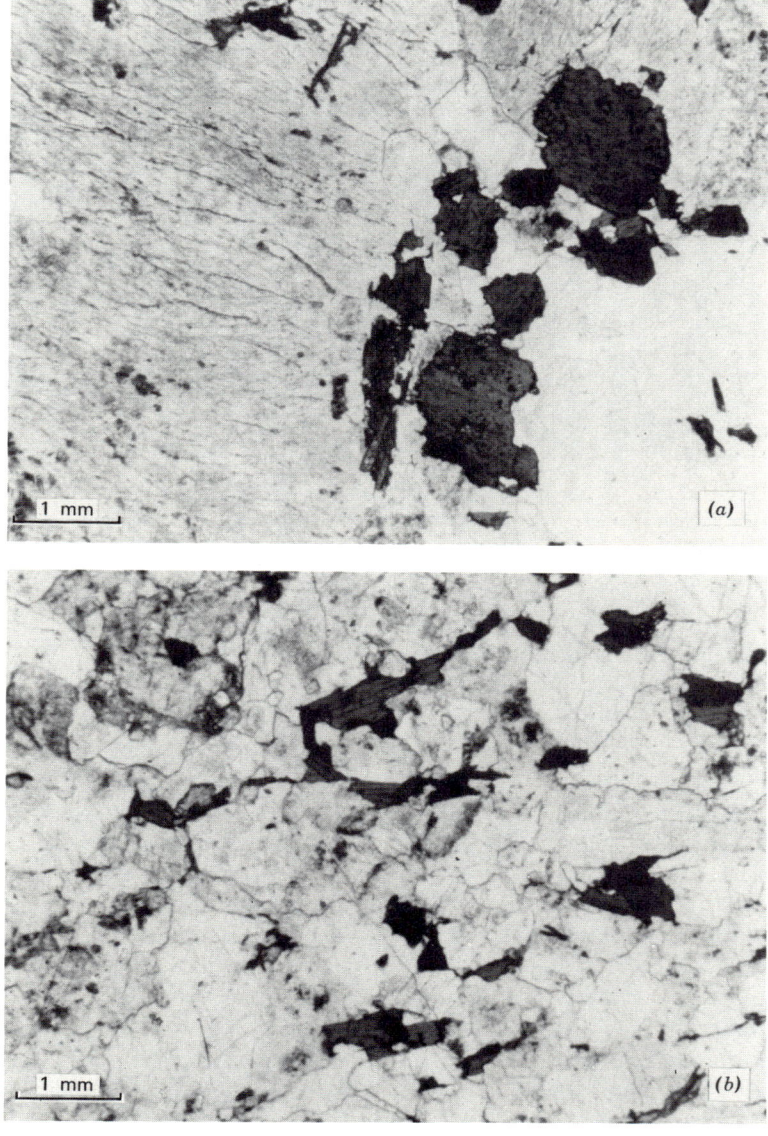

FIGURE 8.6 *Photomicrographs (C. S. Hutchison) in ordinary light:* (a) *Porphyritic biotite granite, Sungei Sok, Kedah, specimen UM 5371;* (b) *biotite granite, Sungei Klau Kechil, Benom, Pahang, specimen 5558.*

xenoliths vary with the composition and texture of the included rock. In the Gunong Jerai area of Kedah, for example, the xenoliths are of pelitic and psammitic metasediments and are characterized by rather elongated tabular shapes, although some assimilation has modified the outlines to be somewhat ellipsoidal. Good examples are to be seen in the granite at Tanjong Jaga to the West of Gunong Jerai, and in the Sungei Batu Pahat at the Merbok reservoir on the southern slopes of Gunong Jerai (Bradford 1965).

In Singapore Island, the granite, near its contact with an earlier gabbro, contains ovoid bodies that are considered to be partly assimilated xenoliths (Figure 8.8b,c) (Hutchison 1964a). The contact between the earlier gabbro and the granite batholith in Singapore and Pulau Ubin provides some interesting examples of assimilation. In describing this contact, Hutchison (1964a, p. 288) wrote as follows:

The granodiorite makes a most irregular contact with the gabbro, and in quarry 3 the intrusive relationship is well displayed [Figure 8.7]. Here dykes of the granodiorite varying from several inches to 15 feet wide anastomose throughout the gabbro. The

FIGURE 8.7 *Granodiorite cutting gabbro as irregular dykes ranging from several centimeters to 5 m thick; from Jurong Quarry 3, Singapore. Photograph C. S. Hutchison.*

dykes make sharp contacts with the gabbro at the present exposed level. They are full of gabbro xenoliths in all stages of assimilation. In some, the outlines are vague and the pyroxene has been completely replaced by coarse hornblende. The same dyke may display advanced assimilation and contain coexisting xenoliths of gabbro that have sharp outlines and preserve their original mineralogy and texture to a large extent. [Figure 8.8a].

The range of assimilation stages present in a single tectonic level indicates that the granite was forcibly intruded from depth. By the time the magma had reached its present level, it had lost most of its assimilative energy through cooling. The well-digested xenoliths have been carried upward by the intrusion, whereas the unassimilated ones have been incorporated at or near the present level.

Contamination by incorporation of xenoliths of the country rock is so common throughout the Malay Peninsula that details of all occurrences cannot be listed. Good examples have been described from the neighborhood of Kuala Dipang and the Sungei Siput South, where the source of the xenoliths has been traced to nearby schist and shale outcrops (Ingham and Bradford 1960).

An interesting granite at Changkat Rembian, near Tapah, Perak, contains abundant prismatic pink andalusite in addition to tourmaline. The andalusite must indicate that the granite has been contaminated by argillaceous sedimentary material.

In the Boundary Range Granite, extensive assimilation of country rock xenoliths along the margin of the intrusion has caused an almost continuous zone of tonalite varying from a fraction of a kilometer to 1.5 km wide (Figure 8.9). The tonalite is medium to fine grained, granular textured, with zoned sodic plagioclase, green hornblende, dark brown biotite, and small amounts of orthoclase. Zircon, apatite, and ilmenite occur as accessories, and chlorite and epidote are secondary. The plagioclase is zoned from labradorite to oligoclase (MacDonald 1968). Outside the tonalite areas, xenoliths are abundant locally; many are of tonalite composition, but some are of sedimentary rock. Contaminated granite is also common on the west side of the Boundary Range Granite near the Sungei Lebir, where it contains ghosts of highly assimilated xenoliths. In such cases, hornblende is invariably present, and hornblende together with biotite may form up to 80% of the rock (MacDonald 1968).

MODIFICATION OF THE GRANITE. Minor late-stage differentiates are common as dykes and veins up to several meters thick. They occur usually within the batholiths, commonly in the marginal zones, but many intrude the surrounding sedimentary rocks. The common types may be described as microgranite, granite porphyry, quartz porphyry, felsite, aplite, pegmatite, and vein quartz. These late-stage differentiates range in origin from magmatic to hydrothermal.

Microgranite. The microgranite is simply a fine-grained modification of the granite in which it occurs. It is found commonly and irregularly throughout the granites of the Malay Peninsula, often abundant in the margins of the batholiths but not restricted to them.

FIGURE 8.8 (a) *Detail of the contact of a granodiorite dyke with gabbro at Jurong Quarry 3, Singapore; note the sharpness of the contact and the advanced assimilation of xenoliths*; (b) *ovoid xenolith in granite, Seng Kee Quarry 16, Singapore [quarry numbers taken from Hutchison (1964a)]*; (c) *photomicrograph in ordinary light of granite–gabbro contact, Pulau Ubin, Singapore, specimen UM 6721. Photographs and photomicrograph C. S. Hutchison.*

Porphyry. Granite porphyry, quartz porphyry, and felsite occur commonly throughout West Malaysia. It is very doubtful, however, if any of these so-called porphyries are of plutonic, or even of igneous origin. All are foliated or sheared, or have a flow structure. Since the majority can now confidently be considered to be metamorphosed rhyolite, they are more appropriately described in Chapters 7 and 9. However, this does not exclude the possibility of some occurrences being truly of plutonic origin.

Aplite. Aplite occurs throughout West Malaysia as irregular veinlike masses within the granite. It also occurs frequently as dykes, which are mainly confined to the marginal zones of the batholiths but also penetrate outward into the surrounding sedimentary rocks. They are fine grained with a saccharoidal granular texture. At Jelapang, Perak, an aplite has been described as containing spherulites of chalcedony (Ingham and Bradford 1960). Usually the aplite is composed of quartz and orthoclase, and may or may not contain sodic plagioclase and muscovite. Cassiterite sometimes occurs in the aplite (Scrivenor 1931).

Aplite is particularly common in the Kinta Valley of Perak, around the granite contact with the sedimentary rocks (Ingham and Bradford 1960).

The aplite veins consolidated slightly later than the granite and they represent a late phase of residual magma of high mobility, more or less contemporaneous with the phases of pneumatolysis and mineralization. The aplite veins, moreover, frequently served as channels for the movement of tin- and tungsten-bearing fluids (Ingham and Bradford 1960).

Pegmatite. Pegmatite occurs mainly as dykes, but occasionally it is observed as irregular veins within the granite. Although usually confined to the margins of the batholiths, the veins frequently extend outward into the country rocks for a limited space. They are typically composed of quartz, feldspar, muscovite, and tourmaline, but biotite is not infrequent in addition to the muscovite. Other minerals that have been found in Malayan pegmatites are: chalcopyrite, pyrite, arsenopyrite, cassiterite, stannite, galena, beryl, wolframite, almandine garnet, and occasionally magnetite, hematite, and torbernite. The last-named mineral has been described as occurring at Gunong Bakau near Bentong, Pahang (Alexander 1968).

In the Kedah Peak area, especially at Tanjong Jaga and in the Sungei Batu Pahat, the pegmatites are characterized by both muscovite and biotite and by abundant small euhedral crystals of almandine garnet (Bradford 1965).

Cassiterite commonly occurs in pegmatites. Pegmatites have been worked as a source of tin ore at Kota Tinggi, Johore, and Ulu Gapis, Pahang (Richardson 1939). In the region around Kuala

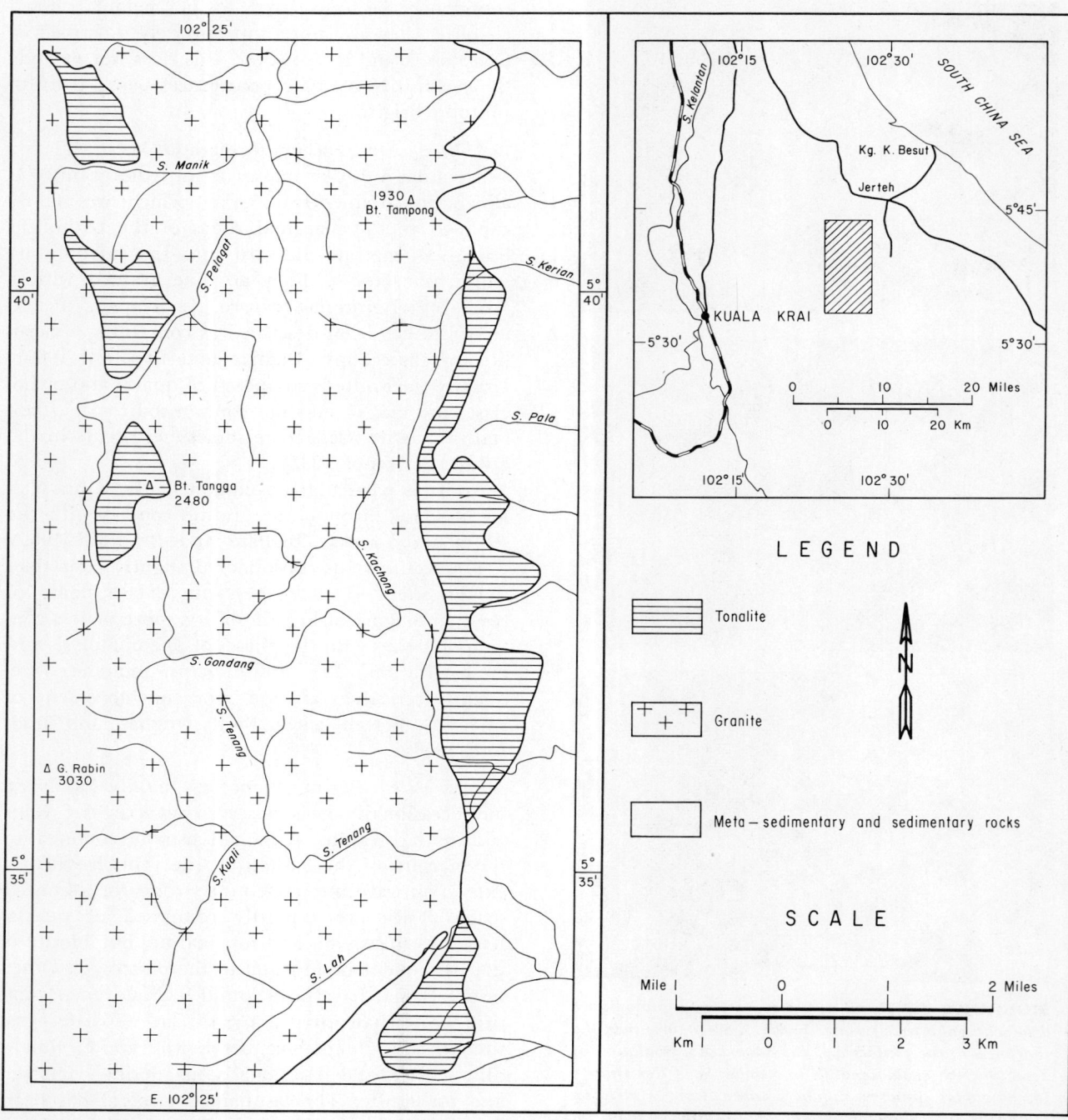

FIGURE 8.9 *Distribution of tonalite at the northern end of the Boundary Range Granite (after MacDonald 1968, fig. 10).*

Kubu Bharu, Selangor, Roe (1951) has described several interesting pegmatite occurrences. A pegmatite in the roof pendent at Ulu Kerling, to the north of Kuala Kubu Bharu, occurs as a dyke 1.8 m wide, and its outcrop can be traced over a kilometer. Cassiterite is present as large crystals associated with muscovite, tourmaline, feldspar, quartz, topaz, and fluorite. A little farther to the north, at Kelumpang, sporadic bunches of coarse cassiterite crystals occur in the pegmatite in two quarries, where the pegmatite cuts both the granite and the sedimentary rocks (Roe 1951). At Gunong Bakau, along the Gap road from Kuala Kuba Bharu, a 9-m pegmatite dyke cuts granite. Along its contact with the granite, a cassiterite-bearing greisen is formed. The pegmatite contains chalcopyrite, arsenopyrite, galena, pyrite, cassiterite, and abundant tourmaline (Roe 1951).

An intimate relation often exists between pegmatite and aplite. Pegmatite occurs sometimes as bands streaked out with aplite and may show a well-developed foliation. The contact between the pegmatite and the granite often contains microgranite, which may be a chilled margin or may result from the shearing of the aplite and pegmatite (Roe 1951). Good examples are to be seen in the quarries along the Gap road.

To the south of Kedah Peak (Gunong Jerai), the pegmatite intrudes sedimentary rocks and constitutes the chief source of cassiterite and columbite-tantalite. A partial analysis of a mica-tourmaline pegmatite from Tanjong Jaga, just west of Kedah Peak gave:

SnO_2 0.06%; $(Ta, Nb)_2O_5$ 0.23%; TiO_2 0.13%; total Fe 2.12% and a radioactivity of less than 0.05% ThO_2 equiv. (Bradford 1965).

The minerals in the following list are found in alluvial deposits in the vicinity of the pegmatite and can safely be assumed to have their source in the pegmatite:

cassiterite, columbite-tantalite, monazite, uraniferous monazite, fergusonite, betafite, pyrochlore-microcline, xenotime, topaz, zircon, ilmenite, rutile, anatase, brookite, allanite, gahnite, garnet, and spinel (Bradford 1965).

Throughout West Malaysia, the pegmatites were intruded or differentiated during a relatively late, pneumatolytic phase of the main granite intrusion. Cross-cutting pegmatite dykes and veins cut the granite, varying in width from 1 cm to several meters. They are often multiple, and some are well zoned. Some pegmatites are affected by later shear movements (Bradford 1965).

Vein quartz. Quartz veins are usually the latest of all the intrusions in the granite, and they cut both aplite and pegmatite dykes. They are characteristically associated with tourmaline greisenization (Hutchinson and Leow 1963). Commonly they occur as swarms, often with approximately parallel alignment, indicating injection along joint directions in the outer portions of a batholith (which had already solidified by the time the hydrothermal solutions rose from the inner, yet molten, areas of the magma body).

The occurrence on coastal exposures immediately to the east of Kuah, Pulau Langkawi, is typical and has been described by Hutchison and Leow (1963).

The mineralogy of the dykes is generally simple. They are composed predominantly of coarse-grained quartz, which commonly contains veins of tourmaline, muscovite, and finer grained quartz. Within the veins, the tourmaline appears to be the earliest formed mineral. The c axes of the tourmaline crystals lie parallel to the veins. The muscovite also shows flow structure and commonly borders the tourmaline. Accessory minerals are pyrite, topaz, and biotite in traces. The mineralogy of these quartz dykes indicates a hypothermal origin (Hutchison and Leow 1963). Quartz veining may also be irregular and may form a dense network of stringers, as in the Kedah Peak area (Bradford 1965).

An interesting occurrence has been described by Alexander (1968) from Gunong Bakau, near Bentong, Pahang. Sill-like bodies of quartz-topaz rock, 5 m thick and with variable amounts of cassiterite and arsenopyrite, occur in the granite. They are feldspar free and sometimes contain mica and tourmaline.

Numerous occurrences of quartz-mica rock have been mapped by Richardson (1939) in the region between Raub and Cheroh, Pahang. In hand-specimen the rock is green or greenish yellow. It is fine grained and strongly brecciated; irregular fissures have been infilled with white quartz, some of it containing gold. The mineralogy is essentially quartz and muscovite, often fine grained; but in places it grades into aplite, containing small subhedral phenocrysts of orthoclase. This rock occurs commonly in the region of the now-abandoned Raub gold mine. It is clear that these rocks are genetically related to the Main Range granite (Richardson 1939).

Elsewhere in West Malaysia, numerous occurrences have been described. Quartz-tourmaline veins in granite, which contain some cassiterite and which have been mined from time to time, have been recorded from Sungei Siput (Savage 1937). In Perak, around Tapah and Telok Anson, the veins carry not only cassiterite but also wolframite (Ingham 1938). In Pahang, gold has been recorded in quartz veins that are cassiterite free (Richardson 1950).

Roe (1951) suggests that the emplacement of quartz occurs synchronously with the period of "tin pneumatolysis" but attains its maximum development after the completion of the cassiterite mineralization. He has also recorded torbernite in quartz-tourmaline veins. The same author (Roe 1953) suggests that greisen and schorl veins carrying cassiterite occur during the main period of "cassiterite pneumatolysis" and also subsequently during a period of hydrothermal injection of siliceous solutions carrying little tin and, in places, tungsten and gold. Tourmalinization and greisenization associated with quartz dykes has been described in the Kinta Valley (Ingham and Bradford 1960).

A completely different type of quartz vein is commonly found in West Malaysia, usually on a very spectacular scale; these are the so-called quartz reefs, which form prominent quartz ridges throughout the country. The largest of these is the Klang Gates Ridge, situated about 13 km (8 miles) northeast of Kuala Lumpur (Plate 18). It has been utilized as the dam site for the Kuala Lumpur water supply reservoir (Alexander and Procter 1955).

The quartz ridge is composed almost entirely of pure white quartz. It has been traced for a lateral distance of more than 16 km (10 miles), and its width varies from only a few meters at some places to as much as 180 m. It was formed by the deposition of hydrothermal quartz along a near-vertical zone of weakness in the granite (Alexander and Procter 1955). As early formed quartz masses and veins cooled and contracted, new fractures were formed and were filled with later veins of crystalline quartz. Repeated quartz injection and deposition, contraction, and fracturing have produced a complex interlacing network of large composite and small simple quartz veins, in many places carrying the remnants of partly or almost wholly altered

PLATE 18 *Quartz dyke intruded into granite and forming a prominent topographic feature. View east along the Klang Gates Ridge, Kuala Lumpur. Photograph D. J. Gobbett.*

granite. Much of the body is of massive quartz, although lenticular vugs with excellent crystalline quartz aggregates are common. This quartz dyke is characteristically formed of pure quartz, with no traces of other minerals. But Scrivenor (1931) reported that one sample was found to contain 0.02% of pyrrhotite and traces of scheelite and cassiterite.

Other prominent quartz reefs occur in Negri Sembilan near Seremban (Willbourn 1922) and in the Sungei Siput area of Perak (Savage 1937). Ingham and Bradford (1960) also recorded quartz reefs in the Kinta Valley. In all these occurrences the quartz reefs are characteristically of pure quartz and barren of ore minerals.

TOURMALINE GREISENIZATION. Zones of metasomatic alteration of the granite host extend outward from the quartz-muscovite tourmaline hydrothermal dykes, commonly as far as 30 m. The amount of metasomatism is strictly in the ratio of the thickness of the dykes or dyke swarms. Very thin dykes commonly show no metasomatism. Tourmalinization of the granite is usually most spectacular on either side of the dykes, although tourmaline can often be found widely throughout the granite and even the neighboring sedimentary rocks, where such dykes occur. Excellent examples are to be seen near Kuah in Pulau Langkawi, where the tourmaline is segregated into subpherical clots up to 5 cm in diameter (Hutchison and Leow 1963). These segregations may show radial arrangement of the tourmaline, they or may be randomly oriented.

Such tourmaline clots are best seen on weathered granite outcrops, where they stand up as prominent, black, resistant projections from the smooth whitish granite surface (Figure 8.10). In zones on either side of the hydrothermal dykes, the granite, as seen in thin section, contains secondary muscovite and quartz, in addition to tourmaline. The muscovite has been introduced interstitially between the primary minerals of the granite. Fine-grained quartz is also associated with the muscovite. The secondary nature of this quartz is demonstrated by its mode of occurrence as blebs in the feldspar as well as by its close association with tourmaline and muscovite. The physical characteristics of the muscovite, tourmaline, and secondary quartz are similar to those of the same minerals occurring in the nearby hydrothermal dykes.

The tourmaline greisenization spread from the dykes into the granite. Greisenization and tourmalinization therefore commonly occur together and the two processes cannot be separated by any major time break. They both occurred before the interior of the granite pluton was fully crystallized (Hutchison and Leow 1963). Tourmalinization and greisenization of the granite is common in West Malaysia, especially in areas adjacent to alluvial tin mining. There is strong evidence to show that all the muscovite in granites is secondary and was introduced as a result of alkali metasomatism causing replacement of the feldspar. Sericitization of feldspar is common in the marginal granites and is often accompanied by kaolinization of the feldspars (Ingham and Bradford 1960).

Tourmaline is also introduced into the country rocks adjacent to the granites to give tourmaline-corundum rocks and tourmaline hornfelses, which

FIGURE 8.10 *Tourmaline clots (dark) formed adjacent to quartz veins in granite, Kuah, Pulau Langkawi. Photograph C. S. Hutchison.*

have been found in the Kinta Valley (Ingham and Bradford 1960).

Catazone

The depth of emplacement of granites in the catazone category usually exceeds 11 km. The plutons are completely conformable with the country rocks, which are characteristically metamorphosed to the amphibolite facies of regional metamorphism. The granites are characterized by a gneissic foliation. Augen gneisses and porphyroblastic granites are common in the larger plutons; these, in turn, are usually sheetlike masses completely conformable with the metasediments, with which they share the same structure and with which they are completely harmonious. Migmatization is common in the catazone (Buddington 1959). The catazone granites cause no contact metamorphism, since the emplacement is synkinematic; they are closely associated with regionally metamorphosed rocks of the almandine amphibolite facies.

Rocks of the catazone are exposed in a belt of country approximately 32 km (20 miles) wide and extending southward from the Kelantan–Thailand border from the region of Gunong Kemahang, some 32 km southwest of Pasir Mas. The area includes the granite referred to as the Kemahang Granite (MacDonald 1968), the Taku Schist, and the so-called Stong Injection Complex (MacDonald 1968), centered on Gunong Stong, Kelantan. The belt can be traced to the south, although catazone rocks are not continuously exposed, only appearing here and there along this belt from under rocks of a higher tectonic level. Exposures to the south of Kelantan are to be found at Bukit Berentin in north Pahang, Bukit Ranjut to the northwest of Kuala Lipis, and in the western foothills of Gunong Benom to the south of Benta and to the east of Raub.

In this belt of country, not all intrusions are of the catazone. Synkinematic harmonious intrusions are characteristic, but postkinematic higher level intrusions have cut through the catazone rocks, giving younger unfoliated granites in association with older foliated rocks, as in the Benom area. It is of interest that throughout this catazone belt, the intrusive rocks are associated with gold mineralization but are apparently free from cassiterite (Richardson 1950, MacDonald 1968).

To the west of this tectonic belt, there are sporadic occurrences of rocks which can be attributed also to the catazone. Near Baling, in Kedah, Burton (1972) has ascribed the Kupang Granite to this level of emplacement and the granites of Gunong Jerai can, in part, be ascribed to this tectonic level also (Bradford 1965).

Throughout the catazone rocks, granite gneisses and pegmatites are developed. These synkinematic igneous rocks cannot be adequately described separately from the metamorphic rocks of almandine-amphibolite facies, with which they are intimately associated. Accordingly they are discussed in Chapter 9.

The Kemahang granite, in north Kelantan, has many features of the catazone but is not completely conformable to the structure of the enclosing metamorphic rocks; therefore it is best considered to be transitional between the mesozone and the catazone. This body of parautochthonous granite occupies the northern part of the Taku Schist outcrop close to the Thai border and forms the hills of Bukit Kemahang and Bukit Kusial. The name was coined by MacDonald (1968).

The predominant rock type is a medium to coarse grained grey rock with large feldspar phenocrysts and abundant biotite. The rock is strongly foliated at many localities by the parallel orientation of the feldspar crystals. The western part of the outcrop is intensely sheared and foliated (MacDonald 1968). The composition is equivalent to granodiorite, except in places, where it is of subordinate syenite. Microgranite occasionally occurs as patches and dykes. The granite is often intimately interfoliated with the schist, and numerous apophyses extend into the Taku Schist along the contact (MacDonald 1968). Quartz veining is common.

The feldspar phenocrysts may be up to several centimeters in length. Clinozoisite is a common accessory. The biotite is brown and pleochroic. Occasionally tremolite and actinolite occur. Pyroxene, allanite, and monazite occur rarely but are very localized (MacDonald 1968).

Many varieties are gneissic; some are cataclastic, even mylonitic. Shearing and foliation increases in intensity toward the west (MacDonald 1968) and augen gneiss is fairly common. In the eastern part of the outcrop, the granite is not foliated but is a porphyritic biotite granodiorite, with quartz, oligoclase-andesine, orthoclase, and biotite. Micrographic intergrowth of quartz and feldspar is very common (MacDonald 1968).

Chemistry

A very large number of whole-rock silicate analyses are available on granites of West Malaysia, and the majority of these are given by Alexander et al. (1964). Since that publication, however, a large

TABLE 8.2 *Chemical analyses and Niggli molecular norms of selected granites, adamellites, and granodiorites** from West Malaysia and Singapore*

	Weight Percentage									
	1	2	3	4	5	6	7	8[a]	9	10
SiO_2	77.52	75.60	74.40	73.90	73.40	72.90	71.58	70.61	67.60	63.70
Al_2O_3	12.28	13.03	13.50	13.20	12.90	13.50	12.96	14.01	15.00	15.30
TiO_2	0.08	0.16	0.23	0.20	0.26	0.24	0.66	0.40	0.45	0.57
Fe_2O_3	0.11	0.11	0.86	1.05	1.03	0.62	0.61	0.45	1.04	0.97
FeO	1.22	1.44	0.81	0.93	1.58	1.22	3.12	2.75	3.07	4.06
MnO	0.02	n.d.	0.08	0.07	0.08	0.02	0.15	0.07	0.09	0.10
MgO	0.20	0.21	0.21	0.09	0.41	0.15	0.35	0.90	0.88	2.72
CaO	0.09	0.84	0.82	0.99	1.70	1.34	1.78	1.85	3.02	4.29
Na_2O	3.22	3.58	3.75	3.62	3.30	3.77	3.39	2.75	3.48	3.35
K_2O	4.65	4.17	4.90	4.99	4.45	5.26	5.03	4.93	3.95	2.90
P_2O_5	0.01	0.06	0.10	0.10	0.04	0.04	0.11	0.17	0.12	0.15
CO_2	0.08	n.d.	0.06	0.03	0.03	0.08	n.d.	0.09	0.07	0.07
H_2O^+	0.35	0.57	0.57	0.60	0.60	1.09	0.36	0.98	1.12	1.66
H_2O^-	0.05	0.13	0.11	0.22	0.26	0.03	0.12	0.12	0.10	0.24
	99.88	99.90	100.40	99.99	100.04	100.26	100.22	100.15	99.99	100.08

[a] $BaO = 0.4$, $F = 0.01$, $S = 0.03$, $Cl = 0.01$; less 0.02 in total for F, S, Cl.

	Niggli Molecular Norms									
	1	2	3	4	5	6	7	8[c]	9	10
Quartz	37.45	33.84	29.94	29.80	30.77	26.13	25.64	27.81	21.88	17.23
Corundum	2.02	1.38	1.01	0.39	—	—	—	1.55	—	—
Orthoclase	28.05	25.12	29.27	30.04	26.89	31.56	30.25	29.82	23.86	17.54
Plagioclase	29.91	36.62	37.12	37.27	37.72	38.81	36.37	32.95	45.91	49.38
Ab_xAn_{100-x}	Ab_{99}	Ab_{89}	Ab_{92}	Ab_{89}	Ab_{80}	Ab_{89}	Ab_{85}	Ab_{76}	Ab_{70}	Ab_{62}
Magnetite	0.12	0.12	0.91	1.12	1.10	0.66	0.65	0.48	1.11	1.04
Ilmenite	0.11	0.23	0.32	0.28	0.37	0.34	0.94	0.57	0.64	0.81
Apatite	0.02	0.13	0.21	0.21	0.09	0.08	0.23	0.39	0.26	0.32
Calcite	[b]	—	0.15	0.08	0.08	0.21	—	0.23	0.18	0.18
Hypersthene	2.33	2.56	1.05	0.80	2.38	0.98	3.63	6.07	6.06	12.08
En_xFs_{100-x}	En_{24}	En_{23}	En_{56}	En_{32}	En_{43}	En_{26}	En_{21}	En_{42}	En_{41}	En_{60}
Diopside	—	—	—	—	0.60	1.22	2.30	—	0.09	1.40

[b] Taken as zero because of insufficient CaO in analysis.

[c] Halite = 0.03, pyrite = 0.08.

* Specimens 1–10 as follows (UM = Department of Geology colln, University of Malaya, Kuala Lumpur. GS = Geological Survey colln, Ipoh):

Rock	Location	Specimen No.
1. Coarse-grained granite	Benom area, Pahang	UM194
2. Adamellite	Singapore granite quarry No. 6, Singapore	UM520; UM525
3. Biotite granite	New Jetty, Kuah, Pulau'Langkawi	GS27422
4. Muscovite-biotite granite	Dusun Tua, Ulu Langat, Selangor	GS27429
5. Porphyritic biotite granite	Tampin quarry, 44.5 milestone Tampin–Gemencheh road, Negri Sembilan	GS27909
6. Biotite granite	Western Road, J. K. R. quarry, Pulau Penang	GS27425
7. Porphyritic granite	Kampar road, near Kuala Dipang, Perak	GS06967
8. Porphyritic granite	Near milestone 52.5 Tranum–Gap road, Bentong area, Pahang	GS15072
9. Biotite adamellite	Ulu Sungei Kemaman, near Kampong Ayer Puteh, Trengganu	GS27423
10. Hornblende granodiorite	Ulu Sungei Sat, 2 miles (3 km) Southwest of Gunong Gagau, Pahang	GS27437

number of silicate analyses have been performed at the headquarters of the Geological Survey at Ipoh. Many of these have been made available by the director. A selection of chemical analyses of granites from West Malaysia and Singapore is given in Table 8.2, together with their Niggli molecular norms as calculated on an IBM 1130 computer using the program of Hutchison and Jeacocke (1971). Chemically the granites are normal as can be seen from the frequency distributions of quartz, albite, orthoclase, anorthite, and plagioclase in the computed norms of more than 100 granites (Figure 8.11). The concentration on the quartz-albite-orthoclase and on the quartz-plagioclase-orthoclase diagrams is extremely high, and the maxima coincide very closely with the 650°C minimum on the isobaric diagram for water vapor pressures of 2000 kg/cm² (Tuttle and Bowen 1958, p. 55). This coincidence of the maxima with the thermal low of the experimental isobaric equilibrium system for low water vapor presures is more than fortuitous. There can be little doubt from this evidence that the granites of West Malaysia crystallized from magmatic fluids. Had they been produced by nonmagmatic mechanisms such as solid diffusion, hydrothermal replacements, or by any other nonliquid mechanism, we would not expect the control on the composition to be as strong as is shown in (Figure 8.11). This must not be taken to mean that the final texture results entirely from crystallization from a magmatic phase. Undoubtedly many euhedral feldspar crystals are part porphyroblast, and it is not uncommon to find such porphyroblasts that have grown in the solid state within xenoliths in the granite. However, this does not affect the basic premise that the bulk composition of the granite results from a magmatic process.

The coincidence between the compositional maxima as found from normative quartz-albite-plagioclase-anorthite frequency distribution diagrams and the thermal low of the laboratory-investigated system $NaAlSi_3O_8$-$KAlSi_3O_8$-SiO_2-H_2O demonstrates only that a liquid phase is necessary to explain the compositional variation of the granite (Figure 8.11); it does not tell us whether the granite magma originated from remelting of salic rocks or by fractional crystallization of a primary magma.

It is of interest to consider the relationship between alkalis and alumina in the granites. If more alumina is present than is required to satisfy all the alkalis and lime as feldspars, it will appear as corundum in the norm. If the alkalis are present in excess of the amount required to combine with the alumina as feldspars, then acmite and/or alkali metasilicates will appear in the norm. Table 8.3 shows these relationships in 147 computed norms of granites from Malaya that contain more than 70% normative quartz + albite + orthoclase. No fewer than 81% of the analyzed rocks contain normative corundum. This datum is to be interpreted as implying that the granites crystallized under conditions that permitted alkalis to be expelled during the final stages of crystallization. The deduction is in agreement with the alkali enrichment in hydrothermal dykes and the commonly occurring alkali metasomatic zones associated with these dykes as products of tourmaline greisenization, described previously.

On the other hand, the presence of normative corundum can result from assimilation of pelitic xenoliths by the granite magma, but this cannot account for the high frequency of granites containing normative corundum. Undoubtedly both assimilation and loss of alkalis together can account for the richness in alumina of the granites.

A selection of chemical analyses and computed Niggli norms of modifications of the granite, including aplite, porphyry, microgranite, and pegmatite is presented in Table 8.4.

TABLE 8.3 *Normative corundum and acmite in 147 salic rocks whose normative $Q + Ab + Or > 70\%$*

	With Acmite	No Corundum, No Acmite	With Corundum
Number of analyzed rocks	1	27	119
Mean normative content	1.53%	0	2.80%

Age

STRATIGRAPHIC EVIDENCE. In at least two widely separated localities in West Malaysia, clasts of an older granite have been found in Paleozoic strata: in metatuffs of probable Carboniferous or Permian age on Pulau Nanas in the Straits of Johore (Scrivenor 1931), and in the Carboniferous to Lower Permian Singa Formation in the Langkawi Islands (Jones in prep.).

The granites of West Malaysia are certainly Post-Silurian, since they intrude both the lower part of the Sungei Patani Formation (now called the Mahang Formation, see Chapter 3) in the Har-

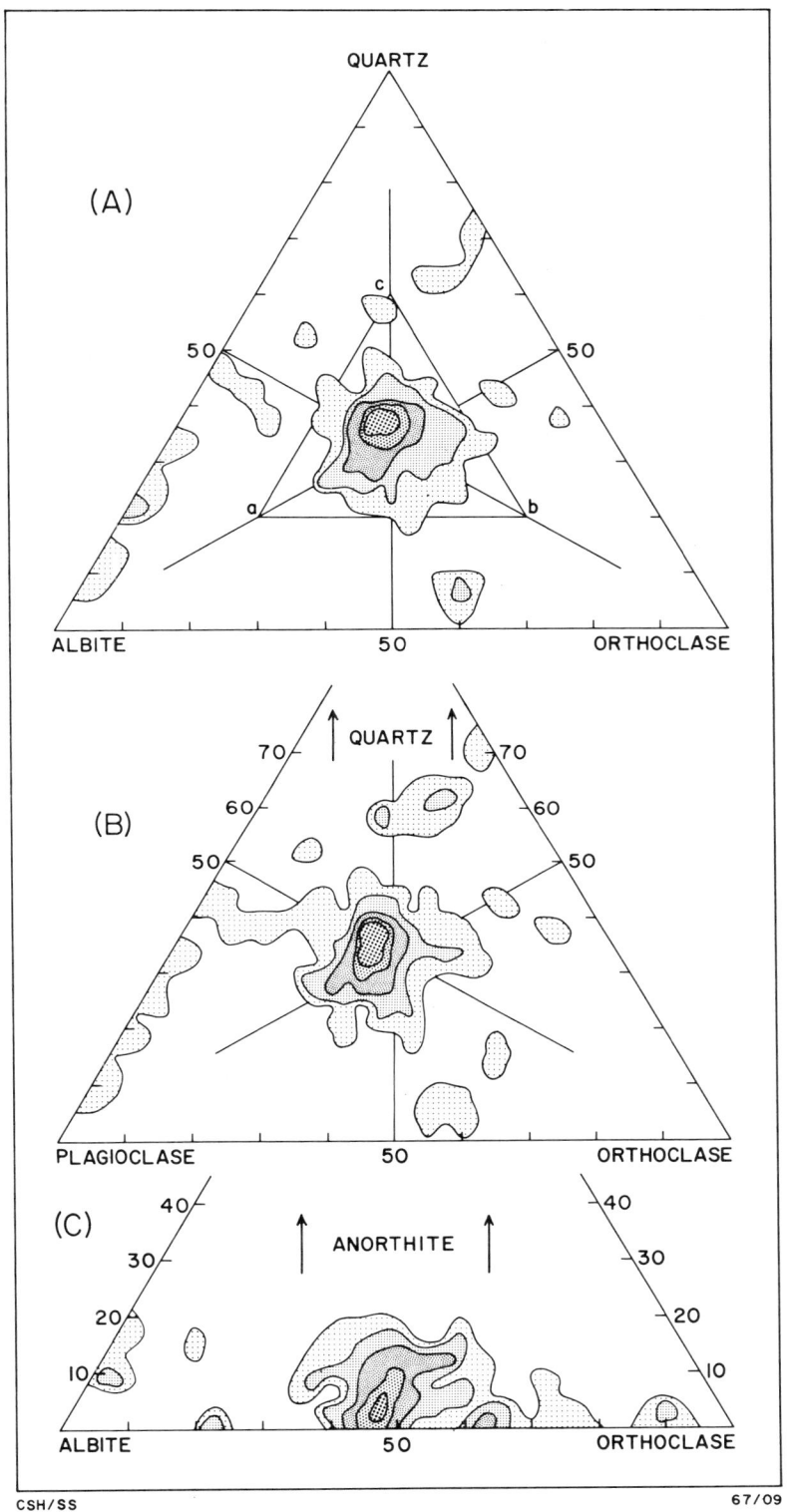

FIGURE 8.11 *Frequency distribution of normative quartz, albite, orthoclase, and plagioclase for 119 analyzed granites and related rocks from the Malay Peninsula containing 80% or more of normative Ab + Or + Q. Contours more than 1, 2, 5, 10, and 15% based on a 0.25% counting area.*

TABLE 8.4 *A selection of chemical analyses and computed Niggli norms of granitic modifications, including aplite, porphyry, microgranite, and pegmatite†*

	Weight Percentages						
	1	2	3	4	5	6	7
SiO_2	76.70	76.55	76.50	75.18	74.26	71.64	65.86
Al_2O_3	12.37	12.78	12.55	12.56	14.83	14.43	15.07
TiO_2	0.23	tr.	0.14	0.38	0.01	0.02	0.61
Fe_2O_3	0.28	0.38	0.08	1.98	0.28	1.13	3.20
FeO	2.76	0.37	1.07	2.58	0.72	0.68	1.90
MnO	0.05	tr.	0.04	0.10	0.02	0.02	0.04
MgO	1.04	0.12	0.08	2.00	0.33	0.23	2.20
CaO	0.13	1.46	0.39	tr.	0.13	2.09	2.65
Na_2O	3.35	2.56	3.55	0.10	4.69	3.08	5.01
K_2O	1.60	5.04	4.35	2.53	4.62	6.02	0.49
P_2O_5	0.07	tr.	0.02	0.07	0.06	0.07	0.12
CO_2	0.34	n.d.	0.14	0.10	0.12	0.10	0.20
H_2O^+	1.28	0.33	0.26	2.65	0.25	0.28	0.68
H_2O^-	0.07	0.22	0.20	0.15	0.07	0.24	0.03
	100.27	99.81	99.37	100.38	100.39	100.03	99.51

n.d. = not determined, tr. = trace.

	Niggli Molecular Norms						
	1	2	3	4	5	6	7
Quartz	45.44	36.93	35.90	60.48	26.05	24.90	25.51
Corundum	5.72	0.51	1.86	11.34	2.20	—	2.51
Orthoclase	9.78	30.51	26.28	16.07	27.22	36.04	2.98
Plagioclase	31.33	30.98	33.54	0.97	42.25	35.09	57.70
Ab_xAn_{100-x}	Ab_{99}	Ab_{76}	Ab_{97}	Ab_{100}	Ab_{99}	Ab_{78}	Ab_{80}
Magnetite	0.30	0.41	0.09	2.23	0.29	1.20	3.33
Hematite	—	—	—	—	—	—	0.08
Ilmenite	0.33	—	0.20	0.57	0.13	0.03	0.87
Apatite	0.15	—	0.04	*	0.13	0.15	0.26
Calcite	*	—	0.36	*	*	0.26	0.52
Hypersthene	6.95	0.66	1.73	8.35	1.84	0.30	6.25
En_xFs_{100-x}	En_{43}	En_{52}	En_{13}	En_{71}	En_{49}	En_{70}	En_{100}
Diopside	—	—	—	—	—	1.22	—

* Taken as zero because of insufficient CaO in analysis.

† Specimens 1–7 as follows (UM = Dept. of Geology colln, University of Malaya, Kuala Lumpur. GS = Geological Survey colln, Ipoh):

Rock	Location	Specimen No.
1. Quartz porphyry	Benom area, Pahang	GS JA.8
2. Aplitic granite	Mandai Quarry No. 13, Singapore	UM968
3. Microgranite	Benom area, Pahang	UM212
4. Quartz porphyry	Benom area, Pahang	GS JA.9
5. Aplite vein	Benom area, Pahang	GS JA.27B
6. Pegmatite	Benom area, Pahang	GS JA.40
7. Pyritized sheared microgranite	Benom area, Pahang	UM192

vard Estate, Kedah (Bradford 1965) and the Kuala Lumpur Limestone at Salak South, near Kuala Lumpur (Gobbett 1965a).

The Bukit Paloh adamellite (No. 23 on Figure 8.1 and Table 8.5) is considered to be typical of the east coast granites. It is definitely intrusive into Lower Carboniferous sediments (Fitch 1952) and is presumed from radiometric evidence to have an Upper Carboniferous age of intrusion (Hutchison and Snelling 1971).

There is strong evidence that the granites of the Main Range are generally Triassic, or Post-Triassic, at least in part. In Kedah, Triassic rocks, dated by fossils, have been mineralized with tin and tungsten ore by the granite and have been enriched in tourmaline (Scrivenor 1931).

However, the general observation that Triassic rocks are folded is not in itself evidence that the granite is Post-Triassic. Scrivenor (1931) attempted to prove a Post-Triassic age for the granite by noting that in Singapore, Triassic fossils are found in beds tilted at a high angle away from the granite. Hutchison (1968c) has shown that this kind of evidence is not acceptable and that, in all probability, the Triassic rocks of Singapore are postgranite.

The simple rule that folded rocks must predate the granite intrusion was established when it was thought that the orogenic evolution of the Thai–Malayan orogen consisted of a single folding period, followed by a single granite intrusion, followed by uplift and erosion, followed by an unconformable postorogenic sequence of undisturbed rocks. Such reasoning can no longer be tenable.

An upper stratigraphic limit is set on some granites of the eastern part of Malaya by the Tembeling Formation. Whose basal conglomerate, the Murau Conglomerate, contains clasts of weathered granite (Koopmans 1968). Postorogenic unfolded conglomerates of the Gagau (Paton 1959) and Panti Formations contain clasts of the underlying granite.

From stratigraphic evidence, then, we know that the major granite intrusions of West Malaysia are definitely Post-Silurian and Pre-Upper Jurassic.

Granite clasts in the Carboniferous Singa Formation in the Langkawi Islands (Jones, in prep.) indicate the existence of a Pre-carboniferous granite, which has never been identified *in situ*. The low thermoluminescence ages of limestones in the Kinta Valley and in the Kuala Lumpur areas, as compared with the thermoluminescence ages of similar limestones in Perlis and Kedah, strongly suggest the existence of Tertiary magmatic events (Hutchison 1968a).

RADIOMETRIC EVIDENCE. Radiometric dates of 43 granites and related rocks from West Malaysia and Singapore (Snelling 1965, 1967, Anon 1966) clearly show that many granites must predate the folded Triassic sedimentary rocks and that some must postdate the gently dipping, relatively undisturbed postorogenic Gagau Formation. It is therefore necessary to recognize that granite emplacements in the Malay Peninsula extended from at least the Upper Carboniferous until the Lower Tertiary. Details of the radiometric determinations age are given in Table 8.5 and the localities referred to appear in Figure 8.1. The frequency distribution of these radiometric dates (Figure 8.12) implies that the intensity of granite emplacement gradually increased to a maximum at the end of the Triassic. This high intensity of granite dates coincides with the important unconformity between the Triassic and the overlying Tembeling Formation (Koopmans 1968, Hutchison 1968c).

A later maximum of granite dates coincides with the unconformity between the folded Tembeling Formation and the unfolded Gagau Formation. However, it is extremly doubtful if these radiometric maxima, correlating with the two spectacular unconformities, actually represent magmatic events. It seems more likely that they result from cooling of the already crystallized granite resulting from uplift associated with the unconformities. Armstrong (1966) has reasoned that radiometric dates of igneous and metamorphic rocks in an orogen would generally be expected to give young values, and that the date determined represents the time that has elapsed since the rock was uplifted, and thereby cooled. A rock will not begin to acquire a potassium-argon age until it has cooled to below about 200°C, the temperature at which it begins to retain its argon in the crystal structures of the micas (Lambert 1964). Hurley et al. (1962) spectacularly demonstrated this by obtaining radiometric dates of only 4 m.y. for ancient metamorphic rocks in New Zealand. What was being dated was not the age of the rocks but the age of their uplift along the Alpine Fault System.

Accordingly, radiometric dates on orogenic granites are more likely to represent phases of uplift than actual magmatic events. It is not, therefore, surprising that the maxima of radiometric determinations on the frequency distribution diagram (Figure 8.12) agree so closely with major unconformities, since unconformities represent uplift and hence cooling.

It would be a mistake to assume a simple correlation between the actual magmatic age of any granite

TABLE 8.5 *Radiometric dates of granites and related rocks from the Malay Peninsula**

Ref. on Figure 8.1	Specimen No.	Nature and locations of samples	Age (m. y.) K–Ar	Age (m. y.) Rb–Sr	Age (m. y.) Other
1	GS 27422 (UM5398)	Biotite granite, Kuah, Langkawi, N6°18′20″ E99°51′20″	B 242 ± 10	WR, KF 218 ± 11	—
2	GS 27919 (UM5371)	Porphyritic biotite granite, Sungei Sok, Kedah, N6°2′15″ E100°48′30″	B 190 ± 10	—	—
3	GS 27413 (UM5392)	†Foliated biotite-muscovite granite, Gunong Jerai, Kedah, N5°44′20″ E100°25′00″	B 47 ± 3 M 59 ± 3 (135 ± 6)	—	—
3A	GS 27414 (UM5393)	Muscovite-tourmaline pegmatite, Gunong Jerai, Kedah, N5°44′20″ E100°25′00″	M 135 ± 6	—	—
4	GS 27415	Muscovite-columbite pegmatite, Semeling, Kedah, N5°43′20″ E100°29′00″	M 135 ± 6		
@ 5	GS 27411 (UM5390)	Biotite-gneiss, Kupang, Kedah, N5°38′30″ E100°51′00″	B 150 ± 8		
6	GS 27425 (UM5399)	Biotite granite, Western Road, Penang Island, N5°26′00″ E100°17′40″	B 67 ± 4	WR, KF 197 ± 10 B 180 ± 6	—
7	GS 27426 (UM5400)	Biotite-muscovite granite Ayer Hitam, Penang Island, N5°22′30″ E100°16′30″	M 188 ± 8		—
8	GS 27914 (UM5370)	Porphyritic pink granite, Penanti, Penang mainland, N5°24′40″ E100°28′20″	B 196 ± 8		
9	GS 27412 (UM5391)	Biotite granite, Karangan, Kedah, N5°24′15″ E100°41′15″	B 190 ± 10 M 180 ± 10		
@10	GS 27073 (UM5408)	Garnet-mica schist, Sungei Kenik, Kelantan, N5°24′45″ E102°05′15″	B 220 ± 8 M 210 ± 8		
@11	GS 27055 (UM5366)	Biotite gneiss, Sungei Anali, Kelantan, N5°24′15″ E102°12′30″	B 215 ± 8		
12	GS 27065	Medium-grained granite, Bukit Yong, Kelantan, N5°44′30″ E102°26′15″	B 140 ± 8		
13	GS AM2	Monazite sand from stream concentrate, Sungei Senang, Trengganu, N5°00′30″ E102°55′30″			U 149 ± 15
13A	GS 27420	Biotite granite, Kuala Brang Trengganu, N5°02′00″ E102°55′45″		B 212 ± 20 KF 225 ± 20	
14	GS 27424	Porphyritic biotite-tourmaline granite, Reservoir quarry, Taiping, Perak, N4°51′45″ E100°45′45″		WR, KF 200 ± 8	
15	GS G3	Galena in limestone, Chemor, Perak, N4°46′00″ E101°06′00″			Pb 144
15	GS G4	Galena in limestone, Chemor, Perak, N4°46′00″ E101°06′00″			Pb 243
16	GS L1	Lepidolite from pegmatite, Gopeng, Perak, B4°28′00″ E101°9′30″		183 ± 8	
17	GS 27430	Porphyritic biotite granite, Kuala Dipang, Perak, N4°23′00″ E101°11′20″	B 232 ± 10		
18	GS AM1	Alluvial monazite, Pulau Attap, Kinta Valley, Perak, N4°18′00″ E101°5′00″			U 175 ± 10
19	GS 27057	Porphyritic biotite granite, Talam quarry, Kampar, Perak, N4°17′30″ E101°10′25″	B 188 ± 8		SS 211

(Continued)

TABLE 8.5 Continued

Ref. on Figure 8.1	Specimen No.	Nature and locations of samples	Age (m. y.) K–Ar	Age (m. y.) Rb–Sr	Age (m. y.) Other
20	GS L2	Lepidolite in pegmatite, Chenderiang, Perak, N4°16′00″ E101°14′00″		186 ± 10	
22	GS 27423	Biotite adamellite, Ulu Sungei Kemanan, Trengganu, N4°17′15″ E103°13′00″	B 260 ± 10		
23	GS 29858 (UM3963)	Nonporphyritic biotite granite, Bukit Paloh, Pahang, N4°03′45″ E103°03′00″	B 240 ± 8	WR 320 ± 60 KF 225 ± 35 B 188 ± 5	
24	GS 27933	Sheared porphyritic granite, Kalumpang, Selangor, N3°38′45″ E101°35′45″	B 142 ± 6		
25	GS 27429 (UM5402)	Muscovite-biotite granite, Dusun Tua, Ulu Langat, Selangor, N3°08′20″ E101°50′00″	B 130 ± 6 M 185 ± 8		
26	GS 27421 (UM5397)	Biotite granite, Bukit Ubi, Kuantan, Pahang, N3°49′00″ E103°19′00″	B 215 ± 8		
27	GS 27436	Sheared granite. Tanjong Ipoh quarry, Kuala Pilah road, Negri Sembilan, N2°44′20″ E102°10′30″	B 180 ± 8		
28	GS 27418 (UM5396)	Pink granite, Bukit Tunggal quarry, Malacca, N2°24′15″ E102°24′45″	B 72 ± 4	KF 87 ± 30 B 35 ± 3	
29	GS 27910 (UM5369)	Pink granite, Selendar–Nyalas road 25 milestone, Malacca, N2°25′40″ E102°26′00″	B 70 ± 5		
30	GS 27428 (UM5401)	Pink granite, Mount Ophir quarry, Johore, N2°21′40″ E102°39′40″	B 52 ± 4		
31	GS 25942 (UM5405)	Porphyritic biotite granite, Bukit Mor quarry, Johore, N1°58′30″ E102°40′30″	B 135 ± 6	WR 291 ± 60 KF 250 ± 30 B 157 ± 6	SS 190
31A	GS 25944 (UM5365)	Muscovite from pegmatite, Bukit Mor quarry, Johore, N1°58′30″ E102°40′30″			SS 191
32	GS 27928 (UM5372)	Granite. Batu Pahat quarry, Johore, N1°49′30″ E102°57′00″	B 196 ± 8		
33	GS 27435 (UM5368)	Hornblende granodiorite, Bukit Batu quarry, Ayer Hitam, Johore, N1°43′30″ E103°27′40″	B 224 ± 9		
34	GS 27416 (UM5394)	Pink granite, Gunong Pulai Estate, Johore, N1°32′40″ E103°34′00″	B 70 ± 4	WR 108 ± 60 KF 83 ± 30 B 14 ± 2	
35	GS 27417 (UM5395)	Granite, Bukit Lanchu, Johore, N1°30′45″ E103°50′45″	B 214 ± 10 B 218 ± 10	WR 230 ± 20 KF 215 ± 10 B 208 ± 6	
36	GS 27929	Granite, Bukit Timah quarry, Singapore, N1°21′40″ E103°46′30″	B 224 ± 9		
38	GS 27437 (UM5404)	Hornblende granodiorite, Ulu Sungei Sat, 2 miles southwest of Gunong Gagau, Pahang, N4°45′10″ 102°37′45″	H 200 ± 8		
40	GS 27074 (UM5407)	Muscovite microgranite, Sungei Baru, Malacca, N2°21′20″ E102°4′15″			SS 201
41	GS 27075 (UM5367)	Porphyritic biotite granite, Simpang Ampat, Malacca, N2°26′30″ E102°12′15″			SS 164
42	GS 27076 (UM5406)	Biotite granite, The Gap, Fraser's Hill, Pahang, N3°43′30″ E101°46′30″			SS 238

For footnotes to table, see page 238.

and the radiometric date determined for one of its constituent minerals. As an illustration of this, the biotite and the muscovite from a foliated granite in Kedah (locality 3 on Figure 8.1 and Table 8.5) gave dates that can be ascribed to the Lower Tertiary. These dates clearly cannot represent a magmatic event, since muscovite from a cross-cutting pegmatite at the same locality gave a date of 135 ± 6 m.y. In this instance, we can conclude that the Lower Tertiary date for the granite results from a Lower Tertiary period of shearing, or diastrophism, which caused the fine-grained mica crystals to lose their radiogenic argon while the large muscovite crystals in the pegmatite retained their argon throughout the diastrophic event (Snelling 1965). Corroborative information of the Lower Tertiary diastrophism comes from the Eocene age of secondary calcite veining in the limestone of Perlis and Kedah, as determined by the thermoluminescence method (Hutchison 1968a).

But is it then scientifically sound to accept the 135 ± 6 m.y. as representing a magmatic event? I feel that there is no basis for so doing. I believe that the main granite magmatism in the Malay Peninsula is represented by the lower maximum of the frequency distribution curve (Figure 8.12), that is, in the Upper Carboniferous to Middle Triassic. This conclusion is justified because, where a potassium-argon date has indicated an Upper Jurassic age (e.g., locality 31 of Figure 8.1 and Table 8.5), the Rubidium-Strontium dates on the same rock strongly suggest a much older age—in this case, even as old as Permian There seems therefore to be a good reason for considering that the majority of granites in the Malay Peninsula were intruded in a period that extended from the Upper Carbonif-

---

Footnotes to Table 8.5

* Symbols used as follows:

Underline	Recommended age, where discrepancy exists in dates.
H	Determination on hornblende.
B	Determination on biotite.
M	Determination on muscovite.
WR	Determination on whole rock.
KF	Determination on alkali feldspar.
†	This foliated granite is cut by pegmatite (No. 3A). It is presumed to have lost argon during a Tertiary tectonic event.
@	Metamorphic rock.
U	Determination by uranium, thorium–lead method.
Pb	Determination by lead method.
SS	Determination by $^{87}Sr/^{86}Sr$ ratio.
GS	Geological Survey colln, Ipoh.
UM	Dept. of Geology colln, University of Malaya, Kuala Lumpur.

Since the compilation of the table, the following radiometric information has been given by J. D. Bignell (personal communication):

Locality	Rock	Method	Age (m. y.)
Papan quarry, Perak, 14.5 km (9 miles) west of Location 16	Granite	K–Ar	203
		Rb–Sr	200
Ampang new village, 13 km (8 miles) north of location 16	Pegmatite	K–Ar	200
Kuala Dipang quarry, Perak, Locality 17	Fine-grained granite	K–Ar	200
	Coarse-grained granite	K–Ar	230
Kampar, Perak, near Locality 19	Granite	K–Ar	180–200
Cameron Highlands road, north-northeast of Locality 21	Granite	Rb–Sr	200
Kuantan, Pahang, near Locality 26	Basalt	K–Ar	1.7
Kuantan, Pahang, near Locality 26	‡Dolerite dyke	K–Ar	110

‡ The dolerite dykes appear to be related to the Cretaceous lavas, which are associated with the Gagau Group, rather than to the Kuantan basalt, which appears to be Quaternary.

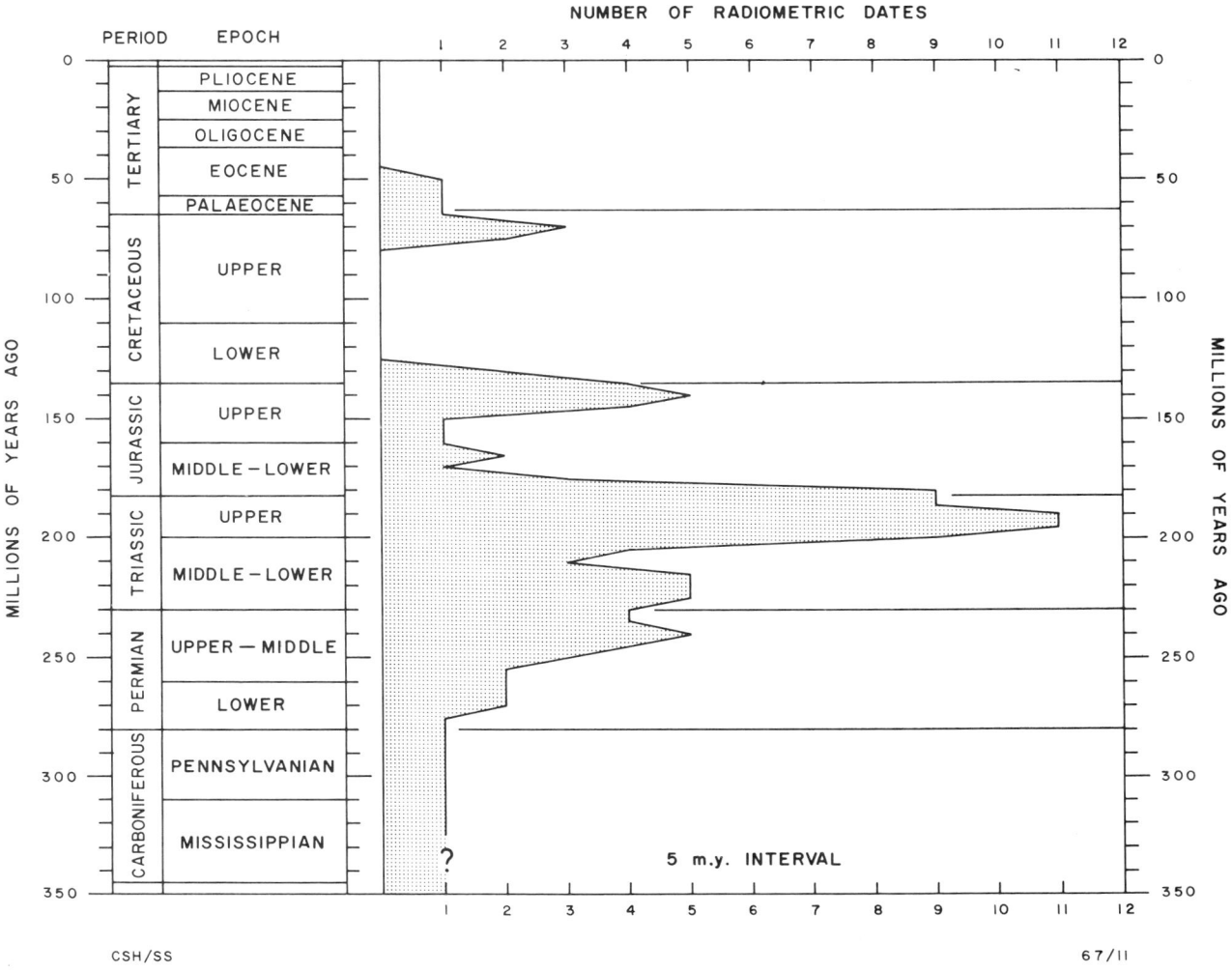

FIGURE 8.12 *Frequency distribution of the radiometric ages listed in Table 8.5.*

erous through the Permian and Triassic, and that the prominent maximum of dates in the Upper Jurassic–Lower Cretaceous actualy represents a major uplift prior to the deposition of the Gagau Formation.

This brings us to the maximum of dates in the Upper Cretaceous–Lower Tertiary (Figure 8.12, Table 8.5). In this instance, there is strong evidence to indicate that the dates represent an actual magmatic event. Similar potassium-argon and rubidium-strontium dates have been determined for the same granite sample (e.g., locality 28 in Figure 8.1 and Table 8.5). Further confirmation of this conclusion is afforded by the consistent petrographic difference between those granites, which have given dates younger than 100 m.y., and all the others. The younger granites are all pink, whereas the older granites are grey. Consistent potassium-argon and rubidium-strontium dates, combined with the dis-

tinctive appearance and the fact that the specimens so determined are confined to outcrops along the extreme western margins of the main batholith system, all conspire to indicate that localized postorogenic granite intrusions did take place in the peninsula in Upper Cretaceous to Lower Tertiary time. The intrusions must be considered to be postorogenic because they are younger than the Gagau molasse rocks. Their consistent potassium-argon and rubidium-strontium dates are also more readily understandable if they are assumed to be postorogenic, having cooled more rapidly in a high-level environment than the syntectonic orogenic granites (Armstrong 1966).

The upper maximum in the frequency distribution diagram (Figure 8.12) is therefore very likely to represent a truly magmatic event, rather than any tectonic event.

In the light of more recent rubidium-strontium

radiometric work, which is not yet published in detail, Snelling et al. (1968) argued that the granite intrusions in Malaya may be fitted into three distinct phases as follows:

1. *Upper Carboniferous,* between 280 and 300 m.y. Fifteen specimens have given ages within this range. Most have come from the east side of the Malay Peninsula, but Late Carboniferous granites occur also in the west and in Thailand. No potassium-argon ages of this magnitude have been determined (see Table 8.5) showing the close relation between potassium-argon ages and tectonism.

2. *Middle Triassic,* between 200 and 230 m.y. Eleven specimens have given ages within this range. Most of the specimens are from the Main Range.

3. *Late Cretaceous,* about 70 m.y. Four specimens have given this age. These appear to be localized postorogenic plutons.

The rubidium-strontium work of Snelling et al. has shown some particularly interesting features:

1. No potassium-argon ages of Upper Carboniferous have survived.

2. No rubidium-strontium whole rock ages of Upper Triassic to Lower Jurassic (ca. 180 m.y.) have been obtained, despite the high concentration of potassium-argon ages in this region (Figure 8.12). Hence this concentration of potassium-argon ages around 180 m.y. is probably related to diastrophism or regional uplift.

3. No rubidium-strontium whole rock ages have been found in the region 135 m.y. Snelling et al. (1968) suggested that this potassium-argon maximum (Figure 8.12) indicates Pre-Gagau uplift. But granite intrusion of this age is not definitely ruled out. It is of interest that two specimens of granite from peninsular Thailand gave ages of 110 m.y., but no granites of this age have been found in Malaya (Figure 8.12).

Orogenic evolution and epeirogenic uplift have caused considerable redistribution of the nuclides of interest in radiometric dating, and thus interpretation of the radiometric evidence is by no means simple.

However, in the general picture that emerges, it would appear that the granites of Malaya east of the Main Range are generally of Upper Carboniferous age, the Main Range granite is generally of a Middle Triassic age, and there are limited amounts of localized postorogenic granite plutons of Upper Cretaceous age.

Correlation with Thailand

As in West Malaysia, granite is by far the most extensive igneous rock outcropping in Thailand. The earliest evidence of granite in Thailand is as clasts in the Carboniferous and Lower Permian Phuket Series. Klompé (1962) suggested it may be of Precambrian age, but it could equally well be Paleozoic. Granite clasts also occur in the Singa Formation of the Langkawi Islands (see Chapter 4). But none of these earlier granites has been definitely identified *in situ.* As in West Malaysia, the granite ranges trend in a general north-south direction, apparently occupying the cores of anticlinal structures, representing the upper parts of a huge batholith (Klompé 1962). Many of these ranges can be traced continuously southward into West Malaysia (Figure 8.13).

There is considerable stratigraphic evidence to indicate that the granite intrusions of Thailand are of two distinct geological ages; indeed, they differ slightly from one another mineralogically (Brown et al. 1951). The older intrusion is commonly of hornblende-biotite granite, with hornblende usually more abundant than biotite. The older granite is associated with gold and is apparently not associated with tin. It has been assigned a tentative Triassic age because it intrudes Paleozoic rocks and is unconformably overlain by rocks of the Khorat Series in at least three widely spaced localities (Brown et al. 1951). The basal conglomerate of the Rhaetic–Jurassic–Cretaceous Khorat Series contains clasts of this granite (Klompé 1962). This granite correlates, in part at least, with the catazone granite of West Malaysia. In particular, the northward extension of the Kemahang Granite into Thailand is shown on the 1:2,500,000 geological map (Brown et al. 1951) as belonging to the older granite.

The younger granite is usually a biotite-muscovite one, with muscovite usually more abundant than biotite. Hornblende is generally present, but of subsidiary importance. Zircon, apatite, and tourmaline are common accessory minerals (Brown et al. 1951). Almost all the tin and tungsten deposits of Thailand are associated with these younger intrusions. They have intruded and metamorphosed the Khorat sedimentary rocks in the Malay Peninsula. This younger granite has therefore been assigned a Late Cretaceous age (Brown et al. 1951), and a recent radiometric date obtained from a muscovite granite in Phuket, southwest Thailand, tends to confirm the Late Cretaceous age, at least for some of the granites (MacDonald 1968).

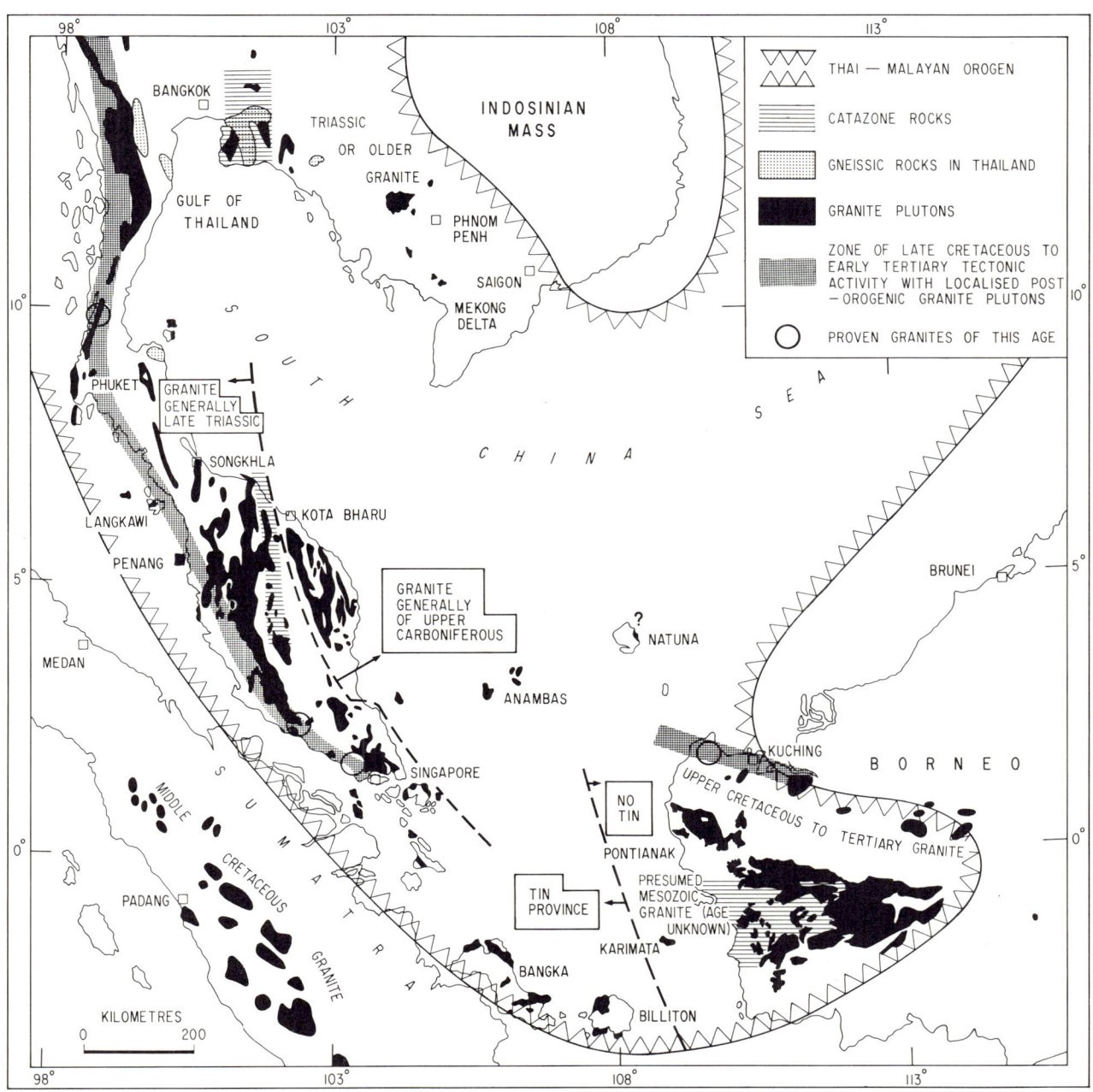

FIGURE 8.13 *Generalized map of Southeast Asia to show the distribution and ages of granite plutons. Note that tin mineralization is absent east of Billiton.*

Generally the granites of central and east Thailand are of hornblende-biotite granite, free of tin, and are considered to be Triassic (Brown et al. 1951); the granites in the peninsula and in western Thailand are considered to be Upper Cretaceous. There are, however, several exceptions to this generalization, as can be seen in the 1:2,500,000 geological map (Brown et al. 1951). It would appear that the age subdivisions of the Thai granites on this map are highly speculative, with the result that no direct correlation of granites can be made across the Malaysian–Thai border. Nevertheless, in general the two ages that have been ascribed to the Thai granites are in exact agreement with those inferred from West Malaysia. From the radiometric and stratigraphic evidence in West Malaysia, however, it seems fairly certain that the postorogenic Upper Cretaceous granites of Thailand have been attributed an excessively wide distribution.

Burton and Bignell (1969) conclude that the granites of the northeastern coast of the Gulf of Thailand, together with those of peninsular Thailand, are Triassic or older. They maintain, on radiometric evidence, that Late Cretaceous to early Tertiary granites are confined to the western margins of the Thai–Malay peninsula. Again they confirm that young potassium-argon dates in this region generally reflect epeirogenic uplift rather than granite intrusion and that magmatism was rather restricted during Upper Cretaceous to Lower Tertiary time.

Lasserre et al. (1968) have reported Upper Permian granite ages for regions of eastern Cambodia, indicating that the granites of this whole region are generally Triassic or older.

Correlation with Indonesia

Granites of similar mineralogical composition, associated with tin deposits, occur on the strike of the granites of the Malay Peninsula in the Indonesian islands of Singkep, Bangka, and Billiton and also on the easternmost parts of Sumatra (Klompé 1962, Hutchison 1968c; Figure 8.1).

In Billiton Island, the plutonic rocks can be classified (Aleva 1960) into: gabbroic rocks, hornblende rich; granodiorite rocks, with biotite and hornblende; adamellitic rocks, conspicuously porphyritic with biotite and hornblende; and granitic rocks (which are the most common), both porphyritic and nonporphyritic, and pegmatitic, associated with greisen. The porphyritic varieties have a fine-grained groundmass, as have the rocks of Singapore to the north (Hutchison 1964a), and the large feldspar crystals are commonly ornamented.

The more reliable of the radiometric dates on the granite in Billiton, as reported by Schürmann et al. (1956, 1957, 1960) are:

180 ± 5 m.y. for biotite.
155 ± 6 m.y. for feldspar.

More recently, Edwards and MacLaughlin (1965) have reported dates of 205 ± 7 and 201 ± 7 m.y. for alkali feldspar and biotite, respectively, from the same rocks. The use of the Billiton granite to set the base of the Jurassic in the time scale (Kulp 1961, Holmes 1959, Harland et al. 1964) has been critized by Hutchison (1968c). The radiometric evidence favors an age for the Billiton granite of Middle to Upper Triassic, rather than basal Jurassic, or even Upper Jurassic, as has been taken by Klompé (1962). The Triassic age is more in agreement with the radiometric dates for the granites of the southernmost part of the Malay Peninsula (localities 35 and 36 of Figure 8.1), to which the Billiton granite is nearest along the regional strike.

Klompé (1962) included the western part of Borneo in the same Indosinian–Thai–Malayan igneous province; with this I disagree for the following reasons:

1. The granite in Borneo is not associated with tin, as it is throughout the province,
2. Radiometric evidence indicates that there are no granites older than Cretaceous in western Borneo (Table 7 in Kirk 1968).

BASIC AND ULTRABASIC ROCKS

Several small outcrops of gabbroic rocks occur in the Malay Peninsula. The two largest known occurrences are in Singapore Island, where they form Bukit Panjang and Bukit Gombak (Hutchison 1964a), and in south Johore in the Senai and Linden Estates, a few miles north of Johore Bahru, near Skudai (Hutchison 1966a). Other small occurrences have been reported on the Tembeling River, in Pahang, below Pasir Siur; near Segamat in north Johore, on the Sungei Segamat; on the Sungei Simat in Negri Sembilan and at Kuala Pasir Alor on the Rompin River in Pahang (Scrivenor 1931).

Small bodies of serpentinite has been found at Bersiah, Upper Perak; Sungei Mas, Pahang; Kuala Siah, Ulu Lipis, Pahang; Kuala Pilah, Negri Sembilan, and in several localities in the country between Kuala Pilah and Tampin; near Durian Tipis, Negri Sembilan; and on the Pertang Road, Negri Semilan (Scrivenor 1931). The main occurrences of serpentinite are as elongated bodies in the eastern flanks of the Main Range. The two largest bodies are in Pahang on the Sungei Telom, approximately 48 km (30 miles) northwest of Kuala Lipis, and in the Sungei Cheroh, the Sungei Chembatu, and the Sungei Batu just to the west of Cheroh village, 13 km (8 miles) north of Raub.

A large body of diorite occurs 5 km (3 miles) east of Kampong Pasir Aka, Trengganu.

The age of these basic bodies and ultrabasic bodies is not precisely known, but they are considered to be preorogenic. A specimen of eucrite from Linden Hill, south Johore failed to give a potassium-argon date on the pyroxene because of the low potassium content. Brown et al. (1951)

described several occurrences in Thailand of basic bodies which are of gabbro, diorite, and pyroxenite with local serpentinization. Their age relations with the granite are uncertain, but they have been ascribed to the Triassic. Klompé (1962, p. 291) has postulated that they ". . . should be considered as representatives of an initial stage of magmatic activity, occurring during the geosynclinal phase of a Mesozoic orogenic cycle." Richardson (1939, 1950) ascribed the variety of igneous rocks in Pahang to assimilation by the granite of xenoliths of these earlier basic and ultrabasic rocks.

The basic and ultrabasic bodies are generally satellitic to the granite batholiths. The gabbro body in Singapore (Hutchison 1964a) is stocklike and is situated just within the edge of the granite batholith. There is good evidence that it is preorogenic and was possibly intruded among the Paleozoic geosynclinal sediments. Alternatively it may be early orogenic. Both interpretations agree with its deduced pregranite age.

Diorite

Rocks of dioritic composition have been described by Dawson (*in* MacDonald 1968) from north Trengganu. They comprise one large stocklike intrusion and numerous smaller bodies. Many, however, have been metamorphosed and are better described as metadiorite. Occurrences are at Bukit Kenuak, Bukit Dara, Bukit Rambutan, Sungei Setiu, and Bukit Titir. Metadiorite is quarried at the J.K.R. quarry at the 59.5 milestone on the main road from Kuala Trengganu to Kota Bharu, at Batang Geraji in the Bukit Rambutan mass. At this locality, metadiorite intrudes chert, rhyodacite, and hornfels (MacDonald 1968).

The diorite contains chalcedony-filled amygdales and occasional small xenoliths of andesite tuff and of porphyritic diorite ranging up to 20 cm in diameter. Porphyritic diorites are fairly common. They have colorless to honey-colored phenocrysts of pyroxene, as well as intermediate plagioclase in varying proportions. The groundmass has a microcrystalline trachytic texture composed of microlaths of plagioclase. None of these rocks are true diorites but are better described as microdiorite, diorite porphyries, or even andesite.

In places, the so-called porphyritic diorite has been reported to contain xenoliths (?) (perhaps lapilli) of yellow glass and xenoliths (perhaps bombs) of microcrystalline andesite. This evidence has been interpreted as suggesting that the dioritic intrusions are high level and in places may have come to the surface as andesite flows. MacDonald (1968) has suggested that the diorites are genetically related to the close-by andesitic lava flows. Therefore, in north Trengganu it appears that it is possible to find a variety of rocks that are transitional from intrusive diorites to extrusive andesites. Their intrusive nature is well authenticated by field evidence, as, for example, at the Sungei Ima and the Sungei Setiu, where they occur as small dykes, sills, and irregular masses intrusive into the country rocks.

Gabbro and Eucrite

In Singapore, the gabbro crops out over an area approximately 1.5 km (1 mile) broad by 5 or 6 km (3 or 4 miles) long (Figure 8.14). It forms the long low hills of Bukit Gombak and Bukit Panjang, which lie parallel to the main Singapore–Johore road between the villages of Bukit Timah and Bukit Panjang. It is well exposed in quarries 1, 2, 3, 10, 11, and 12 (Figure 8.14), which are accessible only from the west side of the hills, either from the Jurong road or the Choa Chu Kang road.

The body is rather variable in composition and changes imperceptibly from gabbro to norite even in the same quarry face. Generally, however, its composition is noritic, gabbro being less abundant and largely confined to quarries 12 and 13.

The western flanks are unmetamorphosed and preserve their original gabbroic texture. Typically the rock has the following modal composition:

Locality: Gammon quarry no. 2
Number of points counted: 5257 Volume %

Labradorite, $Ab_{35}An_{65}$	62
Hypersthene	16
Augite	12
Quartz	4
Magnetite and ilmenite	0.5
Sericite after feldspar	0.3
Biotite after pyroxene	0.2
Actinolite after pyroxene	5

The analysis of this quartz norite indicates that there has been a slight retrogressive metamorphism caused by the granite intrusion or, alternatively, caused by autometamorphism. This has resulted in the partial replacement of the pyroxene by amphibole and mica, and of the feldspar by sericite.

Evidence from quarry 10 suggests that the body is composite. Here there are several sill-like layers approximately 1.8 m thick. Their margins are chilled against the gabbro, and their interiors are

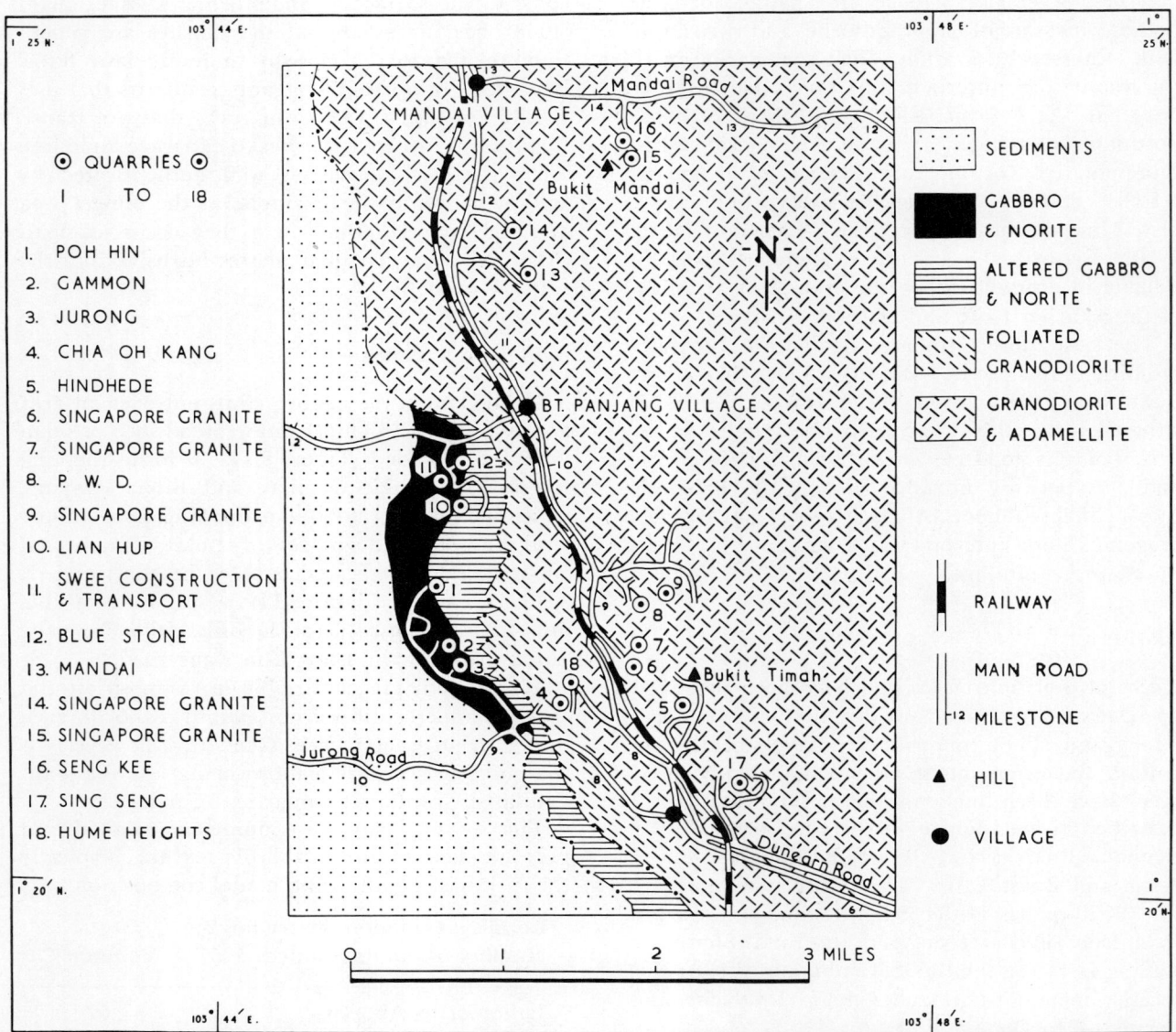

FIGURE 8.14 *Geological map of the southwest margin of the Singapore granodiorite. showing the outcrop of the basic rocks. After Hutchison (1964a, fig. 2).*

petrographically similar to the host rock. These sills have an average east-west strike and dip at 67°N. A major set of joint planes parallel to these sills is seen in quarries 10, 11, and 12. These joints are less common in quarries 1, 2, and 3 to the south, and they are not present at all in the adjacent granite batholith.

Norite and Gabbro occur also near the south coast of Pulau Ubin. About 24 km (15 miles) north of this locality, another satellite body outcrops in the Linden and Senai Estates, south Johore (Hutchison 1966a). Here the rock is considerably more basic and from a mineralogical point of view may be classified as eucrite. The outcrop on Linden Hill is largely free from metamorphism and the following modal composition is typical:

Locality: Linden Hill, south Johore
Number of points counted: 6856   Volume %

Bytownite, $Ab_{12}An_{88}$	50
Hypersthene	8
Augite	29
Olivine	6
Magnetite and ilmenite	1
Hornblende after pyroxene	6

The mineralogical variability of the south Johore eucrite body is shown in the modal analyses variation of olivine, from zero up to as much as 8%, and in the composition of the plagioclase from An_{82} to An_{88}. The relative proportions of augite and hypersthene also vary.

Detailed studies on the other basic bodies in the Malay Peninsula are not available, but the materials are apparently variable in composition from gabbro to norite.

Serpentinite

The largest body of serpentinite extends for a distance of 11 km (7 mles) north-northwest from a locality about 1.5 km (1 mile) north of the confluence of the Sungei Satak Liang and the Sungei Liang, near Cheroh, in the district of Raub, Pahang (Richardson 1939). Throughout this belt, the average width is approximately 1 km (0.75 mile). Several smaller elliptical masses occur close to it. In this area, the serpentinite is associated with schist, phyllite, and metaconglomerate; it is well exposed in the Sungei Chembatu. On its western margins, it is in contact with the granite. Lenses of phyllite and quartz-mica schist occur within the serpentinite.

The serpentinite is well jointed and varies in color from pale yellowish green, to dark green to black. In places the coloring is mottled. The rock is strongly sheared parallel to the regional foliation of the surrounding rocks. The black color is due to an abundance of magnetite. Secondary veining of asbestiform chrysotile is common. Talc is rare. Generally the serpentinite is massive or fibrous and devoid of all traces of an original texture, but obscure relics of completely pseudomorphed olivine and pyroxene occur locally (Richardson 1939).

The accessory minerals include magnetite, ilmenite, chromite, picotite (?), pyrite, pyrrhotite, tremolite, chlorite, and magnesite. The black oxides are euthedral and occur as streaks and lenticles roughly arranged parallel to the foliation of the serpentinite. They cause color banding where they are sufficiently abundant. The serpentinite weathers readily to a characteristic dark brownish-red laterite; good exposures occur in the Raub area in the Sungei Chembatu, the Sungei Cheroh and the Sungei Mas.

Other occurrences of serpentinite in West Malaysia have not been described in detail. The serpentinite at Bersiah in Upper Perak contains relics of olivine (Scrivenor 1931). All known occurrences are well foliated and sheared. It seems clear, therefore, that they have been derived from basic intrusions that have been regionally metamorphosed (Willbourn 1933). This is in agreement with the general conclusion of Miyashiro (1966) that the present mineralogy of serpentinite in orogenic belts appears to be the result of regional metamorphism.

Dyke Rocks

DOLERITE. Dolerite dykes of small dimension are common throughout West Malaysia. In most cases they occur as isolated outcrops and cannot be traced for any distance along the strike. That they are intrusive and hypabyssal is beyond doubt, as can be seen from the numerous coastal and quarry sections. Generally they intrude granite, in which case they are generally unaltered. Often, however, they intrude sedimentary rocks and are in no way related to the granite, in which case they are often altered and usually are better classified as metadolerite or epidiorite. The two distinct modes of occurrence have led several authors to conclude that there are two ages of dolerite dyke intrusion, one pregranite and another postgranite (Alexander 1968).

The finest examples of dolerite dykes are in the Kuantan area of east Pahang, where they seem genetically related to the Kuantan basalt flows and indicate that the flows were of a fissure type (Fitch 1952). Dolerite dykes are most abundant in the granite north of Gambang and near Kuantan town. Very good exposures are to be found along the coast at Tanjong Tembeling and in Telok Champedek (Plate 19). A good locality in Kuantan town is the J.K.R. quarry at Bukit Ubi, where several dykes cut the granite. At Sungei Sengonggoh, Ulu Sungei Bakah, and Sungei Kepau, dolerite intrudes Lower Carboniferous strata.

The dolerite of the Kuantan area is ophitic, nonvesicular, contains no glass, and is coarser in grain than the overlying basalt. Large phenocrysts of plagioclase and pyroxene are present. Some dolerite dykes contain amygdales lined with chlorite or vermiculite and are filled with calcite and rare quartz (Fitch 1952). Titanaugite dolerite and olivine dolerite are equally common. Other minerals include calcic plagioclase, opaque iron oxides and accessory pyrite, chalcopyrite, apatite, biotite, and quartz, but the last three may be xenocrysts from the intruded granite (Fitch 1952). The texture is ophitic and commonly porphyritic, with phenocrysts of plagioclase up to 2 cm in length. Only rarely do the dykes contain xenoliths.

PLATE 19 *Dolerite dykes cutting granite at Tanjong Chempadek, Kuantan. Photograph D. J. Gobbett.*

The field evidence is clear that these dykes represent intrusions of a volcanogenic tectonic level. Wide chilled margins against the granite are occasionally glassy or cryptocrystalline. They are therefore certainly postorogenic and very probably Tertiary or even Quaternary, associated with the basalt flows of the region. However, a radiometric age of 110 m.y. (Table 8.5) indicates that some dykes are Cretaceous.

North of Kuantan, in Trengganu and Kelantan, grey medium-grained dolerites containing lath-shaped labradorite, titanaugite with a 2V of 44 to 48°, chlorite and ilmenite, have been described by J. R. Paton (*in* MacDonald 1968). They are strongly ophitic. Quartz dolerites, which also occur, are subophitic and contain plagioclase of oligoclase-andesine composition. No orthopyroxene has been found to occur. As in the Kuantan area, the dykes have chilled margins of 8 to 10 cm wide, in which feldspar and augite phenocrysts occur in a cryptocrystalline groundmass. J. R. Paton (*in* MacDonald 1968) suggested that the dolerites of the region are genetically related to the lamprophyre dykes that are also common in Trengganu and Kelantan. He suggested further that shortly after the granite crystallizaton, dolerite was intruded, mainly from the south, along channels provided by tear faults that resulted from continuing east-west orogenic pressure. He accounted for the variety of dyke rocks from dolerite to lamprophyre by the assimilation of granitic material by basaltic magma.

Elsewhere in West Malaysia, dolerite dykes occur commonly but sporadically. Alexander (1968) described dykes in the tributaries of the Sungei Kenong, in the Sungei Gombak, Sermi, Rong, and the Bentong near the Shanghai Pahang Estate. At all these localities, the dolerite intrudes granite. At Tras, behind the old court house, dolerite intrudes sedimentary strata. The mineralogy at all these occurrences is rather similar to those on the east coast at Kuantan, except for the frequent replacement of titanaugite by coronas of actinolite and for the presence of epidote. They are fine grained, dark bluish green, and reddish when decomposed. In some localities the dolerite is completely replaced by epidiorite. Alexander (1968) suggested they may be Tertiary in age, but the presence of epidiorites strongly indicates the occurrence of two ages of dolerite, one preorogenic, the other postorogenic.

Occurrences have been recorded sporadically, cutting sedimentary strata and granite in the region north of Kuala Lipis, Pahang (Richardson 1950), and in the Cheroh area north of Raub (Richardson 1939).

Northwest of Serendah, Selangor, in the Sungei Chor and Bukit Muchong Estate and at a few neighboring localities, dolerites have been recorded (Roe 1953). In the Sungei Choh, the outcrop can be traced for 2.5 km (1.5 miles). It cuts conglomerate, schist, phyllite, and sandstone. There is no definite evidence for its age, but Roe (1953) considered it to be preorogenic and distinct from those of postgranite age on the east coast.

Dolerite dykes have also been recorded in Singapore Island (Hutchison 1964a). Earlier, Scrivenor (1931) failed to find boulders of olivine dolerite *in situ* but observed them built into the esplanade wall in Penang.

LAMPROPHYRE. Many lamprophyre dykes occur in the granites of Trengganu. They are common in all granite masses except the Main Range and the Kemahang granite (MacDonald 1968). All the dykes are postgranite and are usually confined to the granite masses, but in a few cases they intrude the surrounding rocks. Unlike the dolerite dykes, the lamprophyres are extremely variable in composition and texture. There appears to be a continuous range of composition of the plagioclase feldspar in the dykes from labradorite, through andesine, to oligoclase. Subdivisions into detailed groups depending on the presence of green hornblende, brown hornblende, or biotite are possible, but a rather artificial grouping results (MacDonald 1968).

Spessartites have an abundance of green amphibole. They are grey, speckled rocks that resemble dolerite. In Trengganu, they vary in thickness from 25 cm to 3 m. The feldspar ranges from oligoclase to labradorite. Augite, hornblende, and biotite also occur. The spessartites are distinguished from the dolerites because they are highly porphyritic (MacDonald 1968). Biotite is always subordinate to hornblende. Ilmenite occurs as large skeletal crystals.

Camptonite dykes contain dark brown hornblende. The feldspar is usually oligoclase but may vary to labradorite. Colorless augite occurs in a few cases. Pyrite occurs commonly. The rock is often porphyritic.

Kersantite dykes are black in appearance, with an abundance of large brown biotite flakes. The biotite is deep reddish brown. Augite, ilmenite, and magnetite occur but amphibole is rare.

In parts of Trengganu, some lamprophyres have an ophitic texture and may be better classified as hornblende dolerite (MacDonald 1968).

Lamprophyres are rare elsewhere in the Malay Peninsula, and known common occurrences are in Singapore Island and south Johore.

In Singapore Island, the last phase of igneous activity is represented by a group of basic dykes. They cut the gabbro and the granite. Variable in composition, they are rich in hornblende and have been much altered. The feldspar varies from albite to labradorite, but it is difficult to ascertain because of advanced saussuritization (Hutchison 1964a). Magnetite is common. The dykes are all fine grained. Some are porphyritic, others equigranular. Many can be termed dolerite, but in view of the advanced deuteric alteration, the albitic nature of many, the high hornblende content, and the tendency to be rich in calcite, a more suitable group term may be hornblende lamprophyre. They vary in thickness from 0.3 to 1.8 m and are generally vertical (Hutchison 1964a).

Scrivenor (1931) has recorded minette occurring in the Muar River near Batu Bersawah; vogesite on Bukit Ulu Chegar, in Kelatan, and at Bukit Bari and Bukit Bari and Bukit Merjar, in Trengganu; kersantite in the Sungei Sokor, in Kelantan; and camptonite at Pasir Dula, Sungei Tehemang, in Trengganu.

Jones (in prep.) has recorded fine-grained to medium-grained lamprophyre dykes, approximately 0.3 m wide, cutting the Setul limestone at the J.K.R.

TABLE 8.6 *Chemical analyses and Niggli Norms of selected mafic rocks* from the Malay Peninsula*

	Weight Percentages								
	1	2	3	4	5	6	7	8	9
SiO_2	59.83	56.20	51.52	49.95	48.59	48.20	44.55	44.34	39.61
Al_2O_3	16.21	14.60	13.29	21.03	24.30	19.80	15.68	9.31	0.76
TiO_2	0.84	0.77	0.58	0.58	3.71	0.30	1.76	1.58	n.d.
Fe_2O_3	1.33	0.94	0.30	—	2.96	0.79	1.47	2.77	6.81
FeO	6.54	6.06	9.91	7.76	9.39	5.60	8.20	6.30	0.64
MnO	0.10	0.13	0.29	0.24	0.25	0.13	0.18	0.14	0.09
MgO	2.43	6.47	12.31	5.95	3.96	7.82	9.29	14.44	37.92
CaO	4.57	8.20	8.77	10.56	7.41	12.42	9.85	14.00	—
Na_2O	3.90	3.26	1.01	2.27	4.05	1.75	2.45	0.73	—
K_2O	2.63	1.17	0.39	0.19	0.82	0.69	1.58	3.75	—
P_2O_5	n.d.	0.18	0.06	0.06	0.88	0.04	0.26	0.12	n.d.
CO_2	n.d.	0.06	n.d.	n.d.	n.d.	0.09	1.52	0.30	0.30
H_2O^+	2.02	1.57	1.54	1.39	3.16	2.63	3.01	1.04	12.57
H_2O^-	0.42	0.22	—	—	0.25	0.27	0.35	0.18	0.65
	99.92	99.83	99.97	99.98	99.73	100.53	100.15	99.60[a]	100.06[b]

[a] Others: $BaO = 0.40$, $F = 0.29$, $S = 0.04$, $Cl = 0.02$, less 0.15 in total for F, S, Cl.
[b] Others: $NiO = 0.27$, $S = 0.06$, $Cr_2O_3 = 0.40$, $Cl = 0.01$, less 0.03 in total for S and Cl.
n.d. = not determined.

	Niggli Molecular Norms								
	1	2	3	4	5	6	7	8	9
Quartz	10.31	5.39	1.79	—	1.14	—	—	—	—
Corundum	—	—	—	—	—	—	—	—	0.87
Orthoclase	16.01	7.02	2.33	1.14	5.13	4.13	9.51	—	—
Plagioclase	55.60	5.180	40.15	67.79	57.98	60.65	49.29	11.09	—
Ab_xAn_{100-x}	Ab_{65}	Ab_{57}	Ab_{23}	Ab_{30}	Ab_{66}	Ab_{26}	Ab_{44}	Ab_0	—
Magnetite	1.43	1.00	0.32	—	3.28	0.84	1.57	2.88	1.78
Ilmenite	1.21	1.09	0.82	0.82	5.47	0.42	2.50	2.19	—
Apatite	—	0.38	0.13	0.13	1.95	0.08	0.55	0.28	—
Calcite	—	0.15	—	—	—	0.23	3.92	0.76	[c]
Hypersthene	12.38	19.06	44.31	24.66	14.40	11.13	—	—	43.91
En_xFs_{100-x}	En_{50}	En_{69}	En_{70}	En_{59}	En_{59}	En_{73}	—	—	En_{100}
Diopside	3.06	14.11	10.16	4.34	10.66	13.50	8.51	43.30	—
Olivine	—	—	—	1.13	—	9.01	23.68[d]	17.87[e]	49.05[f]

[c] Taken as zero because of insufficient CaO in analysis.
[d] $Fo_{73}Fa_{27}$.
[e] Others: Leucite = 12.73, nepheline = 3.82, kaliophilite = 3.69, halite = 0.06, fluorite = 1.22, pyrite = 0.10, $Fo_{87}Fa_{13}$.
[f] Others: chromite = 0.46, hematite = 3.77, pyrite = 0.16, halite taken as zero because of insufficient Na_2O in analysis.

* Specimens 1–9 as follow: GS = Geological Survey coll., Ipoh: UM = Dept. of Geology coll., University of Malaya, Kuala Lumpur.

Rock	Location	Specimen No.
1. Quartz monzonite xenolith in granite	Mandai quarry No. 13, Singapore	UM974
2. Tonalite	Anak Sungei Rek, Kelatan	GS26218
3. Hornblende gabbro	Poh Hin quarry No. 1, Singapore	UM482
4. Norite	Gammon quarry No. 2, Singapore	UM471
5. Microdiorite	Sungei Trengganu, Lata Sauk, Trengganu	GS25465
6. Eucrite	Linden Hill, Johore	GS26485; 3824

quarry at Kaki Bukit and in the Kuala Perlis quarry, Perlis. The dykes are composed of biotite, calcite, and quartz, the original feldspar being completely replaced by calcite.

In describing an important locality of lamprophyre dyke intrusion into the granite of Pulau Ubin and the rhyolite tuff of Pulau Nanas, two islands lying between Singapore and the peninsula in the Straits of Johore, Scrivenor (1931) deduced the following general sequence of igneous events:

1. (Oldest)—clasts of biotite granite in tuff on Pulau Nanas. The granite clasts are enriched in hornblende through metamorphism.
2. Tuffs and lavas, metamorphosed during the Mesozoic orogeny.
3. The Mesozoic granite, usually rich in hornblende resulting from assimilation of earlier gabbroic bodies, and quartz-mica diorite, of Pulau Ubin and Singapore.
4. Hornblende porphyries, hornblende granophyres, and lamprophyres, which cut the older rocks and may be subdivided into
   a. *Fine grained with hornblende and biotite*
      Hornblende porphyry; occurring at Anak Bukit Tinggi; Muncipal quarry 2 (see Scrivenor 1931, Fig. 5, p. 40, for locality map); Che Jevah; Changi.
      Hornblende granophyre: occurring at Wee Cheng Soon quarries.
      Spessartite: occurring in Johore and Seletar.
      Vogesite: occurring at Tanjong Balai and Changi.
   b. *Fine grained with pyroxene*
      Pyroxene microgranite: occurring in the Topham, Jones, and Railton (T.J.R.) quarry
      Pyroxene-bearing rock: occurring at Tanjong Jelutong.
      Enstatite spessartite: occurring at Tanjong Balai; Municipal quarry 2; Changi.
   c. *Coarse grained with pyroxene*
      Quartz biotite gabbro; occurring at Minicipal quarry 1.
      Quartz norite; occurring at Municipal quarry 2.
5. (Youngest). Aplite pegmatite, hornblende granite, and quartz gabbro cutting enstatite spessartite in Muncipial quarry 2. Acid granite cutting hornblende porphyry in Che Jevah quarry. Acid granite cutting enstatite spessartite at Tanjong Balai. Acid Granite cutting hornblende granite at Tanjong Jawa. Granophyre cutting microgranite at the T.J.R. quarry. Hornblende granite aplite cutting hornblende porphyry and enstatite spessartite at Changi. Quartz calcite tourmaline veins cutting lavas and tuffs on Pulau Nanas.

Chemistry

A selection of chemical analyses of basic and ultrabasic rocks, taken from Alexander et al. (1964) is given in Table 8.6, together with their Niggli norms computed by the method of Hutchison and Jeacocke (1971).

PETROGENESIS

The chemical analyses listed in Tables 8.2, 8.4, and 8.6 represent the complete range of plutonic rocks in this particular association of basic satellitic bodies with the granite batholith.

The chemical relations are best displayed on variation diagrams of the cationic percentages of the total alkalis versus silica as a percentage of the total cations, and on a triangular variation diagram of alkalis, total iron, and magnesium (Figure 8.15). The contoured points on the diagrams of Figure 8.15 represent all the chemical analyses of plutonic rocks available from West Malaysia and Singapore (Alexander et al. 1964) and can be taken to represent the relative proportions of the rock types throughout the area. Rocks of the granite kindred are by far the most numerous.

The distribution of the analyses indicates that we cannot consider the granite to have differentiated from the basic and ultrabasic rocks because of the relatively small amount of the latter that was available, compared with the overwhelming amount of granite in the batholiths It is more appropriate to take the position that the basic and ultrabasic rocks have been intruded into the Paleozoic–Mesozoic geosynclinal sediments, giving rise to

TABLE 8.5 Continued

Rock	Location	Specimen No.
7. Hornblende spessartite (lamprophyre)	Sungei Rua, Besut, Trengganu	GS23203
8. Pyroxenite	Sungei Keloi, Gunong Benom foothills, Raub, Pahang	GS13639
9. Serpentinite	Sungei Chembatu, Cheroh district, Raub, Pahang	GS13568

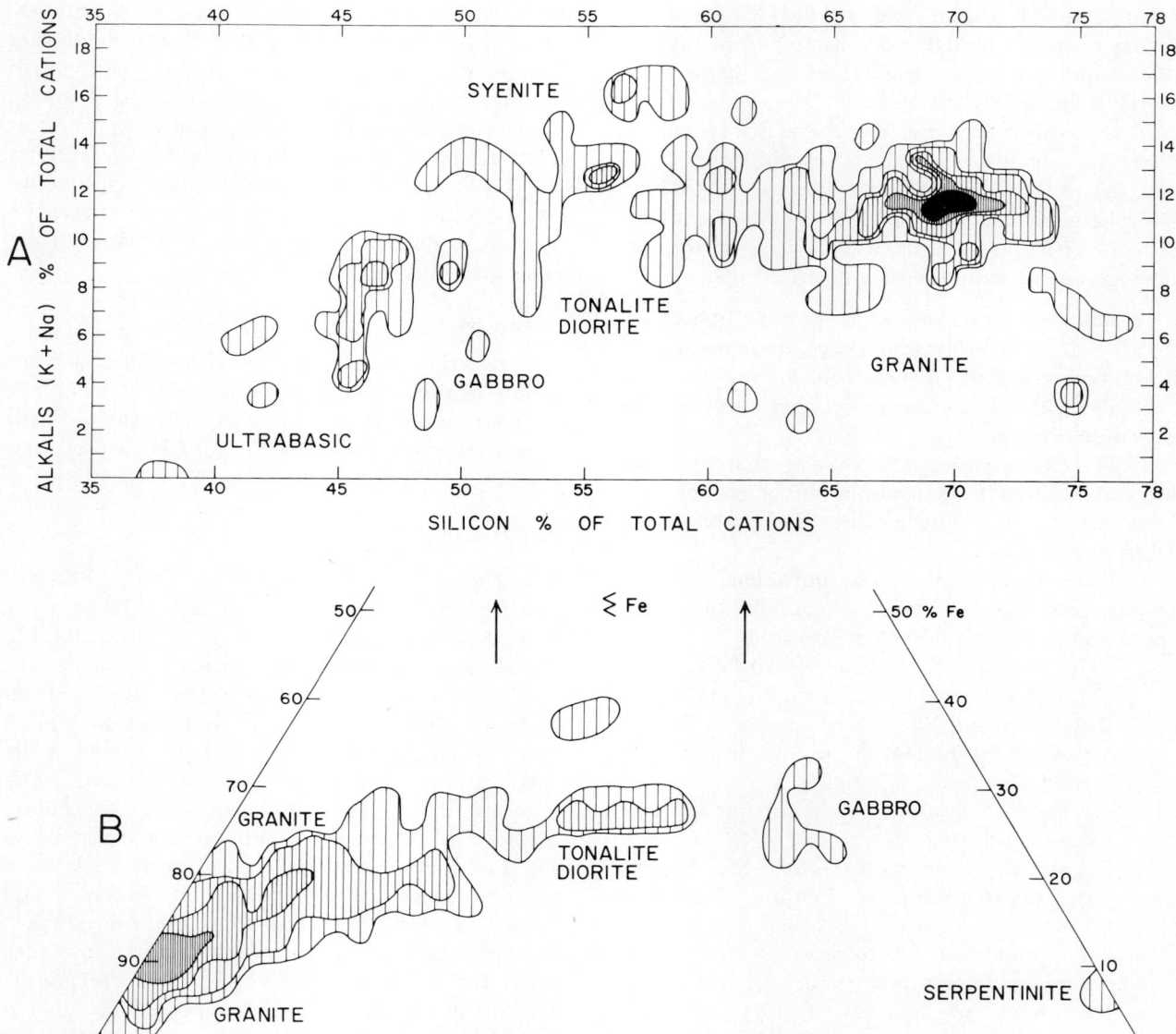

FIGURE 8.15 *Variation diagrams of 192 analyzed plutonic igneous rocks from the Malay Peninsula. Contours at 2, 4, 10, and 20%. Analyses are from Alexander et al. (1964):* (a) *total alkali versus silicon (cationic);* (b) *total iron versus total alkalis versus magnesium (cationic).*

bodies such as the eucrite of Johore, the gabbro of Singapore, and the serpentinites of Pahang. Joplin (1959) reached this general conclusion regarding basic bodies associated with discordant batholiths. In her classification (Joplin 1960), the basic and ultrabasic bodies would be ascribed to the "non-orogenic association," and the granites belong to the "orogenic association."

The field evidence that the basic and ultrabasic bodies predate the granite is on the whole unmistakable. In Singapore, the gabbro has a very pronounced set of east-west joints that are not found in the granite, and granite dykes injecting the gabbro contain up to 60% of gabbro xenoliths (Hutchison 1964a). Elsewhere, the serpentinites of Pahang are foliated concordantly with metaconglomerate, whereas the granite is unfoliated.

It is not necessary that all the basic and ultrabasic bodies were intruded into the sediments of the Paleozoic–Mesozoic geosynclinal pile at the same time. In fact, the evidence favors a divergence of time of emplacement. Some of the basic and ultrabasic bodies have been metamorphosed to amphibolite and serpentinite (e.g. Bukit Larang, Malacca; Cheroh area of Pahang), whereas the gabbro of Johore and Singapore appears to be little affected by regional metamorphism, and these circumstances strongly favor the interpretation that the basic and ultrabasic bodies were emplaced at different times in the devlopment of the geosyncline.

The veining and hybridization of the gabbro bodies of Singapore and Johore by the granite magma does not necessarily imply that the gabbro predates the granite. Three possible age relationships may be proposed to explain the associated acid and basic plutonic bodies in the Malay Peninsula:

1. That the basic intrusions are preorogenic and belong to the ophiolitic or geosynclinal phase (Hutchison 1964a).

2. That the basic intrusions are orogenic and earlier than, but almost contemporaneous with, the granite (Walton 1965).

3. That the basic intrusions are orogenic and followed the granite, which they marginally mobilized (Blake et al. 1965).

Because a basic mass is cut by acid veins, it does not necessarily follow that the granite mass with which the veins connect is younger than the basic mass. Blake et al. (1965) described several areas of intimate association of basic and acid magmas from nonorogenic regions in which basic masses had intruded the granite. Marginally to the basic mass the preexisting acid rock had been mobilized; the mobilized acid material had back-veined the basic material, and on occasions it had veined parts of the acid mass that had remained solid. The veining of the basic mass had produced hybrid characters similar to those described previously from Singapore Island.

In interpreting associations of acid and basic igneous rocks, due regard must be given to the tectonic level of intrusion (Hutchison 1966a).

Basic magma in nonorogenic regions may well have its source in the mantle. Magma is formed along dislocation zones in a manner similar to that observed in the Hawaiian Islands, where dislocations in the mantle migrate toward the surface, accompanied by magma (Eaton 1962). The magma produced is of high temperature, is relatively "dry," and can conceivably be capable of remobilizing acid magma or rock in the manner postulated by Blake et al. (1965).

The conditions pertaining to orogenic belts have been observed to differ in many ways from those in nonorogenic regions. Basic rocks are generally characteristic of the geosynclinal phase (Aubouin 1965). Van Bemmelen (1949), Hutchison (1964a), Joplin (1959), and Reid (1960) have presented many examples. These ophiolitic rocks are characterized by relatively low temperatures of intrusion. The majority of the acid rocks are syntectonic. The water-rich geosynclinal environment results in a "wet" magma. Acid magma penetrates the basic rocks, veining them and incorporating numerous xenoliths, which are in every stage of assimilation. Dioritic hybrids are formed. The basic-acid contacts are sharp or diffuse, depending on the level of intrusion or subsequent erosion level. However, unlike the contacts in volcanogenic levels, the basic rocks do not show chilled margins against the granite, except for the frequently occurring set of postorogenic basic dykes. These, however, are classed as being of the nonorogenic association.

The outcrop distribution of rocks in orogenic belts is radically different from those in volcanogenic levels. Usually a few basic bodies are situated around the edge of the acid batholith in a satellitic manner. The proportions of the two rocks are such that the acid rock generally exceeds the basic bodies by a factor of hundreds of thousands. The mechanism proposed by Blake et al. (1965) could certainly not be applicable to such a distribution and relative proportion. Basic inclusions are extremely common in orogenic granites where the batholith abuts a basic body. Such xenoliths are frequently ovoid and never wisplike as they are in volcanogenic granites as described by Blake et al. (1965) and by Walton (1965).

Joplin (1959) concluded that basic bodies associated with discordant batholiths have been emplaced among the geosynclinal sediments and that the associated granites are orogenic. She explained all intermediate rocks in these associations as products of hybridization. In relation to the Thai–Malay orogenic zone, her analysis seems to be perfectly applicable, as can be seen by reference to the distribution of rock types in Figure 8.14. In the classification of Joplin (1960), the granite and hybridization of West Malaysia and Singapore belong to the "orogenic assocation." The gabbro, and ultrabasic bodies, the metagabbro and the later dolerite and lamprophyre dykes would be included in her "nonorogenic association."

The striking enrichment in hornblende in Singapore and Johore is also found in other unstable regions of the world (e.g., Nockolds 1940, Joplin 1959). In his description of the Garabal Hill igneous complex, Nockolds (1940, p. 461) noted: "Little of the original pyroxene remains in most specimens. More pale green hornblende and biotite (or chlorite) are present." The enrichment in hornblende in orogenic regions, as in the Thai–Malay–Singapore region, can be explained by a volatile-rich environment rising ahead of the main orogenic magma into the geosynclinal materials, bringing silica with it. Reaction between this attenuated magma and the basic rocks of the geosynclinal pile would replace the original pyroxene by hornblende without appreciably altering the rock textures. Quartz-hornblende pegmatites can be formed, as in Singapore and Johore, along the existing zones of weakness in the invaded rocks. The orogenic granites themselves tend to be rich in hornblende, compared with those of stable regions, which can often contain pyroxene (Joplin 1959).

The final phase of dolerite and lamprophyre dyke intrusions is one that is common to orogenic regions. It is rather paradoxical, since the trend has been from basic to acid. This phase can be explained only as belonging to the "nonorogenic association" and resulting from a release of basic magma from deep crustal levels during late-orogenic or post-orogenic faulting.

CHAPTER 9

Metamorphism

C. S. HUTCHISON

Metamorphism can only be effectively described and discussed in relation to the tectonic setting and to the granite batholiths that characterize the orogen. Under the simplest hypothesis, the intensity of regional metamorphism may be expected to increase somewhat uniformly with depth; therefore, together with deformational style, it will afford an indicator of depth. The relationships between the metamorphic and the granitic rocks will also give an indication of the depth of formation within the orogen. These characteristics of metamorphic intensity, deformational style, and nature of granitic plutons can best be generalized in terms of depth zones. As used here, the term zones thus refers in substantial part to intensity zones rather than strictly to depth zones. The subdivisions into zones are threefold: epizone, mesozone, and catazone. These terms have been used frequently over a period of years by a variety of authors to represent a confusing range of depth and metamorphic grades. Hence the terms are strictly defined here before they are used. The definitions generally agree with those of Buddington (1959).

Previous authors, including Buddington, have defined these zones in terms of both temperature and depth within the orogen. Such a procedure is permissible only to a specific orogen where the geothermal gradient is characteristic. Since geothermal gradients may vary from as low as 10°C/km in burial metamorphism terrane, through 20°C/km in Barrovian terrane, to 70°C/km in Abukuma terrane, clearly no single conversion of temperature to depth equivalence can be applied to the crust of the earth generally. Buddington's (1959) zone classification, which is the most suitable to date, is followed and redefined strictly in terms of temperature ranges that are independent of depth in respect of the metamorphic facies. The zones are defined as follows:

Epizone—from surface temperature to maximum 250 to 350°C.
Mesozone—from 250 to 350° range to maximum 500°C.
Catazone—temperature in excess of 500°C.

In these definitions, the temperatures referred to are the regional geothermal temperature, and not those which may be attained in a thermal aureole around a high-level intrusion.

In a later section of this chapter, after a discussion of the metamorphic facies series of the Malayan orogen, some depth equivalence of these temperature zones is attempted.

In the foregoing scheme, the epizone is characterized by a lack of metamorphism, the mesozone by greenschist facies, and the catazone by amphibolite facies metamorphism (see Figure 40 *in* Winkler 1967).

ZONAL CLASSIFICATION OF THE MALAYAN OROGEN

The sedimentary and volcanic formations of the Malayan orogen can be quite conveniently referred to the zones just mentioned.

Epizone

Sedimentary formations and intercalated volcanic rocks of Triassic or younger age belong to the epizone. These include the Jurong formation, the Semanggol formation, the Kerdau formation, the Jelai formation, the Gunong Rabong formation, the Tembeling Formation, the Gagau Group, the Tabak formation, and all the Cenozoic formations. In addition, the Kuantan and Segamat basalt flows belong here. Lower Paleozoic rocks of the miogeosyncline in northwest Malaya have never been deeply buried nor elevated to high temperatures; accordingly, these also belong to the epizone. Included are the Setul and Mahang Formations, in which shale strata have suffered so little metamorphism that graptolites and tentaculites are preserved. In the same areas of northwest Malaya, but including also parts of west-central Malaya, Upper Paleozoic rocks of the Singa, Chuping, and Kubang Pasu Formations are of the epizone, as also are some Upper Paleozoic rocks of the Kinta Valley.

Although devoid of regional dynamothermal metamorphism, the rocks of the epizone may have been intruded by high-level granite plutons associated with thermal metamorphism.

Mesozone

Mesozone rocks outcrop extensively in Malaya (Figure 9.1). They include all the Lower and Upper Paleozoic rocks to the east of the Main Range Granite. In addition, they occur as roof pendants within and as a zone bordering the western margin of the Main Range Granite. These rocks, which are characteristically of the greenschist facies of regional metamorphism, are closely associated with large mesozonal granite batholiths. They are overlain in the axial part of Malaya by Mesozoic epizone formations.

Catazone

The Malayan orogen is not yet deeply enough eroded to expose much of the catazone. The deepest levels have been brought to the surface in north Kelantan as the Taku Schist and Stong Migmatite Complex. Along a zone extending southward from Kelantan, catazone rocks are sporadically brought to the surface (Figure 9.1), the most important areas being around Bukit Berentin and Benta in Pahang. Rocks of the catazone have also been brought to the surface by major faults in restricted areas west of the Main Range, notably in the Gunong Jerai and Kupang areas of Kedah. The catazone rocks are characterized by amphibolite facies and by migmatization. The exact age of the catazone rocks is unknown, but because the mesozone rocks are predominantly of Lower and Upper Paleozoic formations, the catazone probably includes Lower Paleozoic or older rocks.

ROCKS OF THE CATAZONE

High-grade dynamothermally metamorphosed rocks outcrop mainly in a belt of country extending southward from the Thai border in north Kelantan to the region between Gunong Benom and Raub in Pahang. These rocks lie within the Mesozoic belt of axial Malaya and come to the surface through the overlying sedments and low-grade metasediments in discontinuous outcrops. Since the catazone bodes are rather different in character, each is described individually.

Taku Schist

The Taku Schist, an extensive body of pelitic schist in north Kelantan (Figure 9.1), was the first in Malaya to be recognized as a product of deep-seated dynamothermal metamorphism, and it is the only body of metamorphic rock represented on the 6th edition of the geological map of Malaya (Alexander 1965), where it is referred to as "?Pz Pre-Carboniferous undifferentiated." These rocks were first mapped by Savage (1925), who referred to them as the Kelantan schists. He misrepresented them as being part of a thermal aureole, but this was excusable in that only a small number of traverses were made across the schist, and numerous granite intrusions were encountered to the north. The existence of the Taku Schist was forgotten until Macandie and Canavan (1948) recorded that the schist was older than the neighboring ?Carboniferous sedimentary rocks.

The schist body was recognzed as unusual in the geology of Malaya first by MacDonald and later by Slater, as being of deep-seated origin (Alexander et al. 1961). A certain amount of skepticism was encountered from other geologists in the country because it was traditionally held by officers of the Geological Survey that metamorphism in Malaya must be related to the extensive granite masses; accordingly, there was a reluctance to accept regional dynamothermal metamorphism in the country. Paton (in Agocs, 1958–1965, p. 2-J) was the first to formulate his views in writing, in what has

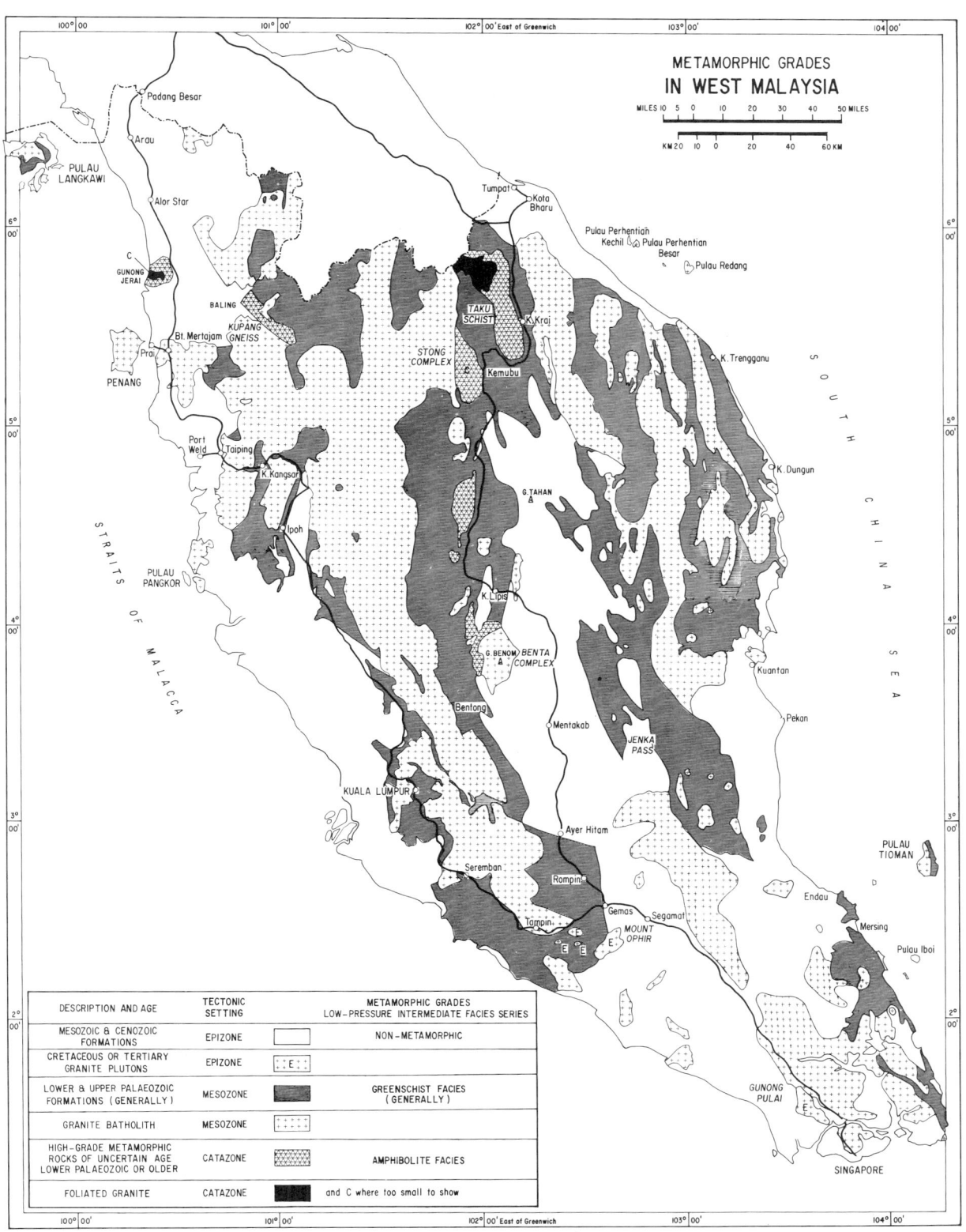

FIGURE 9.1 *Distribution of metamorphic grades and associated granites in the Malay Peninsula.*

become an important landmark in the geological literature:

> The only extensive body of schists discovered occurs in north Kelantan. Apparently it is not related to any of the known Mesozoic granites. It is overlain by shales which are thought to be Carboniferous or Permian, but there are no other pointers to its exact age. The schists may be either the metamorphosed equivalent of some of the Lower Palaeozoic rocks now known to occur in the Langkawi Islands off the west coast, or possibly may form part of the original basement complex.

There then followed a rather premature discussion of the paleogeographic significance of this body of schist (Hutchison 1961, Alexander et al. 1961). Further detailed descriptions are to be found in Aw (1964) and MacDonald (1968).

DISTRIBUTION. The Taku Schist forms an elongate elliptical body about 80 km (50 miles) long and some 8 to 22 km (5 to 14 miles) wide. Its long axis trends north-northwest from the railway line south of the Sungei Galas to a few kilometers north of the Tanah Merah to Nibong road, and thence into Thailand (Figure 9.2). The schist forms a subdued topography, and exposures are largely confined to rivers. Much of the schist is deeply weathered, and outcrops crumble easily to a mixture of clay and quartz grains with a high proportion of muscovite flakes. The best exposures are in the Sungei Taku, the Sungei Sokor, and the Sungei Bertam, in the tributaries of the Sungei Lebir and the Sungei Galas, and at the Tamangan iron mine. The most accessible outcrops are to be observed on the banks of the Sungei Galas between the towns of Dabong and Kuala Krai.

PETROLOGY. The Taku Schist is composed essentially of paraschist, mainly pelitic, with some interfoliated orthoamphibolite bands. The predominant rock type is quartz-mica schist with subordinate quartz-mica-garnet schist, and garnet-mica schist; there are frequent bands of amphibolite and narrow bands of quartz schist and serpentinite. Calc-silicate fels is less common. The most common minerals are quartz, muscovite, garnet, biotite, plagioclase, tourmaline, and kyanite. Chlorite is common as a weathering product of the micas. Pyrite and graphite are locally prominent, and sillimanite and andalusite also occur locally. Magnetite is sometimes locally concentrated, and in one case it occurs in a band of schistose calc-silicate fels. Rutile is very common throughout the schist as an accessory mineral. Quartz veining is extremely common, and in many cases the quartz is gold bearing (MacDonald 1968).

Good exposures of amphibolite are to be found along the Sungei Galas, where the rock is associated with pyroxene schist containing quartz, diopside, titanite, and plagioclase (MacDonald 1968). Clinozoisite and titanite are common accessory minerals in the amphibolite, and epidote occurs locally. Cordierite has been recorded n a quartz-amphibole-garnet schist (MacDonald 1968).

In the Sungei Hau there are occurrences of diopside-tremolite-calcite gneiss and diopside-tremolite-calcite-andesine gneiss, which contain accessory titanite, apatite, magnetite, sericite, and chlorite (Aw 1964).

Small autochthonous granite gneiss bodies occur at a few localities. North of Bukit Kedah, gneiss occurs in a *lit-par-lit* relationship with the schist. Along the south and southeast margin, the schist body is bounded by an interfoliated band of autochthonous foliated granitic gneiss about 1 km wide (Figure 9.2). This gneiss is well exposed in the Sungei Anali.

In the north part of the schist outcrop, there is an extensive parautochthonous granite body that has been named the Kemahang granite (MacDonald 1968). It is formed in part of foliated granite gneiss and in part of cataclastic granite. It contains a number of schist relics, often quite large, which have survived anatexis.

Mica Schist. Mica schist forms the bulk of the Taku Schist. It is strongly foliated. Both muscovite and biotite occur, but the former is more abundant. Quartz and garnet are nearly always present. A chemical analyses is given in Table 9.1, analysis 3.

Quartz usually makes up more than 60% by volume of the rock, as can be illustrated by the following range in modal analyses (Aw 1964):

Quartz	60 to 63%
Muscovite	8 to 24%
Biotite	0 to 10%
Garnet	0 to 32%
Plagioclase	0 to 6% by volume

Muscovite is the predominant mica, but biotite is fairly common both in the garnet-bearing and garnet-free schist. Graphite is commonly associated with the biotite. There appear to be two generations of muscovite. Most crystals lie parallel to and define the schistosity; they are characterized by thin crystals with a thickness to breadth ratio of about 1:8 (Aw 1964). The second type is relatively much shorter and thicker and is devoid of any pre-

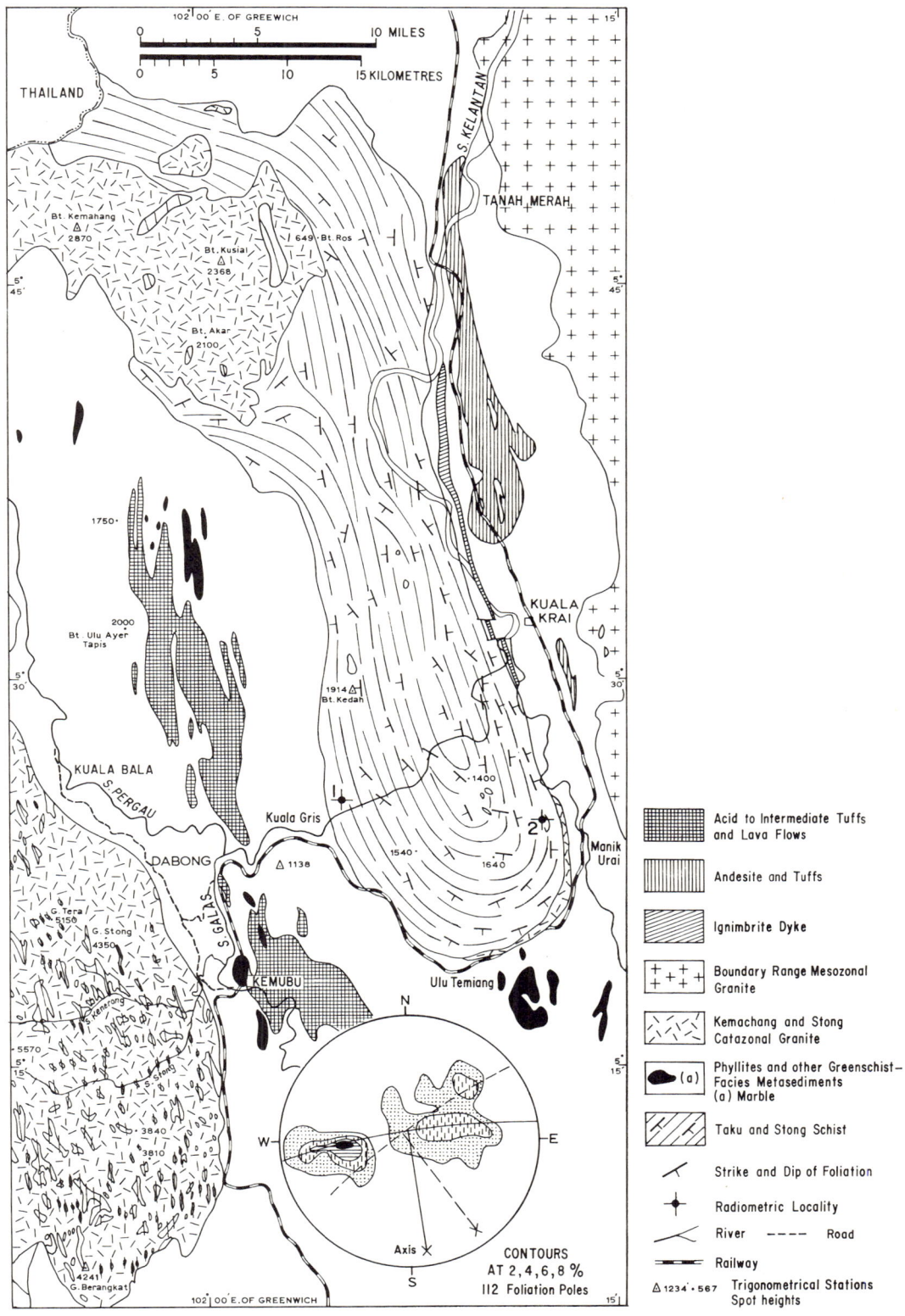

FIGURE 9.2 *Geological map of Ulu Kelantan, showing the outcrop of the Taku Schist and the Stong Migmatite Complex. After Dawson et al. (1968).*

TABLE 9.1 *Chemical Analyses of Selected Metamorphic Rocks* from Malaya*

	Weight Percentages									
	1	2	3	4	5	6	7	8	9	10
SiO_2	93.2	71.5	71.3	59.50	57.8	52.39	48.87	46.7	46.42	8.86
Al_2O_3	3.4	15.7	12.7	11.35	15.2	13.76	10.33	14.2	12.40	78.75
TiO_2	0.14	0.02	0.53	1.68	0.68	1.50	1.25	0.87	0.91	2.06
Fe_2O_3	0.85	0.43	1.68	3.35	1.00	0.76	4.15	1.83	4.36	1.59
FeO	0.42	0.98	3.56	7.45	4.71	9.61	10.86	8.12	6.95	1.72
MnO	0.05	0.03	0.17	0.10	0.15	0.04	0.14	0.18	0.29	0.02
MgO	0.12	0.97	1.62	3.87	6.08	7.28	10.31	9.77	8.73	0.80
CaO	0.04	1.27	0.44	5.12	4.29	5.72	5.32	14.81	14.24	0.40
Na_2O	0.35	4.36	1.39	3.38	1.80	2.05	2.91	1.56	1.47	0.46
K_2O	0.31	3.84	2.79	0.19	6.29	4.08	1.39	0.31	0.27	0.08
H_2O^+	1.08	1.08	1.55	3.41	0.75	} 1.44 {	4.28	1.38	2.04	3.13
H_2O^-	0.14	0.22	0.17	0.67	0.23		0.22	0.07	0.28	0.20
P_2O_5	0.03	0.02	0.13	0.23	0.50	n.d.	nil	0.13	nil	0.13
CO_2	0.01	0.08	0.24	nil	0.04	0.32	0.15	0.20	1.42	0.12
Total	100.14	100.50	98.27	100.24	99.52	99.50	100.20	100.13	99.91	100.22

* Specimens 1–10 as follows [All analyses and references are to Alexander et al. (1964) except those marked†, which are unpublished and by courtesy of the Director, Geological Survey of Malaysia. UM = Dept. of Geology coll., University of Malaya, Kuala Lumpur]:

1. Foliated metaquartzite, Sungei Piah, Lenggong area, Perak (00.3.003).
2. Biotite gneiss, Sungei Anali, Kelantan (01.2.030).
3. Garnet-mica schist (muscovite, minor biotite, quartz, garnet, feldspar, andalusite, sillimanite, graphite), Sungei Kenik, Kelantan (UM5408).†
4. Epidote-albite-chlorite schist, Sungei Tras, Bentong area, Pahang (09.1.005). Analyzed also for ZrO_2 nil; Cl, 0.02; F, nil; S, nil; Cr_2O_3, 0.02; NiO, nil; BaO, nil.
5. Biotite gneiss (biotite, sphene, microcline augen, perthite, quartz, plagioclase, apatite, clinopyroxene), Kupang near Baling, Kedah (UM5390).†
6. Quartz-mica schist, near milestone 3.5 Cameron Highlands road, Perak (09.1.003). Analyzed also for ZrO_2, 0.04; F, 0.34; S, 0.62; less O for F, S, Cl, 0.45.
7. Tremolite amphibolite, Sungei Cheroh, near Cheroh, Pahang (00.4.003). Analyzed also for ZrO_2, nil; Cl, tr.; F, nil; S, 0.04; Cr_2O_3, tr.; NiO, tr.; BaO, nil; Au, nil; Pt, nil; less O for F, S, Cl, 0.02. Density 2.87.
8. Clinozoisite amphibolite, Sungei Galas, Kelantan (09.1.006).
9. Calc-silicate fels, 0.25 mile upstream from confluence of Sungei Kiul and Sungei Sangkau, near Rasa, Selangor (00.1.008). Analyzed also for ZrO_2, nil; Cl, tr.; S, 0.20; BaO, nil; less O for F, Cl, S 0.07.
10. Tourmaline-corundum rock, Khuan Lee Hin Kongsi tin mine, Siputeh, Perak (00.2.004). Analyzed also for ZrO_2, 0.07; Cl, 0.02; F, nil; SO_3, 0.12; S, 0.11; Cr_2O_3, 0.04; BaO, nil; SrO, tr.; LiO_2, nil; B_2O_3, 1.60; As, nil; Sno_2, nil; less O for F, S; Cl, 0.06.

ferred orientation. Crystals of this type are distributed throughout the rock and are also concentrated in the "shadows" of garnet porphyroblasts. Aw (1964) took these crystals to represent a later phase of crystallization after the release of regional pressure. The biotite crystals are shorter and less well formed than the muscovite.

Quartz is often segregated into narrow bands or in low-pressure areas between the mica foliation and garnet porphyroblasts. Quartz veining is also common throughout the schist.

Garnet occurs in two distinct habits. The first forms large poikiloblastic crystals of about 2 mm diameter with good dodecahedral forms; the growth of the crystals has forced the mica schistosity apart (Aw 1964). The porphyroblasts are generally fractured and heavily stained reddish brown. The second habit is that of numerous reddish euhedral crystals of less than 0.25 mm diameter, free from inclusions. These smaller crystals which often occur as inclusions in the feldspar porphyroblasts, grew within the mica crystals and do not deform the schistosity (Aw 1964).

Garnet is characteristically absent where the schist contains graphite. Harker (1956) suggested that the presence of carbonaceous matter in quantity tends to inhibit or delay the formation of garnet. In thin section, the garnet is always completely isotropic. It has been determined as almandine (MacDonald 1968). The crystals are often rotated, cracked and shattered crystals occur locally (MacDonald 1968), and quartz occurs as poikiloblastic inclusions in the garnet.

Plagioclase porphyroblasts of tabular outline, about 3 by 1.5 mm, occur frequently. They are untwinned and of albitic to andesine composition (MacDonald 1968). They are frequently poikiloblastic with inclusions of quartz and garnet.

Pyroxene has been recorded (MacDonald 1968). Magnetite is fairly common, and zircon less common. Zoned tourmaline is common but not abundant. Rutile is widespread, with individual crystals up to 5 cm long on the borders of quartz veins. The tourmaline and rutile probably are derived from the original sediments from which the schist is derived (MacDonald 1968).

Kyanite is of very local occurrence; staurolite is rare; and sillimanite has been recorded locally on the Sungei Hau (Aw 1964). Andalusite-muscovite-biotite schist occurs at the Sungei Siyah and at other localities, near the contact of the schist with granite.

In addition to the quartz veining that resulted from metamorphic segregation, younger quartz veins cut the schist as a result of more recent hydrothermal activity. They contain gold (MacDonald 1968).

Zones of cataclasis have been formed in the schist, for example, on the Sungei Memakah. Thin sections of the schist show mortar structure with angular fragments of quartz in a fine-grained matrix of sericite and granulated quartz (Aw 1964).

Amphibolite. Amphibolite occurs as narrow bands interfoliated with the schist. It attains its maximum development in the south, where it is exposed in the Sungei Galas, particularly at Jeram Berhala, and in the Sungei Sedulek. The amphibolite probably represents metavolcanic rocks that were intercalated with the sediments from which the Taku Schist is derived. Some amphibolite is associated with pyroxene schist (MacDonald 1968). The predominant minerals are hornblende, plagioclase, clinozoisite, quartz, and epidote; tremolite, garnet, and biotite occur in some areas. Titanite is nearly always present, and pyrite and penninite occur locally (MacDonald 1968). The amphibolite that outcrops along the Sungei Galas has been described by Hutchison (1961) as having the following modal composition: hornblende 61% by volume, feldspar 25%, clinozoisite 11%, garnet 2%, and pyrite 1%.

The hornblende has the following characteristics: 2V 80° to 82° $-ve$. $\gamma \wedge c = 14°$ to $15°$, optic plane (010), specific gravity 3.15, $\beta = 1.648$, birefringence 0.02. Following Winchell (1951), the composition can be given as 60% tremolite : 40% ferrotremolite (molecular ratio).

The plagioclase has a 2V that ranges in universal stage determination from 86° $-ve$ to 82° $+ve$. Zoning is wavy and twinning absent. The optical orientation is consistent with a composition in the andesine range.

The clinozoisite has a 2V of 86° $-ve$, with optic plane (010), and crystal elongation parallel to the b axis. It has a perfect (001) cleavage and $\beta = 1.713$. The interference colors are anomalous; birefringence is of the order of 0.01. Following Winchell (1951), the composition may be deduced as a fairly pure clinozoisite with as much as 10% replacement of aluminum by iron. A chemical analysis of the amphibolite is given in Table 9.1, analysis 8.

Serpentinite. A band of serpentinite occurs in the Taku Schist in Ulu Sungei Taku near the contact of the schist with the overlying shale or phyllite. It is dark green and usually strongly foliated with light green and yellowish bands, but in places it is massive. It exhibits a variety of microstructures and is of variable grain size (MacDonald 1968). Antigorite is the main mineral, and calcite, chrysotile, and chlorite are also common. Opaque minerals are magnetite, chromite, and ilmenite. The origin of the serpentinite is in doubt. However, Miyashiro (1966) has demonstrated that serpentinite in orogenic belts can be attributed to the regional dynamothermal metamorphism of basic and ultrabasic lava flows.

Granite and Granite Gneiss. Along the southern outcrop of the schist, a well-foliated biotite gneiss of average grain size 2 mm is concordantly foliated with the schist. The chemical composition of the gneiss is given in Table 9.1, analysis 2. A specimen from the Sungei Anali (no. 5366) has the following modal composition:

Quartz: anhedral, strained granulated and sutured, 24% by volume.

Alkali feldspar: untwinned anhedral microperthitic orthoclase, 30%.

Plagioclase: subhedral, generally untwinned, oligoclase with oscillitary zoning, frequently sericitized and saussuritized, 37%.

Biotite: reddish and pleochroic 9%. Its preferred orientation is very striking in the hand-specimen but generally less obvious in thin section. It occurs as schlieren separating the salic minerals.

AGE. Only two samples of the Taku Schist have been radiometrically dated by the Age Determination Unit of Overseas Geological Surveys (Snelling 1965).

1. Garnet-mica schist, locality Sungei Kenik (1 on Figure 9.2); potassium-argon method. Muscovite gave 210 ± 8 m.y.; biotite gave 220 ± 8 m.y.
2. Biotite gneiss, locality Sungei Anali (2 on Figure 9.2); potassium-argon method. Biotite gave 215 ± 8 m.y.

These dates fall within the range of Middle or Lower Triassic (Kulp 1961) and may well be related to the uplift of these catazone rocks to higher crustal levels; hence we must take them as only an upper limit on the age of the metamorphism (Armstrong 1966). It is unfortunate that no rubidium-strontium determinations are available, since potassium-argon dates by themselves are inconclusive in dating events in the catazone of an orogen.

Since the Taku Schist is largely pelitic, it would be reasonable to conclude that it is the metamorphic equivalent of the Lower Paleozoic rocks of the western part of the Malay Peninsula (the Bentong group and the Mahang Formation), which has been dynamothermally metamorphosed in the tectogene zone of the orogen in the Lower to Middle Triassic or earlier. But this general conclusion may well have to be modified in the light of further radiometric evidence.

GEOCHEMISTRY. Chemcial analyses of the two most common rock types in the Taku Schist, garnet-mica schist and clinozoisite-garnet amphibolite, are presented in Table 9.1 (analyses 3 and 8). A Niggli norm calculation on the amphibolite gave the following results by the method of Hutchison and Jeacocke (1971).

Salic 47%: orthoclase 1.85, plagioclase 45.20 ($Ab_{31}An_{69}$), nepheline 0.12.
Femic 53%: diopside 32.83 ($Wo_{50}En_{36}Fs_{14}$), olivine 16.05 ($Fo_{72}Fa_{28}$), magnetite 1.94, ilmenite 1.23, apatite 0.27, calcite 0.51.

The ACF values corrected for accessory minerals are: A, 27.1; C, 31.7; F, 41.2; which place the rock in the basic composition field as defined in Fyfe et al. (1958), suggesting that the amphibolite is a derivative of basic igneous or pyroclastic rocks.

The granite gneiss, which is interfoliated with the schist along its southern outcrop, has also been analyzed (Table 9.1, analysis 2); its computed Niggli norm is:

Salic 95.34%: quartz 25.02, corundum 2.52, orthoclase 22.79, plagioclase 45.02 ($Ab_{87}An_{13}$).
Femic 4.66%: hypersthene 3.93 ($En_{68}Fs_{32}$), magnetite 0.45, ilmenite 0.03, apatite 0.04, calcite 0.20.

With a normative quartz:albite:orthoclase ratio of 29:45:26, this rock is very close to the granite maximum in Figure 8.5; its composition therefore agrees well with that which would be expected to result from anatexis of sediments during the regional metamorphism that produced the Taku Schist (Winkler 1967, p. 220).

CONDITIONS OF METAMORPHISM. The Taku Schist can be ascribed to the low-pressure intermediate facies series of Miyashiro (1961). The complete outcrop of the schist, together with the amphibolite lenses, is of almandine amphibolite facies.

The minerals that indicate the conditions of metamorphism in the Taku Schist are kyanite, andalusite, sillimanite, staurolite, and cordierite. All these have been recorded in the pelitic schist. Kyanite, sillimanite, and staurolite are characteristic of the Barrovian facies series, whereas cordierite and andalusite are characteristic of the Abukuma facies series (Winkler 1967). The Taku Schist therefore was metamorphosed under conditions intermediate between the two facies series and may be ascribed to the Low-Pressure Intermediate Facies Series of Miyashiro (1961). In the Buchan type of metamorphism of northeastern Scotland, andalusite cordierite and staurolite are associated; in Korea, andalusite, staurolite, and kyanite coexist. Andalusite, kyanite, and sillimanite coexist in Idaho. These terrains are regarded as belonging to the Low-Pressure Intermediate Facies Series (Miyashiro 1961) and are akin to the Taku Schist of north Kelantan.

The common abundance of muscovite, biotite, and almandine garnet is consistent with the allocation to the amphibolite facies of the low-pressure intermediate type.

The mineral paragenesis of the three principal rock types in the Taku Schist is shown by their positions on ACF and $A'FK$ diagrams (Figure 9.3).

The presence of clinozoisite in the amphibolite and of muscovite in the schist set upper limits on the metamorphic temperatures and the presence of andalusite sets an upper limit on the pressure conditions. It can be deduced that the Taku Schist was

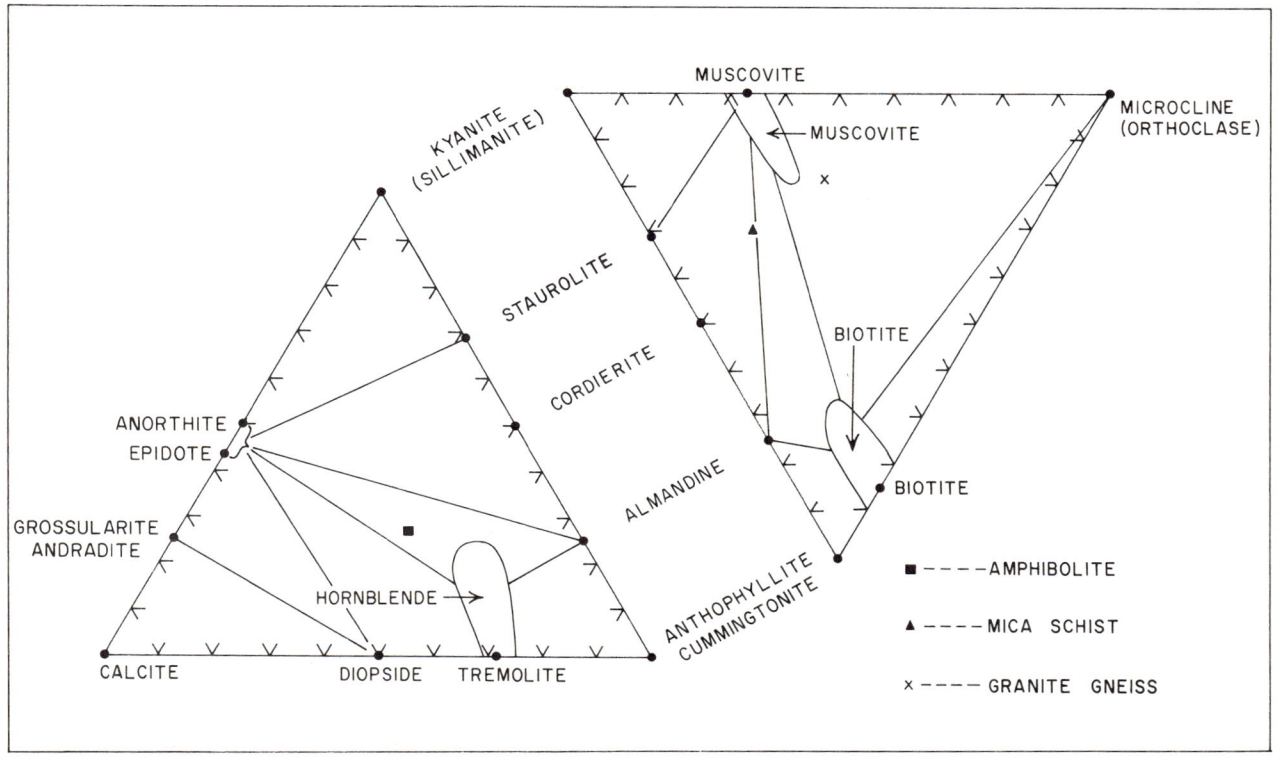

FIGURE 9.3 ACF and A'KF metamorphic facies diagrams that have been deduced for the mineral paragenesis of the Taku Schist. The metamorphic facies series appears to be of the low-pressure intermediate type.

metamorphosed under conditions of temperature around 600°C and a pressure of 6 kb (Figure 40 *in* Winkler 1967). These conditions would be attained in a terrain having a geothermal gradient of approximately 40°C/km. If under this gradient the temperature locally rose as high as 640°C, with a pressure of 6.4 kilobars (kb), then partial anatexis of the salic components of the metasediments would be expected (Winkler 1967). Such local temperature highs can be held responsible for the local production of limited quantities of granitic melt, which on cooling gave rise to the foliated biotite gneiss and granite bodies which locally occur within the Taku Schist, interfoliated with, and in a *lit-par-lit* relationship with, the schist. These locally formed magma bodies did not produce any thermal metamorphism on the enclosing schist because the mineralogy of the schist was in complete harmony with the conditions of pressure and temperature that caused the anatexis.

Not all the granites may have evolved at the tectonic level now exposed; some may have been evolved by anatexis at deeper levels and may have been emplaced at the higher level now represented by the Taku Schist outcrop.

STRUCTURE. Figure 9.2 shows a lower-hemisphere equal-area stereogram of 112 foliation poles that been measured throughout the outcrop of the Taku Schist. The information is of composite origin, taken from Dawson et al. (1968), from Aw (1964), and from several measurements by Hutchison.

The structure can be analyzed as an anticline with its axis plunging 7° to the horizontal in a predominant azimuth of 170° (10° east of south). However, the distribution of the foliation poles indicates that the axial direction is not constant throughout the outcrop and swings from 170°, which represents the majority of the poles, to 145°. This swing to the southeast is largely to be found in the northern part of the outcrop. Despite the swing, the plunge remains remarkably constant (Figure 9.2). MacDonald's (1968) statement that the Taku Schist anticline plunges southward at an angle between 40 and 60° is not supported by the structural data.

The anticlinal structure is clearly asymmetric in that there is a pronounced maximum at 40° for the dip of the foliation planes on the eastern limb of the anticline, whereas the dips on the western limb range evenly from about 10 to 50°.

RELATION TO OVERLYING ROCKS. Along its eastern margins, the schist is overlain by a sequence of rocks that have been described as very micaceous shale and sandstone. Abundant muscovite and sericite flakes lie parallel to the bedding. Aw (1964) ascribed the abundance of mica to denudation of the older mass of the Taku Schist, and he thereby deduced that the overlying sedimentary rocks rest unconformably on the Taku Schist. Nevertheless, on the west and south sides of the Schist, the overlying rocks have all been metamorphosed in the greenschist facies. Hence the abundance of mica in the overlying rocks, which have been described as shale, must be considered to result from greenschist facies metamorphism of the overlying sediments. Furthermore, the prefered orientation of the mica flakes represents the foliation of the rocks that are really phyllite: intense tropical weathering of phyllite in Malaya can be clearly shown in fresh road cuts to quickly destroy the phyllitic nature and render the rock similar to a shale. For this reason many outcrops of phyllite have been mapped and described as shale.

The actual contact between the Taku Schist and the younger Siba formation (Aw 1964) is well exposed in the Temangan iron mine, and along the new road cutting south of the mine. The contact is irregular (Aw 1964). Along the contact are sometimes found lenses of weathered tuff. A striking feature of the contact is the sharp contrast in metamorphic grade, i.e. amphibolite grade in the Schist, greenschist grade in the Siba formation. Another contrast is in the deformational style. The Taku Schist is characterized by a simple asymmetric structure, whereas the lower grade overlying rocks are isoclinally folded and outcrops generally show steep dips.

The contrast shows that the Taku Schist is tectonically of the infrastructure, characterized by high-grade metamorphism but simple deformation, whereas the overlying rocks are of the suprastructure, characterized by greenschist metamorphism, isoclinal deformation, and a phyllitic nature. In such a case, therefore, the contrast shown at the contact between the two formations is more likely to be a tectonic disconformity than a depositional unconformity.

Several bodies of iron ore have been concentrated along this tectonic disconformity between the infrastructure Taku Schist and the suprastructure phyllitic shales, and the biggest of these has been mined south of Temangan (Figure 24 in MacDonald 1968). In this iron mine the contact between the Taku Schist and the overlying rocks is sharp and clearly faulted. The importance of this structural zone, which defines the eastern margin of the Taku Schist outcrop, has already been referred to in other chapters. The zone also includes the long linear feature of the Temangan ignimbrite and is parallel to the major Lebir fault.

The western margin of the Taku Schist is less well known, but in view of the contrast in metamorphic grade and deformational style across the contact, we can assume a tectonic disconformity similar to that in the east. The age of deposition of the Taku Schist sedimentary rocks remains an open question, but it may well be Lower Paleozoic.

The Stong Migmatite Complex

The Stong Migmatite Complex occupies a mountainous region centered on Gunong Stong and Gunong Ayam, lying about 8 km (5 miles) due west of the railway towns of Kemubu and Dabong, which are situated on the high alluvial plain on the east bank of the Sungei Galas in Ulu Kelantan. Some of the best outcrops occur in rapids in the Sungei Kenerang, a tributary of the Sungei Stong, which in turn flows into the Sungei Galas 3 km (2 miles) downstream from Kemubu (Figures 9.2 and 9.6). Access to the eastern margins of the Complex is gained along the numerous streams that cross the road between Kampong Kuala Balah and Kemubu. This road is isolated from the rest of the country because its bridges are commonly swept away by the floods during the northeast monsoon. The road is also accessible from Dabong. Some of the best exposures are on steep eastward-facing cliffs overlooking this road.

The Complex forms an impressive topographic feature, easily identified from the railway line by the reclining spinelike protrusion at the summit of Gunong Stong. The country is accessible only with difficulty because of steep gradients, waterfalls, rapids, gorges, and fast currents in the streams; but the trouble is compensated by good exposures. The streams become torrential during the monsoon months but are easily accessible from July to September. The region has not yet been fully studied; the following general account is taken from a few notes by D. Santokh Singh (in MacDonald 1968) supplemented by petrographic notes by C. S. Hutchison following his several visits to selected parts of the Complex (partly reported in Hutchison 1969).

PETROLOGY. The Stong Migmatite Complex is unique in Malaya. The rocks are typically of the tectogene or root zone of an orogen, characterized by amphibolite facies dynamothermal metamorphism, migmatization, and well-developed ptygmatic folding. The following descriptions of the rocks are after MacDonald (1968) and Hutchison (1969).

The whole Complex is migmatitic to varying degrees, and the following three general types are the most characteristic; however, all transitions exist among them.

1. Agmatite, in which angular fragments and blocks of darker gneiss and schist are surrounded by more or less homogeneous granitic material. It has the appearance of a breccia. The migmatite appears to have formed from the mechanical injection of granitic material into the country rock in the form of dykes and veins of granite, pegmatite, and aplite. The granitic injections occur both concordant and discordant to the foliation planes of the gneiss and schist. Good examples occur in the Sungai Kenerang (Plate VI A *in* MacDonald 1968) and in the Sungai Seladang (UM 6280[1]).

2. Venite, in which discrete layers and patches of granitic material occur in schist. It differs from the agmatite in that individual layers are narrower, closer, and generally concordant to the foliation, giving the rock a distinctly gneissic texture. Good examples occur in the Sungei Semuliang (Plate VII *in* MacDonald 1968) and in the Sungei Kenerang (UM 6357, UM 6358).

3. Nebulite, in which there is a more complete mixing of granitic material and schist. These rocks are considerably more homogeneous than the agmatite or venite migmatites. They are characterized by schlieren, generally enriched in biotite and hornblende (which give the rock a streaky or crude gneissic texture). Good examples occur in the Sungei Kenerang (Plate IX B *in* MacDonald 1968) and in the unnamed river between the Sungei Seladang and the Sungei Nila, 1.5 km (1 mile) south of Kampong Kuala Balah.

The granitic material of the nebulite and of much of the venite is fine grained and nonporphyritic. The granite material in the agmatite and venite may be coarser grained and somewhat porphyritic. Its foliation may vary from excellent to poor. Occasionally the two types of granitic rock can be seen together, either with regular contacts parallel to the foliation or as irregular discordant contacts, and as discrete masses one within the other in the agmatite. Often they are intimately mixed (MacDonald 1968).

The metamorphic parts of the Stong Migmatite Complex are of schist and gneiss rich in biotite, hornblende, and garnet (Figure 9.4); and of marble. A detailed petrological study is not yet available: only a few rock descriptions have been given by MacDonald (1968) after D. Santokh Singh, and the following descriptions appear to represent the most common rock types within the complex.

Hornblende Schist. The hornblende schist is grey, fine to medium grained, finely banded rock, with 1 to 2 cm long augen of quartz and feldspar. Bands rich in quartz, plagioclase, and orthoclase alternate with others rich in hornblende. The hornblende crystals are well lineated and their segregation into bands defines the foliation. The hornblende is dark in thin section with γ deep blue-green. The crystals are imperfectly terminated and are often crowded with crystals of quartz, sphene, allanite, magnetite and ilmenite. Anhedral quartz is the principal constituent of the rock. The plagioclase is oligoclase to andesine in composition, is anhedral and generally untwinned. A colorless mineral that has refractive indices less than quartz and is full of inclusions has been tentatively identified as cordierite (MacDonald 1968).

Biotite-Muscovite Schist. Biotite-muscovite schist is grey, and very fine grained. In thin section the biotite, which is the most abundant constituent, is deep red. Muscovite and quartz are the other constituents. Inclusions of quartz are common in the biotite porphyroblasts. It is possible that a second generation of biotite is represented by smaller flakes. Euhedral crystals of tourmaline are common. Apatite and iron oxide grains are common accessory minerals. Feldspar is apparently absent.

Nebulite. Nebulite is very-fine-grained finely banded migmatite in which dark bands rich in hornblende-biotite alternate with quartz-rich bands. The hornblende crystals are well lineated and have developed with a poikiloblastic texture. Biotite is well foliated. Quartz forms an equigranular mosaic in which the crystals are strained. Plagioclase occurs as small grains associated with the quartz. Small zircon crystals occur in accessory amounts.

Venite. Venite is grey, finely banded gneiss. Granitic segregations form light-colored bands that alternate with the dark, hornblende-rich bands. The granitic segregations vary in width and are commonly lenticular in shape. The hornblende crystals

---
[1] Specimen number, Department of Geology collection, University of Malaya, Kuala Lumpur.

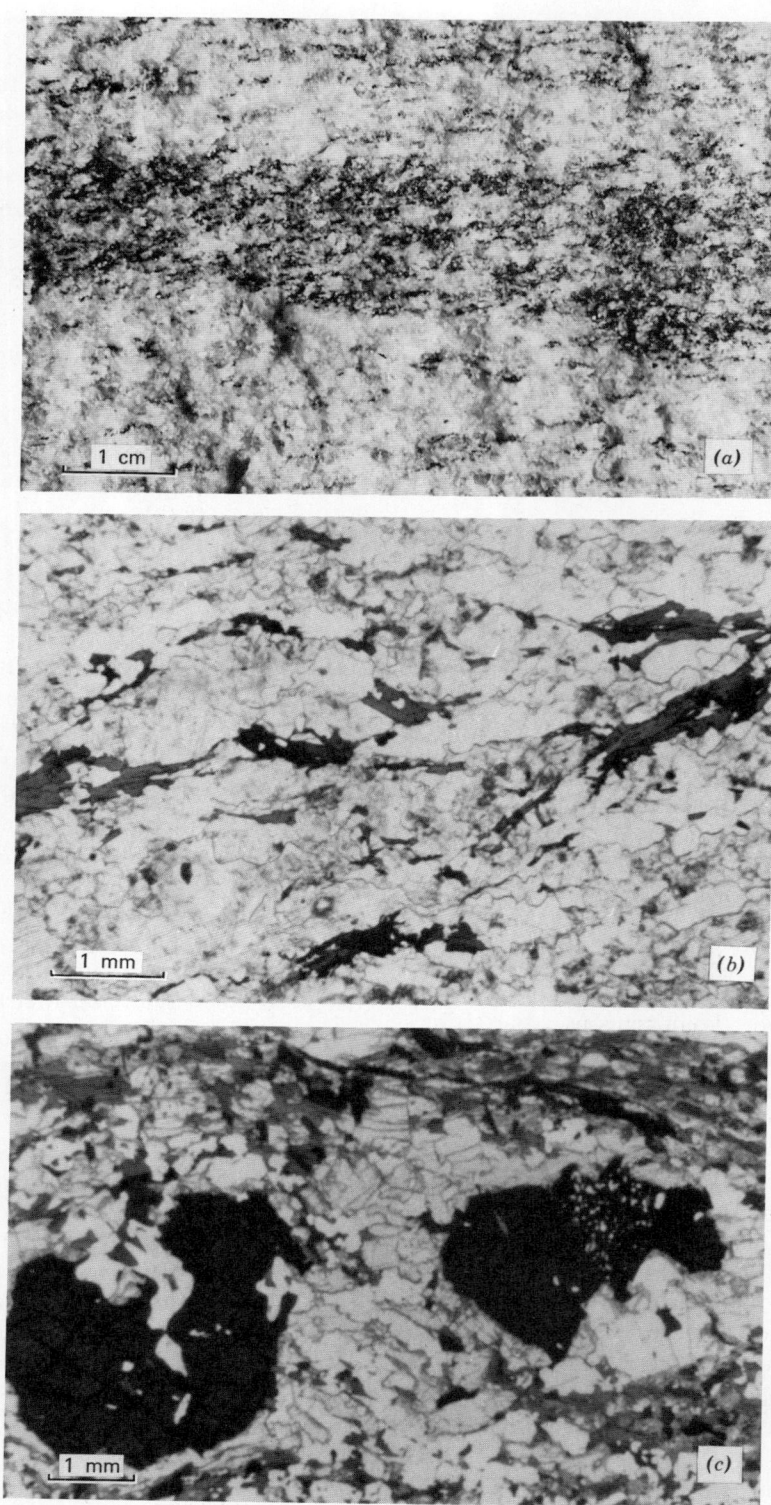

FIGURE 9.4 *Rocks of the Stong Migmatite Complex:* (a) *Biotite gneiss, Dabong, Kelantan, photograph Jaafar bin Haji Abdullah;* (b) *photomicrograph (C. S. Hutchison) of biotite gneiss, Sungei Tuit Besar, specimen UM 5429;* (c) *photomicrograph (C. S. Hutchison) of sillimanite-garnet-biotite gneiss, Sungei Kenerang, specimen UM 6358.*

are blue-green, commonly 1.5 mm long, and crudely lineated. Inclusions of apatite, quartz, and clusters of sphene granules are common in the hornblende crystals. Microcline is the most common feldspar and invariably contains rounded blebs of quartz. Xenoblastic plagioclase commonly occurs as inclusions within the microcline crystals. The plagioclase is of very clear oligoclase, is generally untwinned, and forms local myrmekite with the quartz. The quartz is anhedral and usually strained.

Quartz and Quartz-Feldspar Veins. Veins of quartz and quartz-feldspar of varying thickness are common in the schist and gneiss, especially in the Sungei Kenerang, and are characteristically ptygmatically folded in all degrees of complexity (Plates VIII and IX *in* MacDonald 1968).

Hutchison (1969) described a venite migmatite, which grades to nebulite, from the first rapids on the Sungei Kenrang (UM 6278). It is composed of coarse granitic veins and segregations in a sillimanite-biotite-garnet schist. The mineralogy is as follows: biotite, very abundant and well foliated; garnet, 2-mm diameter porphyroblasts which are perfectly isotropic; quartz, anhedral and more abundant in the granitic segregations; andesine feldspar, strained and generally untwinned; sillimanite, acicular prismatic aggregates of crystals 1 mm long, commonly associated with biotite and often forming euhedral crystals within the plagioclase and cordierite?; muscovite, of minor occurrence. This rock contains some anhedral to rounded crystals that look like cordierite but could not be determined with certainty. However, rounded boulders and pebbles, which are abundant downstream of the Sungei Kenerang rapids, contain abundant cordierite. Specimen UM 6358 of sillimanite-cordierite venite is typical of these.

This rock (UM 6358) is a venite migmatite consisting of narrow granitic veins and segregations varying in width from 2 to 5 mm, separated by schlieren of sillimanite-cordierite-garnet-biotite schist. The biotite is abundant, reddish-brown to yellow, well foliated, and usually of 1 mm maximum dimension. The garnet is subhedral and 3 mm in diameter. Cordierite is abundant as anhedral to rounded colorless crystals. It is characteristically altered along the crystal margins and the imperfect cleavage to sericite aggregates (pinite), giving rise to a laminar texture. Many of the crystals are completely replaced by pinite. The crystals are full of inclusions, characteristically of sillimanite, and some zircon inclusions have pronounced pleochroic yellow to colorless haloes. The 2V is about 80° $-ve$. Plagioclase, which is less abundant, is of andesine composition and is poorly twinned. Quartz is anhedral and strained. Sillimanite is abundant as trains of euhedral acicular prismatic crystals 1 mm long; closely associated with biotite, it characteristically occurs as euhedral inclusions within the cordierite.

Coarsely foliated marble, medium to fine grained and white to pale brown, is interfoliated with the other metasediments in the Stong Complex in the Sungei Stong (Hutchison 1969). It is composed of a crystalline mosaic of calcite with some coarser calcite veining. Pleochroic phlogopite, ranging from colorless to very pale yellowish-brown and having a 2V of about 5°, is very abundant and shows distinct foliation. Diopside occurs as larger colorless porphyroblasts, and quartz and pyrite form accessory amounts (UM 5437). Marble in the Sungei Jelawang (UM 5415), on the eastern slopes of the Complex, also contains phlogopite.

Pyroxene is by no means confined to the marble, and occurs, for example, in a psammitic nebulite at the lower rapids on the Sungei Kenerang (6276). The schist part of the nebulite is fine grained, with average grain size 0.2 mm. It consists of quartz, which is abundant and anhedral; plagioclase, which is untwinned and anhedral; biotite, which is well foliated; and colorless diopsidic augite, which is subhedral to euhedral. Its place is taken by actinolite within the coarser grained granitic segregations. Calcite is also abundant within these segregations. Sphene occurs as an accessory, and some iron oxide crystals are to be found.

On the eastern slopes of the Complex, the rocks are predominantly psammitic. Biotite gneiss is the most common rock type, and the waterfall outcrops on the unnamed river between the Sungei Seladang and the Sungei Nila are typical. Here the predominant rock type is a crudely foliated biotite gneiss with a grain size of 0.7 to 1 mm (UM 5414). The simple mineralogy is as follows: quartz, anhedral and strained; microcline, abundant, anhedral, and generally untwinned or poorly twinned; oligoclase, subhedral with poorly developed twinning, composition around An_{18}; biotite, well foliated but not abundant. The outcrops in this waterfall are predominantly of biotite gneiss, which on close inspection is seen to be nebulite migmatite in which the metamorphic part is psammitic and in which anatexis is advanced. Complex flow folding is common. Elsewhere on the eastern slopes, some of the biotite gneiss is generally well foliated, although locally the foliation may be only poorly deveoped.

In the Sungei Jelawang, the metamorphic parts of the nebulite are of garnet-biotite-muscovite psammite (UM 5416). The garnet porphyroblasts are perfectly isotropic and are poikiloblastic with a maximum diameter of 2 mm. The biotite and muscovite show perfect foliation. Other minerals are plagioclase and quartz.

GEOCHEMISTRY. No chemical analyses of rocks from the complex are available. The only chemical information is that the biotite of the biotite gneiss is characteristically rich in strontium (J. D. Bignell personal communication). This information is of interest because the only other known locality having strontium-rich rocks is in the Benta area of Pahang (Ja'afar bin Ahmad personal communication), and this is also considered to be a catazone metamorphic assemblage (Figure 9.1). No mesozonal or epizonal crystalline rocks have been found to have such high concentration of strontium.

The general aspect of the Stong Complex is that it was derived from a sedimentary sequence that was predominantly arenaceous, with lesser argillaceous and calcareous horizons. The general arenaceous nature has been favorable for large-scale anatexis during the catazone metamorphism.

CONDITIONS OF METAMORPHISM. All the rocks from the Complex so far described can be allocated to the sillimanite-cordierite-muscovite-almandine subfacies of the almandine amphibolite facies of Abukuma-type facies series (the A 2.2 subfacies of Winkler 1967). Of this type of regional metamorphism, Miyashiro (1961, p. 281) wrote:

This type of regional metamorphism appears to be always accompanied by the emplacement of a large amount of granitic rocks. Synkinematic granites are usually abundant in the high-grade parts of the metamorphic terrain.

The mineralogy of the Stong Complex rocks is shown on an *ACF* and *A'KF* diagram (Figure 9.5), which is in fact that of the upper amphibolite facies of the andalusite-sillimanite type of Miyashiro (1961). The presence of cordierite suggests a pressure limit of the Stong Complex during metamorphism of somewhere between 5 and 6 kb, so that the metamorphic facies series is either of Abukuma type or of low-pressure intermediate type. The latter would seem to be the more appropriate choice because that was the deduced facies series for the neighboring Taku Schist. However, the absence of kyanite in the Stong Complex rocks, which are well-exposed in mountainous country, casts doubt on the infrequently recorded identification of kyanite in the rather poorly exposed and somewhat

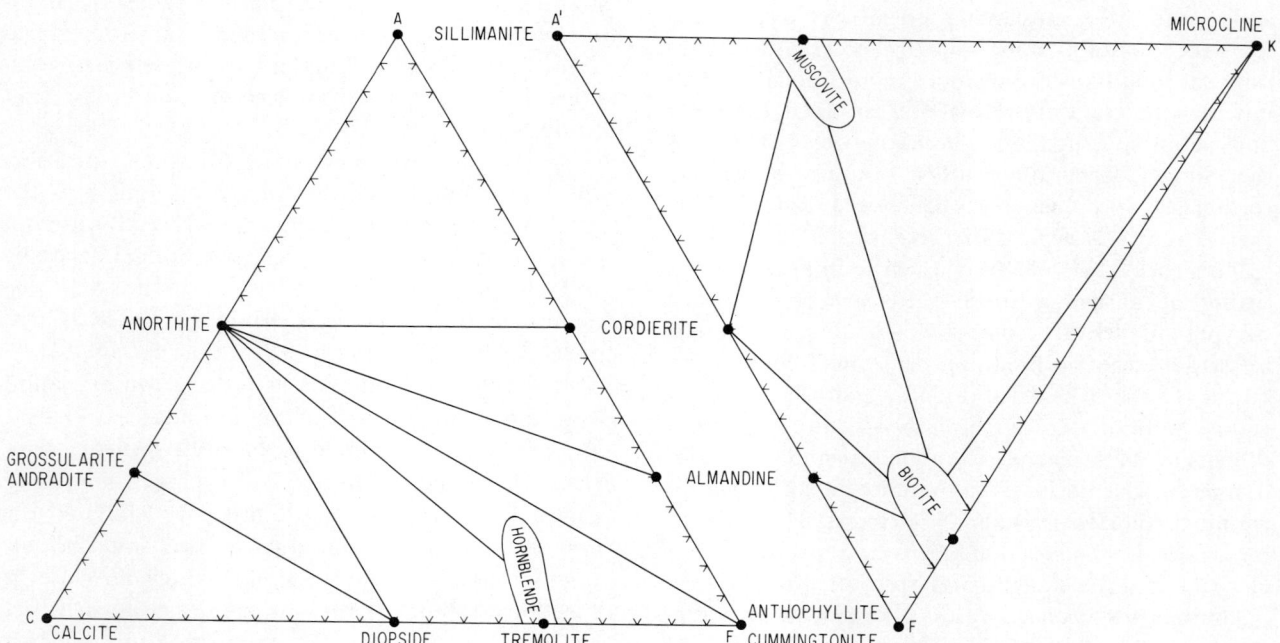

FIGURE 9.5 *ACF and A'KF diagrams to represent the mineral paragenesis of the Stong Migmatite Complex, Kelantan. The mineralogy represents the upper amphibolite facies of the Abukuma Facies Series.*

weathered Taku Schist (see above). Accordingly, we must continue to doubt whether the Stong Complex, together with the Taku Schist, truly belongs to the Abukuma-type or to the low-pressure intermediate facies series. The only certainty is that both Barrovian and burial metamorphism are ruled out for this region, and thus a high thermal gradient of anything betwen 40 and 70°C/km can be confidently deduced.

STRUCTURE. On the geological map of northeast Malaya (Dawson et al. 1968), 38 foliation strike and dip readings are recorded. These, together with four measured by C. S. Hutchison, were plotted and contoured on a lower hemisphere equal-area stereographic net (Figure 9.6). There is quite a scatter at the 1% density level, as revealed by individual poles in the diagram. The 2 and 4% density areas show a crude girdle, for which an axis trending at 60° east of north and plunging at 40° can be deduced. The structure of the Complex is as yet poorly known, and the data of Figure 9.5 may be so incomplete as to be misleading. However, the structural pattern is significantly different from that of the Taku Schist (Figure 9.2), and the present data do not coincide with the prominent NNW-SSE regional trend of the area to the east. The generalization in MacDonald (1968) that the fold axes of the metamorphic rocks have a general northerly trend is misleading.

RELATION TO OVERLYING ROCKS. The Stong Complex is separated from the Taku Schist by a belt of greenschist facies phyllite, marble, and metavolcanic rocks (Figure 9.2). Metamorphism has destroyed any fossils these rocks may have contained, but the similar lithology of less metamorphosed rocks, which occur to the south along the strike, strongly suggests that the greenschist facies rocks belong to the Gua Musang formation and are hence of Permian to Lower Triassic age. Along the eastern margin of the Stong Complex, there is a prominent north-south trending system of vertical faults, usually infilled with pure quartz dykes that occasionally reach 3 m in width. These quartz dykes are clearly visible in the cliff exposures within the river courses to the south of Kampong Kuala Balah. From this evidence the eastern margin of the complex can be taken as a complex fault contact, with the amphibolite facies Stong Complex rocks upfaulted against the greenschist facies Gua Musang formation.

To the south, the Stong Complex is probably represented by gneiss that outcrops in the core of an anticline and is succeeded by greenschist facies metasediments and amphibolites in the Ulu Nenggiri area of Ulu Kelantan (D. J. Gobbett personal communication). As in the east, metamorphism has destroyed any fossil content, but the presence of a prominent marble formation in the sequence suggests a correlation with the Gua Musang Formation that occurs immediately to the south. To the north, the Complex is succeeded by greenschist facies metasediments in the Ulu Pergau region, characterized also by some marble and metavolcanic horizons.

To the west, all published maps show the Complex as being continuous with the Main Range Granite. However, this continuity is not based on field mapping, since the region is difficult of access and is virtually unknown. Furthermore, the Stong Complex is a catazone assemblage of amphibolite facies rocks with migmatite and foliated granite gneiss, whereas the Main Range Granite is generally a coarsely porphyritic batholith of the mesozone. The relation between the Stong Complex and the Main Range batholith are of particular interest but will probably remain unknown for some time because of the inaccessibility of the terrain.

The Benta Migmatite Complex

An interesting migmatite complex outcrops in the region of Benta, Pahang. It occupies the western foothills of Gunong Benom and is well exposed in the Sungei Lipis immediately west of Benta. Scrivenor (1931) described the rocks of the Benta Migmatite Complex as hornblende granite, but he also mentioned exposures of syenite. The first detailed account of the Complex was by Richardson (1939), who identified four major rock types: "perknites" (pyroxenites and amphibolites), "hybrid rocks," "flow-banded rocks," and minor acid-intrusive rocks. He used the name "syenite" for many of the rock types. His view was that the variety of rock types in this vicinity had resulted from the assimilation (hybridization) of earlier ultrabasic rocks by the Benom Granite, which lies immediately to the east.

Khoo (1968) was the first to demonstrate that all the rocks of this region have been metamorphosed, and he suggested a low-pressure facies series. He still maintained the term "syenite." Hutchison (1971) has shown that, in effect, a migmatite complex that is apparently unrelated to the Benom Granite (although intruded by it along its western margins) is involved. He proposed that the terms "hybrid rocks" and "syenite" be abandoned.

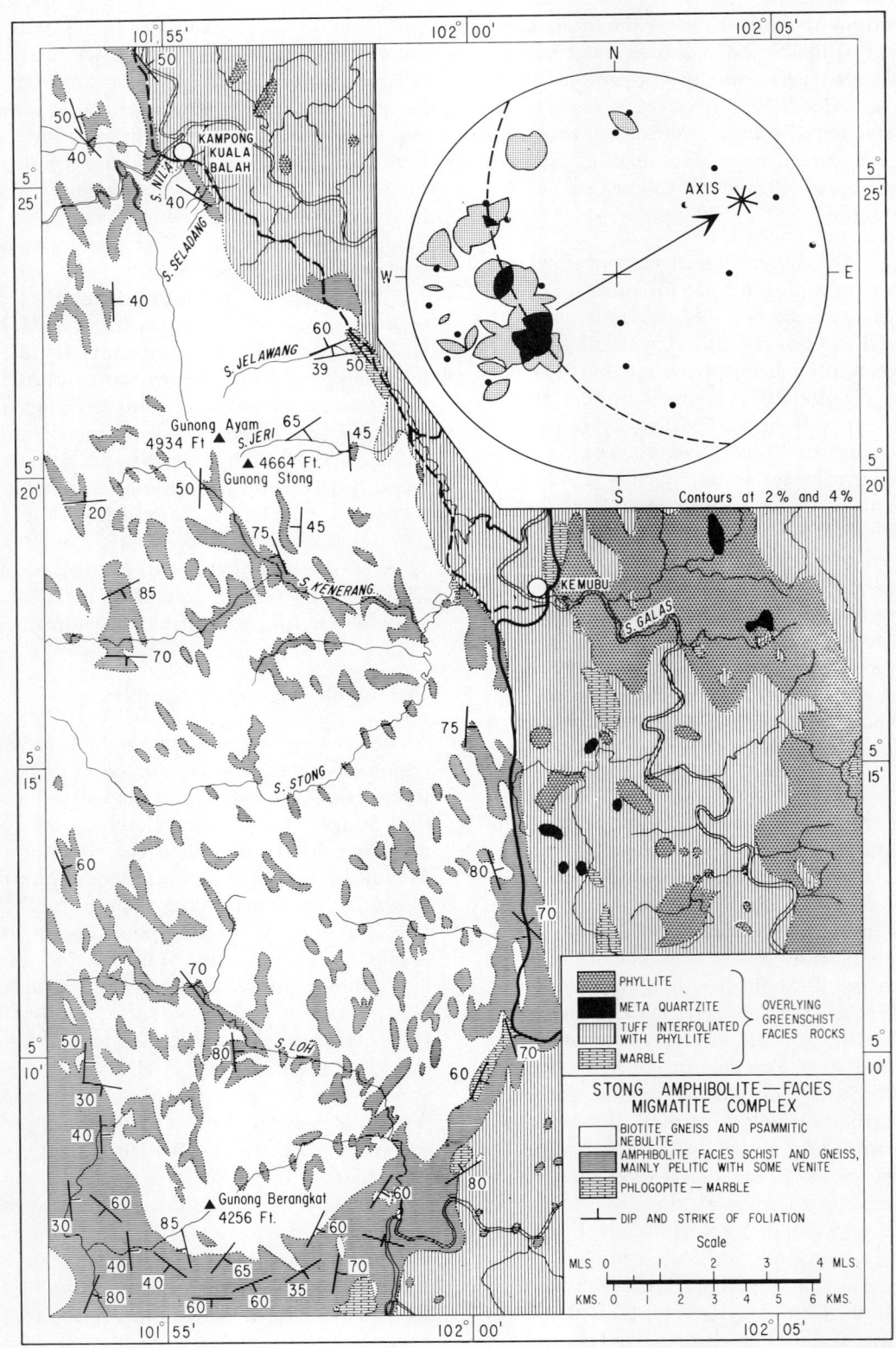

FIGURE 9.6 *Geological map of the Stong Migmatite Complex, Kelantan, after Dawson et al. (1968). The lower-hemisphere equal-area stereogram is of 42 foliation readings collected from the Complex.*

DISTRIBUTION. The belt of the Benta Migmatite Complex extends southward from the immediate neighborhood of Benta to a point about 10 km (6 miles) southeast of Bukit Serdam, in the Raub district of Pahang. This belt of country occupies the land between the main road from Raub to Benta and Gunong Benom. Good exposures are to be found in the Sungei Chin Chin, the Sungei Temurai, the Sungei Cheriong, the Sungei Klau, and the Sungei Teping. The most accessible and the most informative exposures in the Complex are presented by the Sungei Lipis and the J.K.R. quarry, 4 km (2.5 miles) and 3.2 km (2 miles) west of Benta, respectively. These have been described by Hutchison (1971) and are considered below. The geology of these two areas is given in Figure 9.7.

PETROGRAPHY. The Benta Migmatite Complex has five rock types.

1. *Psammite*: a microcline-oligoclase-biotite-quartz-(hornblende) psammitic gneiss. It is fine grained and well foliated (Figure 9.12). The biotite is pleochroic, from light brown to dark green. The alkali feldspar makes anhedral porphyroblasts of up to 4 mm in diameter and full of rounded quartz and plagioclase crystals. Well-formed poikiloblastic crystals of hornblende are prominent. Sphene and pyrite are important accessories. The psammite is apparently the oldest of the rocks in this area.

2. *Foliated gneiss*: a microcline-oligoclase-hornblende-biotite gneiss. It is coarse grained and well foliated. The rock texture is dominated by numerous large tabular to ellipsoidal crystals of alkali feldspar, common dimensions are 3 cm × 3 mm (Figure 9.8). The mafic minerals, quartz, and plagioclase, form the matrix to these K-feldspar porphyroblasts. In hand-specimen the K-feldspar porphyroblasts are pale grey, often with a tinge of pink. The matrix is dark green to black. The most spectacular feature of this rock, which had been called a "flow-banded syenite" by Richardson (1939), is the ellipsoidal to tabular nature and perfect alignment of the K-feldspar crystals.

All the larger crystals are characterized by a simple contact Carlsbad twin. The crystals contain a series of parallel ellipsoidal trains of inclusions of resorbed plagioclase, biotite, and quartz (Figure 9.8). The ellipsoidal ghost ornamentation is characteristic of the larger crystals only. This may be described as a rapakiwi texture (Hutchison 1971). Poorly developed margins of myrmekite occasionally outline the porphyroblasts. The K-feldspar is of cryptoperthitic to microperthitic microcline. As in the psammite, the biotite is green.

3. *Metabasite*: a hornblende-biotite-(epidote)-andesine-orthoclase schist (or gneiss). This rock has been called a perknite by Richardson (1939). It is dark green to black in hand-specimen and is only crudely foliated. It is made up essentially of abundant hornblende and biotite, both of which have a common pleochroic scheme from pale yellow to dark green. K-Feldspar occurs as sieve porphyroblasts.

4. *Monzonite*: a coarse grained porphyritic rock. Monzonite is characterized by large phenocrysts of K-feldspar often up to 3 cm long. (Figure 9.8). Differing from the foliated gneiss, the K-feldspar crystals have no ellipsoidal ghost ornamentation. The rock is generally unfoliated, although locally it may be crudely foliated. The matrix (average grain size 5 mm) is composed of green biotite, hornblende, plagioclase, and quartz. Accessory minerals are sphene, apatite, and minor epidote.

5. *Leuco-microgranite*: average grain size 1 to 2 mm; equigranular intergrowth of quartz, oligoclase, and K-feldspar. There are minor amounts of epidote. Rapakiwi texture is well developed, and the plagioclase may occur either as discrete crystals or as rims mantling cores of alkali feldspar. These rims are of oligoclase and are well twinned.

Mutual relations of the rock types. The psammite occurs as boudinaged concordant inclusions within the foliated gneiss (Figure 9.9) and as larger irregularly shaped inclusions within the monzonite. It is also intimately intruded by the monzonite in the Sungei Lipis section. The psammite and the foliated gneiss share a common foliation (Figure 9.9).

The foliated gneiss and the monzonite have similar mineralogy and similar bulk composition, and the field relations between the two are perplexing. The contacts may be planar, as in the river section, characterized by small-scale block displacements, or they may be both gradational and blocky in the same locality (Figure 9.10).

The monzonite and the metabasite occur together as an agmatite or nebulite, well displayed in the Benta quarry.

The microgranite occurs as a sill, intrusive into the other rocks in the quarry.

CHEMISTRY. The chemical analyses of six rock specimens which are typical of the area are presented in Table 9.2. A notable feature of all these analyses is their high concentration in alkalis, especially potassium; and as seen previously, all the

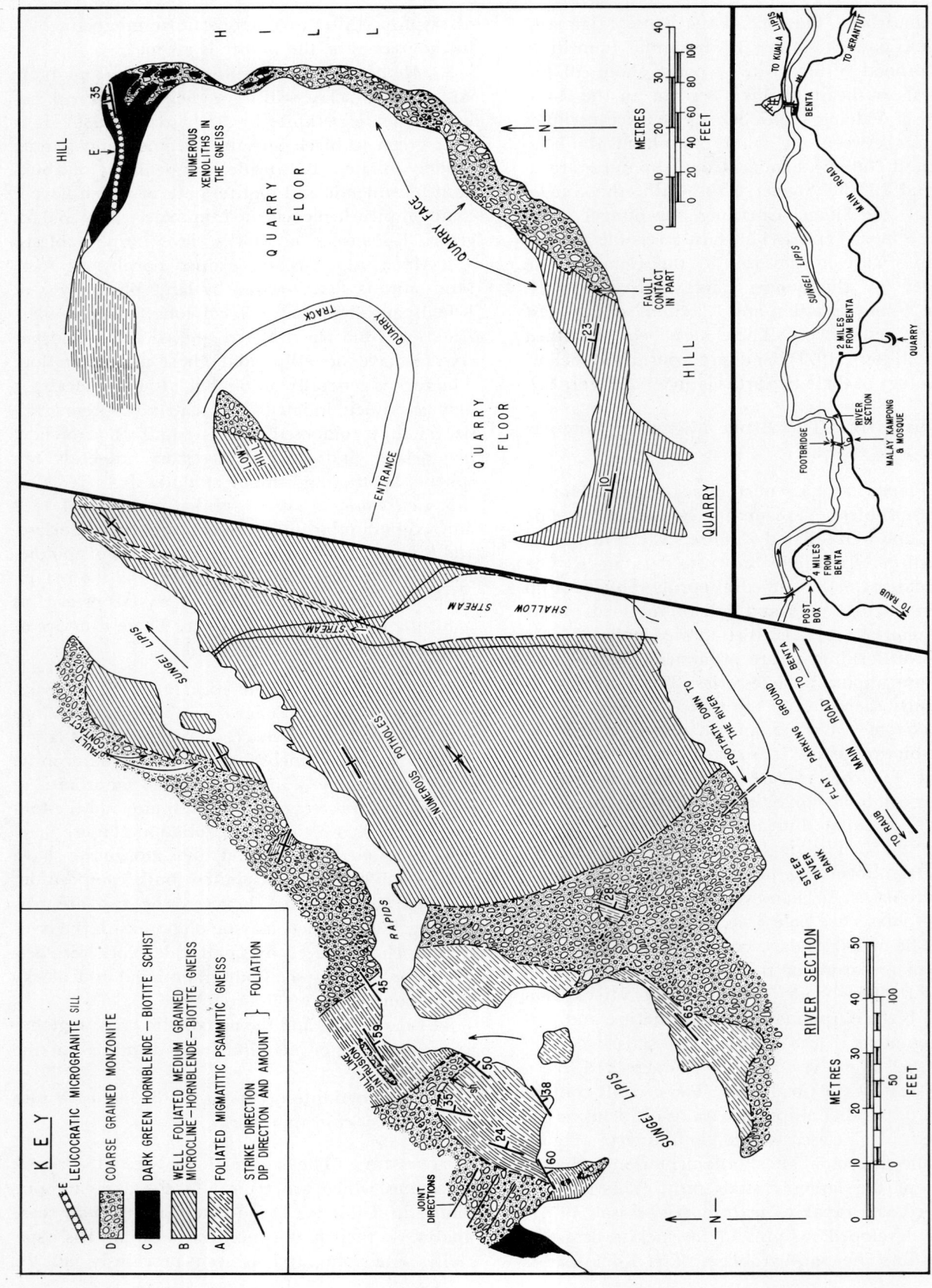

FIGURE 9.7 *Geological map of the Benta migmatite in the Sungei Lipis and in Benta J.K.R. Quarry, Pahang. After Hutchison (1971, Fig. 1).*

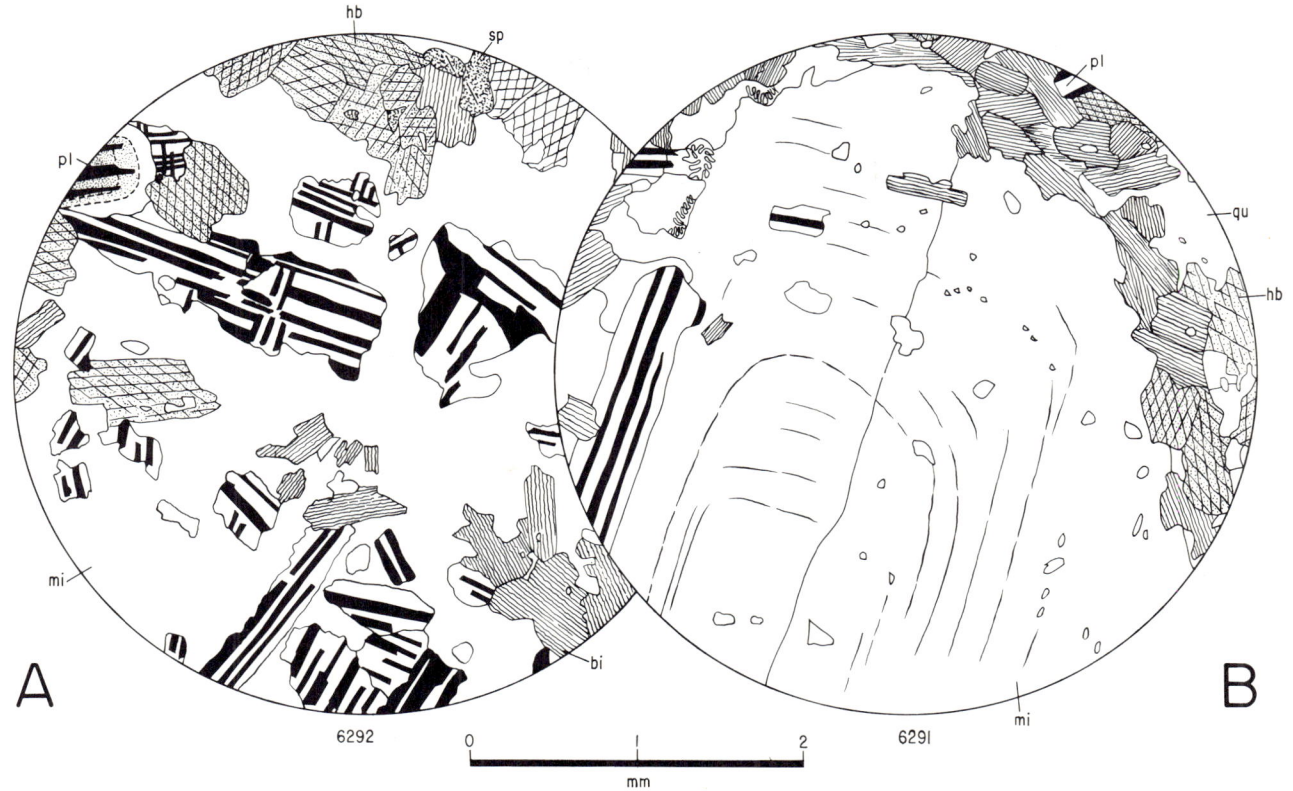

FIGURE 9.8 *Photomicrograph drawings of the Benta migmatite:* (a) *Large K-feldspar porphyroblast in monzonite. It is anhedral and contains poikiloblastic inclusions of plagioclase, quartz, biotite, and hornblende, specimen UM 6292;* (b) *the termination of a large ellipsoidal K-feldspar porphyroblast in foliated gneiss, characterized by ellipsoidal ornamentation and oriented inclusions, specimen UM 6291. qu = quartz, mi = microcline microperthite, hb = hornblende, bi = biotite, pl = plagioclase, sp = sphene.*

FIGURE 9.9 *Foliated gneiss in the Sungei Lipis near Benta, Pahang. The foliation is defined by the alignment of tabular ellipsoidal microcline microperthite porphyroblasts. Note the common oriented inclusions of psammite in conformity with the foliation. Photograph C. S. Hutchison.*

271

TABLE 9.2 *Chemical analyses and Niggli molecular norms of selected rocks* from the Benta area (After Hutchison, 1971, Tables 1 and 2.)*

	Weight Percentages					
	1	2	3	4	5	6
SiO_2	51.04	48.52	48.06	49.66	50.18	52.11
Al_2O_3	16.89	16.24	19.22	13.75	18.76	15.79
TiO_2	1.11	1.15	1.38	1.36	1.46	0.84
Fe_2O_3	1.92	2.47	1.45	2.00	1.22	2.23
FeO	5.17	6.36	6.30	7.10	5.92	5.71
MnO	0.16	0.13	n.d.	n.d.	n.d.	0.11
MgO	5.65	7.40	6.78	9.90	6.23	5.70
CaO	8.19	9.22	8.54	7.60	6.56	7.62
Na_2O	2.18	1.42	1.19	0.29	2.06	1.89
K_2O	5.26	2.90	4.46	5.06	4.76	4.23
H_2O^+	0.59	2.08	n.d.	n.d.	n.d.	2.38
H_2O^-	0.32	0.12	n.d.	n.d.	n.d.	0.13
P_2O_5	0.69	0.64	0.45	0.77	0.58	0.64
CO_2	0.10	0.10	n.d.	n.d.	n.d.	0.10
Total	99.27	99.65	97.83	97.49	97.73	100.07

	Niggli Molecular Norms					
	1	2	3	4	5	6
Quartz	nil	0.83	nil	nil	nil	1.27
Orthoclase	31.49	17.28	26.90	30.77	28.61	25.73
Plagioclase	38.14	42.21	45.56	24.58	47.19	39.96
	$Ab_{45}An_{55}$	$Ab_{29}An_{71}$	$Ab_{24}An_{76}$	$Ab_{11}An_{89}$	$Ab_{40}An_{60}$	$Ab_{42}An_{58}$
Nepheline	1.64	nil	nil	nil	nil	nil
Halite	n.d.	0.22	n.d.	n.d.	n.d.	0.23
Total salic	71.28	60.54	72.46	55.34	75.79	67.19
Magnetite	2.03	2.60	1.55	2.15	1.30	2.40
Ilmenite	1.57	1.62	1.96	1.95	2.07	1.20
Apatite	1.46	1.52	0.96	1.66	1.23	1.55
Fluorite	n.d.	3.91	n.d.	n.d.	n.d.	0.10
Pyrite	n.d.	nil	n.d.	n.d.	n.d.	0.03
Calcite	0.26	0.26	n.d.	n.d.	n.d.	0.26
Diopside	11.94	4.32	4.49	9.39	0.71	9.18
$Wo_{50}En_xFs_{50-x}$	$En_{37}Fs_{13}$	$En_{38}Fs_{12}$	$En_{37}Fs_{13}$	$En_{39}Fs_{11}$	$En_{37}Fs_{13}$	$En_{36}Fs_{14}$
Hypersthene	nil	25.23	2.80	23.91	4.97	18.08
	—	$En_{75}Fs_{25}$	$En_{73}Fs_{27}$	$En_{78}Fs_{22}$	$En_{73}Fs_{27}$	$En_{71}Fs_{29}$
Olivine	11.47	nil	15.77	5.60	13.93	nil
	$Fo_{74}Fa_{26}$	—	$Fo_{73}Fa_{27}$	$Fo_{78}Fa_{22}$	$Fo_{73}Fa_{27}$	—
Total femic	28.72	39.46	27.54	44.66	24.21	32.81

n.d. = not determined.

* Specimens 1–6 as follows:
1. Foliated gneiss (flow-banded syenite) by courtesy of Ja'afar bin Ahmad (JA 97) Geological Survey of Malaysia, Benta area. Contains also Sn, 2 ppm.
2. Metabasite, dark green schist (12457E, Richardson, 1939) (02.3.001, Alexander et al. 1964), Benta quarry. Also analyzed for ZrO_2, nil; Cl, 0.07; F, 0.94; S, nil; BaO, 0.30; Au, nil; MoS_2, nil; C, tr.; less O for F, S, Cl, 0.41.
3. Feldspathized xenolith of metabasite in the monzonite (agmatite) (13104, Richardson, 1939) (02.3.002, Alexander et al. 1964), Benta quarry. Density 2.92.
4. Feldspathized xenolith of metabasite in the monzonite (agmatite) (13105, Richardson, 1939) (02.3.003, Alexander et al. 1964), Benta quarry. Density 2.92.

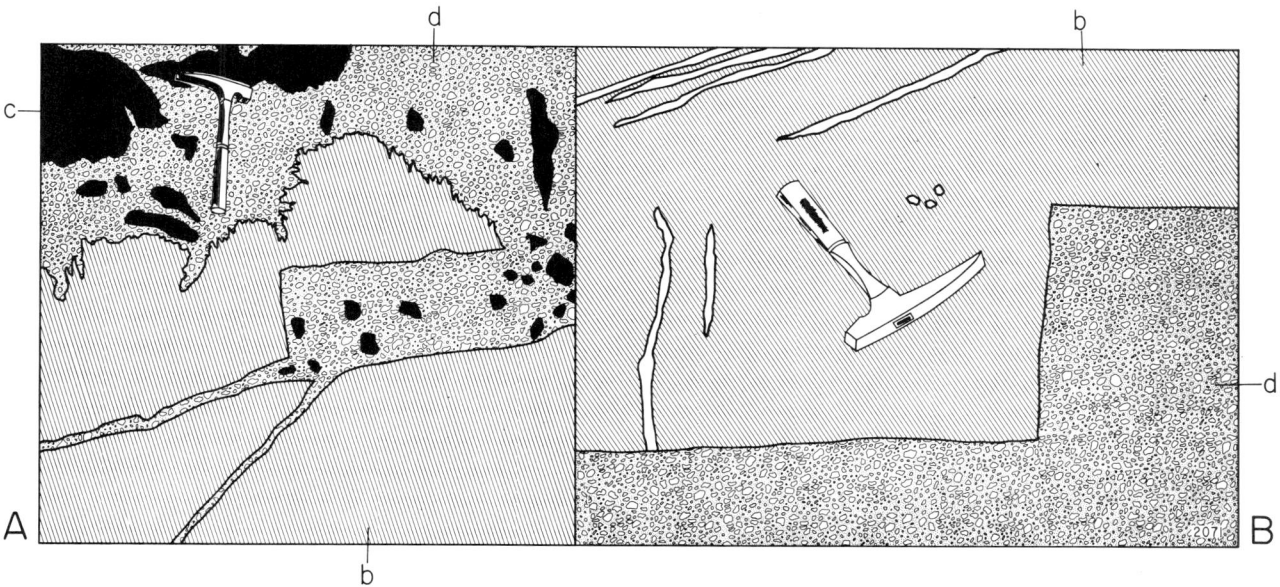

FIGURE 9.10 *Drawings from photographs showing the nature of the contact between the monzonite and the foliated gneiss in the Benta Migmatite Complex. After Hutchison (1971, Fig. 5):* (a) *Quarry exposure showing sharp, joint-controlled, and diffuse contacts;* (b) *exposure on the north bank of the Sungei Lipis showing a joint-controlled contact. Note the premonzonite veining of the foliated gneiss. b = foliated gneiss, c = metabasite, d = porphyritic monzonite.*

rocks contain alkali feldspar porphyroblasts or phenocrysts. The table also includes Niggli norms calculated by the method of Hutchison and Jeacocke (1971).

The difference between the monzonite (6 in Table 9.2) and the foliated gneiss (1 in Table 9.2) is very slight. The normative plagioclase composition is similar (ca. Ab_{42} to Ab_{45}) and the diopside is around En_{36} to En_{37}. The only difference is that the monzonite contains normative quartz, whereas the foliated gneiss is undersaturated.

A triangular plot of cationic percentages of total iron–total alkai–magnesium (Figure 9.11) shows the similarity of the monzonite (analyses 5 and 6) and the foliated gneiss (1). Points 2, 3, and 4 are of the metabasite, which occurs as a paleosome in the monzonitic neosome of the agmatite and nebulite. This rock is of approximate gabbroic composition.

PETROGENESIS. Hutchison (1971) recognized that these rocks were of deep-seated origin and he described their migmatitic nature. He proposed the following sequence of events to account for their present mineralogy, texture, and relationships.

1. The rocks represent an originally sedimentary sequence of lithic sandstone, andesite tuff, and shale in the deeper parts of the Malayan geosyncline.
2. Burial to a depth of about 9 km, in an orogenic region with a geothermal gradient of about 70°C/km.
3. Folding and metamorphism in the sillimanite-cordierite-orthoclase-almandine subfacies of the amphibolite facies of Abukuma type, in which a tectonic temperature of around 650°C and a pressure of about 3 kb was maintained. The tuff became a plagioclase-rich gneiss upon metamorphism, the shale became a schist, and the lithic sandstone became a psammitic gneiss.
4. During the metamorphism, potassium and silica-rich solutions permeated the rock sequence, being squeezed out from the deeper crustal regions undergoing granulite facies metamorphism. The rocks of these deeper regions are not anywhere ex-

---

5. Feldspar-rich nebulite (hybrid between monzonite and dark green schist) (13106, Richardson, 1939) (02.3.004, Alexander et al. 1964), Benta quarry. Density 2.85.
6. Monzonite (quartz-biotite-hornblende syenite) (12456B, Richardson, 1939) (02.2.002, Alexander et al. 1964), Benta quarry. Analyzed also for ZrO_2, nil; Cl, 0.07; F, 0.08; S, 0.01; BaO, 0.39; Au, nil; MoS_2, nil; C, 0.10; less 0 for F, S, Cl, 0.06.

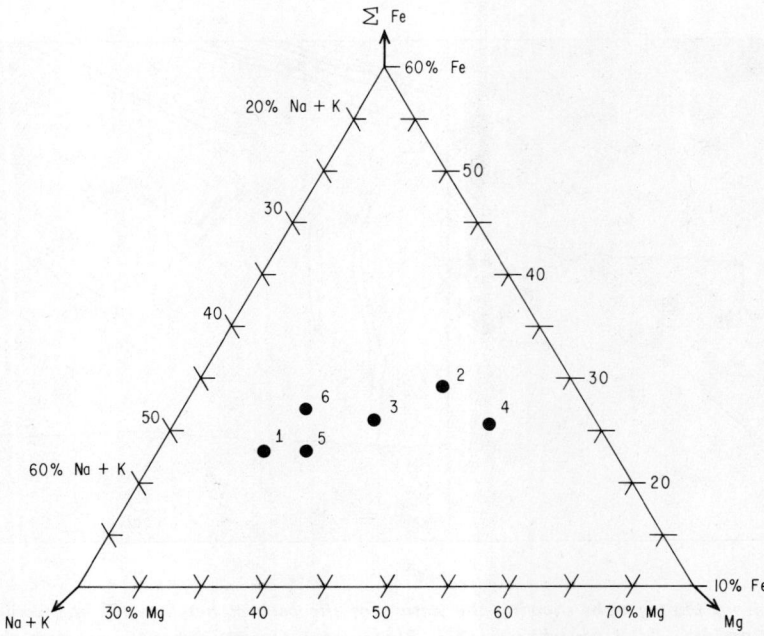

FIGURE 9.11 *Triangular variation diagram of the six rocks whose chemical analyses are given in Table 9.2. The coordinates are of cationic total iron versus total alkalis versus magnesium. After Hutchison (1971, Fig. 10).*

posed. The potassium was preferentially taken up by the abundant plagioclase of the metamorphosed andesite tuff to give the foliated microcline gneiss in which alkali feldspar porphyroblasts show excellent rapakiwi texture, characterized by zonal growth and replacement of other minerals. The alkali metasomatism also introduced K-feldspar into the schist and to a lesser extent into the psammite. During this phase of potassium metasomatism, muscovite was prevented from forming by the prevailing conditions of high temperature and relatively low pressure (not exceeding 3 kb).

5. Orogenic movement in connection with the metamorphism caused the brittle psammite to break, while the interfoliated tuff (now a foliated gneiss) acted in a plastic and pasty manner. The psammite bands therefore became discontinuous and boudinaged (Figure 9.9). This genesis explains the structural and mineralogical conformity of the psammite and the foliated gneiss. The conditions of pressure and temperature must have been sufficient to cause partial anatexis of the more favorable psammite, giving rise to a venite migmatite (Figure 9.12).

6. Following the orogenic folding phase, a period of more static conditions, still under high temperature, witnessed the rheomorphism of the preexisting metamorphic rocks. The rheomorphism of the foliated gneiss was facilitated by a joint system. The metabasite was completely brecciated. The foliated gneiss, being more feldspathic, was more susceptible to anatexis, and large volumes of newly produced melt gave rise to the monzonite. As expected, the monzonite would be somewhat more salic than the paleosome gneiss (Table 9.2). The fraction of the schist that did not melt was also enriched in calcic plagioclase and mafic minerals, forming a metabasite paleosome. The monzonite formed the matrix (neosome) of the agmatitic migmatite, in which are included irregularly shaped brecciated blocks of metabasite and psammite.

7. A microgranite sill, probably anatectic in origin, was intruded. Alkali metasomatism continued, giving rise to rapakiwi textures. But muscovite was prevented from forming by continuing high regional temperature.

8. Uplift and cooling caused retrogressive metamorphism and the formation of epidote, and later zeolite.

It is interesting to compare the foregoing petrogenesis with that porposed by Richardson (1939, p. 61):

1. The flow-banded syenite "(foliated gneiss)" was emplaced first, and, later, veins of feldspar-rock and biotite-feldspar rock were introduced into it.

FIGURE 9.12 *Foliated venite psammite from Benta Quarry, Pahang. It is intruded subconcordantly by monzonite in the upper left-hand corner. The paleosome is of biotite-plagioclase-quartz psammite with rare calc-silicate bands. The neosome is of granitoid K-feldspar and quartz veins and segregations with associated hornblende porphyroblasts. Photograph Ja'afar bin Haji Abdullah.*

2. An ultrabasic rock "(metabasite)", probably pyroxenite, invaded the syenite, and its chilled margin became strongly brecciated.

3. Feldspar-rich rock, comprising orthoclase and labradorite in various proportions "(monzonite)," intrusive into the brecciated pyroxenite, produced a suite of feldspathic hybrids "(agmatite and nebulite)." According as the feldspar added was dominantly orthoclase and labradorite in approximately equal quantities, or dominantly labradorite, so alkalic hybrids were produced in great abundance, next abundant were monzonitic varieties, and calc-alkalic types were the least common.

4. After consolidation of all these rocks, numerous veins of epidote were introduced, of them carrying quartz, calcite, fluorite, and sulphide minerals.

Richardson (1939) implied a genetic relationship to the Benom granite, and as seen from the sequence just quoted, he explained the whole rock complex in terms of magmatic events. Magma was undoubtedly responsible for part of the complex, but the migmatitic nature of many of the rocks is beyond doubt, and hence a deep-seated origin with high-grade metamorphism and migmatization is to be favored.

SOUTHWARD EXTENSION OF THE COMPLEX. The foliated gneiss occurs extensively in the Sungei Ruan area (Khoo 1968), where it is the most abundant rock type (Figure 9.13). The percentage of K-feldspar porphyroblasts varies from 20 to 60% of the rock by volume. The gneiss is intruded by dykes that, like the gneiss, have been metamorphosed. Now amphibolites, they were probably diabase dykes (Khoo 1968). The gneiss contains xenoliths of both psammite and metabasite, which contain K-feldspar porphyroblasts. One important feature is that the alignment of the K-feldspar porphyroblasts in the gneiss is not deflected by the xenoliths (Khoo 1968). Migmatization is well developed in several areas.

The metabasite in the Sungei Ruan area contains both orthopyroxene and clinopyroxene up to 25% total by volume. Khoo (1968) has suggested that these rocks are derived from igneous parents, and in some places he has described relict pyroxene crystals with coronas of actinolite. He has described xenoliths of metapyroxenite in a variety of rock types, including the foliated gneiss, another metabasite, and the Benom granite.

Khoo (1968) has described many indications of potash metasomatism in the rocks of this complex, including the presence of what he interpreted to be relict plagioclase zoning in large K-feldspar crystals, the occurrence of rims of K-feldspar around metapyroxenite xenoliths in metabasite, and the porphyroblastic growth of K-feldspar in a variety of

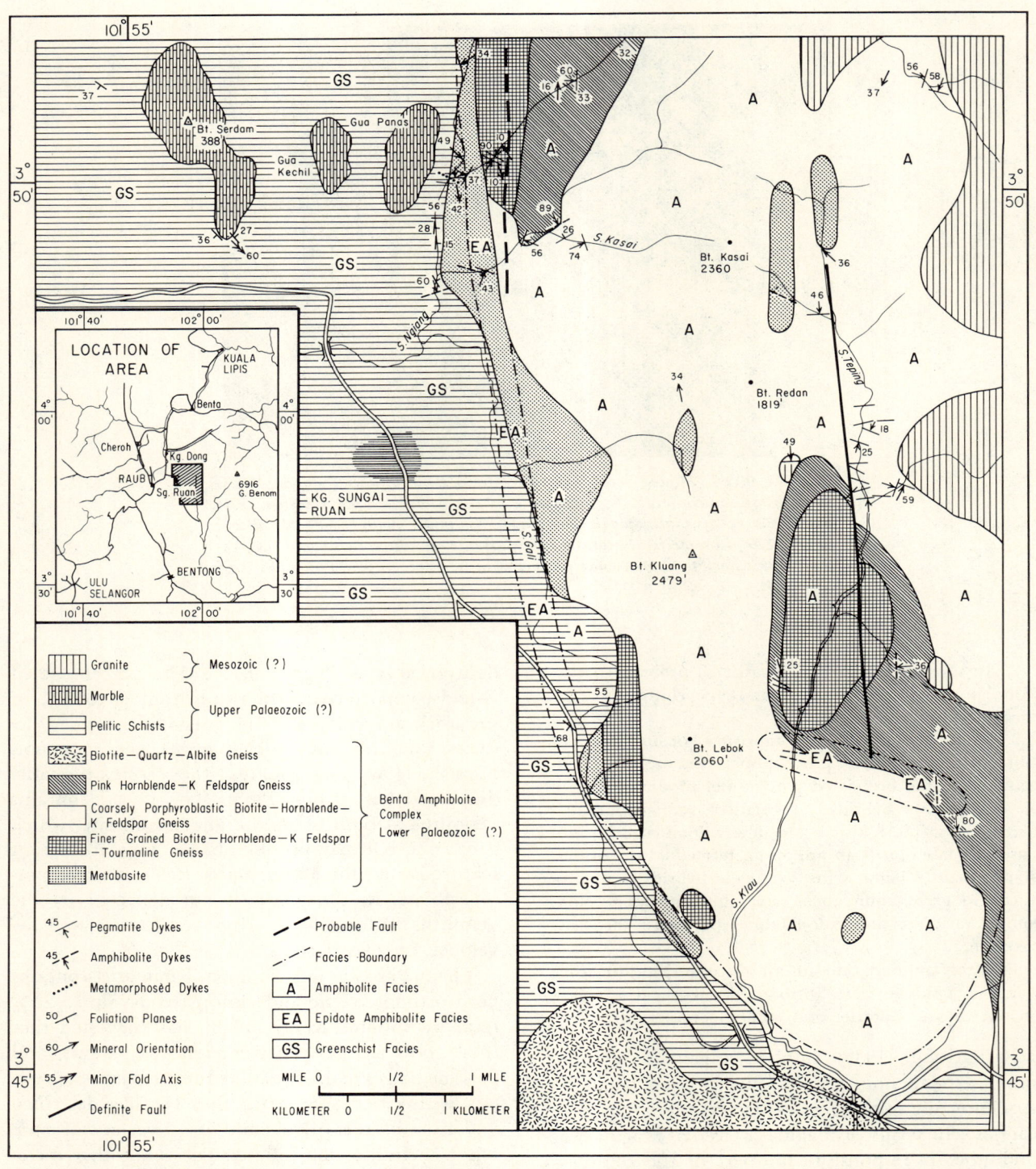

FIGURE 9.13 *Geological map of the Benta Migmatite Complex in the Sungei Ruan area, near Raub, Pahang, showing also the overlying greenschist facies metasediments. After Khoo (1968).*

rocks, including xenoliths and amphibolite dykes. He suggested that the source of potassium was the nearby Benom Granite.

GENERAL CONSIDERATIONS. Richardson was convinced that the metabasites were ultimately of igneous origin and stated that there is "conclusive proof that pyroxene has been converted to amphibole in many rocks of the basic hybrid suite, and the occurrence of perknites in the Sungei Gok, ranging from pyroxenite, through pyroxene hornblendite, to hornblendite suggests that the change has taken place on a large scale" (1939, p. 54). He concluded that the migmatite, in which metabasite xenoliths occur within a monzonitic matrix, was a result of hybridization of an original ultrabasic rock by the Benom granite.

Richardson did not entertain the possibility that the monzonite was derived by partial anatexis of the complete rock mass, thereby splitting into a more basic paleosome (the metabasite) and a more acidic neosome (the monzonite). Accordingly, he proposed the following overall history of the Gunong Benom Granite and the rocks of its western foothills.

Stage 1. Intrusion of ultrabasic magma along a region now occupied by the western foothills of the Gunong Benom Range.

Stage 2. Differentiation of this ultrabasic body into peridotite, pyroxenite, and gabbro, with some assimilation of calcareous sediments.

Stage 3. Solidification of the differentiated ultrabasic intrusion, accompanied by flow banding (producing Richardson's "flow-banded syenite").

Stage 4. Emplacement of the Benom granite. Thermal metamorphism of adjacent sediments and hybridization of the ultrabasic mass.

Stage 5. Differentiation of the granite into coarsely porphyritic, microgranitic, and aplitic varieties.

Stage 6. Solidification of the granite and production of a joint system.

Stage 7. Veining with aplite, pegmatite, quartz-feldspar, and epidote-rich veins.

So we have two opposing interpretations of the Benta Migmatite Complex. One by Richardson (1939), which invokes hybridization by the Benom granite, the other by Hutchison (1971), which does not involve the Benom granite but attempts to explain the whole complex as a result of deep-seated migmatization of the geosynclinal sediments and interbedded tuffs.

The deep seated nature of the Complex is beyond doubt—the venite migmatite illustrated in Figure 9.12 was produced under conditions of high-grade regional metamorphism that led to partial anatexis; furthermore, the Benta Complex lies along the line of country that includes other high-grade metamorphic rocks (Figure 9.1). However, more study is required before the complete history can be interpreted. The Complex is unique in the geology of Malaya in having many features characteristic of the metamorphic and migmatite terrane of, for example, Sweden, in particular a variety of migmatite types and spectacular rapakiwi textures indicative of pronounced and widespread alkali metasomatism. Had this alkali metasomatism had its source in the Benom granite, and had all the processes taken place at a higher tectonic level and corresponding lower temperature, a pronounced enrichment in muscovite would have been expected. Yet the whole complex is characterized by the absence of muscovite and an enrichment only in microcline. The biotite crystals in all the rock types are characteristically green and, in common with the Stong Migmatite Complex, they are rich in strontium (Ja'afar bin Ahmad personal communication). These are the only two known regions in Malaya that have a characteristic high strontium concentration, and this similarity links them both to the catazone of the Malay Peninsula. The Gunong Benom granite is distinctly discordant to the country rocks and is certainly not of the catazone. It is accordingly younger than the Benta Migmatite Complex, but it undoubtedly must have had some modifying effect on the eastern extremities of the Complex.

The Bukit Ranjut Complex

An approximately circular outcrop of rocks similar to the Benta Complex and 6.5 km (4 miles) in diameter, is centered on Bukit Ranjut, about 24 km (15 miles) north-northwest of Kuala Lipis. This region was mapped and described by Richardson (1950, p. 85), who noted that the rocks exhibit "some degree of mineralogical and petrological similarity with the hybrid suite of the Gunong Benom foothills for they range from pyroxene diorite, through pyroxenite, hornblendic and biotitic syenite, to granites." Although Richardson described the complex as entirely of igneous origin, his descriptions do occasionally suggest a metamorphic paragenesis. The position of the Bukit Ranjut complex along a line of country between the Benta Migmatite Complex to the south and the Stong Migmatite Complex to the north suggests that a possible metamorphic origin should also be sought in this case. However, since I have not seen these rocks, I can only report the findings and interpretations of Richardson (1950). The following summary is from his published work.

PETROLOGY. The most important minerals of the Bukit Ranjut Complex are microcline, microcline-microperthite, orthoclase, albite, pargasite, and biotite. Quartz, oligoclase-andesine, myrmekite, augite, and diopside are less important; muscovite is comparatively rare; ilmenite, magnetite, zircon, apatite, pyrite, sphene, and allanite are the most common accessory minerals; leucoxene, limonite, ilmenite, chlorite, sphene, epidote, clinozoisite, zoisite, and rutile are secondary minerals (Richardson 1950). Gold occurs in some of the quartz veins together with traces of ?molybdenite.

The rocks are extremely variable over the Complex, and because outcrops are not extensive, it has been impossible to obtain a clear picture of the relationships among the different types.

Most of the rocks have been classified by Richardson (1950) as syenite, diorite, syenodiorite, and monzonite. Hornblende granite, biotite granite, and leuco-granite also occur. These names, however, have been given by Richardson purely on a mineralogical basis, as they were in the Benta Complex.

Texturally the rocks may vary from fine to coarse to pegmatitic, from feebly to strongly porphyritic, and from unfoliated to strongly schistose. It is clear therefore that the similarity between these rocks and those of the Benta Complex is both mineralogical and textural.

The various rock types found in the complex may be described as follows.

Pyroxene-Bearing Syenite. Pyroxene-bearing syenite is not widely developed. It is medium grained, and the mineralogy includes sodic plagioclase, microcline, biotite, and pale green diopside. Biotite may be replaced by chlorite, epidote, ilmenite and sphene.

Diorite. Diorite forms a dyke across the Sungei Panau. The mineralogy is of intermediate plagioclase, orthoclase, biotite, hornblende, and pyroxene. Ilmenite is an accessory. Biotite has been replaced by chlorite and ilmenite, epidote and leucoxene, and sphene. Richardson (1950) noted that this rock is very similar to some of the rocks in the Gunong Benom foothills.

Hornblende Syenite. The mineralogy of hornblende syenite includes pargasite, orthoclase, subordinate plagioclase, microcline, and epidote. Locally the rocks vary in their relative proportions of amphibole and feldspar. Some localities are coarsely pegmatitic, with microcline-microperthite crystals up to 15×3 cm and amphibole crystals up to 45×7 mm.

Quartz-Hornblende Syenite. The mineralogy of quartz-hornblende syenite consists of microcline, orthoclase, quartz and pargasite. A graphic texture is common.

Quartz-Biotite-Hornblende Syenite. The most abundant rock type in the complex is quartz-biotite-hornblende syenite. It is well jointed and varies in grain size from fine to coarse; some varieties are porphyritic.

The mineralogy includes euhedral and subhedral microcline and microcline-microperthite, some orthoclase, and some albite. The mafic minerals are pargasite and biotite in varying proportions. Quartz is present only in minor concentration. Myrmekite between quartz and plagioclase is common. In some rocks the plagioclase has a crude radial arrangement within the quartz. Apatite is the most abundant accessory, but ilmenite, magnetite, zircon, and sphene are also frequently found.

Occasionally epidote replaces feldspar and amphibole. Rutile commonly has developed within the biotite.

Quartz-Biotite-Hornblende Monzonite. Quartz-biotite-hornblende monzonite is only sparsely developed. The difference between it and the syenite is only in more abundant plagioclase.

Biotite Syenite. Rare in occurrence, biotite syenite may contain quartz. The mineralogy is of orthoclase, sodic plagioclase, biotite, and accessory amounts of apatite. Quite often the biotite is converted to ilmenite, chlorite, sphene, and epidote. Some occurrences are as dykes in the syenite.

Leucocratic microsyenite. The rocks of leucocratic microsyenite are marginal modifications of the main syenite. They are of finely granular aggregates of microcline or orthoclase and sodic plagioclase. Quartz is present in some. Very occasionally minor amounts of biotite may be present.

Hornblende Granite. In the Sungei Kapor the hornblende granite is well foliated and consists of orthoclase and pargasite phenocrysts in a fine-grained matrix of the same minerals and quartz.

A well-foliated granite porphyry occurs near Jeram Star. The foliation is exhibited by bands rich in well-aligned, tabular, white, orthoclase crystals. The amphibole occurs as sieve porphyroblasts with their crystallographic *c* axes lying in the foliation planes. In some rocks the biotite is replaced by epidote.

Biotite Granite. A few localities exhibiting biotite granite have been described. Some are fine grained, others medium grained and porphyritic.

Other Rocks. In the Bukit Ranjut Complex there are minor intrusions of microgranite, granite-porphyry, aplo-granite, vein quartz, and aplite.

PETROGENESIS. Richardson (1950) drew attention to the close similarity of the rock types in the Bukit Ranjut and the Benta Migmatite Complexes. Both complexes contain amphibole as their characteristic mineral, together with pyroxene in some varieties, and both lie along the same tectonic axis that is known to be characterized elsewhere by high-grade metamorphic rocks (Figure 9.1). There is no evidence of migmatization in the Bukit Ranjut Complex, but Richardson proposed that hybridization may have taken place at depth and that the Complex had been subsequently intruded into a higher tectonic level.

Figure 10.7 shows the outcrop of the Bukit Ranjut Complex as an approximately circular area in the lower center of the figure. The Complex seems to have been emplaced as a dome, as witnessed by the concordance of the foliation of the country rocks to the circular outline of the body. The Complex may indeed be purely of igneous origin as proposed by Richardson (1950), or it may be of metamorphic origin as suggested by its enrichment in amphibole and K-feldspar and its similarity to the Benta Migmatite Complex. However, the answer cannot be deduced by the geometry of the body. High-grade gneiss domes can result from the uplift of the metamorphic infrastructure almost in the form of an intrusion through the overlying suprastructure rocks. Many examples have been quoted by Badgley (1965).

Indeed, the Stong Migmatite Complex, which is undoubtedly of very-high-grade metamorphism, is of the form of a gneiss dome that has risen up into higher tectonic levels. Figure 10.7 clearly demonstrates the domelike nature of the Bukit Ranjut Complex, discordant to the regional structure, but modifying it locally. Richardson suggested that the Complex resulted from magmatic differentiation of a quartz-biotite-hornblende syenite (the main rock type of the Complex).

However, its similarity to the Benta Complex, which has now been shown to be a product of high-grade metamorphism and migmatization, must indicate that a restudy of the Bukit Ranjut Complex is necessary before a confident petrogenesis can be proposed.

The Bukit Berentin Complex

The large composite body of granitoid rocks, elliptical in outline, extending northward for 43 km (27 miles) from the Sungei Chekua in Pahang into south Kelantan (the northern body of Figure 10.7) called the Bukit Berentin Complex. It forms the mountainous borderland between Pahang and Kelantan. Richardson (1950) called it an igneous intrusion, but his descriptions indicate that it is on the whole well foliated and the foliations in Figure 10.7 indicate that it is regionally conformable with the surrounding sedimentary and low-grade metasedimentary rocks. Unlike the Bukit Ranjut Complex, it does not appear to be intrusive.

Richardson described the body as a large quartz porphyry intrusion. But the details he provided suggest that the Complex would be better interpreted as a metarhyolite or metarhyolite-tuff. Richardson's descriptions indicated that the metamorphism is high in the greenschist facies, but because of its alignment with other high-grade metamorphic complexes in this region, it is assumed that the Bukit Berentin Complex may have been of higher grade and may subsequently have suffered retrograde metamorphism, as indeed the Benta Migmatite Complex has done. However, it must be stressed that the author's interpretation of the Bukit Berentin Complex is open to question because the only published material available is that of Richardson (1950), from which the following descriptions are taken.

PETROLOGY. The rocks of the Bukit Berentin Complex vary widely. Generally there are three major rock types: aplite and felsite, quartz porphyry and granite porphyry (which make up the bulk of the Complex), and granitic rocks. The phenocrysts in the porphyries may be up to 10 mm in diameter. Most of the rocks, however, are medium grained.

Most rocks are foliated, more or less strongly, but some are unfoliated. Schistosity may vary from very crude to excellent. Many of the well-foliated rocks are rich in chlorite and sericite.

The phenocrysts are of quartz, usually as phenocrysts, which are often corroded, and orthoclase. The latter is the dominant feldspar, but it is sometimes accompanied by oligoclase. The feldspar is often replaced by chlorite and epidote.

The groundmass of the porphyritic rocks is generally cryptocrystalline or microcrystalline and is composed of quartz, feldspar, and sericite. It is generally always well foliated. Chlorite and epidote are also common. In the more schistose varieties the groundmass resembles a quartz-chlorite-mica schist or chloritic phyllite. The individual rock types may be described as follows.

Aplite. Richardson regarded aplite as a marginal fine-grained phase of the granite porphyry. It is usually a pale grey or greenish grey rock. It may be composed of laths of orthoclase, albite, and crystals of quartz in a trachytic groundmass. Some occurrences are rich in muscovite or sericite. Pyrite is common as cubes.

Many of the aplites of Richardson are clearly metamorphic, being rich in chlorite and epidote and often foliated. For example, "aplite" in contact with phyllite in the Sungei Anak–Sungei Kalong is strongly foliated and is composed of lenticles of quartz mosaic, quartz-chlorite aggregate, and quartz-muscovite aggregate, and granules of epidote, zoisite, and ilmenite, all in a quartz-sericite schist base. Chlorite-quartz rock with epidote occurs in several localities in the Sungei Yu.

Quartz Porphyry. Most of the quartz porphyry is more or less strongly foliated, but unfoliated varieties occur in several localities. This is the most abundant rock type in the Complex. In the Sungei Durian it is unfoliated and free of chlorite. The quartz phenocrysts are up to 10 mm in diameter. Corroded quartz phenocrysts occur in a fine-grained quartz-sericite matrix, and pyrite is locally abundant. In the foliated varieties, the matrix is of quartz, feldspar, epidote, and chlorite. Veins of epidote and chlorite are common. In the Sungei Chekua, the quartz phenocrysts are very strongly resorbed by the groundmass.

Granite Porphyry. Generally the granite porphyry is of corroded quartz and feldspar phenocrysts in a foliated groundmass. The groundmass is rich in sericite and chlorite. Sometimes pyrite cubes are abundant. The pyrite is occasionally enclosed in an envelope of muscovite and chlorite flakes.

Tourmaline Granite Porphyry. Tourmaline granite porphyry is of rare occurrence and is found in Ulu Sungei Terisi. It is strongly foliated and contains prisms of pale blue tourmaline and granular aggregates of epidote.

Orthoclase Porphyry. Orthoclase porphyry occurs in the Sungei Sientor. It is unfoliated and is composed of a few orthoclase phenocrysts in a groundmass of feldspar laths and quartz.

Granodiorite Porphyry. Granodiorite porphyry occurs only in the Sungei Yu. It contains phenocrysts of quartz and albite, and some orthoclase, in a groundmass rich in chlorite and sericite with accessory ilmenite.

Microgranite. Microgranite is a common rock type in the Complex, and good exposures are to be found in the Sungei Yu. The phenocrysts are of quartz, orthoclase, and albite. The groundmass is of quartz-orthoclase granophyre and quart-albite myrmekite, chlorite and epidote, ilmenite partly replaced by leucoxene, and pyrite. Some varieties contain no myrmekite.

Granite. Granite is nearly always fine grained, but locally it may be medium grained. Some varieties are well foliated and contain abundant chlorite. The rock is always porphyritic to some extent.

Adamellite. A medium to coarse grained rock, adamellite has been metamorphosed, as shown by the presence of chlorite and epidote.

Several vein rocks cut the foregoing major rocks of the complex. They are as follows.

Epidosite. Bands of fine-grained epidosite are common in the Ulu Sungei Kasai Kechil, where they are associated with a small inclusion of garnet amphibolite in the porphyritic rocks. The rock is composed essentially of epidote and zoisite. In the Ulu Sungei Kasai Besar, alternating bands of epidosite and orthoclase-albite-amphibole-epidote-chlorite rock are found.

Quartz veins. Quartz veins are common throughout the complex. Some may be as wide as 6 m (20 ft). Most are barren, but some contain a trace of gold.

Tourmaline veins. Tourmaline may be present alone or accompanied by quartz in veins. It is a rare mineral in the rocks of the Complex, but it occurs concentrated in veins as acicular crystals.

Surrounding the Bukit Beretin Complex, the sedimentary rocks are cut by swarms of minor granitic apophyses. Richardson was certain that they are associated with the main Bukit Berentin intrusion. They include aplite, microgranite, microgranite-porphyry, quartz porphyry, and granite porphyry.

Gold placers are mined in the valleys of the following rivers or streams: Timah, Merapoh, Kasai, Chadu, Yu, and Chiniau. There can be no doubt that this gold comes from the Bukit Berentin Complex. Indeed, small quartz veins containing gold have been found in a schistose quartz porphyry in a tributary of the Sungei Timah (Richardson 1950).

METAMORPHISM OF ASSOCIATED SEDIMENTARY ROCKS. Richardson (1950) described many localities of metamorphosed sedimentary rocks in the neighborhood of the Bukit Beretin Complex. He has interpreted small outcrops of metasediments within the Complex as "roof pendants." Some examples are as follows. Impure calcareous sedimentary rocks in the Ulu Sungei Kasai Besar and in the Ulu Sungei Kasai Kechil, in the Merapoh area, have been converted to calc-silicate fels. The rock contains yellow-zoned garnet in an acicular tremolite or antinolite matrix. Epidote is also common. Marble associated with the Complex frequently contains abundant epidote. Some "roof pendants" are of phyllite, and mica schist occurs in a tributary of the Sungei Yu and in the Sungei Timah. Around the borders of the Bukit Berentin mass occcur numerous localities of phyllite. Strongly folded white marble, containing contorted and plicated bands of green chloritic marble, and green-grey calcite-quartz-muscovite-chlorite schist, outcrop in the Sungei Serunai near Kuala Lepar, and chloritoid-andalusite schist is exposed nearby.

PETROGENESIS. Richardson (1950) proposed that all the rocks of the Bukit Beretin Complex resulted from the emplacement of granitic magma into a major anticlinorium. The intrusion was followed by further regional diastrophism, which caused the foliation and metamorphism of the granitic rocks of the complex. He also proposed that the metamorphism of the sedimentary rocks around the borders of the complex and within "roof pendants" was because of thermal metamorphism by the intrusive granitic magma.

However, certain features of the Complex are difficult to reconcile with the proposed petrogenesis. First, it is difficult to visualize how such a large intrusion could have cooled so rapidly as to be generally fine grained. This would require that the intrusion was very shallow. Had the intrusion been so shallow, then, it could not possibly have suffered subsequent regional metamorphism. This could only have occurred if the intrusion was at least as deep as the mesozone, in which case the magmatic body would not have cooled to produce this large generally fine grained mass.

Furthermore, the concordance of the foliation of the granitoid rocks and the "roof pendants" and surrounding metasediments (Figure 10.7) is best explained by regional folding accompanied by dynamothermal metamorphism of a generally conformable sequence of igneous-looking rocks and sedimentary rocks. The commonly fine-grained and porphyritic nature and corroded phenocrysts of the major rock types of the Bukit Berentin Complex suggest that the original rocks were volcanic rather than plutonic. It is therefore concluded that the Complex is of metarhyolite or metarhyolite-tuff with some intercalated metasedimentary horizons. Deep burial in the Malayan geosyncline during the main orogenic phase can account for the metamorphism of the whole mass and of some of the surrounding rocks. Although it does appear from Richardson's descriptions that some magmatic intrusions occurred within the Complex and outward along the margins into the surrounding sedimentary rocks, these are of minor significance.

CONDITIONS OF METAMORPHISM. The metamorphic mineralogy of the rocks of the Bukit Berentin Complex includes abundant chlorite, muscovite, and epidote. The associated metasediments have similar mineralogy, but in a few localities actinolite, tremolite, andalusite, and chloritoid occur. The mineralogy is consistent with an allocation of these rocks to the upper greenschist facies (quartz-andalusite- plagioclase-chlorite subfacies) or to the lower amphibolite facies of Abukuma type (andalusite-cordierite-muscovite subfacies). The presence of abundant chlorite and epidote, with the occasional occurrence of garnet, suggests conditions transitional between the greenschist and the amphibolite facies.

It is therefore deduced that the Bukit Berentin Complex is transitional between the mesozone and the catazone. It appears to be related to the Benta Migmatite Complex by the Chegar Perah–Benta fault (Figure 10.11), and it occurs in the zone of country that contains other high-grade, dynamothermally metamorphosed complexes (Figure 9.1).

Unlike the Bukit Ranjut Complex, the Bukit Berentin Complex is not discordant to the regional structure (Figure 10.7). The difference may be explained by regarding the Bukit Beretin Complex as a thermal dome (a region where higher tempera-

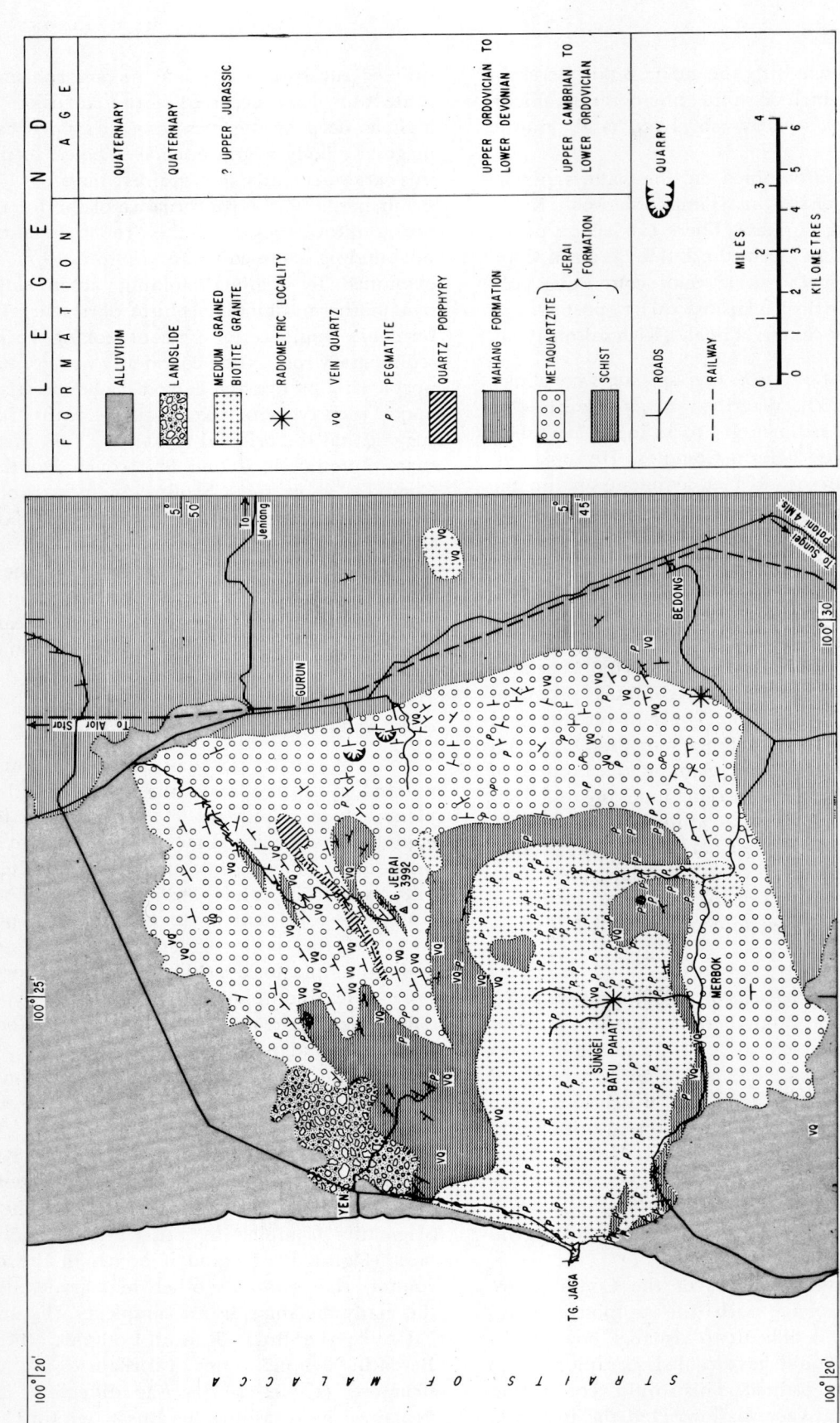

FIGURE 9.14 Geological map of the Gunong Jerai area, south Kedah. After Bradford (1965).

tures were attained during regional metamorphism), whereas the nearby Bukit Ranjut Complex is a gneiss dome, which has risen up from deeper tectonic levels and is both metamorphically and tectonically disharmonious with the surrounding country rocks. The Bukit Beretin mass is metamorphically disharmonious but tectonically harmonious with the country rocks because it has not risen up as a gneiss dome and because it was evolved in a higher level of the crust as a result of a metamorphic temperature somewhat higher than was characteristic of that tectonic level.

The Gunong Jerai Massif

Gunong Jerai, often called Kedah Peak, rises abruptly to an elevation of 1217 m (3992 ft) from the low coastal alluvial plains of south Kedah on the northwest coast of West Malaysia. The peak is anomalous both geographically and geologically, and it provides an impressive landmark for mariners and a mountainous area of good rock exposures in a region generally covered with alluvium.

The geology of the area is illustrated in Figure 9.14. An interfoliated sequence of metaquartzite and schist with some metatuff horizons, collectively known as the Jerai Formation, is intruded by a granite stock. The whole structure is of a dome, with dips of the metasediments generally away from the granite intrusion. The prominent cliff exposures of the summit of Gunong Jerai are of metaquartzite; the granite outcrop does not occupy the center of the peak but, rather, the lower southern slopes.

AGE. The metamorphic rocks of Gunong Jerai are overlain to the north, east, and south by generally unmetamorphosed shales of the Mahang Formation, dated by fossils as Upper Ordovician to Lower Devonian (see Table 3.1). On one of the small islands immediately offshore to the west of the peak, the Pulau Bidan limestone has been dated as Ordovician. Because of the metamorphism of the Jerai Formation schist and metaquartzite, it is assumed that these rocks are of older and presumably Upper Cambrian age. This age is also given on the basis of similar lithology to the Upper Cambrian Machinchang Formation of Pulau Langkawi (see Table 3.1 for correlation).

The granite, which is intrusive into the Jerai Formation, has been dated at radiometric localities 3 and 4 (Figures 8.1 and 9.14). Both localities gave a potassium-argon date of 135 ± 6 m.y. indicating an Upper Jurassic to Lower Cretaceous event. At locality 3, biotite and muscovite in the granite also indicated a Tertiary event of around 47 to 59 m.y. None of these dates may be the age of intrusion, however; they are more likely to represent the ages of major uplift of the peak. The granite may indeed be older than Upper Jurassic.

PETROLOGY

Porphyry. Locally within the Jerai Formation metaquartzite are to be found interfoliated rocks that Bradford (1965) described as "porphyry." He described it as a metatuff or metagrit. This material was described in Chapter 7 under the Lower Paleozoic volcanic rocks. The rock contains varying proportions of blastoporphyritic relict phenocrysts of quartz, microcline, orthoclase, plagioclase, perthite, and biotite in several combinations. The groundmass is of quartz and feldspar, with some muscovite, biotite, chlorite, tourmaline, epidote, clinozoisite, apatite, magnetite, hematite, and pyrite.

Schist. The pelitic rocks of the Jerai Formation include mica schist (Figure 9.15), quartz schist, spotted schist, phyllite, semischist and, rarely, gneiss.

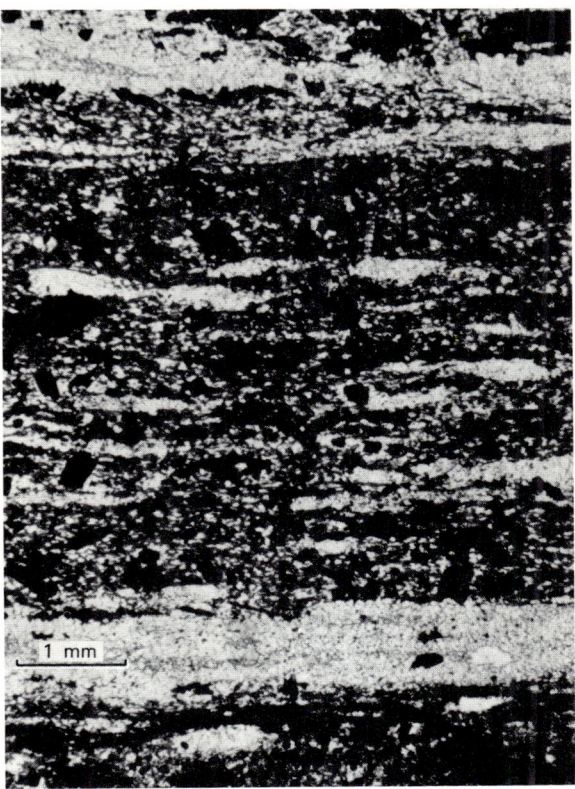

FIGURE 9.15 *Quartz-mica schist from Gurun Quarry, Kedah. Photomicrograph C. S. Hutchison.*

In the Yen area, on the west side of the peak (Figure 9.14), biotite gneiss and muscovite-biotite gneiss occur. They are characterized by augen of quartz and feldspar and contain small amounts of tourmaline and magnetite. Muscovite and biotite are the most common minerals in the pelitic schist, but tourmaline, epidote, hornblende, garnet, diopside, albite, and clinozoisite frequently occur. Actinolite is rare. Micaceous hematite, arsenopyrite, sphene, and cassiterite may also occur. Cordierite showing complex twinning often occurs as fairly large crystals in the biotite-actinolite schist. Schist xenoliths in the granite at Tanjong Jaga commonly contain cordierite and tourmaline (Paramananthan 1964). There are local occurrences of a hornblende schist, a tourmaline schist, and a garnet schist.

Pelitic hornfels is rare, but examples of biotite and quartz-biotite hornfels have been found.

The pelitic schist grades to psammitic schist by way of semischist, metaquartzite, and metaarkose (Bradford 1965).

Paramananthan (1964) noted that a few localities within the schist are distinctly calcareous and contain recrystallized calcite. They contain grossularite garnet, diopside, actinolite, and epidote in varying proportions. The fine-grained grossularite rocks, which are commonly banded, consist of green diopside and grossular garnet that may be granular or massive. Paramananthan (1964) described cordierite as being commonly associated with actinolite, diopside, and epidote. As metamorphic grade increases, the diopside may be seen replacing actinolite, and garnet commonly replaces diopside. The same author described the main occurrence of these calcareous rocks as lying along the eastern margin of the Jerai Massif in the region of Gurun quarry.

Metaquartzite. The most abundant rock type in the Jerai Formation is metaquartzite. The commonest psammite is a pure fine-grained metaquartzite. It is well jointed. The grain size may grade upward to medium grained. The presence of bands of dark pelitic material in the metaquartzite often emphasizes current bedding that has survived the metamorphism. Some of the psammites are feldspar bearing, and pyrite is a common constituent. Micaceous quartzite is common, both with muscovite and biotite. The metaquartzite may also contain tourmaline, garnet, magnetite, hematite, pyrite, pyrrhotite, arsenopyrite, scorodite, pyroxene, actinolite, and epidote.

Excellent exposures of the Jerai Formation occur in and around the J.K.R. road metal quarry near Gurun, at 23.75 milestone on the road from Alor Star to the Sungei Patani. The quarry exposes exceptionally hard metaquartzite (quartz arenite) in the top of the quarry face. It is well foliated and jointed, splitting readily into flat rectangular slabs. The foliation planes strike 175° and the dip varies from 20 to 50° toward the east. Below the metaquartzite, the rock grades into various types of gneiss and schist, containing biotite, muscovite, hornblende, garnet, epidote, clinozoisite, zoisite, pyrite, magnetite, pyrrhotite, and small amounts of bornite and arsenopyrite.

The following description of borehole 1 drilled in 1960 by William Mining Company (at grid reference Y-124187 on map sheet 21/2) is quoted by Bradford (1965):

The section affords:
  a) hornfels with relic phenocrysts perhaps after grit
  b) semi-schist with imperfectly schistose quartzo-feldspathic aggregates enclosing coarser relic sedimentary grains of quartz and feldspar.
  c) schists with distinct foliation
  d) granulites [sic] with or without phenocrysts.

STRUCTURE. Bradford (1965) completed a 1:63,360 geological map of the Kedah Peak and surrounding areas for eventual publication by the Geological Survey of Malaysia (Geological Survey Drawing 59/84). Several of the foliation dip and strike directions have been added to Figure 9.16. These indicate that the metasediments form a dome with the granite at its core. Figure 9.16 is a lower-hemisphere equal-area Schmidt net of 85 poles to foliation planes of the metaquartzite and schist. This diagram confirms the domal structure of the complex but suggests that the dome is asymmetric. The average dip toward the north-northwest is 24°; toward the east, 25°; toward the west, 25°, but toward the south-southeast, the average dip is 40°.

The south and west parts of the dome are not well exposed; the west is under the sea, and the south under alluvium, so that Figure 9.16 is less reliable for southward and westward dips. However, the domal nature of Gunong Jerai seems to be beyond doubt. The dome has not been eroded symmetrically around the granite. It appears that the shallower dips of about 25° for the northern metaquartzite slopes have resister erosion more than the steeper 40° slopes of the southern and western parts of the dome. It can be deduced that the granite underlies at a shallow depth the metaquartz-

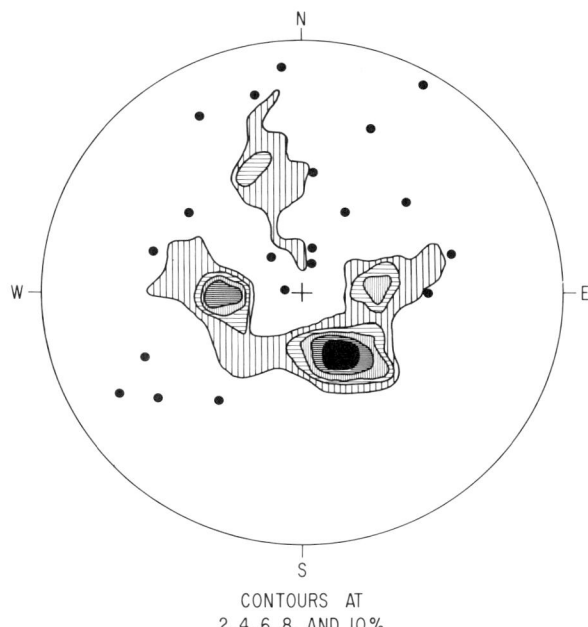

FIGURE 9.16 *Equal-area lower-hemisphere Schmidt net of 85 poles of foliation planes of the Jerai Formation metaquartzite and schist in the Gunong Jerai Massif. Individual poles are shown only outside the 2% density area. Data taken from Bradford (1965).*

ite of the summit of Gunong Jerai and that it has not yet been exposed there by erosion. In a magnetometer survey (Bradford 1965), the width of outcrop of the metaquartzite (Figure 9.14) supports the deduced structure. On the north and east margins, where the dip is 25°, the outcrop width is 3 to 5 km (2 to 3 miles), whereas on the south flank of the dome where the dip is steeper, the outcrop width is reduced to 1.5 to 3 km (1 to 2 miles).

CONDITIONS OF METAMORPHISM. All the rocks of the Jerai Formation and a few isolated patches within the neighboring Mahang Formation have been dynamothermally metamorphosed (Figure 9.17). The distribution of metamorphic minerals throughout this area has been shown by Bradford (1965), and Figure 9.17 is taken from his unpublished thesis. He chose the following isograds, which are incorporated in Figure 9.17: chlorite; muscovite; biotite; epidote, zoisite, and actinolite; hornblende; almandine garnet; diopside, sillimanite, and wollastonite.

Areas left unshaded on Figure 9.17 represent rocks that either have been unmetamorphosed (almost the whole of the Mahang Formation), do not show significant amounts of the index mineral, or are deficient in outcrop.

ROCKS OF THE CATAZONE

Bradford (1965) listed the following factors which control the positions of the isograds: proximity to the granite, degree of tectonism or depth of burial of the sediments during metamorphism, type of original sediment, and local topography.

The highest grades—containing garnet, sillimanite, and diopside—are limited in extent. The high-grade areas on either side of the Sungei Bujang, south of Merbok, are adjacent to the granite. On the other hand, those high-grade areas at the Gurun quarry, near the Kampong Perigi quarry, and in the Sungei Teroi, are not near the granite. Magnetometer surveys, however, point to the granite being shallowly buried underneath these highest grade outcrops (Bradford 1965).

Strong tectonic stress has been responsible for the dynamothermal metamorphism, but higher grades of metamorphism appear to have been caused by the granite dome. Biotite is an abundant mineral in the north, muscovite in the south; the line of subdivision is an east-west line through Gunong Jerai. Bradford (1965) attributed the muscovite around the granite partly to metasomatism from the late pegmatitic phase of the granite.

The establishment of metamorphic facies has been hindered by large extents of psammite containing no metamorphic index minerals.

METAMORPHIC FACIES SERIES. All the rocks of the Jerai Formation are well foliated, and they generally can be allocated to the greenschist facies of regional metamorphism. The localized increase of metamorphic grade around the granite has been quite effectively shown by Bradford, as summarized in Figure 9.17.

It is deduced that the Gunong Jerai granite represents a deep-seated synkinematic stock intrusion into the Cambrian sediments while they were undergoing dynamothermal metamorphism, generally to greenschist facies. The granite stock locally elevated the temperatures in a thermal aureole to raise the metamorphic temperature to that of amphibolite facies.

The facies series developed is that of Abukuma type (Miyashiro 1961). Of this facies series, Winkler (1967) noted that although it is a dynamothermal facies series, it need not be developed on a regional scale and it may be localized in a thermal aureole of a deep-seated synkinematic pluton. The characteristic features of the Abukuma-type facies series is low pressure (only moderate burial), combined with high temperature. High temperature can be attained under relatively low-pressure conditions if

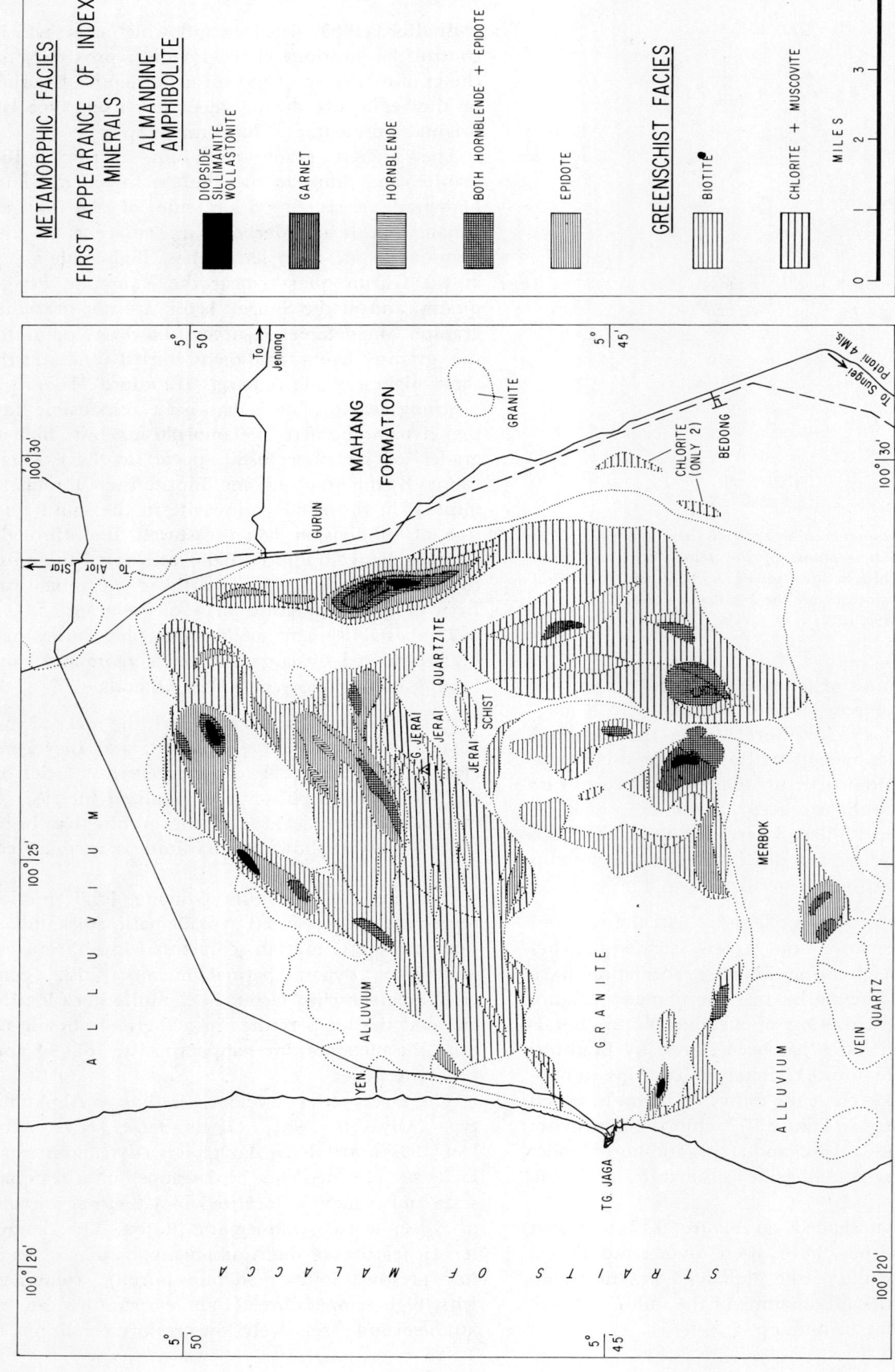

FIGURE 9.17 Distribution of index metamorphic minerals in the Gunong Jerai area. Boundaries are constructed on the first appearance of these minerals. For lithologies see Figure 9.14. Metamorphic boundaries within the granite are based on included xenoliths and roof pendants. Absence of shading does not necessarily mean lack of metamorphism since areas on which the data are insufficient are unshaded as well. The two localities of greenschist facies in the Mahang Formation contain chlorite but no muscovite. All those in the Jerai Formation contain both minerals. After Bradford (1965).

the region had a high thermal gradient of 70°C/km. Some evidence was given in descriptions of the central belt of Malaya that the Malayan Orogen may have been characterized by such a high thermal gradient. In high-level contact metamorphism, the localized thermal gradient around a pluton may be of the order of 100°C/km. This is indeed not much different from the regional thermal gradient of Abukuma Terrane. Hence if a terrane is under greenschist facies metamorphism in an Abukuma-type orogenic belt, it could well be extended to amphibolite facies metamorphism within a thermal aureole of a synkinematic pluton. The regional pressure remained constant, but the temperature was raised locally in the aureole of the granite pluton, and since the thermal gradient of a granite pluton and that of regional Abukuma terrane are of the same order, the facies series developed within the aureole will be the Abukuma facies series.

The only characteristic mineral missing from the rock descriptions is andalusite. But its absence may be accounted for by the tendency of the pelitic rocks to be calcareous, thus taking the mineral paragenesis out of the andalusite field.

In summary, it can be proposed that the Jerai Formation in the region of the peak is characterized regionally by the greenschist facies, probably of Abukuma type. The synkinematic intrusion of the Jerai granite dome locally produced within its thermal aureole rocks that may be ascribed to as high as the sillimanite-cordierite-muscovite-almandine subfacies of the Abukuma-type amphibolite facies, although a low-pressure intermediate facies series is not ruled out with certainty.

Metasomatism of the Jerai Formation by the granite has introduced the following minerals: tourmaline, cassiterite, pyrite, arsenopyrite, chalcopyrite, magnetite, and hematite (Bradford 1965). Tourmaline is widely distributed but is not closely associated with cassiterite.

The Kupang Gneiss

The Kupang Gneiss occupies a rectangular outcrop 32×8 km (20×5 miles) in the Baling region of Kedah (Figure 10.6). It is mainly fault bounded and is intruded by the Damar granite.

This rock was described by Burton (1972). It has a strongly developed gneissic texture and is characterized by large aligned microcline augen, well-developed foliation, and parallel banding. Biotite is abundant and composes from 10 to 40% of the rock. In addition, some clinopyroxene, hornblende, orthoclase, and sodic andesine are the other main constituents. Sphene and apatite are the main accessories. Burton has drawn attention to the similarity of this rock and the foliated Kemahang granite of northeast Malaya, and he suggests a similar evolution in the catazone.

At the locality from which a sample for radiometric dating was obtained (5 in Figure 8.1), the Kupang Gneiss consists of well-foliated micaschist with parallel bands of augen gneiss. In the gneiss bands the augen of feldspar may attain a size of 2×1 cm. The minerals of the matrix to the augen and of the schist bands commonly range from 1 to 3 mm.

Specimen UM 5390 gave the following modal analysis:

> Number of points: 3872 (two-colored staining).
> Quartz: 9%; microcline: 25%; plagioclase: 16%; biotite: 35%; clinopyroxene + amphibole: 13%; sphene: 2%; and apatite: < 1% by volume.

The biotite is extremely well foliated. The pyroxene is a colorless diopside, partly replaced by a colorless to pale yellow amphibole. The plagioclase is poorly twinned, but the occasional albite twinning allowed a determination of An_{38}. The microcline is well twinned. Sphene is anhedral and apatite generally euhedral. The mineralogy of this rock is entirely metamorphic, and there are no relict igneous textures.

A chemical analysis of the specimen is given in Table 9.1, analysis 5. The composition suggests the rock may be a metasediment.

A radiometric date of 150 ± 8 m.y. was obtained by the potassium-argon method for the biotite of this rock. The outcrop of the Kupang Gneiss is, however, bounded by faults (see Figure 10.6); therefore, this date must be interpreted as resulting from tectonic activity rather to indicating the age of metamorphism. It is assumed that the age of metamorphism is considerably older.

The presence of clinopyroxene and microcline and the absence of muscovite indicate that the conditions of metamorphism were high in the amphibolite facies of dynamothermal metamorphism. Hence it is deduced that the Kupang Gneiss has been considerably upfaulted to its present position.

ROCKS OF THE MESOZONE

Greenschist facies phyllitic rocks are widely developed in the Malay Peninsula (Figure 9.1). With the exception of the northwest part of Malaya, where the rocks have never been sufficiently buried, the Lower and Upper Paleozoic formations have been generally isoclinally folded and dynamothermally metamorphosed in the greenschist facies. These rocks are characteristically phyllitic, and the intercalated limestones have been recrystallized to marble.

Geologists in West Malaysia have shown an unfortunate tendency to relate this greenschist facies metamorphism (and especially the formation of andalusite) to the abundant granite batholiths that characterize the mesozone. This theory is likely to be erroneous for two reasons:

1. Phyllitic rocks are widely developed east of the Main Range even in areas where granite batholiths are absent, and andalusite is fairly common throughout Malaya (Scrivenor 1931).

2. Evidence from the catazone has indicated that the Malayan Orogen was characterized by a high thermal gradient, perhaps equivalent to the Abukuma facies series. Hence temperatures high enough to have produced greenschist facies metamorphism with andalusite would have been attained by the regional thermal flow, without any assistance from the granite intrusions. Indeed, the high density of granite emplacement in the Malayan Orogen resulted from the characteristic high thermal gradient of the Orogen. As Miyashiro (1961) noted, this type of regional metamorphism (andalusite-sillimanite) appears to be always accompanied by the emplacement of a large amount of granitic rock, and this may be expected if we accept that granites in orogenic regions are largely of anatectic origin.

Nevertheless, it would be wrong to disregard the granite batholiths entirely. Although they have not been responsible for the high regional temperatures, they have undoubtedly caused some restricted thermal aureoles, and this is displayed where the thermal effect of the granites has been accompanied by metasomatism, especially along the contact with calcareous rocks. The granite plutons have also locally raised the temperatures to cause metamorphic conditions akin to the amphibolite facies or, where pressure was not very high, akin to the hornblende hornfels or K-feldspar-cordierite hornfels facies.

Some attempt will be made to separate the effect of the regional dynamothermal metamorphism of greenschist facies and the thermal effect of the granites that may give rise to higher temperature minerals. However, no clear separation will be possible because the mineral paragenesis of the greenschist facies in Abukuma terrane is identical to that which would be produced around a mesozone granite batholith. This is because the thermal gradients in the two cases are approximately comparable. Only where the thermal effect of the granite is accompanied by metasomatism is it possible to clearly separate the regional from the localized effects. Characteristically, the granite batholiths of Malaya lack contact aureoles. Metamorphism has not been investigated and described in equivalent detail throughout the whole of the Malay Peninsula.

Here the metamorphism is described by regions.

Northwest Malaya

LANGKAWI, PERLIS, WEST KEDAH. The islands of Tuba and Timun, the small islands in between them, and the southeast coast of Langkawi are characterized by intensive deformation with cleavage (Koopmans 1965). The mineral assemblages in the slates of the Lower Paleozoic Setul Formation indicate low-grade regional metamorphism. To the north, toward Pulau Langgon, the influence of regional metamorphism decreases. A mineral lineation also occurs in these rocks. It is nearly always due to the parallel alignment of aggregates of clinozoisite. Adjacent limestone of the Setul Formation has been recrystallized by regional metamorphism. The calcite occasionally shows a noticeable parallel elongation of the grains (Koopmans 1965). This deformation is now considered to be Middle Devonian (Jones 1968).

Elsewhere in the Langkawi Islands are numerous examples of contact metamorphism. Adjacent to the granite, the Machinchang rocks are frequently converted to spotted quartz-biotite hornfels (Jones in prep.). In the Sungei Perangin, the Machinchang Formation is converted to low-grade hornfels. Just below the waterfall of Telaga Tujoh, spotted tourmalinized quartz-biotite-sericite hornfels occurs within beds of purer metaquartzite. The spotting is by segregations of biotite. Incipient chiastolite is also present. Pulau Perak is composed of quartz-tourmaline hornfels. The tourmaline is thought to be a result of metasomatism from the granite (Jones in prep.).

Toward the granite contacts, the Setul limestone frequently recrystallizes to a white, grey, or pinkish marble that may contain tremolite, diopside, wollastonite, garnet, clinozoisite, and vesuvianite (Jones in prep.). A skarn zone is frequently developed at the contact. It may vary from dense, granular, dark-colored diopside-garnet rock to a light cream-colored diopside-tremolite-wollastonite rock of similar texture. Other skarn minerals present in the Langkawi Islands are quartz, chalcedony, corundum, scapolite, apatite, vesuvianite, serpentine, chondrodite, and fluorite. Ore minerals such as hematite, sphalerite, galena, and a variety of copper sulfides also occur near the contacts.

The pelitic bands of the Setul Formation have been converted to spotted hornfels near the granite contacts, and they contain incipient chiastolite and occasionally cordierite (Jones in prep.). East of Kuah, the skarn material is a dense, dark-colored, green-brown, medium-grained granular aggregate of diopside, wollastonite, and garnet containing small amounts of calcite, fluorite, scapolite, and serpentine. The diopside frequently alters to tremolite. East of Kuah and northward toward Gunong Kuang, there occurs a garnet-bearing marble, in which clusters of emerald-green well-formed dodecahedra and trapezohedra of andradite up to 4 cm are set in a coarse-grained bluish-white calcite groundmass.

Of the two hills forming Pulau Bumbon Besar the northern one is of limestone thermally metamorphosed by granite, which forms the southern hill. The commonest contact rock is a tough, medium to coarse grained, cream to white granular rock composed of a dense intergrowth of columnar diopside showing alteration to tremolite. Some of the rock contains quartz and muscovite, and it is closely associated with greisen. There is also a rock composed of talc and a small percentage of apatite and zircon, and there is a distinct zone of well-crystallized columnar pale green vesuvianite. Away from the contact, the impure limestone layers contain granular diopside.

Phyllite occurs on Pulau Tiloi together with metaquartzite that contains biotite.

Adjacent to Batu Puyoh, the rocks of Pulau Tuba are composed of pink garnet associated with minor amounts of diopside and calcite. The marble often contains tremolite. Near Tanjong Peluru on Pulau Tuba, brown medium-grained metaquartzite and some phyllitic hornfels are invaded by granitic veins (Jones in prep.).

Adjacent to the granite in Langkawi and near Jitra in Kedah, the Singa and Kubang Pasu Formations are thermally metamorphosed to spotted hornfels and metaquartzite (Jones in prep). Quartz-muscovite-biotite segregations and incipient chiastolite and cordierite porphyroblasts can be distinguished. Quartz-biotite schist forms the Singa Formation in Dayang Bunting. Thermally metamorphosed roof pendants occur in the Gunong Raja granite and some excellent andalusite and chiastolite crystals have been obtained from metasediments on the eastern side of this pluton.

Spotted phyllite, slate, and quartz-biotite-muscovite-clinozoisite hornfels occur near the Bukit China granite in Perlis at Sungei Bunut.

Near Bukit Berangan, the Kubang Pasu Formation is thermally metamorphosed to hornfels and bears tin. The rocks in this locality are of quartz-biotite and chiastolite-bearing hornfels, containing porphyroblasts of cordierite.

Phlogopite and brucite marble is common in Dayang Bunting in the Chuping Limestone. The contact between the granite and the marble south of Tanjong Telin displays strongly altered muscovite granite- lying against pale green, wollastonite-rich calc-silicate.

EAST KEDAH. In the vicinity of Baling, Lower Paleozoic rocks are strongly folded and invaded by granite. The sequence includes spotted shale, phyllite, metamorphosed siliceous shale, and quartz-biotite hornfels (Burton 1972). Impure metamorphic limestone contains diopside, plagioclase, sphene, hornblende, and epidote. Pure limestone has been metamorphosed to massive or thickly bedded white, dark grey, or black marble. Quartz, tremolite, and wollastonite have been recorded from some of the marble localities.

Calc-silicate rocks are of considerable importance in the Baling area of Kedah. They include hard, brittle, often banded, green, grey, white and yellow, fine-grained hornfels composed of quartz, calcite, garnet, wollastonite, plagioclase, clinozoisite, epidote, hornblende, biotite, tremolite, microcline, zoisite, and sphene. The banding is due to slight differences in original composition of the originally layered sedimentary rock (Burton 1972).

NORTH PERAK. The Lower Paleozoic rocks of the Grik area have in places been metamorphosed to quartz-biotite, and quartz-chlorite-sericite hornfels and schist. In the vicinity of the granite margins are found calc-silicate hornfels composed of diopside, wollastonite, zoisite, clinozoisite, and garnet; and coarsely crystalline marble (Jones 1972b).

In the vicinity of Lenggong, Perak, marble often contains tremolite and flakes of graphite (Jones 1972b).

Chlorite and muscovite in the groundmass of the Lawin Tuff in the Grik area may be of metamorphic origin.

Kinta Valley

CALCAREOUS ROCKS. The calcareous rocks in the Kinta Valley have been everywhere recrystallized (Ingham and Bradford 1960). Regional dynamothermal metamorphism and contact metamorphism combined with metasomatism near the granite bodies have been described.

The grain size of the marble in the Kinta Valley varies generally with distance from the granite contacts. Very coarse marble with grain size of 1 to 2 cm is present near the granite in the Kledang Range near Menglembu, and fine-grained saccharoidal marble is common in the central parts of the Kinta Valley, for example, around Malim Nawar (Ingham and Bradford 1960).

Calc-silicate (Figure 9.18) is developed along the granite contact. In the mines near the Menglembu–Kledang area, large blocks of vesuvianite up to 30 cm in diameter occur. Garnet-vesuvianite rocks are exposed in marble against the granite near Pulai. In these rocks the garnet has been analyzed by Ingham and Bradford (1960) as:

SiO_2, 37.64; Fe_2O_3, 26.57; MnO, 1.74; MgO, 2.46; CaO, 30.90; loss on ignition, 0.50; total 99.81.

And from the same locality, vesuvianite gave:

SiO_2, 36.92; TiO_2, trace; Al_2O_3, 19.08; Fe_2O_3, 4.01; MgO, 0.78; CaO, 36.40; Na_2O, 0.75; K_2O, 0.25; loss on ignition, 0.96; total 99.15.

Garnet occurs in altered limestone at Ulu Piah and Gunong Terendum. Phlogopite occurs at Ampang, Pulai, Ulu Piah, and Selibin, in thin layers within the marble. Phlogopite may also occur along with vesuvianite. Scrivenor noted that lepidolite and small grains of fluorite are common in the residues obtained from dissolving marble (Ingham and Bradford 1960). Diopside has been described from Telok Kruin at Fook Wai, near Ipoh, and a pyroxene rock was exposed in a mine near Kledang. Diopside and tremolite occur at Pulai, and tremolite at Chemor, Ampang, Selibin, and Menglembu. At Pulai, tremolite asbestos occurs along the shear planes in marble. Wollastonite was found at Gunong Datoh and Selibin. Chondrodite and common spinel, in good octahedral crystals, occur near Selibin. Pleonaste has been noted in calcite veins in marble near Chendering near Pulai. It is associated with scapolite. Apatite occurs rarely at Selibin.

Chlorite is developed in sheared marble at Batu Anam and Jelapang; and a 30-cm band of chlorite

FIGURE 9.18 *Calc-silicate hornfels from Tanjong Rambutan, Perak. Photograph Ja'afar bin Haji Abdullah.*

occurred between the granite and marble at Gunong Jala near Ampang (Ingham and Bradford 1960).

Talc occurs at the contact between a pegmatite and marble at Gopeng and at Selibin, and Temoh. The following talc analysis is quoted by Ingham and Bradford (1960):

From Beatrice mine, Selibin: SiO_2, 60.20; Al_2O_3, 0.28; Fe_2O_3, 5.10; MgO, 33.20; CaO, 0.58; loss on ignition, 0.36; total 99.72.

A greenish mineral that may be serpentine or talc is developed along shear planes at Gopeng.

PELITIC ROCKS. Argillaceous strata, interbedded with limestones, have been folded to a greater degree than the limestones. The pelitic bands have been converted to schist, quartz-mica schist, and quartz schist generally over the whole Kinta Valley. Thermal metamorphism, on the other hand, is confined to areas very close to the granite contacts (Ingham and Bradford 1960). Tropical weathering has reduced the schists to clays and the interfoliated quartzites to sandstones, so that the full extent of metamorphism is not readily seen. The Tekka Clay, which occurs on the east side of the Kinta Valley, is considered to be the deeply weathered *in situ* product of underlying schist. Graphitic schists are common around Kinta Kellas Estate and near Batu Gajah (Figure 8 *in* Scrivenor 1931). Carbonaceous schist is widespread and is especially predominant southeast of Chemor. Quartz-mica schist is common near Ulu Piah.

Near the contact with granite, argillaceous beds have been converted to pyroxene schist, amphibole schist, quartz-biotite schist, and tourmaline schist. Pyroxene schist and mica schist, which are best exposed in the area near Pulai and Gunong Terendum, have also been found near Papan. Most of the pyroxene schist consists predominantly of quartz and grey pyroxene. Pale bands of pyroxene-epidote schist and of pyroxene-epidote-zoisite schist occur near the Sungei Choh, Tanjong Rambutan. Minute feldspar crystals may be present, and pyrrhotite and cassiterite are common accessory minerals. More rarely tourmaline is found, and in some of these schists, close to the granite, garnet, vesuvianite, actinolite, and axinite have been developed. Some schists at Gunong Terendum contain prehnite (Ingham and Bradford 1960).

The mica schists are of alternate layers of granular quartz and biotite, muscovite, and quartz, the mica being in minute flakes parallel to the foliation. More rarely grains of pyrrhotite, feldspar, and tourmaline may occur. In some biotite-rich rocks, porphyroblasts of andalusite are found. Accessory minerals may be pyrite, fluorite, and chlorite. Four analyses of pyroxene schist and mica schist from Kramat Pulai are quoted by Ingham and Bradford (1960).

In some places the schist has been tourmalinized by the granite or aplite veins. Tourmaline schist has been recorded from several localities on the east side of the Kinta Valley, including Ulu Piah, Ampang, Pulai, Tekka, and Gopeng, and on the west side of Redhills and Tanjong Tualang.

Andalusite and chiastolite rocks are not widespread, but a grey slate with chiastolite was exposed at Phin Soon Mine near Tanjong Tualang and an andalusite schist, with andalusite partly converted to mica, was found at Rotan Dahan about 1.5 km (1 mile) east of Pusing.

TOURMALINE-CORUNDUM ROCKS. First described by Scrivenor (1910), the tourmaline-corundum rocks appear to be limited in occurrence to the west side of the Kinta River; pure corundum rocks seem to be present only on the eastern side. The tourmaline-corundum rocks are chiefly found as very hard, well-rounded boulders with smooth, often highly polished surfaces; but some *in situ* outcrops have been found (Figure 9.19). In some localities the rock is weathered to a pisolitic clay (Ingham and Bradford 1960). The rock is *always* associated with the schists, and Scrivenor (1910) said that there is a gradual increase of metamorphism from schist to tourmaline-corundum rock. However, corundum appears at temperatures as low as the beginning of the greenschist facies, when bauxites are metamorphosed (Winkler 1967). Diaspore is replaced by corundum at a temperature of 370°C if the pressure is 2000 bars. Such conditions are considered to be the very beginning of the greenschist facies.

Smooth, well-worn specimens of tourmaline-corundum rocks are usually blue-black. Most of the specimens show spheroidal or oval cavities, rings, or bodies of a lighter color varying in size from place to place and averaging 2 mm. The groundmass is very fine grained. The rock is generally very hard and has a density of 3.0.

The rock consists essentially of tourmaline and corundum and/or diaspore, but rutile may be common, and pleonaste has been noted. In thin section, most rocks contain a finely disseminated black dust, which Scrivenor (1910) believed to be magnetite. Other accessory minerals are muscovite, margarite, ?hydrargillite, brown mica, hematite, chalcedony, metallic sulfides (especially pyrite and arsenopyrite), and veinlets of chloritoid. No cassiterite has been found in these rocks, but the nearby schists contain tin (Ingham and Bradford 1960).

The texture is of numerous round to oval bodies in a fine-grained groundmass. The smaller bodies are of corundum alone, but the larger bodies have a shell of small shapeless grains of corundum and a core of irregular tourmaline grains. In a few cases the core is of a single tourmaline grain. Usually in the larger bodies, there is an outer border of corundum grains and an inner border of tourmaline. Sometimes the outer bands are complex and have alternations of tourmaline and corundum layers. The corundum is colorless to pale blue in thin section.

A chemical analysis of a tourmaline corundum rock from the Kinta Valley is given in Table 9.1 (analysis 10).

The question of the origin of the tourmaline corundum rocks is not easy to answer. In Malaya

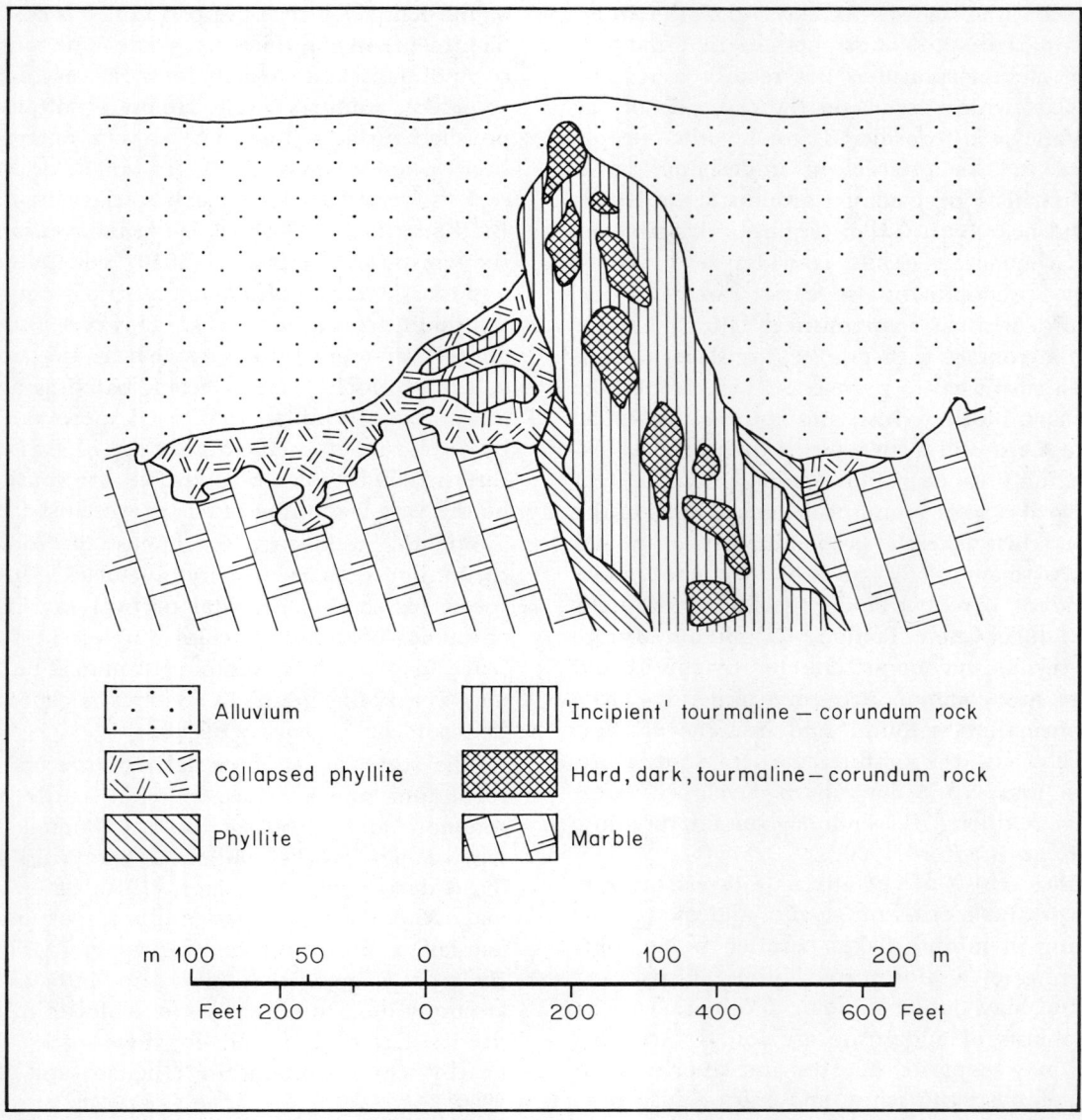

FIGURE 9.19 *Diagrammatic section across the deposit of tourmaline-corundum rock at Kong Chin Mine, Gedong Batu, Kinta, Perak. After Ingham and Bradford (1960, Fig. 7).*

the rocks occur only in the Kinta Valley. The only other described occurrence is in the Pohra area of India. Corundum itself typically occurs in thermally and regionally metamorphosed bauxitic deposits, as in the emery deposits of Samos and Naxos in the Aegean Sea.

The most obvious origin for the Kinta tourmaline-corundum rocks is the dynamothermal metamorphism of an old bauxite combined with tourmalinization from the granite. The tourmaline may have resulted from the combination of boron, introduced from the neighboring granite, with impurities in the bauxitic clay. The high concentration of aluminum gave rise to corundum.

The pisolitic texture of the Kinta tourmaline-corundum rocks suggests metamorphism of a pisolitic bauxite. The pisolites were thus considered to be premetamorphic in origin by Scrivenor (1910).

Willbourn (1931) held that low-grade metamorphism could have produced such tourmaline-corundum bodies. He discarded the theory of origin from pisolitic sediments in favor of an origin from argillaceous bands in limestone, altered by metamorphism to a type of spotted schist, such as is commonly found in metamorphic aureoles.

The only question unanswered by Willbourn's hypothesis is, Where did the enrichment in aluminum come from? Ingham and Bradford (1960) suggested that the spotted schists may have been enriched in aluminum from the neighboring granite. This is unlikely because the granites in Malaya generally gave up their alkalis in preference to aluminum, as witnessed by their richness in normative corundum. Hence we are driven to conclude that the corundum in the Kinta Valley resulted from greenschist facies metamorphism of rocks that probably were pisolitic and certainly were rich in aluminum, i.e. pisolitic bauxitic clays. The tourmaline was introduced as a result of tourmalinization from the adjacent granite bodies. The common association of corundum rocks with schists in the Kinta Valley (Figure 9.19) conclusively indicates the metamorphic origin of these rocks. Not all corundum rocks contain tourmaline, hence it is concluded that the corundum resulted from greenschist facies dynamothermal metamorphism and that tourmaline was introduced only where granite bodies were emplaced nearby.

This origin for the tourmaline-corundum rocks raises the question of the origin of the bauxitic clays. If the interpretation is correct, then there must have existed within the Kinta Valley in Pre-Middle Triassic time some deposits of bauxitic or lateritic clays. This requires an Upper or Lower Paleozoic unconformity with subaerial weathering under tropical conditions. Such an unconformity undoubtedly occurred in northwest Malaya in the Lower Devonian, but there is no evidence for it in Kinta.

Selangor

In Selangor, both regional and thermal metamorphism were detected by Roe (1953). In the region north of Kuala Lumpur, marble, quartzite, graphitic schist, mica schist, and phyllite have been described. Subsidiary pyroxene and amphibole schist and calc-silicate fels also occur. These metamorphic rocks have a foliation trend northwest and dip beneath the marble at Batu and Kanchang.

Roe (1953) particularly noted that the granite in the area north of Kaula Lumpur has no thermal aureole, and most of the metamorphism is dynamothermal. Thermal metamorphism is confined to a narrow contact zone, where hornfels containing biotite and often tourmaline is common. Hornfels is common at Kanchang, the Sungei Udang, and the Sungei Tempayan area east of Ulu Yam.

Mica schist, graphitic schist, calc-silicate hornfels, amphibole schist, and chiastolite and andalusite schist outcrop east of Ulu Yam, in the valleys of the Sungei Rening, Tamu, Batang Kali, and their tributaries. These rocks are quite distinctive, and continue into the areas of south Perak and to the east into the Fraser's Hill area (Roe 1951). The schist is coarse grained, consisting of layers of large muscovite flakes alternating with quartz-rich layers contaning occasional flakes of muscovite and graphite. Graphitic material and pyrite are common, and weathered outcrops are stained brown or yellow. Chiastolite hornfels, which occurs in the Sungei Tamu, is a black, fine-grained siliceous rock containing abundant pyrite. Thin sections show quartz, sericite, chlorite, chiastolite with characteristic carbonaceous inclusions, pyrite and small stumpy crystals of tourmaline (Roe 1953).

In many places limestone in contact with the granite exhibits no contact metamorphism and only patchy thermal metamorphism is shown by limestone and argillaceous rocks (Roe 1953). Hornblende hornfels occurs in the valleys of the Sungei Kanchang. At Bukit Taku, in Templar Park, contact metamorphism is shown at the granite marble contact. The rock contains wollastonite and pale green diopside intergrown with fluorite. It is veined with calcite and quartz. The thickness of the wollastonite band varies from 1 to 6 cm. Below the line of the contact is a pure green pyroxene rock, containing fluorite, quartz, and accessory apatite: nearly colorless epidote occurs embedded in the quartz.

Phyllite is common in the area north of Kuala Lumpur. Good exposures are found along the roads between Rawang and Rasa. The rock weathers readily to a shalelike material, and hence its metamorphic nature is often masked.

Most of the sandstone beds have been dynamothermally metamorphosed to quartzite, with a well-developed foliation striking north-northwest. Exposures are to be found along the valleys of the Sungei Kuang, Kudong, Genil, Manau, and Dungun. Some spotted slate has been described, and biotite hornfels occurs locally. The hornfels consists of quartz, biotite, chlorite, sericite, and iron oxides. Xenoliths of similar composition occur within the nearby granite. Biotite hornfels with mica-rich spheroids, together with spotted siliceous slates, are common near the granite contact in the Kuang Valley and its tributaries. The spheroids are of minute muscovite flakes, and the rock groundmass is of biotite metaquartzite. A similar siliceous spotted slate occurs in the Fraser's Hill area (Roe 1951) and in the Raub area (Richardson 1939).

AMPHIBOLE SCHISTS. The amphibole schists comprise actinolite, tremolite, and pyroxene schist, with associated chlorite schist and some amphibole hornfels. The occurrences in the Rawang area north of Kuala Lumpur are small. The best exposures are east and south of Ulu Yam, where actinolite and tremolite schist and hornfels outcrop in the Sungei Senama, Gapis, Lempor, and a tributary of the Padang. Another occurrence is in the Serendah area between Sungei Choh and the Kampar Bharu Estate. Actinolite, tremolite, pyroxene, and chlorite schists occur as poor exposures in the Sungei Guntong. Other poor exposures occur at the Sungei Tua and the Devon Estate.

These rocks are strongly schistose. They are blueblack to yellow-green and grey in hand-specimen and are fine to medium grained. Accessory mica, quartz, plagioclase, and pyrite commonly occur; epidote, zoisite, and clinozoisite are less common. Some hornfelses contain tourmaline and leucoxene. Preferred orientation of the amphibole is frequent. West of Bukit Muchong, the schist contains 75% pyroxene, partly replaced by amphibole and chlorite. In the Sungei Tua Estate the pyroxene is enstatite. Roe (1953) deduced that the amphibole and pyroxene schists from this area represent metamorphosed dolerite dykes and sills. He did not consider that they were metamorphosed impure calcareous metasediments because northwest of Serendah these rocks occur in an area devoid of calcareous sediments. Also in the Chinchong River, dolerite can be seen partly altered to amphibolite along a shear zone.

To the north and east of Kuala Lumpur, Roe (1951) described metamorphic rocks in the Fraser's Hill area. In this region the metamorphism is predominantly dynamothermal: there is no well-developed thermal aureole around the granite. East of Rasa, good exposures of mica schist, graphite schist, garnetiferous schist, calc-silicate hornfels, and amphibolite occur in the valleys of the Sungei Kiul, Sangkau, Yus, and Kinjai. These well-foliated schists are silvery black when fresh but stained yellow or brown when weathered. Quartz schist grades to mica schist. Garnet mica schist in the Sungei Kiul contains garnet crystals averaging 5 mm in diameter and often largely replaced by limonite. This garnet has been analyzed by Roe (1951) as follows:

SiO_2, 26.18; TiO, 0.50; Al_2O_3, 20.61; Fe_2O_3, 20.03; FeO, 15.50; MnO, 0.52; MgO, 0.98; CaO, 2.00; Na_2O, 0.36; K_2O, 0.37; total H_2O, 10.20; total 97.25.

The calc-silicate fels in the Kiul Valley includes epidote, zoisite, clinozoisite, chlorite, and quartz. A chemical analysis of this rock is given as number 9 in Table 9.1. It is foliated, with greyish-black bands containing graphitic or argillaceous material, alternating with greyish-yellow streaks that are rich in calcareous minerals. Thin sections show epidote, feldspar, calcite, pyroxene, clinozoisite, graphitic material, muscovite, zoisite, and a mineral that is probably wollastonite. Some specimens contain quartz and tremolite (Roe 1951).

Good exposures of interfoliated phyllite, metaquartzite, and schist occur in the Sungei Kerling. The mica is generally muscovite; near the contact with the granite, however, the muscovite is replaced by biotite and tourmaline occurs. Schistose grit containing graphite is exposed in an old quarry near the Kuala Kubu railway station.

Thermal metamorphism is superimposed on the regional dynamothermal metamorphism near the granite contacts as in the Kalumpang J.K.R. quarry where biotite-quartz hornfels and spotted siliceous slate occur.

Here also hornfels xenoliths are common in the granite, and acicular tourmaline is found near the contact. Spotted slate in the Sungei Bernam near Tanjong Malim contains 2-mm diameter spheroids of sericite, and biotite in a quartz-rich rock. Garnet occurs in some roof pendants in the Sungei Perah and in Ulu Kerling.

AMPHIBOLITE. Actinolite schist, tremolite schist, and subsidiary pyroxene schist occur as small exposures west-northwest of Fraser's Hill in the Sungei Sempan and the Sungei Merah and east of Rasa in the Sungei Kiul. They are interfoliated with calc-silicate fels and garnet-mica schist and are thought to result from metamorphism of impure calcareous sediments (Roe 1951). They are strongly foliated and fine to medium grained. The main minerals are actinolite, plagioclase, tremolite, chlorite, and pyrite. In the Sungei Kiul area the mineralogy includes epidote, zoisite, clinozoisite, and garnet. Pyroxene, pyrrhotite, and iron oxides are accessories (Roe 1951).

Kuala Lumpur Area

Gobbett (1965a) described the Lower Paleozoic metasediments of the Kuala Lumpur area. The oldest formation is the Dinding Schist, a well-foliated quartz-mica schist containing variable amounts of muscovite, biotite, and microcline. Quartz-microcline augen give the rock a spotted appearance on weathered outcrops.

Actinolite-diopside-quartz schist occurs along the road from Setapak to Klang Gates, and epidote-actinolite-quartz schist occurs to the west of Bukit Dinding. There is also a quartzite member, containing a small proportion of fine-grained muscovite.

The overlying Hawthornden Schist is strongly graphitic. In places near the granite contact it may contain andalusite.

These schists are overlain by the Kuala Lumpur Limestone, which is a remarkably pure marble. The rock is essentially of calcite, but locally it is a dolostone, massive, and interfoliated with the calcite marble. Occasionally tremolite and brucite occur in the Ampang area; periclase and brucite are common minerals in the metamorphosed dolomite (Hutchison 1968d).

Pelitic rocks in the younger Kenny Hill Formation are distinctly phyllitic.

Pahang, East of Kuala Lumpur

SCHIST SERIES. Immediately east of the Main Range occurs a persistent band of strongly foliated schist, with a regional north-northwest strike. The outcrop varies from 2 to 5 km in width and extends from northern Pahang to Negri Sembilan in the south. It is well exposed in the Bentong area along the main road between Tras and Fraser's Hill and between Bentong and Ginting Sempah. The rocks are Lower Paleozoic and include phyllite, slate, schist, and hornfels (Alexander 1968). Amphibolites occur locally as lenticles.

The dominant rock types are mica schist, quartz-mica schist, and quartz schist, all strongly foliated and contorted. The micas are sericite and biotite, and chlorite may be locally abundant. Tourmaline and black iron oxides are common accessory minerals, and many of the schists are graphitic. Interfoliated quartzite and grit bands are also schistose. Garnet is locally present, and crystals up to 15 mm occur about 1.5 km (1 mile) from the granite contact in the Sungei Bentong and the Sungei Kenong (Alexander 1968).

Interfoliated lenticles of actinolite-bearing schists are also strongly foliated and fine to medium grained. An analysis of one of these appears as number 4 in Table 9.1.

The origin of the amphibole schists is in doubt. Probably they result from regional metamorphism of dolerite dykes which intruded the sedimentary rocks prior to the regional metamorphism.

The phyllite and slate are black, graphitic, and well foliated. They are composed of indeterminate micaceous mineral with minute angular quartz fragments and sericite flakes. Grey spotted siliceous slates sometimes contain andalusite or chiastolite (Alexander 1968).

Biotite and quartz hornfels occur near the granite contact along the region between Bentong and Kuala Lumpur. They are well displayed west of Ginting Sempah, where they occur along with phyllite in roof pendants within the Main Range Granite. Typically they are composed of microcrystalline aggregates of quartz, biotite, chlorite, and iron oxides, although tourmaline may be present. Near the granite they contain pale greenish chlorite or chloritoid, sometimes epidote, graphite dust, and iron oxides (Alexander 1968). Other varieties are rich in biotite. Xenoliths of the hornfels also occur within the granite.

To the east of the Schist Series outcrop, other rock types in the Bentong Group show signs of regional metamorphism. Conglomerate in the Bentong region is foliated and in places has a phyllitic matrix; limestone is schistose in parts.

The oldest rocks (the Schist Series) have, however, suffered the most intense regional metamorphism. Proceeding eastward from Bentong, the grade of metamorphism gradually diminishes as younger and younger rocks are encountered, reflecting the decreasing depths of burial of the original sediments in the Malayan Geosyncline.

Thermal metamorphism resulting from the granite batholiths is of limited importance and is confined to very narrow zones along the contact. In many cases, no contact metamorphism has been detected (Alexander 1968).

Raub Area of Pahang

SCHIST SERIES. The Schist Series continues northward into the Raub area. The rocks here, which were described by Richardson (1939), are mainly of schist and phyllite; hornfels is uncommon except at the Sungei Sempan reservoir, where the rock at the granite contact is a quartz hornfels containing chlorite and epidote. Hornfels xenoliths also occur within the granite. Away from the contact, strongly foliated mica schist, quartz schist, chlorite schist, and graphite schist may be found.

The mica schist has psammitic bands of microcrystalline quartz, chlorite, muscovite and biotite. Arranged in parallel orientation, the psammitic bands alternate with pelitic bands; the latter are often graphitic and of biotite, chlorite, and muscovite. The biotite is strongly pleochroic. Chloritoid is of restricted occurrence (Richardson 1939). Graphite bands are common in most schists and small colorless garnet crystals occur in mica schist near the 3.6 milestone on the Sempan road, together with irregular patches of zoisite aggregate, orthoclase, and epidote.

AMPHIBOLE SCHIST. Amphibole schist occurs commonly within the Bentong Group along the eastern foothills of the Main Range. It is strongly foliated and generally dark green, but epidote-rich varieties are light green. The largest outcrop extends sinuously for 8 km (5 miles) in the north-northwest direction and varies in width up to 1.5 km (1 mile) on the Sungei Lipis. It is flanked by the Main Range Granite on the west. There are also some small elongate lenticular outcrops farther south in the Sungei Cheroh, the Sungei Chembatu, and elsewhere (Richardson 1939).

Both schist and hornfels occur, but hornfels is uncommon and is found only near the granite contact. Epidote and tremolite hornfels is more common than pyroxene hornfels. The most abundant minerals are actinolite, quartz, epidote, zoisite, tremolite, orthoclase, chlorite. The following are less common: plagioclase, clinozoisite, pyroxene, pink garnet, ilmenite, pyrite, calcite, and pyrrhotite.

In the schist the actinolite crystals lie parallel to the foliation. They are strongly pleochroic from

γ dark blue-green
β pale blue green
α yellow green
$\gamma \wedge c = 16°$.

Amphibole forms the bulk of the schist. Epidote forms small lenses. Tremolite is colorless with $\gamma \wedge c = 16°$. A typical epidote schist is exposed in Ulu Lipis 540 m southeast of Kuala Sia. It is a strongly foliated, thinly laminated yellow rock.

Pyroxene schist occurs in the Sungei Semeriong. The pyroxene appears to be intermediate between diopside and hedenbergite (Richardson 1939).

An analysis of the tremolite amphibolite from the Sungei Cheroh is given as number 7 in Table 9.1.

Hornfels is sometimes associated with the schist at the granite contact. In Ulu Keta a hornfels contains quartz, epidote, tremolite, and feldspar. In the Sungei Chembatu, a hornfels consists largely of pale green diopside and tremolite. Large crystals of enstatite and aggregates of quartz and epidote are also present. Tremolite is the most common mineral in the hornfels. Other varieties may contain garnet, epidote, chlorite, pyroxene, and serpentine.

Richardson (1939) concluded that the amphibole schist of this region represents metamorphosed impure calcareous rocks. He particularly pointed out that the amphibole schist is always deficient in zirconium, phosphorus, and barium, and this excluded the possibility of its being derived from basalt or dolerite.

UPPER PALEOZOIC AND TRIASSIC. Khoo (1968) has shown that the Upper Paleozoic, Raub Group sedimentary rocks, which lie to the west of the Benta Migmatite Complex, have been dynamothermally metamorphosed generally in the greenschist facies.

The three hills of Bukit Serdam, Gua Kechil, and Gua Panas, which are prominent features of the country to the east of Raub, are of saccharoidal marble. The marble is essentially of calcite, with small amounts of muscovite and fibrous talc. Epidosite and a gneiss composed of epidote, garnet, quartz, plagioclase, diopside, and calcite occurs in the Sungei Ngiang. From the same river, Khoo (1968) also described in epidote-actinolite-albite gneiss. Amphibolite outcrops on the western foothills of Bukit Lebak and at Ginting Lebak. The mineralogy includes hornblende, quartz, epidote, sphene, and iron oxides. Garnet-andalusite-hornblende-quartz schist and biotite schist containing small amounts of amphibole occur near Bukit Lebak. In the same area biotite-quartz-microcline schist has been described. In places the rock contains andalusite. Phyllite is common near Bukit Serdam and in the Sungei Gali, but on weathering it resembles shale. The presence of andalusite in these rocks has been taken to indicate conditions of Abukuma or low-pressure intermediate facies series (Khoo 1968).

As in the Bentong area, conglomerate occurring east of the Main Range often has a phyllitic groundmass. Mica schist is well exposed in the Sungei Cheroh, and Chembatu. Chlorite and graphite are common constituents. Slate that occurs between the Raub electric power station and the Toon Hing tin mine contains andalusite. Andalusite is widespread throughout the area (Richardson 1939), as indeed it generally is throughout Malaya (Scrivenor 1931).

North Pahang

The Lower Paleozoic Schist Series continues as a belt along the eastern foothills of the Main Range. The rocks here are schistose conglomerate, metaquartzite, quartz schist, muscovite-quartz schist, phyllite, and mica schist (Richardson 1950). Farther east, the overlying shales and mudstones of the Permian to Lower Triassic Gua Musang formation are free of metamorphism. There are, however, localized areas of quartz-epidote-amphibole hornfels in the Sungei Rawa near the contact with igneous rocks. Most of the limestone has been recrystallized to saccharoidal marble. Mica schist occurs in the Sungei Yu. Agglomerate in the Anak Sungei Merapoh is schistose, and chloritoid-andalusite schist occurs in the Sungei Serunai. Around the Bukit Ranjut Complex occur localized pyroxene-bearing hornfelses, and garnet-actinolite hornfels is found in the Sungei Kasai Besar.

The volcanic rocks of the Gua Musang Formation have suffered little metamorphism. However, localized metamorphism has been recorded by Richardson (1950). Rhyolite tuff in the Sungei Tehlong has become a black hornfels containing biotite porphyroblasts. Rhyolite tuff in the Sungei Gapis (Sungei Tegapih) has been altered to a foliated sericite-quartz schist.

Epidiorite occurs in the Sungei Kadjau and in the Sungei Merbau. It is composed of pyroxene, pargasite, intermediate plagioclase, and some orthoclase. Much of the pyroxene and plagioclase has been replaced by epidote, and the amphibole by chlorite. An epidorite dyke 150 m wide intrudes phyllite in the Sungei Jelai Kechil. It is composed of epidote and strongly saussuritized andesine hornblende with a spongy texture.

Northeast Malaya

Argillaceous rocks of general Upper Paleozoic age are commonly converted to phyllite and low-grade mica schist (MacDonald 1968). Biotite-andalusite and calc-silicate hornfels are common in restricted areas close to granite plutons. The former sometimes contains poorly developed cordierite and the latter is characteristically banded (MacDonald 1968).

Calc-silicate hornfels containing quartz, epidote, diopside, tremolite, wollastonite, sericite, and sphene is well exposed in the Ulu Sungei Pergau in north Kelantan. Iron oxides, pyrite, and feldspar are often present in the hornfels, and some varieties containing sphene, chlorite, arfvedsonite, and zoisite have been recorded.

Figure 10.13 shows the distribution of bands of calc-silicate hornfels in central Trengganu. The bands occur within meta-arenaceous rocks and are composed of minute mosaics of quartz, diopside, occasionally garnet and zoisite; rarely, amphibole takes the place of pyroxene. They are hard, tough, banded in white and greenish-white or greenish-black, and are interfoliated with metaquartzite (MacDonald 1968).

The shales between the Stong Complex and the Taku Schist have been metamorphosed to spotted slate, phyllite, hornfels, and mica schist. A green chlorite-rich phyllite occurs in and near the Sungei Jintiang.

Some sandstones are converted to quartzite, but subsequent weathering may make the metamorphic condition less apparent. Near the granites, the quartzite becomes quartz hornfels, quartz schist, and quartz-mica schist. Garnet, epidote, biotite, and andalusite are occasionally developed in the impure sandstones.

All limestones in northeast Malaya are converted to marble, and all show considerable variation in grain size. Contact metasomatism by the granites has given skarn, which in the Ulu Pergau area contains wollastonite and tremolite. Many limestones are schistose and contain augen of calcite in a finer grained psammitic or pelitic groundmass.

An area of south Kelantan, stretching 8 km (5 miles) southward from the Stong Migmatite Complex, from Kampong Setar on the Sungei Nenggiri, was surveyed by D. J. Gobbett (personal communication). The rocks of this area were originally igneous and sedimentary, all metamorphosed in the greenschist facies. The pelitic schists contain quartz, muscovite, chlorite, albite, epidote, and occasionally almandine. The gneisses contain quartz, albite, muscovite, epidote, microcline, and sphene. The calcareous rocks are generally of pure calcite, but occasionally contain epidote and tremolite. Generally the grade of metamorphism increases northward from greenschist facies south of the Sungei Nenggiri to amphibolite facies northward in the Stong Migmatite Complex. Toward the south and east they are succeeded by the unmetamorphosed Gua Musang Formation.

In north Trengganu, at Bukit Bintang, the road up to the new VHF and television transmitting station exposes phyllite and coarse conglomerate. The clasts of the conglomerate are of various rock types. The matrix is phyllitic, and the clasts have been likewise metamorphosed to marble, metaquartzite, schist, and calc-silicate. A rock composed of dark green biotite containing perfectly euhedral red garnet crystals (UM R2800) was found at Bukit Tebak, Kemaman area, Trengganu.

VOLCANIC ROCKS. The matrix of the tuffs has become phyllitic and is frequently chlorite-rich. Often the original tuffaceous nature is destroyed and a green chlorite schist or phyllite takes its place.

In andesites, the pyroxene is replaced by amphibole, especially near granite intrusions, and is eventually replaced by chlorite. In central Trengganu the pyroxene of andesites is replaced by chlorite-amphibole aggregates. The pseudomorphs vary from chlorite with near parallel extinction, a 2V of less than 20°, and a low birefringence to amphibole with a 2V higher than 20°, extinction up to 20°, and high birefringence. All gradations of chlorite and amphibole occur within pseudomorphs within the same rock (MacDonald 1968). Some tremolite occurs in the groundmass. Below Kuala Petuang, tremolite becomes dominant, and some andesites are rich in actinolite.

Near the granite plutons, the original andesitic fabric is destroyed and the rock consists of large prismatic needles of pale green to colorless irregularly oriented tremolite. Biotite is common in meta-andesite at Gunong Kesut.

East Pahang

In the Kuantan region, Fitch (1952) noted that the Upper Paleozoic rocks are often converted to phyllite. Around the Ulu Sungei Reman granite pluton, there occurs a thermal aureole with progressive stages through spotted slate, chiastolite slate, andalusite slate to mica schist which are traversed as the granite is approached. The other granites of the area have caused less metamorphism. Sometimes apparently unmetamorphosed shale abuts a granite.

In the Sungei Panching and the Sungei Kuran, limestones have been metamorphosed to pyroxene hornfels. The arenaceous rocks around Kuantan are apparently unmetamorphosed except at Tanjong Geliga, where there is intense greenschist facies regional metamorphism. At Sungei Darah and Bukit Kuantan, andalusite and chiastolite occur in slate.

At Tanjong Geliga near Chukai, shales have been metamorphosed to mica schist, chlorite schist and, in one locality, muscovite-biotite-pyroxene schist. The argillaceous matrix of conglomerate is converted to schist, and the quartzite pebbles are streaked out in the foliation direction. Phyllite occurs between Bukit Tok Hat and Pintu Gerbang. The headland of Tanjong Gelang, between Kuantan and Chukai, is of a foliated sequence of greenschist facies metaquartzite and phyllite.

Regional metamorphism of rhyolitic rocks at Bukit Minyak has led to the development of white mica in the sheared matrix of tuffaceous rocks.

East Johore

The coastal outcrops southward from Mersing are of interfoliated phyllite, slate, and metaquartzite. Sericite is common in the slate and the metaquartzite. The metaquartzite contains some sedimentary grains of tourmaline. The islands offshore, and especially Pulau Tioman, are composed of metasediments and metavolcanic rocks intruded by granite. Chlorite is abundant in the metasediments. Garnet occurs in some of the rocks at Tanjong Atas. Rhyolite tuff often contains garnet and epidote. Cordierite has been found in the Rompin District of Pahang, accompanied by andalusite (Scrivenor 1931), and in the Sungei Isa near the Pelepah Kanan tin mine, Kota Tinggi (Ganesan 1969).

South Johore

Schist occurs near Tanjong Punggai in southeast Johore. It is composed of black, graphitic muscovite-quartz schist with thin intercalations of slightly coarser quartz-muscovite-cordierite hornfels (Grubb 1968). The schist often weathers to resemble clay or shale.

Rhyolite tuffs have been thermally metamorphosed by the granite to pyrophyllite-quartz rock and diaspore-pyrophyllite rock. The typical diaspore-pyrophyllite-quartz hornfels is of an extremely fine-grained groundmass composed of interlocking quartz grains containing tiny clusters of pyrophyllite and minor muscovite. Within the micaceous growths, diaspore porphyroblasts occur as small crystals. The refractive indices of the diaspore were determined as

α 1.692 to 1.699
β 1.708
γ 1.726 to 1.737
2V about 86° $+ve$.

The pyrophyllite was determined as having $\beta = 1.581$ and 2V varying from 50° to 60° $-ve$ (Grubb 1968).

METABASITES. Along the contact with the granite batholith, the gabbro bodies in south Johore and Singapore Island have been metamorphosed (Hutchison 1964a). Large biotite crystals can be seen replacing pyroxene and hornblende. In some places, relict pyroxene cores occur within large hornblende crystals, but the feldspar of the gabbro is little affected except for the formation of some sericite.

In Singapore Island, a striking feature of the whole gabbro body is the formation of hornblende crystals along joint planes, in veins, and in cracks. Large crystals up to 12 cm long are embedded in a groundmass predominantly of quartz. These hornblende pegmatites are found irregularly throughout the Singapore and South Johore basic bodies. Excellent examples occur in quarries 1 and 11 (Figure 8.14). The following two modal analyses of metamorphosed gabbro from the region are representative of this kind of transformation.

1. Specimen UM 482: Poh Hin quarry 1, Singapore Island (4553 points)—hornblende, 38%; clinopyroxene, 3%; orthopyroxene, 8%; biotite, 10%; plagioclase $Ab_{38}An_{62}$, 37%; quartz, 2%; sericite 2% by volume.

2. Specimen UM 3829: Senai Estate, south Johore (6832 points)—hornblende, 2%; tremolite, 23%; plagioclase $Ab_{18}An_{82}$, 72%; sericite, 1%; magnetite, 2% by volume.

Generally the metamorphic replacement of pyroxene and olivine in the gabbro has been more complete in Johore than in Singapore, where invariably relict pyroxene cores remain within hornblende crystals.

Elsewhere in the Malay Peninsula, basic bodies have been more completely metamorphosed than the south Johore and Singapore occurrences. To the south of Malacca, the Bukit Larang amphibolite is a crudely foliated amphibolite composed essentially of bytownite feldspar, An_{74}, and both actinolite and hornblende. The brown pleochroic hornblende becomes opaque and clouded with magnetite dust as it is replaced by bright green fibrous actinolite. Accessory apatite is common.

The Bukit Larang amphibolite and the Sungei Chembatu hornfels already described from the Raub area of Pahang can be regarded as metagabbro. On either side of the Main Range, amphibolite occurrences in Lower Paleozoic rocks are generally assumed to represent metamorphosed dolerite dykes.

The metabasite and serpentinite bodies farther north in the Malay Peninsula seem to be restricted to the Lower Paleozoic outcrop, which is generally of greenschist facies. Therefore, it seems reasonable to conclude that they are also the products of regional metamorphism. The lenticular nature of these basic and ultrabasic bodies tends to support an ultimate igneous origin, and perhaps they represent ophiolite bodies that were contemporaneous with the Lower Paleozoic metasediments. They might well have been originally of spilitic flows. Miyashiro (1966) concluded that serpentinites in orogenic belts might be regarded as resulting from regional metamorphism of picritic lava flows in the geosynclinal sequence.

Physical Effect of Metamorphism on Calcareous Rocks

Hutchison (1968a,b) investigated the effect of metamorphism and recrystallization of Malayan limestones on their thermoluminescence properties. The thermoluminescence age of calcite is represented by the amount of electrons that have been trapped in crystal imperfections since the crystallization of the mineral, divided by its natural alpha radioactivity. Older limestones have had longer to accumulate trapped electrons and hence will show a greater thermoluminescence age than a young limestone. On the other hand, a highly radioactive limestone will accumulate thermoluminescence more rapidly than a weakly radioactive one because the natural radioactivity gives the electrons sufficient energy to escape from the valence band to the conduction band and thus be in a position to be captured in crystal imperfections. Therefore, the thermoluminescence age is roughly computed as the amount of thermoluminescence divided by the natural alpha radioactivity of a specimen.

The thermoluminescence ages of limestone from the Chuping Formation (Upper Paleozoic) were found to be no smaller than those of limestone from the Setul Formation (Lower Paleozoic). Therefore, the thermoluminescence ages of Malayan limestones are not related to their stratigraphic ages but to the tectonic ages (i.e., to ages of recrystallization during regional metamorphism or to subsequent recrystallization during thermal metamorphism from the granite plutons).

Figure 9.20 is a histogram of thermoluminescence ages of Malayan limestones obtained from 74 determinations. The determinations marked *A* are for specimens far removed from granite intrusions, generally collected from Kedah and Perlis, so that the "age" in these cases can be related to orogenic folding. Those marked *B* were collected from areas close to granite plutons, so that the "ages" determined reflect recrystallization caused by thermal metamorphism.

The younger concentrations of thermoluminescense ages given by Malayan limestones are generally from specimens collected from the Kinta Valley around Ipoh and in the Kuala Lumpur area, all from the western margins of the Main Range Granite batholith. The radiometric ages of Malayan granites indicates that the western margins of the Main Range (Figure 8.1) are characterized by an Upper Cretaceous to Lower Tertiary tectonic event, accompanied in a few localities by small epizonal granite emplacements.

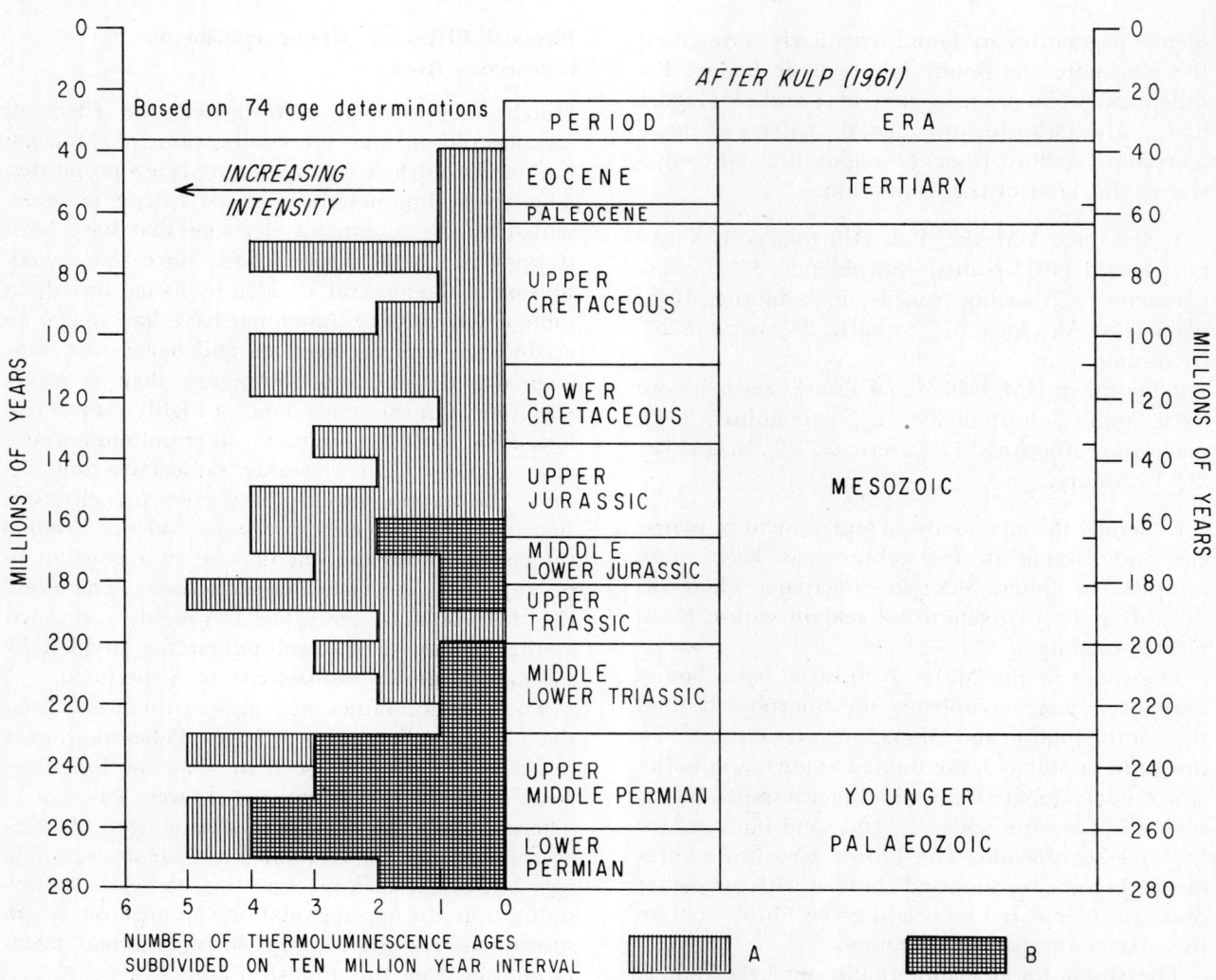

FIGURE 9.20 *Distribution of thermoluminescence ages of marble resulting from orogenic folding* (a), *and from heating or recrystallization by granites* (b). *These results should be compared with the radiometric ages of granites summarized in Figure 8.12.*

The spread of thermoluminescence ages upward from the Triassic period results from this post-orogenic event. The thermoluminescene ages have been reduced from Triassic to younger ages by complete or partial drainage of the earlier accumulation of trapped electrons (Hutchison 1968a). The secondary calcite veining in the marble of Perlis was dated by this method as **Eocene**.

DOLOMITIZATION OF CALCAREOUS FORMATIONS. Hutchison (1968d) showed that the occurrence of dolomite is not related to age of the limestone formations. Dolomite is sporadic and very patchy and has no easily discernible pattern. It seems that in some cases dolomitization was related to the granite intrusions, for dolomite is well represented near the granite contact—for example, at Ampang in Selangor and at several localities along the eastern margin of the Kinta Valley. However, dolomite localities have also been found several miles from the nearest granite. Malayan limestones have a very strong carbonate compositional mode of around 99% cationic calcium with a pronounced mode of 2.5% insoluble residue (Tables 3.2, 4.1, and 5.1; Hutchison 1968d). Because of this high degree of purity, regional and thermal metamorphism has usually resulted in an extremely pure white calcite marble. Black carbonaceous impurity is common in the Lower Paleozoic marble of northwest Malaya, but elsewhere the Lower Paleozoic marble is generally pure white as is younger marble. However, euhedral cubes of pyrite are widespread in Malayan marbles.

ROCKS OF THE EPIZONE

The epizone includes all the rocks of the Malayan Geosyncline that have never been buried deeply enough to have suffered dynamothermal metamorphism. The epizone therefore includes all Middle Triassic, and younger formations.

The Permian to Lower Triassic Gua Musang Formation is apparently intermediate between the epizone and mesozone; in the southern part of its outcrop it is mainly unmetamorphosed, but toward the north in Pahang and Kelantan its grade of metamorphism increases to greenschist facies. This northward increase in metamorphism in north Pahang into Kelantan has been noted by many authors. This is not an actual increase in regional temperature, but rather a reflection of the greater uplift of northern Malaya compared with southern Malaya; the catazone is exposed only in north Malaya.

In northwest Malaya, Upper Paleozoic and Lower Paleozoic rocks have remained in the epizone, but elsewhere all the Paleozoic formations have been buried deeply enough to be generally metamorphosed in the greenschist facies.

The epizone should be characterized by hornfels facies contact metamorphism around the Upper Cretaceous to Lower Tertiary high-level granite plutons.

Metamorphic rocks outcrop extensively on the northern slopes around the reservoir area of Mount Ophir along the granite contact. Their interpretation suggests a thermal aureole, characterised by andalusite porphyroblasts, superimposed on a complex of catazonal metamorphic rocks which includes some migmatite. The thermal epizonal metamorphism from the granite has downgraded the older catazonal metamorphic mineralogy.

Some of the Upper Carboniferous and Permian granites of the East Coast Belt must also be considered as epizonal. The evidence for this is twofold: (a) In the Kuantan area, Upper Carboniferous granite is intruded into Lower Carboniferous sedimentary formations (Hutchison & Snelling, 1971) and (b) they often have distinct contact aureoles, away from which the rocks are fossiliferous and unmetamorphosed.

Dislocation Metamorphism

Postorogenic faulting is pronounced in the Malay Peninsula (see Chapter 10), and intense shearing, fracturing, and mylonitization are often to be found in the fault zones. In the granite quarry east of Kuala Kubu Bharu, Haile (personal communication) has recorded zones of mylonite a few millimeters thick along the fault planes. They include seams of brown tectogenic glass, less than 1 mm thick, associated with stilbite. Along and near the Lebir fault in northeast Malaya Aw (personal communication) has noted fracturing, shearing, and mylonitization. The mylonite is often enriched in sulfide minerals. Shearing in the granite is described in Chapter 8.

SUMMARY OF METAMORPHISM IN THE MALAYAN OROGEN

An upper limit of about 5 kb is set on the deepest exposed parts of the Malayan Orogen by the occurrence of cordierite. Cordierite has been recorded both in catazone amphibolite facies rocks (Gunong Stong) and in mesozone greenschist facies rocks (Kedah Peak, Ulu Pergau, Kota Tinggi, etc.). Hence it can be deduced that high temperatures of about 600°C were attained in the Malayan Orogen under pressure conditions that did not exceed 3 to 5 kb; otherwise, cordierite could not have formed (Winkler 1967). These conditions indicate that the Malayan Orogen was characterized by the Abukuma-type facies series. This facies series represents a metamorphism of regional extent which has come into being under especially low pressure conditions. High temperatures of about 700°C have been reached at a depth of about 10 km through a very high geothermal gradient of 70°C km.

The characteristic mineral of the Abukuma-type upper greenschist and lower amphibolite facies is andalusite, and in combination with cordierite it establishes the andalusite-cordierite-muscovite subfacies of the amphibolite facies. Above temperatures of about 650°C, sillimanite and cordierite (characteristic of the Stong Complex) form the characteristic assemblage of the sillimanite-cordierite-muscovite-almandine subfacies of Abukuma-type amphibolite facies.

The beginning of the amphibolite facies is characterized by such minerals as diopside, grossularite–andradite; if the rock composition is suitable, staurolite and cordierite are also to be expected, and indeed these do occur in the catazone rocks of Kelantan.

The common occurrence of andalusite and cordierite in Malayan rocks must indicate Abukuma-type conditions, and the association of sillimanite with cordierite in the Stong Complex (Hutchison

1969) is conclusive. This allows the tectonic zones of the Malayan Orogen to be subdivided as follows:

	Temperature Range (°C)	Depth Range (km)
Epizone	Surface to 250–350	Surface to 4–5
Mesozone	250–350 to maximum 500	4–5 to 7
Catazone	Temperatures in excess of 500	Depth in excess of 7

Geological features supporting the interpretation that the Malayan Orogen was characterized by the Abukuma-type facies series (high geothermal gradient) are as follows:

1. The cover of Post-Paleozoic sedimentary and volcanic rocks was thin and was deposited in shallow water. Thus the Paleozoic could never have been very deeply buried—the Lower Paleozoic probably lay not deeper than about 7 km. Its subsequent metamorphism to the greenschist facies can only be accounted for by a high geothermal gradient.

2. The high intensity of granitic activity is a feature more characteristic of a region of a high geothermal gradient [e.g., the Abukuma terrane of Japan (Miyashiro 1961) and the Buchan area of northeast Scotland (Read 1960)] than of regions of low geothermal gradient. This is logical if one holds that granites in orogenic belts are largely of anatectic origin. In the deeper parts of the orogen, an Abukuma-type geothermal gradient would give a regional temperature of the order of 700°C and a pressure of about 3 kb at a depth of only 10 km. These conditions are more than adequate for the partial anatexis of the geosynclinal rocks. Moreover, concordance of the Q-Ab-Or maximum for the Malayan granites (Figure 8.11) and the thermal low on the 3000-bar equilibrium diagram for the system SiO_2-Ab-Or-H_2O of Tuttle and Bowen (1960, Figure 25) indicates the strong possibility that the bulk of Malayan granites resulted from anatexis of the infrastructure at depths of the order of 9 or 10 km. Anatexis was possible at such shallow depths because of the high geothermal flow from the underlying mantle.

3. The general absence of thermal aureoles around the large mesozonal batholiths indicates that the country rocks were held at sufficiently high regional temperatures that the granite emplacements caused only very localized heating. The granites did, however, have the ability to raise the temperature enough to cause a higher grade of dynamothermal metamorphism than would have been possible solely as a result of the regional heat flow. The Kedah Peak Dome is clearly a case in which the higher temperatures can be directly related to the granite.

In the past too much importance has been attributed to the granite intrusions in descriptions of metamorphism. Hence the presence of andalusite and cordierite was nearly always attributed to the granites. Dynamothermal metamorphism is clearly more widespread and characteristic of the Malayan Orogen than is contact metamorphism. The general phyllitic nature of the Paleozoic rocks, and the exposures along the central core of Malaya of migmatite complexes, indicate the high geothermal gradient that must have attained in the region and suggest that the large masses of granite result from this rather than cause it.

Pronounced contact metamorphism where granite intrusions are in contact with calcareous rocks gives a skarn that is not entirely thermal but partly metasomatic.

REGIONAL CORRELATION

Thailand

Beds of metasediments occupy a large region along the central axis and west side of peninsular Thailand and extend northward into lower Burma (Brown et al. 1951). These rocks are isoclinally folded and generally have steep dips. Most exposures are of well-foliated schist, slate, argillite, and metaquartzite. Outcrops extend from the vicinity of Phuket in the south to the latitude of Rat Buri (13°30′N), and also in the vicinity of granite intrusions in the Khao Luang mountain chain of eastern peninsular Thailand (Brown et al. 1951). All these metasediments are clearly of low-grade regional metamorphism and represent the northward extension of the mesozonal metamorphism that has been described already from Malaya.

Gneissic and schistose rocks, probably of the catazone, occur in several areas. They occur in an elongate belt of country to the west of Chiang Mai, west of Tak, in a small area northwest of Kanchanburi, in the Hua Hin area to the south of Phet Buri, and west of Surat Thani (Brown et al. 1951).

These gneissic areas are aligned in a general north-northwest direction, which coincides with the gneissic foliation. There are also large outcrops in Changwats Chon Buri and Rayong in the corner of the Khorat Plateau northwest of Loei. The high-grade catazonal rocks of north Kelantan appear to continue into southern Thailand.

At Hua Hin, in Prachuab Province of Thailand, gneiss has evidently been generated by high-grade metamorphism of Lower Permian strata (Burton 1969a). A well-foliated granite gneiss, of more leucocratic and fine-grained nature, occurs around Bang Saen. All these gneissic rocks have been assigned a Prepermian age by Brown et al. (1951) on the grounds that the Kanchanaburi gneiss is overlain by Permian Rat Buri limestone. Burton (1969a), however, suggested that they cannot be older than early Permian.

Indonesia

Low-grade regionally metamorphosed Paleozoic sediments and intercalated metavolcanic rocks, which characterize the eastern parts of Johore, continue into the Indonesian islands to the south of Singapore. These obviously represent mesozonal metamorphism.

Phyllitic formations with metavolcanic intercalations thought to be Permo-Carboniferous, occur on several islands in the Riouw–Linnga Archipelago (van Bemmelen 1949).

The Precarboniferous schists of Songkop are of fine-grained hornblende schist containing hornblende, quartz, epidote, and sphene. The metamorphic basic igneous rocks of the Karimon group indicate regional metamorphism in the mesozone (van Bemmelen 1949). Intercalated phyllitic rocks and metavolcanic rocks occur on Bintan, and van Bemmelen (1949) noted their similarity to the Upper Paleozoic rocks of Johore. Phyllitic rocks also occur in Singkep, and phyllitic and mica schists occur in Bangka. Intensely folded low-grade metasediments are found on Karimata.

The region of West Borneo adjacent to the Karimata Islands has been described by van Bemmelen (1949) as the continental part of Sunda Land, which has not been subjected to Tertiary diastrophism. It contains a basement of crystalline schists, which are the oldest rocks, probably of Precarboniferous age. These schists occur as roof pendants and relics of the batholithic roof of a Triassic granitic and tonolitic plutonic massif. They are of partly migmatized crystalline schist and metabasite, and they contain sillimanite or andalusite, biotite, and cordierite. These metamorphic rocks, apparently of catazonal genesis, are characteristic of the Central Axis or Schwaner Zone of West Borneo (van Bemmelen 1949). The presence of cordierite links them with the Abukuma terrane of Malaya, and apparently indicates that the Thai–Malayan orogen can be extrapolated into West Borneo.

The overlying rocks are assumed to be of Carboniferous or Permian age and are folded and metamorphosed to a low grade. The Matan Complex is comparable with the Permo-Carboniferous volcanic rocks of Malaya. It is closely associated with the Ketapang Complex, which is partly calcareous. Both these complexes are complexly folded and in places have suffered low grade metamorphism. Common accessory minerals in rocks of these complexes are biotite, muscovite, sericite, and hematite. Secondary tourmaline is generally present in the Ketapang Complex, and topaz occurs, but the presence of cassiterite has not been ascertained (van Bemmelen 1949).

Thus although from a structural view point it seems reasonable to extend the Thai–Malayan Orogen into West Borneo, the absence of tin in the large granitic and tonalitic batholiths of West Borneo raises an important question. Does the Thai–Malayan Orogen include West Borneo, or does it terminate somewhere between the islands of Billiton and Karimata? More geological information is required from West Borneo before this question can be answered.

CHAPTER 10

Tectonic History

D. J. GOBBETT AND H. D. TJIA

The overall geological structure of the Malay Peninsula has so far received no definitive discussion or analysis except as a marginal feature of the Indonesian Island Arc System van Bemmelen 1949). Early summaries by Scrivenor (1921, 1923, 1931) described the present structure in very general terms as reflected in the geomorphology. Essentially this is a series of arcuate ridges ("coulisses" of Scrivenor) arranged *en echelon* and running parallel to the axis of the peninsula. These ridges are formed mostly of granite, and between them lie belts of strongly folded sedimentary rocks. This structure continues to the south via the islands of Bangka and Billiton into West Kalimantan. Scrivenor mistakenly continued it northeastward through Kalimantan to the much younger Philippine Island arcs.

Scrivenor (1931) described structural features of some of the Malayan rock formations, but the uncertainty of their stratigraphic sequence, the presence or absence of unconformities, and the relative chronology of events precluded any analysis on a regional scale. The more detailed regional studies published as memoirs by the Malayan Geological Survey described the older (pregranite) sedimentary rocks as strongly deformed into similar folds, typically isoclinal and in places recumbent, by a general east-west compression. The extensive granite plutons were intruded usually in anticlinal axes, and the occurrence of roof pendants and tin mineralization indicated that these plutons had not been deeply eroded. The presence of a major unconformity between the "Calcareous Series" and the "Arenaceous Series" was hinted at by Roe (1951), west of the Main Range Granite, and by Fitch (1952), in the Kuantan area of east Pahang.

In more recent years field studies have shown the presence of an important phase of earth movements in northwest Malaya during the Devonian (Koopmans 1965), the presence of younger Mesozoic terrestrial sediments resting unconformably on older rocks (Paton 1959, Koopmans 1966, Rajah 1969), and major wrench faulting (Burton 1965, 1967e).

THE POSITION OF THE MALAY PENINSULA IN THE STRUCTURE OF SOUTHEAST ASIA

The theory of continental growth by accretion of tectonized zones around ancient massifs has been generally applied to Southeast Asia and Indonesia (Klompé 1961). The position of the Malay Peninsula conforms with this theory. It lies between the Variscan massif of eastern Thailand and Cambodia (Indosinia), which was stabilized in the Middle Carboniferous, and the Tertiary island arc system of the Andaman–Nicobar ridge and west Sumatra (Figure 10.1a). Malaya itself is essentially a Mesozoic structure forming part of the Pacific orogeny of Vialov (1939). Thus it might be expected that there would be a tendency for structures to young southwestward across the Malay Peninsula, and a general geological comparison of the east and west sides gives some support to this.

Van Bemmelen (1949) modified the zonal structure of the Sunda area by postulating two centers

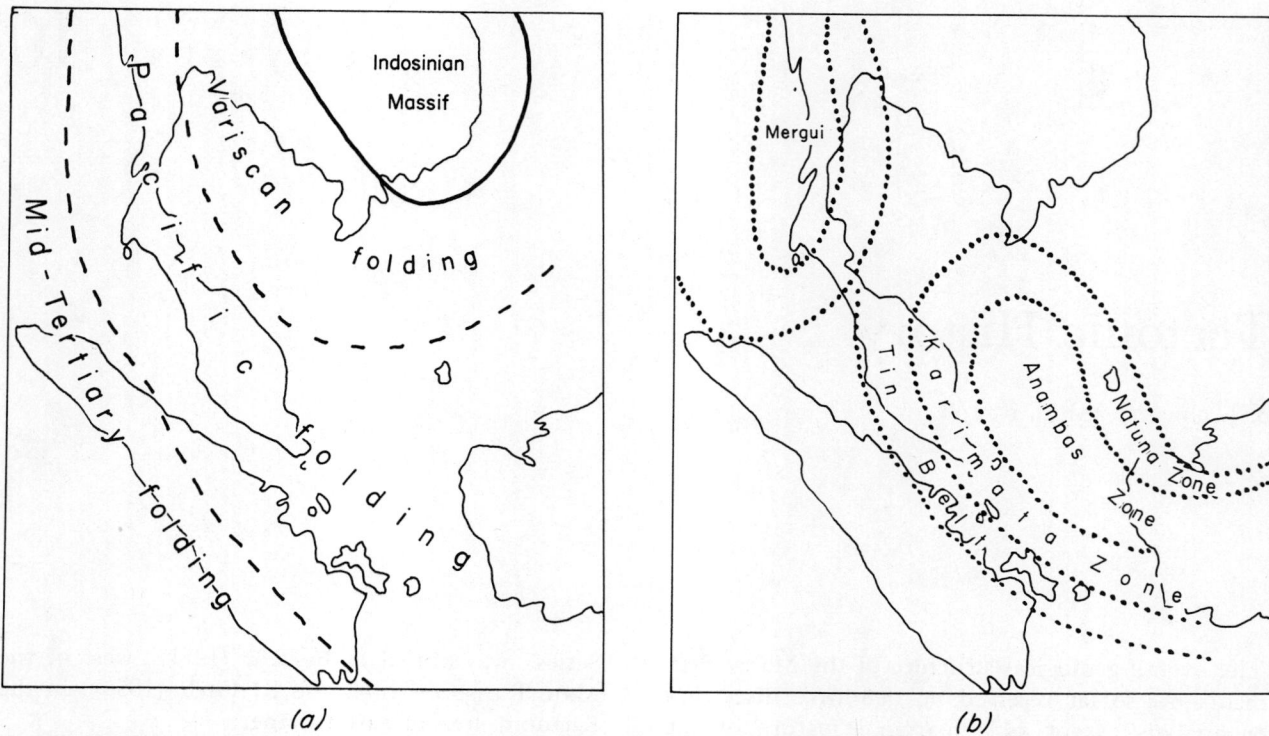

FIGURE 10.1 *Structure of the Sunda area: (a) After Klompé (1961), following Stille, Umbgrove, and Westerveld; (b) after van Bemmelen (1949).*

of continental growth. One of these lay at the Anambas Islands and another at Mergui in peninsular Burma. Most of Malaya lay in two zones of Mesozoic tectonism to the west of the Anambas center (Figure 10.1b). These were successively a volcanic inner arc (Karimata Zone) and a nonvolcanic outer arc (Tin Belt). Perlis and Langkawi were envisaged to belong to an outer zone of the Mergui center. He considered that the Karimata Zone underwent orogeny and plutonism in the late Paleozoic. It was uplifted and supplied Triassic flysch to the west. The second phase of the orogeny produced westerly directed isoclinal folding and thrusting. This was followed by the intrusion of tin-bearing granites, particularly in the Tin Belt.

However, van Bemmelen's thesis was formulated before Lower Paleozoic and Devonian rocks were recognized in the Malay Peninsula, and it now must be reconsidered.

STRUCTURAL OUTLINE OF THE MALAY PENINSULA

Our present knowledge of Malayan geology indicates that the structural grain of the peninsula is not as simple as it has been interpreted to be in the past. Also, its history is longer and more complex. Although the late Triassic movements are dominant, the present structure is a result of superimposed events of mid-Paleozoic, late Paleozoic, late Triassic, later Mesozoic, and Tertiary age. The spatial distribution of these successive earth movements gives little support to the theory of simple accretionary tectonism. An outline of the structure of the Malay Peninsula appears in Figure 10.2. The peninsula is divisible into three structural zones lying parallel to its axis. Regional structural trends are shown by the strikes of bedding and fold axes and by the elongation of plutonic bodies. In the western zone the regional strike is sinuous. Traced southward, it swings from south on the Thai border to southwest in Kedah, back to south in Lower Perak, and to south-southeast and southeast at the southern end of the Main Range.

The eastern zone shows a more uniform south-southeast to south strike, but slight sinuosity of the strike tends to mirror image the arcs of the western zone. The axial zone exhibits uniform regional strike of south-southeast for the Paleozoic and Triassic rocks and includes catazonal metamorphic complexes.

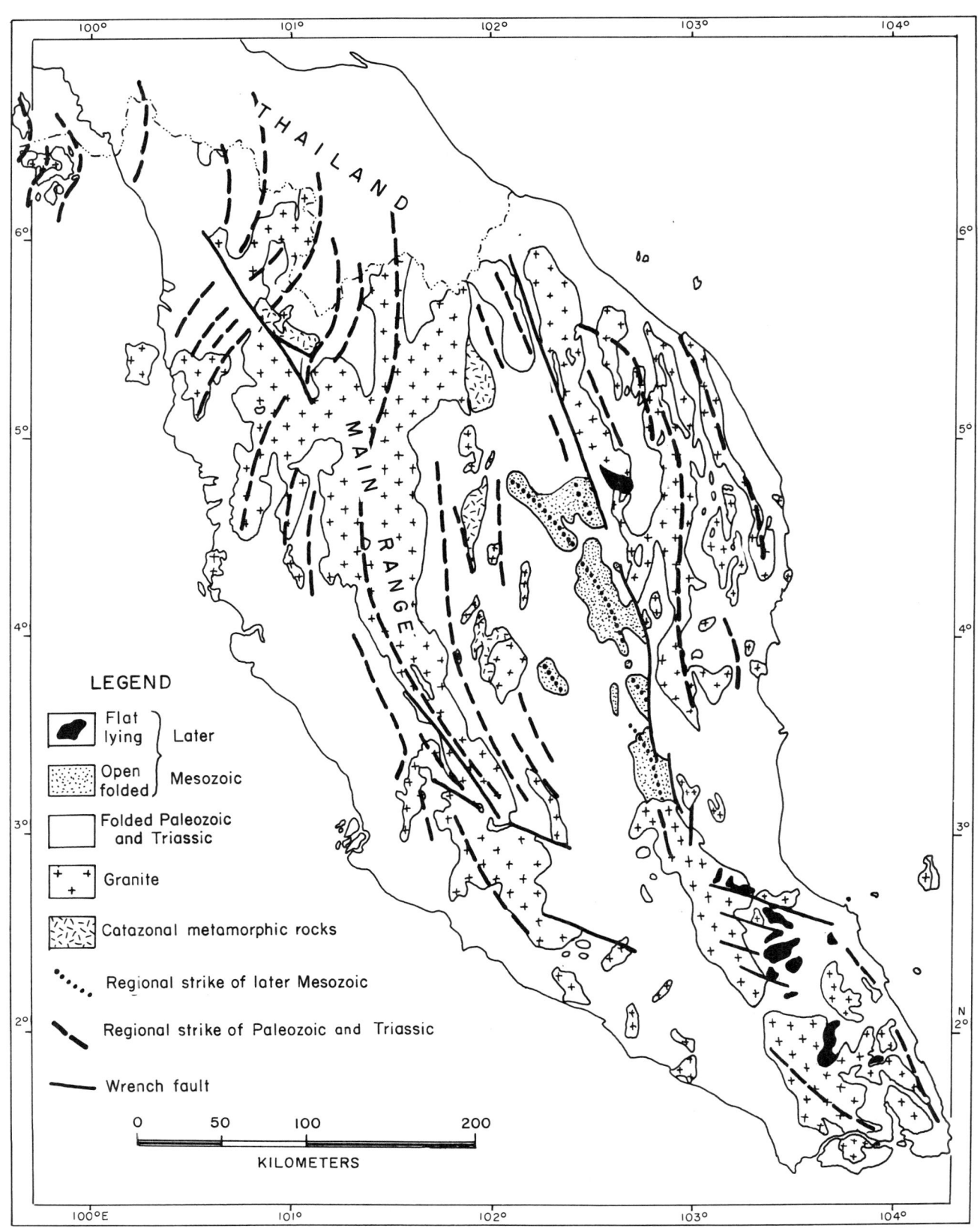

FIGURE 10.2 *Structural outline of the Malay Peninsula.*

Superimposed unconformably on this structure are later Mesozoic strata. These are folded along southeast axes in the north, but to the south the strike swings south. Similar later Mesozoic strata, for the most part flat lying, occur in the eastern zone.

Other prominent elements of the structure are major wrench fault systems. The two most important of these are the Lebir fault, which divides the axial and eastern zones for much of their length, and the Kuala Lumpur–Endau fault, which displaces the structural zones in the southern part of the peninsula sinistrally toward the east.

In summary, Malaya presents a double-sided orogene, with structures convex outward from an axial core. Its tectonic history is now discussed in more detail.

MID-PALEOZOIC TECTONISM

Earth movements of mid-Paleozoic age may have been widespread in Malaya but they are only known with certainty in the extreme northwest (Langkawi and Perlis). In Selangor and west Pahang, cleaved Lower Paleozoic rocks are overlain by uncleaved rocks of Upper Paleozoic and/or Triassic age, but only in the Kuala Lumpur area has an unconformity been clearly demonstrated (Gobbett 1965a, Haile 1970).

In many areas, owing to the effect of compression at the end of the Trias, the strike of the Lower Paleozoic rocks conforms closely to that of the Upper Paleozoic and Triassic. In Langwaki, however, Koopmans (1965) showed structural as well as stratigraphical discordance between the Cambrian to Lower Devonian formations and the overlying Carboniferous and Permian Singa Formation.

Langkawi Islands

The Lower Paleozoic rocks of Langkawi were divided structurally into two units by Koopmans (1965): 1. a zone of recumbent folds occupying the islands Tuba to Timun in the southeast, and 2. an area of less severe deformation to the west and north (Figure 10.3). In zone 1 the Setul rocks have attained a stage of low-grade regional metamorphism, comprising marmorized limestone, slate, phyllite, and quartzite. Axial-plane slaty cleavage is present in the slate, combined with recumbent folding. The fold axes trend approximately north-northeast, and the folds are recumbent toward the west.

A second deformation gently folded the original cleavage planes. Locally the second generation folds show kinking where crenulation cleavage developed as axial-plane cleavage. Characteristic of the kink zones are quartz veins; some of these were folded during the second episode of deformation, and others formed after the folding, as shown by their transections across the crenulation cleavage. The second-generation fold axes are parallel to those of the original folds. Statistically the S_2 planes appear to be slightly steeper than the S_1 planes.

Third-generation structures consist of intensive parallel fracturing and shearing. The fractures dip steeply at about 70° toward the southwest, and they cause knick zones in S_1 and S_2 cleavages. Clinozoisite forms a mineral lineation in the slate and is parallel to the knick zones. Although the lineations associated with the first- and second-generation folds are fairly constant in direction, the variations of pitch directions of the third-generation lineations probably reflect local changes of the stress field.

The massive character of the Setul Limestone considerably reduces the development of small-scale structures. In a few places recumbent folds of the order of 50 m were observed. On Pulau Tuba, granite sills occur within the Setul Formation. The sills vary in thickness from a few centimeters to about one meter and are boudinaged. The long axes of the granite boudins are parallel to the first-generation folds, thus indicating their contemporaneity.

In zone 2, regional metamorphism is not apparent. The Setul Limestone was folded into open anticlines and synclines with vertical axial planes and amplitudes of 30 to 40 m (100-130 ft). Chevron folds occur in the argillaceous beds and are about 50 to 100 m (165-330 ft) high. A statistical study proved the folding to trend northeast. This direction was also observed in the minor folds. Superimposed upon this trend is another fold direction striking southeast.

The Machinchang Formation was deformed into an open anticline of the flexure fold type, striking northeast to north, with a wavelength of several kilometers and a vertical axial plane. Minor flexure folds on the west limb were observed to plunge 40 to 50° to the west. Koopmans (1965) assumed the cross-folding to indicate two different periods of deformation.

The Singa and Chuping Formations present no evidence of regional metamorphism or of the structures described previously; they do, however, show a gentle regional dip toward the east. The deforma-

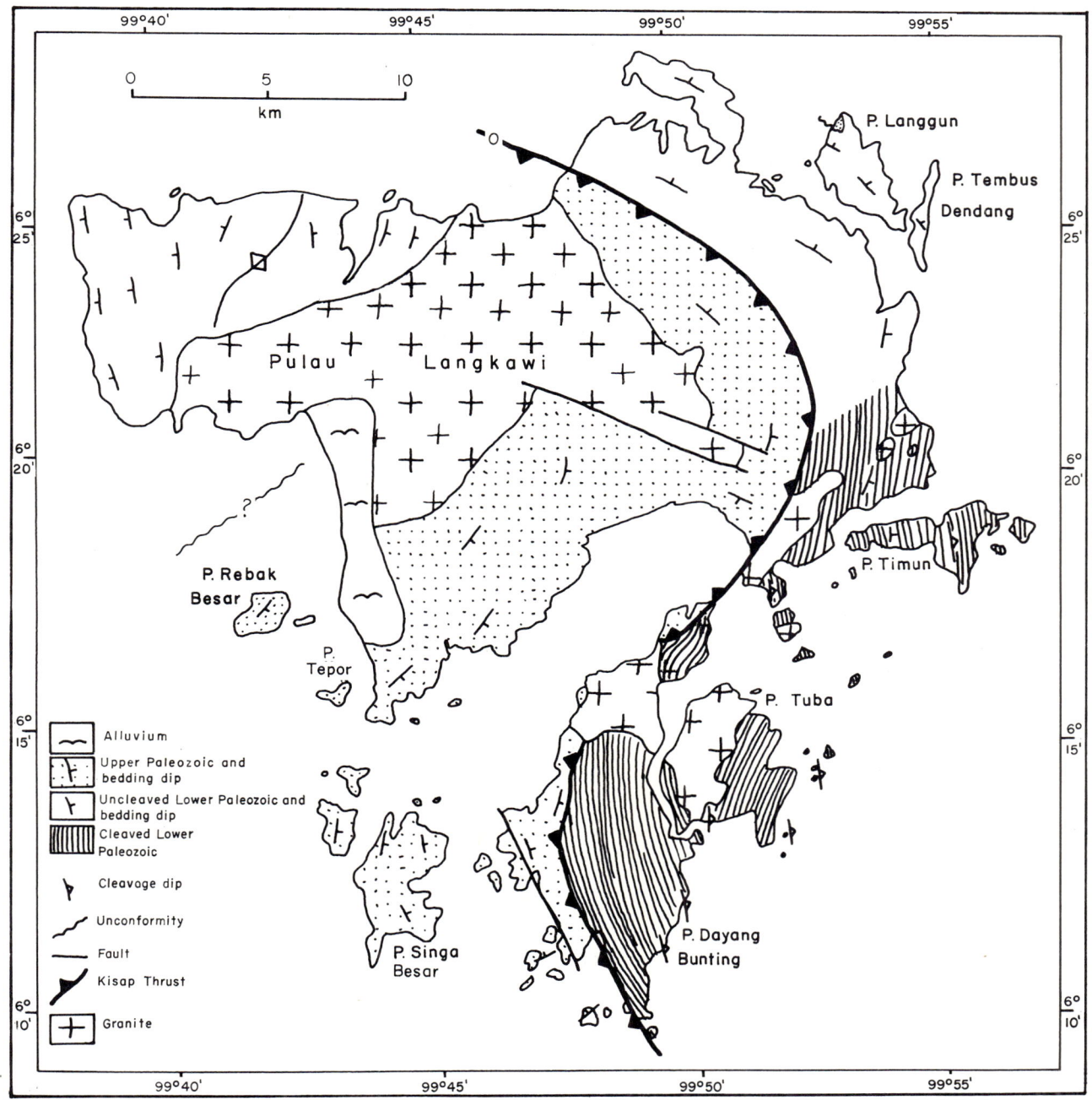

FIGURE 10.3 *Geological sketch map of the Langkawi Islands showing the main structural features. After Jones (1966) and Koopmans (1965).*

tion described earlier can be dated as Devonian, since the youngest beds of the Setul Formation contain graptolites of youngest Silurian or earliest Devonian age, and the oldest beds of the Singa Formation are latest Devonian or earliest Carboniferous. The deformation was accompanied by some acid intrusives found as granitic sills within the Lower Paleozoic of zone 1. Granite pebbles also occur in the Singa Formation.

Kuala Lumpur

Around Kuala Lumpur is a thick sequence of Lower Paleozoic sediments, with Upper Silurian fossils near the top of the succession. These rocks were metamorphosed to schist and marble and deformed into cleavage folds, probably isoclinal, before the deposition of the overlying Kenny Hill Formation. The latter has been but mildly metamorphosed and is gently folded about north-south axes (Figure 10.4). Assumed to be of Upper Paleozoic age (see Chapter 4), the Kenny Hill Formation has so far yielded no distinctive fossils. The Lower Paleozoic rocks form a tectonic basin whose center is occupied by massive to thin-bedded limestones and dolomites, mostly marmorized and commonly showing asymmetrical to recumbent and ptygmatic folds (Gobbett 1965a).

A mineral lineation in the Dinding Schist trends between 230° and 290°. This suggests that the pre-Kenny Hill (?mid-Paleozoic) folding may have produced approximately east-west striking folds in contrast to the north-south striking post-Kenny Hill folds.

North of Kuala Lumpur, ?Lower Paleozoic limestone displays flow and recumbent folds, in contrast to the overlying ?Upper Paleozoic "Arenaceous Series" (Roe 1953). Wong (1971) reported an angular unconformity between an older marble and schist sequence and younger sandstone and phyllite at Rawang, Selangor. On lithology he was able to correlate the older sequence with the Lower Paleozoic of Kuala Lumpur and the younger beds with the Kenny Hill Formation. Further evidence for mid-Paleozoic movements has been provided by Haile (1970), who established the presence of a major angular unconformity in the Ginting Sempah area. Below this unconformity are schist and radiolarian chert of probable Lower Paleozoic age, above it lies the Sempah Conglomerate, which is tuffaceous and is overlain by rhyolite that has been radiometrically dated as late Carboniferous or early Permian.

West Pahang

The north-south striking outcrop of the Bentong Group, running along the eastern margin of the Main Range Granite is probably largely of Lower Paleozoic age (see Chapter 3). In southwest Pahang, Lower Devonian fossils have been found in the upper part of the sequence. The rocks of the Bentong Group, which comprise a typical eugeosynclinal assemblage, are cleaved and in part metamorphosed. The relation of these rocks to those immediately to the east has so far remained obscure, although it seems likely that it is in part tectonic.

In the south, fossiliferous uncleaved Middle Triassic mudstones appear to overlie the Bentong Group. Further north, unfossiliferous shale with lenses of recrystallized limestone is attributed to the Gua Musang Formation of Permian and Lower Triassic age. Thus the mild regional metamorphism of the Bentong Group could have occurred during the middle Paleozoic. However, both the Bentong Group and the Gua Musang Formation were isoclinally folded by the main Triassic earth movements and are now structurally similar.

Plutonism

The evidence for Precarboniferous plutonism in the Malay Peninsula is not conclusive. Granite pebbles in the Singa Formation of Langkawi are accompanied by sedimentary pebbles derived locally from the Lower Paleozoic formations. The granites are well-rounded, however, in contrast to the angular shale and limestone clasts, and they are relatively rare. This suggests that they had been transported from an older granite outside the peninsula. A few radiometric dates give ages older than Carboniferous. Dating by the rubidium-strontium method has indicated microgranodiorite east of Kuala Lumpur to be 430 ± 80 m.y. (Tremadoc to Upper Devonian), and apparent ages of about 355 m.y. (Upper Devonian) have been obtained from granites on the Cameron Highlands Road, Taiping, and East Kedah (J. D. Bignell in correspondence). These Upper Devonian ages correlate with the post-lowest Devonian and pre-uppermost Devonian folding in Langkawi and possible elsewhere. In the Kinta Valley, however, evidence exists for a conformable sequence of Silurian to Permian marine carbonates, and this would preclude the possibility of widespread Devonian movements in the western part of the Malay Peninsula.

UPPER PALEOZOIC TECTONISM

Marine Carboniferous and Permian rocks are widespread in the Malay Peninsula, but the Lower Triassic is confined to central Malaya. Indeed, Triassic rocks appear to be entirely absent in the east and are restricted to northwest Perak and Kedah in the west, where the Upper Ladinian rests unconformably on the Paleozoic (Figure 4.2). Upper Paleozoic earth movements were probably widespread in Malaya, although in many cases they cannot be distinguished with certainty from those

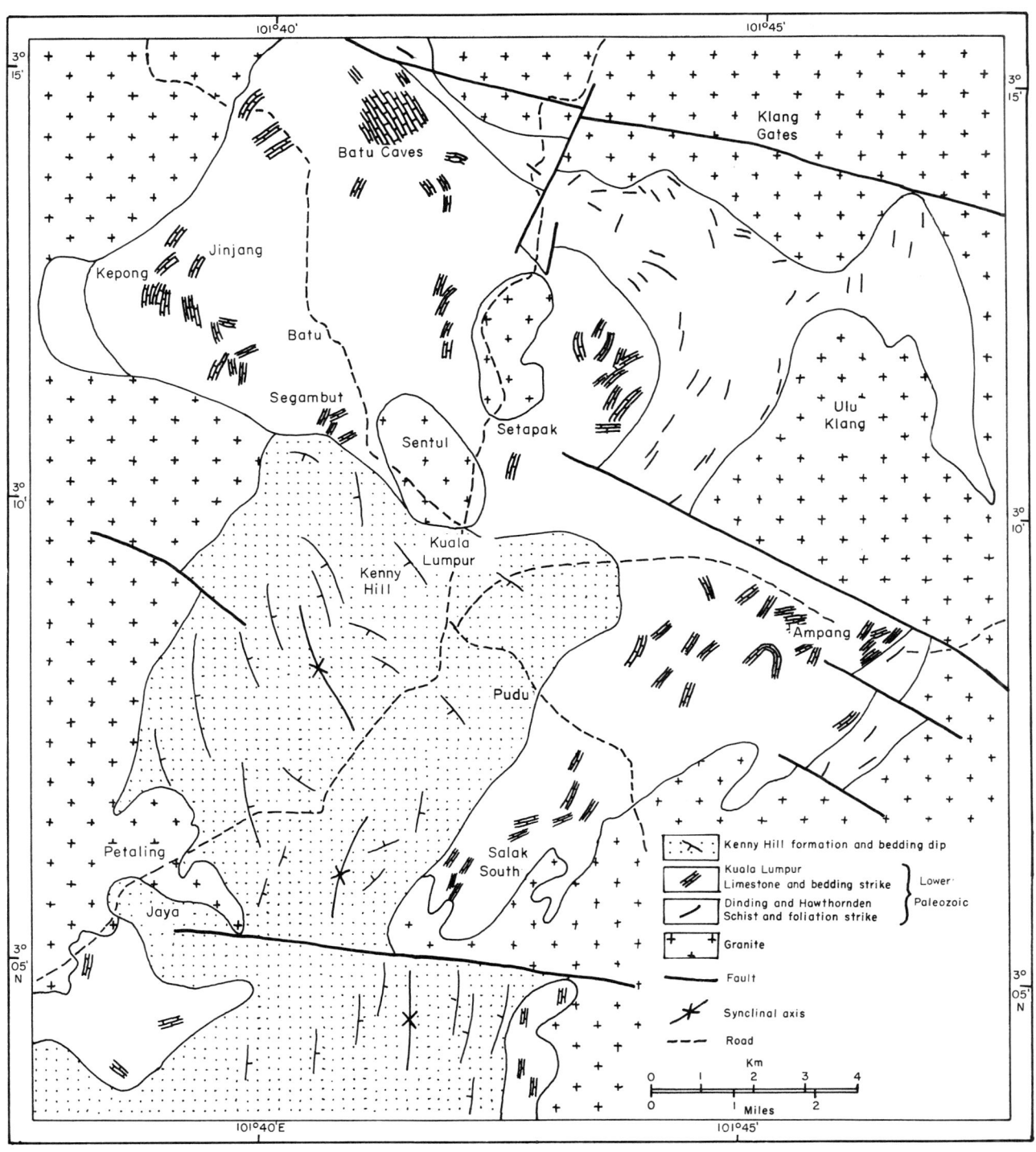

FIGURE 10.4 *Geological sketch map of the Kuala Lumpur area to show the unconformity between the Lower Paleozoic and the ?Upper Paleozoic Kenny Hill Formation.*

superimposed in the late Triassic. They may be correlated with the major Variscan folding and uplift of Indosinia which was initiated in the Middle Carboniferous (Fromaget 1952). The three structural zones defined at the beginning of this chapter were probably initiated in the Upper Carboniferous, and these are discussed in turn below.

Western Zone

LANGKAWI. In Langkawi a low-angle thrust (Kisap Thrust) separates the Lower Paleozoic Setul Formation and the Upper Paleozoic Singa and Chuping Formations. The thrust plane has been arched by the granite intrusion of Gunong Raya and intruded by granites in Pulau Dayang Bunting and southeastern Pulau Langkawi (Figure 10.3). This granite has given radiometric dates of 242 ± 10 m.y. and 218 ± 11 m.y. (Table 8.5) by the potassium-argon and rubidium-strontium methods, respectively, and thus it is likely to be of Upper Permian or Lower Triassic age. The thrusting movements can be dated as Post-Chuping Formation (Middle Permian) and pregranite (Lower Triassic). The Kisap Thrust has been described by Koopmans (1965) and Kimura and Jones (1967) and mapped by Jones (1966). These authors differ in their interpretation of the outcrops of the thrust in the interior of Pulau Dayang Bunting but agree on its general character.

In Pulau Langkawi the thrust plane dips eastward at a low angle, more or less parallel to the bedding planes of the overlying and underlying formations. Sandstone of the Singa Formation occurs in a window to the east of the main outcrop of the thrust. The thrust plane itself is exposed at the northern and southern points of Pulau Dayang Bunting. At the southern outcrop, the Setul Limestone is thrust over phyllite and greywacke of the Singa Formation. Here the thrust plane dips at 15°E. However, at its northern outcrop the thrust, which here separates the Setul Limestone and the marble of the Chuping Formation, dips 50° to the west. Kimuru and Jones interpreted these occurrences as two conjugate thrusts within a more complex zone of thrusting (Figure 10.5b). Koopmans (1965), on the other hand, interpreted the thrust plane as generally lying parallel to the bedding. The inversion of dip at the northern tip of Pulau Dayang Bunting was due to local granite intrusion (Figure 10.5 a).

Evidence of the thrust is also seen near Kisap, where a road cutting exposes mylonized limestone with slickens dipping 35° northeast between the Setul Limestone hills to the northeast and a small outcrop of fossiliferous limestone of the Chuping Formation to the southwest. North of Ayer Hangat, the thrust passes through Pulau Dangli, where the limestone is intensely brecciated, jointed, and limonitized (Koopmans 1965).

A conjugate thrust plane dipping 25° southwest outcrops on Pulau Tembus Dendang and carries massive Ordovician limestone over chevron-folded Lower Silurian flagstones (Kimura and Jones 1967).

MAIN RANGE. The arcuate trend of the Main Range may have been emphasized by later movements, but as a zone of general uplift it was probably important from Upper Paleozoic times onward. Upper Paleozoic granites have been dated from the central and southern parts of the Main Range (Table 8.5 and J. D. Bignell in correspondence).

Axial Zone

In the axial zone of the Malay Peninsula there is no evidence of late Paleozoic or early Triassic folding. Deposition in north Pahang and south Kelantan appears to have been more or less continuous through the Permian and the Trias. However, deep burial and catazone metamorphism of the Taku Schist and the migmatite complexes of Stong, Benta, Bukit Ranjut, and Bukit Berentin may have occurred in the late Paleozoic and early Triassic. Potassium-argon dates on biotite and muscovite of the Taku Schist and related gneiss (Table 8.5) suggest that metamorphism terminated in the Middle Triassic. The difference in metamorphic grade between the Taku Schist and the overlying sediments suggests the presence of an unconformity, but this contact is more likely to be a structural discontinuity, as argued in Chapter 9.

Eastern Zone

In the Kuantan area of Pahang, Fitch (1952) showed that the regional strike of the Lower Carboniferous is north-northeast, but the strike of the overlying "Arenaceous Series" is north-northwest and an unconformity probably lies between the two formations (Figure 4.12). He recorded steep dips and uniform strikes. His sections show the Upper Paleozoic beds symmetrcally folded about vertical axes, with some thinning on the limbs of the folds. The "Arenaceous Series" consists of sandstone, grit, quartz conglomerate, and coaly shale containing

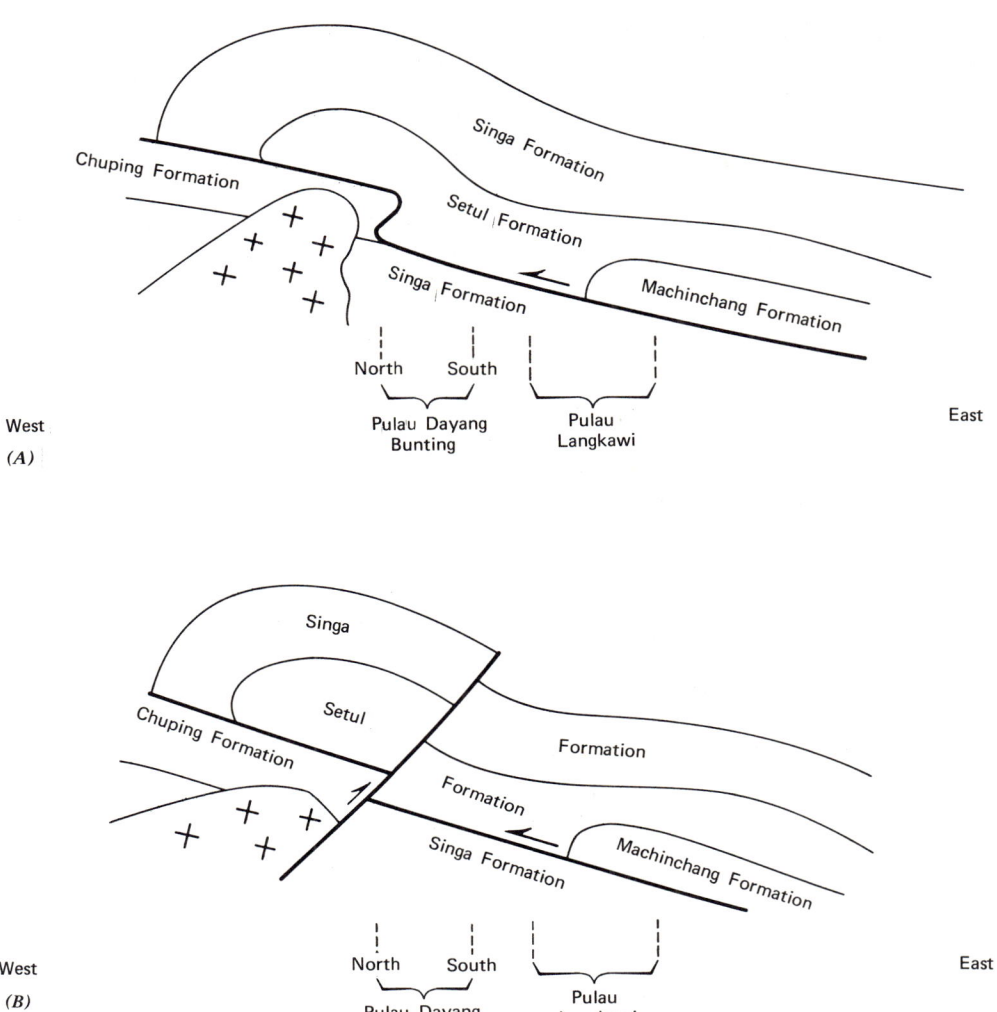

FIGURE 10.5 *Diagrams representing the Kisap Thrust, Langkawi:* (a) *After Koopmans (1965);* (b) *After Kimura and Jones (1967).*

poorly preserved plant remains of Upper Carboniferous or Permian age (Fitch 1952, p. 30). The scattered occurrence of plant fossils in other parts of northeast Malaya indicates the presence of Upper Carboniferous or Permian terrestrial or nearshore marine deposits over this part of the peninsula. It is probable that this area underwent a marked paleogeographical change in Middle to Upper Carboniferous time. This can be accounted for by earth movements correlated with the Indosinian movements and causing a general uplift.

Further evidence for these movements is to be found in the radiometric ages of the granite. J. D. Bignell (in correspondence) has found that Upper Carboniferous and Permian rubidium-strontium ages are widespread in the granites of northeast Malaya. He also noted that the potassium-argon measurements gave ages in excess of 215 m.y. and suggested that the area east of the Lebir fault had been tectonically stable since Lower Triassic time. This contrasts with late paleozoic (Rb-Sr) granites in the west of Malaya, which give potassium-argon ages younger than 215 m.y. If we assume that the Upper Triassic movements have little affected northeast Malaya, then the structure present in that area must have been the result of late paleozoic and perhaps early Triassic (Variscan) earth movements.

LATE TRIASSIC TECTONISM

With the exception of a possible Middle Jurassic fauna from Singapore (Jones et al. 1966, p. 345) and marine fossils from the base of the Tembeling Formation at the Jengka Pass, Pahang (Ichikawa et al. 1966), marine sediments younger than Carnian are unknown in Malaya. The Carnian and older rocks are in places metamorphosed and everywhere intensely folded, except in the extreme northwestern part of the country (Langkawi and Perlis), where the folding is more gentle. The later Mesozoic and Cenozoic is not strongly folded and remains unmetamorphosed.

In the late Triassic a major tectonic event deformed the rocks of the Malay Peninsula and uplifted it above sea level. It has remained essentially a land area ever since. The most extensive intrusive episode in Malaya occurred after the deposition of the Carnian but had ceased by about 200 m.y. ago. The granites intruded were tin-bearing and were most important in the western structural zone, where they have been recognized from Penang, Kulim, the Bintang Range, the Kledang Range, and the Main Range. They are also known from Gunong Benom and Johore and a few from northeast Malaya (Figure 8.1 and J. D. Bignell in correspondence).

Western Zone

In the Baling area of Kedah, Burton (1972) found the Lower (and ?Middle) Paleozoic metasediments of the Baling group isoclinally folded about northeast striking axes with the axial planes of the folds dipping steeply southeast. He mapped the Carnian sediments of south Kedah with a closely similar structure. The Damar granite and the granite masses to the north of Baling show a similar trend (Figure 10.6). The Triassic Kodiang limestone of Bukit Kalong, northwest Kedah, shows a large recumbent fold directed west-northwest (Plate 20).

In the Grik area (Jones 1972b) the regional strike is flexured about a north-northeast trend. Here the metasediments of the Baling group are cleaved and appear to be folded into a series of tight anticlines and synclines with vertical axial planes (Figure 10.6).

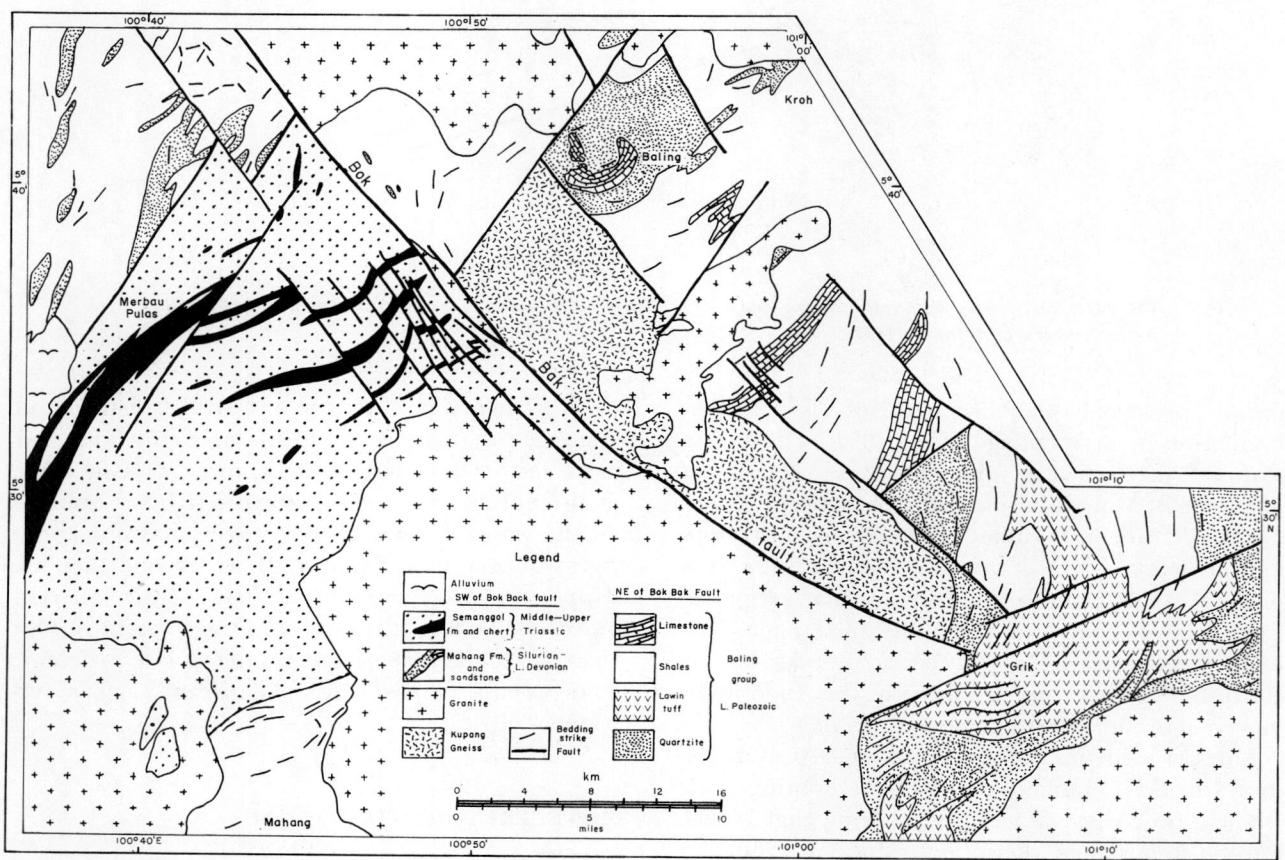

FIGURE 10.6 *Geological sketch map of southeast Kedah and north Perak showing the main structural features. After Burton (1965) and Jones (1970b).*

PLATE 20 *Recumbent fold in Triassic limestone, Bukit Kalong, Kedah. View southwest. Photograph D. J. Gobbett.*

In the Kinta Valley the regional strike is approximately north-south, and Triassic rocks are absent. The Ordovician to Permian Kinta Limestone has been extensively recrystallized and, in places, folded recumbently. In general the beds dip westward, but the structure has been complicated by later faulting and remains obscure.

Ingham and Bradford (1960) recorded complex folding and faulting in the Kinta Limestone in several mines in the central part of the valley and near its boundaries. Isoclinal and recumbent folds occur near the eastern valley margin. Some of the crystalline limestone in the vicinity of Tambun is tightly folded and is accompanied by axial plane shearing (cleavage?). Two schist bands outcrop as the Tambun–Ampang ridge and from Tanjong Rambutan to the Ulu Piah mines; these were interpreted to represent a southward plunging syncline with a north-south strike. This syncline measures 2 km across and is traceable for more than 7 km (4.5 miles) (Willbourn 1936).

It is not at present possible to date the deformation in the Kinta Valley, but it is probably late Triassic. However, the granite of Kuala Dipang, on the southeast border of the Kinta Valley, has been dated as Upper Permian by the potassium-argon method, 232 ± 10 m.y. (Table 8.5).

In Selangor, the Paleozoic rocks strike north-northwest. Locally this strike is deflected by granite intrusion. The ?Upper Paleozoic metasediments are cleaved and isoclinally folded, but no general interpretation of their structure has been attempted (Roe 1953).

Axial Zone

In the Merapoh and Chegar Perah area of Pahang (Richardson 1950), a north-south regional strike is shown by the granite and catazonal metamorphic rocks and also by the Upper Paleozoic and Triassic metasediments (Figure 10.7). Bedding and foliation planes are vertical or steeply inclined. Typically the metasediments are isoclinally folded with axial planes dipping east. Cleavage folding, which is common in intercalated sequences of shales or phyllites with competent beds of crystalline limestone or sandstone, is shown by thickened fold crests with associated attenuated limbs (Figure

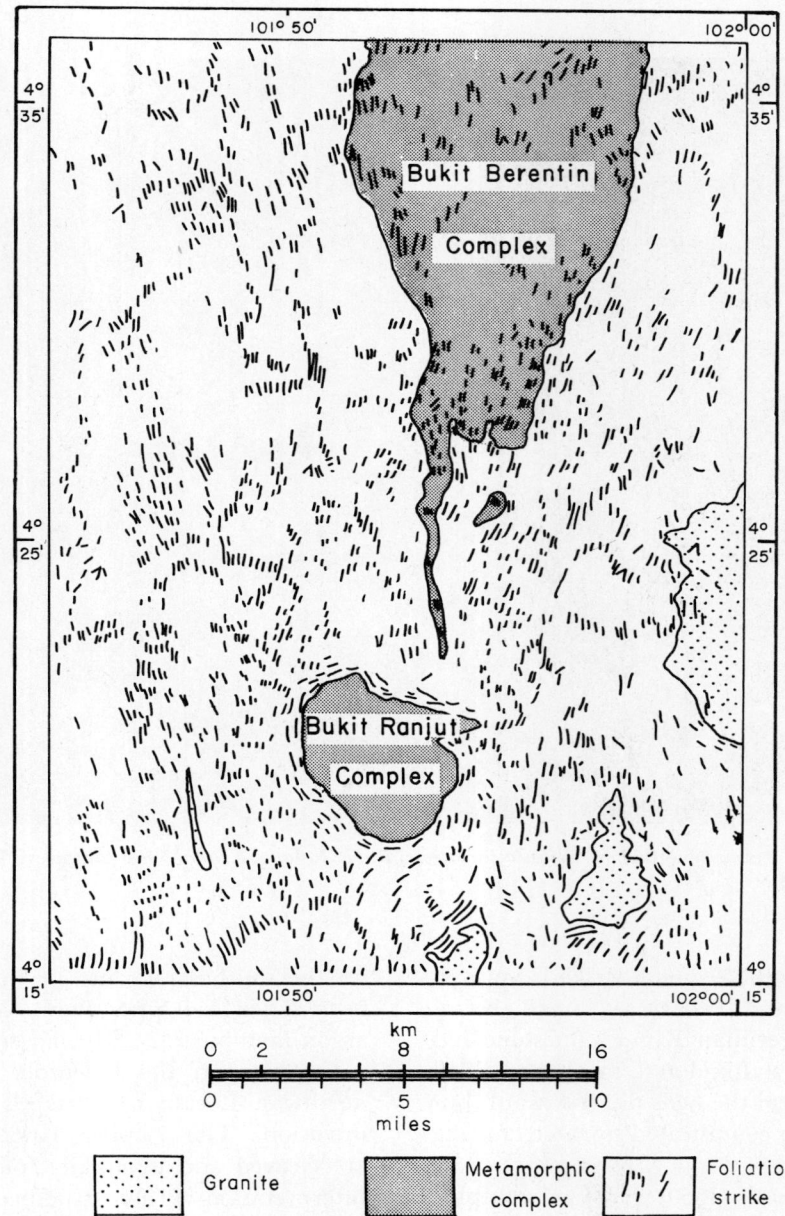

FIGURE 10.7 *Structural map of the Merapoh and Chegar Perah area of north Pahang. After Richardson (1950).*

10.8). A major anticline is cored by the Bukit Ranjut and Bukit Berentin metamorphic rocks. Richardson differentiated four fault sets; according to a geometric classification these are strike faults parallel to the fold axes, dip faults parallel to the dips of the foliation or stratification, and two sets of oblique faults striking north-northeast to northeast and west-northwest to northwest. Auriferous quartz including the Selinsing and Buffalo reefs were reported to follow strike faults. Thinner quartz stringers may occupy dip faults, but quartz emplacements rarely occur along oblique faults. The dispositions of the oblique faults show them to be shear faults formed by the compression that was responsible for the general north-south structures of the region.

Further south, in the Raub area of Pahang, Richardson (1939) described a similar structure. On the east flank of the Main Range Granite the sediments and metasediments are folded into isoclines

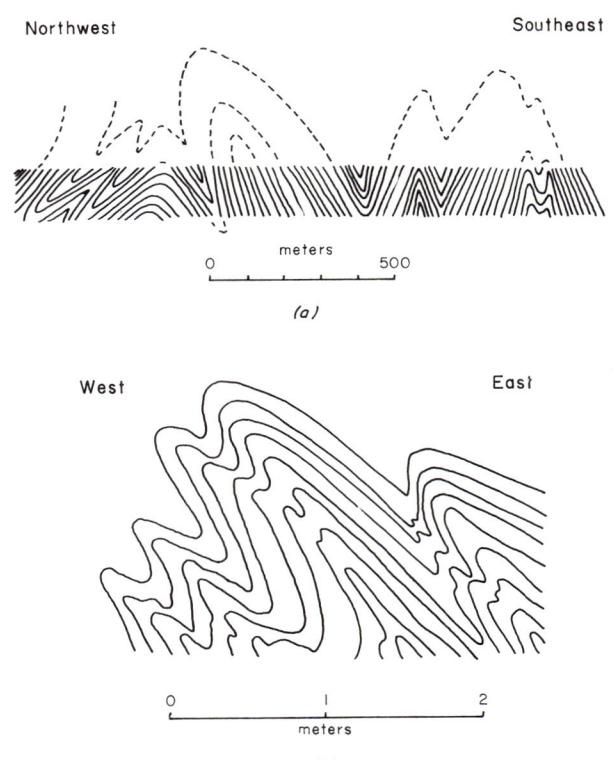

FIGURE 10.8 *Tectonic style in intercalated crystalline limestone and argillaceous rocks in the Sungei Satak, Pahang. After Richardson (1950).*

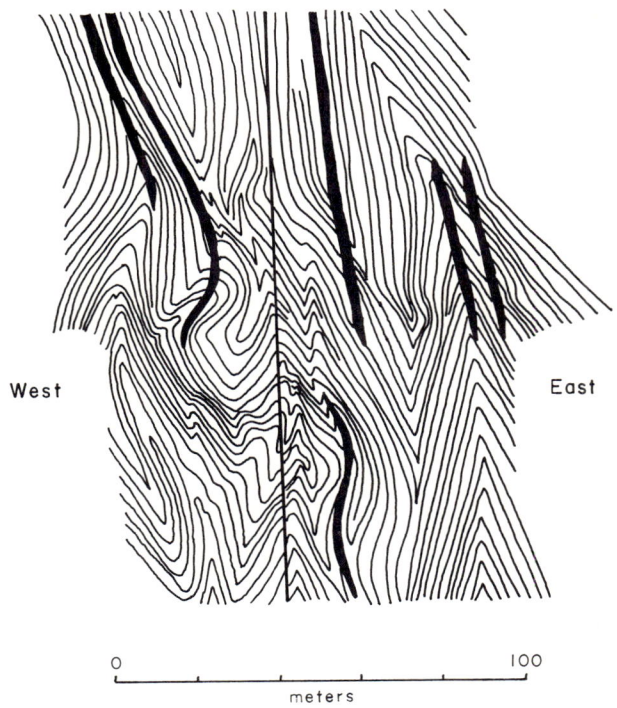

FIGURE 10.9 *Disharmonic folds in the Bukit Jelis shaft, Raub Gold Mine. Lode channels (black) are shown as fillings of tension fractures in calcareous graphitic shales. The eastward inclinations of the lodes and the axial planes indicate that postmineralization deformation was responsible for them. After Richardson (1939).*

with axial planes dipping east at 70°. Richardson (1939) pointed out that the overall structural picture indicated two anticlinoria with axes coinciding with the main Range and Benom granites. The intervening area represented a synclinorium. The smaller folds on the limbs of the anticlinoria were at least 2 km long and pitch at angles between 15 and 20°.

In the Raub Australian Gold Mine, a series of parallel lodes trends north-northwest along the tectonic strike. Postmineralization northeast trending faults caused the lodes to be displaced 100 to 200 m (330-660 ft) in a right lateral sense. The vertical displacements are of similar values. In other words, the true displacement is of a dip-slip nature, pitching at 45° (Figure 10.9).

Two aplite dykes near Cheroh village and the Bukit Kajang granite show marked elongations parallel to the regional structure and therefore appear to have been emplaced in zones of tension fractures. The formations in west Pahang are poorly dated because of the lack of fossils. In south Kelantan, however, cleaved shales contain Ladinian and Carnian fossils that have been distorted by the cleavage folding. Similarly, in the Chegar Perah and Kuala Lipis districts, Carnian *Myophoria* and ammonoids are often strongly deformed. On the other hand, Ladinian and Carnian fossils from south Pahang and Kedah are almost undeformed (Kobayashi et al. 1967).

Eastern Zone

The extent of the late Triassic orogenic phase in east Malaya is conjectural. As pointed out previously, potassium-argon measurements on the granites of east Kelantan and Trengganu suggest stability since the ?Lower Trias, and the area may have formed a stable horst supplying sediment to the west during Triassic time.

POST-TRIASSIC FOLDING

The structure of the Post-Triassic terrestrial sediments of the Tembeling Formation was described by Koopmans (1966, 1968), and its southward extension in South Pahang was shown on geological maps by members of the Geological Survey of Malaysia (Bean et al. 1966). Folding of these rocks is of a flexural character, resulting in open folds with vertical axial planes. The major folds have wavelengths of about 10 km (6 miles) (Figure 10.10). In the incompetent layers, minor folding is also common. A north-northwest structural trend is prevalent with culminations and depressions occurring at regular intervals of 15 km (10 miles) along the strike. A similar structural trend is apparent in the Upper Paleozoic and Triassic rocks underlying the Tembeling Formation. However, these older rocks are isoclinally folded and cleaved.

The Tembeling Formation represents a post-orogenic molasse phase, gently folded during the Jurassic before the deposition of the overlying Gagau Group. It now outcrops extensively only in the axial zone of Malaya; formerly, however, it was probably more widespread. An outlier has been discovered in east Kedah, and it is possible that further outcrops will be found.

The late Jurassic or early Cretaceous Gagau Group and Tebak formation are typically horizontal or slightly tilted, except near faults where locally they may be steeply dipping. At the present day these formations lie on a plane tilted slightly to the south (Figure 5.8).

POST-TRIASSIC FAULTING

A number of major fractures have been recently recognized in the Malay Peninsula. These all show important movement along the strike of the fault plane, as indicated by offset fold axes and granite plutons. They show Post-Triassic movement, and some have offset the Lower Cretaceous Tebak formation; but they appear to be inactive now. Earth tremors recorded from Malaya during the past 150 years have been of low magnitude and have probably been centered on Sumatra or the Andaman Sea. The outcrops of these major wrench faults are shown in Figure 10.11.

Bok Bak Fault, Kedah

In the Baling area of Kedah, two fault sets, one striking north 143° east and the other north 32° east, were interpreted by Burton (1965) as a complementary system of shear faults caused by a north-south compression. Along the north-west set, sedimentary formations were interpreted as laterally displaced in a sinistral sense for about 50 km (30 miles). This sense of relative movement suggests a general east-west compression, presumably subsequent to the initial north-south compression (Figure 10.6). The Bok Bak fault, the largest fault striking northwest, was traced for 75 km (47 miles). Burton extended this fault southward to Johore, based on geological and topographical discontinuities with apparent left lateral offsets. This southward extension was strongly criticized by Procter and Jones (1967), however; they pointed out that there was no field evidence south of the Baling area to support this contention. The Kupang Gneiss is elongated parallel to the Bok Bak fault along its eastern side. This may indicate shearing parallel to the fault or important vertical uplift of the area east of the fault or both.

Lebir Fault Zone

Aw (1967) described an ignimbrite dyke 20 km (12.5 miles) long near Temangan, Kelantan, and Burton (1967a) postulated that this dyke represented the northern end of a much larger fault zone, as indicated by the straight course of the Sungei Lebir southward for a distance of 150 km (96 miles). Mohammad bin Ayob (1968) interpreted, on sedimentological evidence, an important gravity fault along the east side of the Tembeling Formation in the Berantai syncline, Pahang. Provenance studies of the conglomerate horizons pointed to a nearby source east of the syncline, and the various conglomerate beds appeared to indicate intermittent uplifts of the area east of the fault. Topographically and geologically the Temangan dyke–Sungei Lebir–Berantai fault and its southern extension (Chedong fault) seem to form one continuous fault zone extending for a distance of 300 km (188 miles) and striking more or less parallel to the regional structural trend.

Mohammed bin Ayob (1966) interpreted the shear zone to represent block faulting, along which the areas east of the zone gained in elevation with respect to the sedimentary basins west of it. A major shear zone on the east side of the Taku Schist is about 4 km wide in the vicinity of Kuala Krai, and it consists of at least three mylonite zones separated from one another by zones of undisturbed Upper Paleozoic sediments and metasediments (Figure 10.12). The most westerly fault-breccia zone forms the boundary of the Taku Schist and is occupied by iron ore deposits at the Temangan mine and by sheared microgranite. To the east, a second mylonite zone is marked by the Temangan ignimbrite dyke, 22 km (14 miles) long. A third fault zone consists of flaser mylonite of metasediments intermittently outcropping along the Sungei Lebir. In

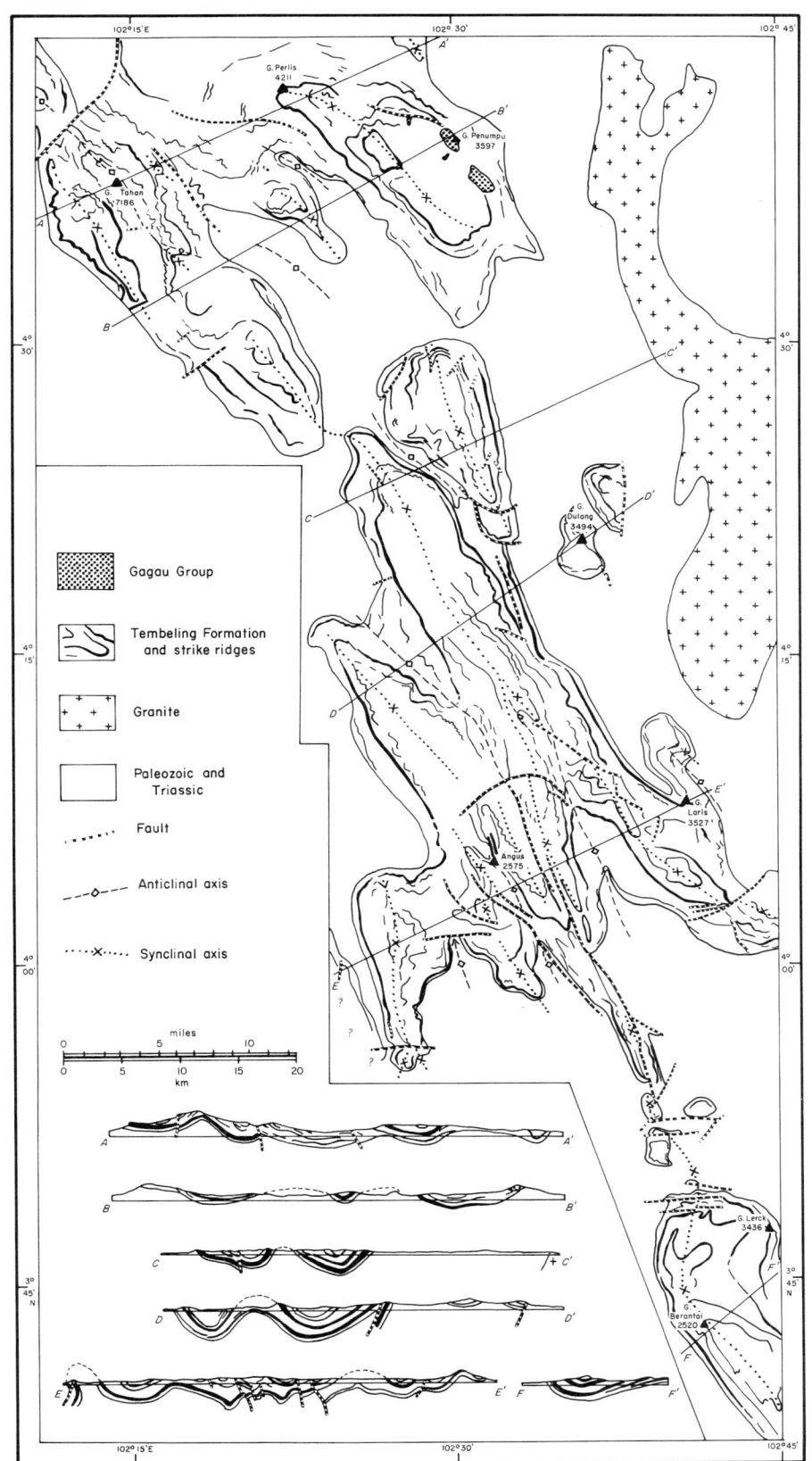

FIGURE 10.10 *Structural map and sections of the Tembeling Formation in north and central Pahang. Heights in feet. After Koopmans (1966).*

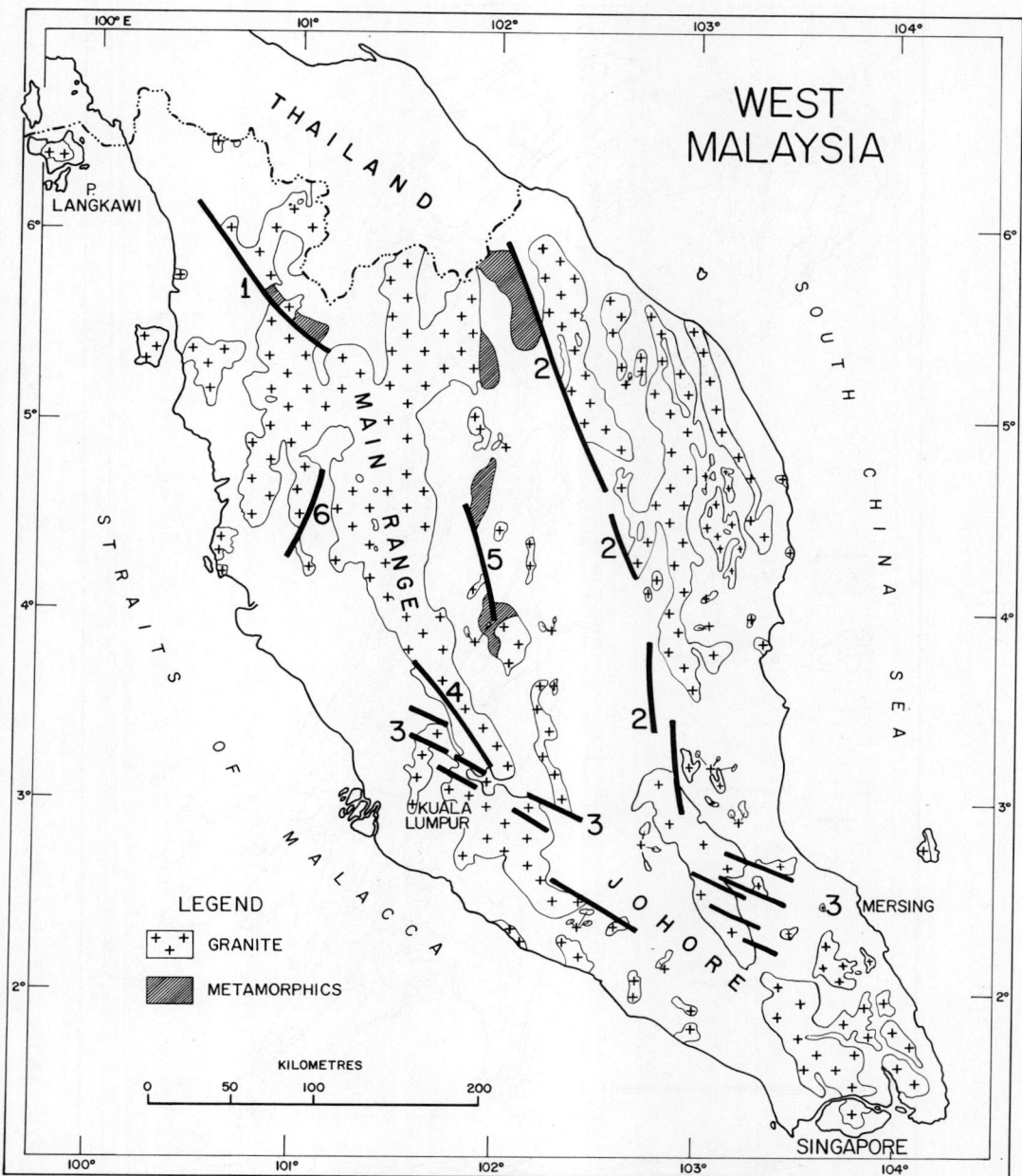

FIGURE 10.11 *Major faults in Malaya: (1) Bok Bak fault, (2) Lebir fault zone, (3) Kuala Lumpur–Endau fault zone, (4) Bukit Tinggi fault, (5) Chegar Perah–Benta fault, (6) Kledang fault.*

this zone the mylonite foliations were measured to strike 160° and are very steep to vertical. The orientation of tension gashes within the mylonite indicates left lateral dislocation. Drag folds also point to sinistral lateral displacements. The rectiinear granite boundary suggests the presence of a fault, although this has not yet been demonstrated in the field.

The Lebir lineament is interpreted here as a major fault zone extending from Temangan for some distance south beyond the Berantai–Chedong fault (Koopmans 1968). Near Kuala Krai, at least three shears mark the fault zone. One shear follows the schist–metasediment boundary, a second shear is indicated by the Temangan dyke, and a third shear follows the straight Lebir River. Tension gashes and dragfolds within and near the shear zone along the river indicate left lateral displacement. The vertical foliation of the mylonite also supports the interpreted transcurrent nature of the

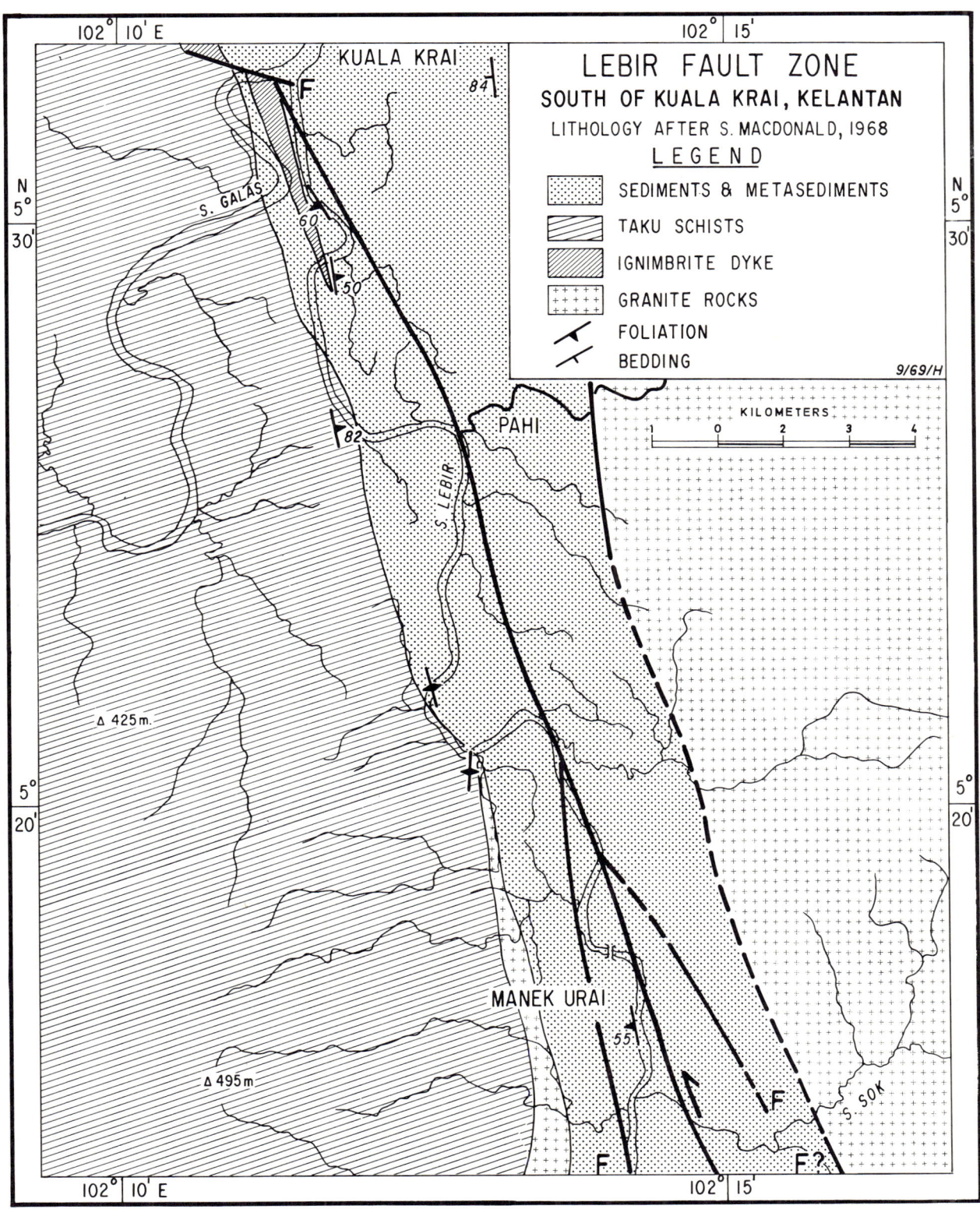

FIGURE 10.12 *The Lebir fault zone near Kuala Krai, Kelantan. Rock units after MacDonald (1968).*

fault. Vertical displacements along the Lebir fault zone could have occurred, but most of the movement was probably lateral.

MacDonald (1968) showed that dolerite dykes in Trenganu were related to joints in the granite (Figure 10.13) and generally strike N75E or N150E. The former trend represents extension. Figure 10.13 also depicts N75E faults with dextral offsets that amount to 600 m (2000 ft) and one sinistral offset of 400 m (1320 ft) along a N105E fault. The horizontal offsets were probably caused by the subsidiary compression that resulted from left lateral shearing along the primary N20W Lebir fault zone.

To the north, the Lebir fault zone disappears below the Kelantan coastal plan, but it may be traced via faults in the Gulf of Thailand (Emery and Niino 1963) into Thailand as the Chao Phraya structural depression. To the south, the Chedong fault, 50 km (30 miles) long, appears to be the continuation of the Lebir fault. It forms the eastern lmit of the Berantai and Gebok synclines (Koopmans 1968).

Another fault, lying east of and parallel to the southern end of the Chedong fault, has been mapped by Bean et al. 1966). The difference in strike of the outcrops on either side of this fault, and the pattern of the outcrops of the argillaceous rocks in the vicinity of the fault, suggest that it extends to the north for at least 5 km (3 miles) and to the south for at least 15 km (10 miles). This evidence, together with the two synclinal axes that appear to be terminated by the fault, is interpreted as due to left lateral dislocation of 18 km (12 miles) (Figure 10.14).

However, an alternative interpretation is that the argillaceous outcrops on the two sides of the fault do not represent the same formation and the two synclines are not parts of the same structure. In this case, movement on the fault may contain an important vertical component. The Lebir fault zone may have been shifted by left lateral displacement along the Kuala Lumpur–Endau fault zone (see below) into the South China Sea immediately east of the Johore shoreline. One indication of sinistral drag is shown by steeply inclined axes of dragfolds in metasediments at Tanjong Punggai (Grubb 1968).

Kuala Lumpur–Endau Fault Zone

A prominent line of dislocation extends from Kuala Lumpur to Ulu Endau in southeast Pahang. It is best known at its western end, where a complex fault pattern has been recognized in the Main Range Granite (Shu 1969, 1970). Left lateral displacement along this fault zone is shown by the Main Range Granite, the Bentong Group, the Trias, the granite masses of Ulu Keratong and Ulu Kemapan, and the Tebak Formation. According to Stauffer (1968), young movements along the Kuala Lumpur fault zone seem to be indicated by a number of geological features. Tertiary beds of Batu Arang in Selangor have steep dips near the northern edge of the basin, where the western extension of the fault zone is believed to be located. In south Pahang, the extensive swampy area of Tasek Bera marks a former course of the Pahang River; stream diversion is thought to have been recently caused by this same fault zone that passes through the south end of the swamps. Shu (1969) reported that field evidence supports a left lateral displacement along the western part of this fault zone. However, it appears that the east-southeast-striking faults are discontinuous and the southeast-striking faults are the major fractures. A study of aerial photographs confirms this.

The main faults at the western end of this zone are shown in Figure 10.15. Southeast and east-southeast-striking faults are dominant and are frequently filled with vein quartz. The largest of these quartz dykes is the Klang Gates ridge, which stands out from the adjacent land surface to a height of 250 m (825 ft) (Plate 18).

Outcrops of the vertical dykes vary from 1 to 2 m to as wide as 200 m (660 ft) and are entirely located in granite. Alexander and Procter (1955) described the detailed structure of this dyke at the site of the Klang Gates dam. The complex interlacing character of deformed and undeformed quartz veins within the dyke, considered with remnants of partly to almost completely altered granite, reflects multiple intrusions. In general, major joints are spaced 30 to 300 cm apart, and minor joints show spacings in the order of a few centimeters to about 10 cm. At the Klang Gates dam area, the major northeast joints were reported to be open fractures. Near-vertical major joints at the dam site indicate the following bearings: 45° to 50°; 105° to 110°; 170° to 180° (Figure 10.16).

According to the 50° to 60° angles between the various joint sets, the entire pattern appears to be related to one fracture system. The east-southeast set parallels the dyke, and the other two sets are diagonal joints. On the other hand, if we relate the joint pattern to the regional structure (striking approximately north-northwest), the following genetic classification may be used. The northeast

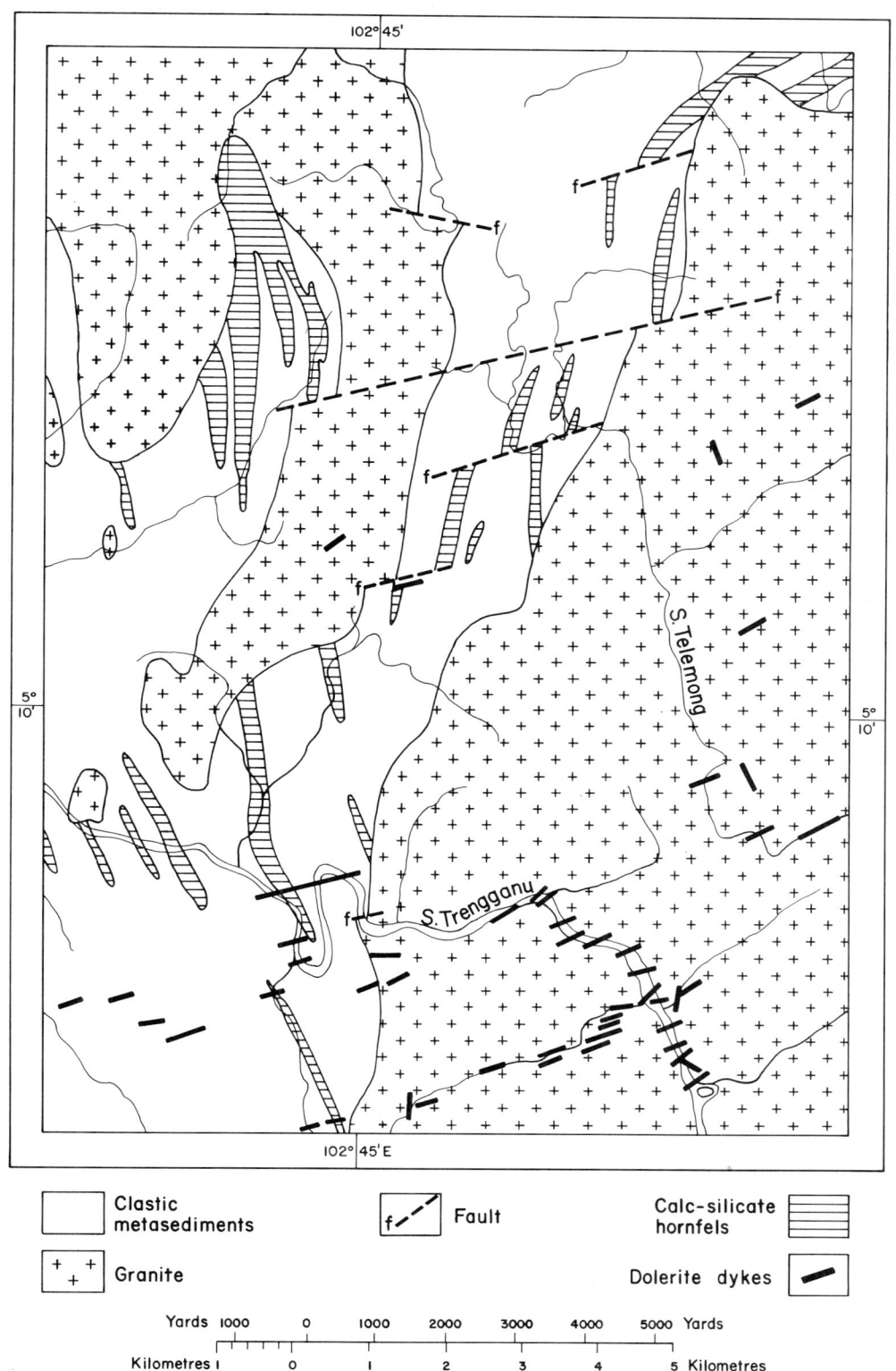

FIGURE 10.13 *Pattern of wrench faults and dolerite dykes in Trengganu. After MacDonald (1968).*

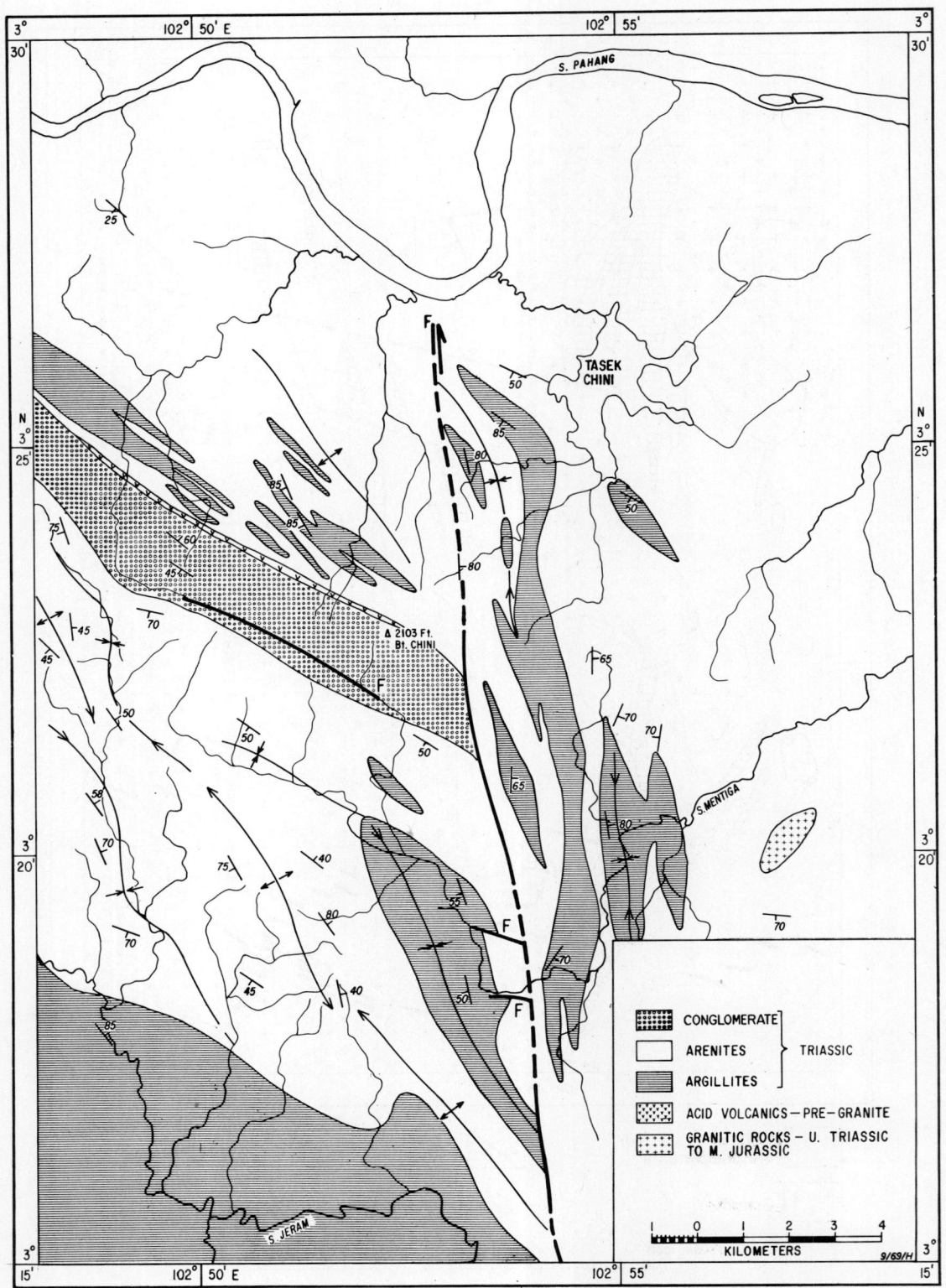

FIGURE 10.14 *Structural map of part of south Pahang. Note the apparent left lateral dislocation along the major fault. After Bean et al. (1966).*

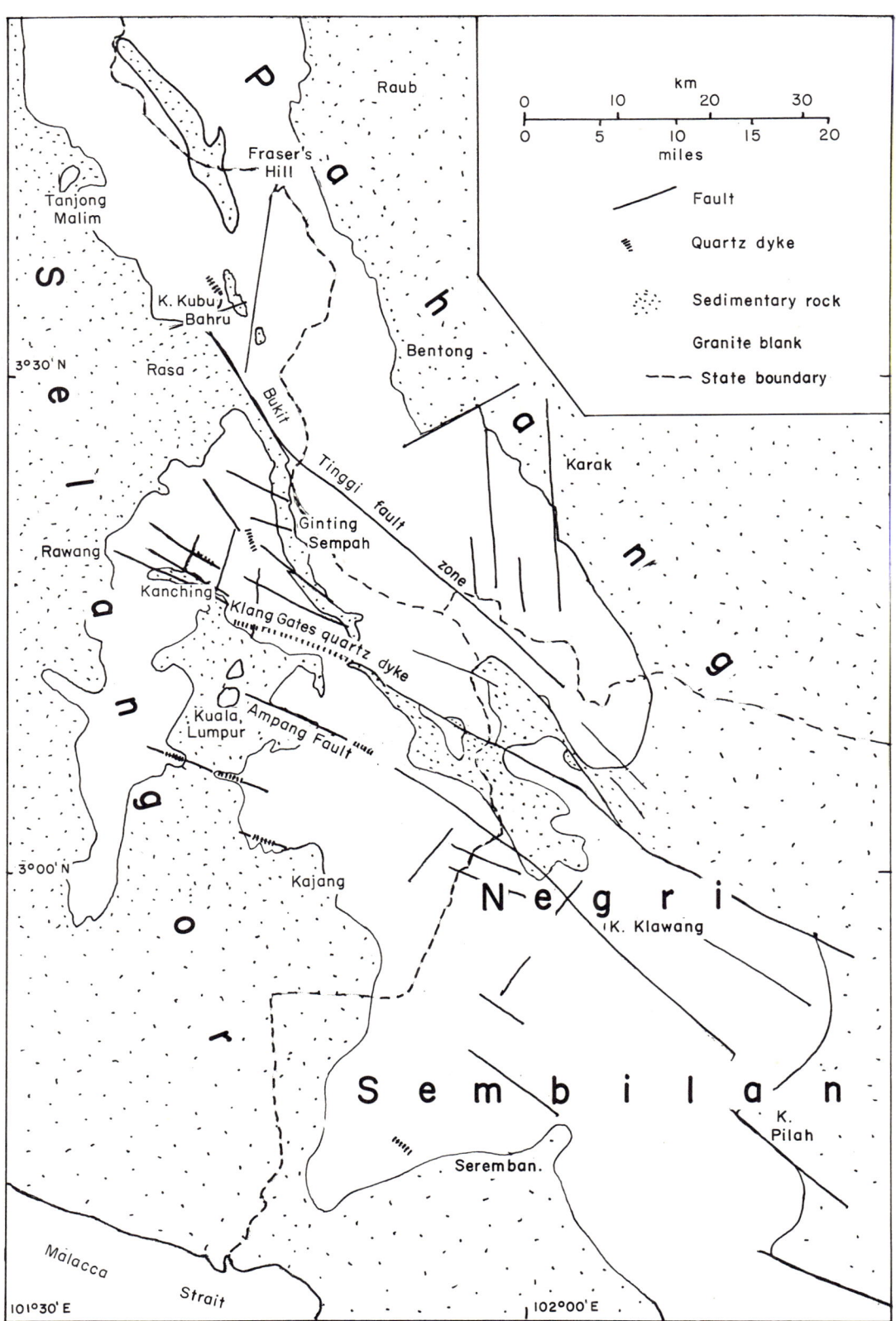

FIGURE 10.15 *Structure of the southern part of the Main Range Granite. Partly after Shu (1969).*

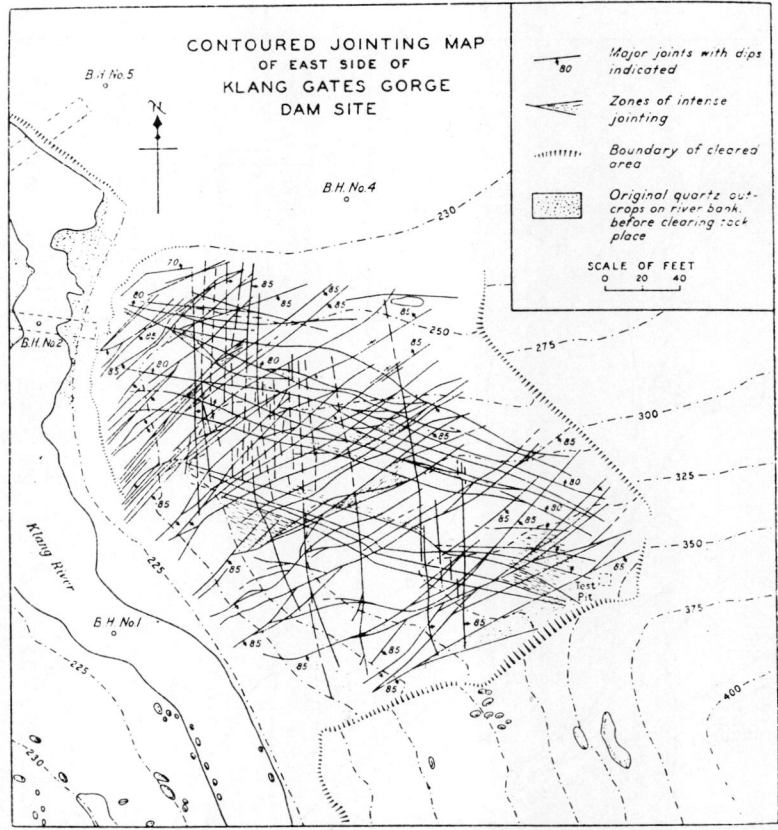

FIGURE 10.16 *Major joints at the Klang Gates dam site, Kuala Lumpur. After Alexander and Procter (1955, Fig. 1). Crown copyright, Geological Survey diagram. Reproduced by permission of the Controller, H. M. Stationery Office.*

joints are extension joints, which agree with their open nature reported previously. The north-south and east-southeast joints are second-order shear joints, respectively, with left lateral and right lateral displacements.

Another lineament runs parallel to the Kuala Lumpur–Endau fault zone from near Johol in southern Negri Sembilan by way of Ayer Kuning in northwest Johore and cuts the northeastern end of the Mount Ophir granite (Figure 10.11). The displacement of outcrops across this line suggests that left lateral movement has occurred along it. The small Tertiary basin of Kepong lies at the southeastern end of this fault.

The eastern end of this fault zone in Ulu Endau has recently been mapped by Chong *et al.* (1970). Here a number of faults trend 290°, lie *en echelon*, and cause the left lateral displacement of the Tebak formation and of the granite of the Gunong Besar area.

Bukit Tinggi Fault

A broad zone of mylonite, fractured granite, and large quartz dykes trends southeast for a distance of 110 km (69 miles), from Kuala Kubu Bharu, Selangor, to the vicinity of Bahau in Negri Sembilan (Shu 1969) (Figure 10.15). This zone may be continued to the north by a lineament separating the Bujang Malaka granite from the Main Range Granite. A left lateral displacement may also be interpreted here from the disposition of the granite bodies on either side of the lineament. The line is parallel to the Kampar Fault (Suntharalingam 1968), but there is no direct evidence for fault movement along it.

Kledang Fault

The evidence for the Kledang fault was discussed in Chapter 4. It can be best interpreted as a normal fault, downthrowing to the east and cut by numerous small wrench faults trending east-southeast.

Chegar Perah–Benta Fault

A major fault zone extending from the southern end of the Bukit Berentin Complex along the Sungei Jolong to the east side of the Benta Migmatite Complex, is indicated by geological and topographical lineaments. Displacement along this N340°-355°E-trending major fault appears to be left lateral and has offset the high-grade metamorphic rocks for at least 40 km (25 miles).

An important lineament, probably a fracture zone, traverses the Benom granite along the Sungei Dong in a N63°E direction. Where it cuts the granite near Kampong Lepar, the apparent left lateral offset is 2 km. However, the lateral offset of the nearby Benom granite boundary is much greater and suggests that the lateral offsets are of an apparent nature, whereas the main displacement was vertical.

Faulting at Sungei Lembing, Pahang

A number of major faults have been mapped in the Sungei Lembing mines (Fitch 1952). Only the Kabang fault, at least 750 m long and striking 60°, could be shown to have a right lateral component of displacement amounting to 600 m (2000 ft) (Figure 10.17). A probable north-northeast-trending fault zone offsets the Kabang fault in a left lateral sense for about 300 m (1000 ft). Horizontal displacement components of the other major faults in the area have been suspected but not yet proved. Dextral offsets along N60°E striking faults, such as the Kabang fault, can be explained

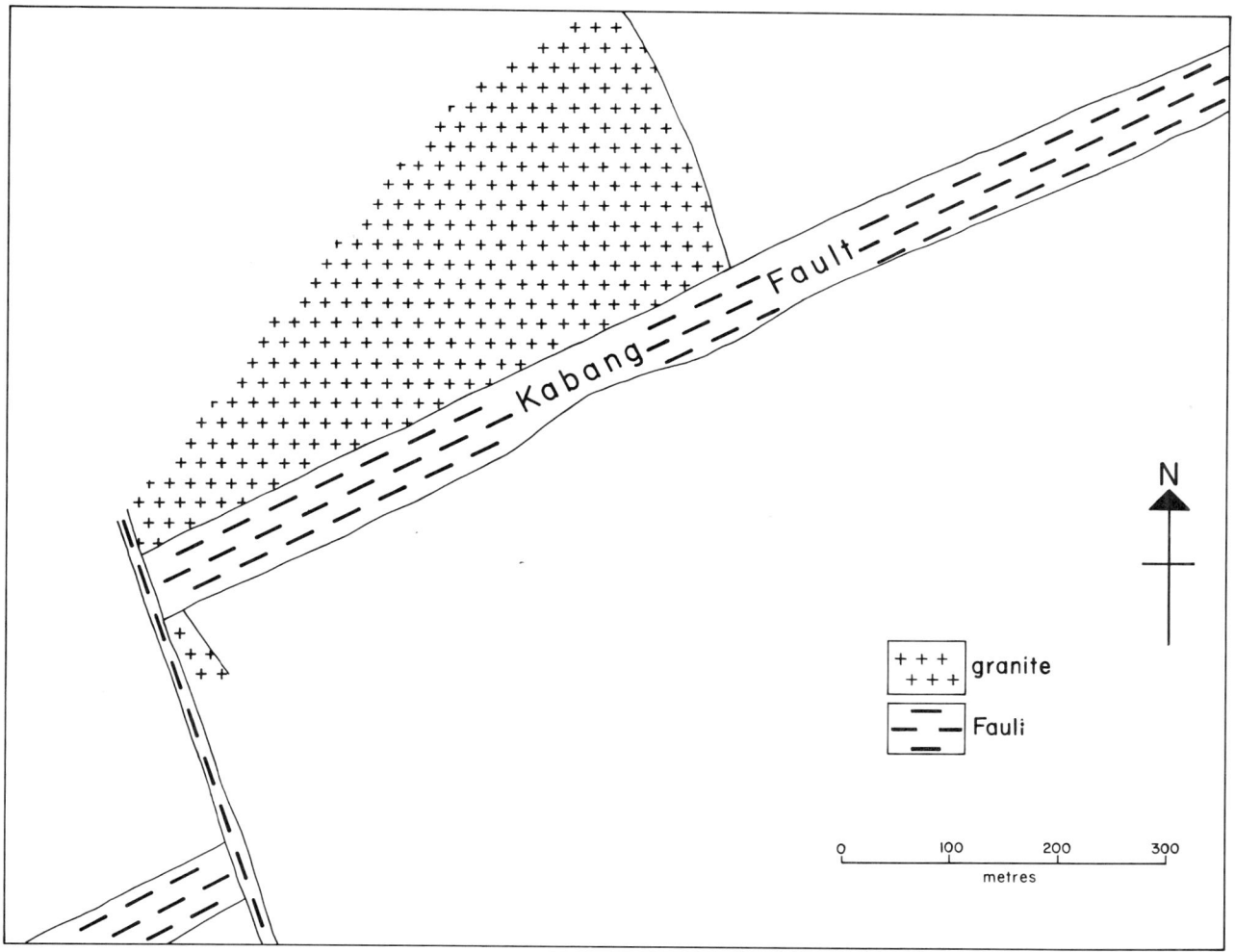

FIGURE 10.17 *The Kabang fault in the 1100 ft Level of the Willink's Mine, Sungei Lembing, Pahang. Dextral horizontal offset along this fault is suggested by the granite–sediment contact. Note also the inferred north-northwest fault that displaced the Kabang fault in a sinistral sense for about 300 m. After Fitch (1952, Fig. 22).*

as a result of left lateral displacement along major transcurrent faults striking south-southeast in the peninsula.

It has been reported that mineralization and a number of tin lodes become more important in the vicinity of the major faults than elsewhere. The very steep dips of the faults and lodes in the Sungei Lembing area suggest their formation by a horizontally acting major compression. The strike of the Lower Carboniferous rocks, in which the majority of faults and lodes occur, is south-southeast. In this respect the fracture directions 60°, 90°, and 120° correspond, respectively, with extension, first-order shear with left lateral displacement, and second-order shear with right lateral displacement. The deformation that accompanied the Upper Carboniferous granite emplacement probably resulted in the opening of existing zones of weakness into which the tin lodes were intruded. Figure 10.18 illustrates the directions of the important tin lodes in this area.

Conclusions

The presence of major left lateral displacements in the Malay Peninsula can be demonstrated for fault systems trending both south-southeast and east-southeast (Figure 10.11). The movement involved totals many tens of kilometers. It can be accounted for by a southeasterly stretching of the peninsula, causing parallel slabs of the crust to move in a counterclockwise direction. The offsets of the Lower Paleozoic rocks are of the same order as those of the granite and the early Cretaceous Tebak Formation. This suggests that movement along the faults is mainly if not entirely post-early Cretaceous. In the Bukit Tinggi fault zone, biotites from cataclastic granites gave Upper Cretaceous potassium-argon ages of 81 to 92 m.y. (J. D. Bignell in correspondence). It seems probable that faulting was associated with the late Cretaceous to Paleogene fold movements that affected Sumatra, and also with the later mid-Tertiary folding. If this was so, it represents Tertiary brittle deformation of the Malay Peninsula complementary to the plastic deformation of the Andaman–Sunda Arc.

STRUCTURE OF THE MAIN RANGE GRANITE

Southern Part

The geology of the Main Range Granite is best known in Selangor, Negri Sembilan, and west Pahang. The granite boundary is here very irregular, and roof pendants of sedimentary rock are prominent. The area is traversed by the Kuala Lumpur and Bukit Tinggi fault zones. In addition to these structures, numerous faults are indicated by the form of the granite outcrop, by zones of foliated and mylonized granite, and by lineations extending in many cases for tens of kilometers and depicted by negative relief. However, many lineations have quartz dykes emplaced along part of their length, giving rise to sharp positive relief. The most important lineaments are indicated on Figure 10.15.

In the Fraser's Hill area, Roe (1951) noticed that the fine-grained varieties of the granite are more intensely fractured than the coarse-grained types. In the latter types, rock strain was relieved through shearing in certain restricted zones. The structural differences presumably resulted in a better development of tin ore in the finer grained granite. However, we suggest that the ore concentration is richer in the finer grained granite because both represent late magmatic phases of the intrusion. The coarse-grained granite should belong to earlier phases of the intrusion.

Granite gneiss, flaser mylonite ("mylonite gneiss"), quartz-sericite schist, and mylonite mark post-con-

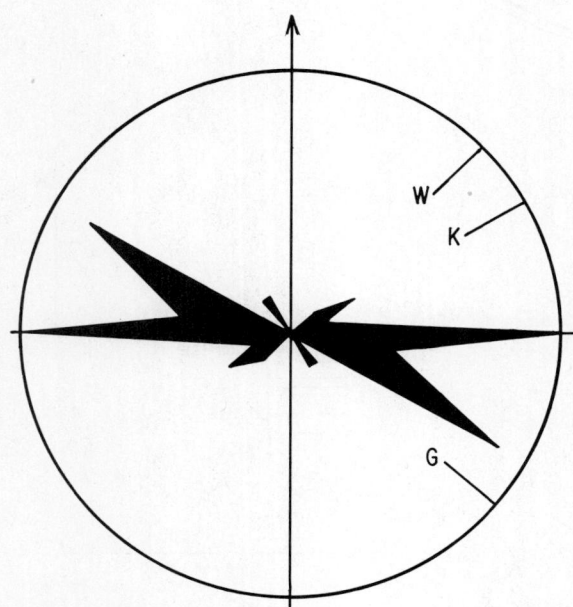

FIGURE 10.18 *Strike frequencies of important tin lodes in the Sungei Lembing mining area, Pahang: W = Willinks fault, K = Kabang fault, G = Gakak fault. Data from Fitch (1952).*

solidation shear zones in the coarse-grained granite. Shear zones and schistosity are usually parallel to the northwesterly regional structural trend. Quartz veins, aplite, and pegmatite intrusions are often displaced up to 6 m by small faults that also affect the sediments. Furthermore, small faults were reported to precede mineralization, whereas in other localities mineralized, slickensided faces indicate contemporaneous faulting. Suitable marker beds have not been recognized, and their absence prevented Roe from interpreting faults of larger dimensions.

Topographic and geological lineaments in the granite, however, are well shown on Roe's map. The longer lineaments are probably faults or fault zones. A lineament 20 km (13 miles) long strikes N10°E from Bukit Kutu toward the Gap. Another lineament strikes from Kuala Kubu Bahru N68°E for a least 8 km (5 miles). Near Kuala Kubu Bahru, this lineament demarcates granite from the sediments and is possibly a fault, downthrowing the southeast block with respect to the northwest block and/or imparting a 3-km left lateral dislocation. Other prevalent lineaments in both the granite and the sediments, with lengths of 4 km or more, trend parallel and perpendicular to the regional N40-44°W strike. These lineaments are genetically related to the regional structure as interpreted in Figure 10.19.

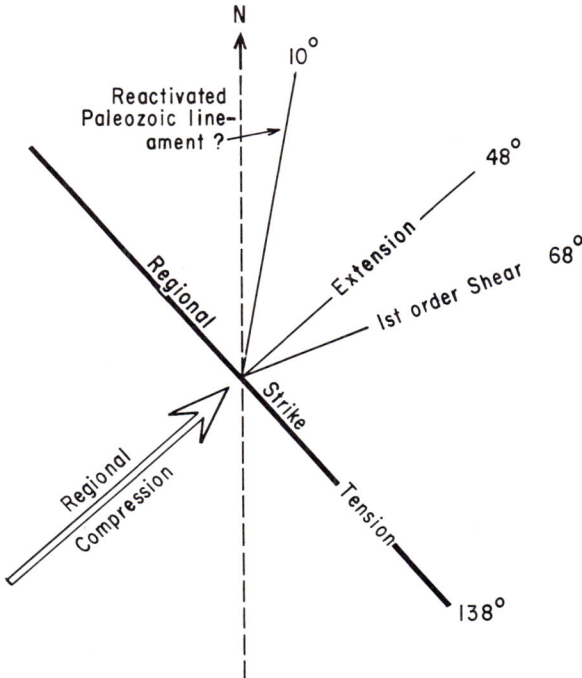

FIGURE 10.19. *Relation between lineaments and the regional strike in the Fraser's Hill area. Data from Roe (1951).*

Northeast of Kuala Kubu Bahru a 1 km long quartz dyke lies in granite and strikes parallel to the regional structure. Another smaller quartz dyke north-northeast of Tanjong Malim occurs in the sediments but is adjacent to and parallels the boundary of the granite stock. Further south, between Kuala Kubu Bahru and Kanching, rectilinear valleys and quartz dykes mark northwesterly and north-northeasterly lineaments. A lineament, more than 10 km (6 miles) long, trends N13°E from Kanching to the north, traversing the granite. Farther to the east, another lineament in granite follows the same trend along the Sungei Batu for almost 10 km. Along the Sungei Ulu Liam a N20–25°W lineament in granite is traceable for 9 km, and a prominent lineation affecting the granite boundary trends N124°E for at least 12 km (7 miles) from Rawang to Kanching and beyond.

Several less conspicuous granite lineaments, 3 to 4 km long, strike parallel or perpendicular to the regional strike, in addition to lineaments in N10°E, N45°E, and N90°E directions. All lineament directions are related to the regional structures, as their distribution pattern (Figure 10.20) reveals. The size of the large lineaments implies that they represent fault zones, presumably with significant displacements. Two of the three quartz dykes 1 to 2 km (1 mile) long through sediments and granite follow the regional strike. The longest dyke, northeast of Kanching, follows a second-order shear direction.

In the Bentong area, fracture systems in the granite usually consist of three sets; two are vertical and the third is more or less horizontal (Alexander 1968). In the northern part of the area, the vertical fractures trend east-southeast to east and north to north-northeast; in the southern part these fractures strike east to east-northeast and south-southeast to south. Other fractures were observed oriented diagonally with respect to the elongation of granite bodies.

On his geological map Alexander (1968) shows three important fault zones. The largest consists of the Bukit Tinggi fault zone trending N126°E for a distance of about 25 km (15 miles). This shows a left lateral discolation of at least 1.5 km. A second fault zone is marked by the Murai and Benus valleys, striking N15°E, with a mappable length of at least 13 km (8 miles). The third fault zone is partly occupied by the Sungei Murai and strikes N51°E with a traceable length of 10 km (6 miles). Field evidence of these fault zones comprise rectilinear valleys, sheared granite, and flaser granite zones.

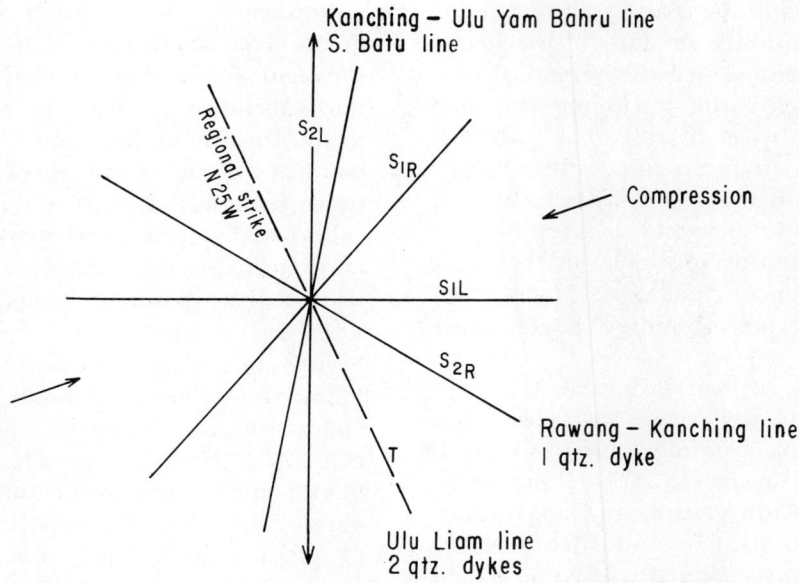

FIGURE 10.20 *Lineament pattern in the Rasa area, Selangor. Data from Roe (1953):* S_{2L} = *second-order left lateral shear,* S_{1R} = *first-order right lateral shear,* S_{1L} = *first-order left lateral shear,* S_{2R} = *second-order right lateral shear,* T = *tensional fracture.*

Batang Padang, Perak

Fractures in the Main Range granite in the area of the Batang Padang hydroelectric scheme were measured in detail by Renwick et al. (1960). The prevalent joint directions appear in Figure 10.21. Joints that strike 270° to 300° are vertical or have low inclinations, varying from horizontal to about 20°. It was noted in the map area that from east to west the number of fractures trending 0° to 40° decreases, whereas fractures trending 220° to 240° increase in importance. Renwick et al. (1960) believed that all joints were initially of tensional origin and that they developed through contraction of the intrusive rock. Quartz and sheared quartz in some of these joints probably indicate that the joints were formed in the early postconsolidation stages of the granite. Shear zones in the granite are indicated by closely spaced foliations, mylonite, and augen gneiss (flaser mylonite) ranging in width up to 30 m (100 ft). These shear zones are vertical or very steeply dipping. Veinlets of hematite, pyrite, epidote, and quartz may occupy such zones. All observed minor faults were determined as reverse in character.

Preece et al. (1962) reported that the prevalent lineaments on aerial photographs strike 10° to 15°, 100° to 120°, 155° to 170°, and 40° to 60°, in order of decreasing frequencies. These lineaments were thought to either represent shear zones or joint zones. The foliation of the metasediments of this area vary in strike between 120° and 140°, but the regional strike appears to be east of north. With respect to the regional strike, the lineaments correspond with fractures to be expected from a stress system with a major compression directed along 100° to 105°.

North Perak

Figure 10.22 shows regional and master joint lineaments as indicated by stream segments in north Perak and west Kelantan. In this area the regional strike is sinuous but is dominantly northeast. The lineaments are divisible into two main sets. One of these strikes northwesterly; the other north-northeasterly. It is likely that many of these lineaments are faults, but only the Bok Bak fault and the Kledang fault are currently recognized following these trends.

SYNTHESIS OF CRUSTAL EVOLUTION IN THE MALAY PENINSULA

C. S. HUTCHISON

The Malayan Orogen comprises a sequence of rocks extending from Upper Cambrian to Cenozoic and now exposes these rocks as a distinct and character-

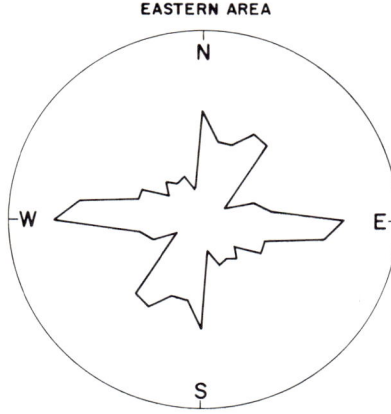

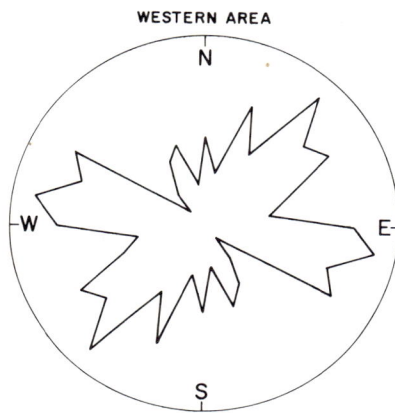

FIGURE 10.21 *Strike frequency diagram of joints in granite in the Batang Padang Hydroelectric Scheme area, Perak. Taken from Renwick et al. (1960).*

istic catazone, mesozone, and epizone (Hutchison 1970). The distribution and structure of the rocks of the three tectonic zones within the Malay Peninsula have been tentatively extrapolated to depth, and the tectonic structure has been schematically displayed in three hypothetical east-west cross sections (Figure 10.23). These sections are hypothetical because no available geophysical data exist on the Malay Peninsula which could assist the deep subsurface extrapolation of the major rock contacts.

The metamorphic facies series is apparently not Barrovian but of a low-pressure or intermediate type, indicating an orogenic high geothermal gradient during the revolutionary phase of the orogen.

The catazone, characterized by amphibolite facies metamorphism and simple deformation style, is generally of Lower Paleozoic rocks and sporadically exposed through the mesozone in a narrow belt of country extending north-south along the east-central axis of the peninsula, shown as the Taku Schist anticline in section A–B and as the Benta Migmatite in section C–D of Figure 10.23. The outcropping catazone belt indicates the risen axial belt of deepest geosynclinal depression. The Stong Migmatite Complex of Kelantan is also of the catazone and is characterized by foliated and migmatitic granite associated with cordierite amphibolite facies metasediments. The Taku Schist can be considered as a mantled gneiss dome, with an unexposed underlying core of rocks similar to the Stong Migmatite, risen up through the mesozonal overlying rocks. The Lower Paleozoic age of these catazonal rocks can hardly be in doubt, since the overlying mesozone contains known Upper Paleozoic formations.

The central belt of catazonal rocks is deficient in tin but is characterized by an association with gold mineralization. Potassium-argon Lower Triassic dates of the metamorphic rocks in the Taku Schist probably indicate the time of uplift of the catazone as gneiss domes in the rising axial zone of the orogen.

The axial zone, having been the zone of deepest geosynclinal depression, is probably underlain by a region of altered mantle. Depression of the mantle under the eugeosynclinal axis is presumed to have caused a phase change in mantle material, leading to acceleration of depression, and the accompanying evolution of heat in this process can be regarded as causing significant anatexis of the Paleozoic metasediments and metavolcanic rocks and of the sialic basement that underlay them.

The mesozone, characterized by greenschist facies metamorphism and an isoclinal folding style, outcrops extensively. Large subcordant granite batholiths occur within greenschist facies Paleozoic metasediments and metavolcanic rocks. It is envisaged that the bulk of the granite magma had been evolved over the tectogene zone, where most heat was generated. During uplift of the axial zone, the granite magma was squeezed upward and outward to the east and to the west as the Main Range and the East Coast Batholith systems, respectively. The batholith margins may have been subsequently modified by metasomatic processes. Granite is generally absent in the axial zone. It is therefore thought that the batholiths are sheetlike intrusions as in Figure 10.23, having been squeezed upward and outward from the axial zone; the Benom granite is one of the few plutons that found its way to higher levels in the rising axial zone.

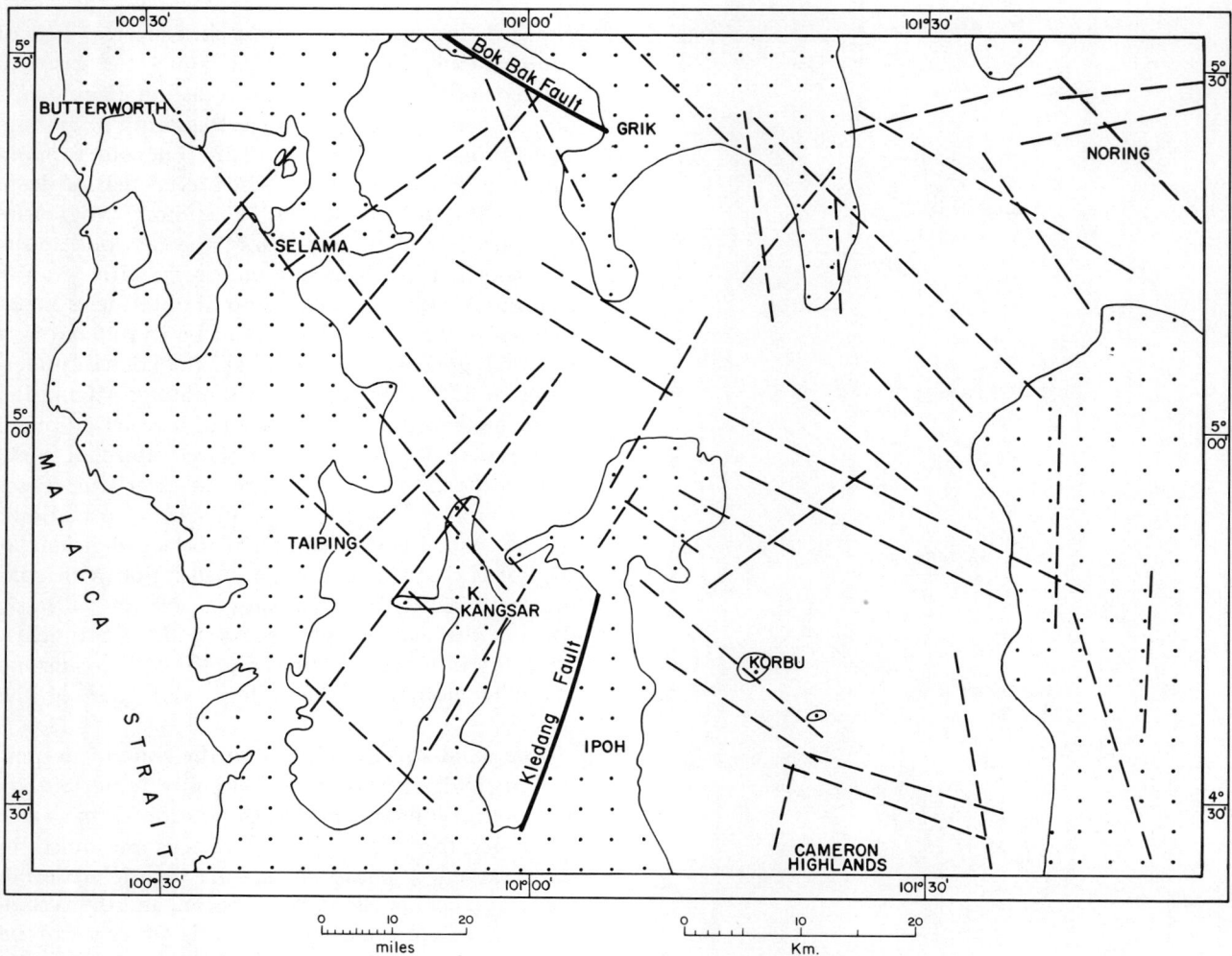

FIGURE 10.22 *Major lineaments indicated by stream segments in north Perak and west Kelantan: blank areas = Granite, dots = sedimentary rocks.*

Rubidium-strontium dates indicate that the granite to the east of the catazone belt is generally of Upper Carboniferous age, whereas that forming the Main Range to the west is generally Triassic. These granites are tin bearing, and during emplacement the tin has been concentrated in the upper parts of the batholiths, especially in cusps that have migrated to higher levels. The granite of the East Coast batholiths system, being older, has been much more deeply eroded than the younger Main Range Granite batholith. The present erosion level of the Main Range is considered to be near the original batholith roof. Hence present outcrops of the Main Range Granite are associated with tin deposits, and those of the East Coast granite are generally devoid of tin.

The rising of the axial zone of the orogen must be considered to have been accompanied by overthrusting toward the foreland to the west (Figure 10.23); the Kisap Thrust in Langkawi is the best example of this type. Within the Malay Peninsula there is no direct evidence of nappe structures. In his analysis of Indonesia, Van Bemmelen (1949) suggested that the Permian and Carboniferous formations of Sumatra form a nappe structure—the Djambi Nappe—derived from a root zone in Malaya. Westward overthrusting, authenticated in the Langkawi Islands, might suggest that westward décollements should be expected, but nothing resembling a nappe structure has yet been recognized in the Malay Peninsula.

A speculation about what underlies the Cambrian Machinchang Formation is in place here. The arenaceous nature of this formation indicates that the basement to the Malayan Orogen must have been continental. Another reason for deducing a

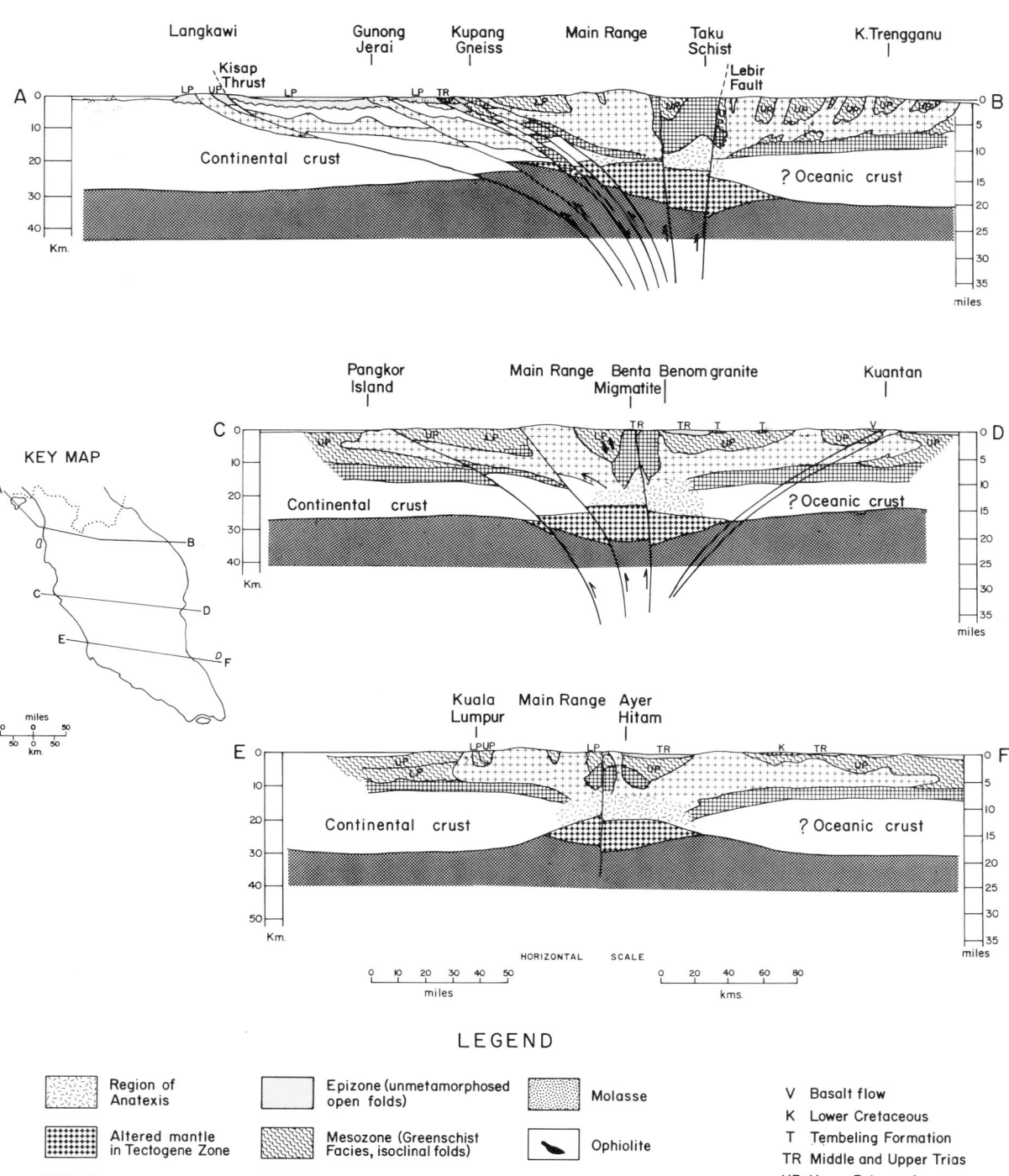

FIGURE 10.23 *Three schematic tectonic sections through Malaya. Cenozoic formations have been excluded. Based only on extrapolation from surface outcrops.*

salic basement is the large volume in granite in the Orogen. It is inconceivable that the immense amount of granite could have been derived from anatexis of the Paleozoic geosynclinal rocks alone—that is, without some considerable contribution from partial melting of an underlying basement, which accordingly must have been salic. Unfortunately, the foreland is nowhere exposed; it must lie at a shallow depth beneath the Langkawi Islands and northwest Malaya generally, however.

The Precambrian foreland was usually conveniently considered to be the Indian peninsula, but Dietz and Holden (1970), in the new global tectonics, would remove India completely from this region during Paleozoic time. The geosynclinal sequence and structure of the Malayan Orogen require a Precambrian landmass immediately to the west of the Malay Peninsula from early Cambrian onward. Whether this was peninsular India is uncertain.

On the other hand, there is no evidence of a landmass directly to the east of the Malayan Orogen either during the geosynclinal or the revolutionary phase. Indosinia lay to the northeast. Therefore, in the constructed cross sections (Figure 10.23) the basement to the east has been shown as oceanic, although this is only an indirect speculation. The presence of ophiolite masses in the Lower Paleozoic sequence immediately east of the Main Range within the eugeosyncline would support the contention that the underlying Precambrian basement in this region was at least in part oceanic. The ophiolite bodies may be regarded as slices of mantle and oceanic crustal material thrust into the Lower Paleozoic geosynclinal sediments. Subsequent mineral equilibration under the prevailing temperature and pressure conditions has largely obliterated the originally igneous aspect of these ophiolites, and they are now of serpentinite and metabasite.

The Malayan Orogen, as deduced in this analysis, is presumed to have evolved by the collision of a continental crustal plate (the western part) and an oceanic crustal plate (the eastern part).

The geosynclinal rocks that were deposited over the salic basement to the west have largely remained in the epizone in an unmetamorphic condition, and this is the only region where unmetamorphosed Lower Paleozoic rocks can be found in Malaya.

East of this region, only the younger late-orogenic formations have remained in the epizone in an unmetamorphic condition. These are the late orogenic Triassic formations and the Mesozoic molasse formations.

Late orogenic epeirogenic uplift and faulting has been elegantly dated by the concordance of potassium-argon dates on granites and the paleontological dates of the unconformities at the base of both the Tembeling Formation and the Gagau-Panti formations. Faulting and mylonitization along the western margins of the Main Range have been dated as late Cretaceous by the frequent potassium-argon late Cretaceous dates of granites in this zone. Minor postorogenic granite emplacements into the epizone have been confirmed along this same zone by rubidium-strontium dates.

The basalt flows of Kuantan and Segamat are postorogenic and form topographic features. Basalt dykes along the east coast apparently range in age from late Mesozoic to Quaternary, and they infill fault zones that must penetrate as deep as the mantle. Some of these dykes have reached the surface to feed postorogenic flows.

CHAPTER 11

> "On a high ridge between the two countries grows a tree that bears red blossoms on the Siamese side and white blossoms on the Perak side. When the red blossoms fall they turn to gold, the white blossoms on the Perak side fall and turn into tin. Could anything be more satisfactory? This theory is worth a whole book on the pneumatolytic genesis of the tin, and on the question whether the gold is of magmatic origin or laterally secreted."
>
> (J. B. Scrivenor, *A Sketch of Malayan Mining* 1928, p. 9).

Primary Mineral Deposits

K. F. G. HOSKING

West Malaysia is the world's major producer of cassiterite, and because most of the tin ore is recovered from placers, by dredging, gravel-pump mining, dry mining, and so on, we tend to overlook the fact that the country is currently exploiting primary (*in situ*) tin and other deposits. At present, for example, cassiterite concentrates are obtained from the large and deep hard-rock mine of Pahang Consolidated, at Sungei Lembing, and this mine also sells a "mixed-sulfide" by-product containing such minerals as chalcopyrite, pyrite, sphalerite, and arsenopyrite. At Rahman Hydraulic Mine, Klian Intan, Perak, and at Sungei Besi Mine, Selangor, bench mining is employed to obtain primary tin ore. In what might be broadly termed the northeast coast mining fields, wolframite and cassiterite are won from veins by mining operations based essentially on shaft sinking or on the driving of adits.

In the past, many types of primary stanniferous deposit have been exploited in West Malaysia by open-cast and underground methods; the Bundi mine, Trengganu and the pipes of the Kinta area may be cited as examples. Also in the past, primary gold deposits were mined by open-cast and underground methods at Raub, Pahang. Scheelite was obtained by the same methods at Kramat Pulai (Perak), and primary iron deposits together with their oxidized portions were, until recently, mined by open-cast methods at Bukit Besi, Trengganu.

In the course of hard-rock mining operations, by-products other than those noted previously have been recovered and sold. Fluorite, which accompanied the scheelite at Kramat Pulai, was stockpiled and eventually shipped during World War II by the Japanese to their homeland, and calcination of the arsenical tin ore from the Beatrice Pipe, Perak, permitted the recovery of salable amounts of arsenious oxide (Ingham and Bradford 1960).

Primary lodes and veins containing cassiterite and/or other mineral species are commonly encountered during placer mining operations. Doubtless, too, significant but unknown quantities of primary cassiterite have been unwittingly recovered by dredges, during gravel-pump mining, and, in the past, during lampanning—that is, whenever the *in situ* ore body is so friable that, as far as the miner is concerned, it behaves as if it were placer material.

Whether special mining methods are employed to exploit a primary tin deposit exposed during gravel-pump mining depends on the size of the "hard-rock" deposit, its grade, and the ease, or difficulty of beneficiating ore derived from it. Some of the stanniferous pipes in the Kinta Limestone, which were discovered by placer mining, were so rich that they yielded handsome profits: the most notable of these was the Beatrice Pipe, which yielded profits of about £750,000 for an outlay of £468 (Ingham and Bradford 1960).

On the other hand, both in Perak and Selangor, lodes that might appear to be of economic interest to the owners of the open-cast mines in which they occur, are neglected because of the considerable sums of money required to test the potential of such bodies and, if they look promising, to mine and beneficiate them in a satisfactory way. Generally there is a considerably greater risk to capital involved in hard-rock than in placer mining.

There are a number of reasons for devoting this chapter to the primary deposits and their oxidized parts. These materials are the source of probably all the cassiterite of the placers and of some of its associates, notably gold and certain tantalum–niobium species. Study of the hard-rock deposits may, therefore, serve to throw further light on some of the characteristics of the placers and may suggest, for example, why the Malaysian tin placers were initially so outstandingly rich.

As the stanniferous placers become increasingly depleted, it is increasingly important to assess the hard-rock mineral potential of the country. This does not simply mean assessing its hard-rock *tin* potential, but its hard-rock mineral and metal potential generally. Hard-rock occurrences of many minerals of economic importance are known; these include, besides those already noted, antimony, lead, zinc, copper, beryllium, barium, manganese, and mercury species. Often the known bodies in which these minerals occur are too small to be of direct economic interest, although small deposits of barite and of manganese are currently being exploited in the Tasek Chini area of Pahang. Their mere presence, however, suggests that the search for similar but larger deposits may be warranted and that, when some of these bodies occur with exploitable tin deposits, their recovery might increase the overall profits made during the mining operation.

Clearly, one important preliminary to the assessment of the country's hard-rock potential is the investigation in some detail, by the employment of modern techniques, of the nature of those deposits which are available for study. Such studies which are currently being carried out in the country have demonstrated that the hard-rock deposits, not solely the stanniferous ones, commonly display a mineralogical complexity considerably greater than was thought to be the case before. For example, before microscopic studies of polished sections of ore in reflected light became routine. Such studies and spectrographic analyses have demonstrated that the Manson Lode (Kelantan), which may well be exploited (primarily for the argentiferous galena which it contains) *may* also yield sphalerite, cassiterite, and even mercury as important by-products.

Certainly these commodities are present; extensive sampling will demonstrate whether they are present in economically important concentrations (Hosking 1969a). During a recent study of certain mineralized veins in the Kampong Pandan area of Selangor (Yeap and Hosking unpublished studies) beryl and helvite—a beryllium species not hitherto recorded from Southeast Asia—were discovered. In conjunction with other beryl occurrences, this discovery serves to highlight the idea that economically important concentrations of beryllium might well occur in some of the greisens and skarns of West Malaysia, and that they should be diligently searched for.

REGIONAL DISTRIBUTION OF PRIMARY MINERALIZATION

Most of the primary mineralization of economic importance, and also those placers which fall into the same category, are closely associated with the north-south-trending Main Range Granite of the western half of the country and the similarly trending discontinuous granitic ranges which parallel the east coast (Figures 11.1 to 11.4). Most of the tin, tungsten, and iron deposits of economic importance occur in these areas. The two granitic belts are separated by the central belt containing minor granitic bodies, generally supposed to be intrusive, and numerous volcanic masses that are, for the most part, acid to intermediate in composition.

Within the central belt is to be found the now-abandoned Raub gold mine, together with many other small gold occurrences, some of which have been worked on a small scale and in a primitive manner. In this little-explored belt, the known mineralization is often so situated with respect to granitic masses that it is easy to feel that a genetic relation probably exists between them. However, there are examples elsewhere in which a spatial relationship between minor acid intrusive and/or volcanic rocks of varying composition is evident. The gold and base-metal deposits of Ulu Ketubong (Kedah), which are associated with metaacid volcanic rocks, tuffs, and quartz porphyry dykes, fall

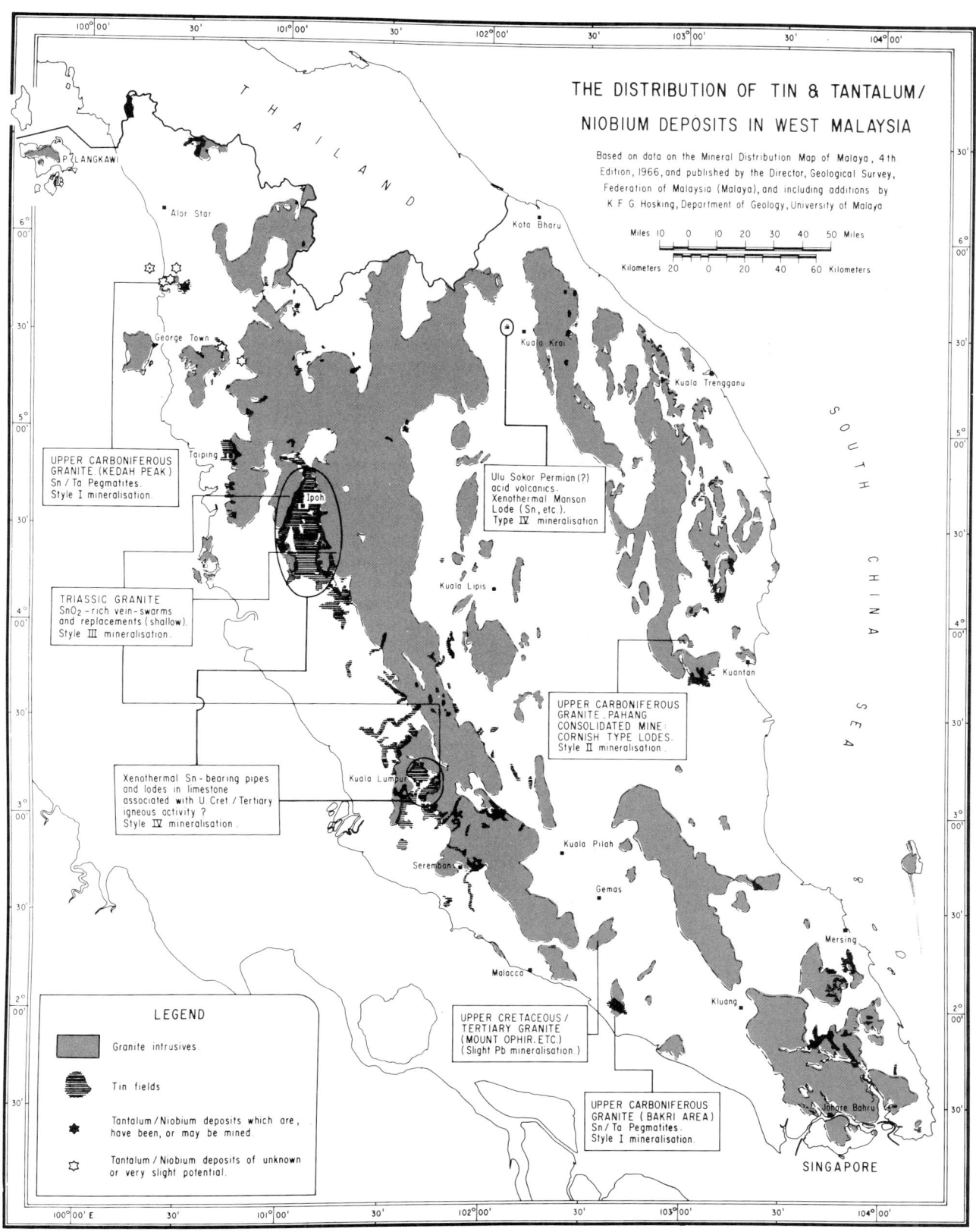

FIGURE 11.1 *The distribution of tin and tantalum–niobium deposits in West Malaysia.*

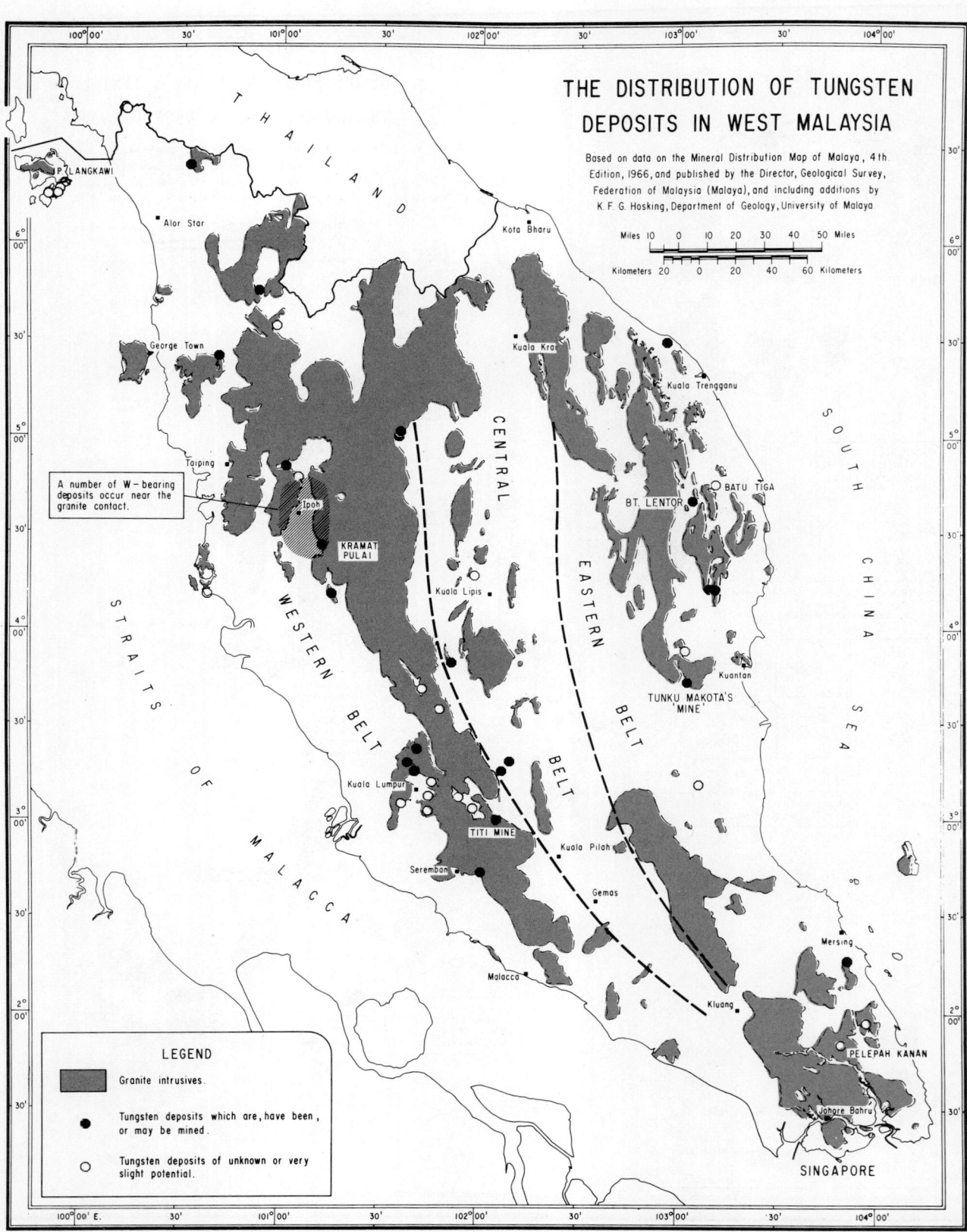

FIGURE 11.2 *The distribution of tungsten deposits in West Malaysia.*

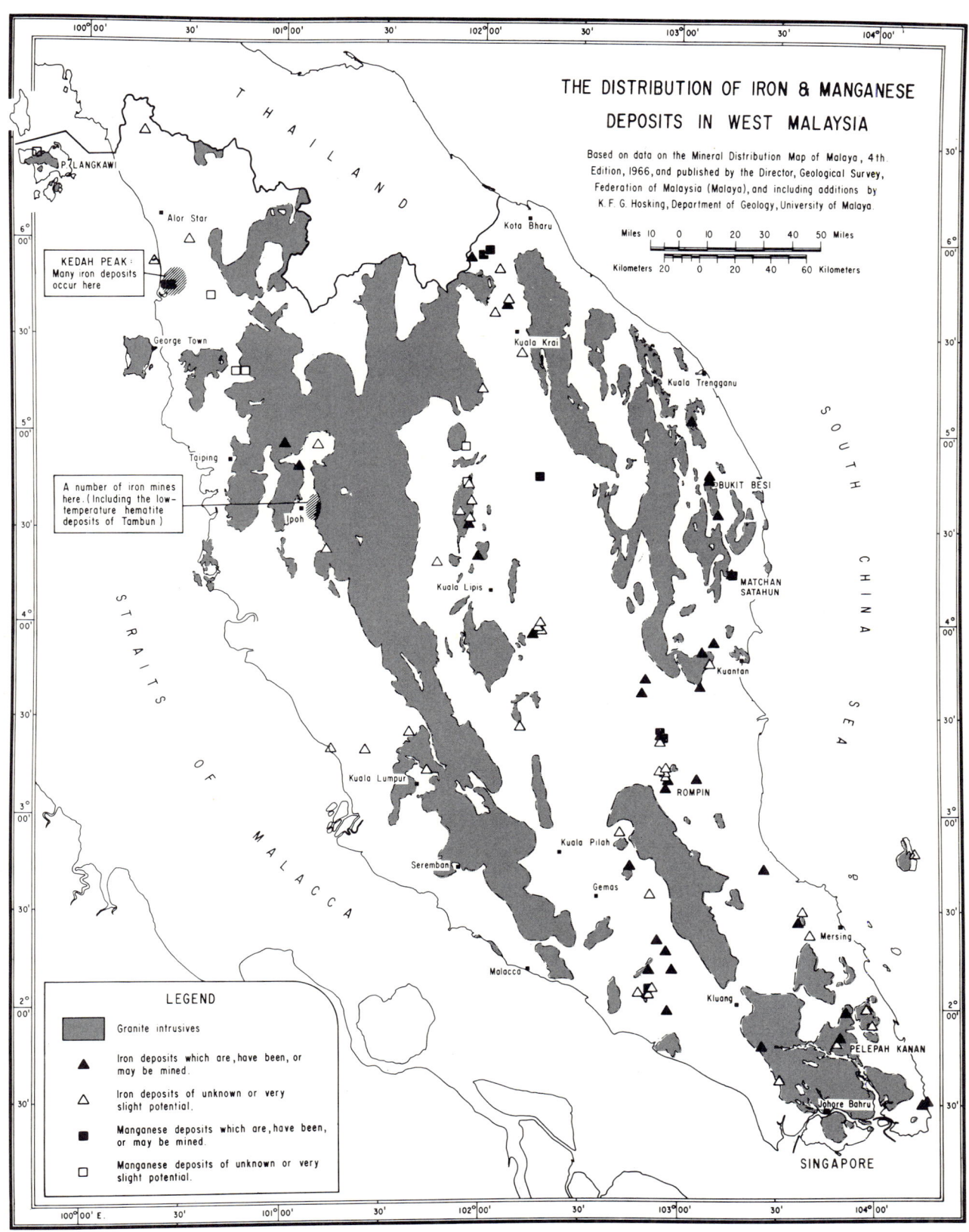

FIGURE 11.3 *The distribution of iron and manganese deposits in West Malaysia.*

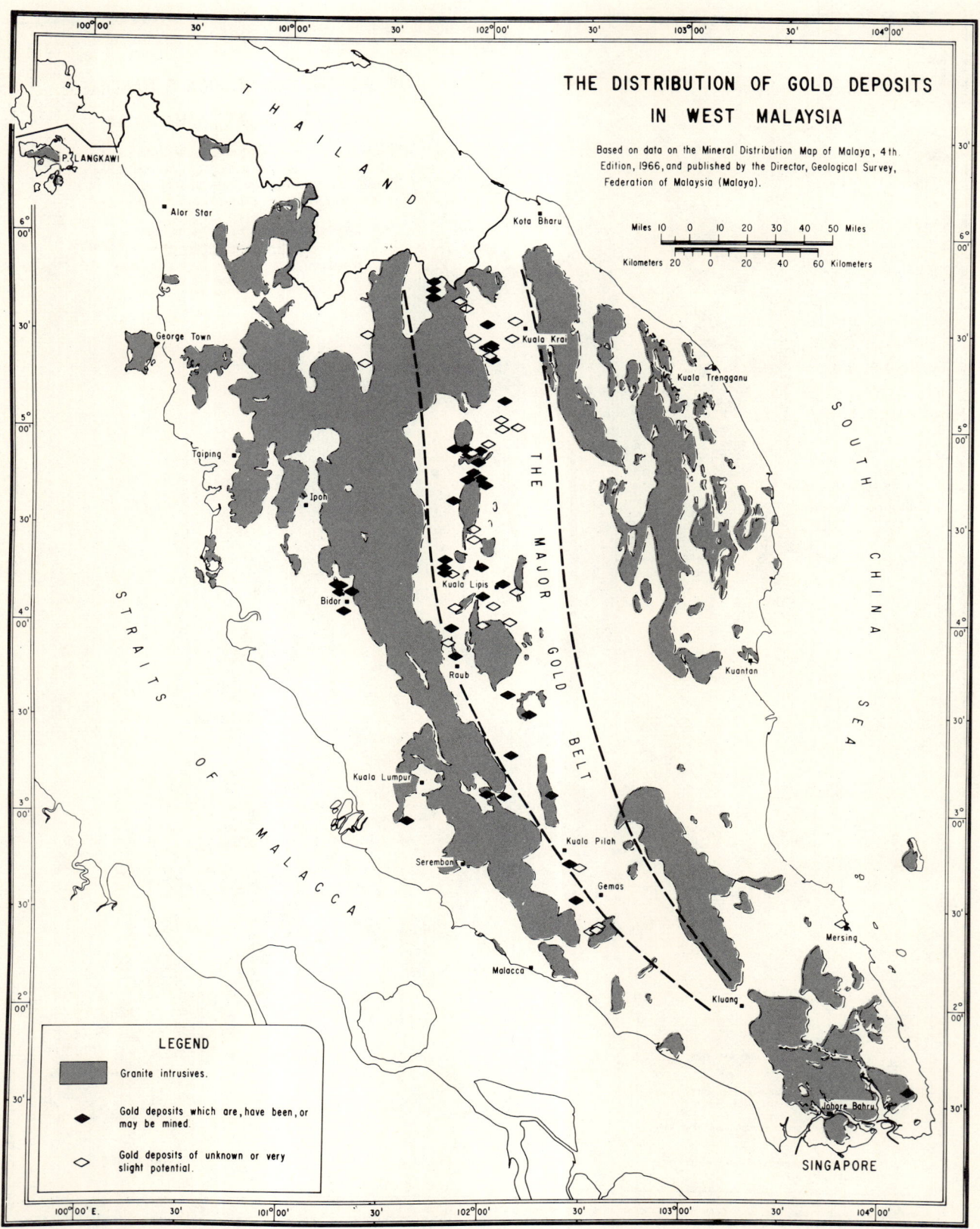

FIGURE 11.4 *The distribution of gold deposits in West Malaysia.*

into this category (Hosking 1969a); so does the copper–lead mineralization at Segamat (Johore), which is found within an andesitic–basaltic pile. The possibility, then, of a genetic relation existing between minor intrusive rocks and volcanic rocks and ore deposits in the central belt has to be considered. It is discussed further in a later section dealing with the whole question of the relation between igneous rocks and mineral deposits.

Finally, as Figures 11.1 to 11.4 indicate, the mineral deposits of the east and west granite belts are disposed in an apparently erratic pattern, but essentially in the vicinity of the granite contact. In the western belt, mineralization is strongest on the west side of the Main Range Granite and reaches its most spectacular development in the Kinta and Kuala Lumpur fields; by comparison, the eastern flanks of this granitic mass are very poorly mineralized. At Pahang Consolidated Mine, in the eastern belt, a plexus of strong and economically most important stanniferous lodes flanks a granite mass. Indeed, nowhere else within the eastern belt have primary tin deposits been discovered which remotely approach those of Pahang Consolidated in economic importance. Are these Pahang Consolidated lodes unique as far as the eastern belt is concerned? And if so, what factors determined their development? We shall return to these and similar questions, but it may be stated here that many questions relating to the distribution patterns of the West Malaysian primary ore deposits, and to the known and possible relation between these deposits and igneous rocks, cannot be answered in a satisfactory manner. This stems in part from our very imperfect knowledge of many aspects of Malaysian geology and in part from our equally incomplete understanding of ore genesis.

MAGMATIC DISSEMINATIONS AND SEGREGATIONS

In order to have a clear understanding of the types of mineral deposits occurring in West Malaysia, and also from an economic point of view, it is most important to differentiate between "magmatic disseminations" and just "disseminations." In the literature this is not always done, by any means.

The phrase "magmatic disseminations of mineral X" means that X is syngenetic with its igneous host rock. It follows, therefore, that there may be a possibility of finding large volumes of the rock which may contain mineral X in such concentrations that they constitute major ore bodies.

In West Malaysia, sparsely disseminated chromite occurs in a few ultrabasite bodies bordering the Main Range Granite on its eastern side, but these bodies are so small and their chromite content so low that they are never likely to be regarded as potential sources of chromite. However, the chromite in them is, beyond reasonable doubt, a true magmatic dissemination.

That magmatic disseminations of cassiterite of possible economic importance occur in West Malaysia has been claimed by some, and because of the important role played by tin in the Malaysian economy it is pertinent to consider this topic. To do this adequately it is necessary to refer to certain reports concerning tin–granite relations in peninsular Thailand. Reference to the Thai work would appear to be in order, since the granites of southern Thailand are essentially extensions of those of West Malaysia.

M. P. Jones (1969) believed that magmatic disseminations of the type under discussion occur in West Malaysia, as he remarks that the biotite granite of the Sungei Besi tin mine (Selangor) "has a tin content of 0.01 percent" and that "the main tin-bearing mineral in these granites is cassiterite and recent work has clearly shown that the cassiterite occurs as a primary accessory constituent." Kaewbaidhoon and Aranyakanon (1961) also held the opinion that locally important concentrations of cassiterite may, in fact, be magmatic disseminations. Indeed, in their classification of the Thai tin deposits they included the following curious group:

Magmatic segregation origin
1. *Disseminated* and aplitic deposits
2. Veins
   a. Pegmatites and aplite veins.
   b. Pegmatitic and hydrothermal quartz veins.

Bradford (1961) was a little more cautious. He remarked that

The question as to what extent the Malayan granite is as a whole tin-bearing has never been investigated fully. A few analyses from ordinary granite suggest that tin-ore is finely disseminated through many of the Malayan granites.

He then gave the tin content of five Malayan granites as: 25, 25, 38, 76, and 200 parts per million (ppm). It would appear that Bradford did not regard such magmatic disseminations as being of direct economic importance, although he believed that cassiterite, liberated from such deposits by deep secular weathering, might concentrate as a worthwhile placer deposit. He also suggested that some of the alluvial deposits in the vicinity of Bukit Yong (Bradford 1961, p. 388) might be "derived from cassiterite disseminated throughout the local granite," since there is a paucity of known stanniferous lodes and stringers in the area.

Sporadic occurrences of crystals of cassiterite that have been recorded in the granite of the Sungei Siput tin field, and the wolframite and cassiterite that is stated to occur in the granite at Ulu Bekor (Bradford 1961, p. 386) are surely epigenetic and probably spatially related to knife-edge feeder-channels that were overlooked.

There is little doubt that tin is a syngenetic component of the Southeast Asian granites. In part it is probably incorporated in the ferromagnesian minerals and in some of the accessory species, and in part it may occur as discrete particles of cassiterite. Certainly ilmenite from the Malaysian stanniferous placers commonly contains from about 1500 to about 2000 ppm tin, and Flinter (personal communication) noted that zircons from the same sources may be highly stanniferous: both the ilmenite and the zircon are derived from granite.

Pitakpaivan (1969) analyzed 80 samples of Thai granite and reported that there was tin in each; but he discovered that the content was somewhat higher in those granites with which tin deposits were associated than in those with which none was known to occur. A summary of his results is as follows:

Percentage of Tin

	Maximum	Minimum	Average
"Tin-bearing granite" (40 samples)	0.52	0.04	0.138
"Tin-barren granite" (40 samples)	0.24	0.03	0.07

It is hardly necessary to point out that large tonnages of rock containing the maximum tin content, which Pitakpaivan recorded even for his so-called barren granite would, in certain circumstances, be of economic interest.

As a whole, by comparison with the Thai granites, the Malaysian granite would appear to be distinctly poor in tin, that is, if the results obtained by the gallein–methylene blue colorimetric method of Stanton and McDonald (1962) can be relied on.[1]

For purposes of comparison with the Thai results, those recently obtained by Yeap Cheng Hock (personal communication) are given in Table 11.1; it should be pointed out, however, that these results have yet to be rechecked to the author's complete satisfaction.

The results accord well with a number of similar unpublished studies carried out in the Department of Geology, University of Malaya, which have also demonstrated that only a few meters away from stanniferous lodes and veins and cassiterite-bearing joints, a concentration of <10 ppm tin is commonly recorded.

Jones's (1969) view of the tin content of the Sungei Besi granite is erroneous because the material he sampled contained epigenetic cassiterite. He was, in fact, sampling tin-anomalous envelopes surrounding a series of extremely rich replacement deposits that have since been exposed and are now in the process of being mined. It is also likely that Pitakpaivan (1969) obtained the high values he recorded in part because he was also investigating, without being aware of it, granite that had been subject to incipient epigenetic mineralization. The writer has demonstrated that on occasion the tin content of apparently fresh granite adjacent to a knife-edge fracture is greatly in excess of that of the granite only a few inches away from the fracture. In part, also, the higher results of Pitakpaivan may be attributed to the analytical method he employed, which was unsuited to the problem he was investigating.

The main reason that Kaewbaidhoon and Aranyakanon (1961) hold the views outlined previously can probably be found in the results of a special study of the primary tin deposits of Haad-Som-Pan, Thailand, that the latter had made earlier (Aranyakanon 1961). In the first study, inverted pear-shaped masses of cassiterite-rich granite, which Aranyakanon believed to be syngenetic, were found to occur beneath an impounding roof of sediments. Such deposits are similar to those recently exposed

---

[1] There is, indeed, good reason for thinking that this method will record, within the limits of the known experimental error, the tin occurring both as cassiterite and in close association with the major components of the granite. It will also record the tin in ilmenite and any occurring as cassiterite inclusions in zircon, provided it has been liberated during grinding. It is probable, however, that, at best, tin in the structure of tourmaline will be only partially released during the analysis.

TABLE 11.1 *Analyses for tin (ppm Sn) of "fresh" vein-free granites by the gallein–methylene blue method**

	Samples from			
	Stanniferous Areas	Vicinity of Major Tin Areas	Vicinity of Minor Tin Areas	Nonstanniferous Areas
Western Belt				
Mean	6.5	6.2	7.1	5.1
Range	2.5–12.5	<2.5–12.5	2.5–12.5	<2.5–12.5
No. of samples	25	21	28	109
Eastern Belt				
Mean	7.5		6	4.2
Range	5–10		<2.5–12.5	<2.5–7.5
No. of samples	2		18	44

* During the computation of the arithmetic mean, samples with less than 2.5 ppm Sn were regarded as being tin-free.

at Sungei Besi and noted previously. The latter deposits developed almost immediately beneath a limestone roof (which has since been removed), but they show an approximately parallel strike direction. Even though in longitudinal section and in plan the deposits have irregular shapes, they are, without doubt, the products of the replacement of the granite by stanniferous agents which ascended into the zone of deposition along fractures and were impounded beneath the limestone.

Both these Thai and Malaysian deposits have much in common with the Cornish stanniferous carbonas, which we know to be epigenetic because their narrow feeder channels have been traced and mapped (Hosking 1969b). The Asian deposits are also similar in a number of respects to some of the tin-bearing deposits occurring in the arkose of the Rooiberg mines of South Africa, but no one would seriously suggest that the latter are syngenetic.

To summarize, then, the writer is of the opinion that no good evidence exists to support the view that in West Malaysia there are magmatic disseminations of cassiterite in the major granitic intrusives that can be classified as ore deposits, or that might in the foreseeable future be so classified. That syngenetic cassiterite has made a significant local contribution to the development of placer deposits in West Malaysia, as Bradford suggested, remains an open question. However, the writer has additional reasons for regarding Bradford's suggestion as one unlikely to be correct. The paucity of tin in the granites is the first. Then, too, we know that some of the tin—perhaps most of it—may not occur as cassiterite but rather in the ionic state in the biotite structure. Furthermore, the work of the writer and his students has provided no evidence that there are always significant differences in the syngenetic tin content of the so-called tin-bearing and tin-barren granites.

PEGMATITES AND APLITES

Pegmatites, aplites, and composite pegmatite–aplite bodies are not uncommon in West Malaysia. With very few exceptions they are small, and all the pegmatites are of the unzoned type. A considerable proportion of them are stanniferous and a few have been mined for cassiterite, as, for example, the pegmatite at Bujang Malaka in Perak and the aplite at Gunong Bakau, Pahang (Scrivenor 1928).

Apart from cassiterite, these granitic bodies do not generally appear to contain any other "heavy" minerals in sufficient amount to make them of direct interest to the miner, or to have allowed them to play a major rôle in the development of placers other than cassiterite placers. There are, however, two notable exceptions and these occur in the vicinity of Gunong Jerai, Kedah, and in the Bakri district of Johore. At Gunong Jerai, cassiterite and columbite–tantalite concentrates have been produced by placer mining operations at Semiling and Bedong, where the economic minerals in question are associated with "tourmaline, zircon, garnet, gahnite, monazite, cheralite, magnetite, haematite and traces of sillimanite." These mines are

... located at the foot of the granite in an area of weathered quartzites which are heavily intersected with pegmatite veins and vein quartz. ... [M]ineralization (placers?) is found in conjunction with the granite and, especially, with the pegmatites which normally carry coarse muscovite, tourmaline, and garnet . . ." (Bradford 1961, p. 384).

At Bakri, the small placers have attracted considerable attention from miners because of the important concentrations of columbite–tantalite associated with cassiterite. According to Bradford (1961, p. 393) "most of the mines work alluvium overlying quartzite and shale which have been intruded by stanniferous pegmatite and quartz veins from the Bukit Mor complex." He also noted that other minerals occurring in the placers of the area included tourmaline, ilmenite, monazite, zircon, garnet, gahnite, tanteuxenite, arizonite, and traces of molybdenite.

Bradford did not state that at Gunong Jerai and at Bakri the columbite–tantalite, and certain other exotic species such as gahnite and cheralite, have been derived from the pegmatites. From the content of his 1961 paper, however, it is reasonable to think that this was, in fact, his opinion. The writer also holds that the pegmatites were probably the source of the minerals in question, although he has not seen any of these species in specimens of pegmatite from these areas nor does he know of anyone who has.

The style of mineralization of these two areas, where the granites have been dated by radiogenic means as Upper Carboniferous, differs appreciably from those of the mining fields associated with the probably largely Mesozoic granites of the Main and Kledang Ranges. Columbite, nevertheless, is quite a common but minor constituent of the heavy mineral concentrates from the placers of the latter areas. Hockin (1957a) demonstrated the common occurrence of columbite in the placers in question and noted in addition that the columbite was sometimes intergrown with rutile, and that niobian rutile was also a not uncommon member of the heavy mineral asemblage.

Previously a Malayan smelting company had shown that columbite was widely but erratically distributed in the stanniferous placers of the western Tin Belt. The source of the columbite and niobian-rutile is still uncertain. It is well known that tantaliferous–niobiferous cassiterite is fairly common in the hydrothermal veins and lodes of West Malaysia, and that these cassiterites often contain columbite–tantalite and possibly tapiolite–mossite exsolution bodies (Hockin 1957a). However discrete crystals or aggregates of columbite have never been recorded in such hydrothermal bodies, and it seems unlikely that they were the source of the free columbite and columbite–rutile intergrowths in the placers.

The writer thinks that such crystals or aggregates have been derived from pegmatites and/or aplites and, because they never occur as large crystals, he favors the latter as the probable source. As far as the author is aware, no one has yet recorded columbite in any of the Mesozoic granitic rocks of West Malaysia. Interim results of studies of the trace-element content of West Malaysian granites have indicated, however, that these rocks are often distinctly niobiferous (see Table 11.2, provided by Yeap Cheng Hock).

TABLE 11.2 *Niobium content (ppm Hb) of West Malaysian Granites determined by the chromatographic method*

	Samples from			
	Stanniferrous Areas	Vicinity of Major Tin Areas	Vicinity of Minor Tin Areas	Non-stanniferrous Areas
Mean	63	48	n.d.	n.d.
Range	25–120	6–90	n.d.	0–48
No. of samples	9	2	n.d.	10

n.d. = not determined.

The evidence now available leaves little doubt that in West Malaysia pegmatites bearing columbite–tantalite are distinctly rare; it is also certain that stanniferous pegmatites are not quite as common in the country as the literature might suggest. This is because deposits have been classified as pegmatites when, clearly, they do not possess such characteristics even when it is acknowledged that today the term "pegmatite" lacks a very precise meaning.

In the writer's view, the mere presence of feldspar in a cassiterite-bearing body does not justify terming the deposit a pegmatite; and yet Bradford (1961) spoke of the Pelepah Kanan (Johore) vein-swarms, which consist largely of quartz, random masses of K-feldspar, cassiterite, arsenopyrite, chlorite, and fluorite, as pegmatites. These veins, like the scores of stanniferous and other feldspar-bearing veins of Cornwall (Hosking 1967), are surely hydrothermal in origin.

Those interested in the genesis of the primary mineral deposits of West Malaysia and in the search for further ore deposits therein, must also ask to what extent the cassiterite, present in the pegmatites and aplites, is syngenetic. Clearly, cassiterite that occurs in a pegmatite was not necessarily deposited at about the same time as the major associated pegmatite silicates, and from the same magmatic fraction. "Pegmatite" is a remarkably readily fractured rock; and so after it had developed, possibly a long time after, it might be preferentially fractured, and ascending mineralizing agents might become impounded within it and might deposit cassiterite, together with an assortment of those other heavy minerals which accompany the tin species in what might be called the "normal" stanniferous ore bodies.

Obviously there is also the possibility that a given pegmatite may contain true pegmatite cassiterite along with later hydrothermal cassiterite. From this it follows that the tin content of pegmatites may vary within wide limits. It is considered that bipyramidal cassiterite is usually a product of the pegmatitic magmatic fraction, since it is not found in deposits other than pegmatites. Cassiterite with such a habit has not been recorded from the West Malaysian pegmatites but it has recently been found, by the writer, in concentrates from placers in Kedah and Johore which are thought to have derived some of their cassiterite from pegmatites. A few such occurrences have also been noted in Peninsular Thailand (Aranyakanan private communication).

Cassiterites of other habits are also found in pegmatites throughout the world and, as far as the writer is aware, there is as yet no certain means of establishing whether some or all of these are also derived from the pegmatite fraction. If, however, the "pegmatite" is distinctly rich in cassiterite, and if the cassiterite tends to be widely distributed in it or displays a spatial relationship to "late" fractures, we can assume with some confidence that the cassiterite is epigenetic with respect to the pegmatite.

Further support of the view that the cassiterite in a given pegmatite was a late arrival would be the presence of a variety of sulfides in the same body and/or the presence of neighboring hydrothermal stanniferous veins. In a given metallogenetic province, pegmatites containing "pegmatitic" cassiterite, columbite–tantalite, wolframite, molybdenite, and so on, and early hydrothermal deposits containing tin, tungsten, molybdenum, and so on, of about the same age, tend to be mutually exclusive. In West Malaysia hydrothermal deposits are overwhelmingly dominant, not the pegmatites.

Thus, according to Bradford (1961, p. 390), at Gunong Gapis, Pahang, cassiterite "occurs disseminated throughout a mass of soft weathered granitic rock and in small irregular bodies distributed in it. These ore bodies . . . consist of quartz, tourmaline, and cassiterite, and of pegmatitic material rich in cassiterite." Although it might be argued that cassiterite was deposited first from the granite magma and later from the pegmatic fraction, the writer thinks it more realistic to consider that essentially all the cassiterite was deposited from hydrothermal solutions that invaded the consolidated granite and pegmatite and subjected them to intense metasomatism.

Again, at Gunong Baku, a pegmatite dyke, about 9 m (30 ft) wide, and occurring in granite, contains (in addition to cassiterite and much tourmaline) pyrite, arsenopyrite, chalcopyrite, and galena (Roe 1951). At Ulu Petai, Perak, on the other hand a pipe that Scrivenor (1928, p. 74) regarded as "a good example of a pegmatitic ore deposit" was followed for 90 m (300 ft) into the granite mountain. This pipe, which occurs in "partly mineralized" aplite, "consists of pink kaolinized orthoclase, tourmaline, cassiterite, metallic sulphides and wolfram, with quartz, muscovite, chlorite, fluorite, and plagioclase feldspar, the last five minerals being recognized only under the microscope."

Ingham and Bradford (1960, p. 62) recorded that a pegmatite intrusive into limestone occurring at Fook Wan Foh Kongsi, Chenderiang, Perak, "is mainly composed of alkali felspar and quartz, and is rich in muscovite, hydromica (probably gilbertite), tourmaline, topaz, beryl, fluorite, zinnwaldite, and cassiterite, and is also associated with various metallic sulphides such as jamesonite and galena." It seems likely that the sulfides in the pegmatites reported by Roe, Scrivenor, and Ingham and Bradford were introduced during a late post-consolidation stage of mineralization, and so it follows that some or all of the cassiterite and wolframite (when present) might also be late introductions.

Within West Malaysia a number of mineralized aplitic sills and dykes are known. In some instances the mineralization is certainly epigenetic. Thus, for example, the so-called Toh Kiri Lode in the Kinta Valley is in fact an aplite body about 180 m (600 ft) long and more than 15 m (50 ft) wide. It was emplaced in schists and phyllite and subsequently strongly fractured and mineralized, and it contains numerous veinlets of cassiterite together with pyrite, arsenopyrite, fluorite, quartz, mica, and kaolin. Adjacent to the aplite the host rocks are also similarly mineralized (Scrivenor 1928, Ingham and Bradford 1960).

The following remarks by Scrivenor (1928, p. 103) concerning the manner of occurrence of cassiterite in granitic veins near Tras, Pahang, further indicate that mineralization of igneous bodies of the type under discussion was, at least on occasion, an epigenetic event:

> The ore occurs disseminated throughout a mass of soft granitic rock, and in small bodies of stone about 8 feet by 6 feet, irregularly distributed in it. The rock . . . might be described as a coarse aplite. The bodies of stone consist, some of quartz, tourmaline, and cassiterite; others of hard white pegmatite with bunches of cassiterite crystals; others again of granular quartz and cassiterite. *Frequently these ore-bodies are traversed by a fissure, on either side of which the cassiterite has collected as an impregnation* . . . (the writer's italics).

At Gunong Bakau, Pahang, however, somewhat more problematical mineralized aplites occur. There, in the granite, sill-like bodies of distinctly stanniferous topaz rock, varying in width from about 2 cm to more than 5 m, are intersected by veins of cassiterite-poor topaz aplite. Alexander (1968) wrote that both sets of topaz-bearing rock are the products of crystallization of injected "molten magma differentiates." This view was expressed earlier by Scrivenor (1928), who also thought that somewhat similar topaz-rick rocks occurring at Ulu Petai, on the side of Bujang Melaka, Perak, developed in the same way.

Although the foregoing view of the genesis of these interesting bodies may be correct, the writer prefers to regard them as granitic bodies whose aplites have been subject to extreme metasomatism, not autometasomatism, particularly since some of the ore bodies were "distinct from the topaz-bearing rocks . . ." (Alexander 1968, p. 128). At Gunong Bakau, the earlier sills are strongly mineralized with cassiterite, whereas the later ones are not; but this does not detract from the writer's view because the sequence of granitic sills, mineralization, granitic sills, mineralization, would seem to be a distinct possibility.

Certainly, from Cornwall and elsewhere, a number of examples have been recorded of granitic dykes intersecting and displacing tin-bearing veins (Hosking 1964). That is to say, three of the four sequential events to account for the phenomena at Gunong Bakau are known to have occurred elsewhere; the second mineralization would seem to be self-evident. Jones (1925c, p. 207), who knew these deposits well, held a position on their genesis different from that expressed by Scrivenor. His view appears to be similar in some respects to that of the writer in that Jones was of "the opinion that the topaz is of secondary origin, being formed from feldspar by the hydrofluoric acid set free during the deposition of the cassiterite. . . ."

As noted earlier, beryl has been recorded from a pegmatite near Chenderiang. At Kramat Pulai, a beryl-bearing pegmatite also occurred, as did another at the Yik Meng Kongsi near the Kledang road, Menglembu, Perak (Ingham and Bradford 1960).

To date neither deposits of beryl, nor indeed of other beryllium-bearing species of commercial value, have been discovered in West Malaysia. However, beryl does occur in some of the West Malaysian pegmatites; it has been recorded in granite and greisen in the Chendai area near Menglembu, and, together with helvite, beryl has been observed in veins in limestone near Kampong Pandan, Selangor (Hosking and Yeap unpublished studies). These data give good reason for searching diligently for worthwhile beryllium deposits. It must be remembered that small crystals of beryllium-bearing species, when associated with other minerals, are easily overlooked both in the field and in the laboratory. Economically interesting concentrations are most likely to be found in the stanniferous, highly greisenized areas, such as occur locally in the Kledang Range: they may also be found in skarns and early vein systems in limestone.

For convenience, the spatial relations existing between the pegmatites and aplites and the major granitic intrusives are discussed later.

PYROMETASOMATIC (SKARN) DEPOSITS

Pyrometasomatic deposits, characterized by possessing a suite of calc- and/or magnesian-silicates, typically develop from rocks that consist essentially of carbonates or from those which are fairly rich in such minerals because heat and chemical substances were supplied to them by the invading igneous body. Usually the igneous body is of granitic composition, although this is not invariably the case and, on occasion, the role of the carbonate-bearing rocks may be played by basic igneous rocks or their regionally metamorphosed equivalents.

The end-products are the same. Rarely, rock types other than those noted previously, may be converted to pyrometasomatic (or skarn) deposits.

For the purpose of this chapter, the pyrometasomatic deposits of West Malaysia may be classified as follows:

Group I —The ferriferous–stanniferous types.
Group II —The stanniferous types.
Group III—Other types.

Before discussing the characteristics of members of each group, it must be made clear that the pyrometasomatic deposits in West Malaysia have only been exploited for their iron oxide and/or cassiterite content. A considerable proportion of iron recovered from these deposits has been from their oxidized portions and, on occasion, as at Matchan Satahun, Trengganu, only the oxidized part has been mined. Scheelite is widespread in the pyrometasomatic deposits of West Malaysia, but such deposits have never been mined for their tungsten content. Primarily this is because the deposits have not been large enough to interest hard-rock miners. A typical example occurs at Templer Park, Selangor. Scheelite occurring in skarns that have been worked for iron or tin, as at Batu Tiga, Trengganu, and Pelepah Kanan, Johore, has never been recovered as a by-product. Sometimes, however, scheelite that has been liberated from skarns by natural processes has been recovered from placers. This has happened, for example, at Sungei Way, where scheelite occurs in both skarn and "normal" hydrothermal deposits.

For reasons given later, the magnificent deposit of scheelite at Kramat Pulai, which is now worked out, but which occurred in marble adjacent to granite, is not classified by the writer as a pyrometasomatic deposit.

Group I: Ferriferous–Stanniferous Types

All known ferriferous skarns that have been mined for their primary iron oxides (magnetite and hematite) or for secondary iron oxides ("limonite," martite, goethite, etc.) occur in the eastern mineralized belt. For the most part they are situated in Trengganu (Bukit Besi and Matchan Satahun) and Pahang (Bukit Bangkong and the Sungei Panching). Some deposits are members of this group, in the writer's view, but differ in several important respects from those found in Trengganu and Pahang; examples of these occur at Pelepah Kanan and Pelepah Kiri in Johore.

All these deposits are stanniferous, and in certain portions of some of them the cassiterite is present in sufficient concentration and amount to make them worth working for their tin content (e.g., at Batu Tiga and Matchan Satahun, Trengganu). On the other hand, the tin content throughout the mass of otherwise potential iron ore has rendered the deposit of no value in the case of the two Johore iron deposits just mentioned.

As far as the writer is aware, only one ferriferous skarn is known in West Malaysia in which no tin has been recorded; that is the small vein, up to 25 cm wide, which occurs at Wang Pisang, Perak. This body has largely developed by replacement of the limestone and consists of magnetite associated with garnet (probably andradite) and epidote (Bean 1969). When it is remembered that andradite has an affinity for tin, even this deposit may be considered to be somewhat stanniferous.

For convenience, the group of skarn deposits under discussion may be divided into the Trengganu–Pahang members and Johore ones. From an economic point of view, the most important member of those of Trengganu–Pahang type is that at Bukit Besi.

Briefly, and in broad detail, the Bukit Besi area consists essentially of shales, some of them originally calcareous, together with minor amounts of quartzite and limestone. These materials, having been subject to mild regional metamorphism and folding, were further deformed and metamorphosed by an invasion of granitic magma that possibly occurred in Upper Carboniferous time. As a result of the invasion by the granite, some of the calcareous beds (marble and calcareous shale) were converted to skarn. Some of the originally argillaceous limestone was altered to marble containing "patches of fractured and broken garnet, and pyroxene which in places is altered to amphibolite" (Bean 1969, p. 135). Elsewhere, near the granite, calcareous bodies have been converted to dark amphibolites and some of these contain magnetite and pyrrhotite.

The iron-ore bodies, which are never far removed from the granite, consist essentially of magnetite and martite that has been derived from it by alteration. According to Bean (1969) some primary hematite is present, together with a large amount of secondary "iron oxide" species that have been derived largely from the oxidation of pyrite and, locally, of pyrrhotite.

These iron ores are formed largely of replacements of shale, which may be irregular in shape and which consist essentially of magnetite, martite, and secondary hematite enclosing fragments of unreplaced country rock; alternatively, the replacements may be tabular bodies whose form has been controlled by the bedding planes of the host rock. These tabular bodies are the major sources of ore, and in the unaltered state they may contain considerable quantites of pyrite (much more than is generally encountered in the irregular deposits) and, of course, dominant amounts of magnetite. On occasion, ore occurs in the granite and is essentially the product of replacement of shale masses engulfed by the igneous rock. Such ore consists of magnetite and hematite together with disseminated pyrite, coarsely crystalline calcite, and fragments of granite and shale.

The skarn ore is composed of massive replacements of magnetite in a gangue of diopside, garnet, and calcite; such bodies are frequently tabular.

Bean (1969) noted that the distribution of the iron ore depended on the presence of favorable host rocks (shales and amphibolites) in close proximity to the invading granite. Distribution also involved the tendency of the "sediments" to dip toward the granite, a structure which "facilitated the ingress of the ore fluids, thus accounting for the numerous tabular ore bodies whose shapes are obviously controlled by the bedding." It is also important to realize that marked embayments at the granite–shale contact provided "physical traps from which further movement of the magmatic effluents was greatly inhibited, thus allowing replacement to be effected on a larger scale than was common elsewhere along the granite contact," and so causing the local development of large concentrations of ore.

Locally at the Sri Bangun and Batu Tiga sections of the Bukit Besi mining area, there occur marked concentrations of cassiterite, due to replacements of the host rocks in the vicinity of fracture zones. At Sri Bangun, the cassiterite is associated with iron oxides in shale. At Batu Tiga, the only part of the area that has been investigated in considerable detail, the cassiterite occurs in a geological setting of some complexity. Thus at the western hill area, which was worked for tin until 1970 and thus gave the maximum opportunity for investigation, the cassiterite was recovered from two rich north-south-trending "lodes." Leong (1970) stated that the mining property is situated in a calcareous horizon representing a shallow-water transition zone between the more arenaceous rocks to the west and the low-lying shales to the east. The mined area marks the extent of a mildly thermally metamorphosed contact aureole (albite-epidote-hornfels facies), probably roofing a granite protuberance. The gently domed country rocks are essentially low-grade calc-silicate hornfels capped by a meta-arenite sequence. These rocks had subsequently been dissected by a series of northerly trending lamprophyric and acid stocks.

The ore consists essentially of early magnetite followed by cassiterite and scheelite; these in turn are followed by chlorite, large quantities of pyrrhotite, a little contemporaneous chalcopyrite, and late pyrite. These were developed in a calc-silicate host, and the disposition of ore was determined by fault zones, which served as channelways for the ore-forming agents, and by the impounding action of the domed meta-arenite. Subsequent to its deposition, the ore was deformed somewhat so that now the cassiterite and scheelite display pull-apart textures, the fractures having been healed subsequently by sulfides, particularly chalcopyrite, which flowed into them (Figure 11.5). In addition, the pyrrhotite was annealed, and triple points are much in evidence in polished section.

Whereas, then, marked concentrations of cassiterite are confined to one or two places at Bukit Besi, a certain amount of tin seems to be present throughout the iron ore. Indeed, according to Bean (1969) the ore contains an average of 0.07% tin and some of this, at least, is rigidly bound in the structure of the magnetite.

This deposit has been dealt with in some detail because it displays a number of features of considerable importance. First, most of the iron ore there cannot be called skarn ore because most of it is not in a calc-silicate hornfels. Second, it seems certain that, although a little tin was deposited contemporaneously with the magnetite, the high concentrations of cassiterite were deposited later from mineralizing agents that moved along faults into the zones of deposition.

Surely we are dealing here with a metasomatic deposit, produced by tenuous hydrothermal solutions, and initially rich in iron but also containing some tin during the initial stages. These reacted equally well with shales and calc-silicate rocks to produce essentially stanniferous magnetite. Subsequently the character of the mineral-depositing agent changed progressively with time, enabling considerable quantities of cassiterite, a little scheelite, and then large amounts of pyrrhotite and other sulfides to be deposited, mainly by replacement processes where favorable passageways and impounding structures existed.

This means that such skarn deposits as exist at the Sungei Besi are essentially hydrothermal de-

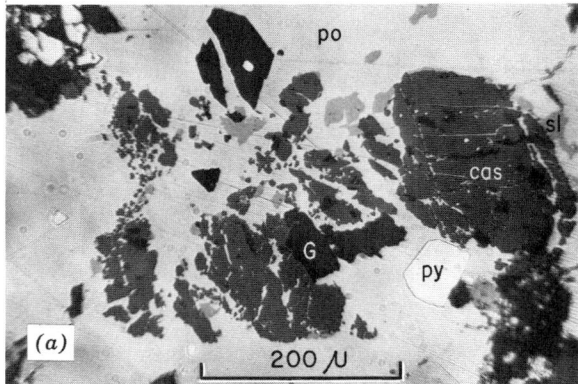

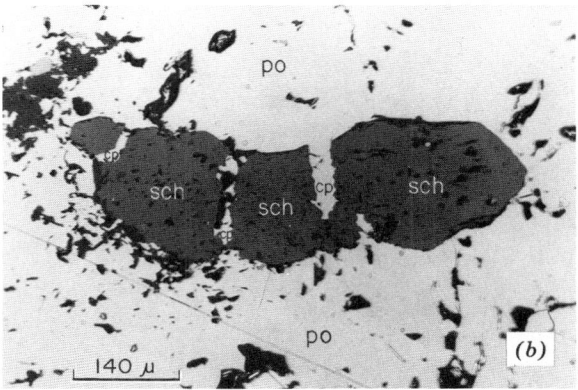

FIGURE 11.5 *Photomicrographs of polished sections of ore from Batu Tiga, Trengganu (in air): (a) Cassiterite (cas) showing pull-apart texture. Matrix essentially of pyrrhotite (po) in which sphalerite (sl) and chalcopyrite occur. Pyrite (py) less common. Near-black component is nonmetallic gangue (G). Photomicrograph K. F. G. Hosking. (b) Scheelite (sch) showing pull-apart texture; matrix essentially of pyrrhotite (po) and minor chalcopyrite (cp), which has healed the fractured scheelite. Near-black components are voids and nonmetallic gangue. Photomicrograph N. H. Chong.*

posits which have developed in a calc-silicate hornfels environment. In the writer's view, all the so-called pyrometasomatic deposits of economic importance are fundamentally the same as those at the Sungei Besi. That is, they are all hydrothermal deposits that happen to have developed within a calc-silicate hornfels host because the latter was favorable to deposition and because an adequate plumbing system could develop in it more readily than in the neighboring rock types. At Matchan Satahun (Trengganu) the mineralization picture is fundamentally much the same as at Bukit Besi. There, a series of hills has been mined for iron, manganese, and one, the South Mine, for tin. The exposed iron–manganese ore, and indeed, all that has been mined, is of the secondary type. However, diamond drilling in and around the South Mine has demonstrated that the primary ore is a magnetite–amphibolite rock. In the same hill, cassiterite occurs in a fracture zone.

At Sungei Panching, diamond drilling has established primary pyrometasomatic ore consisting of magnetite, garnet (probably andradite), and a little pyroxene bordered by skarn consisting of garnet, epidote, diopside, tremolite, hornblende, and wollastonite. In the vicinity of the magnetite, the ore contains some chalcopyrite, pyrrhotite, pyrite, and marcasite. The "major impurity" of the magnetite is tin, which may reach a concentration of 0.244% and which, according to Santokh Singh (*in* Bean 1969, p. 122) is "locked up" in the garnet.

Here is yet another example of a calc-silicate hornfels occurring near the granite to which it owes its existence, proving to be a preferred site for hydrothermal iron, tin, and sulfide mineralization.

A heavily mineralized vein, about 5 cm wide, and occurring in a belt of coarsely crystallized marble in the main pit of Bukit Besi, serves as a good reminder that magnetite–cassiterite–sulfide deposits can develop in an environment in which skarn minerals are barely developed. This vein probably developed by replacement along microfractures that afforded passage ways for incoming mineralizing solutions and were periodically rejuvenated or replaced by new ones during the phase of mineralization.

The vein contains minor amounts of earlier anthophyllite and sericite, but it is not known whether these developed in the vicinity of an early fracture or whether the fractures necessary for the formation of the vein occurred in this zone after the silicates had formed because it was more brittle than the adjacent marble. However, of the ore minerals present, magnetite and cassiterite were the first to be deposited, and these were followed by arsenopyrite, pyrrhotite, sphalerite, chalcopyrite, tetrahedrite, and much galena, in that order (Figure 11.6).

A curious feature is that virtually all the anthophyllite and most of the sericite seen in polished sections of the vein occur in the galena (Hosking et al. 1969).

The best known iron–tin skarn deposit of Johore is that at Pelepah Kanan. The hill on which the iron–tin deposits are found consists essentially of a mass of magnetite, largely altered to martite, which rests on a small roof pendant of quartzite and calc-silicate hornfels; the roof pendant, in turn, is underlain by biotite granite-porphyry.

The calc-silicate rocks which, according to Roe (1941), have been developed by the **metamorphism**

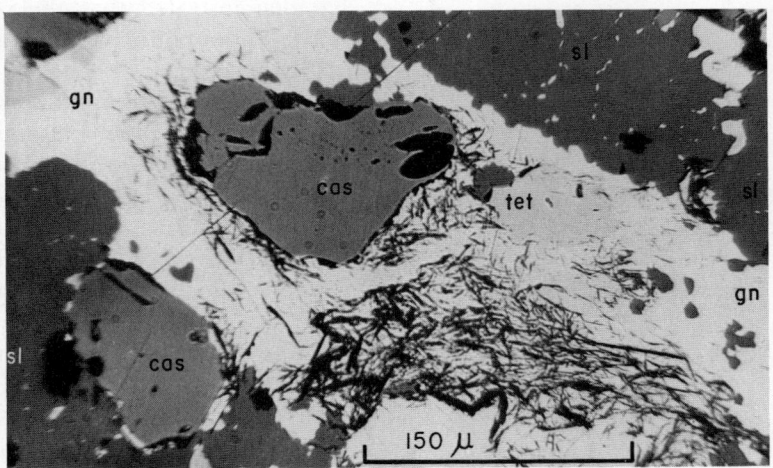

FIGURE 11.6 *Photomicrograph (K. F. G. Hosking) of a polished section (in air) of a galena-rich vein in the marble "reef" of the Main Pit, Bukit Besi, Trengganu. Subangular fragments of cassiterite (cas) and wisplike aggregates of anthophyllite (near black) in a matrix of galena (gn) and sphalerite (sl) in which there are exsolved bodies of pyrrhotite. A single fragment of tetrahedrite (tet) occurs in the galena. Both tetrahedrite and sphalerite have been replaced by galena, but the former shows this more clearly.*

of calcareous shales, siltstones, and mudstones, vary in their mineral content from one spot to another. Confirming Roe's view, Burton (1959) has recognized the following mineral assemblages within them: quartz-epidote-sericite, quartz-diopside-epidote-sphene, and quartz-diopside. In addition, Roe (1941) noted the presence of a garnetiferous hornfels, and Ganesan (1969) held that immediately overlying the granitic rock of the hill there is a "metasomatized porphyroblastic cordierite rock."

Mineralized veins intersect the calc-silicate rocks of the hill and pass into the iron oxide capping. They vary in width from a small fraction of an inch to about a foot, and although most of them are flat dipping, a few are vertical, and small stockworks also occur. These veins vary considerably in their mineral content, as is indicated in Table 11.3. It is important to point out that although some of them contain K-feldspar, these are not pegmatites, and two or possibly three generations of cassiterite are present in some of the members. "The low-attitude veins are invariably parallel to the well-developed banding in the hornfels, and Burton (1959) has shown that they are concordant with the form of a shallow syncline . . ." (Bean 1969, p. 93).

Several different views have been expressed concerning the genesis of this iron–tin deposit (see Bean 1969, pp. 94–95). The writer, however, is of the opinion that the calc-silicate hornfelses, having been developed by the addition of heat and "substance" from the invading and consolidating granitic magma, were fractured. These fractures, which subsequently became the planes along which the mineralized veins developed, were initially the passageways along which iron-rich mineralizing agents migrated from the granite. The latter replaced, virtually completely, some of the roof-rocks in order to provide the massive iron ore that now caps the hill. This marked replacement may well have happened below an impounding body which has since been completely removed by erosion. As the major iron ore body neared complete development, the ascending iron-rich agents were forced to move into the underlying hornfelses, where they locally deposited magnetite.

In addition, the last of the early iron-rich solutions deposited magnetite in the upper portions of some of the channelways. With time, the ascending solutions changed their character and deposited the cassiterite both in the magnetite-bearing hornfels (possibly in part because some of the channelways were restricted in the overlying magnetite body) and in the fissures themselves. The final episode was the unroofing of the magnetite body and its conversion to martite.

Here is yet another example of hydrothermal iron–tin replacement bodies that developed within

TABLE 11.3 *Paragenetic sequence of the minerals of Pelepah Kanan. After Ganesan (1969).*

Mineral	Time: Early → Late
Quartz	———————— Several generations ————————
K-feldspar	— (early)
Biotite–chlorite	—
Cassiterite (coarse)	—
Scheelite?	—
Löllingite	—
Cassiterite (acicular)	—
Ankerite	—
Siderite	—
Hematite	—
Fluorite	——— Several generations ———

Sequence of individual veins listed in the order from Early to Late.
- Blue tourmaline-cassiterite (in quartzite)
- Quartz–feldspar-biotite-fluorite-cassiterite
- Feldspar-chlorite-quartz-fluorite-cassiterite
- Quartz-fluorite-cassiterite-löllingite
- Quartz-fluorite-löllingite
- Quartz-fluorite-löllingite-secondary copper species (cuprite/native copper)

a calc-silicate sequence and with which are associated a swarm of stanniferous veins that would appear to be essentially filled fissures. The genesis of these veins, however, might have involved a great deal more of wall-rock replacement than is readily apparent.

The iron ore of this deposit has proved of little value because of the tin present in it. The weathered "skin" has been mined for its tin content; however, the recovery has been very poor. This is due in part to the crude milling methods adopted, in part to the fine character of much of the cassiterite in the hornfels, and in part to the fact that much of it is magnetic (this interesting property is discussed later).

Group II: Stanniferous Types

Unlike the skarns discussed earlier, the stanniferous skarns are essentially confined to the western mineralized belt of West Malaysia. However, an exception may be provided by the red garnet, green biotite, cassiterite-bearing rock of Bukit Tebak, Kamaman area, Trengganu, a sample of which was given to the writer by Mr. G. Y. L. Lee (Figure 11.7).

In view of the intensity of tin mineralization in the western belt and considering the frequency with which carbonate rocks have been invaded by granite, it might be thought that stanniferous pyrometasomatic deposits would be much in evidence and that their exploitation should have constituted a major feature of the mining industry. However, such is not the case. At present only one such deposit is being mined for its cassiterite content, and that is the pipelike deposit at the Lee Sin Nam Mine No. 2 at Ampang, near Ipoh (Perak). Economically speaking, the most important deposit of the type under review is the Beatrice Pipe (Perak), which has long since been abandoned.

Many other stanniferous pyrometasomatic deposits have been recorded in the western belt, and although some of them have been mined, most of them are small. Others have no economic potential because, apart from their limited size, the tin they contain is not present as cassiterite. Also, the cassiterite content may be too low, or the material may offer too many beneficiation problems for the "small miner."

The paucity of useful stanniferous or, indeed, other pyrometasomatic deposits in the western belt would seem to be primarily due to the purity of the carbonate rocks there. Generally, the sole change that they experienced as a result of invasion in the mesozone by essentially Triassic granitic magma was a general coarsening of their texture in the vicinity of the contact. Only the impure carbonate

FIGURE 11.7 *Photomicrograph (K. F G. Hosking) of a thin section (in plane-polarized light) of a quartz (Q)-plagioclase (plag)-green biotite (biot)-red garnet (gar) rock from Bukit Tebak, Kemaman, Trengganu, in which cassiterite (cas) and pyrite (py) are replacing the silicates.*

rocks contained the necessary ingredients for the development of sizable pyrometasomatic deposits.

Furthermore, even in the mesozone, as soon as the first calc- and/or magnesium-silicate components had developed in the deposits, components from the granite were able to invade the components along intercrystalline boundaries and, somewhat later, along fractures. During this period, and generally during the succeeding phase of hydrothermal mineralization, the hot, deeply buried neighboring pure marble behaved plastically, and forces capable of fracturing granitic and other rock types often simply caused these carbonate rocks to deform by movement along crystal glide planes (Hosking 1970a).

However, veinlike and pipelike pyrometasomatic deposits do occur in pure marble. The former are well developed, for example, in some of the open-cast placer tin mines in the vicinity of Kampong Pandan, near Kuala Lumpur; and the Beatrice and Sin Nam pipes in the Kinta Valley are good examples of the pipelike deposits. Unpublished and as yet incomplete studies by Ong Yeoh Han (private communication) indicate that the Sin Nam pipe has developed at the intersection of two fault systems striking at about 360° and 50°, respectively, and both appear to be of the wrench type. It may be that all the pipes in the Kinta Valley, both pyrometasomatic and early hydrothermal, occur at the intersections of two such wrench systems.

It may also be the case that veinlike skarn bodies in the marble generally develop along tension fractures developed between pairs of wrench faults. Some support for this view is to be seen at the Sungei Lah Section of Chenderiang Tin Ltd., Perak, where a tension fracture between a pair of wrench faults in the limestone has been converted to a vein consisting largely of scapolite. In addition, however, the vein contains a little cassiterite, a very small amount of scheelite, and disseminated metallic species represented largely by arsenopyrite, together with some pyrite. A little native bismuth and tetrahedrite may occur in the arsenopyrite, but the identities of these included species have not been confirmed.

Occasionally granitic veins and dykes, which have invaded the limestone, are enveloped by narrow skarn zones. Thus Scrivenor (1928) recorded that near the Beatrice Pipe there were cassiterite-bearing pegmatite dykes bordered by a thin band of wollastonite. Sometimes such igneous bodies were locally converted to skarnlike rocks, in part owing to reaction between them and the carbonate country rocks. Thus in 1928 at Chebok Mas (Kinta), a pegmatite vein was exposed in the limestone floor of the mine. This pegmatite was locally composed entirely of red garnet and green actinolite (Scrivenor 1928).

MODE OF OCCURRENCE OF TIN IN STANNIFEROUS SKARNS. At some localities during the early stages of skarn development, a certain amount of tin was introduced. This was incorporated in trace or minor amounts in some of the silicates. Thus the andradite in bluish-white calcite that occurs to the east of Kuah in Langkawi contains, according to Sweatman (personal communication), about 2% tin. Ganesan (1969, p. 4) recorded that

A skarn rock in the vicinity of Pelepah Kanan Mines and mainly composed of grossular/andradite, axinite, calcite and hedenbergite, was found to contain about 1.6 percent Sn in the grossular/andradite and 0.3 percent Sn in manganous axinite . . . which occurs in veins and is clearly seen to replace the garnet.

If tin is introduced at a somewhat later stage in the development of the calc-silicate suite, it is deposited as malayaite ($CaO.SnO_2.SiO_2$), which is isostructural with sphene. It may also be deposited as other silicates such as pabsite, arandesite, and stokesite, although none of these has yet been found in West Malaysia or elsewhere in the Tin Belt of Southeast Asia.

Alexander and Flinter (1965) first discovered malayaite in the Batang Padang district of Perak where, in association with cassiterite and varlamoffite (Figure 11.8), it was found in a collapsed pipelike deposit in the limestone floor of an opencast mine. Malayaite fluoresces a characteristic yellow under shortwave ultraviolet light, and specimens from the type area that were so examined by the writer consisted of crystals composed of alternating fluorescing and nonfluorescing zones (Figure 11.9) —that is, of malayaite and possibly stanniferous sphene. (An electron probe study of this material is now being carried out, but the results are not yet to hand.)

Alexander and Flinter (1965, p. 623) recorded that malayaite from this area of Perak is "biaxial negative with a large optic axial angle, and with a mean refractive index of about 1.78." It is of interest to note that in the Batang Padang ore, the varlamoffite, which is probably essentially a hydrous oxide of tin and usually of supergene origin from the oxidation of stannite, here is so intimately related to the cassiterite as to suggest that it has been formed by the hypogene replacement of the cassiterite (Santokh Singh and Bean 1968). The writer finds this view unacceptable, since he has specimens from Batang Padang containing numerous pseudomorphs of varlamoffite after malayaite. In addition, he has seen, in ore from Pinyok, Thailand, malayaite crystals in various stages of alteration to varlamoffite.

At the Hiap Huat Mine, Rawang, Selangor (Hosking et al. 1970), malayaite was found rimming sphene and associated with vesuvianite, diopside, epidote, sphene, alkali feldspar, chlorite, fluorite, calcite, loellingite, and quartz in a tightly folded band of skarn adjacent to a microadamellite dyke 6m thick. From the textural characteristics of the material as seen in thin section, it was concluded that malayaite was one of the latest of the silicates to be deposited and was due to metasomatic processes induced by the invasion of the "protoskarn" by hot stanniferous aqueous agents.

At the Kanching Tin Mine, Templer Park, Selangor (Loganathan 1970, p. 34) malayaite, scheelite, and a little pyrite occur in a band of skarn that consists essentially of vesuvianite, garnet, diopside, wollastonite, hedenbergite, and minor quartz.

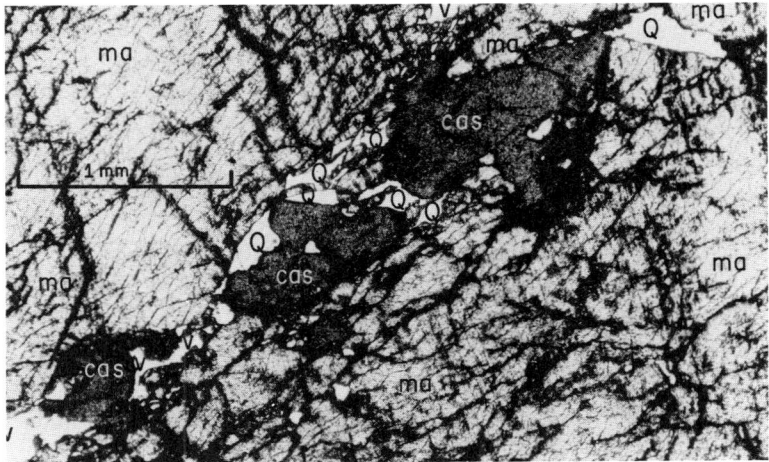

FIGURE 11.8 *Photomicrograph (K. F. G. Hosking) of a thin section (in plane-polarized light) of an aggregate of malayaite (ma) intersected by a vein of cassiterite (cas) and quartz (Q), from Chenderiang, Perak.*

FIGURE 11.9 *Crystal aggregate about 2 in. long from Chenderiang, Perak, under shortwave ultraviolet light. It is composed of zones of fluorescing malayaite (near white) separated by zones of a nonfluorescent mineral, perhaps slightly stanniferous sphene. Photograph K. F. G. Hosking.*

Adjacent to the fault zone are curiously folded and boudinaged skarn bands, of maximum width 4 cm, whose mineralogy is similar to that of the skarn in the fault zone.

At Sungei Gow, Pahang, in a long-abandoned mine, malayaite is locally degraded to cassiterite, calcite, and quartz by hydrothermal agents (Figure 11.10a). Hydrothermal agents also deposited cassiterite directly in the rock (Figure 11.10b), as well as pyrrhotite (which locally replaces the calcite of the degraded malayaite crystals), sphalerite, and and chalcopyrite. In other stanniferous skarns of West Malaysia, there is no evidence that tin was deposited until the hydrothermal stage. Thus in 1926 a small deposit was discovered in the limetone at Ayer Chulit which consisted of calcite, tremolite, cassiterite, pyrite, arsenopyrite, and sphalerite (Scrivenor 1928).

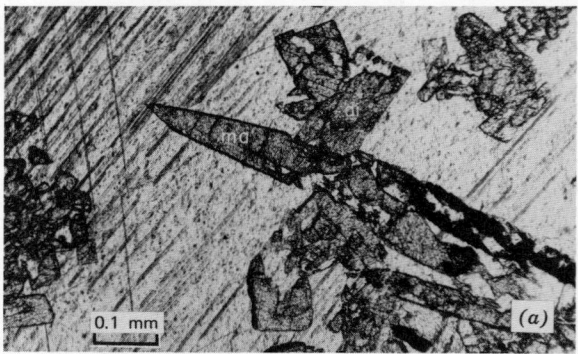

FIGURE 11.10 *Photomicrographs (K. F G. Hosking) of thin sections (in plane-polarized light) of skarn ore from Sungei Gow, Pahang: (a) Elongate crystals of malayaite (ma) partly replaced by calcite and pyrrhotite (black), in a calcite matrix; diopside (di) is associated with the malayaite; (b) cassiterite crystal, about 2 mm wide, showing asymmetric form owing to its development from mineralizing agents that moved in the direction of the arrow. This is also indicated by the thicker deposit of chalcedony on the stoss side of the cassiterite baffle.*

At the Melor Syndicate Mine, Sungei Way, Selangor (Yeow 1969) skarn bands, about 15 cm thick, some of them boudinaged, are intercalated with the marble. These bands, although consisting essentially of vesuvianite and garnet, also contain some epidote, zoisite, calcite, chlorite, quartz, scheelite, and malayaite. Sufides and cassiterite, which are abundant in hydrothermal veins elsewhere in the mine, are completely lacking in the skarns.

Malayaite and scheelite occur at Pudu Ulu Mine I, Selangor, in a skarn fault zone about 1.2 m wide, whose general mineralogic characteristics are similar to those of the Melor Syndicate Mine skarns (Yew Cheng Cheong personal communication).

At Ginting Tua, according to Scrivenor (1928), a vein about 2.5 m wide and occurring in calcareous shales consisted of arsenopyrite, sphalerite, cassiterite, fluorite, pale yellow opaque garnets, iron pyroxene, quartz and tremolite.

There is, of course, the possibility that malayaite may have occurred in the last two mentioned deposits and that Scrivenor did not recognize it. The species was not known until much after his time, and he did not spot that a hitherto unknown mineral (malayaite) occurred at Sungei Gow, even though he examined the material in some detail with the equipment and methods available to him.

However, it is fairly certain that malayaite or other silicates in which tin is an essential component are not always present in the stanniferous skarns. Thus examination of the skarn ore from the Telok Kruen "pipe" has indicated that it consists of cassiterite, arsenopyrite, native bismuth, and scheelite (deposited in the order given) in a gangue of calcite in which there is a small amount of tremolite (Hosking and Yeap, unpublished studies), (Figure 11.11).

In the Beatrice Pipe, the most important stanniferous skarn deposit ever discovered in West Malaysia, the tin occurred largely as cassiterite, but a little stannite was present. According to Willbourn (1932), this pipe consisted essentially of tremolite, whereas the "ore minerals" present, in addition to those noted previously, were arsenopyrite, chalcopyrite, and minor bornite and pyrrhotite. Recently Hosking and Yeap (unpublished studies) observed that the ore also contains a little molybdenite. The nonmetallic gangue minerals, other than tremolite, were fluorite and fluoborite, $Mg_3(BO_3)(F, OH)_3$. Willbourn also recorded that the order of formation of the major components was tremolite, arsenopyrite, cassiterite, and chalcopyrite.

Although this, and all the other cassiterite-bearing skarns noted earlier, can be reasonably classified as pyrometasomatic because of the nature of the silicates present in them, it seems likely that the cassiterite and all the metallic species were deposited during a hydrothermal stage. Indeed, this view was expressed in connection with the genesis of the Group I type of skarn deposit. In any event, the Beatrice Pipe (Figure 11.12), which was exposed on the removal of a placer deposit, extended for about 150 m (500 ft) horizontally and then for a further 90 m (300 ft) vertically before it was faulted out by granite. Its economic importance is indicated by its yield—more than 9000 long tons of cassiterite concentrates. Whether a significant portion of this pipe still awaits discovery is an intriguing thought. All too little is known about the likely vertical extent of either this or other primary tin deposits in the western belt.

As noted earlier, the Sin Nam Lee Pipe at Ampang, Perak, is the only skarn deposit in the country of the type under review that is presently being mined. The country rock in which it occurs is essentially dolomite. In the nonmetallic gangue of dolomite with some tremolite, occur the following ore minerals, deposited in the order given: cassiterite, arsenopyrite, pyrite, stannite, chalcopyrite, jamesonite, and a little tennantite, which occurs

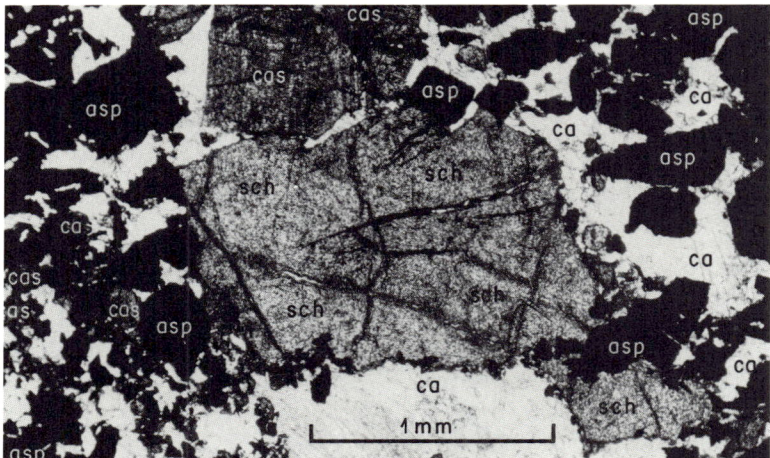

FIGURE 11.11 *Photomicrograph (K. F. G. Hosking) of a thin section (in plane-polarized light) of ore from the Telok Kruen pipe, Perak. Cassiterite (cas), scheelite (sch), and arsenopyrite (asp) occur in a calcite (ca) gangue. The opaque material veining the scheelite is probably native bismuth.*

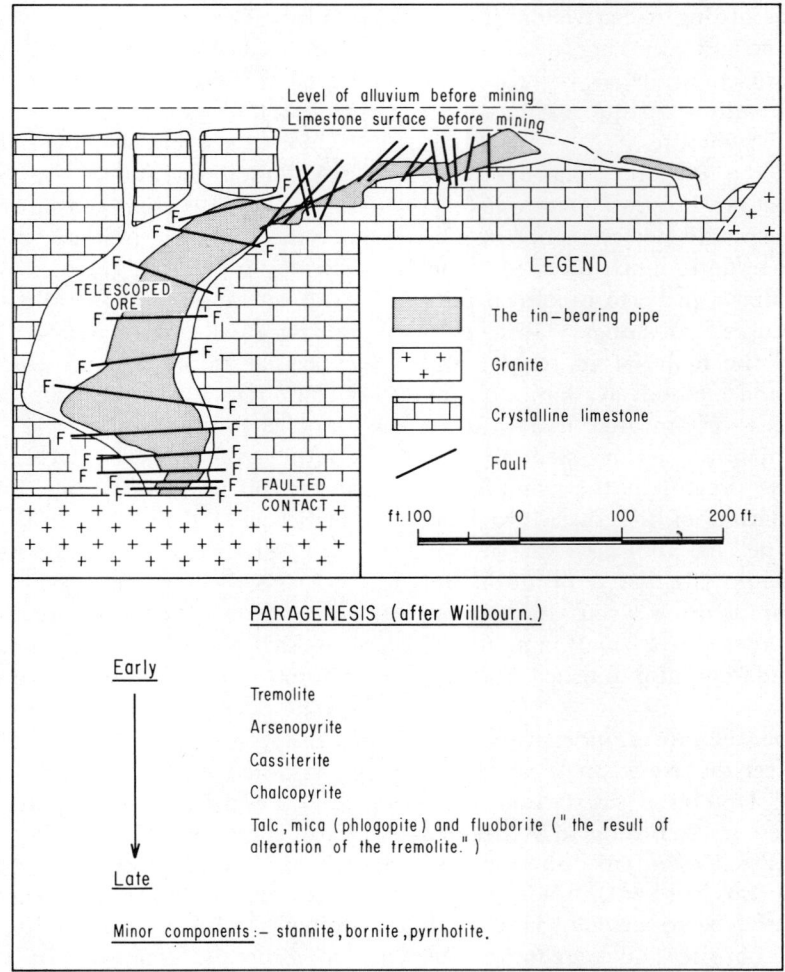

FIGURE 11.12 *Section of the Beatrice Mine, Selibin, Perak, and the paragenesis of the ore. After Willbourn (1926–1927).*

along the cleavage planes of the antimony-bearing species (Yeap and Hosking unpublished studies).

In this, as in all the mineralogically similar skarns and hydrothermal deposits, replacements of one species by another is much in evidence. Both the cassiterite and the sulfides of the Sin Nam Pipe replace both the dolomite and the tremolite, whereas stannite replaces cassiterite.

Group III: Other Types

Although skarns other than the types noted previously do contain minerals of economic interest, none has yet been discovered which is sufficiently large or rich to be exploited.

Hutchison, in Chapter 9, mentions the presence of fluorite-bearing skarns; and, hardly a skarn, particularly on the western belt, and however simple its mineralogy, is completely devoid of scheelite. In Sungei Besi Mine, Selangor, for example, a small vein in the limestone consisting of vesuvianite and red garnet was heavily charged with scheelite (Chan 1970). Some of the Langkawi skarns are scheelite bearing and are best described, from an economic point of view, as tungsteniferous.

Some of the scheelite in the skarns may have been deposited during the late calc-silicate stage (i.e., at about the same time as the malayaite), but when it occurs in considerable amount, it is felt, that most of it was deposited during a later hydrothermal stage and at about the time when cassiterite was being, or would be, deposited. According to Yeap (private communication), in the skarns adjacent to an adamellite dyke at Teh Wan Seng No. 3 Mine, Salak South, Selangor, malayaite has been sometimes partly replaced by scheelite.

Locally, in the Sungei Aring area of south Kelantan a skarn overlies a "ridge" of quartz porphyry. The calc-silicate hornfels, which consists of tremolite, garnet, hedenbergite, calcite, and a little quartz, also contains the following sulfides, which were deposited in the order given: pyrite, sphalerite, chalcopyrite, and galena. The sphalerite is heavily loaded with exsolution bodies of chalcopyrite, and locally there is evidence of slight replacement of both sphalerite and chalcopyrite by galena.

HYDROTHERMAL DEPOSITS

The overwhelming majority of primary mineral deposits of economic importance in West Malaysia are hydrothermal. The expression "of economic importance" is used to cover those deposits which have been exploited, are being exploited, or may be exploited for one or more of the minerals they contain, or which, as a result of weathering processes, have contributed largely to the development of valuable placers.

These primary deposits have developed as a result of fluids ascending from an unknown source along fissures, commonly faults, and depositing minerals in open spaces. More commonly, however, primary deposits develop by replacement of the wall-rock and, in the case of late components, by replacement of earlier deposited ones. Mineralized veins, lodes, and massive replacements are the final result.

The ascending mineralizing fluids must have been very tenuous, since they were capable of migrating through knife-edge fractures, as during the development of the streaky-bacon tin ore (Figure 11.13) of the Kledang Range (Jones 1925). They also changed in composition with time, an earlier fluid depositing, say, cassiterite and a later one, migrating along the same channelways, perhaps depositing galena.

Individual veins, lodes, and replacements of West Malaysia, unlike those of the not dissimilar metallogenetic province of Southwestern England, show virtually no primary zoning. The nearest approach to it is to be seen in some of the lodes of Pahang Consolidated Mine where, possibly, the higher horizons of the essentially tin lodes were somewhat richer in copper than the lower ones. Mineralized veins are sometimes characterized by a varying mineral assemblage when traced along the strike. This is true of some of the multimineralic veins in the limestone bedrock of the mines near Kampong Pandan, Selangor, but such veins can hardly be said to be zoned.

Locally, however, there is some evidence of regional zoning, and this is best displayed in the Kinta Valley, where tungsten occurrences are very close to the granite contacts whereas those of lead–zinc and iron lie in the metasedimentary rocks at some distance from the contact (Figure 11.14). It may, of course, be eventually demonstrated that the Kinta Valley metal distribution pattern is the product of several distinct phases of mineralization, each separated from the others by considerable time.

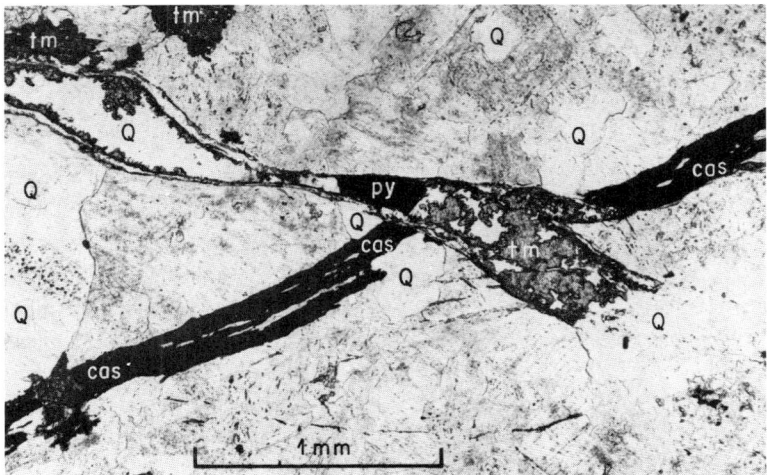

FIGURE 11.13 *Photomicrograph (K. F. G. Hosking) of a thin section (in plane-polarized light) of "streaky-bacon" ore from the Kledang Range, Perak. A veinlet of cassiterite (cas) is intersected and faulted by one of quartz (Q), tourmaline (tm) and pyrite (py).*

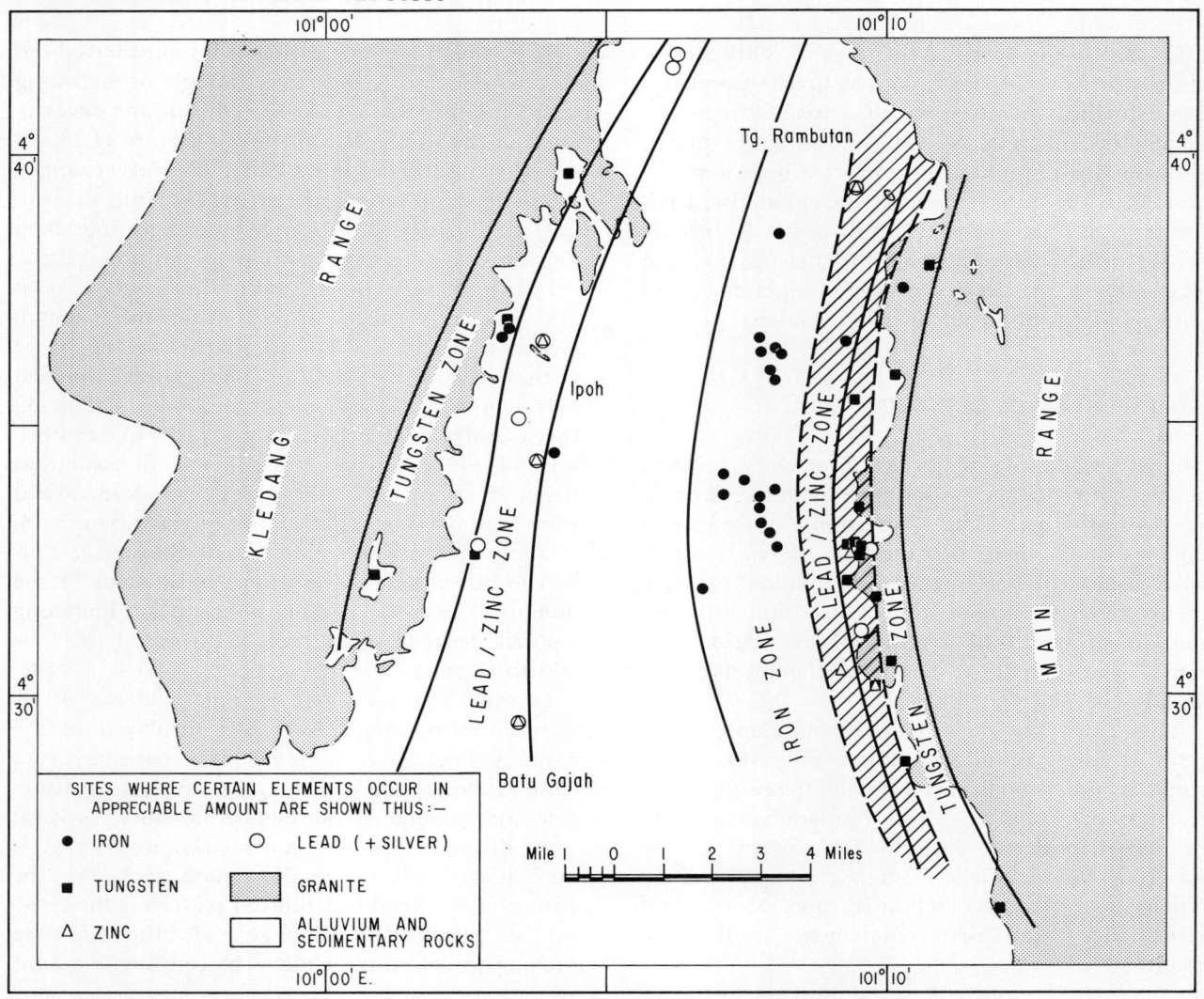

FIGURE 11.14 *Metal zones in the Kinta Valley.*

Although West Malaysia cannot offer a single good example of a vein or lode that displays primary zoning, it can provide scores of beautiful examples of strongly telescoped ones. Some of the most spectacular examples are to be found at the Tekka Mines, Perak and in the mines of the Salak South, Ampang, and Kampong Pandan areas of Selangor. These contain an assemblage of what are usually regarded as high and low temperature minerals; these are intimately spatially related and provide a wide array of most interesting textures. Evidence of replacement is common.

It is pertinent that several of the pyrometasomatic deposits are also markedly telescoped; typical examples are the Telok Kruen and Sin Nam Pipes and some of the cassiterite–sufide-bearing pegmatites noted earlier. This may be regarded as a further indicator that the earlier deposits owe their final character to the deposition of minerals by hydrothermal agents.

The Tekka deposits just mentioned have been termed xenothermal by Riley (1967). This term is best not used unless one believes that the deposits were made near the surface (Park and MacDiarmid 1964, p. 346) and, as will be clear later, this is still a very controversial question. However, the writer completely agrees with Riley's classification. The multimineralic Manson Lode (Kelantan), which is about 11 km (7 miles) from the nearest outcropping granite, and in which cassiterite and cinnabar occur together, can also be classified as xenothermal (Hosking 1969a).

There is good reason for believing that impounding structures commonly played a major role in determining the site of deposition of the hydrothermal deposits.

It is common to find a given mineral deposit limited to a given lithologic unit. Examples noted earlier serve to indicate that mineralization may be restricted to a pegmatite, a band of skarn, or an aplitic dyke. Often mineral deposits in the granite cease at a limestone–granite contact. Thus at the Sungei Besi Mines a carbona-like body, which originally consisted essentially of cassiterite, pyrite, and quartz, developed more or less immediately beneath a limestone roof that has since been removed. At the Melor Syndicate Mines, cassiterite/tourmaline-quartz veins in the granite terminate at the granite–limestone contact. In the Salak South area of Selangor, the stanniferous ore bodies are restricted to the adamellite; they do not persist into the limestone.

That lodes may terminate at the contact between granite and rocks other than limestone is shown by Jones's remark (1925c, p. 204) concerning the behavior of the lodes containing cassiterite and wolframite in the famous Titi Mine (now abandoned) on the eastern side of the Main Range in Negri Sembilan:

The granite-schist contact passes through the property, and in places dykes of quartz-porphyry, very similar to the Cornish elvan, have been encountered in the underground workings. In one part the lodes terminated abruptly against the quartz-porphyry, whereas in other places lodes which have proved productive in the granite have not continued into the adjacent schist, or their tin content has decreased considerably, rendering it unprofitable to follow them into the schists.

At Kramat Pulai, the metasomatic scheelite-fluorite body occurs in limestone, close to the granite, beneath two plunging anticlines of schist (Figure 11.15); there, however, the impounding schist did suffer slight fracture, and the fractures were healed by scheelite which occurred as leak-away "indicator" veins.

Probably it was common for pairs of wrench faults to limit the migration of ore-forming agents

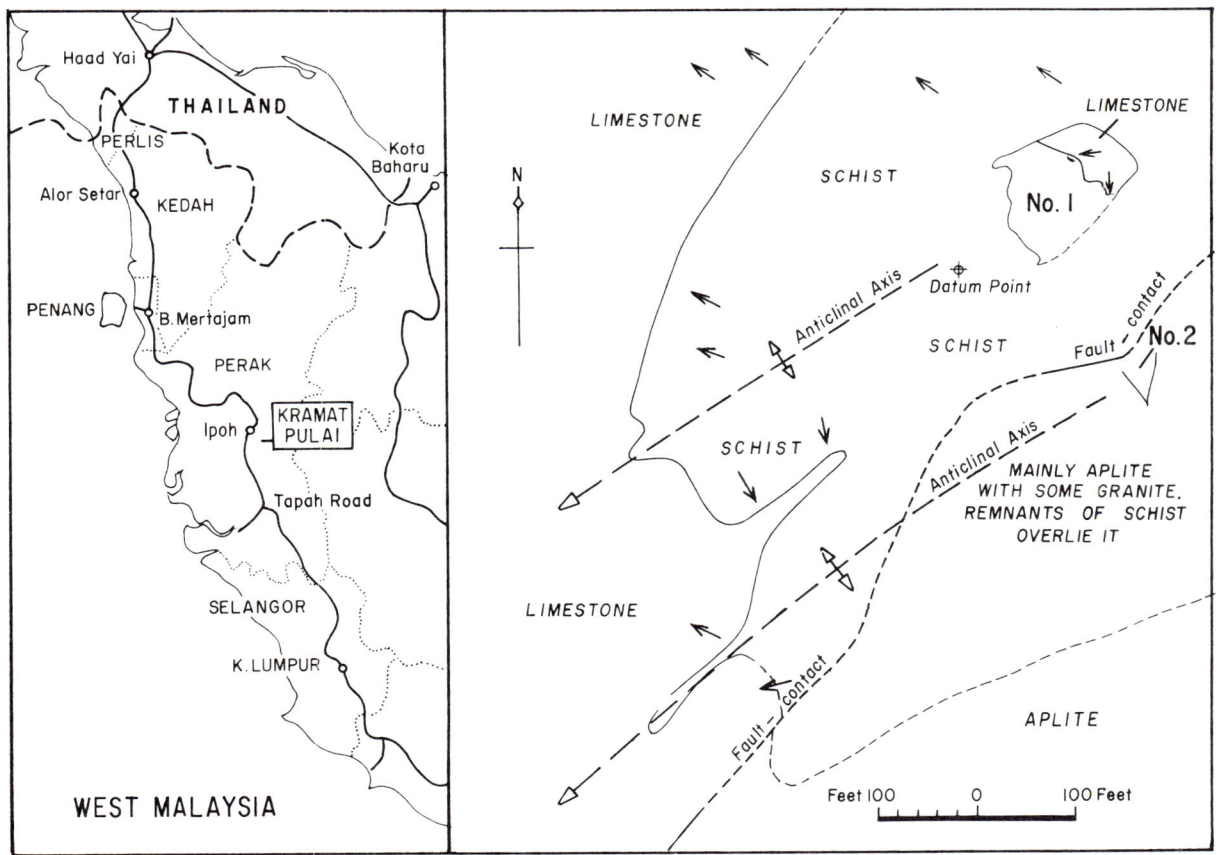

Figure 11.15 *Plan showing the relationship of the structure and ore bodies 1 and 2 at Kramat Pulai, Perak. After Willbourn and Ingham (1933, p. 463).*

to tension fractures which they generated. It seems likely that this was an important factor in determining the disposition of lodes in the Pahang Consolidated Mine area, and the scapolite vein at Chenderiang, noted earlier, is an established example.

The writer believes that the apparently fault-disorganized Gakak No. 3 Lode at Pahang Consolidated (Figure 11.16), may well owe its presence to mineralization along tension fractures formed between a number of pairs of wrench faults. The data available suggest that the lode width in the various segments tends to build up against the faults. This most curious phenomenon is impossible to explain if it is assumed that the Timor Fault zone did not develop until after the Pahang Consolidated lodes had been established. This is not to deny that further movement along the various components of this wrench fault system probably did occur after the lodes had been formed.

Often there is no evidence that an impounding body played a rôle in determining the site of development of a given ore body and/or in controlling its dimensions. It has to be conceded, for example, that the limits of a given ore deposit may have been controlled solely by temperature!

Let us return to the question of the relation between primary mineralization and faulting in West Malaysia. Ever since geologists have worked in the country, they have been aware that the development of lodes and veins was associated with passageways along which the mineralizing agents ascended. Commonly these passage ways were faults and, in the case of the large Sungei Lembing lodes, it was particularly clear that they were essentially mineralized faults that had been repeatedly reopened. (The relation between faults and the large, irregularly shaped iron, scheelite, and cassiterite replacement bodies has never been quite as obvious as in the case of the lodes.) It was also well known long ago that late faults could disrupt primary ore deposits, and commonly the relative ages of the various fault systems, mineralized and otherwise, in a given mine, were established.

Although attempts have been made to explain the lode distribution pattern in a given mining field, and that made by Fitch (1952) to unravel the Pahang Consolidated one is an outstanding example, the efforts were generally unsuccessful. The reasons for this are not far to seek. Until recently, hardly any detailed structural studies of faults in mining fields or elsewhere had been carried out in West Malaysia. However, during the past few years members of the Geological Survey and others have commenced to remedy this situation. Stauffer (1968) and Shu (1969), for example, have already contributed significantly to an understanding of the wrench faults and other faults that may have figured significantly in the development of the mineral distribution in Selangor. The possibility of large wrench faults occurring in the country and the complexity of some of the regional fault patterns are discussed elsewhere in this volume. The continued study of faulting on both a regional and local basis is certain to increase our understanding of why the primary deposits are where they are. It follows that such studies will assist the exploration for primary deposits (and secondary deposits derived from them) on all scales.

In West Malaysia, mineral deposits are commonly spatially associated with Carboniferous, with Triassic and, apparently only to some slight extent, with Tertiary granites. Locally, they are spatially (and so possibly genetically) related to volcanic rocks ranging from acid to basic in composition and of a variety of ages. It may, therefore, be argued that little is to be gained by providing a general paragenesis of the ore deposits of the country. However, such a paragenesis is useful in that it establishes the order of deposition generally encountered. Of course, a given deposit may start its development with any particular mineral in the sequence and finish it with any one above it. The general paragenesis (Table 11.4) is essentially the same as those which have been established in many other broadly similar metallogenetic provinces of the world (Hosking 1964).

Table 11.4 also indirectly indicates the nature of the wall-rock alteration that is encountered. Broadly speaking greisenization, sericitization, and tourmalinization have commonly affected both granitic and metasediment wall-rocks adjacent to deposits that contain significant amounts of the early species, such as cassiterite and wolframite. Even the limestone wall-rocks of veins containing such early species may be sericitized, tourmalinized, and chloritized. Chloritization is somewhat less common than the other phenomena noted, although it is far from rare and is the dominant form of alteration of the metasediments of the Pahang Consolidated Tin Mine. Widespread evidence of silicification, alone or in conjunction with other forms of wall-rock alteration, may be seen in the vicinity of almost all the early primary deposits except those developed in limestone. Late deposits in limestone may be accompanied by little or no wall-rock alteration.

Marked kaolinization of the granite often accompanies tin and tungsten deposits and is clearly

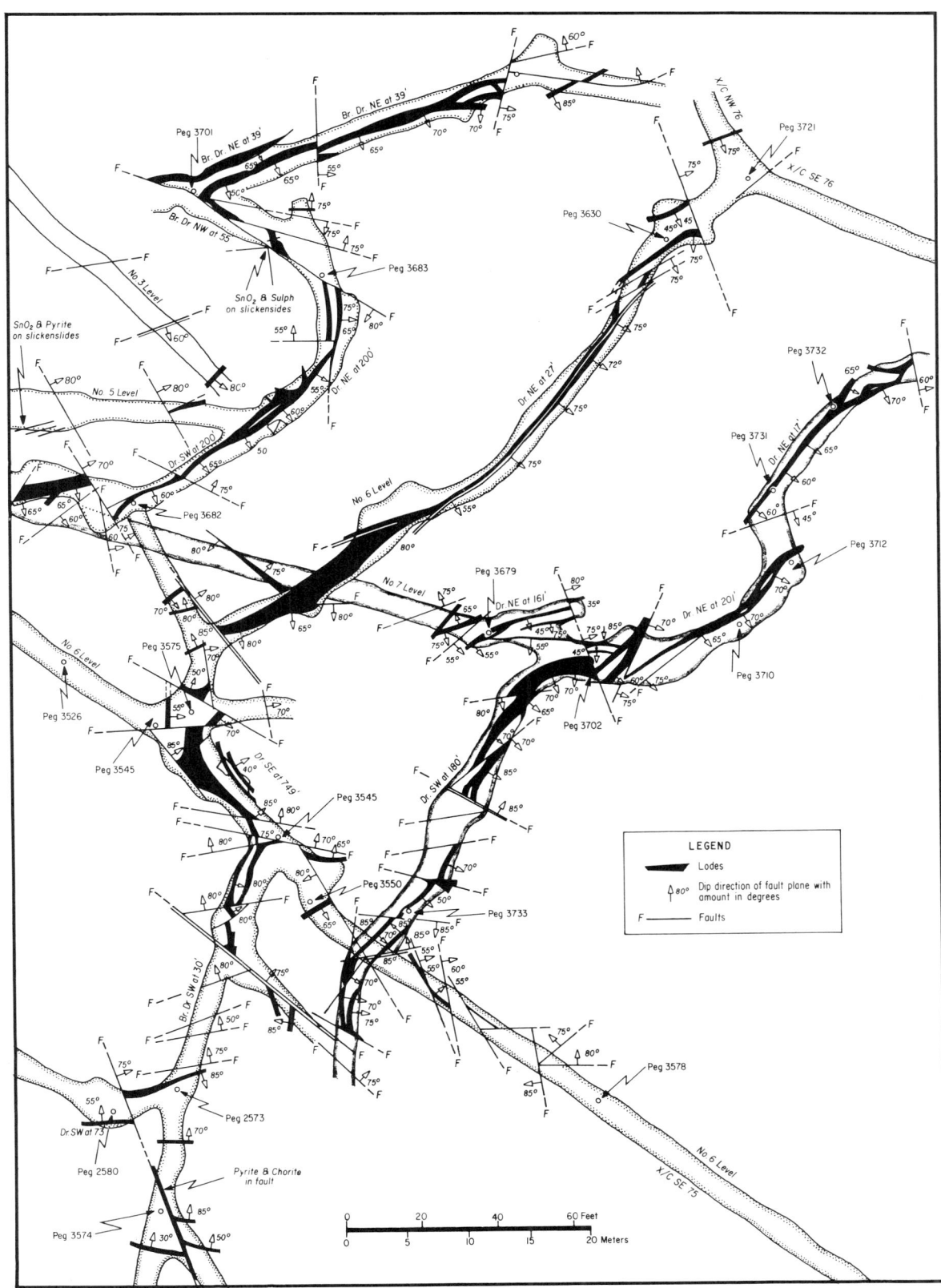

FIGURE 11.16 *Plan of part of the No. 3 lode, Gakak–Gakak Creek Mines, Sungei Lembing, demonstrating the structural complexity of the area. (Published by kind permission of the General Manager, Pahang Consolidated Co. Ltd.)*

TABLE 11.4 *General paragenesis of the mineral deposits of West Malaysia*

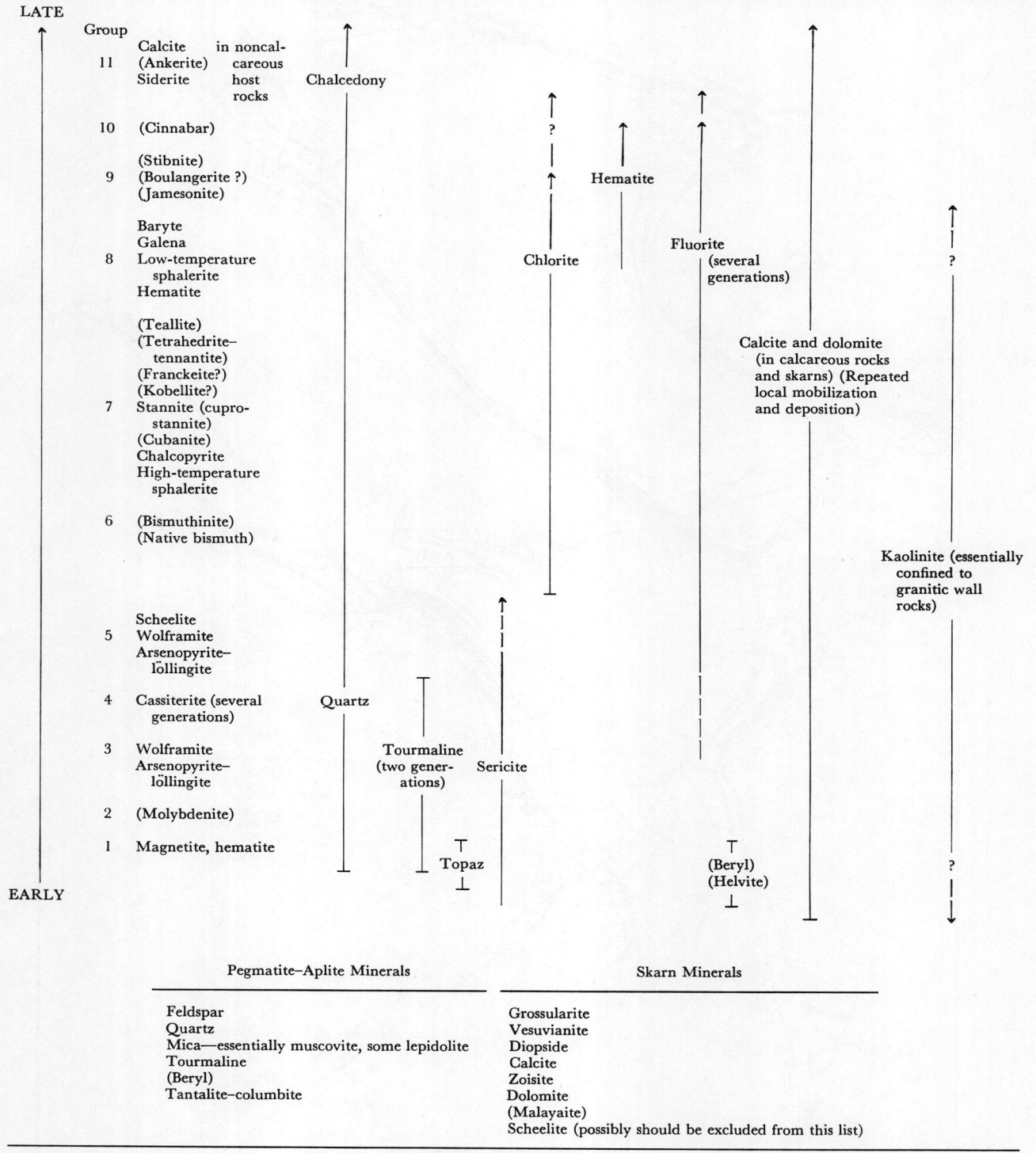

Notes:
1. Some of the vary rare Malaysian species have been excluded fron the table.
2. Species that are of comparatively rare occurrence in West Malaysia are cited in parentheses.
3. A question mark following a species inside the parentheses indicates that the species' identity is still in some doubt.
4. The species within any given group are believed to have been deposited at *about* the same stage during the development of every "ore body" in which they occur.

visible, for example, at Sungei Besi Mine, Selangor. Although some of the granite has been kaolinized by supergene processes, there is little doubt that in areas of "early" mineralization, such as at Sungei Besi, some of the kaolinization is due to hypogene processes.

The problem of differentiating between supergene- and hypogene-kaolinized granite has exercised the minds of geologists in West Malaysia for years (Jones 1915a). Usually visual inspection of a kaolinized mass does not provide a satisfactory answer concerning the genesis of the clay. An easy means of determining this would be welcome to the economic geologist, since hypogene-developed clay is likely to have much greater extension in depth than that developed by supergene processes. Furthermore, hypogene clay existing below the deepest zone of oxidation is likely to be much less contaminated by iron oxides than its supergene counterpart.

At the moment, the only way of assessing the clay potential of a kaolinized granite is to drill it. The results of such an exercise, combined with appropriate laboratory studies of the products of the drilling, will permit not only the indicated tonnage of clay to be assessed but, in addition, will establish the properties of the clay.

In the search for economically important deposits of kaolinized granite, it is possible that resistivity methods, such have been used most successfully in southwestern England, might prove to be useful aids. In Cornwall, open-cast tin mines have become important clay producers: perhaps the same transformation could take place in West Malaysia.

Although some of the primary hydrothermal mineral deposits of West Malaysia can be classified as hypo-, meso-, epi-, or xenothermal deposits, the majority cannot. Most of them contain both early and late species, or what are often regarded (perhaps erroneously) as high-temperature and low-temperature ones. Yet many do not give clear indication of having developed in a near-surface environment, and these cannot be termed xenothermal. Therefore, it is *generally* best not to attempt a genetic subdivision of the hydrothermal deposits. In view of this, the writer proposes to subdivide these deposits on the basis of their metal or mineral content and then to describe in varying details selected members of each subgroup in order to demonstrate variations and to provide illustrative examples of the broad statements made previously.

The subdivisions proposed are as follows:

1. Tin and/or tungsten deposits.
2. Iron deposits.
3. Gold deposits.
4. Base-metal sufide deposits.
5. Barite deposits.

Tin and Tungsten Deposits

In West Malaysia the hydrothermal tin and tungsten deposits are not obviously restricted to any particular rock type. They occur in granite, limestone, clastic sediments and metasediments, and possibly also in some of the volcanic rocks into which, or near which, the granite was emplaced.

Generally the tin and tungsten deposits occur along the flanks of the major granite outcrops. Mineralization along the watershed of the Main Range and Kledang Range is very slight. There is evidence that mineralization may develop preferentially on the flanks of granitic cusps and in the metasediments overlying them. Thus at Tronoh there was an exceedingly rich colluvial cassiterite deposit derived from the disintegration of veins in the metasediments forming thin cover over a granitic cusp (Jones 1925c). Subsequent drilling in 1956 to the north and south of the hill revealed no primary tin mineralization of any consequence (Hosking private report).

Of the granitic islands off Malacca, only Pulau Besar supported an open-cast tin mine. This island is a pronounced granite cusp and the mine was situated immediately over a granite–schist contact that was mineralized and, as a result of weathering, provided minable detrital cassiterite. The wolframite-bearing lodes of Bukit Lentor (Trengganu) and of the Tunku Makota's mine, about 8 km (5 miles) west of Gambang, may be examples of this; other examples may be provided by the cassiterite-bearing vein swarms of Rahman Hydraulic Mine, Klian Intan, Perak. The stanniferous pipes of the Kinta Valley may also overlie granite cusps. However, there is only one example known of a granitic cusp containing a *swarm* of tin- and/or tungsten-bearing lodes or veins such as are commonplace in Burma and not infrequent in Thailand. This occupies a prominent hill near Tapah, Perak, known as Changkat Rembian.

Perhaps some of the West Malaysian deposits of the type under review have developed in the apices of cusps, but erosion has obliterated the evidence. Perhaps also, many of the cusps with which min-

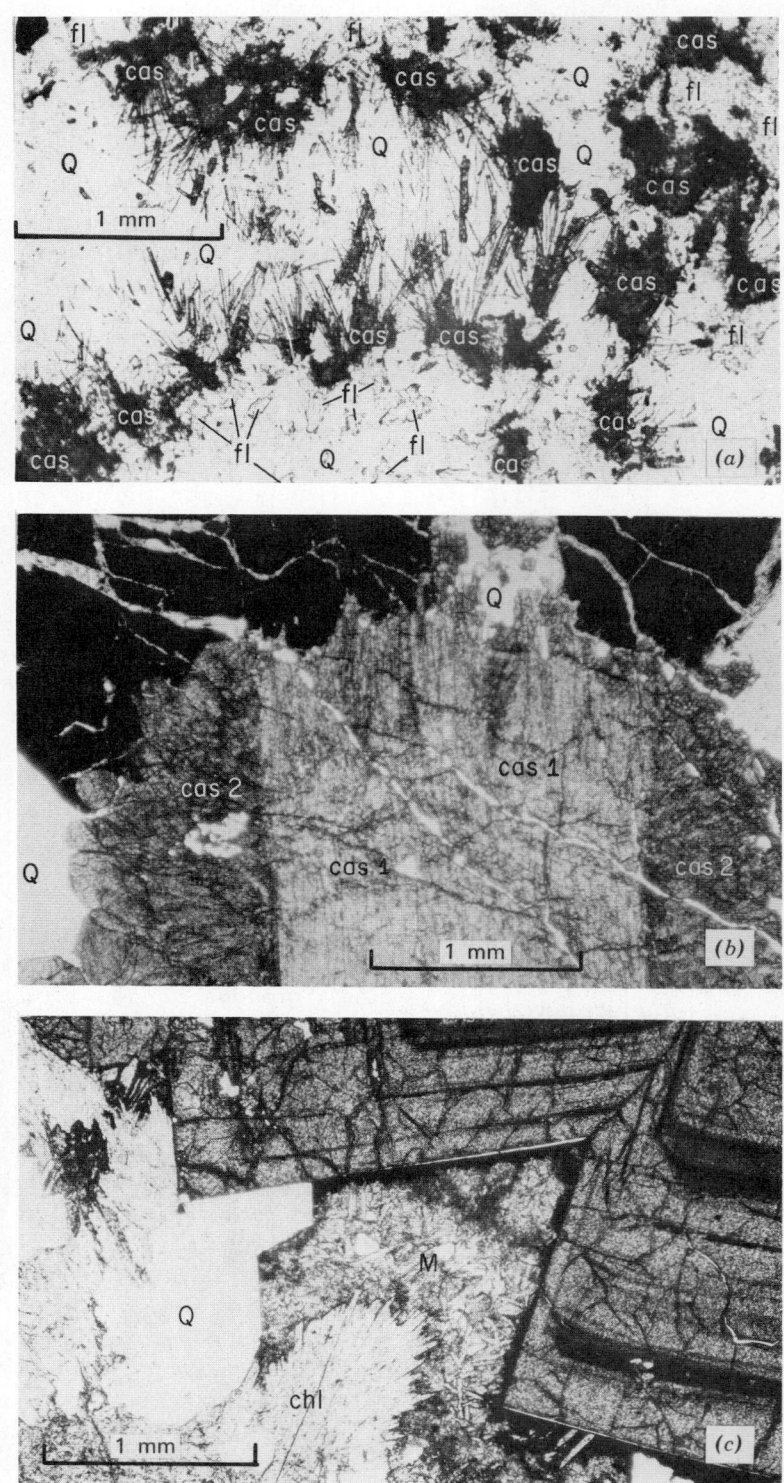

FIGURE 11.17 *Photomicrographs (K. F. G. Hosking) of thin sections (in plane-polarized light) of tin ore: (a) Vein from Hin Fatt No 2 Mine, Salak South, Selangor, showing earlier generation cassiterite (cas) haloed by acicular crystals of cassiterite of a later generation; matrix of quartz (Q) and fluorite (fl). (b) Ore from Thye Sang Mine, Salak South, Selangor, in which a crystal of cassiterite (cas 1) is fringed by prismatic cassiterite (cas 2) of a*

eralization is associated are not obvious because they are composed of "late-phase" granite and have been emplaced within "early-phase" granite. Thus Roe (1951, p. 82), when discussing the mineralization of the Ulu Kul Tin Mine, Selangor, stated that

> The country rock at this mine is mostly grey sparsely porphyritic biotite granite of medium grain, with occasional crystals of tourmaline. Large intrusions of fine-grained granite and granite-porphyry also occur, together with aplite, and pegmatite, dykes and sills. *Cassiterite is carried by schorl rock, and greisen, stringers which are commonest and best developed in the more severely fractured fine-grained granite* (the writer's italics).

TIN AND TUNGSTEN MINERALS. It is convenient at this stage to discuss the tin and tungsten minerals of West Malaysia. Because the writer believes that all the primary tin and tungsten minerals, other than possibly malayaite, are of hydrothermal origin, he feels that it is not necessary to apologize for referring, for example, to cassiterite occurring in pegmatite–aplite and pyrometasomatic deposits.

The tin minerals known with certainty to occur in the primary deposits are: malayaite (which is not discussed further), cassiterite, stannite, cuprostannite, and teallite; possibly franckeite also occurs there. Primary tungsten species are represented by wolframite and scheelite: ferberite occurs at Kramat Pulai, but for no very obvious reasons, it has been regarded by Ingham and Bradford (1960) as secondary. Locally, oxidation of the tungsten species has resulted in the development of tungstic ochre, possibly a little scheelite and, at Kramat Pulai, yttrotungstite and stolzite (Ingham and Bradford 1960). Some of the more interesting characteristics of the primary species noted previously, and of varlamoffite, are recorded below.

Cassiterite. Cassiterite (SnO_2) is the only tin species that is recoverd and sold in West Malaysia.

It shows great variation in habit, although the squat bipyramidal type that occurs in many of the great stanniferous pegmatites of the world [e.g., Bikita (Rhodesia), Kamativi (Zambia), Wodgina (Australia), and also in the Ronphibun and Pak Pak Song areas of peninsular Thailand (Aranyakanan personal communication)] has only been found in the Bakri and Gunong Jerai placers, which are though to be derived partly from pegmatites. Thus, although pegmatitic cassiterite probably exists in West Malaysia, it seems to be of very limited occurrence.

Two generations of cassiterite are not uncommon in the West Malaysian deposits. In the majority of cases, the earlier generation consists of crystals that may be twinned and may occur as aggregates or as isolated crystals, or portions thereof, characterized by short prisms and commonly possessing (111) bipyramidal terminations. The later generation, on the other hand, generally consists of acicular crystals, often but not invariably with acute pyramidal terminations. These later crystals commonly radiate from nuclei consisting of earlier generation cassiterite (e.g., at the Hin Fatt No. 2 Mine in Selangor—Figure 11.17a); or they form a palisade along the faces of the earlier cassiterite (e.g., at the Thye Sang Mine in Selangor—Figure 11.17b).

Wood-tin has never been found in West Malaysia, nor has its presence been established anywhere in the Southeast Asian tin province. Aranyakanon et al. (1969) reported wood-tin in the placers of Huai Jagrao in South Thailand, and the writer was permitted to examine this. Polished sections, while indicating the presence of waterworn, crudely spherical particles of "normal" cassiterite, together with spheroids of colloform limonite, ilmenite, framboidal pyrite, and so on, were devoid of wood-tin. Considering that thousands of samples of tin ore from West Malaysia, and indeed from all the tin-producing countries of Southeast Asia, have been examined by people competent to recognize wood-tin, it seems that the region really is lacking in this particular variety of cassiterite.

The earlier cassiterite, on occasion, is characterized by intense color zoning which, perhaps, is most strikingly exhibited by that in the feldspathic veins of Pelepah Kanan, Johore (Figure 11.17c). Cassiterite needles, on the other hand, never show this feature.

In West Malaysia it is not unusual to find that some of the cassiterite in a given vein displays a strong red-green pleochroism in thin section. It is probable that this peculiar feature is due to appreciable tantalum and/or perhaps niobium in the structure. Polished sections of such cassiterite are char-

---

later generation with long axes at right angles to the faces of the cas 1 crystal. The cassiterite and later arsenopyrite (black) are veined by siderite. Quartz (Q) is also present. (c) Zoned cassiterite fringed by quartz (Q) and sericite mica (M) locally replaced by chlorite (chl). From the feldspathic veins of Pelepah Kanan, Johore.

acterized by the presence of small exsolution bodies that have not been identified with certainty but are probably members of the tapiolite-mossite and/or columbite-tantalite series of minerals. That both tantalum and niobium are often present in appreciable amount in the Malaysian cassiterites is well known, and the tin-smelters' slags contain valuable amounts of tantalum. The tantalum–niobium and other minor and trace-element contents of the Malaysian cassiterites are indicated in Table 11.5.

Such tantaliferous–niobiferous cassiterite occurs in pegmatites (Bakri, Johore, and Kedah Peak) and in hydrothermal veins (Tekka, Perak; and Ulu Langat, Selangor). Where more than one generation of cassiterite occur, the anomalous pleochroism noted previously is exhibited only by the earlier generation cassiterite. Even so, only some crystals display it, and often within these only certain zones do so. That tantalum–niobium may be confined to certain zones in a cassiterite crystals has been confirmed by means of the electron probe (M. P. Jones 1969).

Some of the faces of certain cassiterite crystals in a drusy quartz–cassiterite lode specimen from the western margin of the Kinta Valley fluoresced a dull but quite distinctive orange under shortwave ultraviolet light; cassiterite, which was associated with quartz and pyrite in a specimen from the Puchong area, near Kuala Lumpur, behaved similarly. This phenomenon, which does not appear to have been observed before, may not be uncommon in tin fields generally, and the writer has also noticed it in crystals that are attached to a wolframite crystal from Beralt Tin and Wolfram Mine, in Portugal. It is believed that this phenomenon is due to a thin veneer of a colorless transparent mineral, possibly apatite, activated by a trace of manganese. The presence of such a veneer may explain why certain crystals of cassiterite in a given sample react less readily to the tinning test than do their companions.

Appreciable proportions of the cassiterite from West Malaysia are magnetic, and two different types of magnetic cassiterite are found there. The ferromagnetic type, whose magnetic property is destroyed by heating to 830°C for 15 min (Flinter 1960), occurs in areas in which the cassiterite is in close association with iron ore, as at Pelepah Kanan, Johore. Comparatively recent studies of cassiterite from Pelepah Kanan (Grubb and Hannaford 1966) have demonstrated that the darker the zone, the greater its magnetic susceptibility. This was reconfirmed by Khoo (1969), as Table 11.6 shows.

Grubb and Hannaford conclude "that the magnetization appears to be associated with the presence of hydrated ferrous stannate and that these inclusions are probably accompanied by some local dehydrated regions containing ilmenite-type $FeSnO_3$. This anhydrous ferrous stannate might then be expected to be found in solid solution with any trace of hematite which may have been present— possibly as a result of dehydration of α-FeOOH or γ-FeOOH and form a series of rhombohedral compounds (analogous to $Fe_{2-x}Ti_xO_3$) which, at the appropriate concentrations, should exhibit appreciable ferromagnetism. However, it must be pointed out that the magnetic character of some of the Pelepah Kanan cassiterite crystals is due in part to inclusions of magnetite (Santokh Singh and Bean 1968).

TABLE 11.5 *Chemical analyses of some West Malaysian cassiterites. Based on data of Pryor and Wrobel* (1951).

	Percentage Sn		Percentage of Minor and Trace Elements										
	Spectroscopic Determination	Chemical Determination	Spectroscopic Analysis										
Locality			V	Ta	Nb	Mn	W	As	Ti	Fe	Zr	Si	Sb
Tambun, Kinta	73.1	73.0	tr.	2.2	1.3	0.2	tr.	tr.	0.2	0.9	tr.	0.2	nil
Pulai, Kinta	70.9	71.6	nil	0.2	0.2	tr.	0.1	nil	0.3	0.8	tr.	0.6	nil
Tekka, Kinta	73.8	74.0	nil	0.1	tr.	tr.	0.2	nil	tr.	0.8	tr.	0.5	nil
Kacha, Kinta	75.0	75.2	nil	0.1	tr.	nil	0.1	nil	tr.	0.5	tr.	0.3	nil
Siputeh, Kinta	74.8	75.0	nil	tr.	tr.	nil	0.2	nil	0.2	0.7	tr.	0.2	nil
Sungei Way, Selangor	74.6	74.5	0.3	nil	tr.	tr.	tr.	nil	nil	0.7	tr.	0.5	0.2
Kota Tinggi, Johore	61.2	60.8	nil	6.9	1.1	0.1	tr.	nil	tr.	0.5	nil	0.2	nil

TABLE 11.6 *Cassiterite from Pelepah Kanan in the 100–150 B.S.S. mesh range recovered by the Frantz Isodynamic Separator (forward slope 25°, side slope 15°). After Khoo (1969).*

Current	Weight Percentage Recovered	Maximum Mass Magnetic Susceptibility (10^{-6} c.g.s.)	Relative Abundance of Black, Brown and Colorless Grains	Behavior Towards Hand Magnet
0–0.10	5.72	517.64	All grains black and opaque	Strongly attracted
0.11–0.20	10.48	129.41	↓	
0.21–0.30	5.27	57.52	Mainly black opaque grains with very minor amount of brown grains	
0.31–0.40	4.58	32.35		
0.41–0.50	3.89	20.71	↓	
0.51–0.60	2.71	14.38	Black grains decreasing and brown grains increasing in amount	
0.61–0.70	2.57	10.56		
0.71–0.80	1.53	8.09		
0.81–0.90	1.30	6.39		
0.91–1.00	0.88	5.18	↓	
1.01–1.10	1.00	4.28		
1.11–1.20	0.75	3.59	Mainly brown translucent grains	Weakly attracted
1.21–1.30	0.82	3.06		
1.31–1.40	0.96	2.64		
1.41–1.50	0.74	2.30	↓	↓
>1.50	56.72	—	Virtually all grains brown to colorless	Not attracted

The paramagnetic cassiterite, whose magnetic property is not destroyed by heating, occurs in association with columbite-tantalite and related species, and is found at Bakri, Johore, and at Kedah Peak, Kedah. Bradford (1961) suggests that the magnetic character of such cassiterite may be attributable to the presence of tapiolite in solid solution.

Finally, the overwhelming proportion of the cassiterite apears to have been deposited directly from ascending stanniferous solutions, sometimes in open spaces, but perhaps more often by replacing minerals such as feldspar, mica, and quartz. Nevertheless, sometimes a little cassiterite was developed by the degradation of other tin species. Earlier we noted the development of cassiterite, together with calcite and quartz from malayaite at the Sungei Gow, probably by the action of late hydrothermal solutions. In the stannite in polished sections of ore both from Tekka, Perak, and from the Hock Leong Mine, Ampang, Selangor, the writer has observed veinlets and masses of chalcopyrite and in this latter mine, cubanite also, in all of which are embedded myriads of cassiterite crystals.

The occurrence of "secondary" cassiterite in cubanite in the highly complex Ampang ore was first noted by E. B. Yeap, who pointed it out to the writer. Santokh Singh and Bean (1968) remarked on the same phenomenon in stannite from Serdang, Selangor. This curious feature, which has been observed in a number of tin fields of the world, was thought by Ramdohr (1944) and Edwards (1956) to be due to the spontaneous breakdown of the stannite. The present writer holds much the same view, but he thinks that "late" hypogene agents moving through prefered paths in the stannite may have been responsible for its local degradation.

Stannite and Other Tin-Bearing Sulfides. Stannite, Cu_2S (Fe, Zn) S, SnS_2, is a fairly common member of the West Malaysian primary tin deposits, and systematic examination of polished sections by the writer and his colleagues has already added significantly to the number of occurrences in Malaysia recorded by Santokh Singh and Bean (1968) who also described most of the significant properties of tetrastannite, the only variety they had encountered in the country. Recently, however, Yeap and Hosking (unpublished studies) have found cupro-stannite (mawsonite?) and isostannite in the ore of the Hock Leong Mine, Ampang, Selangor.

The normal (tetragonal) stannite occurs in close asociation with sphalerite and chalcopyrite and would appear to have been deposited more or less contemporaneously with them. As Santokh Singh and Bean (1968) noted, the species is also not uncommonly associated with pyrrhotite, and it may contain exsolution bodies of this sulfide and of any of the companion sulfides mentioned earlier. In two specimens, one from Tekka and the other from Tanjong Tualang, Perak, Hosking and Leow (1970) observed a few sparse silvery exsolution bodies of high reflectivity and birefringence in the stannite. These have been too small to allow an accurate determination of their microhardness and reflectivity, but, provisionally the authors believe the bodies to be kobellite (Figure 11.18a). The species, whatever its identity, is almost certainly the same as that in the Wheal Rock stannite of Cornwall, which was noted but not identified by Ramdohr (1944).

The Malaysian stannite may surround corroded cassiterite in such a way as to suggest that it has developed by replacement of the oxide. When this occurs, as it does, for example, in the Manson Lode (Kelantan), the stannite is separated from the cassiterite by a narrow rim of chalcopyrite (Figure 11.19a). This feature has been noted by Ramdohr (1944) and Hosking (1969b) in samples from other tin fields.

Tetrastannite may also occur in Malaysian ore (and also in ore from elsewhere) in close association with other sulfides rather than with cassiterite, although the latter may also be present in the ore. A typical example is provided by the ore from the Ban Hock Hin Mine, near Kuala Lumpur (Figure 11.18b), which is described in detail later.

Stannite may also occur as exsolution bodies in sphalerite, and examples of this have been noted by the writer in ore from Tekka.

In the recently found occurrences of cuprostannite in the Hock Leong Mine, Ampang, this sufide is present in a multimineralic, telescoped lode in association with cassiterite, tetrastannite, minor isostannite, and teallite. In polished sections of ore from this mine (Figure 11.20), areas rich in cuprostannite, which appears to replace the tetrastannite, are particularly abundant in the vicinity of the veins of chalcopyrite that transect the tin-bearing sulfide area. Also in such sections, silvery teallite [(Pb,Zn,Sn)S] replacement veins, containing relicts of all the varieties of stannite present, together with included masses of chalcopyrite, are commonly much in evidence. That some of the chalcopyrite is strictly contemporaneous with the teallite is suggested by the possibility that reaction between stannite and lead ions might well result in the formation of teallite and chalcopyrite.

In the nearby Hock Aun No. 1 Mine, at Kampong Pandan, teallite also occurs in similar telescoped, multimineralic veins in the limestone. In these, however, cupro-stannite is absent. The teallite there has partly replaced early sphalerite, cassiterite,

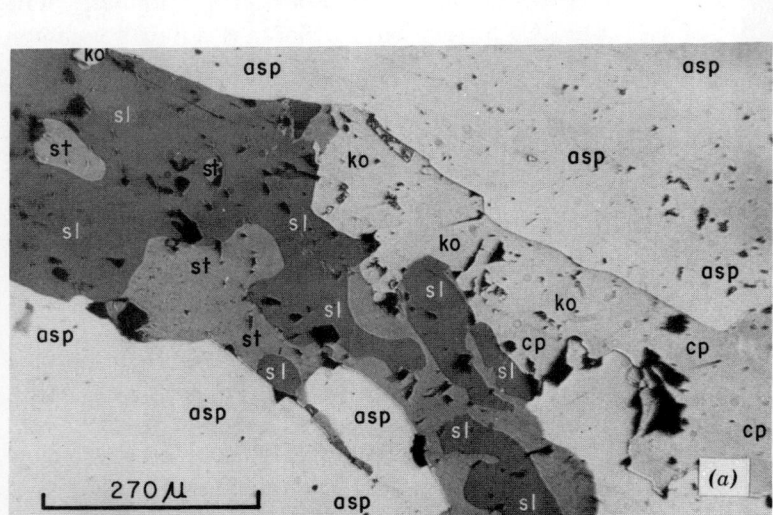 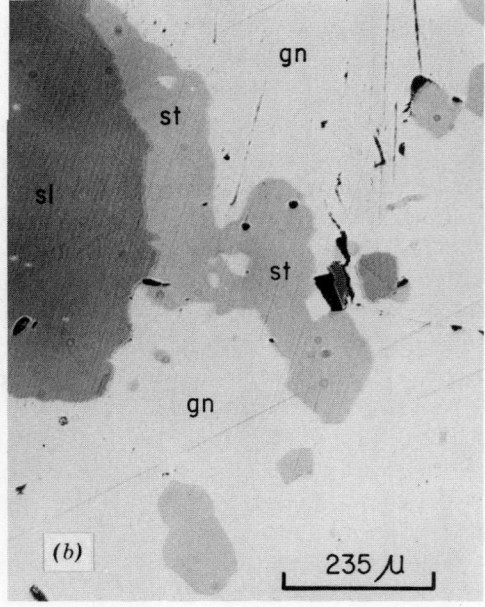

FIGURE 11.18. *Photomicrographs (K. F. G. Hosking) of polished sections (in air) of stanniferous telescoped ores:* (a) *Ore from Tekka, Perak, showing arsenopyrite (asp) locally replaced by later sulfides. Stannite (st) has replaced sphalerite (sl), and the latest species to develop were chalcopyrite (cp) and kobellite (ko). The dark areas are voids.* (b) *Ore from Ban Hock Hin Mine, Selangor, showing stannite (st) rim-replacing sphalerite (sl) and being embayed by galena (gn).*

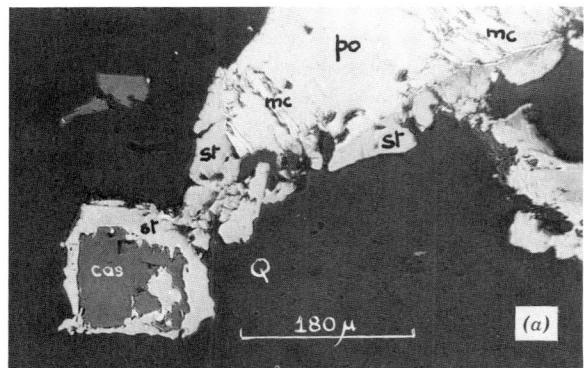

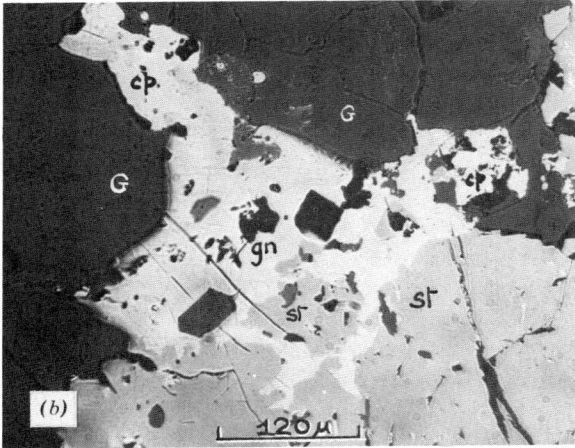

FIGURE 11.19. *Photomicrographs (K. F. G. Hosking) of polished sections (in air) from the Manson Lode, Kelantan: (a) Cassiterite (cas) rim-replaced by stannite (st). Characteristically, a thin rim of chalcopyrite (white) lies between the cassiterite and the stannite. Pyrrhotite (po), marcasite (mc), and quartz (Q) are also present. (b) Galena (gn) locally replacing stannite (st) and chalcopyrite (cp.). The gangue (G) is mainly of quartz.*

and chalcopyrite (Yeap and Hosking unpublished studies).

Hockin (1957b) provisionally identified franckeite ($5PbS.Sb_2S_3.3SnS$) in a sample of ore from Ban Hin Lee Mine, Selangor, which also contained arsenopyrite, boulangerite, cassiterite, chalcopyrite, jamesonite, pyrite, sphalerite, and stannite. The "franckeite" is "commonly found replacing stannite" and is "replaced in turn by jamesonite." This is yet another good example of a telescoped ore.

Wolframite, Ferberite, and Scheelite. By comparison with the primary tin species, surprisingly little can be said about those of tungsten. The writer has no data concerning the trace-element content of Malaysian wolframite [$Fe,Mn (WO_4)$] for example; we do know, however, that tantalum and niobium are both present in the wolframite from Ulu Langat (Gan 1969). Information is lacking on the variations, if any, in the iron-manganese ratio of wolframite in any one ore body, although such studies are underway.

That two generations of wolframite may be present in a single deposit has been demonstrated. Thus polished sections of a small greisen-bordened vein in the granite cliffs to the south of Lumut (which contained cassiterite, wolframite, arsenopyrite, tourmaline, sericite, minor apatite, and relict feldspar in a quartz matrix) showed that most of the wolframite predated the cassiterite (which was tantaliferous) as crystals of the former were completely embedded in the latter (Figure 11.21a). However, the cassiterite was intersected locally by fractures that had been healed by tourmaline and mica, together with small crystals of second-generation wolframite (Hosking 1969a).

As noted earlier, ferberite ($FeWO_4$) was recorded by Ingham and Bradford (1960, p. 300) from Kramat Pulai, where it was found "as black blocks and fragments in a fault-fissure in weathered granite" and where it occurred with yttrotungstite. Although Ingham and Bradford thought that the mineral was secondary, the writer can see no reason for this viewpoint. He believes that ferberite was probably deposited directly from late hydrothermal agents or, possibly, that it was due to the hypothermal replacement of scheelite, although in the latter case discovery might be expected of pseudomorphs of ferberite after scheelite [such as have been found, e.g., in East Africa (Magnee and Aderca 1960) and in the Magdalen Mine, Lanlivery, Cornwall]. The Kramat Pulai ferberite was only present in academically interesting amounts.

In Malaysia, scheelite ($CaWO_4$) may occur in deposits in which there is no wolframite, as in the great scheelite-fluorite replacement body of Kramat Pulai and in many of the skarn deposits. It may also occur, in varying amounts, in association with wolframite in veins, as at Chenderong (where it is also associated with cassiterite) and at Ulu Langat (where it locally cements fractured cassiterite in greisen-bordered veins). Often there is no indication that the scheelite has developed by replacement of wolframite, but the little scheelite present in the dominantly quartz-wolframite veins at Bukit Lentor, Trengganu, and at the Tunku Makota's Mine, Pahang, has been developed by such a process. At the Hock Aun Mine, Kampong Pandan, Selangor, on the other hand, skeletal relics of wolframite in the scheelite of some of the veins in the limestone bear striking witness to the origin of the calcium tungstate (Figure 11.21b).

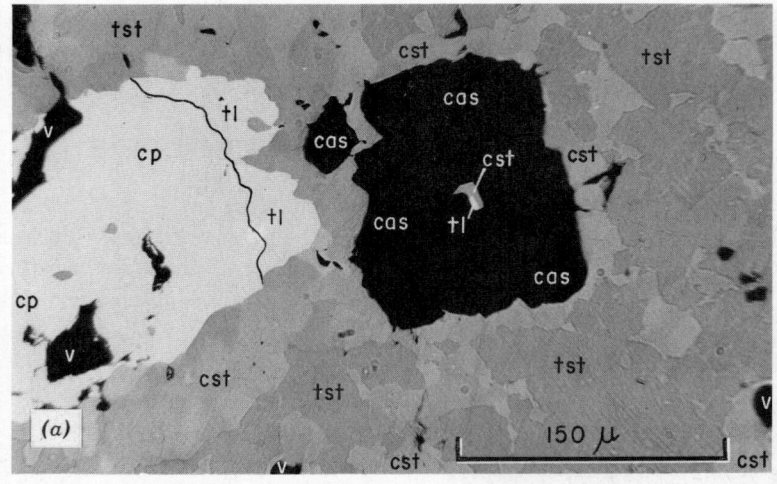

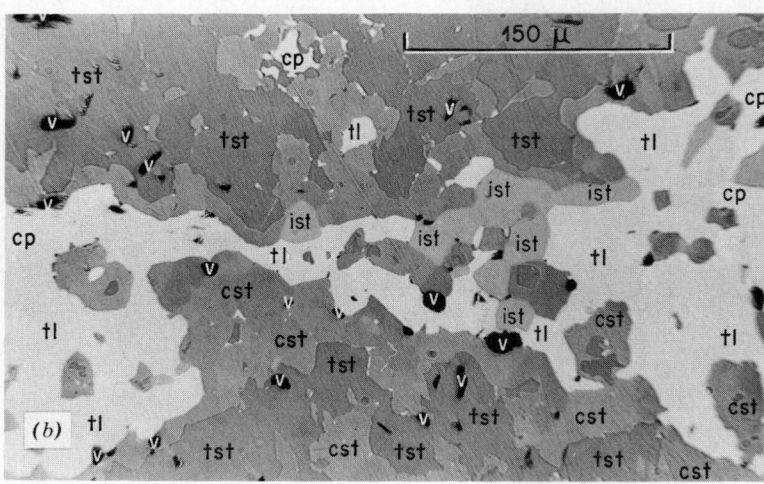

FIGURE 11.20 *Photomicrographs (K. F. G. Hosking) of polished sections (in oil) from the Hock Leong Mine, Ampang, Perak. Dark areas (v) are voids. (a) Cassiterite (cas) locally replaced by tetrastannite (tst) itself partly replaced by cuprostannite (cst). The stannites were later partly replaced by teallite (tl) and chalcopyrite (cp). (b) Tetrastannite (tst), cuprostannite (cst), and some ?isostannite (ist) intersected by a later vein of teallite (tl) and minor chalcopyrite (cp).*

At Batu Tiga, Trengganu, scheelite occurring essentially in pyrrhotite is locally embayed by the latter. Commonly, however, the tungstate displays pull-apart texture and the fractures are healed by pyrrhotite or chalcopyrite that "flowed" during the period of stress. Triple junctions in the pyrrhotite indicate the occurrence of the stress period.

With few exceptions, the West Malaysian scheelite fluoresces a strong blue under shortwave ultraviolet light. However, some of the fragments of the mineral recovered from the placers of Templer Park, Selangor, fluoresced very pale blue, whereas others fluoresced white, indicating that slightly more of the $(MoO_4)^{2-}$ ion occurred in these specimens than was usual.

Finally, it is possible that some of the small grains occurring in skarns, which have been identified as scheelite because they fluoresce blue under shortwave ultraviolet light, may in fact be the barium tin silicate, pabsite, which to date has been recorded only from San Benito County, California. This seems likely, since the Malaysian granites are particularly rich in barium, ranging from 245 to 1250 ppm according to Yeap Cheng Hock (private communication), and tin was certainly available during the late phases of skarn development.

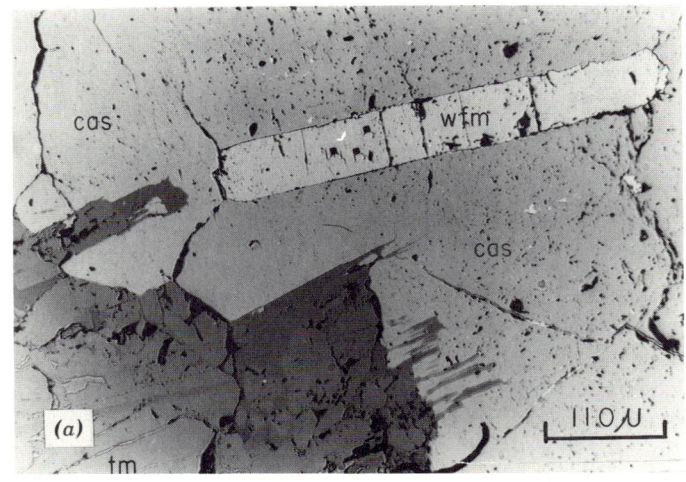

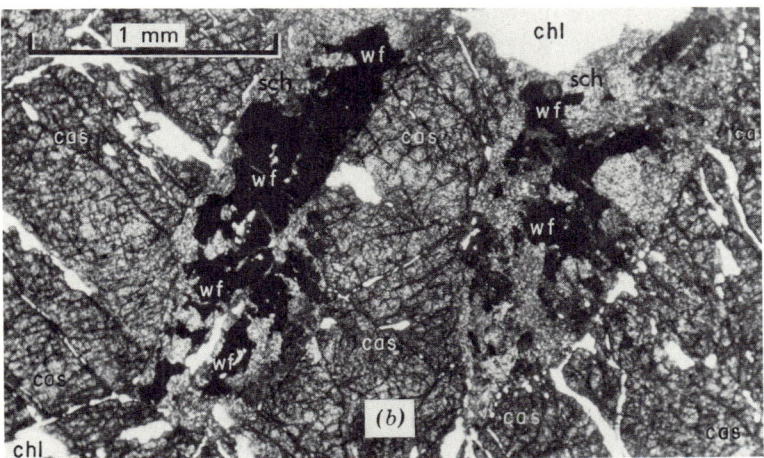

FIGURE 11.21 *Photomicrographs (K. F. G. Hosking) of tin ore: (a) Polished section (in air) of cassiterite (cas) from a greisen-bordered vein at Lumut, Perak, with a crystal of wolframite (wfm). The cassiterite also contains exsolved mossite–tapiolite or tantalite–columbite which appear much paler. Tourmaline (tm) is also present. (b) Thin section (in plane-polarized light) of a vein in limestone from Hock Aun Mine, Kampong Pandan, Selangor. Cassiterite (cas) alternates with bands of scheelite (sch) and wolframite (wf). The scheelite has developed by replacement of the wolframite. Gangue of chlorite (chl).*

MINERALOGICALLY SIMPLE VEINS AND VEIN SYSTEMS. Mineralogically simple veins bearing tin, tungsten, or both are commonplace in the mining fields of West Malaysia, and the stanniferous veins were the major contributions of cassiterite during the development of the world's most important tin placers. The tin members of the group under review may occur singly, but most of those of economic importance are found in swarms that may often be described as stockworks. When the rock in which the stockworks had developed was decomposed and soft, they were mined by open-cast methods. Jones (1925c) provided many examples of such types, and it is notable that they usually occur in granite schists and phyllite but almost never in the limestone. A possible exception was seen at Siputeh (Perak) where, according to Jones (1925c, p. 186) "the tinstone is derived from narrow quartz veins in schists and phyllites that have become completely disorganized . . . and from stanniferous quartz veins traversing the limestone."

Jones (1925c) recorded that locally at Chemor, Perak, "decomposed schist and phyllite, containing cassiterite in small quartz veins, were worked for their tin content" (p. 173). At Tekka, Perak "tinstone occurs in places *in situ* in small quartz veins forming stockworks in the metamorphosed sedimentaries" (p. 176). At Gopeng Consolidated Mines "the schists and phyllites are ramified by narrow (stanniferous) quartz veins, forming in places stockworks on a very large scale" (p. 178).

Ingham and Bradford (1960, p. 135), generalizing about the occurrences of tin deposits in a schist host on the east side of the Kinta Valley (where both Tekka and Gopeng are situated), wrote that such deposits "usually consist of veins of cassiterite associated with pyrite or arsenopyrite and/or tourmaline that have been injected into the schist and that frequently run parallel to its strike. In most instances there is evidence that shearing has taken place in the country rock. . . ." Similar deposits, according to Jones, occur at Lahat, Kacha, Red Hills, Siputeh, and in a number of other places.

Today the Chin Chin Mine, Malacca, is exploiting a deposit that Scrivenor (1928, p. 111) described as "a rich stock of tin-bearing veins, in phyllite with tourmaline." At Rahman Hydraulic Mine, Klian Intan, Perak, a plexus of parallel veins of averaging 1 cm in width (although the width may reach about 60 cm) occupies Gunong Paku; it is composed essentially of schist, phyllite, and shale. Here bench mining has been carried out for years. The mining operation has been facilitated and the milling costs have been kept comparatively low because the ore worked was very decomposed, and very little of it needed to be crushed. The veins, essentially quartz, contain cassiterite and small amounts of pyrite, arsenopyrite, native bismuth, sphalerite, and galena, together with tourmaline. It is understood that these veins extend to depths of at least 210 m (700 ft) beneath the crest of the hill but that the richest ore was found at the highest horizons. As mentioned earlier this vein swarm may halo a buried granite cusp.

Excellent examples of mineralogically fairly simple stanniferous bodies in the granite were encountered on the eastern slopes of the Kledang Range and were mined by underground methods by the Menglembu Lode Mining Company Ltd. There, a swarm of lodes and veins, with strikes varying from 25° to 80° and dipping steeply to the northwest, occurred "en echelon along a zone approximately 1,400 feet long and 100 feet wide, with a general trend of 15° to 20°. . . . Lodes, often pipelike in shape, have formed where veins have been particularly numerous or where they have merged together, and these have been worked to a maximum depth of 560 ft below the head of No. 7 shaft" (Ingham and Bradford 1960, p. 122).

Willbourn recognized two types of lode ocurrence in this mine. One consisted of quartz leaders that were present singly or as two or three parallel veins separated by a few inches or several feet of intervening granite. The leaders and the neighboring rocks were often very rich in cassiterite. The second type consisted of numerous very thin parallel veinlets cut by the first type of lode and penetrating the granite in the form of a lode often exceeding 12 ft in width.

This type was locally known as "streaky bacon" and carried cassiterite, brown biotite, tourmaline, pyrite, magnetite, chlorite and muscovite [Figure 11.13]. The veinlets cut straight through crystals of feldspar and biotite, without apparently displacing the portions of the penetrated crystals and without causing any recognizable alteration of them. There was no sign of a zone of mineralized granite on either side of each veinlet" (Ingham and Bradford 1960, p. 123).

The primary deposits at Kwong Fook Lee Mine, Ulu Langat, are a good example of simple tin-tungsten ores. They consist of a series of greisen-bordered veins that have developed along certain tension joints in coarse-grained porphyritic biotite granite. The writer belives the general order of deposition of minerals in these veins is tourmaline, wolframite, arsenopyrite, cassiterite, scheelite, and marcasite. Quartz of several generations is seen, and the veins proper postdate the development of the associated greisen. Replacement textures are much in evidence. Cassiterite replaces sericite, quartz, and tourmaline; scheelite replaces wolframite, and arsenopyrite replaces tourmaline. Somewhat similar cassiterite, wolframite, scheelite veins occur at Chendrong, Trengganu.

Although wolframite and scheelite may occur in some mineralogically complex bodies (e.g., at Tekka and Kampong Pandan), tungsten bodies that are of economic interest are, without exception, mineralogically simple in the sense that not more than two minerals occur in significant amounts. The fact is that only mineralogically simple tungsten bodies can be exploited profitably, using primative beneficiation methods. Thus the Bukit Lentor lodes consist essentially of quartz and erratically disposed wolframite, although other species are present in small amounts.

The provisional paragenesis of primary minerals of the Bukit Lentor deposits is as indicated in Table 11.7.

TABLE 11.7 *Paragenesis of the Bukit Lentor deposits (K. V. Lee, private communication).*

Mineral	Time (Early → Late)
Quartz	—— Several generations ——
Muscovite	—
Wolframite	—
Scheelite	—
Arsenopyrite	—
Pyrite	— —
Pyrrhotite	—
Sphalerite	—
Chalcopyrite	—
Marcasite	—
Siderite	—
Calcite	—
Chlorite	—— ? ——

The deposit at Najang (Trengganu) consists of wolframite, pyrrhotite, and minor chalcopyrite in a gangue of quartz.

The tungsten deposits of the Tunku Makota's Mine, about 9 km (5.5 miles) west of Gambang, consists of a series of lodes up to 1.2 m wide. The lodes are fringed by quartz–sericite zones, and beyond these are zones of white and black quartz–tourmaline hornfels. The mine is situated on a "bulge" in the andalusite- and cordierite-bearing hornfels that fringes the granite that is exposed further to the east. Immediately to the west, outside the metamorphic aureole, the country rock consists of shale, quartzite, siltstone, and mudstone. The veins' content is essentially quartz and wolframite, but locally a little arsenopyrite, pyrite, molybdenite, native bismuth and bismuthinite occur.

Among the most promising tungsten deposits awaiting further exploitation are those in Kedah, in the Sintok area. Jones (1925c, pp. 208–209) remarked that there, at Bukit Kachi Mine, "wolframite and cassiterite occur . . . in association in large quartz veins, from one to five feet in width, . . . and outcropping on the surface for a distance, in some cases, of well over 1,000 ft. along their strike. The veins are intrusive in schists and phyllites. . . ."

SIMPLE REPLACEMENT DEPOSITS. It may be conceded that most veins, lodes, and pipes are essentially the products of replacement round and about fissures rather than the filling of them. However, we are concerned here with those sizable, rather more equidimensional bodies, which cannot possibly be regarded as having been formed by the filling of open spaces. An example is the Kramat Pulai scheelite-fluorite deposit, whose broad characteristics have already been mentioned.

Another such body is the so-called Lode 4 at the Sungei Besi Mine. This body, which occurs in heavily kaolinized granite and extends over a vertical distance of about 30 m (100 ft), now consists essentially of cassiterite, limonite, and quartz. Within the mass, however, there are islets of sufidic ore—essentially cassiterite, quartz, and pyrite—but containing also minor amounts of chalcopyrite and covellite. These sulfide-rich relicts reveal the nature of the primary ore that developed by the replacement of the granite, possibly as a result of mineralizing solutions' ascending tension fractures, which had developed between two parallel wrench faults. This ore body is accompanied by several similar ones and at least one, the Lee Gossan, differs considerably from the No. 4 Lode. When examined by Chan (1970), the exposure of the latter was roughly circular in plan and about 46 m (150 ft) in diameter. It consisted essentially of cassiterite, fluorite, arsenopyrite, and pyrite (the latter two somewhat oxidized), embedded in quartz.

The Lahat Pipe, Perak, qualifies for inclusion in this section. The upper portions of this irregularly shaped orebody consisted of cassiterite and iron oxide, cemented by secondary calcite; below 90 m (300 ft), however, primary ore consisting of cassiterite, arsenopyrite, and calcite was encountered. This body was worked to a depth of 100 m before working ceased owing to heavy pumping charges (Ingham and Bradford 1960).

MODERATELY COMPLEX VEINS. In West Malaysia, tin- and tungsten-bearing deposits of all degrees of mineralogical complexity exist, and there is little or no relation between the size of the deposit and its mineralogical character. The writer regards mineralogical complexity as being determined not only by the number of species present but also by their relative abundance and, in particular, by their spatial relationship. No really useful rules can be laid down for establishing the degree of mineralogical complexity—it is a personal assessment, and the opinion of the mineral dresser is likely to differ from that of the "pure" mineralogist.

The samples of sulfide-rich ore from Penkalen, which were given to the writer by Mr. McLeod, came from a moderately complex hydrothermal deposit. Thin and polished sections revealed that the ore consisted of early quartz and cassiterite associated with arsenopyrite that was locally rimmed by pyrite, and veined and partly replaced by sphalerite and chalcopyrite. Here and there, large, irregularly shaped masses of sphalerite containing exsolved chalcopyrite occurred. Some of the sphalerite and chalcopyrite had been replaced by galena.

A further example is the Galena Lode on the Pahang Consolidated property (Fitch 1952). A portion of this hitherto undescribed lode was recently given to the writer by Mr. Tay of Pahang Consolidated. It is a symmetrical lode 11 cm wide, and its central band is appreciably richer in sulfides than those on either side of it. Briefly, examination of thin and polished sections established that the body developed within a fault zone that had been repeatedly reopened during a long period of mineralization. Quartz, topaz, and a little pale cassiterite were the first minerals to be deposited, and these were followed by modest amounts of pyrite and fluorite. Later, considerable sphalerite with exsolved chalcopyrite was introduced and this, after fracture, was followed by the deposition of a great deal of chlorite. After further fracture, galena was deposited. The galena effected considerable replacement of the sphalerite, and blebs of the lead mineral, so formed, occur in the zinc species. Locally, relicts of sphalerite occur in a matrix of galena. The galena has, in addition, replaced some of the quartz (Hosking and Yeap unpublished studies).

In the Melor Mine (Selangor) a lode, about 1 m wide and occupying the granite–limestone contact, was characterized by a marked banded structure. It consisted, essentially, of arsenopyrite–cassiterite–quartz bands alternating with essentially sphalerite–chalcopyrite–pyrite bands. Yeow (1969) concluded that a series of parallel fractures developed in an original arsenopyrite–cassiterite–quartz lode and that, within these fractures, sphalerite, chalcopyrite, and pyrite were deposited in the order given. Late chalcopyrite partly replaced the sphalerite and also entered the arsenopyrite-bearing bands, partly replacing the sulfarsenide. Pyrite, the latest of the species to be deposited, is restricted to the sphalerite-bearing bands and locally has replaced the zinc sulfide to a marked degree.

The stanniferous feldspathic veins of Pelepah Kanan, noted earlier are also members of this class.

MINERALOGICALLY COMPLEX PIPES AND VEINS. Most of the stanniferous pipes in the limestone of the Kinta Valley are, again unequivocally hydrothermal; and like many other Malaysian hydrothermal deposits, they display considerable complexity. One of the more complex members was that uncovered in 1929 at Sin Woh Leong Kongsi, Simpang Ampat, Menglembu. It was markedly telescoped and "besides cassiterite, the ore contained pyrite, galena, sphalerite and jamesonite, together with small amounts of arsenopyrite and chalcopyrite ... some calcite gangue was present" (Ingham and Bradford 1960, p. 134).

The strongly telescoped veins of Tekka and of the Kuala Lumpur tin field fall into this class also, as does the Manson Lode of Kelantan. Besides cassiterite, they contain an impressive number of sulfides; in addition, at Tekka and in the Kampong Pandan area, wolframite and scheelite are present. In this chapter, it is only possible to deal in detail with the ore from one of these deposits and to consider that from another more briefly.

Ban Hock Hin Lode. In 1969 a large block of sulfide-rich ore was collected by E. B. Yeap from the old Ban Hock Hin Mine, Setapak, Selangor, which is situated on a granite–limestone contact. The parent lode was not visible but it must have been adjacent. Examination of thin and polished sections (Figure 11.22) permitted the following observations to be made by Hosking and Yeap (1970) concerning the mineralogy. textures, and paragenesis of this remarkable mineral assemblage.

In order of abundance, the "ore minerals" present are as follows: galena, chalcopyrite, sphalerite, stannite, arsenopyrite, cassiterite, chalcopyrrhotite, and tennantite, arsenopyrite, cassiterite, chalcopyrrhotite, and tennantite. In addition, a little quartz and chlorite are present, but these species constitute less than 1% of the ore. The quartz and the chlorite observed were deposited after the cassiterite and arsenopyrite but before all the other species, since the latter have locally replaced both these silicon-bearing minerals. All the species just mentioned may be seen in a single polished section (about 6 cm²), thus emphasizing the telescoped nature of the ore.

Cassiterite, almost colorless and nonpleochroic, and arsenopyrite were the earliest of the "ore-minerals" to be deposited as they are to be seen surrounded and/or replaced by all the other species, but whether cassiterite was deposited before or after the arsenopyrite has not been established.

Rim replacement of the early, initially euhedral minerals is much in evidence; but internal replacement is also found, caused by movement of solutions along fractures deep into them. This internal replacement led to the development of irregularly shaped patches of stannite within the cassiterite, of stannite-chalcopyrite, and of galena within the arsenopyrite

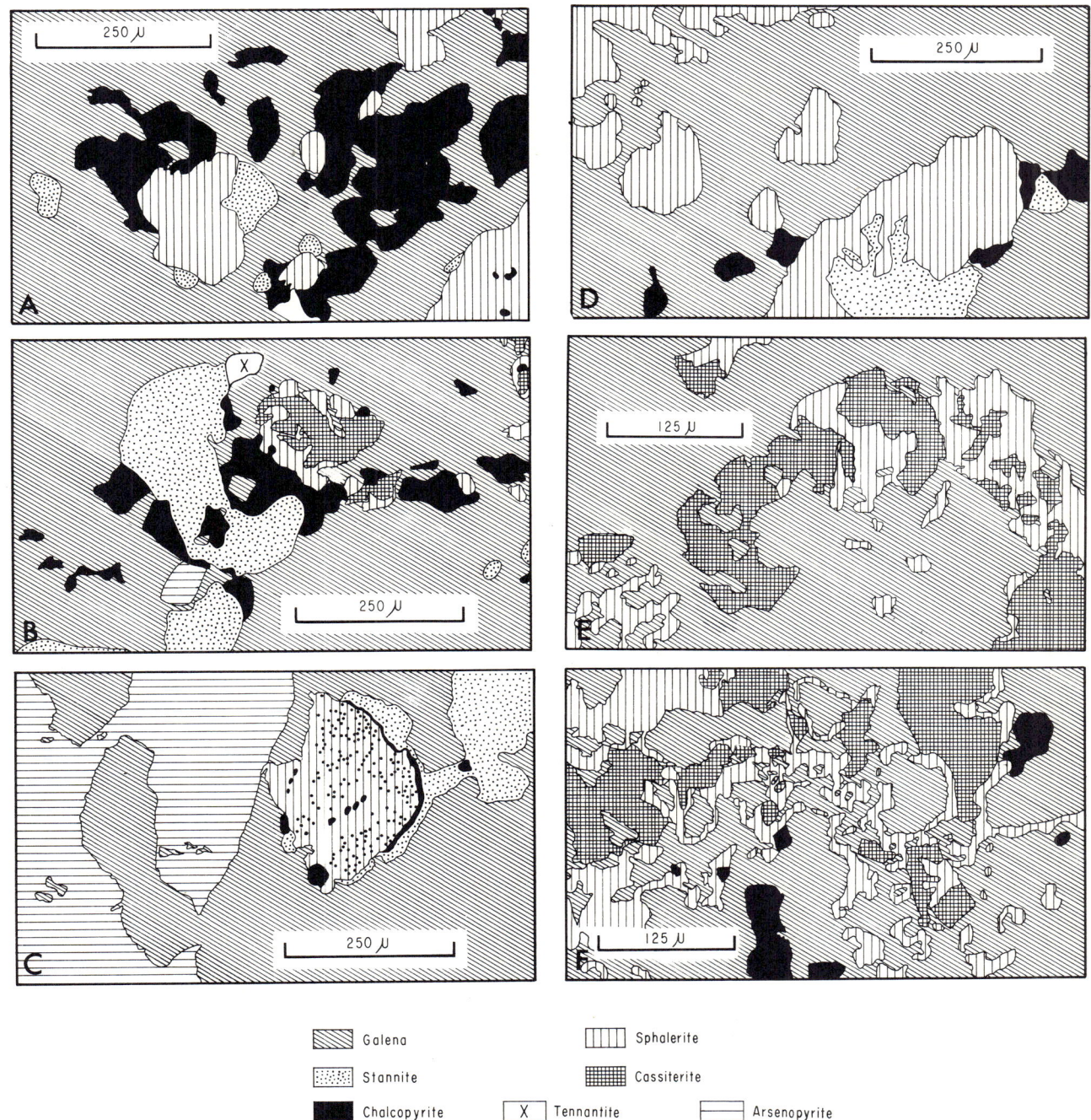

FIGURE 11.22. *Drawings of photomicrographs of polished sections or ore from the abandoned Ban Hock Hin Mine, Setapak, Selangor. For details see text.*

(Figure 11.22c). However, stannite and galena also occur as triangular bodies within the arsenopyrite, which suggests that sometimes replacement was subject to crystallographic control by the host mineral. The rim replacement of elongated euhedral cassiterite by stannite, chalcopyrite, and galena is also of particular interest in that all these sulfides, or rather, the agents that caused their development, seemed to have an affinity for attacking the terminations, commonly ignoring the prism faces.

The arsenopyrite is distinctly more fractured than the cassiterite and, as noted previously, the fractures permitted marked internal replacement of the sulfarsenide to be effected. However, during and after the phase of replacement, these fractures were themselves healed by sulfides, and the manner of their infilling reveals the sequence of deposition of a number of the later species. Commonly the fractures and microfissures in the core of the arsenopyrite are infilled with sphalerite and stannite, whereas those of the intermediate "zone" contain chalcopyrite and those near the surface are healed by galena, which also surrounds the fractured mineral.

Sphalerite occurs as rounded clots and as irregularly shaped relicts, partly or wholly surrounded by galena (Figure 11.22a,d). The largest aggregates of sphalerite are characterized by the presence of emulsion-type exsolution bodies of chalcopyrite and sometimes stannite (Figure 11.22c). In marked contrast, the sphalerite that has rim-replaced the cassiterite is devoid of exsolution bodies (at any rate, none were apparent at the magnifications employed during the examination). Moreover, in thin section this sphalerite is dark brown, whereas that containing exsolved sulfides is yellowish-brown (Figure 11.22f). The dark-sphalerite-generating agents, quite unlike those of chalcopyrite, stannite, and galena, attacked and replaced the cassiterite wherever they come into contact with it, so that commonly irregularly shaped relicts of the tin species are seen completely surrounded by the zinc one.

The dark sphalerite also occurs as irregular blebs in the galena. It is clearly related to neighboring sphalerite, which is rimming cassiterite, and it is difficult not to believe that the texture is due to the local replacement of the cassiterite by galena (Figure 11.22d,e,f).

Stannite rim-replaces some of the sphalerite (Figure 22a,c) and locally deeply embays it.

Some of the clots of sphalerite are patterned by a rectangular grid of microfissures that have been infilled by chalcopyrite, galena, or both. Clearly, the development of these fissures was subject to crystallographic control.

For the most part, stannite, which appears to be the common variety, occurs as rounded clots or irregularly shaped relicts in the galena (Figure 11.22a,b,c). Occasionally this tin-bearing sulfide carries monomineralic exsolution bodies of sphalerite, chalcopyrite, or chalcopyrrhotite; compound bodies of chalcopyrite-sphalerite, and chalcopyrite-chalcopyrrhotite are also present, however. In addition, chalcopyrrhotite also appears to have locally migrated to the stannite rim.

Beyond doubt, the bulk of the stannite was deposited after the sphalerite and before the chalcopyrite, which often veins and replaces the tin mineral.

Most of the chalcopyrite is present as irregularly shaped blebs and masses that either replace one or another of the earlier species or are surrounded by galena and partly-replaced by it (Figure 11.22b).

Locally chalcopyrite replaces sphalerite extensively, and sometimes only wormlike relicts of the zinc species remain in the copper ore, thus giving rise to a most interesting pseudo-eutectic texture.

Tennantite occurs as isolated and comparatively rare small subhedral to euhedral grains that *appear* to have been deposited on either stannite or chalcopyrite, then becoming covered by galena (Figure 11.22b).

Galena, which accounts for more than 50% of the ore and constitutes its matrix, is capable of replacing all the earlier species. However, it shows something of a preference for sphalerite, stannite, and chalcopyrite.

The general sequence of events occurring during the genesis of the ore under review might have been represented in the conventional way by means of a paragenesis table. However, such tables invariably give the impression that the compilers completely understand just how the ores they are dealing with came about. The truth is that rarely, if ever, is it possible to give anything like an adequate account of the genesis of any polymineralic epigenetic ore by examining a few, or even many, so-called representative samples. Even if the ore body has been extensively opened up during mining and is accessible, the limits of the body are never exposed because the body ceases to be economic before the limits are reached. Complete examination, therefore, is still not possible.

In lieu of the usual paragenetic table, the writer has substituted Figure 11.23. The diagram indicates the species present, the order in which they appear to have developed, the major components that must have moved into the site of deposition, the major components liberated when one species was replaced by another, and what might have happened to some of the components liberated during such replacements.

Because the percentages of each of the species present in the entire ore body are unknown and because we cannot determine to what extent an early mineral has been eliminated by subsequent replacements, it is impossible to ascertain whether, say, the tin released when the cassiterite was replaced by sphalerite was sufficient for the formation of all the stannite present, or whether "new" tin had to be added to the system for its formation. The possibility of such new additions, indicated on

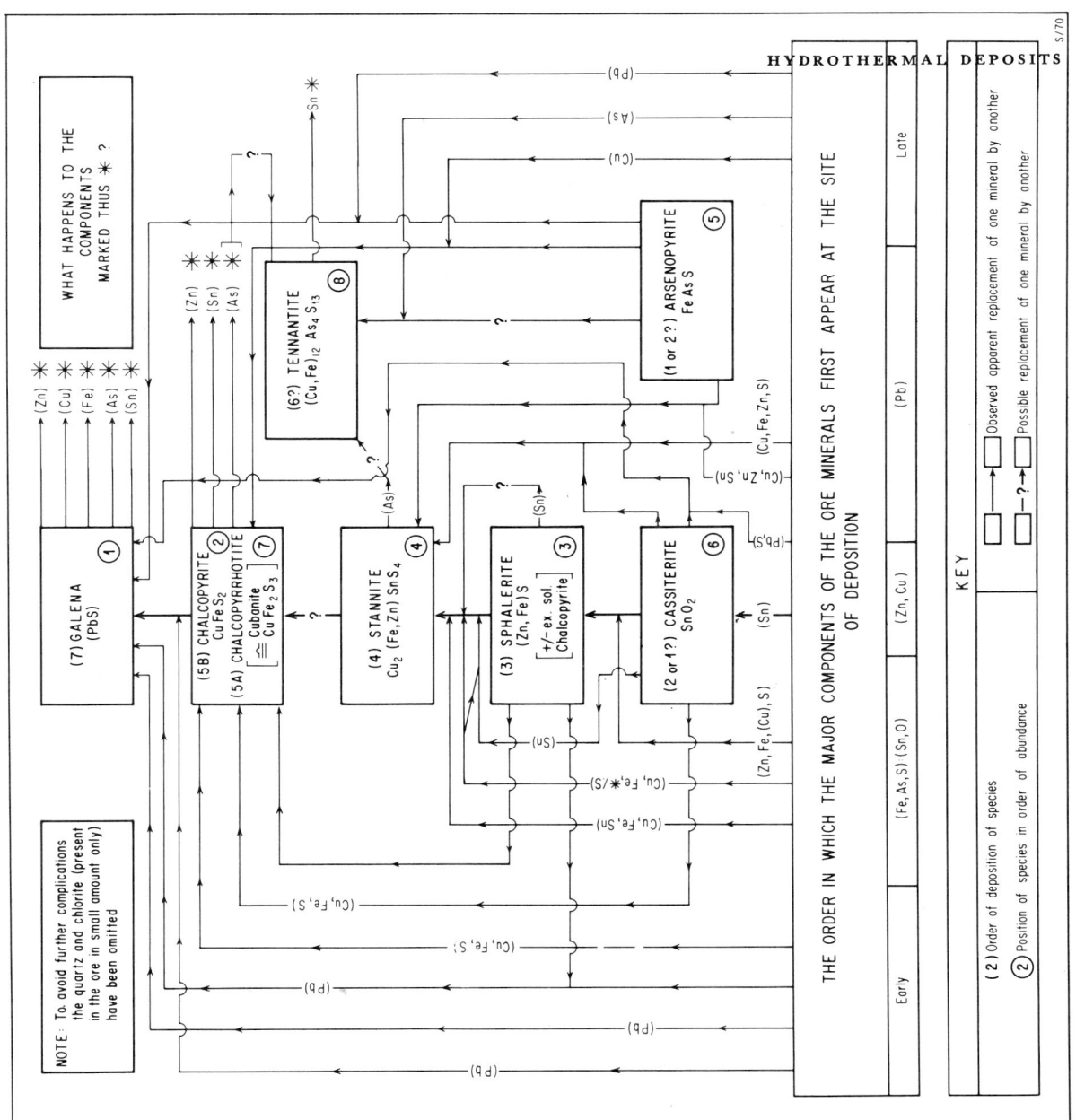

FIGURE 11.23 *A "flow sheet" indicating the actual and possible chemical and mineralogical stages in the evolution of the Setapak telescoped ore.*

the diagram, greatly increases the complexity of the ore-forming picture. It is probably much more realistic to assume that generally each major component was only introduced in quantity once, and that having been deposited as a component of a mineral, some or all of its might later be mobilized and utilized in the development of another species elsewhere in the zone of ore development. Alternatively, it might be incorporated into a new species more or less *in situ* by the little-understood replacement processes.

Figure 11.23 also indicates that replacements, particularly those effected during the later stage of ore genesis, may release components for which no immediate "resting places" are apparent. What happens to these? It might be said that they contribute to the formation of other species elsewhere in the ore body, but this is not a satisfactory answer because "left-over" components will appear everywhere in such a body. In part they may migrate into the wall-rock, thus contributing to the development of the usual envelope of rock characterized by anomalously high concentrations of the ore metals. However, such envelopes are commonly very narrow and their metal content is to be reckoned in parts per million, not in percentage. Thus it is probable that only a fraction of the unused components find a home there.

This only leaves one reasonable solution: namely, that the "left-over" components leak away from the site of deposition by way of fractures. Probably in most cases they ascend by the upward extensions of the fracture systems, which permitted the ore-developing agents to enter the impounding structure in which the ore was deposited, finally becoming dispersed in the groundwater shell. On other occasions, when there is a perfect barrier to such upward migration, they may migrate laterally along the lower contact of the barrier and the ore host rock, or they may become dispersed and absorbed along microfractures and other passageways in the latter. The possibility that sometimes the ore-developing agents might be carried out of the system in the descending portions of convection currents is also worthy of serious consideration. In any event, there must be an adequate plumbing system to cope with the large quantities of solution that enter the site of deposition during the development of epigenetic ore bodies and that, when depleted of their ore-forming components, must be removed.

A further and self-evident requirement is that adequate channelways be maintained within the developing ore body during the period of its development. Probably this is normally achieved by repeated movement within the fault systems with which such bodies are associated. It is difficult to believe that diffusion plays a major role, and it is equally difficult to accept that adequate transfer of material can be effected by movement along interstices between minerals and by way of intercrystalline boundaries, even when it is postulated that the ore-forming agents are very tenuous.

Manson Lode. The Manson Lode, Kelantan, occurs in limestone, indurated shale, phyllite, meta-acid volcanic rock, tuff, and quartz porphyry dykes, all of which may be of Permian age. From the limited information available, it is thought that the lode itself has a hanging wall of tuff and a foot-wall of limestone. The nearest outcropping granite is about 9 km (6 miles) away. A complex series of faults characterizes the area, and the lode itself appears to be a mineralized shear zone that has been repeatedly reopened (Figures 11.19a,b; Figure 11.24).

A tentative paragenesis, based on preliminary studies, appears in Table 11.8, and an indication of the elements present in the body and their relative abundance is suggested by the results of a semiquantitative spectrographic analysis of samples of oxidized and unoxidized ore (Table 11.9). In spite of the apparent abundance of mercury, no mineral in which the element was an essential constituent was noted by the writer. However, Rajah (personal communication via Santokh Singh) reported specks of cinnabar in sphalerite. This suggests that mercury may be somewhat erratically disposed in the ore body (Hosking, Leow, et al. 1970).

All the mineralogically complex ores possess the characteristics of xenothermal deposits. Park and MacDiarmid (1964, p. 346) regarded xenothermal deposits as consisting of "mixed high- to low-tem-

FIGURE 11.24 *Photomicrograph (K. F. G. Hosking) of a polished section (in air) of ore from the Manson Lode, Kelantan. Iron-rich sphalerite (sl) contains abundant exsolution bodies of chalcopyrite (cp) and pyrite (py) locally replaced by galena (gn). Gangue (G) consists of quartz and late calcite.*

TABLE 11.8 *Mineral paragenesis of the Manson Lode, Kelantan. After Hosking, Leow, et al. (1970).*

Mineral	Time
	Early ———————→ Late
Cassiterite	
Arsenopyrite	
Pyrite	
Pyrrhotite	
Sphalerite	
Chalcopyrite	
Stannite	
Mangan-siderite	
Aquilarite (?)	
Galena	
Marcasite	
Quartz	Several generations
Calcite	

TABLE 11.9 *Semiquantitative spectroscopic analyses of oxidized (O) and unoxidized (S) ore from the Manson Lode. After Hosking, Leow, et al. (1970).*

Element	O	S
Calcium	5	5
Cobalt	0	2
Cadmium	2	1
Silver	4	4
Titanium	1	3
Zinc	5	5
Copper	5	5
Molybdenum	1	2
Uranium	1	1
Chromium	0	1
Manganese	0	3
Antimony	2	3
Aluminum	3	5
Bismuth	3	2
Nickel	0	2
Iron	5	5
Indium	3	3
Gallium	0	1
Silicon	5	5
Mercury	5	5
Tin	5	5
Lead	5	5
Arsenic	5	5

Key to the scale of values
0 = Absent.
1 = Very faint indication.
2 = Faint indication.
3 = Element definitely present.
4 = Element strongly indicated.
5 = Element very strongly indicated.

perature ores deposited near the surface," and the writer believes that the West Malaysian deposits just discussed really are xenothermal.

LODES OF THE CORNISH TYPE. In distinct contrast to any of the known primary tin deposits of West Malaysia are those of the Pahang Consolidated Mine at Sungei Lembing (Pahang). These deposits are spatially related to granites that have been dated as Upper Carboniferous (Hutchison and Snelling 1971). Briefly, at Pahang Consolidated there is a series of structurally and mineralogically fairly complex lodes, which are essentially mineralized faults that have been repeatedly reopened; they occur in a suite of slates and quartzites overlying a granite ridge. Mineralization, as at Wheal Vor (Cornwall), does not extend into the granite, and the zone of mineralization occupies a band about 600 m (2000 ft) wide that parallels the granite contact (Figure 11.25).

Although copper was somewhat more abundant in the upper stopes than elsewhere, the ore bodies are essentially telescoped, and cassiterite, chlorite, arsenopyrite, chalcopyrite, sphalerite, pyrrhotite, and pyrite may all occur in a single polished section 6 cm². However, the Sungei Lembing lodes differ from the xenothermal bodies to the west of the Main Range in possessing a less impressive mineral assemblage, displaying less complex textural features, and being very much more persistent both along the strike and in depth. They would appear to have developed along deeply penetrating fault systems which were repeatedly reopened and which constituted passageways for ascending mineralizing agents, whose composition changed with time. Much the same type of mineralization occurs down dip for at least 600 m (2000 ft) which indicates that the temperature gradient was a very gentle one. In this respect the developmental environment of the Sungei Lembing lodes differed markedly from that of the xenothermal deposits.

The lack of obvious primary zoning in individual lodes and veins is a feature of all the Southeast Asian tin deposits, and in this respect they differ markedly from many of their counterparts in southwestern England.

At the Pahang Consolidated Mine, as in Cornwall, cassiterite of several generations can be recognized, but the broad paragenesis appears to be cassiterite, arsenopyrite, pyrite, pyrrhotite, sphalerite (with exsolved chalcopyrite), chalcopyrite, sphalerite (lacking exsolution bodies), and galena. Quartz was introduced on a number of occasions during the phase of mineralization; considerable quantities of

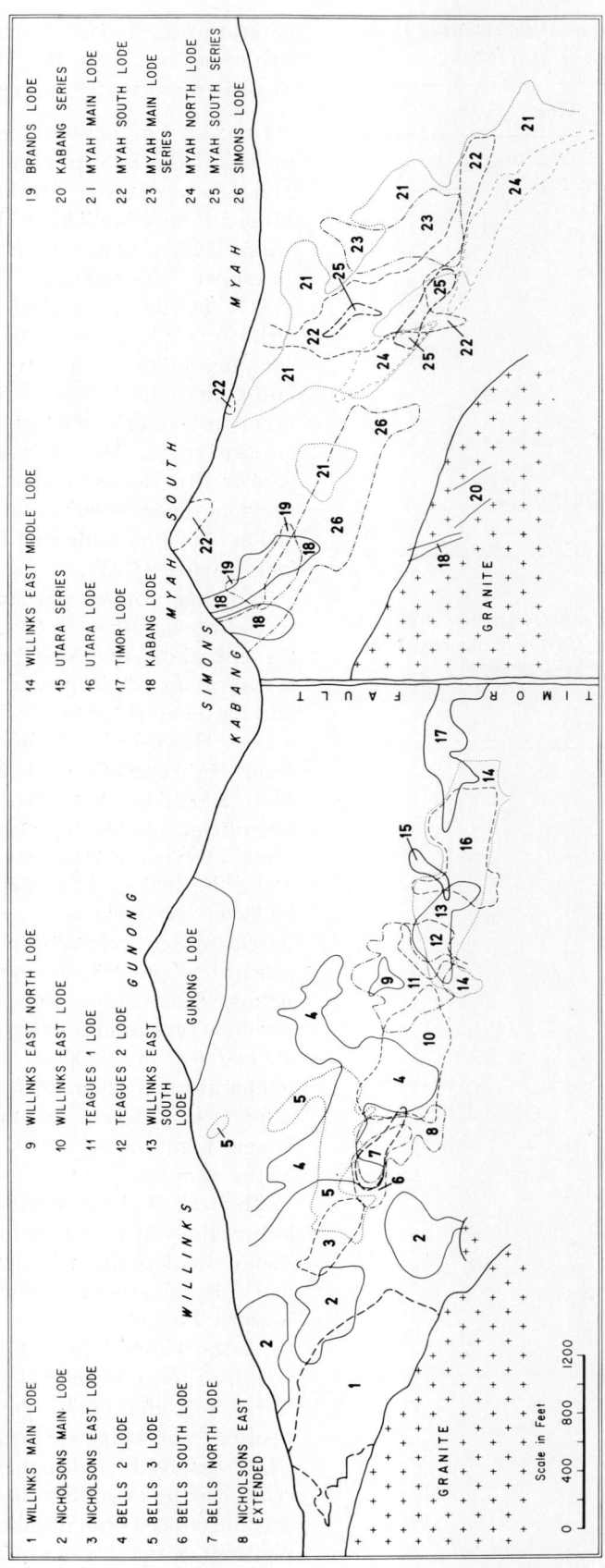

FIGURE 11.25 Section, looking north, through the Pahang Consolidated mines, showing the stanniferous belt that parallels the granite contact. After an unpublished section by G. Smith, by kind permission of the General Manager, Pahang Consolidated Co. Ltd.

chlorite were deposited immediately after the cassiterite (Hosking 1969a), and some was also deposited later. The lodes were repeatedly reopened during their development, and therefore it is not surprising that a brecciated texture is common. Further complications are caused by both prelode and postlode wrench faulting, but the history of the faulting is not understood despite valiant efforts by Fitch (1952) to unravel it.

Iron Deposits

In West Malaysia most of the iron ore that has been mined is the weathered product of an unknown primary ferruginous deposit. However, the nature of the primary deposit which, on oxidation, gave rise to the Mahang Satahun ore, is known, and the primary deposit has been exploited at Bukit Besi. Both in the vicinity of Gunong Jerai and at Tambun, Perak, there are iron deposits of probably hydrothermal origin. Those at Gunong Jerai which have been worked consist mainly of the oxidized portions of the deposits, but some of those exploited at Tambun consists solely of primary ore. At Gunong Jerai "the greatest concentrations of iron ore are to be found in the schists and metaquartzites which surround the granite" (Bean 1969, p. 14).

According to Bradford (1965), magnetite tends to occur as sporadic disseminations along bedding planes or as cross-cutting veinlets, largely in the metaquartzite, occasionally in the schist, and never in the shale. It is also associated with some of the pegmatites. Hematite is also locally present, and Bradford is of the opinion that although some is primary, other portions of it are secondary, having been formed by the removal of silica from ferruginous sediments by supergene processes. Generally, Bradford regards the primary iron oxides as having been formed by mineralizing agents that emanated from the granite, believing that the iron was deposited, at least in the Semiling area, after cassiterite and columbite had been deposited (Bean 1969). The writer thinks that the iron mineralization at Gunong Jerai is equivalent to that of Bukit Besi—both are of hydrothermal origin, and in both the iron mineralization predated the tin ore. The timing of columbite deposition at Gunong Jerai is still an open question—the columbite may be a true pegmatite mineral and may, therefore, have been deposited before the magnetite.

To the east of Ipoh, metasomatic iron ore deposits, probably occupying shear zones, occur in the limestone hills. The ore consists essentially of rosettes of hematite and botryoidal masses of the mineral which have been fractured, at least at Gunong Panjang, and cemented by chlorite. The calcite adjacent to the lodes has been coarsened by recrystallization, and some of it is pink as a result of incipient replacement by hematite. Detrital ore, derived from the weathering of the deposits, occupies areas at the base of these iron-bearing hills and has been exploited. The writer considers that the primary deposits are late, low-temperature, hydrothermal ores that can be classified as mesothermal.

Gold Deposits

Gold in small concentrations occurs in many of the Malaysian stanniferous placers, as it does in similar placers in Cornwall and northern Nigeria. At the present time several hundred ounces are recovered annually as a by-product of tin dredging operations. The most prolific production is from Bidor, where cinnabar, albeit not in economic amounts, also occurs in the heavy mineral fraction. Gold has been won by dredging from placers in Kelantan. In the Ulu Sokor region, gold is found in pyrite in replacement bodies mainly in limestone and calcareous phyllite, and in the free state it occurs intimately associated with pyrite and arsenopyrite, in quartz lodes that are largely confined to andesite and rhyolite lavas (Santokh Singh, personal communication).

In the north-south-trending western half of the central belt of West Malaysia, which is characterized by numerous acid intrusives and effusives, many auriferous areas are known. Some of the placers have been worked there in a small way for gold, and gold in quartz stringers has been reported in a number of localities, for example, in schistose quartz porphyry in a tributary of the Sungei Timah (Richardson 1950).

However, only at Raub has there been a sizable gold mining operation, and the record annual production for the Raub Australian Gold Mining Company in 1938 was 29,789 oz of gold bullion (Richardson 1950).

The "Raub Mine" is situated between the Main Range Granite and the Benom Range, where the country rocks are of fissile calcareous and graphitic shale that has been intensely folded. These rocks have been invaded by apophyses from the neighboring Bukit Kajang granite porphyry. According to Richardson (1939), the resulting gold mineralization was due to hydrothermal solutions emanating from interior reservoirs in the granitic mass and migrating along two fissure zones in the shale (the so-called eastern and western lode-channels) and a third (at the Raub Hole Mine), which lies between the two other channels and was due to tensional faulting. The gold, occurring as shoots, is associated with stibnite, scheelite, arsenopyrite, pyrite, quartz, calcite, and minor chalcopyrite and cerussite (a secondary species). Finally, the area was subject to postmineralization faulting.

Richardson gave the following sequences of deposition for the Raub ore (1939, p. 129):

Stage I. Introduction of translucent, white or pale grey vein quartz, full of opaque dust inclusions, and containing a little sulfide and gold. The quartz invaded the graphitic and calcareous shale, fragments of which were incorporated within it.

Stage II. A. Deposition of sulfides and gold in the quartz. The sulfides were pyrite, arsenopyrite, or stibnite. Stibnite has replaced pyrite in part.

B. Introduction into the quartz of stringers of calcite, sometimes with cryptocrystalline quartz in addition and of scheelite.

Stage III. Metasomatic replacement of stage I quartz by stringers of water-clear cryptocrystalline quartz which has traversed all previous structures and minerals. This quartz commonly occupies the interstices in the quartz of stage I.

Stage IV. Occurrence of minor movements, resulting in slickensiding of the lodes, followed by the deposition of "paint gold."

Table 11.10 indicates the grades of ore mined in the various sections of the Raub mine in the first part of the twentieth century. Richardson believed that the grade generally increased over the years partly because mining became progressively more selective, but also because of an intrinsic increase in the tenor of the ore bodies at greater depths.

Base Metal Sulfide Deposits

Although sulfides are common components of many of the Malaysian primary deposits, comparatively few deposits are known which consist essentially of the lower temperature sulfides and lack such early minerals as cassiterite and wolframite. Only one, the Maran lead prospect, has been seriously investigated recently and, unfortunately, that investigation was not completed. At least one other sulfide occurrence, that in the Segamat volcanic mass, is worthy of examination. However, the

TABLE 11.10 *Production of gold and the average grade of the underground ore from Raub from 1900 to 1934. Based on Richardson (1939, p. 125).*

Section	Period	Tonnage of Underground (sulfide) Ore Mineral	Output of Gold Bullion (troy oz.)	Average Grade of Ore (dwt.) of Bullion per Ton
Bukit Koman Mine	1905–1909	146,377	27,092	3.7
	1910–1919	234,409	49,515	4.2
	1920–1929	109,242	55,223	10.1
	1930–1934	52,463	49,037	18.7
	1905–1934	542,491	180,867	6.6
Bukit Malacca Mine	1900–1904	24,277	8,807	7.2
	1905–1919	549,939*	48,059	1.7
Stope Mine	1905–1909	77,924	23,210	5.9
Anderson Mine	1911–1919	160,650	33,715	4.2
	1920–1929	110,315	46,388	8.4
	1930–1933	35,658	30,526	17.1
	1911–1933	306,623	110,629	7.2
Derrick Mine	1920–1929	75,407	27,796	7.4
	1930–1934	61,864	16,959	5.5
	1920–1934	137,276	44,755	6.5

* Ore from surface operations.

presence of many small showings of sulfides (including stibnite and cinnabar) and gold in the volcanic-rich, little-investigated central belt indicates that it should be prospected for base metal sulfides and gold.

THE MARAN LEAD ORE. Through the good offices of Eastern Mining and Metals Corporation, the writer was permitted to study certain specimens of the lead ore intersected during diamond-drilling operations at Maran, which were carried out in order to assess the mineral potential there. The deposit is probably epigenetic and is associated with a sequence of acid-intermediate volcanic rocks and metasediments. The mineral paragenesis of the ore, of which galena is the dominant component, is indicated in Table 11.11.

TABLE 11.11 *Paragenesis of Maran lead ore. Hosking and Yeap (unpublished results).*

Mineral	Time (Early → Late)
Quartz	—
Pyrite	—
Pyrrhotite	—
Marcasite	—
Chlorite	—
Siderite	—
Hematite	—
Sphalerite	—
Chalcopyrite	—
Galena	—

The relationships between galena and the other components in this essentially late hydrothermal deposit are particularly interesting, as Figure 11.26 demonstrates. Commonly the lead sulfide contains abundant inclusions of chlorite, pyrite, marcasite, and hematite. A considerable proportion of the galena also occurs as microveins in the pyrite and marcasite; elsewhere in the polished sections examined, this sulfide replaces both quartz and sphalerite. High recovery of lead in the form of a clean galena concentrate will offer problems to the mineral dresser.

THE SEGAMAT DEPOSITS. Incipient sulfide mineralization occurs at Segamat, Johore, in a volcanic pile of potassic andesitic basaltic flows (Teng 1970) lying unconformably over the Tenang beds. The Tehang beds consist essentially of a sequence of shale and mudstone with intercalated sandstone bands; their fossil content indicates Triassic age.

Chalcopyrite, bornite, and galena, associated with zeolites, fluorite, and calcite, occupy veins and vesicles. Strontium and barium minerals may also be present (certainly geochemical studies indicate that barium occurs locally in the volcanic rocks, and overlying soil in strongly anomalous amounts), but the writer has not identified them in the material he has examined.

Grubb (1965, p. 344) stated that

"Mineralization in Segamat, chiefly in the form of copper and lead sulfides, was largely synchronous with vesiculation, yet unlike similar deposits in Michigan . . . there was, despite widespread late autometasomatism, little redistribution of the primary sulfides."

Teng (1970, p. 30), on the other hand, saw no field evidence to indicate that the deposition of the sulfides was contemporaneous with the vesiculation of the rocks, and he believed that the sulfides and associated gangue minerals were deposited epigenetically by ascending hydrothermal solutions. In support of his view, Teng mentioned an abundance of sulfides filling "veins and fractures" and that, at Tong Huat Quarry in particular, the degree of sulfide mineralization of the vesicles was a function of the distance they were away from mineralized fracture zones.

The writer strongly supports Teng's view of the genesis of the ore minerals and believes that even though the exposed mineralization is slight, the volcanic area is worth examining by geochemical and geophysical methods. It may be that an economic sulfide deposit is impounded beneath the volcanic pile and that the surface showings are simply due to "leak away" mineralizing agents. In any event, application of the exploration methods noted might indicate targets worthy of investigation by diamond drilling.

Studies of the "ore" in thin and polished section (Figure 11.27) indicate that the nonmetallic species were deposited first and that these were followed by chalcopyrite and then by galena. The bornite, which tends to rim the chalcopyrite, may not be a primary species. Both chalcopyrite and galena locally replace the earlier gangue, but commonly they were deposited in open spaces in the fractures and in the vesicles. Chalcopyrite is also locally replaced by galena.

This late hydrothermal deposit would certainly have been classified by Lindgren (1933) as epithermal, possessing as it does many features in common with the Lake Superior copper deposits.

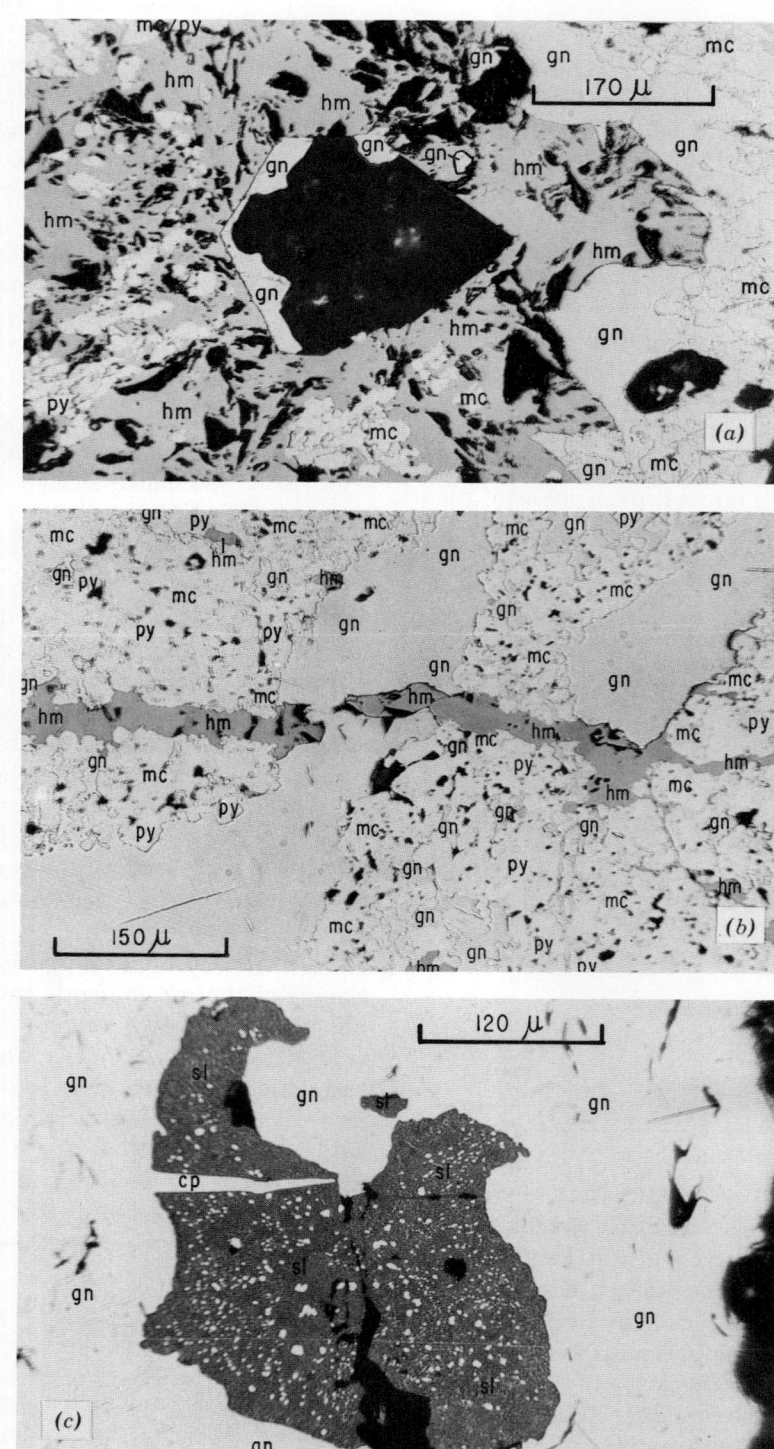

FIGURE 11.26 *Photomicrographs (E. B. Yeap) of polished sections of the Maran (lead) Lode, Pahang: (a) Pyrite (py) and marcasite (mc) partly replaced by hematite (hm) and galena (gn). In the center, euhedral quartz is rim-replaced by galena. Near black areas are mainly voids but some in the hematite are chlorite. In air. (b) Galena (gn) developed by replacement of marcasite (mc) and pyrite (py). Hematite, later than the marcasite and pyrite, has resisted replacement by galena. In air. (c) Relict sphalerite (sl) with exsolved bodies and veins of chalcopyrite (cp) is embayed and enveloped by galena (gn). Near black areas are partly voids and partly gangue. In oil.*

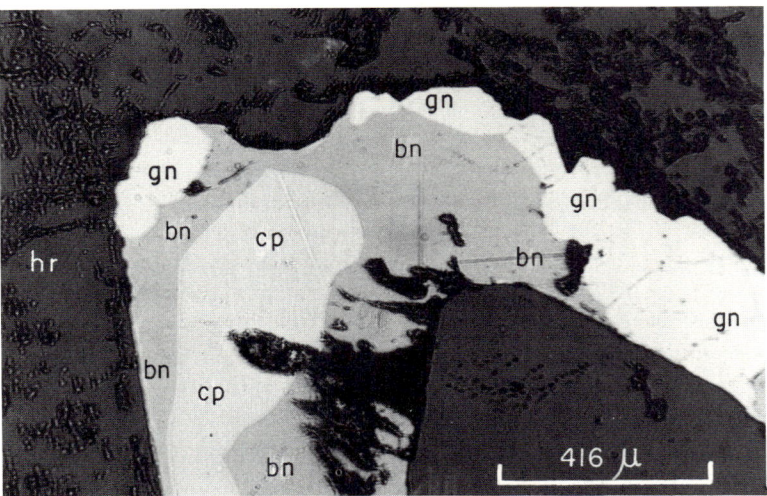

FIGURE 11.27 *Photomicrograph (K. F. G. Hosking) of a polished section (in air) of part of a vesicle in Segamat volcanic rock (hr) which has been infilled with bornite (bn), chalcopyrite (cp), and galena (gn).*

Barite Deposits

At present small amounts of barite are recovered from a hillside exposure at the Good Earth Mine near Tasek Chini, where the mineral is associated with hematite deposits, which are also being exploited. This barite exposure, which is about 3.5 m (12 ft) wide, may be a dip section of a steeply dipping lode-like body, but this is not certain. Locally a little pyrite and occasional small veins of specularite are seen in the essentially dark green ore. Perhaps an epigenetic, late hydrothermal deposit has developed along a fracture zone.

Barite is also known to occur in several localities in Perak, particularly Upper Perak; it has also been found to the north of Jerantut Railway line, in Pahang, and at Bukit Penchuri, Kelantan, where it is associated with hematite. In the Bradwell Estate, south of Kampong Siliau, in Negri Sembilan, an outcrop of barite occurs that is about 9 m (30 ft) wide it strikes east-west, dips vertically, and is exposed over a distance of 19 m (62 ft).

At Kuala Trengan, in Trengganu, barite boulders occur. Santokh Singh (personal communication) wrote that the area.

> ... is underlain by a sedimentary sequence consisting of shales, including carbonaceous shales, siltstones, subgreywacke, sandstones, and conglomerates. Locally the rocks are strongly silicified and numerous quartz veins, infilling tension fractures, have been recorded. Numerous vertical dolerite dykes, striking normal to the regional strike, crop out in the area. ... [W]idespread shearing has occurred in some localities. ...

Santokh Singh also stated that

> The barite is apparently associated with dolerite dykes ... and ... from the few specimens showing the host rock (shales) and barites together, it appears that the barites has been deposited in fissures and fractures in the country rock, but replacement has occurred in a number of places. It is highly significant that the barites is apparently closely associated with a zone suspected of north-south faulting which would have afforded an ideal channel-way for a variety of ore solutions.

Geochemical and other studies carried out at Kuala Trengan by the Geological Survey led Santokh Singh to conclude that there is a distinct possibilitly that in depth this late hydrothermal body may also contain galena and sphalerite.

IN SITU EFFECTS OF WEATHERING OF PRIMARY DEPOSITS

Although strictly speaking, the effect of weathering of the primary deposits is outside the scope of this chapter, it is necessary to explain why primary deposits can be so well studied in West Malaysia where hard-rock mining is relatively rare, and why so little reference has been made to many of the West Malaysian iron deposits, particularly the rich deposits of Rompin, Pahang.

The extent to which a given primary ore body is altered by weathering agents depends on many factors, including the following: the mineralogical composition of the ore body and of the wall-rocks;

the relative proportions of "reactive" and "inert" minerals present; the textural character of the ore; the characteristics of the joints and fractures within both the ore body and the adjacent rocks; the shape of the body; its relationship to the topography; the climatic factors to which it was exposed from the time it was first affected by weathering agents; and the total geologic events with which it has been associated, but particularly those which took place from Tertiary time to the present.

In West Malaysia the present state of the superficial parts of *in situ* ore deposits are due, perhaps above all, to their topographic position immediately before the Pleistocene, to the geologic events and the climate during and subsequent to the Pleistocene, and to the nature of the rock in which the deposit was situated. The mineralogic character of the ore body was also an important factor.

There is little doubt that the lowering of the sea level (of the order of 130 m or more in the Pleistocene) because of the removal of seawater as ice, must have resulted in marked rejuvenation of the rivers and an equally distinct lowering of the water table. Such events would cause increased and rapid oxidation in depth, but because of the speed of the event, deposits rich in sulfides, particularly pyrite, were incapable of being completely oxidized above the water table. Such masses of perched sulfides doubtless were preserved in part by the protective coating of iron oxides. The Manson Lode, Kelantan, which outcrops in the valley side, consists of sulfide protected by a comparatively thin layer of gossan. During this period, many of the iron ore deposits surely developed their present character. The oxidized upper parts of Bukit Besi, which originated essentially from pyrite, is an outstanding example; so, probably, is the Bukit Iban iron oxide deposit (Taylor 1971).

The water table was subsequently raised by the infilling of valleys by alluvium and by marine transgressions, possibly to a height of about 12 m above the present mean sea level. As a result of this, ore bodies that were already partly oxidized, and not in particularly elevated positions, were subject to reducing conditions. One consequence is to be seen at the Sungei Besi Mine, where the Lee Gossan is situated in granite on the side of an old valley. This originally consisted mainly of arsenopyrite and cassiterite, somewhat sparsely distributed in a quartz matrix. It was first subjected to oxidizing conditions owing to the rejuvenation of the river, which caused the arsenopyrite to be partly converted to scorodite (mauve and blue in addition to the usual green) pharmacosiderite, and limonite. Subsequently, as the water table rose because of the raising of the sea level and the infilling of the valley with stanniferous alluvium, the ore body was subject to reducing conditions, which were largely responsible for the development of supergene pyrite on the scorodite. This ore body, with its sparse sulfide content, was only slightly oxidized by comparsion with the adjacent No. 4 ore body, which in its initial state was very pyritic.

In low-lying areas not far from the granitic ridges where the stanniferous placer deposits occur, lodes with little or no signs of oxidation are commonly exposed during open-cast mining, and particularly where the bedrock is limestone. This is the case in many mines along the fringes of the Kinta Valley and also in many mines in the Kuala Lumpur field. The situation must be, in part, related to the lowest water-table level encountered in these areas, but the limited oxidation may have been due to the neutralizing action of the limestone on acids which, in a noncalcareous environment, would be free to attack the sufides. However, sometimes deep preferential weathering of ore bodies took place even in the limestone. This happened in the Lahat Pipe to a depth of 90 m (300 ft) below the limestone floor, and a somewhat similar state of affairs appears to have occurred locally at Telok Kruen, Perak (Ingham and Bradford 1960, p. 278).

The deep secular weathering that has affected the deposits and their host rock on hill sides has been of major economic importance to the country in that it has permitted both nature and man to remove cassiterite easily from the stanniferous ores. Many of the stanniferous lodeswarms that have been mined profitably in West Malaysia by lampanning and gravel-pump methods would not have been exploited if the mining and beneficiation methods needed to treat unweathered rock had had to be employed.

The rejuvenation of rivers during the Pleistocene, accompanied by rapid valley erosion and intense weathering of numerous tin deposits [whose mineralization was condensed into a zone only about 200 m (660 ft) thick] must have been largely responsible for the development of the extremely rich placers in the Kinta Valley and elsewhere west of the Main Range. There is no doubt that such rich secondary deposits would not have developed had the primary deposits been lodes of the Pahang Consolidated type, in which the cassiterite is dispersed over a zone several times as thick.

TIME RELATIONSHIPS BETWEEN THE PRIMARY MINERAL DEPOSITS AND IGNEOUS ROCKS

Recently advances have been made in the understanding of the time relationships existing between the primary ore deposits and the igneous rocks of West Malaysia. The writer's present views concerning this topic stem from the known radiogenic ages of some of the igneous rocks, the zone of emplacement of the granites, the nature of the various primary mineral deposits, and their spatial relationship to the igneous rocks.

According to Bignell (personal communication), the radiogenic dates so far obtained give reason for believing that in West Malaysia there are granites of Carboniferous, Triassic, and Upper Cretaceous to lower Tertiary age.

The granites of Bakri, Johore, and Gunong Jerai, Kedah, with which are associated cassiterite-tantalite-columbite-bearing pegmatites, are tentatively thought to be of Carboniferous age, as is the granite of Pahang Consolidated mine in the eastern belt. Other granites of Carboniferous age occur locally in northeast Malaya and in the Main Range. Many samples of the Main Range Granite are of probable Triassic age, and with this Triassic granite are associated the numerous vein swarms, pegmatites, and occasional lodes from which the major stanniferous placers have been largely derived. Finally, a few isolated granitic masses of Upper Cretaceous to Lower Tertiary age, of which the best known is Mount Ophir, are virtually devoid of associated mineralization, although Bignell (personal communication) noted a small vein of galena in one of these late granites. That late mineralization has occurred, at least locally, is further indicated by the copper and lead sufides found in the probably Late Tertiary or even Quaternary volcanic rocks at Segamat.

In view of the profusion of minor, largely acid, intrusives and extrusives of varying ages in the central belt, it would be surprising if primary mineralization had not occurred on occasions other than those noted previously.

Hutchison has provided strong evidence (this volume) for believing that the Triassic granites of the western belt were emplaced in the mesozone. It is probable, then, that during the phase of early hydrothermal mineralization associated with this granite invasion, ore deposits could develop in the granite, in the brittle noncalcareous metasediments, and in those calc-silicate hornfelses which had been formed from the alteration of impure calcareous rocks. Possibly, also, lodes developed along the faulted granite–pure limestone contacts. However, this is less likely because pure limestone, deeply buried and quite hot, would tend to deform rather than to fracture when subject to stresses capable of fracturing other rock types. Moreover, primary deposits in the granite commonly die out at the contact between granite and pure limestone.

The mesozonal nature of the western belt granite may also account for the absence of the numerous microgranitic dykes which characterize the Cornish tin fields, where the granites were emplaced at a higher level in the crust. It is also probably one of the reasons the western belt lacks those early hydrothermal lodes that, in Cornwall, may have strike lengths of several miles and may be strongly mineralized for several thousand feet down their dips. Clearly, other things being equal, much greater forces would be required to produce the necessary fractures for the development of these dykes and lodes in the western belt of West Malaysia than in Cornwall.

However, strongly telescoped stanniferous veins occur in the limestones of the Selangor field, and similar veins and pipes occur in a similar geologic environment in the Kinta Valley. These deposits, which can be termed xenothermal, must have developed in a near-surface environment. We can only conclude that such deposits are much younger than the Triassic granites and are not genetically related to them. They must have been formed when denudation had brought the limestones sufficiently near the surface for them to behave as brittle bodies. Hence the writer tentatively believes that these primary deposits may be associated with the Upper Cretaceous to Tertiary granite invasion, which until now was thought to have had virtually no part in fashioning the present primary mineral distribution pattern of the country. The numerous hot springs of West Malaysia, and the workable deposits of fluorite in the Quaternary of North Thailand, may represent the last waning phases of this period of mineralization.

The primary mineralization in the granite of the Kledang and Main ranges, occurring mainly on the flanks, is of the type associated in Cornwall and in Portugal with the apices of granite cusps. This is thought to be largely because it is genetically and spatially related to late phase Triassic granite cusps that have developed within an earlier phase. The earlier phase may have caused the inferior mineralization along the Main Range watershed. Some of the flanking deposits in the granite could, of course, be associated with cusps of Late Cretaceous or Lower Tertiary granite. In addition, the mineral content of the earlier deposits may have been increased by additions made to them during the later phase of igneous activity and mineralization (Hosking 1970b). The assumption that the majority of the primary mineral deposits closely associated with the Main Range are linked with cusps of younger granite would go far toward explaining why primary tin deposits are in so much greater profusion along the western side than the eastern side of this great granitic mass.

Largely from the foregoing data, we might reconstruct the following history of primary mineralization in West Malaysia.

Probably geosynclinal conditions were already in evidence in Late Cambrian time in the area now occupied by West Malaysia. By Late Carboniferous time the geosynclinal area was in the process of being converted into a folded belt. Then synorogenic granite was emplaced, for example, in the Pahang Consolidated area, where it is associated with Cornish-type lodes, and in the Bakri area and Kedah Peak, where it is related to stanniferous–tantaliferous pegmatites. Both these types of mineralization are wanting in areas that are dominated by later granites.

If we adopt Bilibin's (1968) views concerning the sequence of events in mobile belts, we would expect that xenothermal tin-sulfide deposits might develop during the late stage of the folded zone and that they would be associated with acid to intermediate minor intrusive and volcanic rocks. The Manson ore body, associated with igneous and other rocks thought to be of Permian age, seems to fit this requirement perfectly.

During the Triassic, the geosyncline was rejuvenated, and when the orogenic phase reached its culmination in Late Triassic time, most of the granite that is now uncovered was emplaced in the mesozone and many tin and related deposits were deposited in those rocks which were capable of fracture. Most of these major deposits may be related to cusps of the later phases of this granite.

During Upper Cretanceous or Lower Tertiary time, renewed acid igneous activity occurred, indicated by the few small granitic plugs in Johore. However, such activity must have affected much of the western belt to the west of the Main Range, and it was probably responsible for the development of the xenothermal deposits, largely in the limestones. This is exactly what Bilibin leads us to expect.

The applicability of Bilibin's theory to West Malaysia is further supported by the nature of the mineralization found at Segamat. As further radiometric dates become available, particularly for the volcanic rocks of the central belt, it should be possible to predict where mineral deposits of a given type are most likely to be found; hence prospecting should be considerably facilitated. In addition, the findings in Malaysia also emphasize that Bilibin's views should not be neglected by anyone concerned with the search for further primary mineral deposits in any part of the Southeast Asian tin and associated metals province.

PROSPECTING

Broadly speaking, the primary mineral deposits can be divided into two classes: those of interest to the private miner and those of interest to large mining groups.

The private miner is only likely to be interested in hard-rock deposits that can be easily mined by wet open-cast methods, by benching, or by adit mining. Furthermore, the deposits must be capable of being beneficiated by simple and cheap methods. The private miner is unlikely to be interested in deposits other than those containing tin, tungsten, or iron. He rarely carries out a systematic search for such deposits, and probably many of the deposits he has worked have been found by local people by chance or as a result of the occurrence of obvious outcrops. Usually he commences mining operations with little or no knowledge of the magnitude of the ore body, of its detailed mineralogical characteristics, and particularly of its overall grade.

As a rule, major mining companies are only likely to be interested in primary deposits if they are large, and it is probable that they would prefer those which can be mined by open-cast methods. In West Malaysia they might be interested in finding low-grade yet high-tonnage tin deposits that would have had little value in the past; because of the present high price of tin, however, and thanks to markedly improved mining and beneficiation techniques, such deposits now constitute viable mining propositions.

A large company might also be interested in exploring the central belt, particularly for base metal sulfide deposits and gold. One important concern, Pahang Consolidated Company, is exceptional in that it operates a large underground tin mine.

Naturally, then, its exploration policy is one in which the search for tin lodes adjacent to its underground workings figures largely. This means that Pahang Consolidated is and will be concerned both with surface and underground exploration; the latter involves cross-cutting and drilling.

Beyond reasonable doubt, fundamentally large-scale, regional exploration is best effected by applying geochemical methods involving the analysis of stream sediments and supported by photogeologic studies and the study of panned concentrates from the streams. Any anomalous areas established by this work should then be further investigated at the surface by detailed photogeologic studies, by classical geological mapping, by detailed mineralogic studies of mineralized outcrops, and by the application of geochemical methods of prospecting involving the analysis of soil samples from spurs and ridges. The search for lode extensions, parallel lodes, and so on, in areas of known mineralization should also be carried out in essentially the way recommended for the examination of anomalous areas established during regional exploration.

It would be impossible to overstress the important rôles which both photogeology and applied geochemistry should play in West Malaysia. Furthermore, continued photogeological researches will surely provide still more aids to the search for primary ore deposits. Recently it has been demonstrated that the hilly area occupied by the wolframite-bearing veins on the Tungku Makota's property to the west of Gambang is characterized by a greater

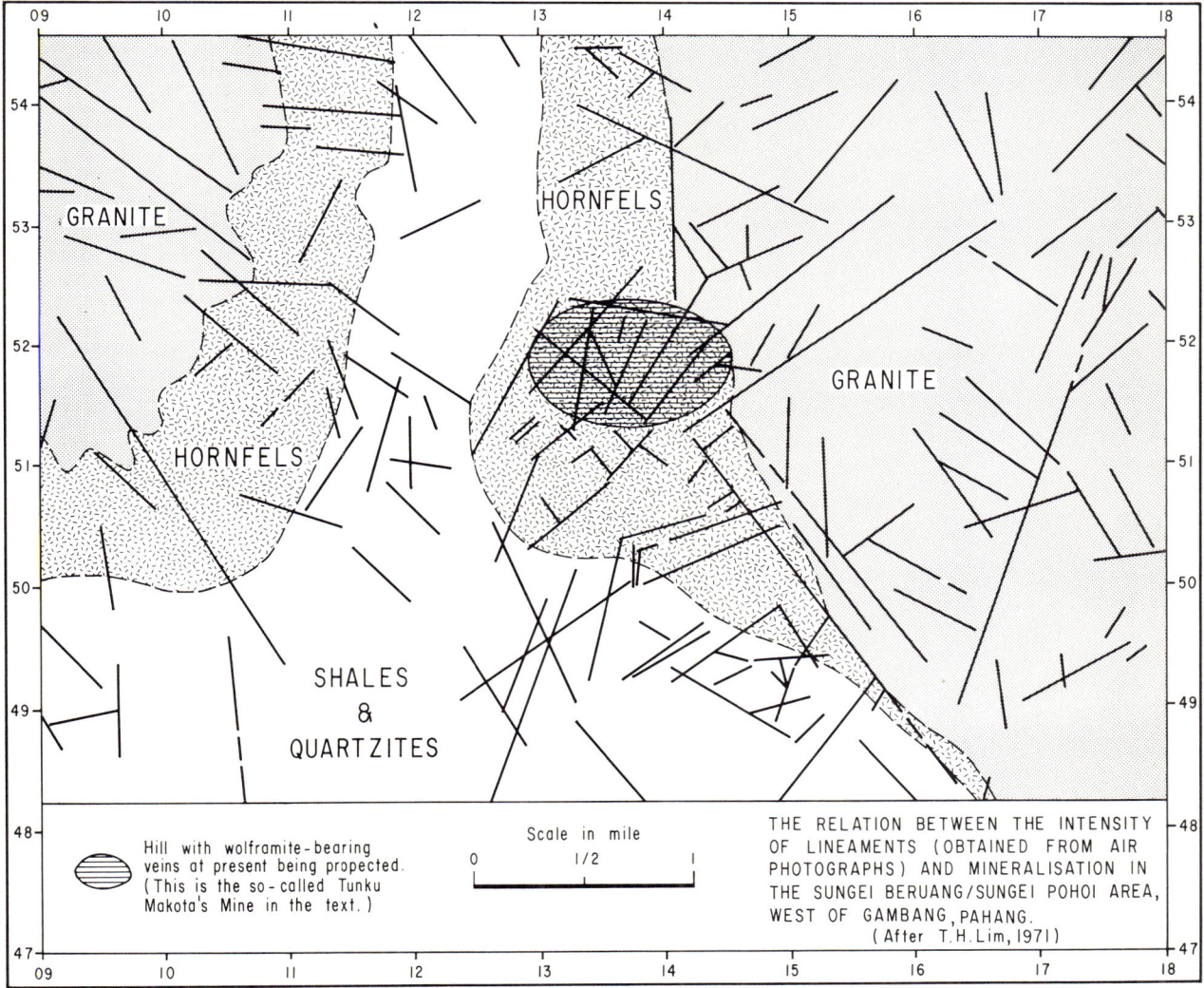

FIGURE 11.28 *Relation between the intensity of lineaments (from air photographs) and mineralization in the Sungei Beruang–Sungei Pohoi area, west of Gambang, Pahang. After T. H. Lim (1971).*

concentration of lineaments than the adjacent terrain (Figure 11.28) (Lim Teong Hua, private communication). Indeed, the strong radial drainage of the wolframite-bearing Bukit Lentor suggests that it has been updomed by a granite cusp that might account for the lodes present. It follows that targets for exploration should now include sedimentary hills with marked radial drainage, which are likely to be spotted more readily on aerial photographs than on the topographical maps, and which are favorably placed with respect to outcropping granite.

Tooms and Kaewbaidhoon (1961) showed that geochemical exploration involving the analysis of soils from spurs and ridges provides an excellent means of locating suboutcropping tin lodes on the property of Pahang Consolidated, and this technique is still in common use by the company. It has been demonstrated by the Geology Department of the University of Malaya, and others, that this technique is also suitable for the detection of suboutcropping deposits containing such elements as tungsten, lead, zinc, copper, and arsenic.

Airborne magnetometer and scintillation surveys were carried out over a considerable proportion of West Malaysia in the 1950s (Agogs 1958–1965), but the results contributed remarkably little to the discovery of further ore deposits. Ground geophysics has proved useful in the search and delineation of iron deposits and also during the search for further sulfide deposits in the vicinity of the Manson Lode. As yet, however, such aids to exploration have not proved particularly useful, except when searching for iron ore.

This is not the place to discuss in detail the best ways and means of prospecting for primary ore minerals in West Malaysia, but it is pertinent to state that library research is of the greatest importance. And not to make use of the data that can be obtained from the Mines Department and the Geological Survey would be sheer stupidity. Armed with such data, one is in a position to do some "discovery thinking" and exploration planning.

Finally, it is a sobering thought that almost all the primary deposits that have been worth exploiting in West Malaysia were found without the aid of photogeology, geophysics, and geochemistry. Often these deposits were discovered by people with no technical training, whose only aids were a dulang and a changkul!

References

Note: Completed yet unpublished items are indicated by (MS) after the date. Items being edited for publication by the Geological Survey Department, West Malaysia, are referred to as "in prep."

Adam, J. W. H. 1960. On the geology of the primary tin-ore deposits in the sedimentary formation of Billiton. *Geol. en Mijnb.* **22**, 405–426.

Agocs, W. B. 1958–1965. Airborne magnetometer and scintillation counter survey. *Econ. Bull. Geol. Survey Dept. W. Malaysia* **1**, pts. 1–6.

——— 1966. *Report on airborne magnetometer and scintillation counter survey of Kedah, Perak, Selangor, Trengganu, Pahang, and Johore, Federation of Malaya*, 340 p. Malaysian Government.

Aleva, G. J. J. 1960. The plutonic igneous rocks from Billiton, Indonesia. *Geol. en Mijnb.* **22**, 427–436.

Alexander, F. E. S. 1950. *Report on the availability of granite on Singapore and the surrounding islands*, 24 p. Singapore: Government Publications Bureau.

——— 1959. Observations on tropical weathering: a study of the movement of iron, aluminium and silicon in weathering rocks at Singapore. *Quart. J. Geol. Soc. London* **115**, 123–144.

Alexander, J. B. 1947. Report on special investigations of coal-bearing formation at Bukit Arang in the state of Perlis during 1941. *Rept. Geol. Survey Dept. Malay. Union* [for 1946], 44–46.

——— 1956. *Lexique stratigraphique international*, **3** (fasc. 6b, 7c), Malaya, 31 p. Paris: International Geological Congress.

——— 1959. Pre-Tertiary stratigraphic succession in Malaya. *Nature, Lond.* **183**, 230–231.

——— 1962. A short outline of the geology of Malaya, with special reference to Mesozoic orogeny. In *Crust of the Pacific basin. Monogr. Am. Geophys. Union* **6**, 81–86.

——— 1965. *Geological map of Malaya*, 6th ed. (Diamond Jubilee), 1:500,000. Geol. Survey Dept. W. Malaysia.

——— 1968. The geology and mineral resources of the neighbourhood of Bentong, Pahang and adjoining portions of Selangor and Negri Sembilan, incorporating an account of the prospecting and mining activities of the Bentong district. *Mem. Geol. Survey Dept. W. Malaysia* **8**, 250 p.

——— and Flinter, B. H. 1965. A note on varlamoffite and associated minerals from the Batang Padang district, Perak, Malaya, Malaysia. *Miner. Mag.* **35**, 622–627.

———, Harral, G. M., and Flinter, B. H. 1964. Chemical analyses of Malayan rocks, commercial ores, alluvial mineral concentrates 1903–1963. *Prof. Pap. Geol. Survey Dept. W. Malaysia* E64.1C, 295 p.

———, MacDonald, S., and Slater D. 1961. The basement rocks of Malaya and their paleogeographic significance in South-East Asia—A discussion. *Am. J. Sci.* **259**, 801–806.

———, and Müller, K. J. 1963. Devonian conodonts in stratigraphic succession of Malaya. *Nature, Lond.* **197**, 681 only.

———, and Procter, W. D. 1955. Investigations upon a proposed dam site at Klang Gates, Federation of Malaya. *Colon. Geol. Miner. Resour.* **5**, 409–415.

American Commission on Stratigraphic Nomenclature. 1961. Code of stratigraphic nomenclature. *Bull. Am. Assoc. Petrol. Geol.* **45**, 645–665.

Anonymous. 1966. Age determination unit. Summary of results from Malaysia. *Ann. Rept. Inst. Geol. Sci.* [for 1965], pt. 2, *Overseas Geol. Surveys*, 44–47.

Aranyakanon, P. 1961. The cassiterite deposit of Haad Som Pan, Ranong Province, Thailand, *Rept. Invest. R. Dept. Mines, Bangkok* **4**.

———, 1969 (MS). *Tin deposits in Thailand*. Preprint of paper read at the 2nd Technical Conference on Tin of the International Tin Council, Bangkok.

Armstrong, R. L. 1966. K-Ar dating of plutonic and volcanic rocks in orogenic belts. In *Potassium Argon Dating*, O. A. Schaeffer and J. Zahringer, Eds. Berlin: Springer-Verlag, 117–133.

Asama, K. 1966. Permian plants from Petchabun, Thailand and problems of floral migration from Gondwanaland. In *Geology and paleontology of southeast Asia*, vol. 2, T. Kobayashi and R. Toriyama, Eds. Tokyo: University of Tokyo Press, pp. 197–237.

———, Iwai, J., Veeraburas, M., and Hongnusonthi, A. 1968. Permian plants from Loei, Thailand. In *Geology and*

Palaeontology of southeast Asia, vol. 2, T. Kobayashi and R. Toriyama, Eds. pp. 82–99.

Aubouin, J. 1965. *Geosynclines*, 335 p. Amsterdam: Elsevier.

Aw, P. C. 1964 (MS). *The geology of the Temangan–Sokor area, Kelantan*. B.Sc. Honours thesis, Department of Geology, University of Malaya.

——— 1967. Ignimbrite in central Kelantan, Malaya. *Geol. Mag.* **104**, 13–17.

Badgley, P. C. 1965. *Structural and tectonic principles*, 521 p. New York: Harper and Row.

Bailey, E. H., and Stevens, R. E. 1960. Selective staining of K-feldspar and plagioclase on rock slabs and thin sections. *Am. Mineral.* **45**, 1020–1025.

Baker, G. 1963. Form and sculpture of tektites. In *Tektites*, J. A. O'Keefe, Ed. Chicago: University of Chicago Press, 1–24.

Barnes, V. E. 1963. Tektite strewn fields. In *Tektites*, J. A. O'Keefe, Ed. Chicago: University of Chicago Press, pp. 25–50.

——— 1967, Tektites. In *International Dictionary of Geophysics*. Oxford: Pergamon Press, pp. 1507–1518.

Batten, R. L. 1972. Permian gastropods and chitons from Perak, Malaysia. Pt. 1. Chitons, bellerophontids, euomphalids and pleurotomarians. *Bull. Am. Mus. Nat. Hist.*, 44 p. **147** (1).

Bean, J. H. 1969. The iron-ore deposits of West Malaysia. *Econ. Bull. Geol. Survey Dept. W. Malaysia* **2**.

——— In prep. Geology and mineral resources of Pulau Tioman. *Bull. Geol. Survey Dept. W. Malaysia*.

———, Bradford, E. F., Burton, C. K., Chung, S. K., Foo, Y. H., Jones, C. R., MacDonald, S. Santokh Singh, D., and Slater, D. 1966. Geological map of Sungei Bera–Tasek Chini–Sungei Jeram, Pahang, 1:63,360. *Geol. Survey Dept. W. Malaysia*.

Becher, H. M. 1893. The gold-quartz deposits of Pahang (Malay Peninsula). *Quart. J. Geol. Soc. London* **49**, 84–88.

Bemmelen, R. W. van. 1949. *The geology of Indonesia*, vol. 1A. *General geology of Indonesia and adjacent archipelagoes*, 732 p. The Hague: Government Printing Press.

Bilibin, Yu. A. 1968. *Metallogenic provinces and metallogenic epochs* (Transl. E. A. Alexandrov). New York: Queens College Press.

Blake, D. H., Elwell, R. W. D., Gibson, I. L., Skelhorn, R. R., and Walker, G. L. P. 1965. Some relationships resulting from the intimate association of acid and basic magmas. *Quart. J. Geol. Soc. London* **121**, 31–49.

Borax, E., and Stewart, R. D. 1966 (MS). *Notes on the Paleozoic stratigraphy of northeastern Thailand*, 26 p. Submitted to the ECAFE meeting in Bangkok, August 1966.

Boucot, A. J., Johnson, J. G., and Jones, C. R. 1966. Silurian brachiopods from Malaya. *J. Paleontol.* **40**, 1027–1031.

Bradford, E. F. 1961. The occurrence of tin and tungsten in Malaya. *Proc. 9th Pacif. Sci. Congr.* **12**, 378–398.

——— 1965 (MS). *The geology and mineral resources of the Gunong Jerai area, Kedah, Malaya*. Ph.D. thesis, University of London.

Brelich, H. 1914. Mining in Trengganu. In *Mining in Malaya*. F. M. S. Information Agency.

Brouwer, H. A. 1925. *The geology of the Netherlands East Indies*. New York: Macmillan Co.

Brown, G. F., Buravas, S., Charaljavanaphet, J., Jalichandra, N., Johnson, W. D., Jr., Sresthaputra, V., and Taylor, G. C., Jr. 1951. Geologic reconnaissance of the mineral deposits of Thailand. Geologic investigations in Asia. 183 p. *Bull. U.S. Geol. Survey* **984**.

Brown, J. C. and Heron, A. M. 1923. The geology and ore deposits of the Tavoy district. *Mem. Geol. Survey India* **44**, 167–354.

Buddington, A. F. 1959. Granite emplacement with special reference to North America. *Bull. Geol. Soc. Am.* **70**, 671–747.

Burton, C. K. 1959 (MS). *A note on the geology of the tin and iron bearing deposits located near Sungei Pelepah Kanan, District of Kota Tinggi, Johore*. Geological Survey Department of West Malaysia.

——— 1964. The older alluvium of Johore and Singapore. *J. Trop. Geogr.* **18**, 30–42.

——— 1965. Wrench faulting in Malaya. *J. Geol.* **73**, 781–798.

——— 1966. Palaeozoic orogeny in north-west Malaya. *Geol. Mag.* **103**, 364–365.

——— 1967a. Graptolite and tentaculite correlations and palaeogeography of the Silurian and Devonian in the Yunnan–Malaya geosyncline. *Trans. Proc. Palaeontol. Soc. Japan*, n.s. **65**, 27–46.

——— 1967b. Dacryonconarid tentaculites in the Mid-palaeozoic euxinic facies of the Malayan geosyncline. *J. Palaeontol.* **41**, 449–454.

——— 1967c. The Mahang Formation: A mid-Palaeozoic euxinic facies from Malaya—with notes on its conditions of deposition and palaeogeography. *Geol. en Mijnb.* **46**, 167–187.

——— 1967d. Wrench faulting in Malaya: A reply. *J. Geol.* **75**, 128–129.

——— 1967e. Ignimbrite in Malaya. *Geol. Mag.* **104**, 397–398.

——— 1969. Neotectonics of the Thai-Malay Peninsula. *Bull. Nat. Inst. Geol. Min. Indonesia* **2** (1), 11–13.

——— 1972. The geology and mineral resources of the Baling area, Kedah and Perak. *Mem. Geol. Survey Malaysia* **12**, 150.

——— In prep. The geology and mineral resources of the area covered by New Series Map Sheet 130, Johore. *Mem. Geol. Survey Dept. W. Malaysia*.

———, and Bignell, J. D. 1969. Cretaceous-Tertiary events in southeast Asia. *Bull. Geol. Soc. Am.* **80**, 681–688.

Cady, W. M. 1950. Classification of geotectonic elements. *Trans. Am. Geophys. Union* **31**, 780–785.

Cameron, W. E. 1924. The deep leads of Kinta Valley. *Min. Mag., London* **31**, 276–285.

——— 1925a. The limestone hills of the Kinta Valley tin-field, Federated Malay States: their geology and physiographic origin. *Geol. Mag.* **62**, 21–27.

——— 1925b. Kinta Valley geology. *Min. Mag., London* **32**, 222–224.

Casey, J. N., and Gilbert-Tomlinson, J. 1956. Cambrian geology of the Huckitta–Marqua Region, Northern Territory. *Int. Geol. Congr.* **20** Mexico: Symposium on Cambrian System, pt. 1, p. 55–74.

Chan, S. C. 1970 (MS). *Geology of the Sungei Besi area with special reference to its primary tin mineralisation*. B.Sc. Honours thesis, Department of Geology, University of Malaya.

Chan, S. H. 1967. Geoelectrical study of some alluvial deposits in West Malaysia. *J. Dept. Eng. Univ. Malaya*, **6**, 95–106.

Chhibber, H. L. 1934. *The Geology of Burma*, 538 p. London: MacMillan Co.

Chin Fatt. 1965 (MS). *The Upper Triassic sediments of Pasir Panjang–Jurong area of Singapore*. B.Sc. Honours thesis, Department of Geology, University of Malaya.

Chong, F. S., Cook, K. H., Evans, G. M., and Suntharalingam, T. 1970. Geology and mineral resources of the Melaka-Mersing area. *Ann. Rep. Geol. Surv. Malaysia* [for 1968]. 89–94.

———, and Yong, S. K. 1967 (MS). *Report on the Jengka Triangle, Pahang 1966*. Proceedings of the Senior Officers' Departmental Conference 1966. Geol. Survey Dept. W. Malaysia.

Choy, K. W. 1970 (MS). *Geology of the western Kuala Lumpur area, West Malaysia*. B.Sc. Honours thesis, Department of Geology, University of Malaya.

Chung, S. K. 1959 (MS). *Geological Investigations of Pulau Bunting*. Geol. Survey Dept. W. Malaysia Rept. GS 58/F/031/33.

Cohen, L. H., Condie, K. C., Kuest, L. J. Jr., MacKenzie, G. S., Meister, F. H., Pushkar, P., and Stueber, A. M. 1963. Geology of the San Benito Islands, Baja California, Mexico. *Bull. Geol. Soc. Am.* **74**, 1335–1370.

Collet, O. J. A. 1903. *L'étain. Étude minière et politique sur les États Fédérés Malais*, 196 p. Brussels: Librarie Falk Fils.

Collings, H. D. 1938. Pleistocene site in the Malay Peninsula. *Nature, Lond.* **142**, 575–576.

Courtier, D. B. 1962 (MS). Note on terraces and other alluvial features in parts of Province Wellesley, south Kedah, and north Perak. *Prof. Pap. Geol. Survey Dept. Fed. Malaya* **E62**.1-T, 6 p.

——— In prep. The geology and mineral resources of the neighbourhood of the Kulim area, Kedah. *Mem. Geol. Survey Dept. W. Malaysia*.

Cox. L. R. 1936. On a fossiliferous Upper Triassic shale from Pahang, Federated Malay States. *Ann. Mag. Nat. Hist.* ser. 10, **17**, 213–221.

——— 1937. On a freshwater shale with *Viviparus* from Johore (Malay States). *Geol. Mag.* **74**, 78–81.

Crawford, J. 1824. Geological observations made on a voyage from Bengal to Siam and Cochin China. *Trans. Geol. Soc. London*, ser. 2, **1**, 406–408.

Croix, J. E., de la. 1882. *Les mines d'étain de Perak*. Paris.

Daly, D. D. 1878. The metalliferous formation of the Peninsula. *J. Straits Br. R. Asiat. Soc.* **2**, 194–198.

——— 1882. Surveys and explorations in the native states of the Malayan Peninsula 1875–1882. *Geogr. J.* **4**, 393–412.

Datta, P. N. 1899. Notes on the Geology of the country along the Mandalay-Kunlon Ferry Railway Route, Upper Burma. *Gen. Rept. Geol. Survey India, 1899–1900*, 96–122.

Dawson, J., MacDonald, S., Paton, J. R., Slater, D., and Santokh Singh, D. 1968. *Geological map of Northeast Malaya*, 1:250,000. Geol. Survey Dept W. Malaysia.

Dietz, R. S. and Holden, J. C. 1970. Reconstruction of Pangaea: break-up and dispersion of continents, Permian to Present. *J. Geophys. Res.* **75**, 4939–4956.

Doyle, P. 1878. Tin mining in Larut. *Min. J. Rail. and Commer. Gaz. London* **48**.

——— 1879a. *Tin mining in Larut*, 32 p. London.

——— 1879b. On some tin-deposits of the Malayan Peninsula. *Quart. J. Geol. Soc. London* **35**, 229–232.

Dzulynski, S. and Walton, E. K. 1965. *Sedimentary features of flysch and greywackes*, 274 p. Amsterdam: Elsevier.

Eaton, J. P. 1962. Crustal structure and volcanism in Hawaii. In *Crust of the Pacific basin. Monogr. Am. Geophys. Union* **6**, 13–39.

Edwards, A. B. 1954. *Textures of the ore minerals and their significance*, 242 p. Melbourne: Austr. Inst. Min. Metall.

Edwards, G., and McLaughlin, W. A. 1965. Age of granites from the Tin Province of Indonesia. *Nature, Lond.* **206**, 814–816.

Edwards, W. N. 1926. Carboniferous plants from the Malay States. *J. Malay. Br. R. Asiat. Soc.* **4**, 171–172.

——— 1933. Triassic wood from the Malay States. *J. Malay. Br. R. Asiatic Soc.* **11**, 236–241.

Elliot, G. F. 1968. Three new Tethyan Dasycladaceae (calcareous algae). *Palaeontology*, **11**, 491–497.

Emery, K. O., and Niino, H. 1963. Sediments of the Gulf of Thailand and adjacent continental shelf. *Bull. Geol. Soc. Am.* **74**, 541–554.

Eyles, R. J. 1966. Stream representations on Malayan maps. *J. Trop. Geogr.* **22**, 1–9.

——— 1967. Laterite at Kerdau, Pahang, Malaya. *J. Trop. Geogr.* **25**, 18–23.

——— 1968a. Stream net ratios in West Malaysia. *Bull. Geol. Soc. Am.* **79**, 701–711.

——— 1968b (MS). *A morphometric analysis of West Malaysia*. Ph.D. thesis, University of Malaya.

——— 1970. Physiographic implications of laterite in West Malaysia. *Bull. Geol. Soc. Malaysia*, **3**, 1–7.

Fawns, S. 1904. Tin-lode mining in Trengganu. *Min. J., London* **76**, 377.

Fermor, L. L. 1939. Coal veins in Malaya. *Geol. Mag.* **76**, 465–472.

Fitch, F. H. 1949. Evidence for recent emergence of the land in east Pahang. *J. Malay. Br. R. Asiat. Soc.* **22**, 115–122.

——— 1952. The geology and mineral resources of the neighbourhood of Kuantan, Pahang. 144 p. *Mem. Geol. Survey Dept. Fed. Malaya*, **6**.

Flathe, H., and Pfeiffer, D. 1965. Grundzüge der Morphologie, Geologie und Hydrogeologie im Karstgebiet Gunung Sewu, Java. *Geol. Jb.* **83**, 533–562.

Flint, R. F. 1957. *Glacial and Pleistocene geology*, 553 p. New York: Wiley.

Flinter, B. H. 1960. The effects of heat, hydrochloric acid and lead chloride on some Malayan mineral grains. *Overseas Geol. Miner. Resour.* **8**, 54–56.

Folk, R. L., and Ward, W. C. 1957. Brazos River bar: a study in the significance of grain size parameters. *J. Sediment. Petrol.* **27**, 3–26.

Fromaget, J. 1941. *L'Indochine Française, sa structure Géologique, ses Roches, ses Mines*. Service Géologique de l'Indochine, Hanoi.

——— 1952. Aperçu de nos connaissances sur la géologie de l'Indochine en 1948. *Int. Geol. Congr. 18.* Great Britain. **13**, 63–77.

Fujii, S., and Fuji, N. 1967. Postglacial sea levels in the Japanese islands. *J. Geosci. Osaka City Univ.* **10**, 43–51.

Fyfe, W. S., Turner, F. J., and Verhoogen, J. 1958. Metamorphic reactions and metamorphic facies, 259 p. *Mem. Geol. Soc. Am.* **73**.

Gan, A. S. 1969 (MS). *Geology and mineralisation of the Ulu Langat area, Selangor. West Malaysia*. B.Sc. Honours thesis, Department of Geology, University of Malaya.

Ganesan, K. 1969. Iron-tin mineralization in the Gunong Muntahak area, Johore. *Newsl. Geol. Soc. Malaysia* **19**, 1–4.

Gilluly, J. 1963. The tectonic evolution of the western United States. *Quart. J. Geol. Soc. London* **119**, 133–174.

Gobbett, D. J. 1965a. The Lower Palaeozoic rocks of Kuala Lumpur, Malaysia. *Fed. Mus. J.* **9**, 67–79.

—— 1965b. The formation of limestone caves in Malaya. *Malay. Nat. J.* **19**, 4–12.

—— 1966. The brachiopod genus *Stringocephalus* from Malaya. *J. Paleontol.* **40**, 1345–1348.

—— 1971. Joint pattern and faulting in Kinta, West Malaysia. *Bull. Geol. Soc. Malaysia* **4**, 39–48.

—— 1972. Geology of the Rebak Islands, Langkawi, West Malaysia. *Newsl. Geol. Soc. Malaysia*, **37**, 1–5.

Gopinathan, B. 1968. *Terrace and alluvial soils in West Malaysia*. Proc. 3rd Malaysian Soils Conf. pp. 45–50.

Gorman, C. F. 1969. Hoabinhian: a pebble-tool complex with early plant associations in Southeast Asia. *Science* **163**, 671–673.

Grabau, A. W. 1924. *Stratigraphy of China*, pt. 1, 528 p. Geol. Survey of China, Peking.

Grubb, P. L. C. 1965. Undersaturated potassic lavas and hypabyssal intrusives in north Johore. *Geol. Mag.* **102**, 338–346.

—— 1966. Distribution and genetic significance of aluminum hydrates in south-east Johore. *Proc. R. Soc. Vict.* **79**, 257–265.

—— 1968. Geology and bauxite deposits of the Pengerang area, southeast Johore, 125 p. *Mem. Geol. Survey Dept. W. Malaysia*, **14**.

——, and Hannaford, P. 1966. Magnetism in cassiterite. *Mineral. Deposita* **2**, 148–171.

Hada, S. 1966. Discovery of early Triassic ammonoids from Gua Musang, Kelantan, Malaya. *J. Geosci. Osaka City Univ.* **9**, 111–113.

Haile, N. S. 1970 (MS). *Rhyolitic and granitic rocks around Genting Sempah, Selangor and Pahang, Malaysia, with a definition of a proposed new formation, the Sempah Conglomerate*. Abstr. Geol. Soc. Malaysia Discussion Meeting, Ipoh, December 1970.

—— In press. Quaternary shorelines in west Malaysia and adjacent parts of the Sunda Shelf. In *Quaternary Shorelines*. International Union for the Study of the Quaternary, special publication.

——, and Mohammed bin Ayob. 1968. Note on radiometric age determination of samples of peat and wood from tin-bearing Quaternary deposits at Sungai Besi Tin Mines, Kuala Lumpur, Selangor, Malaysia. *Geol. Mag.* **105**, 519–520.

Hamada, T. 1960. Some Permo-Carboniferous fossils from Thailand. *Sci. Pap. Coll. Gen. Educ., Univ. Tokyo* **10**, 337–361.

—— 1964. Notes on the drifted Nautilus in Thailand. *Sci. Pap. Coll. Gen. Educ., Univ. Tokyo* **14**, 255–278.

—— 1968. Ambocoeliids from red beds in the Malayan Peninsula. In *Geology and palaeontology of southeast Asia*, vol. 5, T. Kobayashi and R. Toriyama, Eds. Tokyo: University of Tokyo Press, pp. 13–25.

—— 1969a. Late Palaeozoic brachiopods from red beds in the Malayan Peninsula. In *Geology and palaeontology of southeast Asia*, vol. 6, T. Kobayashi and R. Toriyama, Eds. Tokyo: University of Tokyo Press, pp. 251–264.

—— 1969b. Devonian brachiopods from Kroh, Upper Perak in Malaysia (Malaya). In *Geology and palaeontology of southeast Asia*, vol. 7, T. Kobayashi and R. Toriyama, Eds. Tokyo: University of Tokyo Press, pp. 1–13.

Hampton, J. H. 1887. Tin-deposits of the State of Perak, Straits Settlements. *Trans. Min. Assoc. Inst. Cornwall* **1**, 143.

—— 1899. On the occurrence of tin. *Trans. Geol. Soc. A. Afr.* **4**, 37–40.

Harker, A. 1956. *Metamorphism: a study of the transformations of rock-masses*, 362 p. London: Methuen.

Harland, W. B., Smith, A. G., and Wilcock, B., Eds. 1964. *The Phanerozoic Time-Scale*, 458 p. Geological Society of London.

Harris, E. F. 1924. Rahman tin mines, Federated Malay States. *Chem. Eng. Miner. Rev., Melbourne*, 297.

Hayami, I. 1960. Two Jurassic pelecypods from west Thailand. *Trans. Proc. Paleontol. Soc. Japan*, n.s. **38**, 284.

Hill, R. D. 1966. Changes in beach form at Sri Pantai, northeast Johore, Malaysia. *J. Trop. Geogr.* **23**, 19–27.

Hills, E. S. 1963. *Elements of structural geology*, 483 p. New York: Wiley.

Ho, R. 1960. The evolution of the Indo-Malayan Region. *Proc. Centenary and Bicentenary Congr. Biology Singapore, 1958*. University of Malaya Press, pp. 9–20.

Hockin, H. W. 1957a. Tantalum/niobium minerals in Malaya. *Bull. Mines Dept. Fed. Malaya* **2**.

—— 1957b. A note on an occurrence of stannite in Malaya. *Bull. Mines Dept. Fed. Malaya* **3**.

Holmes, A. 1959. A revised geological time-scale. *Trans. Edinb. Geol. Soc.* **17**, 183–216.

Hooijer, D. A. 1963a. Report upon a collection of Pleistocene mammals from tin-bearing deposits in a limestone cave near Ipoh, Kinta Valley, Perak. *Fed. Mus. J.* **7**, 1–5.

—— 1963b. *Rhinoceros sondaicus* Desmarest from the Hoabinhian of Gua Cha rock shelter, Kelantan. *Fed. Mus. J.* **7**, 23–24.

Hosking, K. F. G. 1964. Permo-carboniferous and later primary mineralisation of Cornwall and south-west Devon. In *Present views on some aspects of the geology of Cornwall and Devon*, K. F. G. Hosking and G. J. Shrimpton, Eds., Royal Geological Society of Cornwall, pp. 201–245.

—— 1967. The nature and significance of certain "late" veins of south-west Cornwall which are characterised by the presence of potash feldspar. *Camborne School of Mines Mag.* **67**, 49–59.

—— 1969a (MS). *Aspects of the geology of the tin-fields of Southeast Asia*. Preprint of paper read at the 2nd Technical Conference on Tin of the International Tin Council, Bangkok.

—— 1969b (MS). *The nature of the primary tin ores of the southwest of England*. Ibid.

—— 1970a. Tin deposits and limestone. *Newsl. Geol. Soc. Malaysia* **24**, 3–5.

—— 1970b. The primary tin deposits of Southeast Asia. *Miner. Sci. Eng., S. Afr.* **1**, 24–50.

——, and Leow, J. H. 1970. A further occurrence of malayaite in West Malaysia. *Newsl. Geol. Soc. Malaysia* **22**, 4–5.

———, ———, Chin, L. S. and Wong, Y. F. 1970. The nature and significance of the Manson ore-body, Ulu Ketubong, Kelantan: an interim report. *Malayan Sci.* **5**, 31–41.

———, ———, and Haser, F. E. H. 1969. A magnetite-cassiterite-polysulphide skarn vein at Bukit Besi Mine, Trengganu, W. Malaysia. *Newsl. Geol. Soc. Malaysia* **20**, 3–5.

———, and Stauffer, P. H. 1970. Tektites from the stanniferous placers of eastern Pahang. *Newsl. Geol. Soc. Malaysia* **22**, 1–4.

———, and Yeap, E. B. In press. The nature of the cassiterite-polysulphide ore from Setapak, Selangor, W. Malaysia. *Malayan Sci.*

———, ———, and Wong, T. W. 1970. An occurrence of malayaite ($CaO \cdot SnO_2$) at Rawang, Selangor, West Malaysia. *Newsl. Geol. Soc. Malaysia* **29**, 4–6.

Howell, B. F. 1957. Vermes. in *Treatise on marine ecology and palaeoecology*, vol. 2. N. S. Ladd, Ed., *Mem. Geol. Soc. Am.* **67**, 805–820.

Hsu, K. J. 1964. Cross-laminations in graded bed sequences. *J. Sediment. Petrol.* **34**, 379–388.

Hurley, P. M., Hughes, H., Pinson, W. H. Jr., and Fairbairn, H. W. 1962. Radiogenic argon and strontium diffusion parameters in biotite at low temperatures obtained from Alpine fault uplift in New Zealand. *Geochim. Cosmochim. Acta* **26**, 67–80.

Hutchison, C. S. 1961. The basement rocks of Malaya and their palaeo-geographic significance in Southeast Asia. *Am. J. Sci.* **259**, 181–185.

——— 1963. Interesting coastal exposures east of Kuah, Pulau Langkawi. *Malayan Nat. J.* **17**, 165–169.

——— 1964a. A gabbro-granodiorite association in Singapore Island. *Quart. J. Geol. Soc. London* **120**, 283–297.

——— 1964b. A gabbro-granodiorite association in Singapore Island—reply to a discussion. *Quart. J. Geol. Soc. London* **120**, 547–548.

——— 1966a (MS). *Tectonic and petrological relations within three rock associations of orogenic zones in Malaysia*. Ph.D. thesis, University of Malaya.

——— 1966b. Some relationships resulting from the intimate association of acid and basic magmas—a discussion. *Quart. J. Geol. Soc. London* **122**, 85–86.

——— 1968a. The dating by thermoluminescence of tectonic and magmatic events in orogenesis. In *Thermoluminescence of Geological Materials*. D. J. McDougall, Ed. London: Academic Press, pp. 341–358.

——— 1968b. Dating of Tectonism in the Indosinian–Thai–Malayan Orogen by Thermoluminescence. *Bull. Geol. Soc. Am.* **79**, 375–386.

——— 1968c. Invalidity of the Billiton Granite, Indonesia, for defining the Jurassic/Upper Triassic boundary in the Thai–Malayan Orogen. *Geol. en Mijnb.* **47**, 56–60.

——— 1968d. Physical and chemical differentiation of West Malaysian Limestone Formations. *Bull Geol. Soc. Malaysia* **1**, 45–56.

——— 1969. Some notes on the Stong Metamorphic Complex, Kelantan. *Newsl. Geol. Soc. Malaysia* **21**, 8–11.

——— 1970. The tectonic framework of the southern Malay Peninsula. *Program Ann. Joint Geol. Soc. Am. Allied Soc. Meeting* (Milwaukee, November 1970), vol. 2, no. 7, p. 584.

——— 1971. The Benta Migmatite Complex—petrology of two important localities. *Bull. Geol. Soc. Malaysia* **4**, 49–70.

———, and Leow, J. H. 1963. Tourmaline greisenization in Langkawi, northwest Malaya. *Econ. Geol.* **58**, 587–592.

———, and Jeacocke, J. E. 1971. FORTRAN IV computer programme for calculation of the Niggli Molecular Norm. *Bull. Geol. Soc. Malaysia* **4**, 91–95.

———, and Snelling, N. J. 1971. Age Determination on the Bukit Paloh Adamellite. *Bull. Geol. Soc. Malaysia* **4**, 97–100.

Ichikawa, K., Ishii, K., and Hada, S. 1966. On the remarkable unconformity at the Jengka Pass, Pahang, Malaya. *J. Geosci. Osaka City Univ.* **9**, 123–130.

———, and Yin, E. H. 1966. Discovery of Early Triassic bivalves from Kelantan, Malaya. *J. Geosci. Osaka City Univ.* **9**, 101–106.

Igo, H. 1964. Permian fossils from northern Pahang, Malaya. *Jap. J. Geol. Geogr.* **35**, 57–71.

——— 1966. Some Permian fusulinids from Pahang, Malaya. In *Geology and Palaeontology of southeast Asia*, T. Kobayashi and R. Toriyama, Eds. vol. 3. Tokyo: University of Tokyo Press, pp. 30–38.

———, and Koike, T. 1966. Ordovician and Silurian Conodonts from the Langkawi Islands, Malaya. *Ibid.* pp. 1–29.

———, and ———. 1968. Carboniferous conodonts from Kuantan, Malaya. *Ibid.*, vol. 5, pp. 26–30.

———, Yin, E. H., and Koike, T. 1965. Triassic conodonts from Kelantan, Malaya. *Mem. Mejiro Gakuen Women's Junior Coll.* **2**, 5–19.

Ingham, F. T. 1938. The geology of the neighbourhood of Tapah and Telok Anson, Perak, Federated Malay States, with an account of the Mineral deposits. *Mem. Geol. Survey Dept. Fed. Malaya* **2**, 72 p.

——— 1948. *Geological map of Malaya*, 1:760,320. Geol. Survey Dept. Fed. Malaya.

——— 1949. *Rept. Geol. Survey Dept. Fed. Malaya* [for 1948]. 29 only.

———, and Bradford, E. F. 1960. The geology and mineral resources of the Kinta Valley, Perak. 347 p. *Mem. Geol. Survey Dept. Fed. Malaya* **9**.

Ishii, K. 1966. On some fusulinids and other foraminifera from the Permian of Pahang, Malaya. *J. Geosci. Osaka City Univ.* **9**, 131–136.

———, Ichikawa, K., and Hada, S. 1966. Notes on the geology and palaeontology of Malaya. *J. Geosci. Osaka City Univ.* **9**, 89–91.

———, and Nogami, Y. 1966. Discovery of Triassic conodonts from the so-called Palaeozoic limestone in Kedah, Malaya. *J. Geosci. Osaka City Univ.* **9**, 93–98.

Iwai, J. Asama, K., Veerabus, M., and Hongnusonthi, A. 1966. Stratigraphy of the so-called Khorat Series and a note on the fossil-bearing Palaeozoic strata in Thailand. *Jap. J. Geogr.* **39**, 21–35.

Izokh, E. P., Le Din Khyu, and Nguen Van Tien. 1964. Nove danne o magmatizme Severnogo Vietnama [New information on Magmatism in North Vietnam.] *Dokl. Akad. Nauk SSSR* **155**, 1321–1324.

Jack, W. 1822. Extract from a letter from Mr. William Jack to H. T. Colebrooke, Esq. V.P.G.S., etc., containing a notice respecting the rocks of the islands of Penang and Singapore. *Trans. Geol. Soc. London*, ser. 2, **1**, 165–166.

Ja'afar bin Ahmad. 1965a (MS). 1963. Progress Report of Geological Survey Work in Western Pahang. *Prof. Pap. Geol. Survey Dept. Fed. Malaya* E65.1-G, 11–14.

—— 1965b (MS). Annual report on regional mapping, 1964, in Pahang. *Prof. Pap. Geol. Survey Dept. Fed. Malaya* E65.2-G, 10–12.

—— 1966 (MS). *Geology of the Benom-Karak area*. Senior Officers' Departmental Conference 1965, Geol. Survey Dept. W. Malaysia.

—— In prep. The geology and mineral resources of the area between Gunong Benom and Karak, South Pahang, map sheets 3C/1, 3C/5, 3C/9. *Mem. Geol. Survey Dept. W. Malaysia*.

Jaeger, H. 1959. Graptolithen und Stratigraphie des jungstens Thuringer Silurs. *Abh. Deut. Akad. Wiss. Berlin (Kl. Chem., Geol. Biol.)* 2, 1–197.

—— 1962. Das Silur (Gotlandium) in Thuringen und am Ostrand des Rheinischen Schiefergebirges (Kellerwald, Marburg, Giessen). *Symposium-Band der 2nd International Arbeitstagung Über die Silur Devon-Grenze und die Stratigraphie von Silur und Devon*, pp. 108–135.

Jefferies, R. P. S. 1962. The palaeoecology of the *Actinocamax plenus* subzone (lowest Turonian) in the Anglo-Paris basin. *Palaeontology* 4, 609–647.

——, and Minton, P. 1965. The mode of life of two Jurassic species of "*Posidonia*" (Bivalvia). *Palaeontology*, 8, 156–185.

Jones, C. R. 1959. Grapolites recorded from Malaya. *Nature, Lond.* 183, 231–232.

—— 1961. A revision of the stratigraphical sequence of the Langkawi Islands, Federation of Malaya. *Proc. 9th Pacif. Sci. Congr.* 12, 287–300.

—— 1965. The limestone caves and cave deposits of Perlis and north Kedah. *Malay Nat. J.* 19, 21–30.

—— 1966. *Geological map of New Series Sheet 150 (Pulau Langkawi)*, 1:63,360. Geol. Survey Dept. W. Malaysia.

—— 1967a. Graptolites of the *Monograptus hercynicus* type recorded from Malaya. *Nature, Lond.* 215, 497 only.

—— 1967b. Fauna from the Foothills Formation at Tuan Estate, Karak, Pahang and its significance. *Newsl. Geol. Soc. Malaysia* 7, 1–3.

—— 1968. The Lower Paleozoic rocks of the Malay Peninsula. *Bull. Am. Assoc. Petrol. Geol.* 52, 1259–1278.

—— 1970a. On a lower Devonian Fauna from Pahang, West Malaysia. *Bull. Geol. Soc. Malaysia* 3, 63–74.

—— 1970b. The geology and mineral resources of the Grik area, Upper Perak, 144 p. *Mem. Geol. Survey Dept. W. Malaysia* 11.

—— In prep. The geology and mineral resources of Perlis, North Kedah and the Langkawi Islands. *Mem. Geol. Survey Dept. W. Malaysia*.

——, Gobbett, D. J., and Kobayashi, T. 1966. Summary of fossil record in Malaya and Singapore 1900–1965. In *Geology and palaeontology of southeast Asia*, T. Kobayashi and R. Toriyama, Eds., vol. 2. Tokyo: University of Tokyo Press, pp. 301–359.

Jones, M. P. 1969. The tin industry. *Geol. en Mijnb.* 48, 451–466.

Jones, T. R. 1905. Note on a Triassic *Estheriella* from the Malay Peninsula. *Geol. Mag.* 42, 50–52.

Jones, W. R. 1915a. *Clays of economic importance in the Federated Malay States*. Kuala Lumpur: Government Press.

—— 1915b. Mineralization in Malaya. *Min. Mag., London* 13, 195–202, 322–330.

—— 1916. The origin of topaz and cassiterite at Gunong Bakau, Malaya. *Geol. Mag.* 53, 255–260.

—— 1917. The origin of the secondary stanniferous deposits of the Kinta district, Perak (Federated Malay States). *Quart. J. Geol. Soc. London* 72, 165–197.

—— 1925a. The tin-deposits of Kinta district. *Min. Mag., London* 32, 26–31.

—— 1925b. Tin deposits of Kinta district. *Min. Mag., London* 33, 83–89.

—— 1925c. *Tin-fields of the world*. London: Mining Publications Ltd.

Jongmans, W. J. 1951. Fossil plants of the island of Bintan. *Proc. K. Ned. Akad. Wet.*, ser. B 14 (2).

Joplin, G. A. 1959. On the origin and occurrence of basic bodies associated with discordant bathyliths. *Geol. Mag.* 96, 361–373.

—— 1960. On the tectonic environment of basic magma. *Geol. Mag.* 97, 363–368.

Judd, W. R., Ed. 1964. *The state of stress in the earth's crust*, 732 p. New York: Elsevier.

Kaewbaidhoon, S., and Aranyakanon, P. 1961. Tin and tungsten deposits of Thailand. *Proc. 9th Pacif. Sci. Congr.* 12, 400–404.

Kee, T. M. 1966 (MS). *Geology of Gunong Panti area, southern Johore, West Malaysia*. B.Sc. Honours thesis, Department of Geology, University of Malaya.

Keller, G. H. 1967. Pleistocene–Recent boundary in the Malacca Strait, Southeast Asia. *Abstr. 7th Int. Sediment. Congr. Gt. Britain*.

Khoo, H. P. 1969 (MS). *Mineralisation at Pelepah Kanan, Kota Tinggi, Johore, West Malaysia*. B.Sc. Honours Thesis, Department of Geology, University of Malaya.

Khoo, T. T. 1968 (MS). *A petrological study of the Sungai Ruan area, Raub, West Malaysia*. B.Sc. Honours thesis, Department of Geology, University of Malaya.

Kimura, T., and Jones, C. R. 1967. Geological structures in the northeastern and southern parts of the Langkawi islands, northwest Malaya. In *Geology and Palaeontology of southeast Asia*, T. Kobayashi and R. Toriyama, Eds., vol. 3, Tokyo: University of Tokyo Press, pp. 123–124.

King, W. B. R. 1937. Cambrian trilobites from Iran (Persia). *Palaeontologica Indica*, n.s. 22, (5), 1–22.

Kirk, H. J. C. 1968. Igneous rocks of Sarawak and Sabah. *Bull. Geol. Survey Dept. Borneo Region, Malaysia* 5, 210 p.

Klompé, T. H. F. 1961. Pacific and Variscian orogeny in Indonesia: a structural synthesis. *Proc. 9th Pacif. Sci. Congr.* 12, 76–115.

—— 1962. Igneous and structural features of Thailand. *Geol. en Mijnb.* 41, 290–302.

—— 1965. *Geological map of Indonesia*, 1:2,000,000. U. S. Geol. Survey, Washington, D.C.

——, Katili, J., Johannas, and Soekendar. 1961. Late Palaeozoic–Early Mesozoic volcanic activity in the Sunda Land area. *Proc. 9th Pacif. Sci. Congr.* 12, 204–217.

Kobayashi, T. 1957. Upper Cambrian fossils from peninsular Thailand. *J. Fac. Sci. Tokyo Univ.* 10, 367–382.

—— 1958. Some Ordovician fossils from the Thailand–Malayan borderland. *Jap. J. Geol. Geogr.* 29, 223–231.

———— 1959. On some Ordovician fossils from Northern Malaya and her adjacence. *J. Fac. Sci. Tokyo Univ.* **11**, 387–407.

———— 1963a. On the Triassic *Daonella* beds in central Pahang, Malaya. *Jap. J. Geol. Geogr.* **34**, 101–112.

———— 1963b. *Halobia* and some other fossils from Kedah, northwest Malaya. *Jap. J. Geol. Geogr.* **34**, 113–128.

———— 1964. Geology of Thailand. In *Geology and Palaeontology of Southeast Asia*, vol. 2, T. Kobayashi and R. Toriyama, Eds. Tokyo: University of Tokyo Press, pp. 1–16.

————, Burton, C. K., Tokuyama, A., and Yin, E. H. 1967. The Daonella and Halobia facies of the Thai-Malay peninsula compared with those of Japan. *Ibid.* vol. 3, pp. 98–122.

————, and Hamada, T. 1966. A new proetid trilobite from Perlis, Malaysia (Malaya). *Jap. J. Geol. Geogr.* **37**, 87–92.

————, and ————. 1970. A cyclopid-bearing Ordovician faunule discovered in Malaya with a note on the Cyclopygidae. In *Geology and Palaeontology of Southeast Asia*, vol. 8, Kobayashi, T. and Toriyama, R. (eds.). Tokyo: University of Tokyo Press, pp. 1–18.

————, and Igo, H. 1965. On the occurrence of graptolite shales in northern Thailand. *Jap. J. Geol. Geogr.* **36**, 37–44.

————, Jones, C. R., and Hamada, T. 1964. On the Lower Silurian shelly fauna in the Langkawi Islands, Northwest Malaya. *Jap. J. Geol. Geogr.* **35**, 73–80.

————, and Tamura, M. 1968a. *Myophoria* (s.l.) in Malaya with a note on the Triassic Trigoniacea. In *Geology and palaeontology of southeast Asia*, T. Kobayashi and R. Toriyama, Eds., vol. 5. Tokyo: University of Tokyo Press, pp. 88–137.

————, and ————. 1968b. Upper Triassic pelecypods from Singapore. *Ibid.* pp. 138–150.

Kon'no, E. 1963. Some Permian plants from Thailand. *Jap. J. Geol. Geogr.* **34**, 139–159.

———— 1966. Some younger Mesozoic plants from Malaya. In *Geology and Palaeontology in southeast Asia*, T. Kobayashi and R. Toriyama, Eds., vol. 3. Tokyo: University of Tokyo Press, pp. 135–164.

———— 1968. Addition to some younger Mesozoic plants from Malaya. *Ibid.*, vol. 4, pp. 139–155.

————, Asama, K., and Rajah, S. S. 1970. The late Permian Linggiu Flora from the Gunong Blumut area, Johore. Malaysia. *Bull. Nat. Sci. Mus., Tokyo*, **13**, 491–580.

Koopmans, B. N. 1964. Geomorphological and historical data of the lower course of the Perak River (Dindings). *J. Malay. Br. R. Asiat. Soc.* **37**, 175–191.

———— 1965. Structural evidence for a Palaeozoic orogeny in Northwest Malaya. *Geol. Mag.* **102**, 501–520.

———— 1966. A structural map of north and central Pahang. *J. Trop. Geogr.* **22**, 23–29.

———— 1968. The Tembeling Formation—a lithostratigraphic description (West Malaysia). *Bull. Geol. Soc. Malaysia* **1**, 23–43.

Kulp, J. L. 1961. Geologic time scale. *Science* **133**, 1105–1114.

Kummel, B. 1960. Anisian ammonoids from Malaya. *Breviora* **124**, 1–8.

Lacroix, A. 1932. Les tectites de l'Indochinie. *Archives du Museum d'Histoire Naturelle*, 6th ser. **8**, 139–240.

Lake, H. 1894. A journey on the Sembrong River from Kuala Indau to Batu Pahat. Pt. II. Topography and geology. *J. Straits Br. R. Asiat. Soc.* **26**, 19–24.

Lambert, R. St. J. 1964. The relationship between radiometric ages obtained from plutonic complexes and stratigraphical time. In *The Phanerozoic Time Scale*, W. B. Harland, A. G. Smith and B. Wilcock, Eds. London: Geological Society of London, pp. 43–52.

Lamoreaux, P. E., Charaljavanphet, J. Jalichan, N., Phan Na Chiengmai, P., Bunnag, D., Thavisri, A., and Rakprathum, C. 1958. Reconnaissance of the geology and groundwater of the Khorat plateau, Thailand. *U.S. Geol. Survey Water-Supply Pap.*, 62 p. **1429**.

Laserre, M. M. M., Lacombe, P., and Saurin, E. 1968. Étude pétrographique, chimique et géochronologique du granite de Bo Khâm (Cambôdoge Oriental). *C.R. Acad. Sci. Paris* **267**, 2073–2076.

La Touche, T. H. D. 1899. Preliminary report on the geology of the northern Shan States. *Gen. Rept. Geol. Survey India*, *1899–1900*, 74–95.

Law, W. M. 1961 (MS). *The Batu Arang coalfield, Selangor*. B.Sc. Honours thesis, Department of Geology, University of Malaya.

Leong, L. S. 1970 (MS). *Geology and mineralisation of the Western Hill area, Batu Tiga, Bukit Besi, Trengganu, West Malaysia*. B.Sc. Honours thesis, Department of Geology, University of Malaya.

Leow, J. H. 1962. A glimpse of the sedimentary structure of Singapore. *Malay. Nat. J.* **16**, 54–60.

Lim, T. H. 1971 (MS). *Geology, mineralization and geochemical studies in the western Yambang area, Pahang, West Malaysia*. B. Sc. Honours thesis, Department of Geology, University of Malaya.

Lindgren, W. 1933. *Mineral deposits*. New York: McGraw-Hill.

Logan, J. R. 1847a. On the local and relative geology of Singapore, including notices of Sumatra, and Malay Peninsula, etc. *J. Asiat. Soc. Bengal* **16**, 519–557.

———— 1847b. The rocks of Pulo Ubin. *Verh. Batav. Genoot. Kunst. Wet.* **22**, 3–43.

———— 1848a. Sketch of the physical geography and geology of the Malay Peninsula. *J. Indian Archipel.* **2**, 83–138.

———— 1848b. Notices of the geology of the east coast of Johore. *J. Indian Archipel.* **2**, 625–631.

———— 1851. Notices of the geology of the Straits of Singapore. *Quart. J. Geol. Soc. London* **7**, 310–344.

Loganathan, P. 1970 (MS). *Geology and geochemical study of the Templer Park area, Selangor, West Malaysia*. B.Sc. Honours thesis, Department of Geology, University of Malaya.

Low, J. 1833. Observations on the geological appearances and general features of portions of the Malayan Peninsula, and of the countries lying betwixt it and 18° north latitude. *Asiat. Res.* **18**, 128–162.

———— 1847. Notes on geological features of Singapore and some of the islands adjacent. *J. Indian Archipel.* **1**, 84–100.

Lu, K. U. 1950. *Palaeogeographic atlas of China*, 118 p. Peking.

Macandie, A. G., and Canavan, F. 1948 (MS). *The iron-ore and manganese–ore deposits of Malaya*, 78 p. Melbourne: Broken Hill Pty. Co. Ltd.

McBride, E. F. 1962. Flysch and associated beds of the Martinsburg Formation (Ordovician), central Appalachians. *J. Sediment. Petrol.* **32**, 39–91.

McCall, T. L. 1922. Description of the coal field at Batu Arang. *Bull. Can. Inst. Min. Metall.* **123**, 834.

MacDonald, S. 1968. The geology and mineral resources of north Kelantan and north Trengganu. *Mem. Geol. Survey Dept. W. Malaysia*, 202 p. **10**.

Magnée, I. de, and Aderca, B. 1960. Contribution à la connaissance du Tungsten-belt ruandais. *Mém. Acad. r. Sci. d'outre-mer Cl. Sci. nat. méd. 80*, n.s. **11**, fasc. 7, 1–56.

Marks, P. 1956. *Lexique stratigraphique international*, **3** (fasc. 7a) *Indonesia*, 241 p. Paris: Centre National de la Recherche Scientifique.

Marriott, G. 1927. *The mining geology and structure of the Kinta Valley*, 17 p. Penang.

Medway, Lord. 1969. *The wild mammals of Malaya and offshore islands including Singapore*, 127 p. Kuala Lumpur: Oxford University Press.

Mitchell, A. H. G., Young, B., and Jantaranipa, W. 1971. The Phuket Group, Peninsula Thailand: a Palaeozoic ?geosynclinal deposit. *Geol. Mag.* **107** [for 1970], 411–428.

Miyashiro, A. 1961. Evolution of metamorphic belts. *J. Petrol.* **2**, 277–311.

—— 1966. Some aspects of peridotite and serpentinite in orogenic belts. *Jap. J. Geol. Geogr.* **37**, 45–61.

Mohammed bin Ayob. 1965 (MS). *Studies in bedrock geology and sedimentology of Quaternary sediments at Sungei Besi Tin Mines, Selangor*. B.Sc. Honours thesis, Department of Geology, University of Malaya.

—— 1968 (MS). *Stratigraphy and sedimentology of the Tembeling Formation in the Gunong Berantai area, Pahang*. M.Sc. thesis, University of Malaya.

—— 1970. Quaternary sediments at Sungei Besi, West Malaysia. *Bull. Geol. Soc. Malaysia* **3**, 53–61.

Mohr, E. C. J., and Van Baren, F. A. 1954. *Tropical soils, a critical study of soil genesis as related to climate, rock and vegetation*, 498 p. The Hague: van Hoeve.

Molengraaff, G. A. F. 1921. Modern deep-sea research in the East Indian Archipelago. *Geogr. J.* **58**, 95–121.

—— and Weber, M. 1919. Het verband tusschen den plistoceenen ijstijd en het ontstaan der Soenda-Zee en de invloed daarvan op de verspreiding der koraalriffen en op de land- en zoetwater fauna. *Versl. Gew. Verg. Wis.- en Nat. Afd. K. Akad. Wet.* **28**, 497–544.

Morgan, M. J., de, 1886. *Note sur la géologie et sur l'industrie minière du Royaume de Perak et des pays voisins (Presqu'ile de Malacca)*. Paris.

Muir-Wood, H. M. 1948. *Malayan Lower Carboniferous fossils and their bearing on the Viséan palaeogeography of Asia*. 77 p. London: British Museum (N.H.).

Newell, R. A. 1971. Characteristics of the stanniferous alluvium in the southern Kinta Valley, West Malaysia. *Bull. Geol. Soc. Malaysia* **4**, 15–37.

Newton, R. B. 1900. On marine Triassic lamellibranchs discovered in the Malay Peninsula. *Proc. Malacol. Soc. London* **4**, 130–135.

—— 1906. Notice on some fossils from Singapore discovered by John B. Scrivenor, F.G.S., Geologist to the Federated Malay States. *Geol. Mag.* **43**, 487–496.

—— 1923. On marine Triassic shells from Singapore. *Ann. Mag. Nat. Hist., Ser. 9* **12**, 300–321.

—— 1925. On marine Triassic fossils from Kedah and Perak. *Geol. Mag.* **62**, 76–85.

—— 1926. On *Fusulina* and other organisms in a partially calcareous quartzite from the Malayan-Siamese frontier. *Ann. Mag. Nat. Hist., Ser. 9* **17**, 49–64.

Nikiferova, O. I., and Obut, A. M. 1965. *Stratigraphy of the U.S.S.R.: The Silurian System*, 529 p. Moscow: National Academy of Sciences.

Nockolds, S. R. 1940. The Garabal Hill-Glen Fyne Igneous Complex. *Quart. J. Geol. Soc. London* **96**, 451–510.

Noetling, F. 1890. Field notes from the Shan hills (Upper Burma). *Rec. Geol. Survey India* **23**, 78–79.

—— 1891. Report on the coalfields in the northern Shan States. *Rec. Geol. Survey India* **24**, 99–119.

Nossin, J. J. 1961a. Relief and coastal development in north-eastern Johore (Malaya). *J. Trop. Geogr.* **15**, 27–39.

—— 1961b. Occurrence and origin of clay pebbles on the east coast of Johore, Malaya. *J. Sediment. Petrol.* **31**, 437–447.

—— 1962. Coastal sedimentation in northeastern Johore (Malaya). *Z. Geomorph.*, n.s. **6**, 296–317.

—— 1964a. Beach ridges on the east coast of Malaya. *J. Trop. Geog.* **18**, 111–117.

—— 1964b. Geomorphology of the surroundings of Kuantan. *Geol. en Mijnb.* **43**, 157–182.

—— 1965a. The geomorphic history of the northern Pahang delta. *J. Trop. Geogr.* **20**, 54–64.

—— 1965b. Analysis of younger beach ridge deposits in eastern Malaya. *Z. Geomorph.*, n.s. **9**, 186–208.

Oakley, K. P. 1940. The organic content of Recent rhyolitic ashes in Malaya. *Geol. Mag.* **77**, 289–294.

Ong, S. S. 1969 (MS). *Geology of the Muda Dam area, Kedah, West Malaysia*. B.Sc. Honours thesis, Department of Geology, University of Malaya.

Opik, A. A. 1956. Cambrian Geology of the Northern Territory. *Int. Geol. Congr. 20, Mexico: Symposium on Cambrian System* **1**, 25–54.

Panton, W. P. 1956. Types of Malayan laterite and factors affecting their distribution. *6th Int. Congr. Soil Sci.* E, 419–423.

Paramananthan, S. 1964 (MS). *The geology of the Gunong Jerai massif, south-west Kedah*. B.Sc. Honours thesis, Department of Geology, University of Malaya.

—— 1966 (MS). *Petrographic studies of granite in Malacca and Negri Sembilan*. Department of Geology, University of Malaya.

Park, C. F., and MacDiarmid, R. A. 1964. *Ore deposits*. San Francisco: Freeman and Co.

—— 1969 (MS). *Tin bearing granite and tin barren granite in Thailand*. Preprint of paper read at the 2nd Technical Conference on Tin of the International Tin Council, Bangkok.

Pascoe, Sir E. H. 1959. *A manual of the geology of India and Burma*, 3rd ed., vol. 2. Calcutta: Government of India Press.

Paton, J. R. 1959. Jurassic/Cretaceous sediments in Malaya. *Nature, Lond.* **183**, 231–232.

—— 1960 (MS). *The intermediate, extrusive and dyke rocks of Central Trengganu, Malaya*. M.Sc. thesis, University of Leeds.

—— 1964. The Origin of the limestone hills of Malaya. *J. Trop. Geogr.* **19**, 134–147.

—— 1966. Geology—Federation of Malaysia. In W. B. Agocs, *Report on airborne magnetometer and scintillation counter survey of Kedah, Perak, Selangor, Trengganu, Pahang and Johore, Federation of Malaya*, Malaysian Government. Chapter 2, pp. 33–97.

Penrose, R. A. F. 1903. The tin deposits of the Malay Peninsula with special reference to those of the Kinta district. *J. Geol.* 11, 135–154.

Pettijohn, F. J. 1957. *Sedimentary rocks*, 2nd ed. 718 p. New York: Harper.

Pitakpaivan, K. 1965. The fusulinacean fossils of Thailand, pt. I Fusulines of the Rat Buri limestone of Thailand. *Mem. Fac. Sci. Kyushu Univ., Ser. D* 17, 1–69.

Preece et al. 1962 (MS). *Report on Batang Padang hydro-electric scheme. 3, Geological investigations.* London: Preece, Cardew and Rider, and Binnie and Partners.

Price, N. J. 1968. *Fault and joint development in brittle and semi-brittle rock*, 176 p. Oxford: Pergamon Press.

Prior, G. T. 1927. Tektites. *Nat. Hist. Mag.* 1, 8–13.

Procter, W. D., and Jones, C. R. 1967. Wrench faulting in Malay: A discussion. *J. Geol.* 75, 127–128.

Pryor, E. J., and Wrobel, S. A. 1951. Studies in cassiterite flotation. *Trans. Inst. Min. Metall.* 60, 201–237.

Raguin, E. 1965. *The Geology of Granite.* 2nd ed. (English transl.). New York: Wiley-Interscience.

Rajah, S. S. 1967 (MS). *Geology and Mineral Resources of sheet 125—Johore.* Proceedings of the Senior Officers Conference, 1966, pp. 41–47. Geol. Survey Dept. W. Malaysia.

——— 1969. Younger Mesozoic sedimentary rocks, State of Johore, West Malaysia. *Bull. Am. Assoc. Petrol. Geol.* 53, 2187–2194.

Ramdohr, P. 1944. Zum Zinnkiesproblem. *Abh. Preuss. Akad. Wiss. math. naturwiss. Kl.* 4, 1–30.

Rastall, R. H. 1927. The geology of the Kinta valley. *Min. Mag. London* 36, 329–338.

Read, H. H. 1949. A contemplation of time in plutonism. *Quart. J. Geol. Soc. London* 105, 101–156.

——— 1955. Granite series in mobile belts. In *The crust of the earth, Spec. Pap. Geol. Soc. Am.* 62, 409–430.

——— 1960. Aspects of Caledonian magmatism in Britain. *Liverpool Manchester Geol. J.* 2, 653–683.

Reed, F. R. C. 1906. Lower Palaeozoic fossils of the Northern Shan States of Burma. *Palaeontol. Indica*, (n.s.) 2 (3), 1–154.

——— 1920. Carboniferous fossils from Siam. *Geol. Mag.* 57, 113–120, 172–178.

——— 1949. The Straits Settlements and Protected Malay States. In *The geology of the British Empire.* London: Arnold, pp. 526–532.

Renwick, A., and Rishworth, D. E. H. 1966. *Fuel resources (coal, lignite, and petroleum) in Malaya*, 123 p. Ipoh: Geological Survey Department of West Malaysia.

———, Santokh Singh, D., and Ledgerwood, E. 1960 (MS). *Preliminary geological report on the area of the Batang Padang hydro-electric scheme.* Geological Survey Department, Federation of Malaya.

Richardson, J. A. 1939. The geology and mineral resources of the neighbourhood of Raub, Pahang, Federated Malay States, with an account of the geology of the Raub Australian Gold Mine, 166 p. *Mem. Geol. Survey Dept. Fed. Malay* 3.

——— 1941. The coal veins of British Malaya. *Geol. Mag.* 78, 451–462.

——— 1946. The stratigraphy and structure of the Arenaceous Formation of the Main Range Foothills, F. M. S. *Geol. Mag.* 83, 217–229.

——— 1947a. The origin of the amphibole schist series of Pahang, Malaya. *Geol. Mag.* 84, 241–249.

——— 1947b. Facies change and lithological variation in the Permocarboniferous formation of north-west Pahang and south-west Kelantan, Malaya. *Geol. Mag.* 84, 281–288.

——— 1947c. An outline of the geomorphological evolution of British Malaya. *Geol. Mag.* 84, 129–144.

——— 1950. The geology and mineral resources of the neighbourhood of Chegar Perah and Merapoh, Pahang, Malaya, 162 p. *Mem. Geol. Survey Dept. Fed. Malaya* 4.

Riley, G. 1967. The cassiterite-stannite occurrence at Tekka Mines, Kinta. *Newsl. Geol. Soc. Malaysia* 7, 10–12.

Rishworth, D. E. H. In prep. A. The geology and mineral resources of Malacca and the south part of Negri Sembilan. *Bull. Geol. Survey Dept. W. Malaysia.*

——— In prep. B. The upper Mesozoic continental Gagau Group of Malaya. *Bull. Geol. Survey Dept. W. Malaysia.*

Roe. F. W. 1941 (MS). *Visit to Pelepah Kanan Mines, Johore.* Geological Survey Department of the Federation of Malaya.

——— 1949. Progress report on geological work in Selangor. *Rept. Geol. Survey Dept. Fed. Malaya* [for 1948], 24–37.

——— 1951. The geology and mineral resources of the Fraser's Hill area, Selangor, Perak and Pahang, Federation of Malaya, with an account of the mineral resources, 138 p. *Mem. Geol. Survey Dept. Fed. Malaya,* 5.

——— 1953. The geology and mineral resources of the neighbourhood of Kuala Selangor and Rasa, Selangor, Federation of Malaya, with an account of the Batu Arang Coalfield, 163 p. *Mem. Geol. Survey Dept. Fed. Malaya* 7.

Roux-Brahis, J. 1910 *Étude du district stannifère de Tekkah.* Bordeaux: Gounouilhou.

Ruedemann, R. 1934. Paleozoic plankton of North America. 114 p. *Mem. Geol. Soc. Am.,* 2.

Rumbold, W. 1906. The tin-deposits of the Kinta valley, Federated Malay States. *Bull. Am. Inst. Min. Eng.* 11, 755–765.

Rutten, L. M. R. 1938. Le paysage des Indes Orientales et Occidentales. *C. R. Congr. Int. Geogr., Amsterdam 1938,* 1, 58–77.

Sakagami, S. 1963. Bryozoa from Pulau Jong, the Langkawi Islands, north-west Malaya. *Jap. J. Geol. Geogr.* 34, 205–209.

——— 1966a. Three Carboniferous species of bryozoa from Khao Noi, Central Thailand. In *Geology and Palaeontology of Southeast Asia,* T. Kobayashi and R. Toriyama, Eds., vol. 2. Tokyo: University of Tokyo Press, pp. 9–16.

——— 1966b. The Permian bryozoan fauna of Ko Muk, Peninsular Thailand with the description of the Cylostomata. *Ibid.,* vol. 2, pp. 255–272.

——— 1968. Permian bryozoa from Khao Phrik, near Rat Buri, Thailand. *Ibid.,* vol. 4, pp. 45–66.

Santokh Singh, D., and Bean, J. H. 1968. Some general aspects of tin minerals in Malaysia. *Pap. Tech. Conf. Tin, London 1967* 2, 457–478. London: International Tin Council.

Sato, T. 1963. Ammonites du Trias de la Malaisie. *Jap. J. Geol. Geogr.* 34, 93–99.

Savage, H. E. F. 1925. A preliminary account of the geology of of Kelantan. *J. Malay. Br. R. Asiat. Soc.* 3 (1), 61–73.

––––– 1937. The geology of the neighbourhood of Sungei Siput Perak, Federated Malay States, with an account of the mineral deposits, 46 p. *Mem. Geol. Survey Dept. Fed. Malaya* **1**.

––––– 1950. Triassic fossils from near Kuala Lipis, Pahang. *Colon. Geol. Miner. Resour.* **1**, 76–77.

Saurin, E. 1956. *Lexique stratigraphique international*, **3** (Fasc, 6a) *Indochine*, 140 p. Paris: Centre National de la Recherche Scientifique.

Schürmann, H. M. E., Bot, A. C. W. C., Steensma, J. J. S., Suringa, R., Eberhardt, P., Geiss, J., von Gunten, H. R., Houtermans, F. G., and Signer, P. 1956. Second preliminary note on age determinations of magmatic rocks by means of radioactivity. *Geol. en Mijnb.* **18**, 312–330.

––––– 1957. Third preliminary note on age determinations of magmatic rocks by means of radioactivity. *Geol. en Mijnb.* **19**, 398–413.

––––– 1960. Further preliminary note on age determination of magmatic rocks by means of radioactivity. *Geol. en Mijnb.* **22**, 93–104.

Scrivenor, J. B. 1907. *Geologists report of progress, Federated Malay States, September 1903–January 1907*, 44 p. Kuala Lumpur: Government Press.

––––– 1908. Note on the sedimentary rocks of Singapore. *Geol. Mag.* **45**, 289–291.

––––– 1909. Obsidianites in the Malay Peninsula. *Geol. Mag.* **46**, 411–413.

––––– 1910. The tourmaline-corundum rocks of Kinta (Federated Malay States). *Quart. J. Geol. Soc. London* **66**, 435–449.

––––– 1911a. *The geology and mining industries of Ulu Pahang*, 61 p. Kuala Lumpur: Government Printing Office.

––––– 1911b. Report on the Rantau Panjang coal measures, 4 p. *F.M.S. Fed. Council Pap.* **4**.

––––– 1912. The Gopeng beds of Kinta. *Quart. J. Geol. Soc. London* **68**, 140–163.

––––– 1913a. *The geology and mining industry of the Kinta district, Perak, Federated Malay States, with a geological sketch map*, 90 p. Kuala Lumpur: Government Press.

––––– 1913b. The geological history of the Malay Peninsula. *Quart. J. Geol. Soc. London* **69**, 343–371.

––––– 1913c. *Report on a visit to Perlis.*

––––– 1914a. The junction of the Malayan Gondwana clays with the Mesozoic granite of the Malay Peninsula. *Geol. Mag.* **51**, 309–311.

––––– 1914b. The topaz-bearing rocks of Gunong Bakau (Federated Malay States). *Quart. J. Geol. Soc. London* **70**, 363–381.

––––– 1916. Mineralization in Malaya. *Min. Mag., London* **14**, 94–95.

––––– 1917. Report on the Enggor coalfield, 3 p. *F.M.S. Govt. Gaz., Suppl.*, Feb. 2.

––––– 1918. The origin of the clays and boulder clays, Federated Malay States. *Geol. Mag.* **55**, 157–168.

––––– 1919. Topaz as a rock constituent. *Geol. Mag.* **56**, 190–191.

––––– 1921. The physical geography of the southern part of the Malay Peninsula. *Geogr. Rev.* **11**, 351–371.

––––– 1923. The structural geology of British Malaya. *J. Geol.* **31**, 556–570.

––––– 1924. The geology of Singapore Island. *J. Malay Br. R. Asiat. Soc.* **2**, 1–8.

––––– 1925. Geology of the Kinta valley. *Min. Mag., London* **33**, 156–157.

––––– 1926. The palaeontology of British Malaya. *J. Malay. Br. R. Asiat. Soc.* **4**, 173–184.

––––– 1927. The geology of Malacca, with a geological map and special reference to laterite. *J. Malay. Br. R. Asiat. Soc.* **5**, 278–287.

––––– 1928. *The geology of the Malayan ore-deposits*, 216 p. London: Macmillan.

––––– 1930. A recent rhyolite-ash with sponge-spicules and diatoms in Malaya. *Geol. Mag.* **67**, 385–393.

––––– 1931. *The geology of Malaya*, 217 p. London: Macmillan.

––––– 1946. Sea levels in Malaya. *Geol. Mag.* **83**, 103–104.

––––– 1949. Geological and geographical evidence for changes in sea-level during ancient Malayan history and late prehistory. *J. Malay. Br. R. Asiat. Soc.* **22**, 107–115.

–––––, and Jones, W. R. 1919. *The geology of south Perak, north Selangor and the Dindings*, 196 p. Kuala Lumpur: Government Press.

–––––, and Willbourn, E. S. 1923. The geology of the Langkawi Islands, with a geological sketch map. *J. Malay. Br. R. Asiat. Soc.* **1**, 338–347.

Serra, C. 1968. Sur quelques empreintes Mesozoique de Malaisie. *Arch. Geol. Viet-Nam* **11**, 43–51.

Service, H. 1949. Progress report. *Rept. Geol. Survey Dept. Fed. Malaya* [for 1948], 17–19.

Shu, Y. K. 1969. Some NW trending faults in the Kuala Lumpur and other areas. *Newsl. Geol. Soc. Malaysia* **17**, 1–5.

––––– 1970. Geology of the Jelebu District, new series map sheet 95. *Ann. Rep. Geol. Surv. Malaysia* [for 1968], 94–99.

Sieveking, A. de G. 1962. The Palaeolithic industry of Kota Tampan, Perak, Malaya. Pt. II. Archeology. *Proc. Prehist. Soc.* **28**, 107–139.

Sivam, S. P. 1968. Radiocarbon dates in the Kinta Valley. *Newsl. Geol. Soc. Malaysia* **15**, 1 only.

––––– 1969 (MS). *Quaternary alluvial deposits in the north Kinta Valley, Perak*. M.Sc. thesis, University of Malaya.

Smiley, C. J. 1970a. Later Mesozoic flora from Maran, Pahang, West Malaysia. Pt. 1: Geologic considerations. *Bull. Geol. Soc. Malaysia* **3**, 77–88.

––––– 1970b. Later Mesozoic flora from Maran, Pahang, West Malaysia. Pt. 2: taxonomic considerations. *Ibid.* 89–113.

Snelling, N. J. 1965. Age Determination Unit. Summary of results from Malaysia. *Ann. Rept. Overseas Geol. Surveys* [for 1964], 32–33.

––––– 1967. Age Determination Unit. Summary of results from the Pacific and S.E. Asia. *Ann. Rept. Inst. Geol. Sci.* [for 1966], 142–147.

–––––, Bignell, J. D., and Harding, R. R. 1968. Ages of Malayan Granites. *Geol. en Mijnb.* **47**, 358–359.

Snow, A. B. 1902. *Mineral resources of Johore*, 11 p. Singapore: Straits Times Press.

Stanton, R. E., and McDonald, A. J. 1962. Field determination of tin in geochemical soil and stream sediment surveys. *Trans. Inst. Min. Metall.* **71**, 27–29.

Stauffer, P. H. 1968. The Kuala Lumpur fault zone: A proposed major strike slip fault across Malaya. *Newsl. Geol. Soc. Malaysia* **15**, 2–4.

—— 1970 (MS). *Quaternary volcanic ash in West Malaysia.* Abstracts of papers, Annual General Meeting Geological Society of Malaysia, January, 1970.

Stephens, F. J. 1901. Mineral features of Pahang, Malay Peninsula. *Trans. Inst. Min. Metall. London* **9**, 419–424.

Stok, J. P. van der, 1922. Maritieme geologie en getijden. In *De Zeeen van Nederlandsch Oost-Indie*. Leiden: Brill, pp. 182–212.

Sujkowski, S. L. 1957. Flysch sedimentation. *Bull. Geol. Soc. Am.* **68**, 543–554.

Sun, Y. C. 1946. The Sino-Burmese geosyncline of early Palaeozoic times with special reference to its extent and character. *Bull. Geol. Soc. China* **25**, 1–7.

—— 1947. On the occurrence of Fengshanian (the Late Upper Cambrian) trilobite faunas in West Yunnan. *Bull. Geol. Soc. China* **27**, 29–34.

Suntharalingam, T. 1968. Upper Palaeozoic stratigraphy of the area west of Kampar, Perak. *Bull. Geol. Soc. Malaysia* **1**, 1–15.

Swan, R. M. W. 1900. Extracts from notes on prospecting in Ulu Pahang. *Ann. Rept. Pahang, F.M.S.* [for 1899], appendix C, x–xii.

Swinney, A. J. G. 1891. *Report on work done etc., on the Rumpen river during the year 1890*. London: Crowther and Goodman.

Tamura, M. 1968. *Claraia* from north Malaya, with a note on the distribution of *Claraia* in southeast Asia. In *Geology and palaeontology in southeast Asia*, T. Kobayashi and R. Toriyama, Eds., vol. 5, Tokyo: University of Tokyo Press, pp. 78–87.

Taylor, D. 1971. An outline of the Bukit Ibam Orebody, Rompin, Pahang. *Bull. Geol. Soc. Malaysia* **4**, 71–89.

Teng, H. C. 1970 (MS). *General geology and geochemical study of the Segamat volcanics area*. B.Sc. Honours thesis, Department of Geology, University of Malaya.

Tenison-Woods, J. E. 1884a. On the stream tin-deposits of Perak. *J. Straits Br. R. Asiat. Soc.* **13**, 221–240.

—— 1884b. Geology of the Malay Peninsula. *Nature, Lond.* **30**, 76 only.

—— 1885. Report on the geology and physical geography of the State of Perak. *Proc. Linn. Soc. N.S.W.* **9**, 1176–1203.

Thomas, H. D. 1963. Silurian corals from Selangor, Federation of Malaya. *Overseas Geol. Miner. Resour.* **9**, 39–46.

Thomas, H. D., and Scrutton, C. T. 1969. Palaeozoic corals from Perak, Malay, Malaysia. *Overseas Geol. Miner. Resour.* **10**, 164–171.

Thompson, J. T. 1851. Description of the eastern coast of Johore and Pahang, and adjacent islands. *J. Indian Archipel.* **5**, 83–92, 135–154.

Tilley, C. E. 1922. Density, refractivity, and composition relations of some natural glasses. *Miner. Mag.* **19**, 275–294.

Tjia, H. D. 1968. Fluvial correlates of "Daly levels." *Z. Geomorph.* **12**, 194–199.

—— 1969a. Slope development in tropical karst. *Z. Geomorph.* **13**, 260–266.

—— 1969b. Sunda-Land bauxites: related to Late Cenozoic see level? *Newsl. geol. Soc. Malaysia* **18**, 1–2.

—— 1970a. Monsoon-control of the eastern shoreline of Malaya. *Bull. Geol. Soc. Malaysia* **3**, 9–15.

—— 1970b. Quaternary shore lines of the Sunda Land, Southeast Asia. *Geol. en Mijnb.* **49**, 135–144.

Tokuyama, A. 1961. On some Triassic pelecypods from Pahang province, Malaya. *Trans. Proc. Palaeontol. Soc. Jap.* (n.s.) **44**, 175–178.

Tooms, J. S., and Kaewbaidhoon, S. 1961. Dispersion of tin in soil over mineralisation at Sungei Lembing, Malaya. *Trans. Inst. Min. Metall.* **70**, 475–490.

Toriyama, R., Sato, H., Hamada, T., and Komalarjun, P. 1965. *Nautilus pompilius* drifts on the west coast of Thailand: *Jap. J. Geol. Geogr.* **36**, 149–161.

Tuttle, O. F., and Bowen, N. L. 1958. Origin of granite in the light of experimental studies in the system $NaAlSi_3O_8$–$KAlSi_3O_8$–SiO_2–H_2O. *Mem. Geol. Soc. Am.*, 153 p. **74**.

Tweedie, M. W. F. 1965. *Prehistoric Malaya*, 44 p. Singapore: Donald Moore.

Umbgrove, J. H. F. 1929. The amount of the maximum lowering of the sea level in the Pleistocene. *Proc. 4th Pacif. Sci. Congr.* **2A**, 105–113.

U.S. Navy Hydrographic Office. 1960. Summary of oceanographic conditions in the Indian ocean. *Spec. Publ. U.S. Navy Hydrogr. Office* **53**, 1–142.

Vialov, O. S. 1939. Le plissement mesozoique (pacifique) en Asie. *Int. Geol. Congr. 17, Moscow* **2**, 563–570.

Walker, D. 1956. Studies on the Quaternary of the Malay Peninsula. Pt. I. Alluvial deposits of Perak and the relative levels of land and sea. *Fed. Mus. J.* **1–2**, 19–34.

—— 1962. The Palaeolithic industry of Kota Tampan, Perak, Malaya. Pt. I. Geography and geology. *Proc. Prehist. Soc.* **28**, 103–107.

Walton, B. J. 1965. Sanerutian appinitic rocks and Gardar dykes and diatremes, north of Narssarssuaq, South Greenland. *Medd. Grønland*, 66 p. **179** (9).

Ward, D. E., and Bunnag, D. 1964. Stratigraphy of the Mesozoic Khorat Group in northeast Thailand. *Rept. Investigation, Dept. Min. Res. Bangkok, Thailand*, 95 p. **6**.

Ward, T. 1833. Short sketch of the geology of Pulau Pinang and the neighbouring islands, with a map and sections. *Asiat. Res.* **18**, 149–168.

Waterhouse, J. B., and Piyasin, S. 1970. Mid-Permian brachiopods from Khao Phrik, Thailand. *Palaeontographica*, **135A**, 83–197.

Weir, J. 1925. On some specimens of fossiliferous sandstone from Pahang. *Geol. Mag.* **62**, 347–350.

Wilford, G. E. 1964. The geology of Sarawak and Sabah Caves. *Bull. Geol. Survey Dept. Borneo Region, Malaysia*, 181 p. **6**.

—— 1965. Penrissen area, West Sarawak, Malaysia, 195 p. *Rept. Geol. Survey Dept. Borneo. Region, Malaysia*, **2**.

—— 1969. Sunda-Land bauxites and sea levels. *Newsl. Geol. Soc. Malaysia* **19**, 5.

Willbourn, E. S. 1917. The Pahang Volcanic Series. *Geol. Mag.* **54**, 447–462.

—— 1922a. *An account of the geology and mining industries of south Selangor and Negri Sembilan*, 115 p. Calcutta: Baptist Mission Press.

—— 1922b. A general account of the geology of the Malay Peninsula and the surrounding countries, including Burma, the Shan States, Yunnan, Indo-China, Siam, Sumatra,

Java, Borneo and other islands of the Dutch East Indies. *J. Straits Br. R. Asiat. Soc.* **86**, 237–256.

——— 1923. *Report of the Geologist, F.M.S.* [for 1922], 7 p.

——— 1926. The geology and mining industries of Kedah and Perlis. *J. Malay. Br. R. Asiat. Soc.* **4**, 289–332.

——— 1926–1927. The Beatrice Mine, Selibin, Federated Malay States. *Min. Mag., London* **35**, 329–338; **36**, 9–15.

——— 1928. The geology and mining industries of Johore. *J. Malay. Br. R. Asiat. Soc.* **6** (4), 5–35.

——— 1931. The occurrence *in situ* of corundum-bearing rocks in British Malaya. *De Mijning. Bandung* **10**, 170–176.

——— 1933. *Report of the Geological Survey Department, F.M.S.* [for 1932], 18 p. Kuala Lumpur.

——— 1934. *Report of the Geological Survey Department, F.M.S.* [for 1933], 19 p. Kuala Lumpur.

——— 1936. A short account of the geology of those tin-deposits of Kinta that are mined by alluvial methods. *J. Eng. Assoc. Malaya* **4**, 255–264.

——— 1937. *Report of the Geological Survey Department, F.M.S.* [for 1936].

——— 1938. *Report of the Geological Survey Department, F.M.S.* [for 1937].

——— 1940. *Report of the Geological Survey Department, F.M.S.* [for 1939].

——— 1941. *Report of the Geological Survey Department, F.M.S.* [for 1940].

———, and Ingham, F. T. 1933. Geology of the Scheelite Mine, Kramat Pulai Tin Limited, Kinta, Federated Malay States. *Quart. J. Geol. Soc. London* **89**, 449–479.

Winchell, A. N., and Winchell, H. 1951. *Elements of optical mineralogy: an introduction for microscopic petrography, Part II: Description of minerals*, 4th ed., 55 p. New York: Wiley.

Winkler, H. G. F. 1967. *Petrogenesis of metamorphic rocks*. Rev. 2nd ed. (Transl. N. D. Chatterjee and E. Froese), 237 p. Berlin: Springer-Verlag.

Wong, I. F. T. 1960 (MS). *The Agglomerate of Bukit Kepayang, Pahang*. B.Sc. Honours thesis, Department of Geology, University of Malaya.

Wong, T. W. 1971 (MS). *Bedrock stratigraphy of the Rawang area, Selangor*. Abstr. Discussion Meeting Geological Society of Malaysia, February 1971.

Wood, A., and Smith, A. J. 1959. The sedimentation and sedimentary history of the Aberystwyth Grits (Upper Llandoverian). *Quart. J. Geol. Soc. London* **114**, 163–196.

Wray, L. 1886. *Notes on Perak, with a sketch of its vegetable, animal, and mineral products*, 33 p. London: W. Clowes and Sons.

——— 1893. Alluvial tin-prospecting. *Perak Museum Notes* **1** (2), 1–114.

——— 1894–1898. Some account of the tin mines and the mining industries of Perak. *Perak Museum Notes* **1** (3), 1–24; **2**, 81–88.

Yeap, E. B. 1970 (MS). *Geology of the Petaling Jaya–Salak South area, Selangor, West Malaysia*. B.Sc. Honours thesis, Department of Geology, University of Malaya.

Yeow, B. C. 1969 (MS). *Studies in geology and mineral resources of the Sungei Way area, Selangor*. B.Sc. Honours thesis, Department of Geology, University of Malaya.

Yin, E. H. 1965a (MS). 1963 progress report on geological work done in the area of the new series sheet 45 in southern Kelantan. *Prof. Pap. Geol. Survey Dept. Fed. Malaya* **E65.1G**, 7–10.

——— 1965b (MS). 1964 progress report on geological survey work done in the area of sheet 45 in south Kelantan. *Prof. Pap. Geol. Survey Dept. Malaya* **E65.2G**, 20–24.

——— In prep. Geology and mineral resources of the Kuala Lumpur area, Malaya. *Mem. Geol. Survey Dept. W. Malaysia*.

Yochelson, E. L., and Jones, C. R. 1968. *Teiichispira*, a new early Ordovician gastropod genus. *Prof. Pap. U.S. Geol. Survey*, 15 p. **613-B**.

Young, B., and Jantaranipa, W. 1970. The discovery of Upper Palaeozoic fossils in the "Phuket Series" of peninsular Thailand. *Proc. Geol. Soc. London* **1662**, 5–7.

Zeuner, F. E. 1959. *The Pleistocene Period*, 447 p. London: Hutchinson.

Index

Numbers in boldface type indicate the pages on which the major references appear. Pages followed by "F" represent illustration entries; "T" table entries.

Aberystwyth Grits, 402
ablation, 18
abrasion, 18, 19
　marine, 18
Abukuma amphibolite facies, 273
　facies series 260, 266, 266F, 281, 285, 287, 288, 296, 301
　terrane, 253, 302, 303
accretion, 305
ACF diagram, 261F, 266F
acid and basic magma associations, 392, 395
　plutonic rocks, 28F
acmite, 183
　normative, 232, 232T
acodus, 35
acontiodus, 35
acrochordiceras sp., 112
actinocamax plenus subzone, 396
actinoceras sp., 36, 38F, 40
actinoceras perlisense KOBAYASHI, 36
actinolite, 51, 53, 221, 265, 281, 284, 352
　optical properties, 296
ADAM, J. W. H., 105, 391
adamellite, 71, 101, 185, 220T, 221, 215, 231T, 244F, **280**
　biotite, 231T, 237T
　　porphyritic, 220T
　hornblende, 103
　hornblende, pink, 220T
　leuco, 220T
　porphyritic, 220T
ADERCA, B., 369, 398
adze heads, 175
Aegean Sea, 292
aegirine, 211
aequipecten sp., 127
aequitradites sp. (cf. *aequitradites* cf. *verrucosus*), 136
aerial photographs, 9, 15, 73, 330, 389F
age, carbon-14, 143
　determination, 400
　rubidium-strontium, 123, 216F
Age Determination Unit, 260, 400
ages, granite, 9
　lead method, 216F

　metamorphic rocks, 216F
　potassium-argon, 123, 141, 143, 153, 204, 216F, 235
　radiocarbon, 168, 174, 167, 173, 170
　radiometric, 71, 78, 87, 90, 121, 122, 141, 204, 216F, 217, 235
　radiometric locality, 217F
agglomerate, 81T, 82, 103, 131, 187, 188, **190**, 197, 200, 209, 211
　andesite, 83, 185, 192
　fragments, 197
　mineralogy, 190
　rhyolite, 190
agglomeratic tuff, 199
aggradation, 15, 19, 22, 168
　of bays, 19
agmatite, **263**, 269
AGOCS, W. B., 8, 254, 390, 391
airborne magnetometer, 390
A'K F diagram, 261F, 266F
albite, 188, 200
　-epidote hornfels facies, 348
　normative, 233F
albitization, 199
ALEVA, G. J. J., 242, 391
ALEXANDER, F. E. S., 8, 13, 24, 108, 109, 110, 111, 127, 391
ALEXANDER, J. B., 7, 8, 15, 23T, 34T, 37, 39, 51, 54, 61, 71T, 75, 84, 102T, 103T, 105, 111, 114, 115, 120, 123, 128, 134, 144, 151, 154, 178, 184, 185, 186T, 187, 197, 202T, 210F, 212F, 216F, 225, 227, 228, 230, 245, 247, 249, 250F, 254, 256, 258T, 272T, 295, 322, 326F, 329, 346, 353, 391
algae, 68
　dasycladacean, 78
alkali enrichment, 273
　expulsion, 232
　feldspar, 180, 182, 220T, 350
　　ellipsoidal, 269, 271F
　　modal, 221F
　　pink, 218
　　radiometric dates, **236-238T**

　　sieve porphyroblasts, 269
　feldspar-cordierite hornfels facies, 288
　feldspar porphyroblasts, 271F, 274, 275
　metasilicate, normative, 232
　metasomatism, 274, 277
　total, 274F
　variation, 250F
allanite, 227, 230
alluvial, 5
　deposits, 12
　in caves, 171
　flat, 23T
　plains, 18
　soils, 394
　terraces, 20
　tin mines, 37
alluvium, 3, 9, 22, 23T, 64F, 153, 155, 282F, 286F, 292F, 314F
　coastal, 205F
　collapse, 157
　fluviatile, 22
　high level, 158, 160
　post Pliocene, 3
　Quaternary, 71
　Recent, 5
　river, 196F
　rudaceous, 166F
　thickness, 160, 165
　tin-bearing, 12
　tin-bearing in caves, 171
almandine garnet, 225, 259, 261F, 266, 285
Alor Star, 31F, 64F, 68, 99F, 145F, 173, 284
Alpine Fault System, 235
alpine forms, 112
Alps, 112, 121
aluminium enrichment, 292
　hydrates, 394
amang, 6
American Code of Stratigraphic Nomenclature, 106
　Commission on Stratigraphic Nomenclature, 391
AMIES, A. C., 7
ammonoids, 11

Anisian, 397
  Lower Triassic, 101
  Middle Triassic, 7
  Triassic, 394
ammonites, Triassic, 399
ambocoeliids, 394
amorphognathus, 35
Ampang, 290, 311F, 358
Ampang Fault, 325F
Ampang new village, 238T
  Perak, 74F, 182
  Selangor, 4, 220T, 290, 300
amphibole, 161, 185, 220T, 243
  replacement of augite, 191
  sieve porphyroblasts, 279
Amphibole Schist Series, 7, 399
amphibole, sodic, 200
amphibolite, 53, 185, 250, 256, **259**, 261F, 294, 295, 296, 347
  clinozoisite, 258T
  dykes, 275
  facies, 253, 254, 263, 285, 287, 301, 331
  garnet, 280
  mineralogy, 260
  modal analysis, 259
  tremolite, 258T, 296
amphipora, 76
amygdales, 204, 245
amygdalophyllum sp., 69F, 86
anaerobic conditions, 40
Anak Bukit Tinggi, 249
Anak Sungei Belat, 204, 206
Anak Sungei Merapoh, 297
Anak Sungei Pengau, 202T, 203
Anak Sungei Rek, Kelantan, 248T
Anak Sungei Tembeling, 202T
analcites, 106
Anambas, **213**, 241F
  Islands, 306
  Zone, 306F
anatase, 161, 166
anatectic granite, 288
anatexis, 256, 260, 277, 331, 333F
  partial, 261, 274
andalusite, 161, 180, 256, 260, 281, 287, 288, 289, 295, 296, 298, 301, 302, 303, 291
  -cordierite-muscovite subfacies, 301
andalusite in granite, 224
  porphyroblasts, 301
Andaman-Nicobar ridge, 305
  Sea, 2F, 55, 57F
  -Sunda Arc, 328
Anderson Mine, 382T
andesine, 103
  saussuritised, 297
andesite, 4, 10T, 81T, 82, 86, 87, 106, 177, 184, 187, **188**, 190, **191**, 195, 196F, 197, **199**, 201, 202T, 203, 209, 210F, 211, 213, 214, 243, 257F
  agglomerate, 179T, 195
  augite, 190, 212
  flows, 243
  hornblende, 202T
  hornblende-hypersthene, 214

andesite lava flows, 178F
  microcrystalline, 243
  mineralogy, 191, 197, 199, 211
  old, 213
  Permian, 191
  pillow, 199
  porphyritic, 195, 202T
andesite porphyry, 184
  scoriaceous, 199
  vesicular, 191, 199
  xenoliths, 199
andesitic breccia, 214
  material, 32
andradite garnet, 289, 347, 349
  tin-bearing, 353
angiopteris erecta, 149
ankaramite, potassic, 206
Anisian, 11, 81T, 101, 102, 106, 112
ankerite, 351T
annonaceae, 149
ANONYMOUS, 9, 104, 122, 391
anorthoclase, 206
anthophyllite, 349, 350F
anticlinal arch, 33
  axis, 319F, 359F
  crest, 18
  fold, 30, 33
  structure, 261
anticline, 6, 196F, 308, 314, 316, 359
  plunging, 261
anticlinorium, 11, 73, 317
antigorite, 206, 259
antimony, 336
apatite, 149, 180, 183, 220T, 240, 366
ape bones, 175
aphanitic nature, 182
apheoorthis sp., 32
aplite, 222, **225**, 263, 279, **280**, **343-346**, 359F
  mineralized, 345, 346
  mineralogy, 225
  pegmatite, 249
  spherulitic, 225
  topaz, 346
aplite veins (veining), 225, 341
aplitic granite, 234T
applied geochemistry, 389
arandesite, 353
Aranyakanon, P., 341, 342, 345, 365, 391, 396
Arau, 31F
arca, 110, 173
archaeology, 171, **175**
arctoceras sp. cf. *A. blomstrandi* (LINDSTROM), 101
arcuate form, 11
Arenaceous Formation, 7, 399
Arenaceous Series, 7, 84, 87, 108, 305, 312
Arenaceous Series (*Myophoria*), 106
arenite, 29T, 41, 46, 48, 53, 115, 116, 117, 118, 119, 324F
  cross-laminated, 135
  immature, 110, 113
  micaceous, 117
  quartz, 33

arfvedsonite, 297
argillaceous facies, 80, 84
  rocks, 317F
argillite, 29T, 50, 53, 115, 324F
  black, 102
  carbonaceous, 33, 41
  graptolitic, 33
  red, 138
arizonite, 344
arkose, 31
armonoceras chediforme KOBAYASHI, 36
ARMSTRONG, R. L., 217, 235, 239, 260, 391
arpadites, 112
  cf. *cinensis,* 112
  sp. ex. gr. *A.* Szaboi, 112
arsenious oxide, 335
arsenopyrite, 161, 227, 335, 345, 355F, 356F, 364F, 365, 368F, 372, 374, 375F, 377F
Artinskian, 94
ASAMA, K., 94, 391, 395, 397
asaphus, 32
ash, 184
  aeolian, 207
  age, 155
  analysis, 154
  beds, volcanic, 150
  falls, 119
  rhyolite, **153-155**, 207
  rhyolitic, 143, 153
    origin of, **154-155**
  stratigraphic position, 155
  thickness, 154
  wind-borne, 154
ashy tuff, 199, **200**
Asia, 36, 43
Asian continent, uplift, 140F
Assam, 55
assimilation, 223, 225F, 232, 249, 267
  by basalt, 247
  stages, 224
association, nonorogenic, 250, 251
  orogenic, 250, 251
astarte guthriensis sp. nov., 110
  scrivenori sp. nov., 110
atrypella sp., 37, 38F, 39
AUBOUIN, J., 97, 134, 139, 251, 392
augen gneiss, 230, 330
  microcline, 287
augite, 182, 188, 191, 197, 198, 204, 206, 211, 243, 244
  optical properties, 197
  uralitized, 183
aureole, 34
Australia, 56
Australian tektites, 175
Austrian Tyrol, 111
autometamorphism, 243
auto-metasomatism, 207
AW, P. C., 194, 256, 259, 261, 262, 318, 392
axe heads, 175
axial Malayan Province, Mesozoic, **106-114**
axial plane cleavage, 308

axinite, 291
  tin-bearing, 353
Ayer Chulit, 354
Ayer, Hangat, 312
Ayer Hitam, 99F, 236T
Ayer Kuning, 326
AYOB, see under MOHAMMED bin AYOB

back veining, 251
Badong Conglomerate, **129-130**, 130F, 136, 137, 139
BADGLEY, P. C., 279, 392
Bagan Datok, 172, 173
Bahau, 326
BAILEY, E. H., 220, 392
Bajocian, 110
BAKER, G., 174, 392
Bakri, 337F, 344, 365, 366, 367, 387, 388
Bakri district, Johore, 343
Baling, Kedah, 31F, 41, 47, 104, 230, 289, 314F
Baling area, 314, 392
Baling Formation, 10T, 46
Baling Group, 29T, 31F, 40, 46, **47-49**, 48F, 54, 58, 314F
Baling Limestone, 34T
Baling region, 287
BALL, H. W., 112
Baltic States, 59
Bangka, 1, 215, 241F, 242, 303, 305
Bangkok, 46, 50, 93F, 94
Bang Saen, 303
Ban Hin Lee mine, Selangor, 369
Ban Hock Hin lode, **374-378**
Ban Hock Hin mine, Setapak, Selangor, 368F, 375F
Ban Na, 45
Bantam Tuffs, 214
bar, offshore, 22
BAREN, F. A. van, 13
Barisan Highlands, 214
  Range, 214
barite, 336
  deposits, **385**
Barito Basin, 2F
barium, 336
BARNES, V. E., 174, 392
Barrovian facies series, 260
  terrane, 253
basalt, 131, 133, 205F, 210F, 213, 214, 238T
  alkali, 203, 206
  chemistry, **207**, 208T
  feldspathoid-bearing, 204
  flat lying, 153
  flows, 143, 153, 177, 203, 209, 333F, 334
  mineralogy, 204, 206
  nepheline, 208T, 209
  olivine, 204, 214
    vesicular, 208T, 209
  petrogenesis, **206-207**
  petrology, **204**, 206
  potassic, 383
  postorogenic, 178F, 213

at Segamat, **206-207**
  thickness, 204
  vesicular, 153
basalt flows, thickness, 153
  flows, submarine, 54
  highest point, 204
  Late Cenozoic, 145F
basaltic lava, 12, 145F, 153
basaltic volcanicity, 139, 214
basanite, 204
  leucite, 206
  nepheline, 204
basement, anatexis, 334
  oceanic, 334
  Precambrian, 30F, 56, 57F, 334
  rocks of Malaya, 391, 395
base metal sulfide deposits, **382-385**, 388
basic bodies, 215, 396
  bodies, satellitic, 243
  magma, 396
  rocks, **242-249**
    chemistry, **248**
    map, 244F
basin, 27
  deposits, 27, **40-46**, 144
  deposits, miogeosynclinal, 33
  down-faulted, 133
  facies, 37
  intrageosynclinal, 27
  of sedimentation, 152
  Tertiary, 152
Batang Geragi, 243
  Padang, Perak, 330, 353
  Padang district, 353
  Padang hydroelectric scheme, 330, 331F, 399
Batang Malaka granite, Malacca, 217, 217F, 220T
batholith, 215, 219, 240
  discordant, 250
  granite, 253
  subcordant, 219
  tonalite, 303
bats, 171
BATTEN, R. L., 76, 392
Batu, 293
Batu Anam, 290
Batu Arang, 24, 144, 145F, 147F, **146-150**, 158F, 397
  Arang, fossils, 149
    stratigraphic column, 148F
    thickness, 146
  Arang beds, 10T, 147F, 148F
  Arang coal field, 4, 397
  Arang Tertiary Basin, 147F
Batu Bersawah, 247
Batu Besi, 85F
Batu Caves, Selangor, 34T, 37, 311F
  Gajah, 4, 7, 72, 76, 74F, 78, 164, 164F, 165F, 167, 168, 175, 291, 358F
  Hitam, 204
  Kurau, 49
  Pahat, 397
  Pahat quarry, Johore, 220T, 237T
Batu Puyoh, 289

Tiga, Trengganu, 338F, 347, 348, 349F, 370, 397
bauxite, 143, **156-157**, 199, 291, 401
  alluvial, 157
  deposits, 13
  old, 293
  metamorphism, 292
  pisolitic, 157
  pisolitic, metamorphism, 293
  residual, 157
  and sea levels, 401
Bawdwin, 209, 210
  Volcanics, 58
Bawdwin Volcanic Series, 209
Bay of Bengal, 55
beach deposits, 111
  deposits, marine, 19
beaches, Johore, 394
  raised, 20
beach ridges, 9, 19, 21F, 22, 23T, 143, **173-174**, 398
  ridge deposits, 398
  ridges, older, 173
    younger, 174
  ridge zones, 21F, 22
beach sand, 173
  sand, white, 173
BEAN, J. H., 201, 318, 322, 324F, 347, 348, 349, 350, 353, 366, 367, 381, 392, 399
Beatrice mine, Selibin, 290, 356F, 402
Beatrice Pipe, Perak, 335, 351, 352, 355
  Pipe, Mineralogy, 355
  size, 355
Bebar River, 22
BECHER, H. M., 3, 4, 392
bedding, cross, 89, 136
  current, 33
  flaser, 89
  graded, 33, 89, 118, 167
  horizontal, 134
  lenticular, 109, 163, 167, 170
  massive, 166
  prolapsed, 117
  structures, Old Alluvium, 164F
Bedok, Singapore, 166F
Bedong, 41, 282F, 286F, 343
bedrock, 9
  exposure, 3
beds, detrital, 33
Belakang Mati, 109
BELLAMY, H. F., 4
bellerophon, 68
Beluh, 214
BEMMELEN, R. W. van, 2F, 94, 105, 111, 138, 213, 214, 251, 303, 305, 306F, 332, 392
Bendang Riang formation, 29T, **48**
Benom area, Pahang, 5, 231T, 234T
  granite, 111, 112, 275, 277, 331, 333F
Benom-Karak area, 396
Benta, 230, 254, 266, 267, 270F
Benta-Gunong Benom, 207
Benta Migmatite Complex, 255F, **267-277**, 271F, 273F, 296, 331, 333F, 327, 395

Migmatite Complex, chemistry, **269**, 273
  distribution, **269**
  map, 270F, 276F
  petrogenesis, **273-275**
  relation among the rocks, **269**
  southward extension, **275-277**
Benta quarry, 269, 270F, 272T, 273, 273F, 275F
Bentong, 7, 15, 39, 51, 52, 53, 99F, 154, 183, 255F, 295, 391
  area, 295, 329
Bentong-Ginting Sempah road, 186T, 187
Bentong Group, 10T, 16, 29T, 39, 51, **51-53**, 52F, 53, 54, 56, 58, 185, 296, 310
Bentong road, Pahang, 220T
benthonic fossils, 26
Benut Sandstone, 67F
Beralt Tin, 366
Berantai, 127
Berantai-Chedong fault, 320
Berantai syncline, 318
Bernam River, 172, 173
Bersiah, 242, 245
Bertam, 191
beryl, 336, 346
beryllium, 336
bibos cf. *bubalus* sp., 171
Bidor, Perak, 159F, 340F, 381
BIGNELL, J. D., 91, 153, 204, 238T, 242, 266, 310, 312, 313, 328, 387, 392, 400
Bikita, Rhodesia, 365
BILIBIN, Yu. A., 388, 392
Billiton, 1, 215, 219, 241F, 242, 303, 305, 391
  granite age, 242
  Granite, Indonesia, 395
  Island, 242
Bilut Valley road, 53
Bintan, 396
  formation, 138
  island, 140F, 141
Bintang-Gunong Bubu granite, 49
  -Gunong Bubu Range, 49
Bintang Range, 47, 121
biotite, 47, 117, 154, 180, 182, 185, 183, 189, 192, 198, 220, 220T, 247, 256, 259, 261F, 266F, 289F, 295
  bent crystals, 183
  chloritised, 222
  granite, 200
  green, 269, 352F
  radiometric determination, **236-238T**
  tin in structure of, 343
bioturbation, 90
bismuth, native, 352, 355F
bismuthinite, 373
bipyramidal habit, 365
bivalves, 11
bivalves, early Triassic, 395
bivalve molluscs, 4
Black River-Trenton Group, 59
  River-Trenton Limestone, 36
BLAKE, D. H., 251, 392
Blue Stone Quarry, 244F
Bok Bak Fault, Kedah, 314F, 318, 320F, 330, 332F

Bak fault extension, 318
bombs, 181, 192, 243
bone fragments, 167
BORAX, E., 138, 139, 392
bored surfaces, 173
boreholes, 32, 37, 150, 172
  depths, 172
Borneo, 7, 241F, 242
Borneo (Kalimantan), 1
Borneo, west, 303
  granite age, 242
bornite, 383, 385F
boron metasomatism, 293
BOT, A. C. W. C., 400
BOUČEK, B., 43, 45, 49
BOUCOT, A. J., 8, 37, 392
boudinage, 94
boudinaged inclusions, 269
boudinage, psammite, 274
boulangerite, 369
Boulder Beds, 9, 24, 129, 145F, 146, 147F, 148F, 152, 153, 158F, **158-159**, 160
Boulder Beds, age, 159
Boulder Clays, 5, 6, 158, 400
boulders, core, 155, 156F
boulder gravels, 144, 158
boulders, granite, 60
Boundary Range, 185
  Range Granite, 97, 104, 128, 133, 191, 192, 195, 218, 222, 224, 226F, 257F
BOWEN, N. L., 221, 232, 302, 401
brachiopods, 34, 35, 42, 63, 68
  amboceliid, 75
  chonetid, 68
  Devonian, 394
  dielasmid, 76
  late Cambrian, 32
  late Paleozoic, 394
  marginiferid, 83
  orthid, 30
  Silurian, 392
brachy-synclinal areas, 16
BRADFORD, E. F., 8, 33, 37, 41, 49, 61, 72, 73, 74F, 122, 159, 160, 172, 173, 180, 182, 222, 223, 225, 227, 228, 229, 230, 235, 282F, 283, 284, 285, 285F, 286F, 287, 290, 291, 292F, 315, 335, 341, 344, 345, 346, 365, 367, 369, 372, 373, 381, 386, 392, 395
Bradwell Estate, 385
Brazos River bar, 393
breccia, 263
  andesitic, 109
  intraformational, 52
  volcanic, 86
BRELICH, H., 4, 392
British Museum, 150, 174; *see also* Museum
bronze age, 175
brookite, 161
BROUWER, H. A., 141, 392
BROWN, G. F., 33, 39, 46, 50, 144, 212F, 213, 217, 240, 242, 302, 392
BROWN, J. C., 92, 392
brucite, 289, 295

Brunei Shell Petroleum Ltd., 136
bryantodus, 35
bryozoans, 35
Buchan, 302
  type of metamorphism, 260
BUCHANAN, 156
buchania sp., 77F
BUDDINGTON, A. F., 215, 218, 219, 230, 253, 392
Buffalo Reef, 316
building stone, 156
Bujang Malaka, Perak, 156F, 343, 346
  Malaka granite, 362
Bukit Arang, 144, 145F, 391
  Arang Beds, 10T, 144
  Arang-Betong, **144-146**
  Arang Coal Beds, 144
  Baling, 47
  Baloh, Sungei Linchong, Pahang, 220T
Bukit Bangkong, 347
Bukit Bangsal Kunyit, 200, 202T, 203
  Bari, 247
  Batu Bertelor, 41
  Batu quarry, Ayer Hitam, Johore, 237T
  Berangan, 289
  Berentin, 230, 254
Bukit Berentin Complex, **279-283**, 316F, 327
  Berentin Complex, petrogenesis, **281**
    petrology, **279-281**
  Besi, Trengganu, 339F, 335, 347, 349, 350F, 381
    petrology, 347
Bukit Betong, 190
  Betong railway bridge, 188
  Bintang, 297
  Biwah, 84, 86, 85F
  Bopeng, 199
  Buloh, J. K. R., quarry, Kelantan, 220T
  Chabang, 18F
Bukit Charas, Kuantan, Pahang, 71T, 84
  Cherating, 195
  Chermingat, 123, 126
  China, 222
  China granite, 289
  Chini, 324F
Bukit Chuping, Perlis, 18F, 71T, 173, 174
  Dada Ayam, 118
  Dara, 243
  Dinding, 182, 295
  Gasing, 88F, 90
Bukit Gerai, 195
Bukit Goh forest reserve, 205F
  Gombak, 242, 243
  Ibam ore body, Rompin, Pahang, 401
  Ibam iron oxide deposit, 386
  Jelis shaft, 317F
  Jelutong granite, 118
  Jerneh, 18F
Bukit Kachi mine, 373
  Kajang granite, 317
    porphyry, **198**, 381
  Kajang range, 198
  Kalong, Kedah, 120, 315F
  Kasai, 276F

Kechil, 120
  Kodiang, Kedah, 102T
  Kedah, 256
Bukit Kemahang, 230
  Kenuak, 243
  Kepala, Kedah, 19F
  Kepayang, Pahang, 197, 402
  Keriang, 19
  Ketri, Perlis, 18F, 72F
  Kluang, 276F
  Kodiang, Kedah, 120F
Bukit Koman mine, 382T
  Kuantan, 298
  Kusial, 230
  Kutu, 329
  Labohan, J. K. R. quarry, Trengganu, 220T
Bukit Lagi quarry, Perlis, 34T, 36
Bukit Lanchu, Johore, 237T
Bukit Lanjan, 183
  Larang, Malacca, 250
Bukit Larang amphibolite, 299
Bukit Lebak, 276F, 296
  Lentor, 338F, 363, 369, 372, 390
    mineral paragenesis, 373T
  Malacca mine, 382T
  Mandai, 244F
  Merah, 115, 117, 117F, 120
  Merjar, 247
  Minyak, 195, 298
  Mor complex, 344
  Mor quarry, Johore, 237T
  Muchong, 294
  Muchong Estate, 247
Bukit Pakir Terbang range, 118
  Paloh, Pahang, 237T
  Paloh adamellite, 235, 395
  Panjang, 242, 243
  Panjang village, 244F
Bukit Pelali, 140F, 141
  Penchuri, Kelantan, 385
  Peradong, 135
  Raja, 199
  Raja Muda, 82
  Rambutan, 243
  Rambutan mass, 243
Bukit Rangin, 195
  Ranjut, 230, 277
  Ranjut Complex, **277-279**, 297, 316F
    mineralogy, 278
    petrogenesis, **279**
    petrology, **278-279**
    structure, 279
Bukit Redan, 276F
  Resam, 109
  Resam clastic member, 104
  Resam member, 103, 108, **109**, 107T
  Saiong, 126
  Sagu, 84, 195
  Senggeh granite, Malacca, 217, 217F
  Serdam, 198, 269, 276F, 296
  Setongkol, 195
Bukit Taku, 293
  Takun, Selangor, 18F, 39, 40F
  Tampong, 226F

Tanah Merah, 199
Tangga, 226F
Tebak, Kemaman area, Trengganu, 297, 351, 352F
Temiang, Perlis, 68, 64F, 70F, 128
Tengku Lembu, 64F, 68
Timah, 243, 244F
Timah quarry, Singapore, 237T
Bukit Tinggi, 153, 204, 205F
Bukit Tinggi Fault, 320F, *326*
  Tinggi Fault, radiometric age, 328
  Tinggi Fault Zone, 325F, 329
Bukit Titir, 243
Bukit Tok Hat, 298
  Tunggal quarry, Malacca, 237T
  Ubi, Kuantan, Pahang, 87, 195, 205F, 237T, 245
  Ubi quarry, 204
  Ulu Chegar, 247
  Wang Mu quarry, Perlis, 34T
Bukit Wang Pisang, 64F, 68
  Yong, Kelantan, 236T, 342
Bundi mine, Trengganu, 335
BUNNAG, D., 138, 139, 397, 401
BURAVAS, S., 392
burial metamorphism terrane, 253
Burma, 1, 25, 30, *39*, 46, **45-46**, 50, 56, 91, 97, 121, **209-210**, 211, 212F, 302
  geology, 393
burrows, 89, 89F, 90
BURTON, C. K., 8, 15, 22, 23T, 24, 37, 40, 41, 42, 46, 55, 58, 59, 68, **97-141**, 101, 104, 110, 112, 115, 116F, 117F, 120, 126, 129, 134, 139, 150, 151, 152, 159, 166, 168, 170, 172, 195, 200, 222, 230, 242, 287, 289, 303, 305, 314, 314F, 318, 350, 392, 397
BUSH, BILL, 167
Butterworth, 20, 172, 332F
bytownite, 244

CADY, W. M., 134, 392
calcareous coatings, 171
  facies, **80-82**, 84
Calcareous Formation, 7, 61, 102
  Series, 7, 61, 72, 81T, 84, 102, 305
  Series (*Daonella*), 106
calcilutite, 102
calcite, 47, 193, 354F, 355F
  gangue, 378F
  glide planes, 352
  recrystallized, 299
  secondary, 192
calc-silicate, 47, 290, 346; *see also* Hornfels
Calc-silicate, analysis, 294
Calc-silicate fels, 256, 258T, 281, 294
calc-silicate, mineralogy, 294
calc-silicate hornfels, 293, 297, 348, 349; *see also* Hornfels
  mineralogy, 297
calc-silicate rock, 41, 284, 289
  mineralogy, 350
Caledonian magmatism, 399
caledonite, 206
Cambodia, 211, 212F, 213, 242

granite, 397
Cambrian, 9, 10T, 25, 26F, 30, 55-58, 57F
  Late, *9*, 25
Cambrian, Upper, 8, 29T, 181, 283
CAMERON, W. E., 5, 6, 392
Cameron Highlands, 15, 215, 332F
  Highlands road, 238T, 258T, 310
camptonite, 247
CANAVAN, F., 254
cancrinella, 76
  cf. *cancrini* (VERN), 68
capellinella, 37
carbon, 40, 41
carbona, 359
carbonaceous matter, 148
carbonas, Cornish, 343
carbonate deposition, 11
Carboniferous, 6, 10T, 41, 49, 51, 61, 65F, 72, 75, 76, 80F, 91, 92T, 187, 239F, 387
  fossils, 399
  glacial deposits, 158
Carey's Island, 172
carlsbad twinning, 269
Carnian, 11, 80, 110, 112, 114, 119, 122
carpolithes sp., 132, 136
CASEY, J. N., 56, 392
cassianella sp., 110
  cf. *tenuistria*, 110
cassiterite, 144, 153, 160, 165, 166, 170, 171, 172, 221, 225, 226, 227, 229, 287, 335, 336, 342, 343, 346, 348, 349F, 350F, 351F, 352F, 353F, 354F, 355F, 356F, 357F, 358, 359, 360, 364F, **365-367**, 369F, 370F, 371F, 374, 375F, 377F
  bipyramidal, 345
  chemical analyses, **366T**
  color zoned, 365
  corroded, 368
  crystallography, 365
  detrital, 5
  disseminations, 341
  deposition in pegmatite, 345
  epigenetic, 342
  epigenetic in pegmatite, 345
  ferromagnetic, 366
  floatation, 399
  fluorescent, 366
  in granite, 341
  hydrothermal, 345
  magmatic segregation, 341
  magnetic, 351, 366, 394
    susceptibility, 367T
  paramagnetic, 367
  in pegmatite, 345
  pleochroic, 365
  pneumatolysis, 228
  replacement, 374
    by galena, 376
  several generations, 379
  tantaliferous, 369
  tantaliferous-niobiferous, 344, 366
  twinned, 365
  zoned, 366

cataclasis, 222
cataclastic zones, 259
catazone, 141, 215, 253, 254, 255F, 281, 302, 331, 333F
  definition, 253
catazonal metamorphism, 11, 301, 312
  plutonism, 11
catazone rocks, 241F, 254-287
catfish, 167
Cathaysia, 27, 55, 56F, 57F
caves, 16, 401
  alluvium, tin-bearing, *171*
cave deposits, **171**
  limestone, 19F
cement, calcitic, 131
  ferruginous, 129
  siliceous, 135
central belt, 338F, 341, 381, 388
central Malaya, Geological map, 79F
  Malaya, Upper Paleozoic, **78-84**
Cenozoic, **143-176**
  formations, 254
  sedimentary rocks, distribution, 145F
cephalopods, 34, 116
ceratiocaris sp., 53
cervus (rusa) sp., 171
cerussite, 161
chalcedony, 354F
  cement, 154
  spherulitic, 115
chalcopyrite, 204, 335, 345, 348, 349F, 356F, 357, 367, 368F, 369F, 370F, 374F, 375F, 377F, 378F, 383, 384F, 385F
chalcopyrite-chalcopyrrhotite, 376
chamosite, 157
CHAN, S. C., 356, 373, 392
CHAN, S. H., 9, 393
Changi, 249
Changkat Rembian, 224, 363
changkul, 390
Changwats Chon Buri, 303
channeling, 167
channels, 170
channel deposits, 170
channels, erosional, 149
  feeder, 343
Chao Phraya structural depression, 322
CHARALJAVANAPHET, J., 392, 397
Chaung Magyi beds, 209
Chebok Mas, Kinta, 352
Chedong fault, 318, 322
Chegar Perah, 7, 101, 106, 185, 152, 315
  Perah, Pahang, fossils, 100F
  Perah area, structural map, 316F
  Perah-Benta Fault, 327, 320F
  Perah and Merapoh, Pahang, 399
Chemendong volcanic rocks, 200, **200-201**
chemical analyses, 34, 37, 186T, 201, 202T, **208T**, 231T, 232, 234T, 248T, 249, 258T, 269, 272T, 287, 366T
  analyses, granodiorite, 103T,
    limestones, 34T, 71T
    rhyodacite, 103T
    Triassic limestones, 102T

of Malayan rocks, 391
chemical decomposition, 155
Chemor, Perak, 37, 49, 71T, 72, 73, 74F, 182, 236T, 290, 291, 371
  Limestone, 29T, 37, 49, 59
Chendai area near Menglembu, 346
Chenderiang, Perak, 222, 237T, 290, 346, 353F, 354F, 360, 369
Chendrong, Trengganu, 372
cheralite, 344
Cherating, 195
Cheroh, 198, 227, 242, 245, 317
Cheroh area, Pahang, 250, 247
Chersonesia, 1, 2F
chert, 11, 33, 43, 51, 58, 68, 115, 120, 198
  bedded, 82, 115
    Semanggol formation, 116F
  breccia, 53
    carbonaceous, 52
  member, 52, **115-116**
  origin of, 119
  pebbles, 109
  radiolarian, 4
Chert Member, Semanggol formation, 115
chert nodules, 68, 70
  nodules, black, 68
Chert Series, 4
cherty beds, 33
  rock, carbonaceous, 51
  rocks, 26, 41, 42, 43, 50
chevron folding, 312
Chia Oh Kang quarry, 244F
Chiang Mai, 302
chiastolite, 288, 289, 291, 295, 298
CHHIBBER, H. L., 92, 211, 393
chief minister, 1
CHIENGMAI, P., 397
Chiengrai, 138
chilled margins, 204, 206, 218, 246
chilling, 206
Chin Chin mine, Malacca, 372
CHIN FATT, 108, **109**, 111, 393
CHIN, L. S., 395
China, 35, 55, 56, 59, 104, 174, 394, 397
Chinchong River, 294
Chinese continents, 59
Chinese troughs, 56
Chindrass gold mines, 4
Chintamani, 39
chlamys valoniensis, 110
  valoniensis DEFRANCE, 106
chlorite, 32, 51, 54, 89, 165, 180, 182, 183, 189, 280, 281, 286F, 295, 348, 364F, 374, 383, 384F
  gangue, 371F
chloritization, 360
chloritoid, 281, 291, 295
Choa, Chu Kang road, 243
chondrodite, 289, 290
CHONG, F. S., 195, 196F, 201, 326, 393
CHONG, N. H., 349F
CHOY, K. W., 88F, 89, 90, 393
chromite, 54, 245, 259
  disseminations, 341
chrysotile, 259

asbestiform, 245
Chuchoh siltstone, 67F
Chukai, Trengganu, 23T, 298
Chumphon, 138
Chumphon River, 138
CHUNG, S. K., 180, 392, 393
Chuping Formation, 10T, 18F, 63, 64F, 65F, 65, 67, **68-70**, 70F, 71T, 71, 73F, 92T, 299, 313F
Chuping limestone, 66F, 67F
cigar-shaped bodies, 117
cinnabar, 358, 378, 381
circulina sp., 127
circum-Pacific, 112
claraia, 11, 101, 102, 401
  griesbachi concentrica (YABE), 100F
classification, environmental, 27
classopolis classoides PFLUG, 127
  sp. cf. *classopollis classoides*, 136
clasts, chert, 118
  dyke rocks, 166
  granite, 9, 20, 103, 133, 158, 166
clast imbrication, 170
clasts, limestone, 133
  lutite, 136
  metamorphic rock, 109
  mud, 163
  quartzite, 158
  rhyolite, 131
  unweathered, 168
  weathered, 158
clay, 146, 148, 148F, 159, 165, 172
  blue, 173
clays and boulder clays, 5
clay, carbonaceous, 148
  economic, 396
  Gondwana, 5
  grey, 160
clay minerals, 160
clay pebbles, 398
  peaty, 163
  plastic, 144
clay slate, Paleozoic, 3
clay, tight, 165F
cleavage, 314
  crenulation, 308
cleiothyridina, 76
cliffs, 16, 35
  steep, 19
climacograptus, 45
  rectangularis (McCOY), 42
  scalaris (HISINGER), 38F, 39, 42, 43, 49
  scalaris normalis ELLES and WOOD, 43
  yangtzeensis HSÜ, 43
clinopyroxene, 275, 287
clinozoisite, 34, 47, 49, 180, 230, 256, 259, 260, 283, 284, 288, 289, 294
  opitcal properties, 259
coal, 109, 111, 136, 137, 146
  analysis, 149
  fragments, 150
coal laminae, 149
  lenticular, 146, 150
  lenses, 131
  lignite and petroleum, 399

lignitic, 143, 144, 146, 150, 151
measures, 11, 24, 146
noncoking, 149
seams, 136, 144, 147F
streaks, 148F
sub-bituminous, 149
thickness, 149
veins, 160, 393, 399
coast, drowned, 19
ria, 19
coastal accretion, 20, 22
coastal development, 398
coastal features, **19-23**
coastal flats, narrow, 22
plain, 16, 19, 22, **171-174**
south Perlis, 20F
plain deposits, thickness, 172
plains, thickness, **172**
wide, 22
sedimentation, 398
cobaltite, 161
coelenterates, 34
COHEN, L. H., 119, 393
COLLET, O. J. A., 3, 393
COLLINGS, H. D., 154, 393
columbite, 161, 344
columbite-tantalite, 227, 343, 366
colluvium, 15F, 156F, 173
colluvial deposits, 143, 165
material, 155
Colombo Plan, 8, 9
color banding, 190, 245
columnar jointing, 204
composita sp, 65, 68
compression, major, 330
comptonite dykes, *247*
conchidium, 37
cf. *triangulum* KHODALEVICH, 38F, 39
conchostracan remains, 106, 112
concretions, 13, 109
iron oxide, 13
CONDIE, K. C., 393
conduction band, 299
conglomerate, 4, 11, 16, 30, 31, 32, 50, 51, 84, 86, 108, 109, 113, 115, 124F, 129, 148, 148F, 324F
basal, 123, 136
continental, 118
foliated, 295
intraformational, 52, 52F, 59, 133
member, 52, *118*
phyllitic, 297, 298
polymictic, 123, 126, 129, 137
quartz, 196F
sandstone, 135
schistose, 54
sheared, 52
Conglomerate Member, Semanggol formation, **115**
coniferales, 132, 136
conites spinulosus nov., 132
conodonts, 34, 75, 91, 120
Carboniferous, 395
Devonian, 37, 75, 391
Lower Namurian, 84

Ordovician, 395
and Silurian, 35
Triassic, 395
Upper Triassic, 70
zones, 35
conostichus, 65
contact, aureoles, 218, 301, 348
granodiorite/gabbro, 225F
metamorphism, 217, 288, 301
monzonite-gneiss, 273F
continental, 144
basement, 332
conditions, 133
deposits, 152
drift, 123, 393
formations, Mesozoic, 79F, 85F
convolute laminations, 117F
COOK, K. H., 393
copper, 336
native, 161
sulfides, 289
corals, amplexiform, 75
lophophyllidiid, 82
Paleozoic, 401
rugose, 65
Silurian, 401
tabulate, 75
waagenophyllid, 86
cordierite, 183, 256, 260, 261F, 263, 265, 266F, 284, 289, 297, 298, 301, 302, 303
amphibolite facies, 331
rock, 350
cordyolus, 35
core boulders, granite, 15F
coreanocephalus, 32, 58
coreanocephalus planulatus KOBAYASHI, 32, 38F
Cornish-type lodes, 388
Cornwall, 344, 368, 381
correlation chart, major rock formations, 10T
corundum, 5, 6, 161, 172, 289, 291
-bearing rocks, 402
metamorphic conditions, 291
normative, 201, 232, 232T, 293
pebbles, 5
rocks, *291-292*
costatoria, 118, 120
chegarperahensis KOBAYASHI and TAMURA, 100F
cf. *Myophoria* (BOETTIGER), 100F
pahangensis KOBAYASHI and TAMURA, 100F
quinquicostata KOBAYASHI and TAMURA, 100F
costatoria singapurensis KOBAYASHI and TAMURA, 100F
coulisses, 305
COURTIER, D. B., 20, 23T, 41, 42, 112, 114, 157, 173, 174, 393
covellite, 373
COX, L. R., 7, 101, 106, 108, 110, 112, 150, 393
craton, 57F

cratonization, 122
CRAWFORD, J., 3, 393
Cretaceous, 9, 10T, 13, 16, 107T, *122-141*, 239F, 300F
early, 11
Lower, 127, 132, 136
tectonism, 241F
-Tertiary radiometric dating, 392
Cretaceous, Upper, 216F
crinoids, 4, 42, 43
columnals, 76
cups, 76
ossicles, 82, 83
stems, 37, 65
CROIX, J. E. de La, 3, 393
cross bedding, 109, 116, 119, 125, 129, 131, 133, 135, 138, 149, 160, 163, 164F, 167, 170, 171
bedding, coarse, 133
bedded, fluvial sandstone, 125F, 126F
Semanggol formation, 117F
cross lamination, 65, 395
cross sections, 57F
crust, continental, 333F
oceanic 333F
crustal evolution, summary, **330-334**
plate, continental, 334
oceanic, 334
crystals, angular, 184
crystal ash, 192
ash, feldspathic, 192
crystals, bent, 222
crystal fragments, 189
imperfections, 299
crystals, intratelluric, 193
ornamented, 220
crystal lapilli, 192
crystal tuff, 180
cubanite, 367
cucullaea sp., 106
scrivenori sp. nov., 110
cuestas, 134
cummingtonite, 199
cupressinocladus acuminifolia nov., 132
cupro-stannite, 365, 367, 368, 370F
current-bedding, 30, 48, 52, 55, 284
current, longshore, 20, 22
monsoon-controlled, 23
cuspidoria sp., 110
cycadales, 132
cyclopigidae, 41, 43, 49
cyclopygids, Ordovician, 43
cymbidium, 37
cyrena, 173
cyrtograptus sp., 46
lapworthi TULLBERG, 42
lundgreni TULLBERG, 42
cyrtonotella thailandica HAMADA, 38F
cyrtosymbole, 68
perlisensis, 91
KOBAYASHI and HAMADA, 69F
cystids, 34, 35

Dabong, Kelantan, 101, 191, 256, 257F, 262, 264F

410 INDEX

dacite, 103, **192**, 197, 200, 203, 211, 214
    mineralogy, 192, 197, 211
dacryonconarid tentaculites, 392
dalmanella, 37
dalmanitina, 43
    malayensis KOBAYASHI and HAMADA, 43
DALY, D. D., 4, 24T, 393
Daly levels, 24, 401
Damansara, 90
damar, 167
Damar granite, 287, 314
daonella, 11, 105, 112, 114, 118, 121
    biofacies, 80
    facies, 121
    and *halobia* facies, 397
    indica, 119
    pahangensis KOBAYASHI, 100F
    cf. *pichleri*, 119
dasycladaceae (calcareous algae), 393
DATTA, P. N., 393
dating, radioisotope, 8
DAWSON, J., 135, 190, 243, 257, 261, 267, 268F, 393
Dayang Bunting, 289; *see also* Pulau
décollements, 332
deep water conditions, 71, 121
deer, 167
deformation, Middle Devonian, 288
    postmineralization, 317F
    soft sediment, 119
    style, 253
deltaic aspect, 55
    beds, 31
    conditions, 127
    deposits, 26, 27, 91, 57F
deltas, cuspate, 19
    extuarine, 19
delthyris, 37
demirastrites convolutus, 45
    convolutus (HISINGER), 42, 45
    decipiens (TORNQUIST), 42
denudation rates, 13
    rates, Europe, 13
        Java, 13
        North America, 13
deoxygenated conditions, 121
deposition, deep water, 26
    marine, 20
    rapid, 68
    shallow water, 26
depositional basins, 139
    subcycles, 139
deposits, channel, 163, 168
    lenticular, 159
    overbank, 163
    unconsolidated, 166
depressions, solutional, 15
derbyia, 70, 76, 83
Derrick mine, 382T
detrital member, Setul Formation, **43-45**
detrital sedimenta, 57F
deuteric alteration, 247
Devon Estate, 182, 294
Devonian, 9, 10T, 45, 49, 61, 72, 75, 75-76, 63, 65F, 76, 92T
    early, 25, 57F, **59-60**
    Lower, 9, 29T, 33, 53, 61, 63
    Upper, 45
diagenetic changes, 34
diallage-enstatite rock, 54
diamond drilling, 383
diaspore, 291
    optical properties, 298
    porphyroblasts, 298
    -pyrophyllite rock, 298
diatoms, 154
dictyotidium, 43
dielasma, 65
DIETZ, R. S., 334, 393
differentiates, late-stage, 224
    marginal, 6
differentiation, 207, 277
    gravity, 206
dimorphograptus malayensis JONES, 38F, 39, 43, 45
Dinding Schist, 10T, 29T, 53, 55, 180, 182, 294, 311F
Dindings, Perak, 5, 20, 23T
diopside, 34, 47, 49, 256, 265, 266F, 284, 286F, 287, 288, 289, 293, 301, 354F
    columnar, 289
    -garent rock, 289
    normative, 273
    -tremolite-wollastonite rock, 289
diopsidic augite, 265
diorite, 213, 243, 250F, **278**
    amygdales, 243
    high-level, 243
    porphyries, 243
    porphyritic, 243
dip, gentle, 129
dip slopes, gentle, 131
diplograptus, 45
    cf. *magnus* H. LAPWORTH, 50
    modestus, 43
        LAPWORTH, 38F, 39
    thuringiacus EISEL, 42
discoceras (*hardmanoceras*) *chrysanthimum* KOBAYASHI, 35
    (*hardmanoceras*) *laeviventrum* KOBAYASHI, 35
dislocation metamorphism, **301**
displacement, sinistral, 318
disseminations, 341
distamodus, 35
distribution diagram, 221F
district offices, 7
Djambi, 93F
    Nappe, 332
    region, 212
dolerite, 50, 51, 55, 208T, 209, 245-247
    dykes, 185, 238T
        metamorphic, 294
    hornblende, 247
    mineralogy, 204, 245, 246
    petrology, **204-206**
    titanaugite, 245
    two ages, 247
dolines, 15, 16, 68

dolomite, 34T, 71T, 75, 76
    saccharoidal, 37, 49
dolomitic, 34
dolomitization, 82, *300*
dome, 279, 283
    asymmetric, 284
    geometry, 284
    granite core, 284
    thermal, 283
downwarping, 152
    Quaternary, 24
DOYLE, P., 3, 393
drag folds, 24
drainage divides, 14F
    patterns, rectangular, 15
    radial, 15
    system, 13
    trellis, 16
drepanodus, 35
drill hole location, 147F
    hole log, 148F
drilling, 144
dripstone, 73
    coating, 68
drowned coastlines, 22
    topography, 20
DRUMMAND, P. V. O., 131, 134, 135, 136, 137, 166, 167, 168
duboisia santeng (DUBOIS), 171
dulang, 390
Dunearn road, 244F
Durian Chondong, 150
    Tipis, Negri Sembilan, 185, 242
Dusun Tua, Ulu Langat, Selangor, 183, 231T, 237T
dwarfs, 118
dyke rocks, 245-249
    rocks of central Trengganu, 398
dykes, 51, 243, 276F
    age of, 246
    aplite, 49
    dolerite, 55, 153, 204, 245, 246F, 323F
    granite, 90, 201
    granodiorite, 224F
    ignimbrite, 194
    in granite, 245
    lamprophyre, 246
    microadamellite, 353
    quartz, 16, 73, 74F, 90
        porphyry, 195, 214
    two ages of, 245
dynamothermal metamorphism, 254; *see also* Metamorphism
DZULYNSKI, S., 121, 393

earth movements, 9
    pillars, 15F
East Africa, 369
east coast railway, 189
East Indian Archipelago, 398
    Indies, 399
        geology, 392
East Malaya, geological map, **85F**
    Malaya, Upper Paleozoic, **84-87**
eastern belt, 338F

belt granite, 343T
Eastern Clays, 5
eastern mineralized belt, 347
Eastern Mining and Metals Co., 383
EATON, J. P., 251, 393
EBERHARDT, P., 400
echinocoelia, 49
EDWARDS, A. B., 367, 393
EDWARDS, G., 242, 393
EDWARDS, W. N., 7, 83, 106, 393
Eifelian, 43
electrical resistivity methods, 9
element migration into wallrock, 377
    mobilization, 377
elephant teeth, 167, 170
ELLIOT, G. F., 78, 393
elluvial, 5
Elphinstone, 211
elvan, Cornish, 359
ELWELL, R. W. D., 392
embayments, 19
EMERY, K. O., 322, 393
Endau, 145F, 178F, 185, 201
endoceras sp., 35
Enggor, 144, 145F, **146**
    beds, 10T
    coalfield, 400
    district, Perak, 207
enstatite, 294
entolium quotidianum, 106
Eocene, 204, 300F
eomidion chumphonensis, 138
eoorthis sp., 32
eophyton, 65
eosaukia buravasi KOBAYASHI, 32
eotomaridae gen. and sp. nov., 77F
epeirogenic movements, 78F, 123
    uplift, 242, 334
epidiorite, 245, 247, 297
epidote, 47, 51, 53, 103, 180, 182, 183,
    188, 192, 193, 197, 199, 200, 220T,
    247, 256, 259, 274, 278, 280, 281,
    284, 286F, 293, 296, 298
    veins, 275
epidosite, 296
    veins, **280**
epizone, 11, 215, 253, **254**, 255F, 302, 331,
    334, 333F
    definition, 253
    metamorphism, **301**
equisetales, 132
equisetites sp., 127
    burchardti, 132
ericaceae, 149
erosion, 11, 71
    differential, 6
    intraformational, 26
    lateral, 19
    level, 80F
    mature stage, 16
    stage, late youthful, 15
    subaerial, 18
    surface, 29T, 134, 164
    vertical, 19
erosional history, 15

Erzgebirge, 5
escarpment, 129
estheria mangalensis, 110
estheriella, Triassic, 396
estuarine conditions, 111
    environment, 24, 168
eucrite, 242, **243-245**, 244, 248T, 250
    mode, 244
eudoceras sp., 40
eugeanticlinal ridge, 98F
eugenia, 149
eugeosynclinal affinities, 30, 37, 45
    axis, 331
    basin, 58, 121
eugeosynclinal furrow, 98F
    rocks, 28F, 29T, 30F, 39, **50-55**, 51
    sequence, 52, 53, 54
    trough, 9, 58, 59
    zone, 97
eugeosyncline, 27, 104, 137
euomphalus, 68
    (*philoxene*) sp., 37
euphobiaceae, 149
Eurasian landmass, 1
Europe, 13, 43, 45, 53, 59
euxinic affinities, 41
    aspect, 58
    deposition, 46, 57F
    facies, 40, 58
    sediments, 59
    sequence, 27
    type, 26
EVANS, G. M., 393
exploration in Malaya, 393
    for primary deposits, 360
explosive activity, 193
    submarine andesitic activity, 197
    volcanism, 200
exsolution bodies, 368
EYLES, R. J., 15, 156, 393

facies, 27
    almandine amphibolite, 56, 104, 180,
        230, 260, 266, 286F
    amphibolite, 255F, 276F
    arenaceous, 41, 47
    argillaceous, 41, 47
    basin, 29T
    calcareous, 41, 47
    changes, rapid, 51
    deltaic, **30-33**, 27
    diagrams, 260
    epidote amphibolite, 218, 276F
    greenschist, 183, 218, 255F, 276F, 286F
    low pressure intermediate, 260, 261F
    lower amphibolite, 281
    mixed, 27, 29T
    orthoquartzite-carbonate, 58
    quartz-andalusite-plagioclase-chlorite, 281
    siliceous, 41
    upper amphibolite, 266F
    variation, 50, 80
    volcanic, 29T
FAIRBAIRN, H. W., 395
fanglomerate, 137

fault, Bok Bak, 314F
    Lebir, 139, 399
    offset, 327
fault bounded, 133, 287
fault breccia zone, 318
    Chegar Perah-Benta, 281
    contact, 267
    control, 348
    displacement, 317, 329
fault(ing), 15, 24, 33, 44F, 64F, 73, 74F,
    83F, 88F, 90, 97, 141, 143, 153,
    276F, 309F, 311F, 314F, 315, 319F,
    323F, 324F, 327F, 333F, 334, 356F,
    361F, 400
    age, 334
    en echelon, 326
    extension, 141
    Kinta Valley, 394
    left-lateral, 324F, 328
    normal, 11, 326
    oblique, 316
    post-orogenic, 252, 301
    post-Triassic, **318-328**
    shear, 318
    sinistral, 11
    strike, 316
    strike-slip, 73, 207
    submarine, 83
    Sungei Lembing, Pahang, **327-328**
    tensional, 382
    transcurrent, 11
    thrust, 90
    wrench, 45, 90, 114, 118, 123, 133, 195,
        305, 307F, 308, 318, 323F, 360, 392
    zones, 15, 329
fault lineaments, 329
fault-mineralization relationship, 360
fault reopening, 379
    scarp, 133, 134, 137, 139
    scarp, contemporary, 139
faults, major, 320F
    mineralized, 379
fault system, 267
fault zone, Kuala Lumpur-Endau, **322-326**
fauna, Artinskian, 92
    benthonic, 11, 108
    Lower Devonian, 8
    Viséan, 61
    Lower Carboniferous, 7
    nektonic, 11
    Lower Silurian, 8
faunal breaks, 41
faunal gaps, 42
favosites sp., 37, 38F, 39, 49
FAWNS, S., 4, 393
Federation of Malaya, 1
feldspar, 5, 52, 154, 182
    microlites, 190, 194
    ornamented, 242
    staining, 392
    structural state, 218
felsite, 180, 279
ferberite, 365, 369
FERMOR, L. L., 160, 393
filicales, 132, 136

fissure, 194
FITCH, F. H., 7, 22, 23T, 84, 153, 160,
    172, 173, 177, 185, 195, 203, 204,
    205F, 208T, 209, 222, 235, 245, 298,
    305, 327F, 328F, 360, 374, 381, 393
flaser granite zone, 329
  mylonite, 328
FLATHE, H., 19, 393
flexure fold, 308
FLINT, R. F., 15, 23, 393
FLINTER, B. H., 103, 342, 353, 366, 391,
    393
flora, Cathaysian, 94
  late Permian, 397
  Upper Permian, 87
flow-banded rhyolite, 203F
flow-banding, 82, 188, 199, 200, 277
flow-banded rocks, 267
flow-folding, 265
flow texture, 192
fluoborite, 355, 356F
fluorescence, ultraviolet, 354F
fluorine, 5
  introduction, 5
fluorite, 5, 161, 220T, 221, 226, 290, 293,
    335, 364F, 374, 387
flute casts, 117
fluvial, 144
fluvial deposition, 167, 170
  deposits, 143
  environment, 136, 163
  origin, 152, 165, 166, 167
  terrace deposits, *157*
fluviatile, 11
  -deltaic-lacustrine, 123
  environment, 127
fluvio-glacial, 5
  -lacustrine deposition, 144
flysch, 67, 110, 112, 113, 122
  beds, 397
  conditions, 55
  deposition, 92T
  facies, 59, 119
flyschlike aspect, 121
flysch sedimentation, 401
  sedimentary features, 393
  type affinities, 51
  type deposits, 113
  type detrital sediments, 50, 71
  type rocks, 58
  type sequence, 51
fold, plunges, 16
fold(ing), 9, 10T, 11, 33, 59, 63, 73, 115
  chevron, 308
  cleavage, 315, 317
  Devonian, 70, 63, 305, 309
  disharmonic, 317F
  drag, 320
  flexure, 11, 63, 125
  flow, 71
  gentle, 90, 318
  isoclinal, 68, 80, 201, 302, 305, 310, 314,
      315, 316
  late Triassic, 306
  Mesozoic, 401

mid-Tertiary, 306F
post-Triassic, **317-318**
ptygmatic, 104, 310
Quaternary, 24
Triassic, 315
Variscan, 312, 313
folded tight, 52
fold(ing) recumbent, 65, 308, 310, 314,
    315F, 315
  second generation, 308
fold limbs, attenuated, 316
folds, open, 318
foliation, gneissic, 230
  strong, 182
FOLK, R. L., 163, 393
Foo Brothers Hydraulic Gold mine, 187
FOO, K. Y., 117, 118
FOO, Y. H., 392
Fook Wai, 290
Fook Wan Fah Kongsi, Chenderiang, Perak, 345
Foothills Formation, 29T, 51, 111, 180,
    182, 185, 396
foraminifera, 42, 115
foreland, 9, 26, 30F, 40, 332, 334
  Precambrian, 334
fossil deformation, 317
fossils, 4
  Carboniferous, 8, 86, 69F
    locality, 74F, 85F
  in cave deposits, **171**
  current-oriented, 121
  deformed, 114
  Devonian, 8, 37, 69F
  Devonian, locality, 74F
  diminutive forms, 118
  first record, 3
  Gagau Group, 132F
  Lower Paleozoic, 8, 38F
  Permian, 8, **77F**, 395
    locality, 74F, 79F, 85F
    and Carboniferous, 4
  Silurian, locality, 74F
  trace, 65, 90
  Triassic, 4, 6, 8, **100F**, 398
  Silurian, 8
  Upper Paleozoic, 6
  Viséan, 398
    locality, 79F, 85F
fossil impressions, 90
  locality, 124F
  plants, 124F
fossil record, summary, 396
fossiliferous sandstone, 401
fracture analysis, 15
  conchoidal, 198
  fillings, 16
  healing, 376
  knife-edged, 342
fracture patterns, 15
  system, 322
fractures, open, 322
  tension, 317F, 330F, 360
fracture system, element migration via, 378
franckeite, 365, 369
Frantz isodynamic separator, 367T

Fraser's Hill, 178F, 184, 293, 294, 295
  Hill area, 328, 329F, 399F
frecherites sp., 101
frenelopsis malaiana nov. KON'NO, 132,
    132F, 136
  malaiana parvifolia KON'NO nov., 136
frequency distribution diagram, 233F
FROMAGET, J., 211, 212F, 213, 312, 393
FUJI, N., 23, 393
FUJII, S., 23, 393
fusulina, 398
fusulinacean fossils of Thailand, 399
fusulinella konnoi, 68
fusulinids, 91
  Lower Permian, 87
  Permian, 395
FYFE, W. S., 260, 394

gabbro, 197, 223, 225F, 243, 244F, 243-
    245, 248T, 250, 250F
  age, 243
  cut by granite dykes, 224F
  -granite relationship, 251
  -granodiorite association, 395
  metamorphosed, 298
    modal analysis, 299
    mineralogy, 299
  mode, 243
  pregranite, 243
  quartz-biotite, 249
  stocklike, 243
gabbroic composition, 273
  rocks, 242
Gagau, 235
  Formation, 128, 235, 239
  Group, 10T, 16, 99F, 107T, 122, 126,
      127, **128-134**, 136, 137, 179T, 201,
      209, 238T, 254, 318, 319F, 399
    age, **131-133**
    flat lying sandstone, 128F
    fossils, **132F**
    lithology, 129
    palaeontology, **131-133**
Gagau Group, thickness, 131
  Group, summary, **133-134**
  summit profile, 140F
  plateau, 130F
Gagau-Panti formations, 334
Gagau upland, 129
gahnite, 344
Gakak fault, 328F
  -Gakak Creek Mines, Sungei Lembing,
      361F
Gakak No. 3 Lode, 360
galena, 161, 289, 345, 349, 350F, 357,
    368F, 369F, 374, 375F, 376, 377F,
    378F, 383, 384F, 385F, 387
  argentiferous, 336
  in limestone, 236T
Galena Lode, 374
gallein-methylene blue colorimetric method,
    342
Gambang, 204, 218, 245, 363, 389F
  Valley, 175
Gammon quarry, 244F

Gammon quarry, Singapore, 248T
GAN, A. S., 369, 394
GANESAN, K., 298, 350, 351T, 353, 394
gangue, 348, 349F
Gap, Fraser's Hill, Pahang, 221, 237T, 329
Gap road, 222
Garabal Hill-Glen Fyne, 398
  Hill igneous complex, 251
garnet, 34, 47, 49, 161, 172, 180, 201, 256,
    259, 285, 286F, 288, 290, 294, 295,
    296, 298, 344, 352, 354, 356, 357
  analysis, 290, 294
  -biotite-cassiterite rock, 351
  -biotite rock, 297
  in pegmatite, 225
  pink, 289
  porphyroblasts, 258
garnet-pyroxene rock, 54
  red, 352F
  rotated, 259
  tin-bearing, 349, 353
  -vesuvianite rock, 290
  zoned, 281
gastropods, 34
  bellophontid, 76
gastropod-cephalopod association, 35, 36
gastropod fossils, silicified, 109
  fresh-water, 150
gastropods, murchisoniid, 75
  Permian, 392
geanticlinal barrier, 119
  margin, 54
  mixed facies, 49
  ridge, 9, 30, 37, 40, 41, 42, 46, 104
  rocks, 28F, 29T, 30, 30F, **46-50**
  upwarp, 58
geanticline, 27, 40, 97, 122, 137
Gedinnian, post, 71
  Stage, 42, 45, 53
  Upper, 42
Gedong Batu, Kinta, 292F
GEISS, J., 400
Gemas, 99F, 174, 185, 217F, 340F
Genting Sempah, Selangor, 394
geochemical methods, 389
geoelectrical study, 393
geographyphysical studies, 8
geological controversy, 5
geological history, **55-60**
  history, Upper Paleozoic, **94-95**
  investigations, history of, **1-9**
    modern, **8-9**
    pre-1870, **1-3**
  investigations 1870-1903, **3-4**
  investigations 1903-1930, **4-6**
  investigations 1930-1955, **7-8**
  map, 3, 5, 391, 392, 393, 395, 396
    of Indonesia, 396
  observations, 393
Geological Society of London, 5
  Society of Malaysia, 9, 12, 50
  Survey, 231T, 232, 238T, 248T, 258T,
    326F, 390
  Survey Department, 4, 7, 8
geology, physical, 6

summary correlation chart, 10T
Geology Department, 8
geology of Malay Peninsula, summary, **9-12**
geology of Malaya, 6
geology of the Malayan ore deposits, 6
geomorphological evolution, 399
geomorphology, 3, 9, **13-24**, 173
  eastern coastal plains, **22-23**
  highland areas, **15-16**
  limestone areas, **16-19**
  lowland areas, **16-23**
  Malacca Straits coast, **19-20**
  southern coastal plain, **20-22**
geomorphometry, 15
geophysical methods, 383
Georgetown, 155F
geosynclinal basin, 55, 59
  belt, 25
  deposits, 25
  organization, Ordovician to Upper
    Triassic, 98F
  phase, 251
  sedimentation, 9
geosynclines, 9, 392
  Cenozoic, 1
geosyncline, epicontinental, 27, 55
  intercratonic, 27, 55
  Lower Paleozoic, cross-section, 30F
  Malayan, cross-section, 57F
  marginal, 27
geotectonic elements, 392
geothermal gradient, 253, 261, 267, 273,
    287, 301, 331
  temperature, 253
GERTH, 114
gervillia, 110, 118, 120
  hanitschi sp. nov., 110
  scrivenori sp. nov., 110
gibbsite, 157
GIBSON, I. L., 392
GILBERT-TOMLINSON, J., 56, 392
GILLULY, J., 97, 394
Ginting Bidai, 183, 184, 186T, 187
  Highlands, 183, 184
    road, 184
  Lebak, 296
  Sempah, Selangor, 178F, 179T, 180, 183,
    184, 295, 310, 394
    road, 183
  Tua, 355
girtyoceras, 86
Givetian, 43, 75
glabellae, swollen, 41
glacial origin, 5
  periods, 24
  theory, 5
glaciations, Pleistocene, 23
glass, devitrified, 192, 211
  fragments, 154
  lapilli, 192
  natural, 401
  shards, 194, 200
    tricuspate, 194
  wisps, 192
glassy margins, 246

gleichenites gagauensis KON'NO pinna of.,
    132F
  pantiensis KON'NO pinna of., 132F
gleichenites stenopinnula KON'NO pinna
    of., 132F
gleichenoides gagauensis nov., 132
  pantiensis KON'NO nov., 136
  serratus nov., 132
  stenopinnula nov., 132
glyptograptus lunshanensis HSÜ, 42
  persculptus (SALTER), 43
gneiss, 53, 263, 283, 284
  Archean, 3
  biotite, 236T, 258T, 264F, 265, 268F,
    284
  biotite-hornblende-K feldspar, 276F
    -tourmaline, 276F
  biotite, radiometric age, 260
    -quartz-albite, 276F
  catazone, 302
  diopside-tremolite-calcite, 256
    -andesine, 256
  foliated, *269*, 271F, 272, 273F
  granite, 256, **259-260**, 261F, 303
  microcline-hornblende-biotite, 270F
  microcline-oligoclase-biotite-quartz-(horn-
    blende), 269
  microcline-oligoclase-hornblende-biotite,
    269
  mineralogy, 259, 260, 263, 265, 287, 296,
    297
  modal analysis, 259, 260, 287
  muscovite-biotite, 284
  pink hornblende-K feldspar, 276F
  psammitic migmatitic, 270F
  sillimanite-garnet-biotite, 264F
gneiss dome, 279
  dome, mantled, 331
  uplift, 331
gneissic rocks, Thailand, 241F
  texture, 263
GOBBETT, D. J., **1-12**, 8, 15F, 16, 37, 39F,
    40F, 55, **61-95**, 53, 67F, 72F, 75, 87,
    90, 182, 228F, 235, 246F, 267, 294,
    308, 315F, **305-330**, 396
goethite, 347
gold, 161, 227, 240, 259, 278, 280, 388
  belt, 340F
  deposits, **381-382**
    distribution, **340F**
  dydrothermal, 381
  mineralization, 198, 230, 331
  mining, 3, 4
  placers, 281, 381
  production, 382T
  in pyrite, 381
  -quartz deposits in Pahang, 392
  theory of genesis, 335
Gombak, 183
Gondwana Clays, 5
  continent, 5
  ice, 5
  rocks, 5, 105, 115, 123
Gondwanaland, 9, 27, 55, 56F, 57F, 141
goniatites, 68, 91

goniomya scrivenori sp. nov., 110
  singaporensis sp. nov., 110
gonodon sp., 109, 110
Good Earth mine, 385
Gopeng, Perak, 74F, 236T, 290, 291
  Beds of Kinta, 5, 6, 9, 158, 160, 400
Gopeng Consolidated Mines, 372
GOPINATHAN, B., 157, 394
GORMANN, C. F., 175, 394
gossan, 386
Government geologist, 4
GRABAU, A. W., 27, 55, 58, 394
graben, 73
graded bedding, 117, 130, 133, 169F
gradient, 141
granite, 6, 7, 8, 11, 47, 49, 57F, 64F, 74F,
  73, 78, 88F, 108, 205F, 215, 215-
  242, 225F, 226F, 250, 255F, 259-
  260, 263, 276F, 280, 286F, 307F,
  309F, 311F, 314F, 316F, 320F,
  319F, 321F, 323F, 324F, 327F,
  332F, 333F, 337F, 338F, 339F,
  340F, 356F, 358F, 380F, 389F,
  399F
  accessory minerals, 221
  acid, 249
  age, 232-240, 235, 332, 400
    radiometric evidence, 235-240
    rubidium-strontium, 238
    stratigraphic evidence, 232-235
  ages, frequency, 239F
  anatectic, 302
  apophyses, 230, 280
  autochthonous, 256
  barium content, 370
  batholiths, subcordant, 331
  belt, 341
  biotite, 222, 223F, 231T, 236T, 279,
    282F, 365
  boulders, 5
  biotite, porphyritic, 231T
  biotite-muscovite, 236T, 240
  biotite-tourmaline, 236T
  Carboniferous, 360
  catazone, 230
  chemical analyses, 218, 231T, 230-232
  clasts, 232, 235, 240
    weathered, 158F
  clasts in tuff, 249
  coarse, 231T
  contamination, 222-224, 224
  correlation with Indonesia, 242
  correlation with Thailand, 240-242
  cusps, 332, 363, 390
    mineralization around, 363
  cusp apices, 387
  cut by dykes, 246F
  definition, 215
  distribution, 216F
  dome, 285
  dykes, 223, 250
  emplacement, 60, 105, 392
  epizone, 215-218
  fine-grained, 238T
  foliated, 230, 236T
  fragments, 116
  graphic, 222
  gneiss, 328
  gneissic, 230
  grey, 222
  grey porphyritic, 219
  high-level, 239
  hornblende, 5, 278-279
  hornblende-biotite, 240
  intrusion, 10T, 92T, 95
  kaolinized, 363
  Late Cretaceous, 217, *240*
    petrography, *218*
  -limestone contact, 374
  magmatic age, 238
  magmatic origin, 232
  marginal modifications, 185
  marginal phase, 184
  medium grained, 236T
  Mesozoic, 5, 27, 249, 400
  mesozone, 218-230, 388
  middle Cretaceous, 141
  middle Triassic, 240
  mineralogy, 218, 219-221, 222, 230
  modal analyses 218, 220T, 221
  modifications, 224-230
    analyses, 234T
    chemical analyses, 234T
  muscovite-biotite, 231T
  mylonitic, 230
  niobiferous, 344
  niobium content, 344T
  microporphyritic, 222
  normative composition, 233F
  origin, 401
  Paleozoic, 27
  parautochthonous, 230, 256
  pebbles, 137, 309, 310
  Permian, 218
  petrography, 219-222
  pink, 220T, 222, 236T, 237T, 239
  plutons, 241F
  porphyroblastic, 230
  porphyritic, 219F, 221, 222, 231T, 236T,
    237T
    weathered, 219F
  porphyritic-biotite, 223F
  porphyry, 180, 186T, 187, 198, 215, 221,
    279, **280**
    biotite, 349
  post-orogenic, 239, 334
  pre-Carboniferous, 235
  radiometric age, 283, 310, 312, 313, 315,
    317
    determinations, 236T, 237T, 238
  relicts, 322
  replacement, 373
  rocks, age dating, 400
  series, 399
  -shale contact, 348
  sheared, 237T, 325F
  sheetlike, 230
  sills, boudinaged, 60
  staining, 220
  Tertiary, 360
  Thailand, radiometric dating, 240
  tin-barren, 342
  tin-bearing, 314, 342
  tin content, 341, 343T
  Triassic, **121**, 235, 241, 337F, 360
  Upper Carboniferous, 240, 218, 301, 313,
    337F
  Upper Cretaceous, 217F, 239, 337F
  Upper Permian, 60
  wash, 160, 163, 164F, 166
  weathering profile, 155F
granite stock, 283
  system, thermal low, 232
    thermal valley, 302
  tectonic level of intrusion, 215
granitic rocks, 90, 242
granitoid rocks, 279
granodiorite, 215, 220T, 225F, 244F, 242
  dykes, 224F
  greisenized, 220T
  hornblende, 231T, 237T
  porphyry, 222, *280*
granophyre, 222
  hornblende, 249
  sheared, 185
  quartz-orthoclase, 280
granulite facies, 273
graphite, 49, 256, 289
graptolites, 8, 33, 43, 396, 397
  Silurian, 42, 396
graptolite-tentaculite association, 42
graptolites, Upper Wenlockian, 42
gravel, 144, 159, 160, 162F, 165, 168, 169
  -pump methods, 386
  weathered, 161F
Great Tenasserim River, 211
Greenland, 401
greenschist facies, 11, 253, 254, 262, 267,
  279, 285, 288, 291, 296, 297, 298,
  299, 301, 331
  facies metamorphism, 194
  facies phyllitic rocks, 287
greisen, 220, 242, 336, 346, 365
  -bordered veins, 369, 371F, 372
  cassiterite, 226
greisenization, 228, 360
greywacke, 41, 112, 116, 117
  polymict, 118
Grik, 27, 42, 46, 47, 48, 54, 55, 58, 151,
  177, 178F, 181, 207, 209, 289, 314,
  332F
  area, Upper Perak, 48F
  siltstone, 29T, **48**
  tuffs, 31F
grit, 32, 51, 84
  feldspathic, 31
  schistose, 51
grooves, basal, 18
groove casts, 117
grossularite-andradite, 301
grossularite garnet, 284
ground geophysics, 390
groundmass, 220-221
  aphanitic, 184
  black, 183

chloritic, 188
cryptocrystalline, 194, 198, 246, 280
devitrified glass, 194
glassy, 184, 188
microcrystalline, 180, 192
micrographic, 198
GRUBB, P. L. C., 13, 153, 156, 177, 198, 199, 204, 206, 208T, 209, 222, 298, 322, 366, 383, 394
Gua Bama, 82
Gua Cha, Kelantan, 171, 175
Gua Kechil, 276F, 296
Gua Keh, 67, 67F
  Keh quartzite, 67F
Gua Musang, 82, 99F, 101, 103, 113, 123, 79F, 185
  Musang Formation, 10T, 78, 97, 99F, **101-103**, 102T, 104, 113, 107T, 191, 187, 267, 297, 301, 310
  Musang, Kelantan- fossils, 100F, 101
Gua Panas, 276F, 296
Gua Panjang, 101
Gua Sai, 4
Gua Sae, 82
Guadalupian, 94
guano deposits, 171
  mining, 171
Gubir beds, 115
gugup, 155
Gulf of Thailand, 215, 242, 322
  sediments, 393
Gunong Ayam, 262, 268F
  Badong, 130
  Bakau, 5, 225, 226, 227, 343, 346
Gunong Baku, 345
  Benom, 97, 99F, 121, 222, 230, 267, 396 granite, 277; *see also* Benom granite
Gunong Berangkat, 268F
  Berantai, 99F, 125
  Beremban, 134, 141, 140F
  Besar area, 326
Gunong Blumut, 99F, 200
  Datoh, 290
  Dulang, 17F
  Gagau, 8, 99F, 128, 139, 141, 140F
    fossils, 132F
    plateau, 128F
  Gapis, Pahang, 345
  Hutan Haji, 64F, 68
  Irong, 140F
Gunong Jala, 290
  Jerai, 27, 175, 178F, 180, 223, 236T, 227, 230, 254, 255F, 282F, 283, 333F, 343, 381, 387
  Jerai area, Kedah, 392
    map, 282F, 286F
  Jerai, conditions of metamorphism, **285-287**
    petrology, **283-284**
    structure, **284-285**, 285F
  Jerai dome, 33
  Jerai massif, **283-287**, 398
    age, **283**
  Jerai placers, 365
Gunong Kanthan, 75

Gunong Kemahang, 230
  Kendrong, 48F, 48
  Keriang, Kedah, 71T, 64F, 173
  Kesut, 298
  Kuang, 289
  Laris, 17F
  Ledang, Malacca, 217F
Gunong Lesong, 99F, 135, 135F
  Muntahak area, Johore, 394
  Paku, 372
  Panjang, 381
  Panti, 99F, 134, 135, 136, 137, 140F
    Johore, fossils, 132F
  Panti area, 396
Gunong Panti West, 136, 141
  Pelangei, 17F
  Penumpu, 99F, 126, 128, 127, 139
  Perlis, 128
  Pondok, 49
  Pulai, 101, 178F, 185, 255F
    Estate, Johore, 237T
    granite, 217
Gunong Pulai Member, **103-104**, 107T, 108
Gunong Pulai Volcanic Member, 179T, 200
Gunong (Putus) Semanggol, 114
Gunong Rabong, 99F, 113, 123
  Rabong formation, 10T, 99F, 106, 107T, 110, **113-114**, 127, 201, 254
  Raja granite, 289
  Rapat, Perak, 161F, 162F, 164
Gunong Raya, 71, 312
  Sah, 112
  Semanggol, 118, 120
  Serudom, 140F
  Sinyum, 79F, 112
  Stong, 82, 97, 101, 230, 262, 268F
  Sumalayang, 99F, 140F, 141
Gunong Tahan, 13, 14F, 99F, 113, 123, 138
  Tembat, 191
  Tempurong, 158
  Terendum, 290, 291
GUNTEN, H. R., 400
Gunz glaciation, 168
Gurun, 282F, 284, 286F
  quarry, 180, 283F, 284, 285
gymnosperms, 136

Haad Som Pan, Ranong Province, Thailand, 342, 391
HADA, S., 8, 80, 82, 101, 394, 395
HAILE, N. S., 151, 156, 167, 168, 173, 174, 176, 184, 301, 308, 394
halobia, 11, 105, 112, 118, 121, 397
  aotii, 119
  charlyana MOJSISOVICS, 100F
halysites sp., 37, 49
HAMADA, T., 49, 63, 68, 92, 112, 397, 394, 401
hamletella, 68
HAMPTON, J. H., 3, 394
HANNAFORD, P., 366, 394
HANITSCH, 110
HARDING, R. R., 400
hardpan, 13

HARKER, A., 258, 394
HARLAND, W. B., 242, 394
HARRAL, G. M., 391
HARRIS, E. F., 4, 394
Harvard Estate, Kedah, 232, 235
HASER, F. E. H., 395
hauyne, 206
Hawaii, 393
Hawaiian islands, 251
Hawthornden Schist, 29T, 45, 58, 179T, 295, 311F
HAYAMI, I., 138, 394
headland, 20
hedenbergite, 296, 357
helicotoma sp., 40
  jonesi KOBAYASHI, 35, 38F, 39
helicotoma costata KOBAYASHI, 35
heliolites, 49
  aff. *barrandei* var. *spongodes* LINDSTROM, 37
helix, 150
helminthopsis, 65
helvite, 336, 346
hematite, 157, 161, 172, 180, 289, 347, 348, 384F, 381, 385
  micaceous, 284
  replacement, 381
  rosettes, 381
  veinlets, 330
HERON, A. M., 92, 392
Hiap Huat Mine, Rawang, Selangor, 353
high-level alluvium, 6
  -level intrusion, 217
high thermal gradient, 288
HILL, R. D., 22, 174, 394
hill, summit profile, 140F
  towerlike, 68, 82
hillocks, sinoid, 19
HILLS, E. S., 394
hills, fault-bounded, 5
  flat-topped, 135F
Himalayan, 56
  trough, 58, 59
Himalayas, 30, 55, 112, 121
Hin Fatt mine, Salak South, 364F, 365
Hindehede quarry, 244F
Hindu temples, 175
hippopotamus sp., 171
HO, R., 394
Hoabinhian, 394
  age, 175
  culture, 175
Hock Aun mine, Kampong Pandad, Selangor, 368, 369, 371F
Hock Leong mine, Ampang, Selangor, 367, 368
  Ampang, Perak, 370F
HOCKIN, H. W., 344, 369, 394
HOLDEN, J. C., 334, 393
HOLMES, A., 242, 394
holmia fauna, 55
Holocene, 9, 143, 157
homo erectus (*pithecanthropus*), 175
homoclinal valley, 17F
HONGNUSONTHI, A., 391, 395

HOOIJER, D. A., 167, 171, 394
hormotoma sp., 35
hornblende, 47, 53, 103, 182, 184, 188, 191, 197, 200, 221, 220, 222, 244, 247, 259, 261F, 266F, 286F, 298
  coarse grained, 298, 299
  enrichment, 251
  fibrous, 183
  green, 224
  hornfels facies, 288
  lineated, 263
  optical properties, 259
  pegmatites, 299
  poikiloblastic, 263, 269
  radiometric determination, **236-238T**
hornblendite, 277
hornfels, 47, 51, 52, 284, 293, 295, 389F
  andalusite-cordierite, 373
  biotite, 295
    -andalusite, 297
    -quartz, 294
  calc-silicate, 48, 290F, 323F, 387
  chiastolite, 289, 293
  chondrodite-bearing, 49
  epidote, 296
  garnet-actinolite, 297
  hornblende, 293
  mineralogy, 295, 296
  phyllitic, 289
  pyroxene, 297, 298
  quartz-biotite, 47, 289
    -muscovite-clinozoisite, 289
  quartz-chlorite-epidote, 295
    -sericite, 48, 289
  quartz-epidote-amphibole, 297
  quartz-feldspathic, 180
  quartz-tourmaline, 288, 373
  quartz-tremolite-zoisite, 53
  spotted, 45, 289
    quartz-biotite, 288
    tourmalinized quartz-biotite-sericite, 288
  tourmaline, 229
  xenoliths, 295
horridonia, 83
horsts, 18
Hose Mountains, 214
HOSKING, K. F. G., 174, 343, **335-390**, 336, 344, 349F, 350F, 352F, 353F, 354F, 355F, 357F, 364F, 368F, 369F, 370F, 371F, 374F, 378F, 379T, 385F, 394, 395
hot springs, 387
HOUTERMANS, F. G., 400
HOWARTH, 106
HOWELL, B. F., 113, 395
HSÜ, K. J., 119, 395
Hua Hin, 302, 303
Huai Jagrao, 365
Huat Choe village, 110
Huckitta-Marqua region, Northern Territory, 392
Hué, 93F
HUGHES, H., 395
Huin Fatt, Salak South, Selangor, 34T

Hume Heights, 244F
HURLEY, P. M., 235, 399
Hutan Haji, Perlis, 69F
HUTCHISON, C. S., 8, 19F, 34, 35F, 55, 56, 103T, 122, 166F, **177-214**, 181, 185, 193F, 201, 203F, **215-252**, 218, 219F, 220, 221, 222, 223F, 223, 224F, 225F, 227, 229F, 232, 235, 242, 244F, 247, 249, **253-303**, 256, 259, 261, 262, 264F, 270F, 271F, 273F, 273, 274F, 283F, 301, **330-334**, 356, 379, 387, 395
hybrid rocks, 267
hybridization, 251, 275, 277
hydrated ferrous stannate, 366
hydrodictya cylix HALL and CLARKE, 53
Hydrographic office, U. S. Navy, 20, 21F
hydropotes, 167
hydrothermal action, 220
  activity, 259
  alteration, 153
  deposits, 348, 349, **357-385**
  deposition in skarn, 349
  quartz, 228
    veins, 341
  solutions, 348
  tin and tungsten deposits, **363-381**
hyolithids, 35
hypersthene, 199, 206, 211, 243, 244
hypabyssal rocks, petrology, **206**
hypogene kaolinization, 363
  processes, 363
hypothermal origin, 227
hystrichospheres, 42, 43

ice rafting, 68
ICHIKAWA, K., 83, 101, 102, 118, 123, 124F, 125, 127, 314, 395
Idaho, 260
igneous activity, mid-Paleozoic, **60**
  intrusions, 281
  origin, 219
  rock, ages, 387
  rocks, 85F
ignimbrite, 16, 184, 193F, 200, 207, 257F, 392
  dyke, 318, 321F
  mineralogy, 194
  Temangan, **194-195**
IGO, H., 8, 35, 84, 86, 101, 105, 395, 397
ilmenite, 160, 166, 170, 172, 243, 244, 245, 344, 366
  tin-bearing, 342
Imperial Institute, 6
impounding body, 350, 360
incertae sedis, 32, 132, 136
India, 5, 55, 334
  and Burma, 398
Indian Ocean, 401
Indochina, 55, 94, 104, 105, 108, 112, 121, 138, 213, 211, 400
  geology, 393
Indochinites, 174
Indo-Malayan region, 394
Indonesia, 11, 58, 121, 150, 211, 303, 392, 398
Indonesian, 174
  Archipelago, 24
  granite, 242
  age, 393
Indosinia, 11, 60, 95, 305, 312, 334
Indosinian massif, 2F, 241F, 306F
  movements, 313
Inferior Oolite, 110
infrastructure, 262, 302
INGHAM, F. T., 7, 8, 49, 51, 61, 72, 73, 74F, 105, 111, 115, 120, 122, 123, 129, 134, 146, 151, 159, 160, 182, 222, 225, 229, 227, 290, 291, 292F, 315, 335, 345, 346, 359F, 365, 369, 372, 373, 386, 395, 402
inland valley fill, **157-171**
inliers, faulted, 75
insoluble residue, 300
intercratonic pattern, 56
interglacial, 15
  periods, 24
Institute of Geological Sciences, 8
introduction, **1-12**
intrusions, hypabyssal, 51, 194
  multiple, 322
  sheetlike, 331
Ipoh, 15, 49, 74F, 169, 232, 290, 337F, 338F, 339F, 332F, 358F
  area, 215
Iran, 58
  trilobites, 396
iron, 336, 358F
  predating tin mineralization, 381
  -rich mineralizing agents, 350
  -tin skarn deposit, 349
  total, 274F
  variation, 250F
  zone, 358F
iron age, 175
  deposits, 381
    distribution, **339F**
  hydrothermal, 381
  ore, 8, 262, 397
    cassiterite-bearing, 348
    metasomatic, 381
    primary, 347
    stanniferous, 347
  ore deposits, 392
  oxide films, 89
  oxides, 149, 165, 182, 183
  pan, 143
isculites, 101
ISHII, K., 8, 70, 101, 120, 395
island arc, 54, 56F, 59
  arc system, Indonesia, 305
    Philippines, 305
islands, offshore, 19
isoclinal folding, 262, 288, 331
isocrinus sp., 127
isopic deposits, 58
  zones, geosynclinal, 55
iso-stannite, 367, 370F
isotelus, 49
IWAI, J., 138, 391, 395

IZOKH, E. P., 60, 395

JAAFAR BIN AHMAD, 8, 39, 51, 54, 55, 111, 112, 266, 272T, 277, 396
JAAFAR BIN HAJI ABDULLAH, 203F, 219F, 264F, 275F, 290F
Jabor Valley Estate, 205F
JACK, W., 1, 395
JAEGER, H., 42, 396
JALICHANDRA, N., 392, 397
Jalong Tinggi Estate, 182
jamesonite, 345, 369
JANTARANIPA, W., 91, 398, 402
Japan, 23
Japanese islands, 59
  occupation, 7
Jasin Volcanics, 201
Java, 1, 13, 19, 93F, 175, 204, 214, 393
  Sea, 141
JEACOCKE, J. E., 103T, 185, 201, 249, 232, 273, 395
JEFFERIES, R. P. S., 76, 106, 108, 110, 112, 396
Jelai Formation, 10T, 99F, **106-109**, 106, 108, 109, 111, 107T, 254
Jelapang, Perak, 182, 225, 290
Jelebu, 185
Jemaluang, 200
Jengka Pass, 79F, 83, 125, 126, 127, 314
  Pass, exposed section, 83F
    Permian fossils, **77F**
    road cutting, **83**
    unconformity, 395
  Pass formation, 123
  Triangle, Pahang, 393
    geological map, 185, 196F
Jeniang, 41
Jenut Batu Papan, 188
Jerai Formation, 29T, 31F, **33**, 37, 56, 177, 179T, 180, 282F, 283, 286F
  Formation, thickness, 33
Jerai Quartzite, 286F
  Schist, 286F
Jeram Berhala, 259
Jeram Gading, 184
  Kuantan Estate, 204
  Limau, 185
Jeram Star, 188, 279
Jerantut, 106, 125, 195, 196F, 178F, 201
  railway line, 385
Jinjang, 311F
Jitra, 64F, 289
JOHANNAS, 396
JOHNSON, J. G., 392
JOHNSON, W. D. jr., 392
Johore, 3, 6, 8, 10T, 11, 15, 23T, 26F, 28F, 61, 62F, 63, 84, **87**, 101, 103, 108, 109, 122, 134, 138, 156, **166-167**, 177, 179T, 185, 215, 218, 221, 222, 242, 397, 402
  Bahru, 24, 101, 109, 166, 145F, 201, 242
  east, 10T
Johore, fossils, 100F
  geology, 392

north-east, 23T
and Pahang, 401
River, 22
Strait, 200
Johol, 326
joints, 15, 90, 330
  extension, 326
  in granite, 322
  Kinta Valley, 394
  major, 322, 326F
joint pattern, 188
  planes, 19
joints, second order shear, 326
  strike frequency, 331F
  tensional, 330
  columnar, 153
  intense, 326F
JONES, C. R., 8, 18F, 20F, 25-60, 30, 32F, 32, 33, 35, 36F, 37, 42, 43, 46, 48F, 48, 51, 52F, 53, 55, 60, 61, 62F, 63, 65, 68, 83, 84, 86, 70, 71, 73F, 75, 76, 80, 81T, 101, 105, 106, 110, 112, 113, 114, 118, 120, 120F, 133, 134, 135F, 144, 151, 152, 170, 171, 172, 173, 181, 219F, 232, 235, 247, 288, 289, 309F, 312, 313F, 314F, 318, 392, 396, 397, 399, 402
JONES, M. P., 341, 342, 363, 366, 396
JONES, T. R., 396
JONES, W. R., 5, 6, 149, 151, 158, 173, 346, 357, 371, 396, 400
JONGMANS, W. J., 138, 396
JOPLIN, G. A., 139, 250, 251, 252, 396
Jordan, 121
JUDD, W. R., 396
Jurassic, 1, 9, 10T, 11, 16, 107T, **122-141**, 239F, 300F
  -Cretaceous sediments, 398
  late, 11
  Lower, 124F, 110
  Upper, 124F, 132
Jurong, 108
  area, **109**
  Formation, 10T, 99F, 101, 103, 106, 107T, 109, **108-109**, 126, 137, 200, 254
  quarry, Singapore, 224F, 225F, 244F
  road, 243, 244F

Kabang fault, 327F, 328F
Kacha, Kinta, 366T, 372
KAEWBAIDHOON, S., 341, 342, 390, 396, 401
Kajang, 170, 325F
Kaki Bukit, Perlis, 64F, 65, 171, 249
kaksa, 169
Kalimantan, 91, 93F, 94, 212F, **213**, **214**, 305
Kalumpang, Selangor, 237T
  J. K. R. quarry, 294
Kamativi, Zambia, 365
Kampar, Perak, 49, 69F, 72, 74F, 75, 75F, 76, 160, 164F, 172, 238T, 401
  Bharu Estate, 294
  fault, 326

Kampar road, near Kuala Dipang, 231T
Kampong Awah, 79F, **83**, 178F, 179T, 185, 195, **197**
  Awah, fossils, 77F
  Batu Hampar, 219F
  Belimbing, 101
  Bukit Kepong, 150
  Dong, 154
  Jabi, 206
  Kermoi, 187
  Kuala Balah, 262, 263, 267, 268F
  Kampong Lepar, 198, 327
  Liang, 188
  Pahit, 49
Kampong Pandan, near Kuala Lumpur, 195, 336, 346, 352, 357, 358, 368, 372, 374
  Pasir Aka, Trengganu, 242
  Peling, 202T
  Perigi quarry, 285
  Sena Formation, 68, 115
  Setar, 297
  Siliau, 385
  Sungei Kroh, 76
Kampong Sungai Ruan, 276F
Kanchanburi, 302
Kanchanaburi gneiss, 303
  Series, 46, 50, 91, 94
Kanching, Selangor, 158F, 293, 329, 330F
  Tin mine, Templar Park, Selangor, 353
  Valley, 158
Kangar, 36, 64F, 172
Kanthan, 74F
  hills, 37
  quarry, Perak, 71T
kaolin, 160
kaolinite, 157
kaolinization, 229, 360
Kapal granite, 104
Kapar Bharu Estates, 182
Karak, Pahang, 51, 52, 52F, 53, 112, 111, 198
  Formation, 51, 53
Karangan, Kedah, 236T
Karimata, 241F, 303
Karimata islands, 303
  Zone, 306F
Karimondjawa, 214
karst, 15
  features, 16
  kegelkarst, 68
  mogote, 15
  old-age, 68
  plain, 157
  sinoid, 19
  topography, 34, 47, 48
  tower, 15
  window, 171
KATILI, J., 396
Kechau Valley, 81T
Kedah, 3, 6, 8, 19, 20, 25, 26, 26F, 27, 28F, 29T, 30F, 31F, 31, 33, 37, 41, 45, 57F, 58, 61, 62F, 63, 64F, 68, 70, 114, 115, 117, 120, 144, 171, 172, 173, 174, 179T, 236T, 310, 402

418 INDEX

Kedah, north, 65F
Kedah Peak, 283, 284, 337F, 339F, 366, 367
   Peak dome, 302
   River, 19
   southeast map, 314F
KEE, T. M., 135, 136, 396
kegelkarst, 68
Kelantan, 6, 8, 10T, 11, 16, 25, 26F, 27, 28F, 50, 56, 61, 62F, 63, 79F, 86, 97, 101, 114, 122, 123, 128, 171, 174, 177, 179T, 185, 218, 246, 399, 402
   north, 82-83
Kelantan River, 19, 22, 194
   schists, 254
   Upper Paleozoic, **80-82**
KELLER, G. H., 23T, 396
Kelumpang, 226
Kemachang granite, 104, 133, 230, 240, 247, 256, 257F, 287
Kemubu, 191, 257F, 262, 268F
Kenny Hill, 87, 311F
   Hill Formation, 10T, 63, **87-91**, 95, 295, 311
      age, **90-91**
      distribution, 87
      geological map, **88F**
      lithology, **87-90**
      sediment origin, **91**
      thickness, 90
Kepong, 144, 145F, **150-151**, 311F
Kerdau formation, 10T, 99F, 103, 106, 108, 107T, 110, **111-114**, 119, 201, 254
Kerdau station, 111
kersantite, 247
   dykes, 247
Kertam, Pahang, 86F
Ketapang Complex, 303
ketaphyllum aff. *turbinatum* (LINNAES), 37, 38F, 39
Khao Luang, 302
Khao Noi, central Thailand, 399
Khao Phrik, near Rat Buri, Thailand, 399, 401
KHOO, H. P., 366, 367T, 396
KHOO, T. T., 222, 267, 275, 276F, 296, 396
Khorat Group, 138, 139, 141, 401
Khorat Plateau, 93F, 94, 213, 303, 397
   Series, 240, 395
      basal conglomerate, 240
Khuan Lee Hin Kangsi tin mine, Siputeh, Perak, 258T
Kim Loong No. 1. beds, 75, 75T
   Loong No. 3 beds, 75T, 76
KIMURA, T., 312, 313F, 396
KING, W. B. R., 58, 396
Kinta, 3, 62F
   district, 400
   Kellas Estate, 291
   Limestone, 9, 10T, 69F, 71T, 92T
   Limestone, fossils, 77F
   River, 291
   Valley, Perak, 3, 4, 5, 7, 8, 9, 15, 18, 24, 37, 46, 61, 63, 78, **71-78**, 78F, 92, 92T, 94, 145F, 157, **160-165**, 162F, 171, 182, 229, 300, 398, 399, 400, 395
   Valley, geological map, 74F
   Valley geology, 392
   Valley, Lower Permian cross-section, 78F
      ore, 392
      Upper Paleozoic rocks, 254
KIRK, H. J. C., 212F, 213, 214, 242, 396
Kisap, 312
   thrust, Langkawi, 309F, 312, 313F, 332, 333F
Klang, 145F
   Gates, 295, 311F
   Gates dam site, 322, 326F, 391
   Gates quartz ridge, 16, 228F, 228, 322, 325F
   Gates reservoir, 184
   River, 326F
Klau Forest Reserve, 197
Kledang, 290, 387
   beds, 162F
   fault, 320F, **326**, 330, 332F
   Range, Perak, 49, 73, 74F, 78F, 121, 164, 182, 290, 344, 346, 357F, 358F, 363, 372
   scarp, 73
KLOMPÉ, T. H. F., 212, 212F, 213, 215, 240, 242, 243, 305, 306F, 396
Kluang, 151, 185, 200
Kluang-Niyor, 144, 145F, **151**
klukia sp., 127
knickpoints, 15, 16
knick zones, 308
knife-edge fractures, 357
KOBAYASHI, T., 8, 32, 35, 36, 43, 45, 49, 58, 68, 94, 97, 106, 110, 111, 112, 113, 114, 115, 118, 119, 141, 152, 317, 396, 397
kobellite, 368F
kockerella, 35
Kodiang, 64F, 70, 172
   area, 121
   limestone, 10T, 70, 102, 102T, 107T, 120F, *120-121*
   quarry, Kedah, 102T
   railway station, 120
KOIKE, T., 35, 84, 395
KOMALARJUN, P., 401
Ko Muk, Peninsular Thailand, 92, 399
Kong Chin Mine, 292F
KON'NO, E., 8, 87, 94, 127, 132, 136, 397
KOOPMANS, B. N., 8, 20, 22F, 23T, 60, 63, 65, 70, 79F, 113, 122, 123, 124F, 125, 128, 133, 137, 288, 308, 309F, 312, 313F, 317, 319F, 320, 322, 305, 397
Korbu, 332F
Korea, 58, 260
   north, 59
Kota Baharu, 145F, 172, 216F, 243
Kota Jin, 79F
   Tampan, Perak, 145F, 155, 157, 400, 401
   Tampan Estate, 154
Tinggi, Johore, 225, 366T
Tinggi road, 168
Krabi Series, 144
Kramat Pulai, Perak, 291, 335, 338F, 346, 347, 359F, 365, 369, 373
   Pulai ore bodies, 359F
Kroh, 47, 314F
Kuah, Langkawi, 35F, 219F, 227, 229F, 229, 236T, 289, 353, 395
Kuala Bala, 257F
   Brang, Trengganu, 236T
   Dipang, Perak, 74F, 78, 224, 236T
   Dipang, J. K. R. quarry, 156F, 238T
   Dungun, 216F
   Gris, 257F
   Jelai, 185
Kuala Kangsar, 46, 49, 154, 157, 332F
Kuala Kelawang, 325F
   Keluong, 125
   Kemaman, 85F
   Krai, 145F, 178F, 191, 192, 194, 226F, 256, 257F, 318, 320, 321F, 340F
   Kubu Bharu, Selangor, 221, 222, 225, 226, 301, 325F, 326, 329
   Kubu railway station, 294
   Kubang badok, 33
Kuala Lepar, 281
   Lipis, 4, 7, 8, 79F, 80, 82, 99F, 101, 105, 106, 108, 114, 145F, 157, 178F, 185, 230, 242, 247, 277, 340F
   Lipis, fossils, 100F
   Lumpur, 8, 16, 18, 45, 54, 87, 88F, 93F, 99F, 145F, 171, 180, 183, 216F, 228, 255F, 293, **309-310**, 311F, 322, 325F, 337F, 338F, 394
   Lumpur area, **165-166**, 215, 393, 402
      map, 311F
Kuala Lumpur-Endau fault, 308
   Lumpur-Endau fault zone, 320F, 322
   Lumpur fault zone, 401
   Lumpur Limestone, 10T, 29T, 34T, 37, 37-39, 39F, 40F, 45, 59, 87, 88F, 90, 235, 311F
   Lumpur Limestone, fossils, 38F, 39
   Lumpur lowlands, 157
   Medang, 188, 190
   Pahang, 23T
Kuala Pasir Alor, 242
   Perlis, 32, 172
   Perlis quarry, Perlis, 249
Kuala Petuang, 298
   Pilah, Negri Sembilan, 185, 242, 325F, 340F
   Selangor and Rasa, Selangor, 399
   Seli, 186T
   Sia, 296
   Siah, 242
   Sungei Chempedak, Kuantan, 208T, 209
Kuala Tamang, 182
   Tembeling, 112
   Trengan, Trengganu, 385
   Trengganu, 23T, 85F, 191, 216F, 243
Kuan On beds, 75T, 76
Kuantan, Pahang, 7, 12, 22, 23T, 84, 85F, 125, 143, 153, 160, 172, 173, 174,

178F, 185, 195, 203, 208T, 209, 216F, 218, 238T, 298, 301, 333F, 334, 393, 398
area, Upper Paleozoic, **84-86**
basalt, 10T, 145F, 179T, 238T, 254
basalt, geological map, 205F
basalt flows, 245
Kuantan Group, 10T, 51, 61, 84
  limestone, 71T
  road, 52
Kubang Pasu Formation, 10T, 63, 64F, 65F, 67, **68**, 69F, 70F, 71, 91, 92T, 115, 254
Kuching, 93F
Kueichou, 112
KUEST, L. J. Jr., 393
Kulim area, Kedah, 393
Kulim granite, 41, 114
KULP, J. L., 242, 260, 300F, 397
KUMMEL, B., 112, 397
Kupang, Kedah, 104, 236T, 254, 258T
Kupang Gneiss, 255F, **287**, 314F, 318, 333F
  Gneiss, radiometric age, 287
  Granite, 141, 230
Kwong Fook Lee mine, Ulu Langat, 372
kyanite, 256, 259, 260

labradorite, 183, 243
LACOMBE, 397
LACROIX, A., 174, 397
lacuna, 42
lacustrine, 144
  basins, 143
  deposits, 128, 151
  sediments, 11
Ladinian, 106, 108, 112, 114, 121
  Upper, 119
lagoonal mud, 173
lagoons, 173
Lahat, 74F, 372
  Pipe, Perak, 373, 386
Lain Seng mine, 162F
LAKE, H., 4, 397
Lake Superior copper deposits, 383
Lake Toba, 207, 212F
  Toba caldera, 154
LAMBERT, R. St. J., 235, 397
lamellibranchs, 42, 115, 116
lamellibranch shells, 118
lamellibranchs, Triassic, 398
laminations, 135
  convolute, 116, 117
  cross, 117, 135
  horizontal, 116
LAMOREAUX, P. E., 138, 397
Lampang, 138
lampanning, 386
Lampong Tuffs, 214
lamprophyre, **247-249**
  dykes, 251
  hornblende, 247
  mineralogy, 247
  stocks, 348
Lanchang, 178F, 181F, 185

village, **198**
landforms, 15
landslide, 282F
Lang mudstone, 115
Langat River, 20
Langkawi island(s), 6, 8, 19, 20F, 23T, 25, 26, 27, 29T, 31F, 32, 36F, 43, 46, 55, 56, 59, 61, 62F, 63, 64F, 65F, 66F, 68, 70, 71, 91, 93F, 94, 144, 171, 216F, 232, 240, 255F, 283, **308-309**, 312, 313F, 396, 397, 400
  islands, map, 309F
  islands, structure, 309F
  southwest, geological map, 66F
  island, southwest, 67F
  -Perlis area, 58
Laos, 139, 211, 212F, 213
Lapan Utan, Selangor, 23T
lapiés, 16, 18F
lapilli, 181, 189, 201, 243
  hydrated glass, 192
Larut, 3, 4
  tin mining, 393
LASSERRE, M. M. M., 60, 242, 397
Later Mesozoic flora, 400
laterite, 5, 13, 143, **156**, 393, 398
  aluminous, 156
  bedrock, 156
  brownish-red, 245
  elevation, 156
  fossil, 156
  hilltop, 156
  at Kerdau, Pahang, 393
  pebbles, 173
lateritic pellets, 156
laterization, 3, 13, 200, 206
  iron metasomatism, 3
laterized horizons, 206
LA TOUCHE, T. H. D., 397
lauraceae, 149
laurinea, 149
lavas, potassic, Segamat, 394
  subaerial, 113
LAW, W. M., 146, 147F, 148, 149, 397
Lawin, 145F, **151-152**, 178F, 181, 207
  Tuff, 29T, 54, 179T, 181, 181F, 289, 314F
Lawit granite, 104
Layang Layang, 145F, **152**
LE DIN KHYU, 395
lead, 210, 336, 358F
  method, radiometric determinations, **236-238T**
  ore, epigenetic, 383
  zinc zone, 358F
LEAMY, M. L., 156
leaves, 148
  coriaceous, 149
  zerophytic, 149
Lebir fault, 133, 139, 213, 262, 301, 308, 313, 333F
  fault, extension, 318
  fault zone, **318-322**, 320F, 321F, 322
  Valley, 192
LEDGERWOOD, E., 399

Lee Gossan, 373, 386
LEE, G., 351
Lee, H. S., beds, 75T, 76
LEE, K. V., 373T
Lee Sin Nam mine No. 2, Ampang, 351
Lenggong, Perak, 49, 207, 289
lenses, conglomerate, 52
Leonardian, 78
  of North America, 78
LEONG, L. S., 348, 397
LEONG PAK CHEONG, 103T
Leong Wah mine, 162F
leopardstone, 75
LEOW, J. H., 221, 227, 368, 378, 379T, 394, 395, 397
lepidodendron, 84, 86
lepidolite, 290
lepidolite from pegmatite, 236T, 237T
lepidopleurus sp., 77F
leptodus sp, 11, 77F, 82, 83, 102
Lesong sequence, 134
lesueurilla zonata KOBAYASHI, 35
leucite, 206
  -tephrite, 208T, 209
leucogranite, 222
  leucoxene, 161, 188, 199
Lexique stratigraphique international, 391
Lian Hup quarry, 244F
library research, 390
Liesegang rings, 161
lignite, 146, 151
  high-grade, 149
  seams, 148
  veins, 6
lignitic coal, Miocene, 150
  coal seam, 148F
LIM, T. H., 389F, 390, 397
lima, 110
Lima Kedai, 109
limestone, 4, 6, 7, 26, 34T, 34, 41, 46, 50, 52, 57F, 70F, 74F, 76, 83F, 102T, 196F, 314F, 359F, 371F
  argillaceous, 76
  basal Ordovician, 30
  bedrock, 165F
  bedrock, solution, 164, 172
  bedrock surface, 164F
  bioclastic, 75T, 76, 86
  biohermal, 75T, 76
  biogenic, 71
  biostromal, 76
  bituminous, 83
  breccia, 171
  brecciated, 312
  calcitic, 68, 75, 80
  cancrinella, 70F
  carbonaceous, 75, 102T
  Carboniferous, 3, 7, 61
  caves, 394, 396
  chemical analyses, 68, 71T, 300
  crinoidal, 70, 75T
  crystalline, 317F, 356F
  current-bedded, 34
  Devonian, 94
  dolomitic, 34T, 71T, 102T

facies, 79F, 85F
footwall, 378
formations, 395
  -granite contacts, 359
grey, 49, 70
hills, 6, 16, 18F, 74F
hill-forming, 84
hills of the Kinta Valley, 392
hills of Malaya, 398
hills, notches, 18
impure, 351
intrasparitic, 120
Kinta, 3
magnesian, 34T, 71T
marmorized, 71
micritic, 120
mylonitized, 312
notches, 173
oolitic, 34, 37, 76
Ordovician, 32
Permian, 61, 82
plateau, 92
pinnacles, 164
replacement, 347
shelf, 27, 33-40, 58
shelly, 26, 27
silicified, 30, 32
solution, 5, 73, 157, 160
stylolitic, 33, 37
thermoluminescence age, 235
thick-bedded, 36F
Triassic, 120F, 315F
Viséan, 80, 94
limonite, 157, 161, 172, 294, 347, 386
  colloform, 365
Linau Balui Plateau, 214
Linden Estate, 242, 244
  Hill, 242, 244, 248T
LINDGREN, W., 383, 397
lineament, 327, 329F, 389F
  Lebir, 320
lineaments, major, 332F
  master joint, 330
lineament patterns, 15, 330F
lineaments, structural, 152
lineations, 73, 328
Lingga Archipelago, 123, 138
Linggiu Formation, 87
lingula, 84
linograptus aff. *posthumus* (RICHTER), 53
linoproductus, 76, 83
  kokdscharensis (GRÖBER), 69F
Lipis district, 54
  Group, 105, 114, 123, 137
lit par lit, 256, 261
lithophagus, 173
lithostrotion, 86
litsea, 149
Llandeilian, 35
Llandovery, 29T, 33, 42
  late, 36
  Lower, 43
  Middle, 43, 47, 50
  Upper, 49, 59
  zones, 42

load-cast structures, 136
load-structure, 125
  -structures, Singa Formation, 67F
lode, mineralogy, 374
lode channels, 317F
lodes, 317, 361F
  Cornish type, **379-381**
  pipelike, 372
  stanniferous, 380F
  telescoped, 368
lode tin deposits, 335
Loei, 93F, 94, 303
loellingite, 351T, 353
LOGAN, J. R., 3, 397
LOGANATHAN, S. P. K., 158F, 353, 397
logs, 163
  trapped in alluvium, 169F
lombong siam, 175
Lone Pine hotel, Penang, 220T
longshore current movement, 21F
lopha cf. *montis-caprilis*, 110
lophospira sp, 40, 43
Lotong Sandstone, 127, **130-131**, 133, 136, 137, 139, 201
LOW, J., 3, 397
Lower Badong Conglomerate, 129
Lower Detrital Member, 29T, **33**, 35, 43, 45, 44F
  Detrital Member, thickness, 43
  Devonian fauna, 396
  Indosinias, 211
  Myophorian Sandstone, 106
  Paleozoic, 9, 65F, 396
    classification, **25-27**, 29T
    cross-sections, 57F
    distribution, **25-27**, 26F, 31F
  Paleozoic formations, 64F, 79F
    thickness, 29T
Lower Paleozoic rocks, lithofacies, 28F
  Paleozoic, undifferentiated, 26F
  Paleozoic volcanic rocks, chemical analyses, **185-186**
    east and south of the Main Range, **185**
Lower Paleozoic volcanic rocks, Pahang, **182-185**
Lower Paleozoic volcanic rocks, Selangor, **182-185**
  Paleozoic volcanic rocks, Sungei Siput North, **182**
  Upper Perak-Grik area, **181-182**
  West Kedah, **180-181**
  Paleozoic volcanism, **177-185** 178F
  Permian, cross-section, Kinta Valley, 78F
Lower Redbeds, 138
  Scythian, 11
  Setul Limestone, 29T, 34,
loxonoma sp., 37
LU, K. U., 55, 397
Lubok Liang, 188
Lubok Sukam, 112
lucina sp., 110
Ludlow, 29T, 36, 37, 50
Lumut, Perak, 369, 371F
lutite, 89, 109, 116, 117
  black, 121

clasts, 136
red, 126
lyrodictya sp., 53
lytospira rectangularis KOBAYASHI, 36, 38F, 39

MACANDIE, A. G., 254, 397
macaranga, 149
McBRIDE, E. F., 116, 397
McCALL, T. L., 4, 148, 397
MAC DIARMID, R. A., 358, 378, 398
McDONALD, A. J., 342, 400
MACDONALD, S., 86, 102, 172, 173, 185, 190, 191, 192, 193, 218, 222, 224, 230, 236F, 240, 243, 247, 254, 256, 259, 262, 263, 297, 298, 321F, 322, 323F, 391, 392, 393, 398
Macdonnell Ranges, 56
Machang, 216F
Machinchang Formation, 10T, 29T, 31, 31F, **30-33**, 48, 55, 63, 65F, 66F, 283, 313, 332
  Formation, fossils, 38F
  thickness, 30-31
  Formation quartzite, 32F
  Hills, 30
MACKENZIE, G. S., 393
MACLAUGHLIN, W. A., 242, 393
McLEOD, 373
Mae Moh basin, 213
mafic rocks, chemical analyses, **248T**, 249
Magdalen Mine, Lanlivery, Cornwall, 369
magma, attenuated, 252
magmatic differentiation, 279
  disseminations, 341
  and segregations, **341-346**
magmatic event, dating of, 238
  Tertiary, 235
magnesian — silicates, 346
magnesite, 54, 245
MAGNEE, I. DE., 369, 398
magnesium, variation, 250F, 274F
magnetite, 54, 161, 166, 180, 183, 191, 198, 243, 244, 245, 259, 291, 347, 348
  amphibolite rock, 349
  deposition, 350
  stanniferous, 348
  tin-content, 349
  titaniferous, 204
magnetometer, 391
and scintillation counter survey, 398
  survey, airborne, 8
  surveys, 285
Mahang, 41
  Basin, 45, 58
  Formation, 10T, **41-43**, 41, 42, 46, 58, 59, 29T, 31F, 65F, 232, 254, 282F, 285, 286F, 314F, 392
  Formation fossils, 38F, 39
  Satahun ore, 381
Main Range, 4, 47, 121, 137, 78F, 164, 179T, 185, 215, 216F, 219, 235, 307F, 312, 314, 333F, 320F, 358F, 363, 387, 388

batholith, 267
Granite, 25, 28F, 50, 114, 124F, 180, 182, 184, 217, 217F, 218, 322, 325F, 326, 336
Main Range Granite batholith, 122
Granite, erosion level, 332
Maingay's Island, 211
Malacca, 3, 5, 6, 8, 10T, 11, 13, 27, 29T, 50, 53, 58, 156, 215, 216F, 363, 399, 400
and Negri Sembilan, 398
Strait, 1, 13, 14F, 22F, 23T, 152
Malay Peninsula, 1, 2F, 13, 23, 397, 400, 401
Malaya, 400
Axial, Mesozoic, 107T
East, 92T
east-geological map, 85F
Central, 92T
central geological map, 79F
geology, 397
North-West, 10T, 30F, 57F, 92T
northwest, geological map, 64F
northwest-Mesozoic, 107T
prehistoric, 401
south-Mesozoic, 107T
malayaite, *353F*, 354F, 365, 394, 395
alteration, 353
fluorescence, 353, 354F
optical properties, 353
replacement by scheelite, 356
unrecognized, 355
Malayan Collieries, 146, 150
mineral granis, 393
ore-deposits, 400
Malayan Orogen, 287, 301
Malayaspira sp., 36
rugosa KOBAYASHI, 35, 38F, 39
Malaysia Waterfall, 180
Malim Nawar, 74F, 76, 290
Mammals, wild, 398
Mandai Quarry Singapore, 234T, 248T, 244F
Road, 244F
Village, 244F
Mandalay — Kunlon Ferry, 393F
manganese, 336
deposits, distribution, **339F**
Manek Vrai, 321F
Manson Lode (Kelantan), 336, 358, 368, 369F, 374, **378-379**, 378F, 386, 390
mineral paragenesis, 379T
ore spectroscopic analysis, 379T
Manson ore body, 388, 395
Mantle, 333F
altered, 331, 333F
slices, 334
maps, ancient, 22
general geological, 6
historical, 20
mineral distribution, 6
Maran, 99F, 123, 125, 127, 195, 196F, 197, 383
Kuantan road, 83, 125F
lead ore, **383**

paragenesis, 383
lead prospect, 382
Maran Lode, Pahang, 384F
Maras-Jerong granite, 104
Marattia ceae, 149
marble, 34, 39, 47, 49, 83, 162F, 257F, 263, 267, 268F, 276F, 288, 290, 292F, 293, 295, 297, 300, 300F
argillaceous, 47
carbonaceous, 47
chloritic, 281
crystalline, 49
epidote, 281
foliated, 265
garnet, 289
grain size, 290
mineralogy, 289
phlogopite, 268F, 289
saccharoidal, 296, 297
tremolite, 289, 297
Upper Silurian, 3
Marcasite, 161, 369F, 372, 383, 384F
Marginifera, 68
Marginirugus, 65
marine clastic deposits, 23T
clay, 173
deposits, 173
environment, 83-84
erosion notch, 19F
transgressions, 55
MARKS, P., 105, 398
marl, sandy, 50
MARRIOTT, G., 6, 398
martite, 347, 350
alteration to, 349
Mata Ayer, 67F
Matan Complex, 303
Matchan Satahun, 339F, 347, 349
matrix, andesitic, 197
cryptocrystalline, 182, 195
felsitic, 194
fine grained, 182
glassy, 190, 199
limestone, 197
quartzo-feldspathic, 200
maturity index, 139
mawsonite?, 367
Mawson Series, 40, 46
Medan, 93F
Mediterranean area, 23
medusoid impressions, 118
MEDWAY, LORD, 171, 398
Meekospira sp., 77F
Megalomphala? sp., 43
MEISTER, F. H., 393
Mekong Delta, 2F
Melor Syndicate Mine, Sungei Way, Selangor, 354, 359, 374
Menam, 211
Mengkarak, 154
Menglembu, Perak, 74F, 290, 346
Menglembu-Kledang area, 290
Lode Mining Co. Ltd., 372
Mentakab, 111, 112, 145F, 222
Merapoh, Pahang, 7, 102, 102T, 315

Pahang, fossils, 100F
area, 281
structural map, 316F
Merbok, 282F, 285, 286F
Reservoir, 223
Merchang, 22
mercury, 336, 378
meretrix, 173
Mergui, 212F, 306F
center, 306
islands, 211
Series, 92, 211
Mersing, 62F, 185, 201, 298
area, 393
district, 87
mesas, 134
mesolithic, age, 175
Mesozoic, 6, 86, 97-141, 177
fossils, 400
orogeny, 391
sedimentary rocks, 399
synthesis, **137-141**
rocks, 307F
distribution, **99F**
volcanism, 178F
mesozone, 11, 215, 253, 254, 255F, 281, 302, 331, 333F, 387
-catazone transition, 230
definition, 253
metamorphic rocks, **287-300**
mesozonal metamorphism, 302, 303
mesozone metamorphism, East Johore, 298
East Pahang, **298**
Kinta Valley, **290-293**
Kuala Lumpur area, **294-295**
Northeast Malaya, **297-298**
North Pahang, **296-297**
Northwest Malaya, **288-289**
Pahang east of Kuala Lumpur, **295**
Raub area of Pahang, **295-296**
Selangor, **293-295**
South Johore, **298-299**
Upper Paleozoic and Triassic **296**
mesozonal granite batholiths, 254
meta-agglomerate, 201
meta-andesite, 201, 211, 298
mineralogy, 211
meta-argillite, 41
metabasite, **269**, 273F, 275, 276F, **298-299**, 344
brecciation, 274
metacoceras sp., 77F, 78
meta-conglomerate, 245
metadiorite, 243
metadolerite, 245
metagabbro, 251, 299
metagrit, 283
metal zones, Kinta Valley, 358F
metalliferous formation, 393
metallogenetic province, 360
metallogenic epochs, 392
metamorphic belts, 398
complex, 316F
conditions, 261
facies, 253, 394

facies series, **285**, 331
formations, 320F
grade, 53, 255F
index minerals, **285**, 286F
infrastructure, 279
intensity zones, 253
isograds, 285
petrology, 402
rocks, 5, 27
  catazonal, 307F
  chemical analyses, 258T
  high grade, 79F, 85F
  temperature, 301
metamorphism, 33, 200, 253-303, 394
  amphibolite facies, 230
  dynamothermal, 54, 191, 201, 302
  northwards increase, 114
  regional, 11, 33, 63, 84, 92T, 218, 281
  regional correlation, **302-303**
  retrogressive, 243, 279
  summary, **301-302**
  thermal, 254, 281
metapyroclastic, 180
metapyroxenite, 275
metaquartzite, 45, 268F, 282, 283, 284, 298
  foliated, 258T
  mineralogy, 284
meta-rhyolite, 201, 279, 281
metasediments, 10T, 84, 87, 201, 226F, 321F, 323F
  greenschist, 257F
  low grade, 303
  Paleozoic, 147F, 148F
metasomatic, 348
  replacement, 382
metasomatism, 229, 285, 288, 302, 346
  alkali, 229, 232
meta-trachyte, 201
meta-tuff, 283
metavolcanic rock, 101, 180, 182, 259, 267
mica, 89, 170, 180, 182, 356F
  schist, modal analysis, 256; *see also* Schist
michelina, 70, 83
micrite, 104
microcline, 47, 166, 180, 188, 189, 219, 261F, 266F, 278
  -microperthite, 222, 278
  -microperthite porphyroblasts, 271F
microdiorite, 243, 248T
microfelsitic, 192
microgranite, 180, 183, 207, 221, 224, 234T, 279, **280**
microgranite-leuco-, **269**, 270F
microgranite, muscovite, 237T
  porphyry, 211
  pyroxene, 183, 184, 249
  sill, 269
micrographic intergrowth, 222
microperthitic, 220
microkarst, 164
  topography, 164F
microparia, 49
microsparite, 102
microspherulitic texture, 183
microsyenite, leucocratic, **278**

Midai, 214
Middle Indosinian, 138
migmatite, 267
  complexes, 11, 10T
migmatite, venite, 274
migmatization, 104, 230, 254, 263, 275, 279
Milazzian, 24T
Mindel, 170
  glaciation, 168
Mines Department, 390
Mine levels, 361F
mines, open-cast, 172
mine pits, 147F
mineral deposits, 397, 401
  deposits, mineralogy, 362T
  paragenesis, **362T**
  primary, **335-390**
mineral lineation, 288, 308, 310
  paragenesis, 376
  paragenesis table, **362T**
  resources, Johore, 400
minerals, heavy, 163
  tin, **365-370**
  tungsten, **365-370**
mineralization, Cornwall, 394
  epigenetic, 346
  over granite cusps, 363
  impounding, 343
  in Malaya, 396, 400
  primary and igneous rocks, relationship, **387-388**
  relation to igneous rocks, 336, 341
mineralizing fluid migration, 357
  solutions migration, 378
mineralogical complexity, 336, 373
mineralogically simple veins, **371-373**
mineralogy, high temperature, 218
  optical, 402
minette, 247
mining exposures, 153
  gravel pump, 335
  Malayan, 393
  tin, 3
  in Trengganu, 392
MINTON, P., 110, 112, 396
Miocene, 143, 151
miogeanticlinal ridge, 98F, 121
miogeosynclinal basin, 26, 33, 40, 45, 41
  conditions, 45, 63
  furrow, 98F
  ridge, 97
  rocks, 29T
    basin facies, 28F, 30F
    shelf facies, 28F, 30F
  trough, 9, 35, 37, 58, 59
  zone, 97
miogeosyncline, 26, 37, 43, 104, 137
  northwest Malaya, 254
  Lower Paleozoic, **27-46**
missellina claudiae, 78
MITCHELL, A. H. G., 91, 398
MIYASHIRO, A., 245, 259, 260, 266, 285, 288, 299, 302, 398
mobilization, marginal, 251

modiolopsis gonoides, 110
  gonoides HEALEY, 106
modiolus, 106
  cf. *nachamensis*, 110
Magok Gneiss, 56
MOHAMMAD bin AYOB, 124F, 125, 127, 153, 165F, 165, 166, 167, 168, 318, 394, 398
MOHR, E. C. J., 13, 398
molasse, 119, 137, 138, 333F
  facies, 111, 122, 123, 127
  formations, 334
  oxidised, 134, 138
molasse type sedimentation, 11
MOLENGRAFF, G. A. F., 24, 141, 398
molluscs, 63, 68
  marine, 172
molybdenite, 278, 344, 345, 373
Monastirian, 24T
monazite, 6, 149, 160, 166, 171, 172, 227, 344
  alluvial, 236T
  sand, 236T
Mongolia, 55
Mongolian Sea, 59
monocarpia marginalis (*Annonaceae*), 149
monoclimacis flumendosae (GORTANI), 42
  vomerina (NICHOLSON), 46
monocline, 318
monograptus distans (PORTLOCK), 45
  dubius (SUESS), 46
  formosus BOUČEK, 45
  aff. *flemingii* (SALTER), 42
  aff. *flexilis* ELLES, 42
  hemiodon JAEGER, 42
  hercynicus, 43,
  hercynicus group, 42, 45, 46
  hercynicus PERNER, 53
  langgunensis C. R. JONES, 45
  marri PERNER, 42
  praecedens BOUČEK, 49
  praehercynicus, 53
monograptus cf. *praehercynicus* JAEGER, 53
  cf. *riccartonensis* LAPWORTH, 46
  sedgwicki, 45
monograptus sedgwickii (PORTLOCK), 42, 45
  thomasi JAEGER, 42
  cf. *uniformis* PRIBYL, 45
monograptus yukonensis JACKSON and LENZ, 42
montmorillonite, 131, 200
monzonite, **269**, 271F, 272T, 273, 273F, 274, 275F, 278
  coarse grained, 270F
  quartz-biotite-hornblende, 278
MORGAN, B. E., 127
MORGAN, M. J. De, 3, 398
morphologic stage, mature, 16
morphometric analysis, 393
Morse Road, 110, 111
mortar structure, 222, 259
Moscovian, 94

mossite-tapiolite, 371F
Moulmein, 93F
  limestone, 92
mountain-building, Mesozoic, 60
Mount Faber, 110, 111
  Guthrie, 109, 110, 111
  Ophir, 4, 217F, 255F, 301, 337F, 387
  Ophir granite, 10T, 217, 326
  Ophir quarry, Johore, 220T, 237T
mountain ranges, 215
movement, counter-clockwise, 328
  post-early Cretaceous, 328
Muar River, 16, 247
mud, 162F, 165F
  organic, 160
  sandy, 165F
Muda, Dam area, Kedah, 398
Muda River, 19
mudstone, 27, 30, 31, 46, 66F, 68, 87, 108, 116, 120, 124F, 135, 196F
  basal red, 65F
  bentonitic, 131
  black, 63
  carbonaceous, 43
  conglomeratic, 45
  graptolitic, 37
  lenses in sandstone, 89F
  Lower Devonian, 49
  pebbly, 11, 65
  red, 125
  red bedded, 45, 63
  sandy, 148F
  siliceous, 41
MUIR-WOOD, H. M., 7, 61, 84, 398
MÜLLER, K. J., 8, 37, 75, 391
muntiacus, 167
Murai, 214
Murau Conglomerate, 107T, 125, 126, 127, 137, 235
  Conglomerate, bedded, 125F
  Conglomerate Member, 125, 127
murchisonia sp., 69F
murchisoniacea, 76
muscovite, 5, 52, 117, 154, 165, 220, 220T, 227, 256, 261F, 266F, 281, 286F
  absence, 277
  from pegmatite, 237T
  radiometric determination, 236-238T
  secondary, 229
  two generations, 256
museum, 4
Museum, British, 4, 6, 7, 150, 174
  Perak, 4
  Raffles, 174
  Sedgwick, 46
Myah, south, 380F
mylonite, 326, 330
  flaser, 318
  gneiss, 328
mylonitized granite, 328
mylonite, left-lateral displacement, 320
  zones, 318
mylonitization, 222, 301, 334
myoconcha sp., 110
myophoria, 102, 105, 110, 121, 317

biofacies, 106, 111
bittneri, 110
facies, 121
cf. *goldfussi,* 110
goldfussi lipisensis sp., 106
inaequicostata, 110
in Malaya, 397
myophoria newtoni KOBAYASHI and TAMURA, 100F
ornata, 110
Myophorian Sandstone, 105, 106, 122, 111
myophoriopsis sp., 110
cf. *carinata,* 110
myrmekite, 265, 269, 278
  growths, 218
myrtaceae, 149

nageiopsis sp., 132
Najang, Trengganu, 373
nakamuranaia, 133
Nam Loong beds, 75T, 76, 78
Namhsin Series, 50
Nami, Kedah-fossils, 100F
Nam Phong Formation, 138
Napeng Beds, Burma, 106, 108
nappe structures, 332
  Upper Djambi, 94
Na Suan, 46, 50
natrolite, 206
Natuna, *213*
  Archipelago, 174
  Zone, 306F
Naungkangyi Series, 50
  Stage, 40
nautiloids, 35, 37
  onocerid, 75
nautilus, 112
  pompilius, 401
  in Thailand, 394
Naxos, 292
near-surface deposition, 379
  environment, 387
nebulite, **263**, 269
  feldspar-rich, 272T, 273
nebulite, psammitic, 265
Negeri Sembilan, 10T, 6, 25, 27, 29T, 39, 50, 53, 54, 58, 62F, 177, 185, 229, 295, 326
neichiashan fauna, 35
nektoplanktonic, 112
nektoplanktonic habit, 110
Nenggiri Valley, 83
neoqene, **11**
neolithic midden, 175
neoschizodus laevigatus, 108
neoschizodus laevigatus elongatus PHILIPPI, 106
neoschwagerinids, 101
Neoschwagerina cheni SHENG, 77F
neosome, 273, 275F, 277
neotectonics, 392
nepheline, 153, 204, 206, 211, 213
  normative, 207
nepionic individuals, 118
Neritacea, 76

neritic, conditions, 108
  environment, 111
neuropteris, 86
New Jetty, Kuah, Pulau Langkawi, 220T 231T
New Zealand, 235
Newark type, 134
Newer Arenaceous Series, 105, 123
NEWELL, R. A., 153, 160, 161, 163, 164F, 398
NEWTON, R. B., 4, 6, 105, 106, 108, 109, 111, 110, 112, 118, 122, 68, 398
NGUEN VAN TIEN, 395
Nha Trang, 211
Nibong road, 256
Nieuwenhuis Mountains, 214
Nigeria, 381
Niggli molecular norms, **185-186**, 202T, 208T, 218, 231T, 232, 234T, 249, 260, 272T, 395
NIINO, H., 322, 393
NIKIFEROVA, O. I., 55, 59, 398
niobium, 365, 369
Niobium analyses, 344T
Niut, 214
  stock, 214
Niyor, 151
Niyor and Kepong beds, 10T
  Railway Station, 151
NOCKOLDS, S. R., 251, 398
NOETLING, F., 398
NOGAMI, Y., 70, 120, 395
nonmarine conditions, 133
Norian, 110, 119
Noring, 332F
norite, 243, 244F, 248T
  quartz, 243, 249
Norms, **248T**
norm, Niggli, 103T, 186T, 201
North America, 13, 36, 59
  Borneo Geosyncline, 2F
  Kelantan and north Trengganu, 398
  Vietnam magmatism, 395
northern Pahang, 398
  Territory, 398
  Territory of Australia, 56
northwest Malaya, **63-71**
  Malaya, geological map, 64F
  relationships between Upper Palaeozoic formation, 65F
NOSSIN, J. J., 9, 22, 23T, 173, 174, 398
notches, 16
  wave-cut, 19
Nowakia, 43, 50
  acuaria, 43
  acuaria (RICHTER), 43
  cf. *acuaria Richter,* 49
nuclide redistribution, 240
Nucula peelii, 106
Nucula perlonga, 106
Nyalas, 217F

OAKLEY, K. P., 154, 398
Obsidianities, 400
OBUT, A. M., 55, 59, 398

ochre, tungstic, 365
Ogygiocaris, 49
oil companies, 9
Oistodus, 35
Old Alluvium, 9, 16, 22, 24, 143, 145F, 158, 159F, **159-168**, 160, 161F, 163F, 164F, 165F, 166, 169F
  Alluvium, age **167-168**
Old Alluvium, mineralogy, 161
  Alluvium, origin, **167-168**
    rodaceous, 166F
    stratigraphic sections, 162F
older alluvium, 392
Older Arenaceous Series, 7, 8, 29T, 51, 59, 61
Olenellus, 55
oligoclase, 183
oligoclase rims, 269
olivine, 54, 204, 206, 244
  normative, 201, 207, 209
  relics, 245
ONG, S. S., 114, 115, 117, 118, 123, 124F, 126, 137, 398
ONG, Y. H., 352
opaque minerals, 220T
open-cast methods, 388
operculum, gastropod, 35
Ophiolite, 29T, **54-55**, 28F, 30F, 57F, 251, 333F
ophiolite bodies, 299
  masses, 334
ophiolitic rocks, 27, 251
  suite, 25, 50, 60
ophitic texture, 245, 247
OPIK, A. A., 56, 398
Orbiculoidea sinensis MANSUY, 53
Ordovician, 25, 29T, 41, 54, **58-59**, 72, 10T, 26F, 57F
  early, 32
  fossils, 397
  geosynclinal organization, **98F**
  Limestone, 44F
  Lower, 34, 35, 61, 181
  Upper, 34, 40
ore, alluvial, 3
  body, structural control, 348
  paragenesis, 356F
  telescoped, 356F
ore deposits, 398
ore mineral textures, 393
origin, fluvial, 160
ormoceras sp., 40
  langkawiense KOBAYASHI, 35, 36, 38F
ornamentation, ellipsoidal, 269
orogene, double-sided, 308
orogen, Malayan, 122, 330
orogene, Thai-Malayan, 25
orogenesis, Lower Triassic, 121
orogenic axial zone, 331
orogeny, Late Triassic, 138
  mesozoic, 219
  Palaeozoic, 392, 397
Orogeny, Rocky Mountain, 97
  Thai-Malayan, 97
  B Variscan, 396

orogenic activity, Lower Triassic, 104
  uplift, 25
orogeny, Late Triassic, **11**
orthoamphibolite, 256
Orthoceras Beds, 40, 46
orthoceratids, 53
orthoclase, 103, 180, 183, 188, 189, 216, 219, 166
  -albite-amphibole-epidote-chlorite rock, 280
  microlites, 188
  microperthite, 222
  normative, 207, 233F
orthoclase porphyry, **280**
orthogeosyncline, 137
Orthograptus vesiculosus (NICHOLSON), 43
orthopyroxene, 183, 275
orthoquartzite, 41, 131
ostracods, 63
ostrea, 173
otoliths, 167
Otozamites gagauensis (nov.), 132
outlier, faulted, 70
  tectonic, 18
overfolding, 125
Overseas Geological Surveys, 260
overthrusting, 332
Owenites carpenteri SMITH, 100F, 101
  koeneni HYATT and SMITH, 101
Ozarkodina, 35

pabsite, 353, 370
Pachycardia sp., 110
Pacific basin, 56
  folding, 306F
Padang, 93F
  Besar, 67, 145F
  Highlands, 212
Pagodia, 32, 56, 58
  thaiensis KOBAYASHI, 32
Pahang, 3, 8, 10T, 11, 16, 22, 26F, 28F, 25, 27, 29T, 50, 59, 61, 63, 51, 54, 80, 84, 87, 97, 109, 122, 123, 128, 134, 62F, 79F, 85F, 174, 177, 179T, 185, 207, 295
  Central, 80F, 83
  Consolidated Co. mine, Sungei Lembing, 335, 337F, 341, 357, 360, 361F, 379, 388, 386
  Consolidated Mines, sections, **380F**
  north, 80F
  north, structural map, 316F
  northeast, 17F
  River, 16, 111, 157
  south, Structural map, 324F
  structural map, 319F
  Upper Palaeozoic, **80-82**
  Volcanic Series, 4, 6, 7, 61, 81T, 177, 179T, 187, 401
  west, 30F, 57F
  west, Limestone, 39
Pahi, 321F
"paint gold," 382
Pak Lay, 211

Pak Pak Song, 365
Palaeocardita cf. *crenata,* 110
Palaeocardica singularis, 106
Palaeoloxodin namadicus, 167
Palaeomphalus giganteus KOBAYASHI, 35
Palaeoneilo, 106
Palaeoneilo curvirostris HEALY, 106
Paleogene, *11*
paleocurrent, 136, 164
paleodrainage, 164
Paleogeography, 27, **55-60**, 119, 256
  Late Cambrian, 56F
  Lower Palaeozoic, 56F
  Ordovician and Silurian, 56F
Paleolithic, 175
paleontologists, Japanese, 8
palaeontology, 6, 395, 400
  Laboratory, 136
paleosome, 273, 274, 275F, 277
Paleozoic, 333F
  Lower, 25-60, 51, 54, 177, 309F
  rocks, 319F
    folded, 307F
Paleozoic, upper, **61-95**, 177, 309F
  Upper-distribution, 62F
  Upper, map, 64F
  Upper-Southeast Asia distribution, 93F
paludal conditions, 139, 144, 152
Panderodus, 35
Panghsa-pye Series, 46, 50
panned concentrates, 389
Pantai, 90
Panti Formations, 235
  -Lesong area, 139
  sandstone formation, 134
PANTON, W. P., 156, 398
Papan, 291, 74F
  quarry, Perak, 220T, 238T
Papulut Forest, 48
  quartzite, 29T, 48, 56
Paraceratites sp., 101, 112
  cf. *Frecherites?* sp., 101
  trinodosus, 112
  cf. *trinodosus,* 101, 106
  trinodosus (MOJS), 100F, 101
parafusulina, 86, 101
paragenesis, 360
paragenetic sequence, 349
paralic environment, 127, 138
Parallelodon sp., 78, 106, 77F
PARAMANANTHAN S., 217F, 218, 284, 398
Paranannites aspensis HYATT and SMITH, 100F, 101
Paranowakia, 45
Paratrachyceras? sp., 114
  regledanus (MOJS), 100F
pargasite, 278, 297
Parit, 49
PARK, C. F., 113, 358, 378, 398
PASCOE, SIR E. H., 40, 46, 50, 55, 105, 209, 211, 212F
Pasir Dula, 247
Pasir Mas, 230
Pasir Panjang conglomerate, 111

-Panjang-Jurong area of Singapore, 393
Pajang formation, **108-109**
Pajang member, **109-111**, 109, 111, 126, 127, 107T, 137
Pasir, Siur, 242
Pasoh gold mine, 185
passage beds, 30, 67F
Patalung, 92, 93F
PATON, J. R., 8, 18, 25, 104, 122, 128, 190, 218, 235, 246, 254, 305, 393, 398
"Paton conglomerate," 128
Paya Besar, 195
Peat, 23T, 159, 160, 162F, 165, 172, 173
peat and wood age, 394
pebble beds, 136
  clay, 149
  granite, 60, 65
  lithology, 65
  spreads, 65, 66F, 91
Pecopteris beds, 10T
Pecten, 110, 118
pegmatite, 222, **225-227**, 234T, 238T, 263, 278, 282F, 341, **343-346**, 344
  -aplite bodies, 343
  -aplite minerals, 362T
  beryl-bearing, 346
  chemical analysis, 227
  definition, 344
  dyke, 226
  hornblende, 221
  mica-tourmaline, 227
  mineralogy, 225-226, 227, 344, 345
  muscovite-columbite, 236T
  muscovite-tourmaline, 236T
  quartz, hornblende, 252
  unzoned, 343
  veins, 352
pegmatitic phase, 242, 285
Pekan, 145F
pelecypods, Triassic, 397, 401
Pelepah Kanan, Johore, 338F, 339F, 334, 347, 349, 364F, 365, 366, 367T, 374, 392, 396
Pelepah kanan, paragenetic sequence, 351T
Pelepah Kanan tin mine, 298, 353, 399
Pelepah Kiri, 347
Pelourdea cf. *megaphylla* (PHILLIPS), 132
Penang, 1, 20, 40, 155F, 216F, 247, 395, 401
  Island, 19, 157, 236T
  state, 41
Penanti Penang mainland, 236T
peneplain, Cretaceous, 156
  drowned, 24
  Neogene, 157
Pengerang, 178F
  area, southeast Johore, 394
Peninsular Divide, 14F
Penkalen, 373
Penoh, 15
Penrissen area, West Sarawak, 401
PENROSE, R. A. F., 3, 399
Perak, 3, 4, 6, 8, 24, 25, 26, 27, 29T, 31, 37, 26F, 28F, 30F, 31F, 41, 42, 46, 50, 54, 55, 57F, 58, 62F, 114, 115, 177, 179T, 182, 402
  geology, 398, 401
  lower, 10T
  mines, 393
Perak River (Dindings), 19, 20, 54, 157, 172, 397
  migration, 22F
Perak River valley, 154
periclase, 295
peridotite, 50, 54
  and serpentinite, 398
perknites, 267, 269, 277
Perlis, 6, 8, 19, 20F, 20, 23T, 25, 26, 27, 29T, 32, 33, 35, 36, 40, 61, 62F, 63, 64F, 65, 67, 68, 70, 94, 144, 171, 172, 173, 308, 400
  north, 67F, 65F
  north Kedah and the Langkawi Islands, 396
permatang, 173, 174
Permian, 6, 9, 10T, 11, 49, 61, 71, 72, 75, 76-78, 92T, 65F, 80F, 102, 187, 216F, 239F, 300F
  fossils, **77F**
  Japan, 78
  Lower, 65, 68
  Middle, 63, 101
  road section, Jengka Pass, *83F*
Permocarboniferous formation, 399
Pernerograptus revolutus (KURCH), 45
Pernerograptus revolutus (TÖRNQUIST), 38F, 39
Pernerograptus revolutus austerus (TÖRNQUIST), 42
Pertang, 185
  Road, Negeri Sembilan, 242
perthite, 180
Petaling, 87, 88F
Petaling Jaya, Selangor, 15F, 87, 88F, 90, 311F
  -Salak South, 402
Petalolithus tenuis (BARRANDE), 42, 45
Petchabun, 93F, 94
PETTIJOHN, F. J., 40, 58, 111, 119, 134, 399
PFEIFFER, D., 19, 393
PHAN NA, 397
pharmacosiderite, 386
phenoclasts, 52
  elongated, 52
  granite, 123
phenocrysts, 219-220
  alignment, 219F
  angular, 183
  corroded, 198, 280, 281
  feldspathoid, 207
  flow alignment, 219
  embayed, 198
  feldspar, 183
  pyroxene, 184, 191
  quartz, 194, 198
  relic, 283
Phet Buri, 302
Philippine, 174
Phin Soon Mine, 291
phlogopite, 265, 290
photogeologic studies, 389
Photomicrographs, 181F, 193F, 203F, 223F, 225F, 264F, 271F, 283F, 349F, 350F, 352F, 353F, 354F, 355F, 357F, 364F, 368F, 369F, 370F, 371F, 375F, 378F, 384F, 385F
Phra Wihan Formation, 138
Phricodothyris sh., 69F
Phu Kadung, 141
Phu Kadung Formation, 138
Phuket, 93F, 302, 241F
Phuket Group, Peninsula Thailand, 33, 91, 398F
Phuket Island, 33, 91
Phuket Series, 240, 402
Phu Phan Formation, 139
phyllite, 3, 33, 37, 43, 45, 48, 51, 53, 89, 94, 101, 183, 218, 245, 257F, 259, 262, 268F, 281, 283, 289, 292F, 293, 295, 297, 298, 303
  carbonaceous, 52
  chloritic, 280
  Gondwana, 5
  spotted, 289
physical geography, 400
phytoplankton, 40
picotite, 190, 245
picritic lava, 299
Pigeonite, 191, 213
Pilai river, 22
pillow-lava, 184
pillow texture, 199
pinching, 116
Pindaya Beds, 50, 210
  Formation, 40
pinite, 265
pinnacles, 18
PINSON, W. H. JR., 395
Pintu Gerbang, 298
Pinyok, Thailand, 353
pipe, 345
  deposit, 351
  deposits, Kinta Valley, 335
pipes at fault intersections, 352
Pipes, mineralogically complex, **374-379**
pipe, mineralogy, 345
  tin-bearing, 356F
pipelike pyrometasomatic deposits, 352
pisolitic clay, 291
pisolitic texture, 292
PITAKPAIVAN, K., 94, 342, 399
pitting, Test, 144
PIYASIN, S., 94, 401
Placers, 386
placer deposits, source of, 336
  tin deposits, 335
plagioclase, 47, 53, 180, 182, 183, 189, 197, 220T, 233F, 259, 352F
plagioclase exsolution lamellae, 220
plagioclase laths, 184
plagioclase modal, 221F
  normative, 273

optical properties, 259
saussuritized, 197
untwinned, 259
zoned, 220
zoning, relict, 275
plankton, Palaeozoic, 399
planktonic organisms, 26, 41
Planolites, 65
Planoshirina sp., 77F
plants, Carboniferous, 393
plant fossils, early Cretaceous, 8
fragments, 84, 148
plants, Mesozoic, 397
plants, modern, 149
Permian, 391, 397
remains, 121, 167
remains liquitized, 166
Upper Carboniferous, 63, 94
plastic deformation of marble, 352
plateau, remnant, 15
"Plateau Sandstone," 138, 139
Platyceratacea, 76
Platystrophia, 3
Plectodonta, 49
Plectospathodus, 35
Pleistocene, 1, 16, 22, 24, 155, 158, 143, 173, 207, 402
age, 151, 167
geology, 9, 393
mammals, 394
Middle, 167
-Recent boundary, 396
pleochroic haloes, 265
pleonaste, 161, 290, 291
Pleuromya sp., 110
Pleurotomariacea, 76
Pleurophorus elongatus?, 106
Pliocene, 24, 143
plutonic activity, 215-252
rocks, 79F
plutonic rocks, petrogenesis, 249-252
plutonism, 8, 9, 399
Pre-Carboniferous, 310
synorogenic, 54
Podocarpus imbricatus type, 167
Podozamites cf. *lanceolatus,* 110
Poh Hin quarry, Singapore, 244, 299, 248T
Pohra, 292
poikiloblastic crystals, 258
Pokok Sena, 64F
Poleumita cf. *discors* (SOWERBY), 37
scamnata CLARKE and RUEDEMANN, 37
poljes, 15
pollen, 149
Polyalthia, 149
polyzoa, fenestellid, 68, 82
Pong Onn, Ampang, Selangor, 34T
porcellanite, 116
porphyritic texture, 222
porphyry, 180, 225, 283
hornblende, 249
mineralogy, 283
porphyroblasts, 232
Port Swettenham, 172, 145F

Portugal, 336
Posidonia sp., 68, 69F, 105, 106, 109, 110, 112, 114, 118, 121
biofacies, 106
(Bivalvia), 396
cf. *japonica,* 112
cf. *kedahensis,* 114
cf. *wengensis,* 112
postdepositional changes, 89
postgeosynclinal phase, 134
postkinematic intrusions, 230
posttectonic, 217
potassic ankaramite, 208T, 209
potassium metasomatism, 274, 275
permeation, 273
pot hole, 165F, 270F
pottery, 175
Prachuab Province, 303
Prai, 172
Precambrian age, 3
Precambrian landmass, 334
Pre-Tertiary stratigraphic succession, 391
PREECE *ET. AL,* 330, 399
prehnite, 291
Prenkites? sp., 101
PRICE, N. J., 399
primary deposits, weathering of, 385-386
iron deposits, 381
mineralization, regional distribution, 336-341
PRIOR, G. T., 174, 399
Pristiograptus argutus (LAPWORTH), 45
concinnus (LAPWORTH), 43
cyphus (LAPWORTH), 43
gregarious (LAPWORTH), 42
meneghini (GORTANI), 42
Pristiograptus nudus (LAPWORTH), 42
regularis (LAPWORTH), 42
regularis (TÖRNQUIST), 45
sandersoni (LAPWORTH), 45
Proboliolina, 83
PROCTER, W. D., 106, 112, 228, 318, 322, 326F, 391, 399
proetid trilobite, 397
progradation, 174
progress reports, 396, 399
Prolaria sp., 110
Promathildia colon, 110
prospecting, 388-390
Prosphingites austini HYATT and SMITH, 101
Prospondylus comtus, 110
protoquartzite, 31, 48, 41, 47, 52, 116, 131
Province, Atlantic, 55, 59
Australo-Asian, 55, 58, 59
Pacific, 55, 59
Wellesley, South Kedah, 157, 173, 174, 393
West Malayan, 114-121
PRYOR, E. J., 366T, 399
psammite, 269
garnet-biotite-muscovite, 266
Pseudodoliolina, 68
Pseudofusulina gobbetti, 77F, 160
krafti, 76

Pseudomonotis (Claraia) aurita HAUER, 101
pseudomorphs, 191, 194, 197, 298
Pseudosageceras multilobatum NOETLING, 101
Pseudoschwagerina, 94
Pseudozygopleura sp., 77F
"*Pteria,*" 110
Pteroperma malayensis NEWTON, 100F
Ptilophyllum sp., 127
cf. *pterophylloides* (YOKOHAMA), 132F, 136
Ptychites sp., 112
cf. *rectangularis* (KRAUS), 100F
ptygmatic folding, 263, 265
Puchong, 366
Pudu, 311F
Ulu Mine I, Selangor, 354
Pugnax asiaticus MUIR-WOOD, 69F
Pulai, Kinta, 290, 366T
Pulau Attap, Kinta Valley, Perak, 236T
Ayer Chawan, 110
Beras Basah, 64F, 65, 66F
Besar, 363
Pulau Bidan, 37, 31F
Bidan limestone, 29T, 283
Bintan, 138
Bumbon Besar, 289
Bunting, 178F, 180, 393
Dangli, 312
Dayang Bunting, 64F, 66F, 68, 309F, 312, 313F
Jong, Langkawi Islands, 66F, 70, 73F, 399
Kentut Besar, 66F
Kueh Besar, 65
Kuning, 68
Langgun, 33, 35, 43, *43, 45,* 45, 63, 65, 68, 288, 309F
Langgun, geological map, *44F*
Melaju Series, 213
Nanas, 103, 200, 232, 249
Pangkor, 145F
Panjang, 68
Perak, 288
Rebak, Langkawi, 19, 394
Rebak, Besar, 63, 66F, 309F
Rebak Kechil, 63
Rhu, 125
Senang, 111
Singa Besar, 63, 65, 67F, 70, 64F, 66F, 309F
Singa Kechil, 65, 66F, 70
Pulau Tanjong Dendang, 43
Tembus Dendang, 312
Tepor, 65, 66F, 309F
Terutau, 30, 31F, 32, 32F, 39, 56, 64F
Tiloi, 289
Timun, 35, 43, 45, 288, 308, 309F
Tioman, 8, 87, 185, 201, 255F, 298, 392
Tuba, 35, 45, 43, 65, 288, 289, 308, 309F
Ubin, 3, 223, 225F, 244, 249, 397
Ujol, 87, 185
Ular, 65
pull-apart texture, 348, 349F, 370

pumice, 154, 174, 207, 211
Punctospirifer pahargensis (MUIR-WOOD), 69F
Punjab, 55
PUSHKAR, P., 393
Pusing, 291
P. W. D. Quarry, Baling Town, Kedah, 34T
P. W. D. quarry, Singapore, 244F
pyrite, 37, 54, 117, 165, 166, 172, 180, 182, 183, 188, 191, 148, 161, 227, 280, 284, 293, 300, 345, 348, 335, 349F, 352F, 357F, 359, 372, 378, 384F
  cubes, 89
  framboidal, 161, 162, 365
  gold-bearing, 185
  oxidation, 347
  pyroclastic, 186T
  andesite, 11
  deposits, 207
  fragments, 197
  material, 58
  member, 52
  origin, 181
  rocks, 4, 25, 51, 54, 82, 78, 84, 86, 95, 177, 185
    petrology, **188-190, 192-194**
    correlation chart, 179T
    occurrence, 178F
pyrometasomatic (Skarn) deposits, **346-357**
pyrophyllite-quartz rock, 298
pyroxene, 54, 161, 230, 252, 259, 349
  granodiorite porphyry, 186T
  -microgranite, 183, 186T, 187
  optical properties, 191, 206
  phenocrysts, 183
  relic, 275, 299
  replacement, 297, 298
  rock, 290
  uralitized, 199
pyroxenite, 50, 54, 243, 248T, 249, 275, 277
pyrrhotite, 54, 191, 229, 284, 347, 348, 349F, 350F, 354F, 367, 369F, 370

quartz, 16, 47, 53, 103, 154, 160, 165, 180, 182, 183, 188, 189, 200, 220T, 227, 243, 256, 259, 352F, 353F, 357F, 359, 364F, 369F, 384F
  augen, 182
  -calcite-tourmaline veins, 249
  corroded, 103, 192, 195
  diorite, 213
  dolerite, 246
  dyke, 228F, 267, 322, 325F, 326, 328, 329, 330F
  gangue, 378F
  gold-bearing, 256, 316
  keratophyre, 213
  -mica rock, 227
  microcrystalline, 115
  modal, 221F
  monzonite, 248T
  normative, 201, 233F
  porphyry, 177, 180, 182, 183, 198, 207, 210F, 210, 213, 215, 234T, 279, **280**, 282F, 359, 381
  Dong, **198**
  Gali, **198**
  phyllitic, 182
  purple, 183
  schistose, 74F
  pure, 229
  reefs, **228-229**
  ridges, **228-229**
  rounded, 182
  secondary, 13, 229
  -sericite rock, 185
  -sericite zones, 373
  smoky, 180
  stringers, 43
  stringers, gold-bearing, 381
  -topaz rock, 227
  -tourmaline veins, 227
  veins, 129, 229F, **280**, 322, 372
    mineralogy, 372
  veining, multiple, 228
quartzite, 7, 30, 46, 50, 51, 54, 148F, 205F, 293, 314F, 389F
  micaceous, 51, 284
  Triassic, 7
Quaternary, 9, 12, 13, 23, 143, 159, 177, 204, 401
  alluvial deposits, **153-175**, 400
  sediments, distribution, 145F
  shorelines, 394
  volcanic ash, 401
  volcanic rocks, 387
  volcanism, 178F

radial drainage, 390
radioactivity, alpha electron trapping, 299
radiocarbon dates, 400
radiogenic argon diffusion, 395
  argon retention, 238
radiolaria, 42, 43, 119
radiolarian tests, 115
radiolarite, 58
  black, 53
  carbonaceous, 41
radiometric ages, 397
  age determination, 391
  age frequency, 239, 239F
  dating, 11, 400
  maxima, 235
rafesquina komalarjuni HAMADA, 38F
Raffles Museum, 174
RAGUIN, E., 219, 399
Rahman Hydraulic mine, Klian Intan Perak, 4, 335, 363, 372, 394
railway, East Coast, 101
rains, torrential, 13
RAJAH, S. S., 134, 136, 200, 305, 378, 397, 399
RAKPRATUM, C., 397
RAMDOHR, P., 367, 368, 399
Ranau, 214
ranges, granitic, 15
  mountain, 13
Rangoon, 93F

Rantau Panjang, 146
  Panjang coal measures, 400
  Panjang Forest Reserve, 146
rapakiwi texture, 269, 274, 277
Rapat beds, 162F
rapids, 15, 270F
Rasa, 293, 294
  area, Selangor, 330F
RASTALL, R. H., 6, 158, 159, 160, 166, 399
rastrites hybridus LAPWORTH, 42
  maximus CARRUTHERS, 42
Rat Buri, 93F, 94, 138, 302
  Buri limestone, 91, 94, 303
Raub, 7, 54, 79F, 99F, 154, 167, 207, 227, 242, 245, 269, 335, 340F, 381, 399
  area, 293
  Australian Gold Mine, 227, 317, 317F, 336
  Australian Gold Mining Company, 381
Raub Gold Mine, paragenetic sequence, 382
  Group, 61
  Hole Mine, 382
  Series, 4, 8, 61, 105, 111
  shales, 6
Rawang, 182, 183, 293, 294, 329, 330F
  area, Selangor, 402
Rayong, 303
READ, H. H., 251, 302, 399
Recent, **11**
recrystallization, 82, 89
red beds, 127
  bed association, 134
Red Hills, 291, 372
redlichia, 55
Redlichia Gulf, 55
Redlichia Sea, 55
REED, F. R. C., 46, 92, 399
references, **391-402**
regional metamorphism, 277; *see also* Metamorphism
regional tectonic setting, **305-306**
regolith, 155
rejuvenation, 11
remanié assemblage, 49
RENWICK, A., 144, 146, 150, 151, 330, 331F, 399
replacement, crystallographic control, 375
  deposits, simple, **373**
  processes, 377
  textures, 372
residual deposits, 153, **155-157**
resin from forest trees, 167; *see also* Damar
resistivity, 363
reticulatia, 70
retiolites (pseudoretiolites) perlatus (NICHOLSON), 45
reworking, 111, 125
rhabdosome, graptolite, 42
Rhaetic, 106, 110, 138
rhaphidograptus törnquisti (ELLES and WOOD), 43
Rhenish succession, 53
rheomorphism, 274
rhinoceros, 167

INDEX

rhinoceros sondaicus (DESMAREST), 171, 394
Rhio Archipelago, 1
rhodea, 86
rhyodacite, 101, 135, 185, 191, 192, 200, 202T, 210F
   mineralogy, 200
   porphyritic, 203
   flow, 200
rhyolite, 10T, 11, 82, 87, 121, 177, 180, 183, 185, **187-188**, 190, **191-192**, 195, 200, 202T, 203, 209, 210F, 211
   aphanitic, 189
   ash, 153-155, 179T, 398, 400
     origin, **154-155**
     thickness, 154
   banded, **199**
   contemporaneous, 198
     felsitic, **199**
   flows, 210
   flow-banded, 201, 203F
   groundmass, 188
   mineralogy, 199, 211
   non-banded, **199**
   phenocrysts, 188
   porphyritic, 181F, 197
   porphyry, 213
   -pumice, 207
   schistose, 182
   spherulitic, 185, 187
   tuff, 178F, 180, 186T, **189**, 210F
     foliated, 178F, 179T, 186T, 187
     sheared, 181F
Rhythmite Member, **117-118**
   Semanggol Formation, 115, 117F
rhythmite sequence, 115
RICHARDSON, J. A., 7, 13, 51, 54, 61, 80, 81T, 82, 101, 102, 105, 111, 123, 154, 160, 167, 175, 177, 185, 187, 188, 198, 207, 225, 227, 230, 243, 245, 247, 267, 269, 272T, 274, 277, 279, 281, 293, 295, 296, 315, 316F, 317F, 381, 399
RICHTER, 113
ridge, geanticlinal, 27
   quartz, 16
   sandy, 20
   strike, 16
   submarine, 26
ridges, arcuate, 305
RIDLEY, H. N., 149
riebeckite, 206
RILEY, G., 358, 399
rim replacement, 374
Riouw Archipelago, 138
   -Lingga Archipelago, 111, 212-213, 303
ripple marks, 117, 121, 125, 130
ripples, 136
RISHWORTH, D. E. H., 53, 123, 126, 127, 128, 128F, 129, 130, 130F, 131, 133, 134, 139, 144, 146, 150, 151, 399
Riss glaciation, 168
Riss-Würm, 170
river alluvium, 173
   anticlinal, 16
   capture, 16
   deposits, 172
   mouth deflection, 20
   mouth displacement, 21F
   rejuvenation, 386
rivers, subterranean, 16
   tidal, meandering, 20
river terrace deposits, 145F
robsonoceras, 35
rocks, miogeosynclinal, **27-46**
ROE, F. W., 7, 23T, 24, 51, 54, 55, 122, 146, 147F, 148, 148F, 149, 174, 182, 184, 226, 227, 247, 293, 294, 305, 310, 315, 328, 329F, 330F, 345, 349, 399
Rompin, Pahang, 339F, 385
   district, Phang, 298
   River, 242
Ronphibun, 365
roof pendants, 50, 180, **182**, 254, 281, 303, 328
   pendants, Lower Paleozoic, 184
Rooiberg mines, South Africa, 343
Rotan Dahan, 291
ROUX-BRAHIS, J., 4, 399
Royal Dutch Shell Group, 149
rudaceous rocks, 108
rudite, 119, 129
RUEDEMANN, R., 113, 399
Rui Valley, 49
RUMBOLD, W., 4, 399
Rumpen River, 401
Russia, Asiatic, 59
Russian, 46
rutile, 149, 161, 172, 256, 259, 291, 344
   niobian, 344
RUTTEN, L. M. R., 13, 399

S-Planes, 308
Sadao, 144
Sagenopteris sp., 127
Saigon, 93F
Saiong beds, 107T, 118
SAKAGAMI, S., 8, 70, 92, 94, 399
Sakmarian, 76
Salak, Perak, 167
   South, 88F, 90, 235, 311F, 358, 359
Sallya sp., 77F
Sambo, 109
Samos, 292
San Benito County, California, 370
   Islands, Baja California, 393
sand, 146, 159, 162F, 165F
   feldspathic, 144, 171
   grains, 20
   muddy, 165
   unconsolidated, 168
   weathered, 166
   white, 173
sandstone, 4, 16, 26, 32, 30, 50, 66F, 86, 84, 87, 109, 124F, 148, 148F
   arkosic, 124F, 138
   brown, 70F
   current-bedded, 31
   deltaic, 27, 55
   facies, 85F
   feldspathic, 68, 125
   flaggy, 134, 135
   immature, 106, 108
   lithic, 63, 129, 273
   pebbly, 31
   quartz, 63, 125
   red, 63
   tuffaceous, 53, 83, 83F, 125, **190**, 196F, 197
   weathered, 106
sanidine, 199
SANTOKH SINGH D, 190, 262, 263, 349, 353, 366, 367, 378, 381, 385, 392, 393, 399
Sao Khua Formation, 138
Saravanamuthu Pillai Estate, 195
Sarawak, 56, 91, 94, 93F, 212F, 213, 214
   igneous rocks, 396
Satak Valley, 82
satellitic basis bodies, 251
SATO, J., 8, 101, 110, 112, 399, 401
Satun, 25, 39
Saukiella, 32
Saukiella terutaoensis KOBAYASHI, 32
SAURIN, E., 105, 397, 400
saussurite, 193, 194
saussuritization, 247
SAVAGE, H. E. F., 6, 7, 49, 75, 108, 167, 177, 182, 227, 229, 254, 399, 400
saxonite, 54
scapolite, 289, 290, 352, 360
scarp, 131
   retreat, 141
scheelite, 161, 229, 335, 347, 348, 349F, 353, 355F, 356, 365, **369-370**, 371F, 382
   fluorescence, 370
   -fluorite, 359, 369
   Mine, Kramat Pulai, 402
schist, 6, 37, 45, 51, 52, 54, 162F, 245, 263, 268F, 282F, **283-284**, 290, 295, 359F
   actinolite, 51, 54, 55, 294, 295
   -diopside-quartz, 295
   amphibole, 33, 51, 53, 54, 55, 182, 291, 293-294, 296
   andalusite, 53, 291, 293
   -muscovite-biotite, 259
   biotite, 53
   -muscovite, **263**
   calcite-quartz-muscovite-chlorite, 281
   carbonaceous, 37, 45, 291
   chlorite, 53, 295, 298
   chloritoid-andalusite, 281, 297
   cordierite-bearing, 43
   crystalline, 303
   dark green, 272T
   hornblende-biotite, 270F
   diopside, 54
   epidote, 54
   -actinolite-quartz, 295
   -albite-chlorite, 258T
   -chlorite-actinolite, 54
   garnet, 51, 284

garnet-andalusite-hornblende-quartz, 296
   -bearing, 33, 53
    -mica, 236T, 256, 258T, 294
  -granite contact, 359
graphite, 51, 53, 54, 291, 293, 295
hornblende, 54, **263**, 284
   -augite, 183
   -biotite-(epidote)-andesine-orthoclase, 269
member, 52
mica, 108, 256, 261F, 283, 293, 295, 296, 297, 298
mineralogy, 263, 284, 291, 294, 295, 296, 297
muscovite-bearing, 53
origin of, 296
pelitic, 29T, 45, 53, 72, 74F, 76, 254, 276F
psammitic, 29T, 45, 53, 59, 74F
pyroxene, 256, 259, 291
   -epidote, 291
quartz, 51, 53, 256, 283
   -amphibole-garnet, 256
   -biotite, 43, 53, 289
   -chlorite-actinolite, 54
   -chlorite-mica, 280
   -chlorite-sericite, 53
   -mica, 54, 245, 256, 258T, 283F, 291, 294, 297
   -mica-garnet, 256
   -mica, lateritized, 3
   -muscovite, 53
radiolarian graphitic, 53
relics, 256
sericite-chlorite, 194
  -feldspar-quartz, 198
  -quartz, 297
Series, 7, 29T, 51, 53, 185, **295-296**
sillimanite, 53
  -biotite-garnet, 265
  -cordierite-garnet-biotite, 265
Silurian, 3
  and Devonian, 3
spotted, 283
tourmaline, 53, 284
weathered, 256
Schizodus sp., 77F, 78, 76
Schizophoria, 84
schlieren, 263, 265
schorl rock, 111, 365
SCHÜRMANN, H. M. E., 242, 400
Schwagerina cf. *granum-avenae*, 68
Schwagerina cf. *gumbeli* (DUNBAR and SKINNER), 77F
Schwaner Zone, 303
scintillation counter survey, 8, 390, 391
Scolopodus, 35
scoriaceous rhyolite, 181
scorodite, 284, 386
Scotland, 260
Scour channel, 164F
scree, 83F
SCRIVENOR, J. B., 4, 4-6, 5, 6, 13, 16, 23T, 50, 61, 54, 68, 72, 73, 90, 105, 109, 110, 111, 114, 115, 120, 122, 123, 129, 134, 149, 151, 153, 154, 155, 166, 156, 158, 160, 172, 173, 174, 175, 177, 144, 146, 181, 183, 184, 185, 195, 200, 207, 218, 219, 221, 225, 229, 232, 235, 242, 245, 247, 249, 267, 288, 290, 291, 292, 296, 298, 305, 335, 343, 345, 346, 352, 354, 372, 400

SCRIVENOR, retirement, 7
SCRUTTON, C. T., 49, 75, 401
Scythian, 81T, 82, 101, 102
sea arches, 19
  cliffs, 18
  level(s), 15, 386, 393, 400, 168
  level changes, 174
  levels, Daly, 24
  level changes, recent, 3
  levels, former, 9
  level fluctuations, 173, 176
  level, high, 160
  -levels, Japan, 393
  level lowering, 401
  -level notches, 18
  -level rise, Quaternary, 19
  levels, Quaternary, 15, 23T, 23, 23-24, 156, 176
  level stability, 157
  level stillstand, 170
  stacks, 19
  straits, 19
"seat-earths," 149
seaway narrow, 56
secondary iron deposits, 381
sedili volcanic rocks, **200**
sedimentary pullaparts, 117
sediments, deep-water, 27
  shallow-water, 9
  tuffaceous, 51
  unconsolidated, 172
sedimentary rocks, 332F, 399
Segamat, Johore, 143, 153, 99F, 178F, 206, 208T, 209, 216F, 217F, 218, 204, 242, 334, 341, 387, 388
  basalt, 145F, 179T, 254
  vesicle, 385F
  deposits, **383**
  volcanics area, 401
  volcanic mass, 382
  Kuala Lumpur, 39F, 311F
seismic traverses, 172
Selama, 41, 49, 332F
Selangor, 10T, 5, 7, 6, 20, 23T, 24, 30, 30F, 27, 29T, 25, 37, 45, 47, 50, 54, 55, 58, 63, 57F, 62F, 177, 179T, 399, 401
  Coal Measures, 146
  River, 172
Selat Pulau Peluru, 33
Selendar-Nyalas road, 237T
Selensing gold mine, 82
Selenoharpes, 49
Seletar, 249
Selibin, 290
Selinsing, 316
Semabang Trachyte Member, 213

Semambu Estate, 208T, 209
Semanggol Formation, 10T, 47, 99F, 110, 114, **114-120**, 141, 65F, 107T, 124F, 254, 314F
Semanggol formation, bedded chert, 116F
  formation, palaeontology, **118-119**
Semangko, 214
Semantan, 214
Semeling, Kedah, 236T
semi-schist, 283, 284
Semiling, 343, 381
Sempah Conglomerate, 310
Sempan road, 296
Senai, 242
Senai Estate, south Johore, 244, 299
Seng Kee Quarry, Singapore, 225F, 244F
Sentol quarry, 195
Sentul, 311F
sequence, alternating, 26
  structural, 80
Serau Valley, 81T, 82
Serberuang, 213
Serdang, Selangor, 367
Seremban, 4, 174, 229, 325F
Serendah, 182, 247, 294
Serian Volcanic Formation, 213
sericite, 117, 220T, 265, 295, 349, 364F
sericitization, 229, 360
"Series," 4
serpentine, 289, 290
  minerals, 54
serpentinite, 50, 54, 185, 190, 242, 248T, 249, 250, 250F, *245*, 256, 259, 334
  age, 243
  foliated, 250
  metamorphosed, 245
  sheared, 245
  yellow-green, 245
serpentinization, 243
serpulids, 116
Serpulites, 113, 118, 121
SERRA, C., 127, 400
SERVICE, H., 7, 101, 106, 400
Setapak, Selangor, 295, 311F, 395
  ore, mineralogical changes, 377F
Setul — Boundary Range, 33, 35, 40, 171
  Formation, 29T, 30, 10T, 32, 33-36, 36, 37, 46, 31F, 92T, 63, 44F, 65F, 254, 299, 313F
  thickness, 33-34, 43
  Formation fossils, 38F, 39
  Limestone, 34-35, 34T, 35F, 36F, 39, 43, 67F, 247
  bedded, 43, 45
Shale, 4, 11, 16, 46, 51, 53, 81T, 87, 109, 113, 83F, 124F, 148, 148F, 205F, 314F, 389F
  bituminous, 149, 150
  black, 83, 113, 83F
  carbonaceous, 112
  brown, 149
  calcareous, 46
  graphite, 317F
  carbonaceous, 4, 27, 46, 48, 41, 84, 109, 116, 150, 151

facies, 79F, 82F, 85F
fissile, 26, 43, 82, 86, 148
flaggy, 30, 33, 43, 41
Lower Devonian, 54
oil, 144
pellets, 125
phyllitic, 31, 76
pyritiferous, 75T
red, 33, 125, 128
replacement by iron-ore, 347
sandy, 124F
siliceous, 52
silty, 63, 66F
spotted, 47
tentaculites, 43, 46, 50
tuffaceous, 82, **189-190**
Upper Triassic, 393
shallow water, 111, 121
water conditions, 105
-water deposition, 84, 134, 138
-water environment, 136
Shan hills (Upper Burma), 398
Shan States of Burma, 9, 40, 46, 50, 59, 92, 93F, 209, 397, 399
States coalfields, 398
States, Northern, 40
Southern, 40
Shanghai Pahang Estate, 247
shards, glass, 155
shear(ing), 181, 301, 308, 328, 329F
shear directions, 330F
sheared quartz, 330
shear, second-order, 329
zone, mineralized, 378
zones, 330
shearing, axial plane, 315
Granite, 222, 329
radiometric age, 238
regional, 47
shelf, 27, 33
area, 97
continental, 26
Deposits, 27, **27-40**
Facies, 29T, 37, 42
Shelf limestone, **33-40**, 35, **36-40**
stable, 78
shells, aragonitic, 76
shonkinite, 206
sill, 206
shore progradation, 172
shore lines, correlation chart, 24T
shoreline, elevated, 23T
former, 173
Malaya, east coast, 24T
west coast, 24T
Monsoon control, 401
Shorelines, Quaternary, 24T, 401
Quaternary, Ages, 24T
Shoreline, Singapore, 24T
of submergence, 22
warped, 24T
shoshonite, 206
dykes, 206
SHU, Y. K., 45, 322, 325F, 326, 360, 400
sialic basement, 331

Siamese mine, 175
Siba formation, 262
siderite, 161, 172, 192, 364F, 365
Siegennian Stage, 9, 42, 53
Siegenian, lower, 42
SIEVEKING A de G., 154, 400
SIGNER, P., 400
Sik, Kedah, 49
siliceous rocks, 27
silification, 82
silicon variation, 250F
sillar, 194
sillimanite, 256, 259, 260, 265, 266F, 285, 286F, 301, 303
-cordierite-muscovite-almandine subfacies, 287
-orthoclase-almandine subfacies, 273
sills, 51, 243
andesite, 87, 185
boudinaged, 308
granitic, 71
-layering, 243
silt, 159
siltstone, 26, 30, 41, 63, 113, 129, 83F
black carbonaceous, 33
carbonaceous, 43, 48
flaggy, 66F
Micaceous, 32
red, 131
tuffaceous, 131
Silungkang Volcanic Series, 212
Silurian, 9, 10T, 25, 26F, 29T, 49, 53, 57F, 59-60, 72
limestone, 44F
Lower, 34, 45, 54
Upper, 30, 37
silver, 210, 358F
Simpang Ampat, Malacca, 237T
Sin Nam pipe, 352, 358
Sin Nam Lee pipe, 355
Sin Nam Lee Pipe mineralogy, 355
Sin Woh Leong Kongsi, Simpang Ampat, Menglembu, 374
Sing Seng Quarry, 244F
Singa Formation, 10T, 33, 44F, 45, 60, 63, **63-68**, 64F, 65F, 65, 66F, 67F, 71, 92T, 235, 240, 232, 289, 313F
Formation, cross-sections, 67F
geological map, 66F
thickness, 63
Singapore, 1, 3, 4, 6, 8, 10T, 15, 22, 108, 109, 122, 138, 56F, 145F, 168, **166-167**, 215, 216F, 218, 221, 222, 313, 395, 397
-fossils, 100F
geologic map, 244F
geology, 391, 397
Granite Quarry, 244F, 231T
Island, 13, 23T, 23, 160, 172, 223, 242, 247, 298, 400
Singapore-Malaya Company Mine, 162F
sedimentary rocks, 400
structure, 397
Singkep, 215, 242, 303
Sino-Burmese geosyncline, 401

Sinopora dendroides (YOH), 68, 70
Sinophyllum, 70
Sintok, 373
Siphonophyllia, 76
Siphonophyllia cf. *cylindrica,* 76
Siphonophylia aff. *gigantea,* 76
Siputeh, Kinta, 366T, 372
Sitiawan, Perak, 23T, 24
SIVAM, S. P., 153, 159, 160, 161, 162F, 163, 167, 168, 169, 170, 400
skarn, 288, 295, 302, 336, 347
bands, boudinaged, 354
deposits, **346-357**
other types, **356-357**
stanniferous types, **351-356**
diopside-garnet, 34
-tremolite-wollastonite, 34
ferriferous-stanniferous types, 347, 347-351
fluorite bearing, 356
minerals, 362T
mineralogy, 349, 353, 354
ore, 348, 354F
rocks, 5
stanniferous type, 347
vein, Bukit, Besi, 395
SKELHORN, R. R., 392
skewing, 163
Skudai, 242
slate, 45, 51, 52, 53, 94, 218, 289, 288, 293, 295
andalusite, 296, 298
spotted, 43, 294, 298
SLATER, D., 190, 193, 391, 392, 393
slickensides(ing), 153, 222, 329
slope break, 23T
mogote, 19
slump(ing), 24, 41, 46, 52, 59, 159, 161, 165, 167
slump bedding, 11
features, 119
slump folds, 67F, 89
folds, Carboniferous rocks, 86F
structures, 84
slumping, Old Alluvium, 161F
Semanggol formation, 117F
submarine, 65
SMILEY, C. J., 127, 400
SMITH, A. G., 394
SMITH, A. J., 122, 402
SMITH, C., 380F
SNELLING, N. J., 8, 78, 71, 87, 104, 121, 122, 141, 218, 235, 238, 240, 260, 301, 379, 395, 400
SNOW, A. B., 4, 400
SOEKENDAR, 396
soft sediment deformation, 89F
soils, 153, **155-156**, 143
profile on granite, 155F
samples, 389
sole marks, 117, 119
Survey, 155
solution, 18
of limestone bedrock, 167
pressure, 89, 135

INDEX

Songkhla, 25, 39
sorting, poor, 163, 165, 166, 169, 171
South China Sea, 1, 9, 13, 14F, 19, 22
South China Sea Basin, 2F
South East Asia geology, 401
South Johore, 299
   Mine, 349
   Perak, 400
   Vietnam, 211, 213
Southeast Asia, granite, 241F
   Asia, granite ages, 241F
     structural elements, 2F
      Upper Palaeozoic distribution, 93F
      volcanic rocks, 212F
   Asian Archipelago, 55
Specularite, 385
spessartite, **247, 248T**, 249
   enstatite, 249
Sphaerium, 133
sphalerite, 166, 289, 335, 336, 349F, 350F, 367, 368F, 374, 375F, 376, 377F, 378, 384F
   iron-rich, 378F
   relicts, 376
   replacement, by chalcopyrite, 376
     by stannite, 376
sphene, 47, 182, 220T, 265
   stanniferous, 353, 354F
   -tin, 353
Sphenolepis cf. Kurriana, 132
cf. *sphenophyllum,* 85
Sphenopteridium?, 86
spherasters, 154
spheroidal weathering, 206
spherules, mineralogy, 188
spherulites, 194
spicules, 154
spilite, 201
spilitic flows, 299
   pillow lava, 108, 109
spines, cidaroid, 78
spinel, 290
Spirifer, 67, 76
   scrivenori MUIR-WOOD, 69F
Spiriferina, 83
   cf. *fragilis,* 110
Spirograptus grobsdorfiensis (HEMANN), 49
   minor (BOUČEK), 42, 49
   spiralis (PERNER), 38F, 39
   spiralis contortus (PERNER), 42
spit-end, 23
Spondylus dubiosus, 110
sponges, 34
sponge spicules, 115
sporomorphs, 136
SRESTHAPUTRA, V., 392
SRI BANGUN, 348
St. Cassian beds, 108, 111
stagnant conditions, 40, 122
staining, sodium cobaltinitrite, 206
stalactites, 171
stalagmites, 171
stanniferous alluvium, 398
   deposits, 396

stannite, 353, 355, 365, 367, 368F, 369F, 374, 375F, 377F
   breakdown, 367
   in Malaya, 394
   replacing cassiterite, 356
STANTON, R. E., 342, 400
STAUFFER, P. H., 15F, 18F, 86F, **87-91**, 89F, 67F, 125F, 154, 155F, 156, 158F, 159F, 161F, 163F, 164F, 165F, 169F, 174, **143-176**, 322, 360, 395, 401
staurolite, 166, 260, 259, 261F, 301
STEENSMA, J. J. S., 400
Stenocisma, 76
Stenopareia sp., 43
STEPHEN, F. J., 3, 401
stereogram, 261, 285F
stereographic net, 267
stereoplasmoceras(?) sp., 36
STEVENS, R. E., 220, 392
STEWART, R. D., 138, 139, 392
stibnite, 382
Stigmaria, 84
stilbite, 301
stille, 306F
stockworks, 350, 371
STOK, J. P. VAN DER., 23, 401
stokesite, 353
stolzite, 161, 365
Stomatograptus cf. grandis (SUESS), 49
stone bracelets, 175
stong complex, conditions of metamorphism, **266-267**
   complex, geochemistry, **266**
     structure, 267
     petrology, **262-266**
     relation to overlying rocks, **267**
   Migmatite Complex, 101, 104, 191, 192, 230, 255F, 254, 257F, **262-267**, 264F, 265, 297, 331, 395
     map, 268F
     mineral paragenesis, 266F
Stope Mine, 382T
Straits of Johore, 232, 249
   of Malacca, 1
   Settlements, 399
   of Singapore, 397
Straparollus sp., 69F
stratigraphic code, 391
   nomenclature, 12
   nomenclature code, 12
Stratigraphic units, 27
stratigraphy, 8
streaky-bacon tin ore, 357F, 372
stream capture, 13
stream, deposition, 171
   net ratios, 393
   representations, 393
   sediments, 389
streams, consequent, 13
   homoclinal, 16
   north Perak, 332F
   subsequent, 13
   synclinal, 16
   trunk, 13

Streptograptus lobiferus (MCOY), 42
   nodifer (TÖRNQUIST), 38F, 39, 49
stress in the earth's crust, 396
strike control, 16
   regional, 329F
   ridges, 16
Stringocephalus, 76, 394
   perakenis, 75
Stringocephalus perakensis GOBBETT, 69F
Stromatoporoids, 75
strontium, 266
   concentration, 277
Strontium 87/86 ration radiometric determination, **236-238T**
structural discontinuity, 312
   Geology, **305-334**, 392, 394, 400
   map, 397
   outline, Malay Peninsula, 307F, **306-308**
   sequence, Upper Palaeozoic, 81T
   studies, 360
   zone, axial, 306, **312**, **315-317**
     Eastern, 306, **312-313**, 317
     Western, 306, **312**, **314-315**
   zones, 306
structure, 6, 8
structures, impounding, 348, 358
structure, Kenny Hill Formation, **90**
   Main Range, Granite, **328-330**
structures, penecontemporaneous, 89
structure, synclinal, 144
   Upper Palaeozoic and Triassic, 80F
struverite, 6
STUEBER, A. M., 393
sturia sansovini, 112
styliolina sp., 49
   fissurella (HALL), 43, 45
styliolinids, 43
stylolites, 32, 34, 35F
subaerial eruption, 206
subarkose, 131
subgreywacke, 30, 41, 43, 46, 47, 48, 51, 52, 113, 116
submarine ridge, 58
submerged character, 20
   regions, 21F
subvolcanic rock, 184
subsidence, rapid, 134
subsurface deposits, **172-173**
Suid, 167
SUJKOWSKI, S. L., 119, 401
Sukadana, 214
sulphide mineralization in vesicles, 383
   in pegmatite, 345
Sultan, 1
Sulu Basin, 2F
Sumatra, 1, 9, 56, 56F, 91, 93F, 94, 143, 154, 173, 212F, 212, 213-214, 241F, 242, 328, 332
   Highlands, 94
summit gradation plane, 16
   profile, 140F, 141
SUN, Y. C., 55, 58, 401
suncracks, fossil, 131
Sunda, 23
   Arc, 2F, 11

area, 305
   structure, 306F
   zonal structure, 306
Sundaland, 1, 176, 213, 214, 303
Sundaland, volcanism, 396
Sunda peneplain, 141
   region, 157
   Sea, 398
   Shelf, 24
   Shield, 97
Sung, 175
Sungei Anak Sungei Kalong, 280
Sungei Anali, Kelantan, 236T, 256, 258T, 259, 260
   Aring area, south Kelantan, 357
   Badong, 129
   Baru, Malacca, 237T
   Batang, 293
   Batu, 195, 242, 329
      Pahat, 21F, 223, 225, 282F
   Bebar, 21F
   Bentong, 247, 295
   Benus, 329
   Benut, 67F
      Valley, 67
   Bernam, Selangor, 21F, 23T, 24, 294
   Bertam, 256
   Beruang, 389F
   Besi, Selangor, 90, 165, 165F, 168, 398
      area, 392
      granite, 342
      Tin mines, Selangor, 165, 335, 341, 356, 359, 363, 373, 386, 398
Sungei Besut, 21F
Sungei Bilut, 197
   Bisek, 185, 187, 189
   Biwah, 86
   Bruas, 21F, 22F
   Bujang, 285
   Buloh, 183
   Buloh village, 22F
   Bunut, 289
   Chadu, 189, 281
   Chekua, 188, 279, 280
   Chembatu, Cheroh district, 242, 245, 248T, 249, 294, 296, 299
   Chenerai, 196F
   Cheriau, 111
   Cheriong, 269
   Cheroh, 242, 245, 296
   Chiku, 83, 79F, 114
   Chin Chin, 269
Sungei Chiniau, Pahang, 187, 189, 190, 202T, 281
   Choh, 182, 247, 291, 294
   Chuchoh, 67F
      Valley, 67
   Darah, 298
   Dindings, 21F, 22F
   Dong, 198, 327
   Dungun, 293
   Durian, 280
   Endau, 21F, 22
Sungei Enggor, 146
Sungei Galas, Kelantan, 194, 256, 257F, 258T, 259, 262, 268F, 321F
   Gali, 276F, 296
   Gapis, 294, 297
   Genil, 293
   Gok, 154, 277
   Golok, 21F
   Gombak, 247
   Gow, Pahang, 354F, 355, 367
   Guntong, 294
   Hau, Kelantan, 193F, 256, 259
   Henderik, 189
   Ima, 243
   Ingsor, 187
   Isa, 298
   Jelai, 106, 185, 187, 188, 189, 190
   Jelai Kechil, 297
   Jelawang, 265, 266, 268F
   Jempol, Pahang, 125, 126F
   Jengka, 83, 195, 196F
   Jeram, 126, 324F
   Jeri, 268F
   Jintiang, 297
   Johore, 21F, 167
   Jolong, 327
   Kachong, 226F
   Kadjau, 297
   Kali, 293
   Kanchang, 293
Sungei Kapor, 278
Sungei Karang, 208T, 209
   Kasai, 276F, 281
   Kasai Besar, 297
   Kasai Kechil, 188
   Kau, 189
   Kechau, 187
   Kedah, 21F
   Kelantan, 13, 14F, 21F
   Keloi, Gunong Benom foothills, 248T, 249
   Kelubi, 114
   Kemaman, 21F
   Kemasih, 21F, 22
   Kenerang, 262, 263, 264F, 265, 268F
   Kenik, Kelantan, 236T, 258T, 260
   Kenong, 247, 295
   Kepau, 245
   Kerling, 294
   Kerum, 17F
   Ketial, 17F
   Kinjai, 294
   Kiul, 294
   Kiul and Sungei Sangkau, Rasa, Selangor, 258T
   Klang, 21F, 184
   Klau, 269, 276F
   Klau, Kechil, Benom, 223F
   Krian, 21F
   Kroh, 74F
   Kuang, 293
   Kuantan, 21F, 22
Sungei Kudong, 293
Sungei Kuran, 298
   Lah, 188, 226F
      section, Chenderiong Tin Ltd., 352
   Langat, 21F, 183
   Lawin, 151, 219F
   Lebak, 114
   Lebir, 133, 192, 195, 224, 256, 318, 321F
   Lembing, area, 328
      faults, 327F
      lodes, 360
      lodes, mineralogy, 379
      mine, 361F
      mine, sections, **380F**
      mineral paragenesis, 379
      Pahang, 3, 7, 61, 62F, 69F, 84, 85F, 87, 328F, 335
Sungei Lempor, 294
   Lepar, 84, 85F
   Liang, 245
   Linggi, 21F
   Lipis, 157, 267, 269, 270F, 271F, 273F, 296
   Loh, 268F
   Lotong, 130
   Luit, 84, 85F
   Malim, 189
   Manau, 293
   Maran, 21F
   Mas, 242, 245
Sungei Memakah, 259
Sungei Mentiga, 324F
   Merah, 185, 294
   Merapoh, Pahang, 102T, 281
   Merbau, 297
   Merbok, 21F
   Merchang, 21F
   Mersing, 21F, 22
   Mesah, 190
   Muar, 14F, 21F
   Muda, 14F, 21F, 47, 114, 118
   Murai, 329
   Nak Sap, 186T, 187
   Naning, 186T
   Nenggiri, 297
   Neriap, 17F
   Ngiang, 276F, 296
   Nila, 263, 265, 268F
   Padang, 294
   Pahang, 13, 14F, 19, 21F, 196F, 324F
   Panau, 278
   Panching, 298, 347, 349
   Paka, 21F
   Panjang, 20, 21F
   Patani, 31F, 33, 37, 41, 284
      formation, 41, 232
      limestone, 29T, 31F, 37
   Payong, 200
   Penor, 21F
   Perah, 294
Sungei Perak, 13, 14F, 21F, 22F, 151, 146, 181
Sungei Perangin, 288
   Perenggan, 101
   Pergau, Kelantan, 193F, 257F
   Perlis, 21F
   Pertang, 130
   Pesasi, 185
   Piah, Lenggong area, 186T, 187, 258T

Plus, 207
Pohoi, 389F
Pontian, 21F, 175
Pulai, 21F
Rawa, 297
Reman, 195
Rening, 293
Rompin, 21F
Rong, 247
Rua, Besut, Trengganu, 248T, 249
Ruan, 275
   area, Raub, 276F, 396
Rui, 49
Sala, 21F
Sangkau, 294
Sapetang, 21F
Sat, 130, 202T, 203
Satak, Pahang, 317F
   Liang, 245
Sedili Besar, 21F
Sedulek, 259
Sungei Segamat, 242
Sungei Seladang, 263, 265, 268F
   Selangor, 21F
   Seli, 184
   Selor, 189
   Semeriong, 296
   Sempan, 294
     basin, 184
     reservoir, 295
   Semuliang, 263
   Senama, 294
   Senang, Trengganu, 236T
   Senggarang, 21F
   Sengonggoh, 245
   Serau, 185, 188
   Serembun, 188, 189, 190
   Sermi, 247
   Serunai, 281
   Setiu, 243
   Setul, 21F
   Sientor, 280
   Simat, 242
   Siput, Perak, 7, 49, 54, 177, 179T, 207, 209, 229, 400
Sungei Siput north, 74F
   Siput south, 74F, 224
   Siput tin field, 342
   Siyah, 259
   Soi, Kuantan, 22, 23T, 172
   Sok, Kedah, 220T, 223F, 236T, 321F
   Sokor, 247, 256
Sungei Som, 112
   Stong, 262, 265, 268F
   Taku, 256
   Tamu, 293
   Tebak, 134, 136
   Tehemang, 247
   Tekai, 17F, 123
   Telemong, 323F
   Telom, 242
   Tembeling, 17F, 79F, 123
   Temengor, 181
   Tempayan, 293
   Tempenis, 200

Temurai, 269
Tenang, 226F
Teping, 269, 276F
Terap, 205F
Termus, 17F
Teroi, 285
Timah, 281, 381
Tinggi, 187
Tras, Bentong area, Pahang, 258T
Trengganu, 21F, 323F
   Lata Sauk, Trengganu, 248T
   (Lata Terap), 202T
Sungei Trong, 21F
Tua, 112, 182, 294
   Estate, 294
Tui, 187, 190
Sungei Tuit Besar, 264F
Sungei Tulang, 204
   Udang, 293
   Ulu Liam, 329
     Sedili, 200
   Way, Selangor, 347, 366T
   Way area, Selangor, 402
   Yih, 125
   Yu, 188, 280, 281, 297
   Yu Halt, 189
   Yus, 294
SUNTHARALINGHAM, T., 8, 63, 72, 74F, 75, 75T, 326, 393, 401
superficial deposits, 143
supergene, hypogene clay, distinction, 363
   kaolinization, 363
suprastructure, 262
Surat Thani, 39, 302
surficial deposits, 173
SURINGA, R., 400
swamps, 127
   coastal, 20
swamp deposits, 172, 173
swamps, fresh-water, 22
swamp water, 18
swampy conditions, 137
SWANN, R. M. W., 3, 401
SWEATMAN, T. R., 353
Sweden, 277
Swee Construction quarry, 244F
swelling, 116
SWINNEY, A. J. G., 3, 401
syenite, 7, 250F, 267, 278
   augite, 5
   biotite, 278
   flow-banded, 269, 274
   hornblende, 278
   pyroxene, 278
   quartz-biotite-hornblende, 278
   -hornblende, 278
syenodiorite, 278
syncline, 47, 87, 88F, 90, 153, 196F, 308, 314, 315, 350
   axis, 311F, 319F
   Berantai, 322
   Mata Ayer, 67
synclinal trough, 48
   valley, 17F
synclinorium, 11, 73, 97, 317

syngenetic tin, 342
synkinematic emplacement, 60, 230
   pluton, 287
   stock, 285

Tabak formation, 254
Tahan coulisse, 123
   Range, 123, 126
tailings, mining, 162F
Taiping, 3, 49, 99F, 117, 236T, 332F
Tak, 302
Taku Schist, 10T, 16, 25, 56, 97, 104, 133, 230, 254-262, 255F, 318, 321F, 331, 333F
   Schist, chemistry, **260**
   distribution, 256
   map, 257F
   mineral paragenesis, 261F
   petrology, **256-259**
   radiometric age, **260**
   relation to overlying rocks, **262**
   structure, **261**
Talam quarry, Kampar, 236T
talc, 290, 356F
   analysis, 290
talus deposit, 13
   slope, 131
Tamangan iron mine, 256
Tambun, Kinta, Perak, 74F, 315, 339F, 366T, 381
   -Ampang ridge, 315
Tampanian, 154, 175
Tampin, 185, 217F, 242
   quarry, 231T
TAMURA, M., 80, 101, 106, 110, 397, 401
Tanah Merah, 194, 256, 257F
Tanglir-Benus valley, 15
Tanjong Atas, 298
   Balai, 249
   Batu Hitam, 205F
   Batu Pak Mok, 195
   Bulat, 199
   Chempadek, Kuantan, 205F, 246F
   Chinchin, 30
   Gelang, 21F
   Geliga, 298
   Ipoh quarry, Kuala Pilah road, Negri Sembilan, 237T
   Jaga, 223, 225, 282F, 284, 286F
   Jelutong, 249
   Kling village, Singapore, 202T
   Langgun, 44F
   Malim, 294, 329
   Murau, 123, 126
   Pasang, 199
   Peluru, 289
   Perak, 154
     Estate, 154, 207
   Punggai, 298, 322
   Rambutan, Perak, 74F, 145F, 151, 161F, 163F, 169F, 169, 290F, 291, 315, 358F
   Sepang, 199
   Tanjong Telin, 289
   Tembeling, Kuantan, 204, 206, 208T,

209, 245
Tokong, Penang, 15F
Tualang, Perak, 291, 368
tantalite-columbite, 371F
tantalum, 365, 369
  -niobium deposits, distribution, 337F
  -niobium minerals, 394
tanteuxenite, 344
Tanum Valley, 81T
Tapah, Perak, 7, 224, 363, 395
tapiolite, 367
  -mossite, 366
  -mossite exsolution bodies, 344
Tarim, 59
Tasek Bera, 14F, 16
  Bera diversion, 322
  Chini, 324F, 336, 385
Tasmania, 174
Tavoy, 211, 212F
Tavoy district, geology, 392
Tawar, Kedah, 116F
Tawng-peng, 209
TAY, 374
TAYLOR, D., 386, 401
TAYLOR, G. C. Jr., 392
teallite, 365, 368, 370F
Tebak formation, 10T, 16, 107T, 134-137, 135F, 139, 203, 326
  formation, age, 136
  fossils, 132F
  lithology, 134-136
  nomenclature, 134
  occurrence, 134
  outliers, 137
  paleontology, 136
  summary, 136-137
  summit profile, 140F
  thickness, 134
Teblas Mine, 169, 169F
tectogene, 27, 60, 333F
  zone, 331
tectogenic glass, 301
tectonic ages, 299
  cross-sections, 331, 333F
  disconformity, 262
  framework, 395
tectonic harmony, 283
  history, 305-334
  pressure, 273
  relations, 395
  setting, 253
  stability, 175, 317
  style, 317F
  temperature, 273
  zones, defined, 302
tectonism, late Triassic, 313-317
  mid Paleozoic, 308-310
  Upper Paleozoic 310-313
teeth, antelope, 171
Teh Wan Seng No. 3 mine, Salak South, Selangor, 356
teiichispira kobayashi, 35, 402
Tekka, Kinta Valley, 291, 366T, 366, 367, 368F, 372, 374
  Clays, 6, 8, 9, 158, 291

Mines, Perak, 358, 399
tektites, 174-175, 392, 395, 397, 399
  age, 175
  in alluvium, 175
Telaga Tujoh, 288
telescoped ore, chemical changes, 377F
  mineralogical changes, 377F
telescoped stanniferous ore, 368F
telescoping in veins, 358
telescopium, 173
Telok Anson, 7, 22F, 171, 395
Telok Champedek, 204, 245
  Kruen, Perak, 290, 358, 386
  Kruen Pipe, Perak, 355F
  Kubang Badak, 30
  Memplam, 44F
Temangan, 191, 262, 318
Temangan Ignimbrite, 178F, 193F, 194-195, 262, 318
Temangan iron mine, 262, 318, 256
  -Sokor area, Kelantan, 392
Tembeling Formation, 10T, 16, 17F, 99F, 107T, 113, 114, 118, 122, 123-128, 137, 235, 254, 314, 317, 319F, 333F, 334, 397, 398
  Formation, cross-bedded fluvial sandstone, 125F, 126F
  cross-sections, 319F
  east Johore, 124F
Tembeling Formation, Jengka Pass, 124F
  Formation, lithology, 123-127
    nomenclature, 123
    northeast Kedah, 124F
    occurrence, 123
    paleontology, 127
    stratigraphic sections, 124F
    structural map, 319F
    Sungei Tekai type section, 124F
    Sungei Yih, 124F
    summary, 127-128
    thickness, 125
  River, 123, 242
  Series, 4, 123
Temerloh, Pahang, 80, 83, 99F, 106, 108, 111, 79F, 178F, 195, 196F, 198
  Pahang, fossils, 100F
Temerloh-Kuantan road, 86F
  -Maran, 101
Temoh, Perak, 71T, 290
temperature-depth equivalence, 253
Templar Park, Selangor, 158, 293, 347, 370
  Park area, Selangor, 397
Tenang beds, 383
Tenasserim, 92, 93F, 211
TENG, H. C., 383, 401
tennantite, 374, 375F, 376, 377F
TENNISON-WOODS, J. E., 3, 401
tension gashes, 320
tentaculites, 8, 33, 42
tentaculites beds, 40
  dacryonconarid, 42, 45
  lower Devonian, 50
tentaculites matlockianus CHAPMAN, 45
tephrite, leucite, 206

Terbat Formation, 94
  limestone, 93F
terrace(s), 15, 157
terrace deposits, 143
  deposits, fluvial, 157
  soils, 157
terrorist activities, 7
Tertiary, 9, 10T, 11, 12, 16, 23, 177, 239F
  basins, 145F
  denudation, 11
  deposits, 64F
    thickness, 144, 146, 150, 151
  sedimentary deposits, summary, 152-153
  sedimentary rocks, 143-153
  volcanism, 178F
Terutau, 27
Tethyan affinities, 112
Tethys, 68, 108
tetradenia (neolitsea), or Lindera, 149
tetrahedrite, 350F, 352
tetrastannite, 367, 368, 370F
texture, brecciated, 381
  ophitic, 204
  subophitic, 204
Thai-Malay Peninsula, 25
Thai-Malayan border survey, 8
  -Malayan Orogen, 241F
Thailand, 1, 25, 26F, 28F, 30, 31F, 33, 35, 39, 45-46, 56, 58, 64F, 91, 94, 97, 104, 114, 138, 144, 145F, 210, 211-212, 212F, 213, 302-303
  Cambrian fossils, 396
  fossils, 394
  geological map, 241
  geology, 392, 397
  granite, 240-242
  igneous features, 396
  northeast, 392
  Ordovician fossils, 396
  pelecypods, 394
  tin and tungsten, 396
thailandium solum KOBAYASHI, 32, 38F
thamnopora, 75
THAVISRI, A., 397
thecia swinderniana (GOLDFUSS), 37
thermal aureole, 285, 288, 294, 298, 301, 373
  aureoles, absence of, 302
thermoluminescence, 238, 395
  ages, 235, 299, 300F
  histogram, 299
  of limestones, 299-300
  young age, 300
thiara, 151
  variabilis, 175
THOMAS, H. D., 8, 37, 49, 75, 401
THOMPSON, J. T., 3, 401
thorite, 172
thrust(ing), 10T, 11, 33, 63, 71
  age, 312
thrust, geometry, 312
Thrust, Kisap, 92T, 313F
thrust, low-angled, 312, 333F
  plane, conjugate, 312
thrusts, two conjugate, 312

Thung Song, 39, 50, 138
  Song limestone, 39, 40
Thye On beds, 75, 75T
  On mine, 76
  San mine, 36
  Sang mine, Salak South, 364F, 365
Tibetia, 55
Tien Shan, 55, 59
TILLEY, C. E., 174, 401
tillite, 5
tilting, 16
time scale, 394, 397
Timor, 112
  Fault, 360, 380F
tin, 240, 336, 392
  absence of, 303
  additions, 376
  analysis, 400
  bearing, 332
    granite, 398
Tin Belt, 306F
  western, 344
tin in cassiterite, 342
tin concentration in lodes, 342
  content, syngenetic, 343
  chemical analyses, 343T
  deposits, 242, 393, 399
    distribution, 337F
    Kinta district, 396, 402
    and limestone, 394
    Perak, 394, 401
    pregranite, 5
    Southeast Asia, 394
    Thailand, 391
  dispersion in soil, 401
  dredge, 4
  in ferromagnesian minerals, 342
  fields, 337F
    Southeast Asia, 394
    the World, 396
tin, hydrothermal, 354
  industry, 396
  lode mining, Trengganu, 393
  lodes, 390
    -fault relationships, 328
    strike, 328F
  minerals in Malaysia, 399
  mineralization, 11, 90, 219, 215
  mine, open cast, 159F
  mines, 71
tin mining, 3, 4, 39F
  occurrence, 394
  ore, 328, 364F
    alluvial, 3
    lode, 3
  ores, South-west England, 394
  placers, 157, 165
  pneumatolysis, 227
tin prospecting, 402
Tin Province, 241F
tin-pyrite problem, 399
tin-rich zone, 169
tin in skarns, 352-356
  theory of genesis, 335
  in tourmaline, 342

titanaugite, 204, 246
  replacement of, 247
titanite, 256; *see also* Sphene
Titi Mine, 338F, 359
TJIA, H. D., **13-24**, 19, 23, 24, 157, 174, 176, 195, **305-330**, 401
Toba, 214
Toh Kiri Lode, 345
TOKUYAMA, A., 8, 106, 108, 118, 397, 401
tombolo, 19
tonalite, 222, 224, 226F, 248F, 250F
  mineralogy, 224
Tong Huat quarry, 383
Tonkin, 106, 108
tools, paleolithic, 154
tools, Tampanian, 157
TOOMS, J. S., 390, 401
Toon Hing tin mine, 296
topaz, 5, 161, 166, 172, 226, 227, 374, 400
  -bearing rocks, Gunong Bakau, 400
  and cassiterite, 396
  -quartz rocks, 5
  rock, stanniferous, 346
  secondary origin, 346
topography, limestone, 15
  scarp, 30
torbernite, 225, 227
TORIYAMA, R., 112, 402
Toufangian Limestone, 36
tourmaline, 5, 89, 149, 160, 165, 166, 170, 172, 180, 182, 189, 198, 220, 221, 224, 225, 227, 240, 256, 259, 263, 284, 287, 293, 298, 345, 344, 351T, 357T, 359, 365, 371F, 372
  clots, 229
  clots in granite, 229F
  -corundum boulders, 5
  -corundum rocks, 3, 72, 74F, 229, 258T, **291-293**, 292F, 400
  -corundum rock, analysis, 291
    mineralogy, 291
    petrology, 291
  granite porphyry, **280**
  greisenization, 221, 227, **229**, 232, 395
  veins, 221, **280**
tourmalinization, 72, 228, 291, 360
Tournaisian, 94
toxic conditions, 113
trachyandesite, 82, 134, 187, **188**, 197
  mineralogy, 188
trachyceras (paratrachyceras) cf. *regelodanum*, 110
trachyceratidae gen. et. sp. indet. *orestites* sp., 110
trachyte, 4, 82, 131, 177, 179T, 187, **188-192**, 201, 202T, 203, 210F
  mineralogy, 192
  -rhyolite tuff, **190**
trachytic lava flows, 129
  texture, 243
Trang, 92
Tranum-Gap road, Bentong area, 231T
transgression, Lower Silurian, 59

of basement, 31
Upper Devonian, 59
Transjordan, 108
Tras, 247, 295, 346
tree trunks, 136
trellis drainage, 16
tremolite, 34, 47, 49, 53, 54, 161, 259, 281, 288, 295, 356F
  asbestos, 290
Trengganu, 4, 7, 8, 10T, 61, 62F, 63, 84, 85F, 86, 97, 122, 125, 128, 177, 179T, 185, 218, 246, 247, 297, 323F
  Upper Paleozoic, **86-87**
Traing, 106, 108, 154
Triassic, 4, 6, 7, 9, **9-11**, 10T, 11, 16, 63, 70, 80F, 84, 120, 107T, 235, 239F, 300F, 333F, 387
Triassic *daonella* beds, 397
  formations, 64F
  fossils, 100F, 400
  geosynclinal organization, 98F
  Lower, **97-105**, 187
    nomenclature, **97-101**
    occurrence, **97-101**
  Middle and Upper, **105-122**
    nomenclature, **105-106**
    occurrence, **105-106**
    synthesis, **121-122**
  structure, central Malaya, 80F
  Upper, 61, 110, 124F, 216F
tributaries, major, 13
trichonodella, 35
triclinicity, 218
trigonia cf. *zlambachensis*, 110
trigonodus sp., 110
trilobites, 33, 35, 42, 63
trilobite impressions, saukid, 30
trilobites, Late Cambrian, 32
  Ordovician, 49
trilobite pygidia, 84, 91
trilobites, saukids, 31, 32, 58
trinodus, 49
Trong, 20
Tronoh, 363
tropical karst, 401
tropical soils, 398
tropical weathering, 391
TSUYAMA, H., 152
Tuan Estate, 53
tubes, 89, 89F, 90
  sand filled, 65
tuff, 54, 81T, 84, 185, **188-189**, 195, **199-200**, 202T, 205F, 210, 211, 268F
  acidic, 109, 193F, **196F**, 195, 257F
  andesitic, 82, 102, 195, 212, 273
  andesite-trachyte-rhyolite, **190**
  agglomeratic, **200**
  calcitic, 194
  crystal, 182, 189
    constituents, **189**
  -lithic, 82, 103, 190, 195, 196F, 200
  -vitric, 42
  dactic, 181
  flow, 194
  foliated, 182

INDEX   435

foliated rhyolitic, 181
greenish, 182
hanging wall, 378
hornblende dacite, 212
jointing of, 188
lapilli, 192
lenticular, 181
lithic, 189, 194, 200
    constituents, **189**
metamorphic constituents, **189**
metamorphosed, 192, 210
mineralogy, 195
phyllitic, 297
rhyodacite, 103
rhyodacitic crystal, 50
rhyolite, 187, **189**, 192, 197, 201, 214, 249
    bedded, 46
    crystal, 54
rhyolitic, 42, 48, 50, 53, 82, 81T, 112
rock inclusions, **189**
    spherulitic, 189
trachyte-rhyolite, **190**
trachytic, 82, 102
vitric, 42, 194
volcanic, 185
water deposited, 189
weathered, 262
welded, 103, 184, 200
    rhyolite, 194
tuffaceous rhyolite, 195
    shale, 189-190
Tui Valley, 79F, 81T, 82
tulang mawas ("ape bones"), 175
Tungku, 123
tungsten, 240, 336, 358F, 392
    deposits, 398
        distribution, 338F
    ores, 6
    zone, 358F
Tunku Makota's Mine, Pahang, 338F, 363, 369, 373, 389F
turbidites, 122
turbidity flow, 41, 46, 59, 122
TURNER, F. J., 394
turtle shells, 167
TUTTLE, O. F., 221, 232, 302, 401
TWEEDIE, M. W. F., 175, 401
twinning, 218
type section, 123
Tyrrhenian, 24T

ulrichodina, 35
ultrabasic bodies, 215
    complex, 206
    rocks, 50, 54, **242-249**, 250F
Ulu Bekor, 342
Ulu Besut, 85F
    Endau, 103, 134, 136, 201, 203F, 322, 326
        Johore, fossils, 132F
    Gapis, Pahang, 225
    Jempol, 195
    Jurong village, 202T
    Kelantan, 1, 257F, 262

Kenong, 108
Kerling, 226, 294
Keta, 296
Ketubong, Kedah, 336
Klang, 178F, 180, 182, 183, 184, 311F
Klang road, 182
Kul tin mine, Selangor, 365
Ulu Lakit, 113, 114
    Langat, 90, 184, 394, 366, 369
    Lebir, 79F
    Liam, 330F
    Lipis, Pahang, 242, 296
    Nenggiri, 50, 97
    Nenggiri area, 267
    Pahang, 1, 4, 5, 61, 400, 401
Ulu Pedu, 119
    Perak, 1
    Pergau, 267, 297
    Petai, Perak, 345, 346
    Piah, 290, 291, 315
    Rompin, 3
    Sedili, 62F, 87
    Sokor, 337F, 381
    Sungei Atok, Pahang, fossils, 77F
Ulu Sungei Bakah, 245
    Sungei Kasai Besar, 280
    Sungei Kasai Kechil, 280
Ulu Sungei Kemaman, Trengganu, 87, 237T, 231T
    Sungei Kenong, 106
    Sungei Meledu, 189
    Sungei Pergau, 83, 297
    Sungei Reman, 298
    Sungei Sat, 231T, 237T
    Sungei Taku, 259
    Sungei Terisi, 280
Ulu Tanum Valley, 81T
    Triang, 53
    Yam, 182, 293, 294, 330F
UMBGROVE, J. H. F., 23, 306F, 401
uncinunellina, 82
unconformity, 9, 36, 45, 52, 61, 63, 68, 83, 87, 90, 92, 95, 108, 124F, 126, 127, 133, 137, 139, 144, 146, 150, 152, 159, 161, 167, 168, 169F, 170, 201, 235, 262, 293, 305, 308, 309F, 310, 311F, 312
    angular, 67, 129, 310
    Badong Conglomerate on Permian? shale and quartzite, 130F
unconformities, dating of, 235
U.S. Navy Hydrographic Office, 401
U.S.S.R. stratigraphy, 398
United States tectonics, 394
University of Malaya, 8, 220T, 231T, 234T, 238T, 248T, 258T, 263, 390
uplift, 11, 71, 95, 114, 134, 138, 141, 153, 156, 174, 301, 312
    age, 334
    Cenozoic, 15
    dating of, 235, 240
    epeirogenic, 143
    Late Cretaceous, 141
    Late, Triassic, 314
    Permian, 104

post-Tembeling, 139
radiometric age, 239, 283
Upper Cretaceous, 387
    Cretaceous mineralization, 387
    Detrital Member, 29T, 33, 35, 43, 44F, 45
    Carboniferous, 11
    Devonian sediments, 44F
    Gondwanas, 123
Upper Indosinian, 138
Upper Lotong Sandstone, 129
Naungkangyi Stage, 40
Upper, Paleozoic, 7, 25, **61-95**
    clastic rocks, 78F
    correlation, **91-94**
    formations, inferred relationship, 65F
    formations, Kinta, thickness, 75T
    geological history, **94-95**
    limestone, 70F
    rocks, distribution, 62F
    Southeast Asia distribution map, 93F
    structure, central Malaya, 80F
    volcanism, 178F
    Zone 1: northwest Malaya, 61
    Zone 2: Kinta Valley, Perak, 63
    Zone 3: central Malaya, 63
    Zone 4: east Malaya, 63
    Redbeds, Indosinian, 138
    Sandstone, Indosinian, 138
    Setul Limestone, 29T, 33, 34
Ural geosyncline, 59
uranium, thorium-lead radiometric determinations, 236-238T
Usun Apau Plateau, 214
uvalas, elongated, 15
Uzbekistan, 59

vaccinium scortechinii, 149
vaginatenkalk, 59
valence band, 299
valley, dead-end, 16
    in filling, 158
    inland, fill, 157-171
    rectilinear, 15, 329
VAN BAREN, F. A., 398
VAN DER STOK, 21F
variation diagram, 209, 210F, 232, 249, 250F, 274F
Variscan folding, 306F
    massif, 305
varlamoffite, 353, 365, 391
VEERABURAS, M., 391, 395
vein, galena-rich, 350F
    mineralized, 349
    mineralogy, 372
    quartz, 182, 222, 227-229, 282F, 286F
veins, Cornwall, 394
    moderately complex, 373-374
    quartz-feldspar, 265
    simple mineralogy, 371-373
    telescoped, 374, 387
vein swarms, 227, 371
    swarm around granite cusp, 372
venite, 263, 265
    psammite, foliated, 275F
    sillimanite-cordierite, 265

verbeekina sp., 82, 101
verbeekina verbeeki (GEINITZ), 77F
verbeekinid limestone, 10T
VERHOOGEN, J., 394
vermes, 395
vermiculite, 204, 245
vertebrate, 167
vesicles, 204, 206
   mineralized, 385F
vesicular horizons, volcanic, 153
vesuvianite, 288, 289, 290, 291, 353, 354, 356
   analysis, 290
VIALOV, O. S., 305, 401
Vietnam, 211, 212F
   north, 30
Viséan, 63, 80, 83, 84, 94
viviparus, 150, 151
   from Johore, 393
   willbourni, 151
vogesite, 247, 249
volcanic activity, 119, 186T, 177-214
   activity, explosive, 83
      summary, **207-209**
   arc, 46, 56F, 58
   ash, 209
   breccia, 108, 184, 210
   deposits, Quaternary, **153-155**
   eruption, 154
   facies, 28F, 30F, **54**, 60, 79F, 80, *82*, 85F
   glass, 154
   islands, 95
volcanic member, 101
   origin, 184, 198
   rocks, 42, 57F, 113
      acid, 324F
      Bentong area, **197-198**
      central Johore, **200-201**
      chemistry, **201**
      correlation chart, 179T
      east Johore, **201**
      east Pahang, **195-197**
      Jengka Triangle, **195-197**
      metamorphism, **297-298**
      north Kelantan and north Trengganu, **190-195**
      occurrence, 178F
      Permian, 187
volcanic rocks, petrology, **187-190**, 191-192
   rocks, Raub area, **198**
      South Johore, **198-200**
      Triassic, 201
      west Pahang, **197-198**
   vents, 58
volcanism, 9, 11, 40
   active, 214
   Cenozoic, **203-207**, 213-214
   central and north Pahang, **187-190**
   contemporaneous, 27, 187
   Jurassic and Cretaceous, **201-203**
   Kuantan, **204-206**
   Lower Carboniferous, 195
   Lower Paleozoic, **177-185**, 209-211
   middle to upper Triassic, 201

Permian, 195, 213
regional correlation, **209-214**
submarine, 187, 195
Upper Paleozoic to Lower Mesozoic, **185-201**, 211-213
volsella cf. *compressa*, 110
vugs, 229

waagenoconcha, 76
WALKER, D., 9, 23T, 24, 151, 152, 153, 154, 155, 158, 159, 160, 168, 401
WALKER, G. L. P., 392
wall-rock alteration, 360
wall-rock replacement, 351
WALTON, B. J., 251, 401
WALTON, E. K., 121, 393
wangs, 16
Wang Kelian, 67F
   Pisang, Perak, 347
   Tangga, 171
WARD, D. E., 138, 139, 401
WARD, T., 3, 401
WARD, W. C., 163, 393
waterfalls, 15
WATERHOUSE, J. B., 94, 401
watersheds, 14F
water table fluctuation, 386
wave action, 19
Wealden flora, 132
weathered feldspar, 89
weathering, 3, 5, 155, 290
   advanced, 161F
   chemical, 13
   deep, 386
   depth of, 155
   factors, 386
   granite, 15F
   in situ, 160
   intense, 262
   profile, granite, 155F
   subaerial, 13
   tropical, 13, 141
WEBER, M., 24, 398
WEIR, J., 106, 108, 401
Weng Valley, 47
Wenlock, 29T, 50
wentzelella, 70, 83
   malayensis IGO, 77F
west coast, Malaya, 23T
West Malaysia, 1, 93F, 212F
west Pahang, 310
west Yunnan, 401
western belt, 338F, 341
   belt granite, 343T, 387
Western Boulder Clays, 6, 8, 9, 158
Western Clays, 5
western mineralized belt, 351
Western Road, Penang, 236T
   Road J. K. R. quarry, Penang, 231T
WESTERVELD, 306F
Westphalian age, 86
Wheal Vor, Cornwall, 379
Wheel Rock, 368
whole rock radiometric determination, 236-238T

WILCOCK, B., 394
WILFORD, G. E., 16, 94, 157, 401
WILLBOURN, E. S., 6, 7, 23T, 24, 54, 87, 115, 118, 120, 150, 154, 158, 174, 177, 180, 182, 183, 184, 185, 204, 207, 229, 245, 292, 315, 355, 356F, 359F, 372, 400, 401, 402
William Mining Company, 284
Willinks, 380F
   fault, 328F
Willink's Mine, Sungei Lembing, 327F
WINCHELL, A. N., 259, 402
WINKLER, H. G. F., 253, 260, 261, 266, 285, 291, 301, 402
Wodgina, Australia, 365
Wolfram Mine, 366
wolframite, 227, 335, 345, 360, 365, **369-370**, 371F, 389F
   relicts, 369
   two generations, 369
wollastonite, 34, 47, 49, 285, 286F, 288, 289, 290, 293, 352
WONG, I. F. T., 197, 402
WONG, T. W., 310, 395, 402
WONG, Y. F., 395
WOOD, A., 122, 402
wood fragments, 163
   fragments in alluvium, 159F
   lignitised, 159
wood tin, 365
   Triassic, 393
worm burrows, 118
   casts, 136
worthenia, 65
WRAY, L. Jr., 4, 402
wrench faulting, 399; *see also* Faulting
WROBEL, S. A., 366T, 399
Würm glacial period, 168
Würm interglacial, 174

xenocrysts, 245
xenoliths, 90, 199, 222, 225F, 243
xenolith, assimilation, 221, 222
   feldspathized, 272T
xenoliths in dolerite, 206
   gabbro, 224, 250
   ghosts, 224
   in granite, 199
   metasediment, 223
   ovoid, 223, 251
xenoliths, psammite, 271F
   serpentinite, 206, 207
xenothermal, 358, 387
   deposits, 358, 378, 388
   tin-sulfide deposits, 388
xenotime, 161, 172, 227
X-ray diffraction, 218

yabeina asiatica ISHII, 77F
YANCEY, T. E., 63
YEAP, C. H., 342, 370
YEAP, E. B., 88F, 89, 90, 336, 346, 355, 356, 367, 369, 374, 383, 384F, 395, 402
YEE, C. C., 354

Yen, 282F, 286F
  area, 284
YEOW, B. C., 354, 374, 402
Yew Ling Quarry, Lanchang, 181F
Yik Meng Kongsi, 346
YIN, E. H., 87, 90, 97, 101, 102, 112, 113, 123, 127, 180, 395, 397, 402
YOCHELSON, E. L., 35, 402
YONG, S. K., 195, 196F, 393
Yong Peng, 110
YOUNG, B., 91, 398, 402
Young Alluvium, 9, 143, 159, 160, 169F, **168-171**
  Alluvium, age, 170, **170-171**
    criteria, 168
    drainage pattern, 170
    origin, **170-171**
Younger Arenaceous Series, 7, 105, 123
yttrotungstite, 161, 365, 369
Yuen Wan Foong mine, 162F
Yunnan, 9, 25, 56, 58, 59, 97, 105, 108, 112
  -Malaya geosyncline, graptolites and tentaculites, 392

zamites sp., 127
zaphrentites, 76
Zebingyi Series, 40, 46, 50
zeolite, 153, 197, 204, 274
zeolitization, 213
ZEUNER, F. E., 24, 168, 402
zig-zag, ridge, 16, 17F
zinc, 336, 358F
zinnwaldite, 345
zircon, 89, 91, 148, 149, 160, 165, 166, 170, 172, 240, 259, 263, 344
  tin-bearing, 342
zoisite, 47, 49, 161, 280, 289, 296
zonal classification of Malayan Orogen, 253-254
zone, *acodus mutatus-acontiodus hamari*, 35
  acodus similaris-drepanodus altipes, 35
  cyrtograptus lundgreni, 42
  demirastrites convolutus, 45, 46, 47
  demirastrites triangulatus, 46
  glyptograptus persculptus, 43
  meekoceras gracilitatis, 101
  monoclimacis griestoniensis, 49
  monograptus flexilis, 42
  monograptus hercynicus, 42, 53
  monograptus praehercynicus, 42, 53
  monograptus sedgwicki, 45, 46
  neoschwagerina, 94
  orthograptus vesiculosus, 43, 46
  panderodus unicostatus, 35
  pristiograptus cyphus, 42, 43, 46
  pristiograptus gregarius, 47, 50
  pseudofusulina ambigua, 78
  pseudoschwagerina, 94
  rastrites linnaei, 42, 49
  rising axial, 333F
  scolopodus staufferi-Scolpodus giganteus, 35
  streptograptus crispus, 49
  spirograptus minor, 42, 49
  spirograptus spiralis, 42, 49
  stomatograptus grandis, 42, 49
  verbeekina-neoschwagerina, 83
  Viséan-*dibunophyllum*, 84
zones, breccia, 90
  crush, 90
  shear, 90
  Upper Paleozoic, 92T
zoning, primary, 357
  regional, 357